Multilevel Block Factorization Preconditioners

Panayot S. Vassilevski

Multilevel Block Factorization Preconditioners

Matrix-based Analysis and Algorithms for Solving Finite Element Equations

Panayot S. Vassilevski
Center for Applied Scientific Computing
Lawrence Livermore National Laboratory
Box 808, L-560
Livermore, CA 94550
vassilevski1@llnl.gov

ISBN: 978-1-4419-2448-3 e-ISBN: 978-0-387-71564-3
DOI: 10.1007/978-0-387-71564-3

Mathematics Subject Classification (2000): (Primary) 65-02, 65F10, 65N22, 65N55;
(Secondary) 65F50, 65N30, 65Y20, 65H10

Printed on acid-free paper

9 8 7 6 5 4 3 2 1

springer.com

To Tanya, Virginia, and Alexander

Preface

The purpose of this monograph is to offer a comprehensive presentation of methods for solving systems of linear algebraic equations that typically arise in (but are not limited to) the numerical solution of partial differential equations (PDEs). Focus is on the finite element (or f.e.) method, although it is not presented in detail. It would, however, help the readers to be familiar with some basic knowledge of the finite element method (such as typical error estimates for second-order elliptic problems). There are a number of texts that describe the finite element method with various levels of detail, including Ciarlet [Ci02], Brenner and Scott [BS96], Braess [B01], Ern and Guermond [EG04], Solin [So06], and Elman, *et al.* [ESW06]. The presentation here utilizes matrix–vector notation because this is the basis of how the resulting solution methods are eventually implemented in practice. The choice of the material is largely based on the author's own work, and is also aimed at covering a number of important achievements in the field that the author finds useful one way or another. Among those are the most efficient methods, such as multigrid (MG), especially its recently revived "algebraic" version (or AMG), as well as domain decomposition (DD) methods. The author found a common ground to present both as certain block-matrix factorizations. This framework originates in some more classical methods such as the (block-) approximate (or incomplete) LU (or block-ILU) factorization methods. This led to the somewhat unusual title of the book. The approach, as well as the specific topics covered, should offer a different view on topics covered in other books that deal with preconditioned iterative methods.

This book starts with a motivational introductory chapter that describes the class of matrices to which this book is mainly devoted and sets up the goals that the author tries to achieve with the remainder of the text. In particular, it describes sparsity, conditioning, assembly from local element matrices, and the Galerkin relation between two matrices coming from discretization of the same PDE on coarse and fine meshes (and nested finite element spaces). The introduction ends with a major strong approximation property inherited from the regularity property of the underlining PDE. A classical two-grid method is then introduced that is illustrated with smoothing iterations and coarse-grid approximation. The motivational chapter also contains some basic facts about matrix orderings and a strategy to generate a popular nested dissection ordering arising from certain element agglomeration algorithms. The element agglomeration is later needed to construct a class of promising algebraic multigrid methods for solving various PDEs on general unstructured finite element meshes. Also discussed is the important emerging topic in practice of how to generate the f.e. discretization systems on massively parallel computers, and a popular mortar f.e. method is described in a general algebraic setting. Many other auxiliary (finite element and numerical linear algebra) facts are included in the seven appendices of the book.

The actual text starts with some basic facts about block-matrices and introduces a general two-by-two, block-factorization scheme followed by a sharp analysis. More specific methods are then presented. The focus of the book is on symmetric positive definite matrices, although extensions of some of the methods, from the s.p.d. case to nonsymmetric, indefinite, and saddle-point matrices, have been given and analyzed. In addition to linear problems, the important case of problems with constraints, as well as Newton-type methods for solving some nonlinear problems, are described and analyzed. Some of the topics are only touched upon and offer a potential for future research. In this respect, the text is expected to be useful for advanced graduate students and researchers in the field. The presentation is rigorous and self–contained to a very large extent. However, at a number of places the potential reader is expected to fill in some minor (and obvious) missing details either in the formulation and/or in the provided analysis.

Specific comments due to Yvan Notay, David Silvester, Joachim Schöberl, Xiao–Chuan Cai, Steve McCormick, and Ludmil Zikatanov are gratefully acknowledged. Special thanks are due to Tzanio Kolev for his comments and for providing numerous illustrations used throughout the book.

The author is thankful to Arnold Gatilao for his invaluable help with editing major part of the text.

Finally, the help of Vaishali Damle, Editor, Springer is greatly appreciated.

Portions of this book were written under the auspices of the U.S. Department of Energy by University of California Lawrence Livermore National Laboratory under Contract W-7405-Eng-48.

Livermore, California
March 2007

Panayot S. Vassilevski

Contents

Part II Block Factorization Preconditioners

Part III Appendices

Part I

Motivation for Preconditioning

1

A Finite Element Tutorial

This introductory chapter serves as a motivation for the remainder of the book. In particular, we illustrate the type of matrices that we focus on (but are not limited to) and describe the need for methods for fast solution of associated linear systems of equations. In particular, this chapter provides a brief finite element tutorial focusing on a matrix-vector presentation.

1.1 Finite element matrices

To be specific, consider the Poisson equation here

$$-\Delta u \equiv \frac{\partial^2 u}{\partial x^2} + \frac{\partial^2 u}{\partial y^2} + \frac{\partial^2 u}{\partial z^2} = f, \tag{1.1}$$

posed on a polygonal domain $\Omega \subset \mathbb{R}^d$. Here, $d = 3$, but we often consider the case $d = 2$. To be well posed, the Poisson equation needs some boundary conditions, and to be specific, we choose

$$u = 0 \text{ on } \partial\Omega. \tag{1.2}$$

Norms of functions in Sobolev spaces

In what follows, we consider functions that have derivatives up to a certain order (typically, first- and second-order) in the L_2-sense. The formal definitions can be found, for example, in Ciarlet [Ci02] and Brenner and Scott [BS96]. We use the following norms

$$\|u\| = \|u\|_0 = \left(\int_\Omega u^2 \, dx \, dy \, dz \right)^{1/2} \quad \text{and } \|u\|_1 = \left(\|u\|_0^2 + \|\nabla u\|_0^2\right)^{1/2},$$

P.S. Vassilevski, *Multilevel Block Factorization Preconditioners*,
doi: 10.1007/978-0-387-71564-3_1,

where

$$\|\nabla u\|_0 \equiv \left(\left\| \frac{\partial u}{\partial x} \right\|^2 + \left\| \frac{\partial u}{\partial y} \right\|^2 + \left\| \frac{\partial u}{\partial z} \right\|^2 \right)^{1/2}.$$

The latter expression is only a seminorm and is frequently denoted by $|u|_1$. Finally,

$$\|u\|_2 = \left(\|u\|_1^2 + |u|_2^2\right)^{1/2},$$

where $|u|_2^2$ stands for the sum of the squares of the L_2-norms of all second derivatives of u.

In general, we may want to explicitly denote the domain $\tau \subset \Omega$, for example,

$$\|u\|_\tau = \left(\int_\tau u^2 \, dx \right)^{1/2}.$$

If the domain is omitted, it is assumed that the integration is taken over the given domain Ω.

The spaces of functions that are complete in the above norms give rise to the so-called Sobolev spaces of the given order.

Also, sometimes we use the L_2-inner product of functions denoted by $(\cdot,\ \cdot)$.

The construction of finite element spaces

The popular finite element method consists of the following steps.

- Partition the domain Ω into a number of simply shaped elements τ in the sense that they cover Ω and have the property that two adjacent elements can share only a vertex, a face, or an edge (in 3D). In other words, two elements cannot have a common interior, partial face, or part of an edge only. In what follows, in two dimensions (2D), we consider triangular elements τ. Denote the set of these elements by $\mathcal{T}$. The elements are often assumed to be quasiuniform in the sense that their diameter is proportional to a characteristic mesh-size h. We denote this property by $\mathcal{T}_h = \mathcal{T}$. A 3D tetrahedral mesh is illustrated in Figure 1.4.
  The goal is to select an h small enough to obtain a suitable approximation to the continuous (infinitely dimensional) problem (1.1)–(1.2).
- Construct a finite element space $V = V_h$. For this purpose, we introduce a set of nodes $\mathcal{N}_h = \{x_i\}_{i=1}^n$, typically the vertices of all elements $\tau \in \mathcal{T}_h$ in the interior of Ω (because of the boundary condition (1.2)). With each vertex $x_i \in \mathcal{N}_h$, we associate a basis function ψ_i, which is supported in the union of the triangles that share vertex x_i. The function ψ_i restricted to any of the triangles τ is linear. Also, $\psi_i(x_i) = 1$ and, by construction, $\psi_i(x_j) = 0$ for any $x_j \in \mathcal{N}_h \setminus \{x_i\}$, so the basis $\{\psi_i\}$ is often called nodal or Lagrangian. Some nodal basis functions in 2D are illustrated in Figures 1.1 to 1.3.

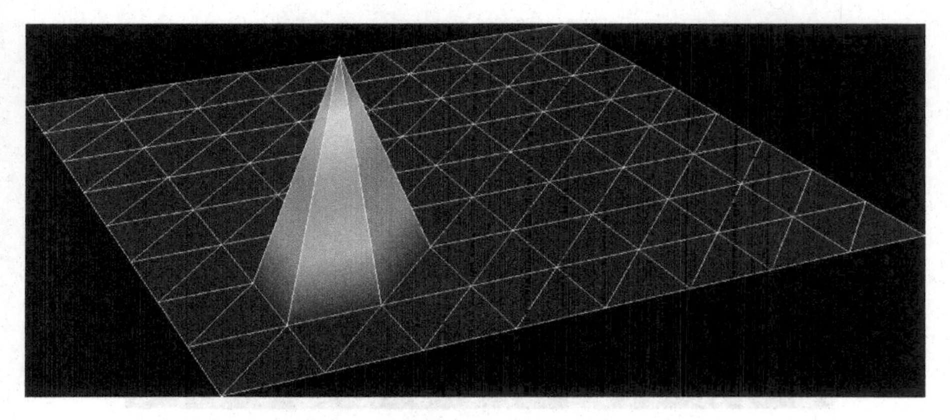

Fig. 1.1. *Fine-grid piecewise linear basis function.*

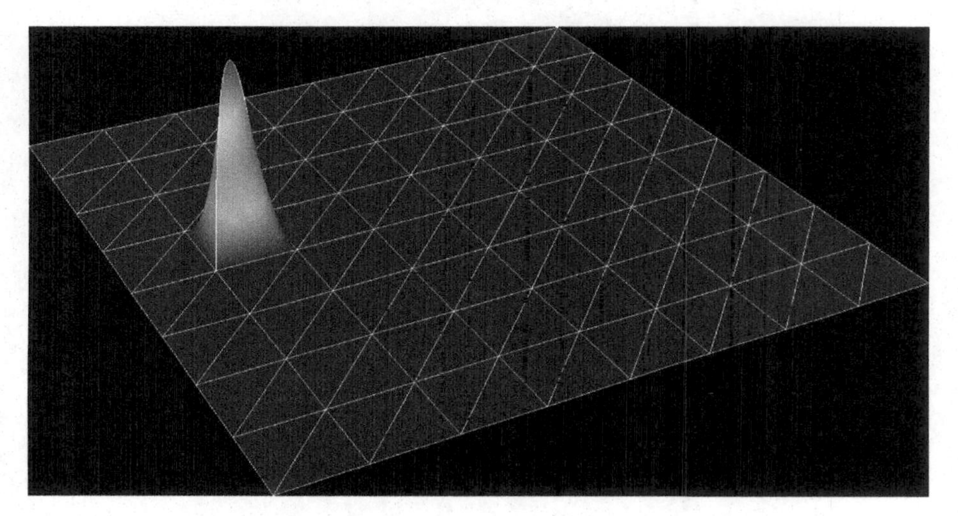

Fig. 1.2. *Fine-grid piecewise quadratic basis function associated with a midpoint of an element edge.*

Then, any function in the finite element space V_h (by definition) takes the form

$$v = \sum_{i=1}^{n} v_i \ \psi_i .$$

Because ψ_i is a Lagrangian basis, we see that $v_i = v(x_i)$. Thus, there is a one-to-one mapping between $v \in V_h$ and its coefficient vector $\mathbf{v} = (v_i)_{i=1}^n$ represented by the nodal values of v on $\mathcal{N}_h$.

In what follows, we adopt the convention (unless otherwise specified) that the same letter is used for the f.e. function and in boldface for its coefficient vector with respect to a given Lagrangian basis.

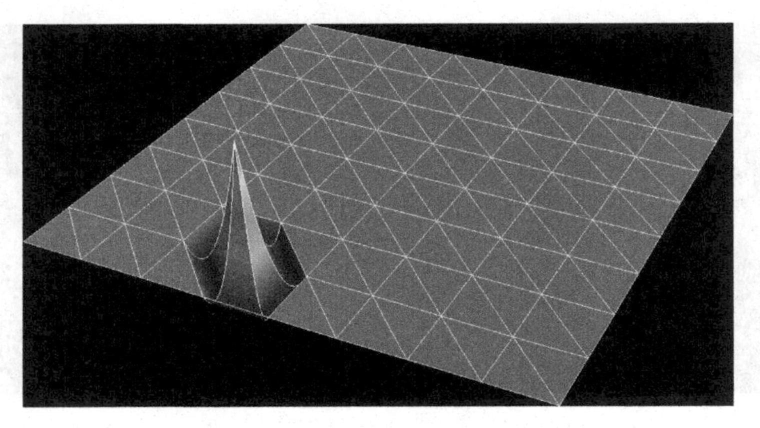

Fig. 1.3. *Fine-grid piecewise quadratic basis function associated with a vertex node.*

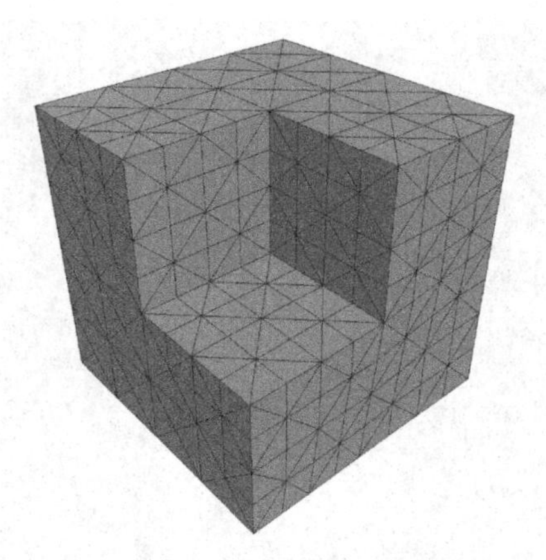

Fig. 1.4. *A 3D mesh.*

Weak form and Galerkin finite element discretization

To derive the finite element approximation to the continuous problem (1.1)–(1.2), we use its "weak formulation". The PDE is multiplied (tested) by functions φ and integrated over Ω, so that after using integration by parts, we end up with the desired "weak form":

$$\begin{aligned} a(u, \varphi) \equiv \int_{\Omega} \nabla u \cdot \nabla \varphi \, dx - \int_{\partial\Omega} \nabla u \cdot \mathbf{n} \, \varphi \, d\varrho = \int_{\Omega} f \varphi \, dx \\ = \int_{\Omega} \left(\frac{\partial u}{\partial x}\frac{\partial \varphi}{\partial x} + \frac{\partial u}{\partial y}\frac{\partial \varphi}{\partial y} + \frac{\partial u}{\partial z}\frac{\partial \varphi}{\partial z} \right) dxdydz \\ - \int_{\partial\Omega} \nabla u \cdot \mathbf{n} \, \varphi \, d\varrho = (f, \varphi) \equiv \int_{\Omega} f \varphi \, dx. \end{aligned} \tag{1.3}$$

Here, $\mathbf{n}$ stands for a unit vector normal to $\partial\Omega$ (pointing outward from Ω). Note that the integrals above make sense even for functions that are only once piecewise differentiable. The latter fact is used by the finite element method. Assuming that $\varphi = 0$ on $\partial\Omega$, we end up with

$$a(u, \varphi) = (\nabla u, \nabla\varphi) = \int_{\Omega} \nabla u \cdot \nabla\varphi \, dx = (f, \varphi) = \int_{\Omega} f\varphi \, dx. \tag{1.4}$$

The finite element discretization of (1.4) is obtained by the Galerkin method (equivalent to the Ritz method in the present setting); we approximate u with $u_h \in V_h$ determined from

$$\int_{\Omega} \nabla u_h \cdot \nabla\psi_i \, dx = \int_{\Omega} f\psi_i \, dx, \quad \text{for all } i = 1, \ldots, n.$$

Because $u_h = \sum_{j=1}^{n} u(x_j)\psi_j$, we get n equations for the n unknowns $u_h(x_j)$, $j = 1, \ldots, n$:

$$\sum_{j=1}^{n} u_h(x_j) \int_{\Omega} \nabla\psi_j \cdot \nabla\psi_i \, dx = \int_{\Omega} f\psi_i, \, dx, \quad \text{for } i = 1, \ldots, n.$$

Introducing $A = (a_{i,j})$ with

$$a_{i,j} = \int_{\Omega} \nabla\psi_j \cdot \nabla\psi_i \, dx = \sum_{\tau:\, x_i,\, x_j \in \tau} \int_{\tau} \nabla\psi_j \cdot \nabla\psi_i \, dx,$$

the vector of unknowns $\mathbf{x} = (u_h(x_i))_{i=1}^{n}$, and the r.h.s. vector $\mathbf{b} = \left(\int_{\Omega} f\psi_i \, dx\right)_{i=1}^{n}$, we end up with the discrete problem of our main interest,

$$A\mathbf{x} = \mathbf{b}.$$

Degrees of freedom (dofs)

It is customary, in finite elements, to use the notion of "degrees of freedom" or simply dofs. In our setting dofs can be identified with the vertices x_i. In general, dofs are equivalent to unknowns. In some situations we may have several degrees of freedom associated with a vertex. This is the typical case for systems of PDEs, such as elasticity equations. Then, in 2D, we have two dofs associated with every vertex of the mesh.

Properties of f.e. matrices

S.p.d.

The finite element method guarantees (by construction) that A is symmetric positive definite (s.p.d.). Symmetry follows because, from the relation between A and the bilinear form $a(.,.)$ defined in (1.3), we have

$$\mathbf{w}^T A\mathbf{v} = a(v, w) = a(w, v).$$

We also have that $\mathbf{v}^T A \mathbf{v} = a(v, \ v) \geq 0$ and that $\mathbf{v}^T A \mathbf{v} = a(v, v) = 0$ implies $\nabla v = 0$. The latter means that $v =$ const and, because $v = 0$ on $\partial\Omega$, that $v = 0$, hence, $\mathbf{v} = (v(x_i)) = 0$. Thus, A is also positive definite.

Sparsity

Another important property is that A has a bounded number of nonzero entries per row. Notice that row i of A has nonzero entries $a_{i,j}$ only for vertices x_j such that x_i and x_j belong to a common element τ. That is, the number χ_i of nonzero entries in row i of A, equals the number of edges of the triangles that meet at vertex x_i. This number is bounded by a topological constant $\chi \geq 1$ (depending on the triangulation $\mathcal{T}_h$), which can stay bounded when $h \mapsto 0$ if, for example, the minimal angle of the triangles is bounded away from zero.

For the matrix A corresponding to the f.e. Laplacian on a uniform mesh $h = 1/(n+1)$ and $\Omega = (0, \ 1)^2$, it is well known that every row's nonzero entries, up to an ordering, equal $(-1, \ -1, \ 4, \ -1, \ -1)$ (some off-diagonal entries are missing for rows that correspond to vertices near the boundary of Ω).

Matrix diagonal and matrix norm estimate

We need the next result in what follows.

Proposition 1.1. *Let χ_i be the number of nonzero entries of row i of A, and let $D = \mathrm{diag}(a_{ii})$ be the diagonal of A. Then,*

$$\mathbf{v}^T A \mathbf{v} \leq \mathbf{v}^T \chi D \mathbf{v},$$

where χ is either the diagonal matrix $\mathrm{diag}(\chi_i)$ *or simply the constant* $\max_i \chi_i$.

Proof. We first use the Cauchy–Schwarz inequality in the A-inner product for the coordinate vectors $\mathbf{e}_i$ and $\mathbf{e}_j$: $a_{ij}^2 = \left(\mathbf{e}_j^T A \mathbf{e}_i\right)^2 \leq \mathbf{e}_i^T A \mathbf{e}_i \mathbf{e}_j^T A \mathbf{e}_j = a_{ii} a_{jj}$. Then the sparsity of A, and one more application of the Cauchy–Schwarz inequality confirm the result:

$$\begin{aligned}
\mathbf{v}^T A \mathbf{v} &= \sum_i v_i \sum_{j:\ a_{ij} \neq 0} a_{ij} v_j \\
&\leq \sum_i |v_i| \sum_{j:\ a_{ij} \neq 0} |a_{ij}|\ |v_j| \\
&\leq \sum_i \sum_{j:\ a_{ij} \neq 0} a_{ii}^{1/2} |v_i| a_{jj}^{1/2}|\ |v_j| \\
&\leq \left(\sum_i \sum_{j:\ a_{ij} \neq 0} a_{ii} v_i^2 \right)^{1/2} \left(\sum_i \sum_{j:\ a_{ij} \neq 0} a_{jj} v_j^2 \right)^{1/2} \\
&= \sum_i \chi_i a_{ii} v_i^2 \\
&\leq \max_i \chi_i \ \mathbf{v}^T D \mathbf{v}. \qquad (1.5)
\end{aligned}$$

□

Remark 1.2. We can actually prove (see Proposition 1.10) a more accurate estimate of the form (1.5) with $\chi_i = \max_{\tau \in \mathcal{T}_h:\, i \in \tau} |\tau|$, where $|\cdot|$ stands for cardinality; that is, $|\tau|$ is the number of dofs that belong to τ. Thus, for linear triangular elements, $\chi_i = 3$.

We also have the following property of A.

Proposition 1.3. *The norm of the f.e. matrix A exhibits the behavior that:*

$$\|A\| \simeq h^{d-2},$$

which is asymptotically (for $h \mapsto 0$) sharp.

Proof. The proof follows from the estimate (1.5), which reads

$$\mathbf{x}^T A\mathbf{x} \le \mathbf{x}^T \chi D\mathbf{x} \le \max_i \chi_i a_{ii} \; \|\mathbf{x}\|^2.$$

Recall that χ_i is the number of nonzero entries of A in row i. Let Ω_i be the support of the ith basis function ψ_i. Note that Ω_i is the union of a bounded number of elements. Due to quasiuniformity of $\mathcal{T}_h$ (i.e., $|\tau| \simeq h^d$) we have that Ω_i has measure $|\Omega_i| \simeq h^d$. Then $a_{ii} = \|\nabla\psi_i\|^2 \simeq |\Omega_i|\; h^{-2} \simeq h^{d-2}$; that is, $\|A\| \simeq h^{d-2}$. This estimate is sharp asymptotically inasmuch as

$$\|A\| \ge a_{ii} = \|\nabla\psi_i\|^2 \simeq h^{-2}\; |\Omega_i| \simeq h^{d-2}. \qquad \square$$

1.2 Finite element refinement

Consider now two nested finite element spaces $V_H \subset V_h$. Let $V_H = \text{Span}\,(\psi_{i_c}^{(H)})_{i_c=1}^{n_c}$ and $V_h = \text{Span}\,(\psi_i^{(h)})_{i=1}^{n}$ with their respective nodal (Lagrangian) bases. Because each $\psi_{i_c}^{(H)} \in V_H \subset V_h$, we have the expansion

$$\psi_{i_c}^{(H)} = \sum_{i=1}^{n} \psi_{i_c}^{(H)}(x_i)\; \psi_i^{(h)}.$$

Interpolation matrix

Consider the coefficient (column) vector $\boldsymbol{\psi}_{i_c} = (\psi_{i_c}^{(H)}(x_i))_{i=1}^{n}$. The matrix $P = (\boldsymbol{\psi}_{i_c})_{i_c=1}^{n_c}$ is referred to as the *interpolation matrix*. It relates the coefficient vector $\mathbf{v}_c \in \mathbb{R}^{n_c}$ of any function $v_c \in V_H$, expanded in terms of the coarse basis $\{\psi_{i_c}^{(H)}\}$, to the coefficient vector $P\mathbf{v}_c$ of $v_c \in V_h$, expanded in terms of the fine-grid basis $\{\psi_i^{(h)}\}$. The finite element bases are local, thus the $n \times n_c$ rectangular matrix P is sparse. The number of nonzero entries of P per column depends on the support of each $\psi_{i_c}^{(H)}$, namely, on the number of fine-grid basis functions $\psi_i^{(h)}$ that intersect that support. That is, the sparsity pattern of P is controlled by the topology of the triangulations $\mathcal{T}_H$ and $\mathcal{T}_h$.

Finite elements often use successive *refinement*, which refers to the process of constructing $\mathcal{T}_h$ from $\mathcal{T}_{2h}$ by subdividing every element (triangle) of $\mathcal{T}_{2h}$ into four geometrically similar triangles of half the size. Then, by construction, $V_{2h} \subset V_h$, because if a continuous function v is linear on a triangle $T \in \mathcal{T}_{2h}$, it is linear on the smaller triangles $\tau \subset T$, $\tau \in \mathcal{T}$. The interpolation mapping P in this case is linear. Its columns have the form

$$\boldsymbol{\psi}_{i_c} = \begin{bmatrix} 0 \\ \frac{1}{2} \\ \vdots \\ 0 \\ \vdots \\ 1 \\ \vdots \\ 0 \end{bmatrix}.$$

The coefficients $\frac{1}{2}$ appear at rows j of $\boldsymbol{\psi}_{i_c}$ for which x_j is a midpoint of an edge $(x^c_{i_c}, x^c_{j_c})$ of a coarse triangle T (such that one of its endpoints is $x^c_{i_c}$). We note that the coarse nodes $x^c_{i_c}$ are also fine-grid nodes. That is, $x^c_{i_c} = x_i$ for some i. The latter means that the coarse indices i_c are naturally embedded into the fine-grid indices $i_c \mapsto i(i_c)$. The entry 1 of $\boldsymbol{\psi}_{i_c}$ appears exactly at the position $i = i(i_c)$. All remaining entries of $\boldsymbol{\psi}_{i_c}$ are zero.

Galerkin relation between A and A_c

Based on V_H and its basis, we can compute $A_c = (a(\psi^{(H)}_{j_c}, \psi^{(H)}_{i_c}))^{n_c}_{i_c, j_c=1}$. Similarly, based on V_h and its basis, we can compute $A = (a(\psi^{(h)}_j, \psi^{(h)}_i))^n_{i,j=1}$. We easily see that $a(\psi^{(H)}_{j_c}, \psi^{(H)}_{i_c}) = \boldsymbol{\psi}^T_{j_c} A \boldsymbol{\psi}_{i_c} = (P^T AP)_{i_c, j_c}$, which yields the variational (also called Galerkin) relation

$$A_c = P^T AP. \tag{1.6}$$

1.3 Coarse-grid approximation

The fact that geometrically smooth functions can accurately be represented on coarse grids is inherent to any approximation method; in particular, it is inherent to the f.e. method. Some illustrations are found in Figures 1.7 to 1.10.

We summarize the following fundamental finite element error estimate result (cf., e.g., Ciarlet [Ci02], Brenner and Scott [BS96], and Braess [B01]). Our goal is to prove at the end of this section a "strong approximation property" in a matrix–vector form.

Because $a(u - u_h, \varphi) = 0$ for all $\varphi \in V_h$, we have the following estimate,

$$\|\nabla(u - u_h)\|^2 = a(u - u_h, u - u_h) = a(u - u_h, u - \varphi) \leq \|\nabla(u - u_h)\| \|\nabla(u - \varphi)\|.$$

It implies the following characterization property of the finite element solution u_h.

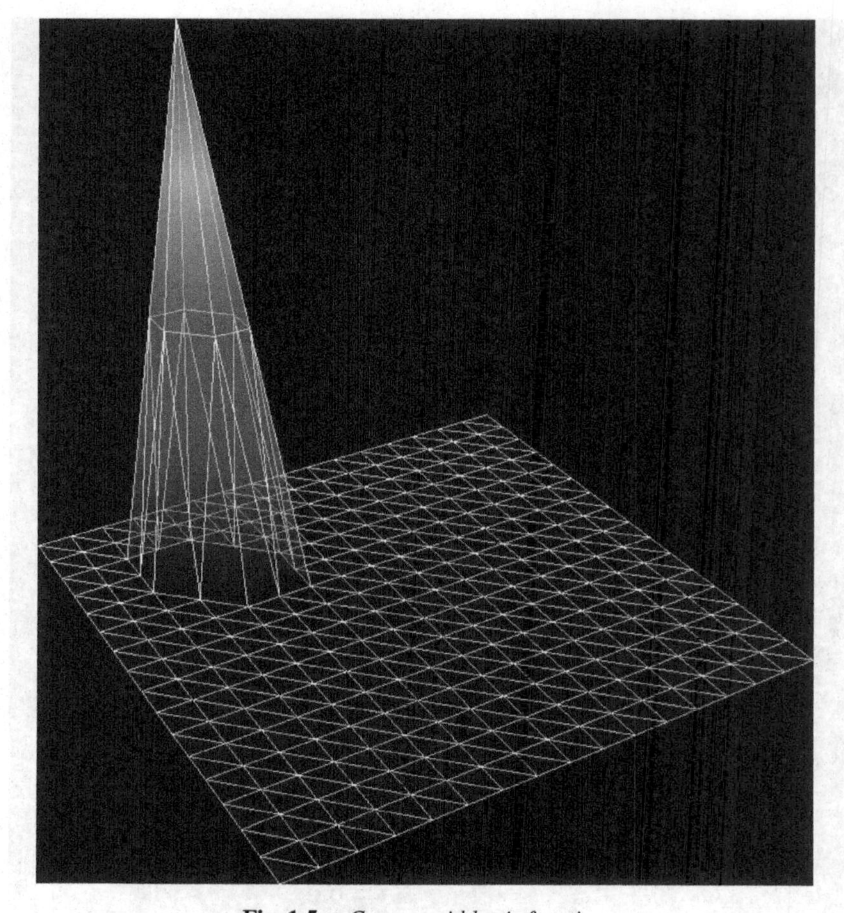

Fig. 1.5. *Coarse-grid basis function.*

Proposition 1.4. *The finite element solution u_h is an $a(\cdot, \cdot)$-orthogonal projection of the PDE solution u on the f.e. space V_h. In other words, we have the characterization,*

$$\|\nabla(u - u_h)\| = \inf_{\varphi \in V_h} \|\nabla(u - \varphi)\|. \tag{1.7}$$

Assuming now that u has two derivatives in $L_2(\Omega)$, we immediately get the first-order error estimate

$$\|\nabla(u - u_h)\| \leq Ch\|u\|_2.$$

To be more precise, we first form a nodal interpolant $I_h u = \sum_i u(x_i)\psi_i$. Then, based on (1.7) we have $\|\nabla(u - u_h)\| \leq \|\nabla(u - I_h u)\|$. Therefore, to estimate the latter term, splitting it over every triangle $\tau \in \mathcal{T}_h$, yields

$$\|\nabla(u - I_h u)\|^2 = \sum_{\tau \in \mathcal{T}_h} \int_\tau |\nabla(u - I_h u)|^2 \, dx \leq \sum_{\tau \in \mathcal{T}_h} C_\tau \, h^2 \|u\|_{2,\,\tau}^2 \leq Ch^2\|u\|_2^2.$$

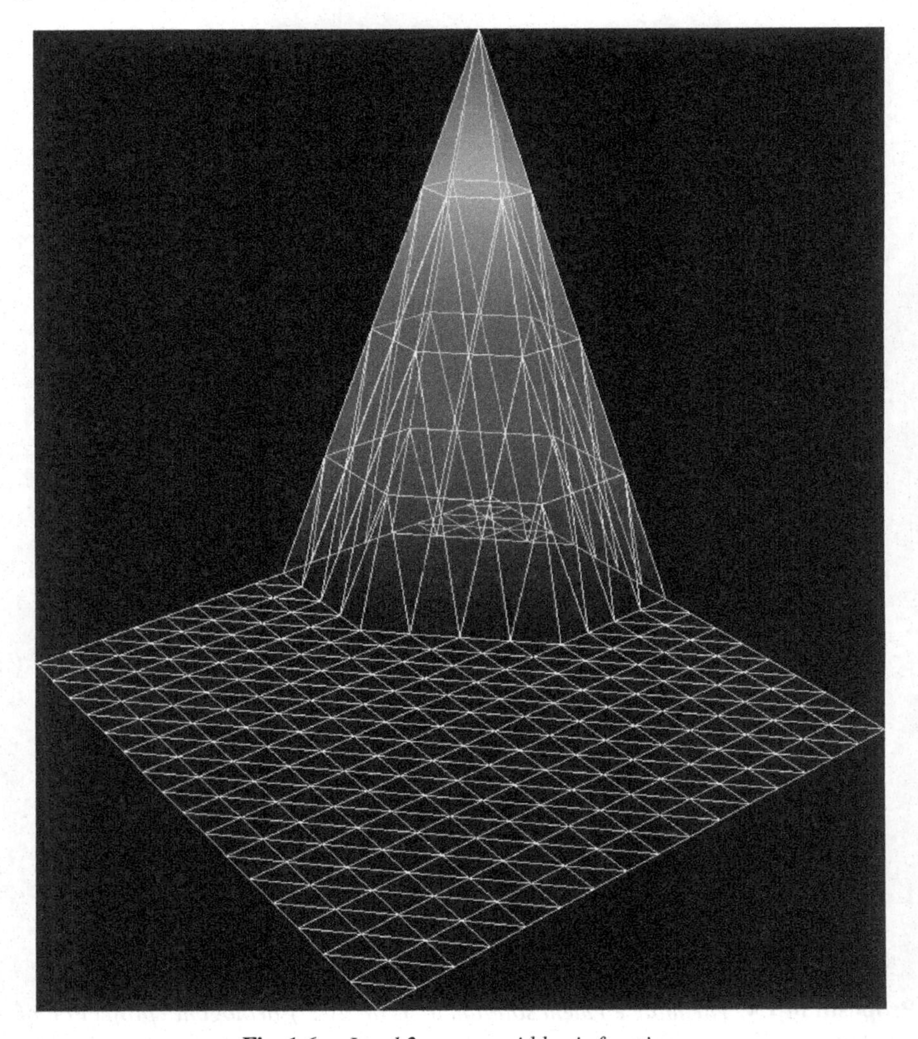

Fig. 1.6. *Level 2 coarse-grid basis function.*

Here, we use the Taylor expansion on every triangle τ and the fact that the triangles are geometrically similar to a fixed number of an initial set of coarse triangles. Hence, C_τ will run over a fixed number of mesh-independent constants. The estimate above shows that for smooth functions u (e.g., having two derivatives) the finite element approximations on grids $\mathcal{T}_H$ will give approximations u_H such that the error $u - u_H$ behaves as $H \; \|u\|_2$.

For a given f.e. function u_h, consider now its coarse finite element projection u_H defined from $a(u_h - u_H, \; \varphi) = 0$ for all $\varphi \in V_H$. We want to measure the coarse-grid approximation, that is, to estimate $u_h - u_H$. The preceding argument is not immediately applicable because u_h does not have two derivatives. To overcome this difficulty we introduce the f.e. function $A_h u_h \in V_h$ defined on the basis of the matrix

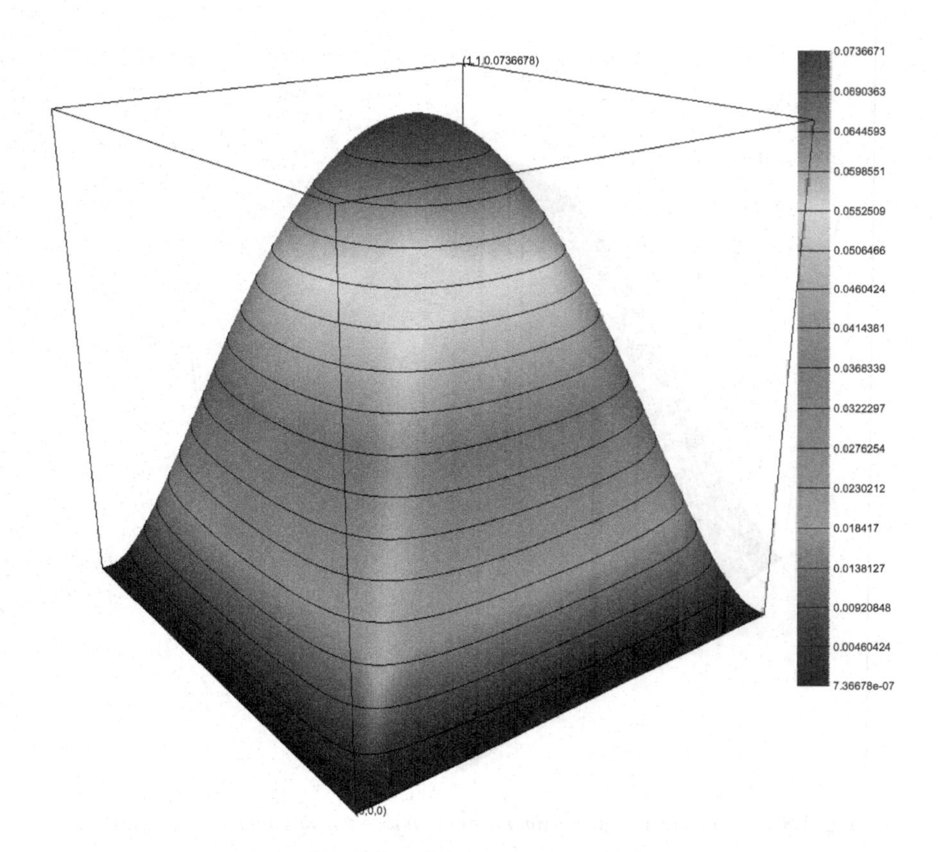

Fig. 1.7. *Solution to* $-\Delta u = 1$.

$A = (a_{i,j})$ and the coefficient vector $\mathbf{x} = (u_h(x_i))$ of u_h. Let $A\mathbf{x} = (c_i)$. Define then,

$$A_h u_h = \sum_i \frac{c_i}{(1,\ \psi_i)}\ \psi_i.$$

Because

$$c_i = (A\mathbf{x})_i = \sum_j a_{i,j}\ u_h(x_j) = \sum_j a(\psi_j,\ \psi_i) u_h(x_j) = a(u_h,\ \psi_i),$$

we have

$$A_h u_h = \sum_i \frac{a(u_h,\ \psi_i)}{(1,\ \psi_i)}\ \psi_i.$$

Introduce now the quasi-interpolant $\widetilde{Q}_h : L_2(\Omega) \mapsto V_h$ defined as follows,

$$\widetilde{Q}_h v = \sum_i \frac{(v,\ \psi_i)}{(1,\ \psi_i)}\ \psi_i.$$

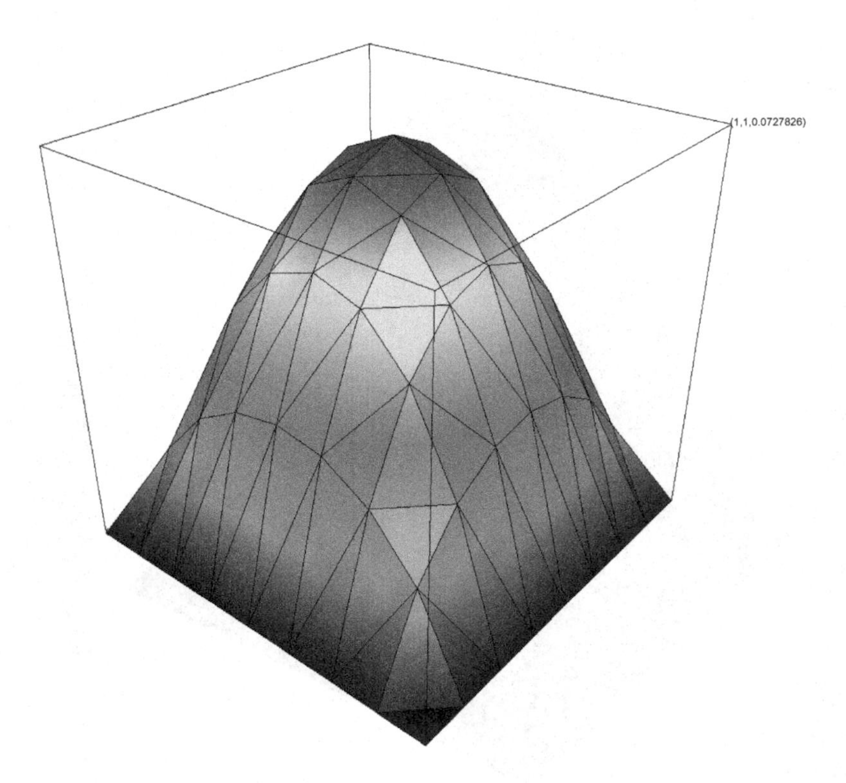

Fig. 1.8. *Finite element approximate solution to* $-\Delta u = 1$ *on a coarse mesh.*

We then have, for any $v \in L_2(\Omega)$,

$$(A_h u_h,\ v) = a(u_h,\ \widetilde{Q}_h v).$$

It is easy to see the following result.

Theorem 1.5. $\widetilde{Q}_h\ :\ V_h \mapsto V_h$ *is a symmetric and uniformly coercive operator; that is,*

$$\begin{aligned} (\widetilde{Q}v,\ w) &= (v,\ \widetilde{Q}w), && \textit{all } v,\ w \in L_2(\Omega), \\ (\widetilde{Q}\psi,\ \psi) &\geq \delta\ \|\psi\|^2 && \textit{for all } \psi \in V_h. \end{aligned}$$

Proof. Consider the basis $\{\psi_i,\ x_i \in \mathcal{N}_h\}$ of V_h. Recall that $\mathcal{N}_h$ is the set of degrees of freedom (the vertices x_i of the elements $\tau \in \mathcal{T}_h$ in the interior of Ω). We also consider $\overline{N}_h$, which is $\mathcal{N}_h$ augmented with the vertices of $\tau \in \mathcal{T}_h$ on the boundary of Ω. The coefficient vectors $\mathbf{v} = (v_i)$ of functions $v \in V_h$ (that vanish on $\partial\Omega$) are extended with zero entries whenever appropriate; that is, we let $v_i = v(x_i) = 0$ for $x_i \in \overline{N}_h \cap \partial\Omega$. Similarly, we use basis functions ψ_i associated with boundary nodes $x_i \in \overline{N}_h$ (whenever appropriate).

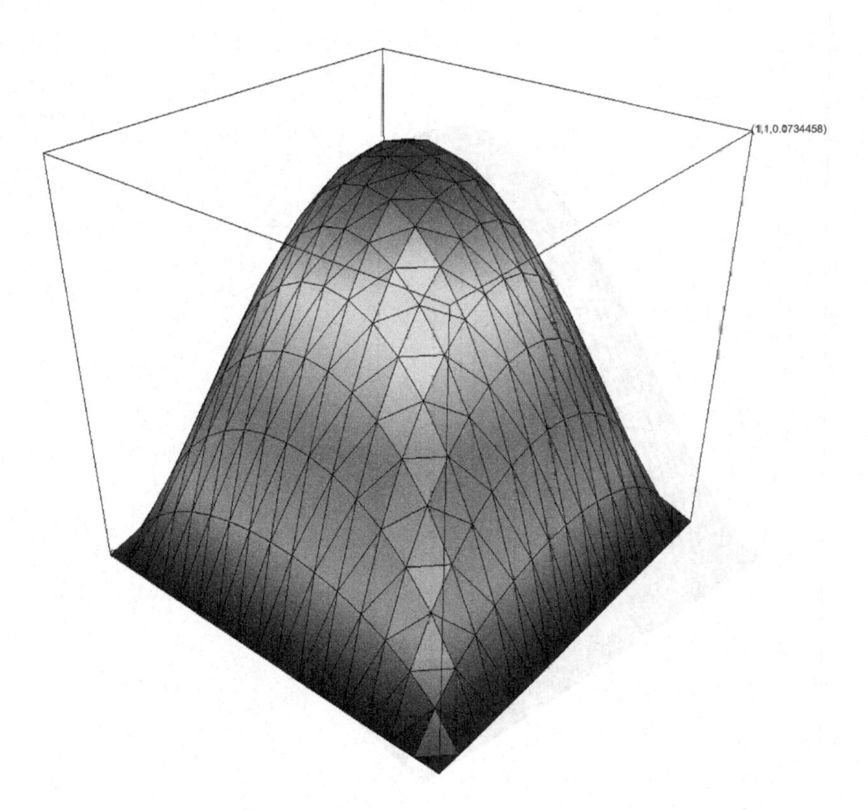

Fig. 1.9. *Finite element approximate solution to $-\Delta u = 1$ on a refined mesh.*

The symmetry is trivially seen because

$$(\widetilde{Q}v,\ w) = \sum_i \frac{(v,\ \psi_i)(w,\ \psi_i)}{(1,\ \psi_i)},$$

which is a symmetric expression for v and w.

We prove the uniform coercivity of $\widetilde{Q}_h$ in the following section. □

1.4 The mass (Gram) matrix

To prove the uniform coercivity of $\widetilde{Q}$ on V_h, introduce the Gram (also called mass) matrix $\overline{G} = \{(\psi_j,\ \psi_i)\}_{x_i, x_j \in \overline{\mathcal{N}}_h}$. Due to the properties of $\mathcal{T}_h$, it is easily seen to be uniformly well conditioned. Similarly, to estimate (1.5), we prove that (recalling that χ_i is the number of elements τ sharing the node x_i)

$$\mathbf{v}^T \overline{G} \mathbf{v} \leq \max_{x_i \in \overline{N}_h} (\chi_i \|\psi_i\|^2) \mathbf{v}^T \mathbf{v} \simeq h^d \ \|\mathbf{v}\|^2. \tag{1.8}$$

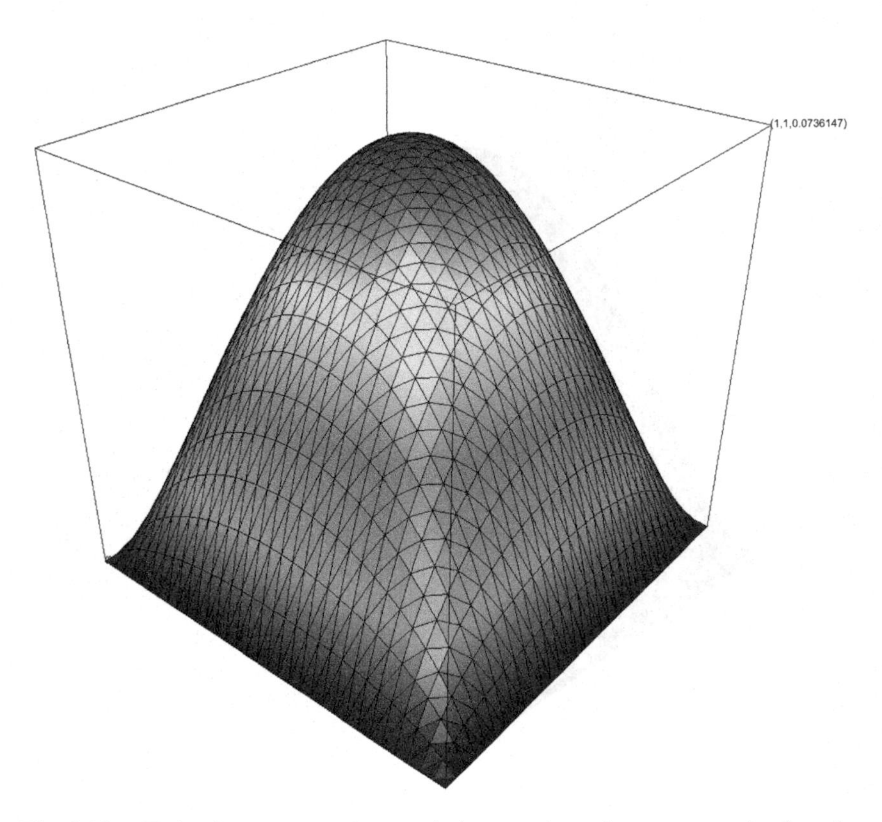

Fig. 1.10. *Finite element approximate solution to $-\Delta u = 1$ on a more refined mesh.*

For any vector $\mathbf{v}$, define its restriction $\mathbf{v}_\tau = (v(x_i))_{x_i \in \tau}$ to every element τ. We then have,

$$\mathbf{v}^T \overline{G} \mathbf{v} = (v,\ v) = \sum_{\tau \in \mathcal{T}_h} \int_\tau v^2\, dx = \sum_{\tau \in \mathcal{T}_h} \mathbf{v}_\tau^T \overline{G}_\tau \mathbf{v}_\tau.$$

Here, $\overline{G}_\tau = \left(\int_\tau \psi_j \psi_i\ dx \right)_{x_i,\ x_j \in \tau}$ is the so-called element mass matrix. For the particular case of triangular elements τ and piecewise linear basis functions ψ_i, we get

$$\overline{G}_\tau = \frac{|\tau|}{12} \begin{bmatrix} 2 & 1 & 1 \\ 1 & 2 & 1 \\ 1 & 1 & 2 \end{bmatrix}.$$

Because the minimal eigenvalue of this matrix is $|\tau|/12$, we get the following estimate from below,

$$\mathbf{v}^T \overline{G} \mathbf{v} \geq \sum_{\tau \in \mathcal{T}_h} \frac{|\tau|}{12}\, \mathbf{v}_\tau^T \mathbf{v}_\tau.$$

In general (under the assumption of quasiuniform $\mathcal{T}_h$), we get an estimate

$$\mathbf{v}^T\overline{G}\mathbf{v} \geq Ch^d \; \|\mathbf{v}\|^2. \tag{1.9}$$

Both estimates (1.8) and (1.9) imply the following result.

Theorem 1.6.

$$(v, v) = \mathbf{v}^T\overline{G}\mathbf{v} \simeq h^d \sum_{x_i \in \overline{\mathcal{N}}_h} v_i^2 = h^d\mathbf{v}^T\mathbf{v}, \quad \text{all } v = \sum_{x_i \in \overline{\mathcal{N}}_h} v_i\psi_i, \; \mathbf{v} = (v_i)_{x_i \in \overline{\mathcal{N}}_h}. \tag{1.10}$$

In other words the scaled inner product h^{-d} $(\mathbf{v}, \mathbf{v})$ is bounded above and below by the coefficient vector inner product $\mathbf{v}^T\mathbf{v}$ uniformly w.r.t. $h \mapsto 0$. Alternatively, we have

$$Cond\,(\overline{G}) = \mathcal{O}(1).$$

As a corollary, consider G, the principal submatrix of $\overline{G}$ corresponding to the nodes in $\mathcal{N}_h$ (i.e., in the interior of Ω). We then have $\lambda_{\min}[G] \geq \lambda_{\min}[\overline{G}]$.

Finally, under the assumption that the number of elements in $\mathcal{T}_h$ which share a given node is kept bounded, it follows that $\overline{G}$ is uniformly sparse, namely, that the number χ_i of nonzero entries of any row i of $\overline{G}$ is bounded by a fixed number $\chi = \max_i \chi_i$, independent of $h \mapsto 0$.

Proof (of the uniform coercivity of $\widetilde{\mathcal{Q}}_h$). Consider the coordinate unit vectors $\bar{\mathbf{e}}_i = (\delta_{i,j})_{x_j \in \overline{\mathcal{N}}_h}$, $x_i \in \overline{\mathcal{N}}_h$. Similarly, let $\mathbf{e}_i = (\delta_{i,j})_{x_j \in \mathcal{N}_h}$ be the unit coordinate vectors corresponding to $x_i \in \mathcal{N}_h$ (the boundary nodes excluded). It is clear then, that the following matrix–vector representation holds,

$$(\widetilde{\mathcal{Q}}_h v, v) = \sum_{x_i \in \mathcal{N}_h} \frac{(\mathbf{v}^T G\mathbf{e}_i)^2}{\mathbf{1}^T\overline{G}\bar{\mathbf{e}}_i}, \qquad \mathbf{1} = \sum_{x_i \in \overline{\mathcal{N}}_h} \bar{\mathbf{e}}_i.$$

Based on the decomposition for any $v \in V_h$ (i.e., vanishing on $\partial\Omega$), $G\mathbf{v} = \sum_{x_i \in \mathcal{N}_h}((G\mathbf{v})^T\mathbf{e}_i)\mathbf{e}_i$, and $\mathbf{v} = \sum_{x_i \in \mathcal{N}_h}((G\mathbf{v})^T\mathbf{e}_i)G^{-1}\mathbf{e}_i$, we get

$$\begin{aligned}
(v, v) &= \mathbf{v}^T G\mathbf{v} \\
&= \Big(\sum_{x_i \in \mathcal{N}_h} (\mathbf{v}^T G\mathbf{e}_i)\mathbf{e}_i\Big)^T G^{-1}\Big(\sum_{x_i \in \mathcal{N}_h} (\mathbf{v}^T G\mathbf{e}_i)\mathbf{e}_i\Big) \\
&\leq \lambda_{\max}[G^{-1}] \sum_{x_i \in \mathcal{N}_h} (\mathbf{v}^T G\mathbf{e}_i)^2 \\
&\leq \lambda_{\max}[\overline{G}^{-1}] \sum_{x_i \in \mathcal{N}_h} (\mathbf{v}^T G\mathbf{e}_i)^2.
\end{aligned}$$

Therefore, the following estimate is obtained.

$$\frac{(\widetilde{Q}_h v, v)}{(v, v)} \geq \lambda_{\min}[\overline{G}] \min_{x_i \in \mathcal{N}_h} \frac{1}{\mathbf{1}^T \overline{G} \overline{\mathbf{e}}_i}.$$

At most χ terms $\overline{\mathbf{e}}_j^T \overline{G} \overline{\mathbf{e}}_i$ in the first sum below are nonzero (these indices j define the set $\mathcal{I}(i)$), therefore we get

$$\begin{aligned}
\mathbf{1}^T \overline{G} \overline{\mathbf{e}}_i &= \sum_j \overline{\mathbf{e}}_j^T \overline{G} \overline{\mathbf{e}}_i \\
&\leq \sum_{j \in \mathcal{I}(i)} \left(\overline{\mathbf{e}}_j^T \overline{G} \overline{\mathbf{e}}_j\right)^{1/2} \left(\overline{\mathbf{e}}_i^T \overline{G} \overline{\mathbf{e}}_i\right)^{1/2} \\
&\leq \lambda_{\max}[\overline{G}] \sum_{j \in \mathcal{I}(i)} \|\overline{\mathbf{e}}_j\| \, \|\overline{\mathbf{e}}_i\| \\
&\leq \lambda_{\max}[\overline{G}] \, \chi.
\end{aligned}$$

That is, the desired uniform coercivity estimate takes the final form,

$$\frac{(\widetilde{Q}_h v, v)}{(v, v)} \geq \frac{1}{\chi \operatorname{Cond}(\overline{G})} = \mathcal{O}(1). \qquad \square$$

1.5 A "strong" approximation property

In what follows, we proceed with the proof of the following main result.

Theorem 1.7. *Assume that the Poisson equation $-\Delta u = f$, posed on a polygonal region Ω with $u = 0$ on $\partial\Omega$, satisfies the full regularity estimate $\|u\|_2 \leq C \, \|f\|$. Let $V_H \subset V_h$ be two nested f.e. spaces equipped with Lagrangian f.e. bases. Assume that h is a constant fraction of H. Let the corresponding finite element matrices be; A, the fine-grid matrix and $A_c = P^T A P$, the coarse-grid matrix, where P is the interpolation matrix that relates the coefficient vector of a coarse f.e. function expanded first in terms of the coarse basis and then in terms of the fine-grid basis. Then, the following strong approximation property holds, for a constant $\eta_a \simeq (H/h)^2$.*

$$\|\mathbf{x} - P A_c^{-1} P^T A \mathbf{x}\|_A^2 \leq \frac{\eta_a}{\|A\|} \, \|A\mathbf{x}\|^2.$$

The matrix $\pi_A = P A_c^{-1} P^T A$ is the so-called coarse-grid projection. Equivalently, we have

$$\|A^{-1}\mathbf{x} - P A_c^{-1} P^T \mathbf{x}\|_A^2 \leq \frac{\eta_a}{\|A\|} \, \|\mathbf{x}\|^2.$$

For less regular problems (see (1.14)*), we have that for an $\alpha \in (0, 1]$, the following weaker approximation property holds.*

$$\|\mathbf{x} - \pi_A \mathbf{x}\|_A^2 \leq \frac{\eta_a}{\|A\|^\alpha} \, \|A^{(1+\alpha)/2}\mathbf{x}\|^2.$$

Here $\eta_a \simeq (H/h)^{2\alpha}$.

Proof. First, estimate the L_2-norm of $A_h u_h \in V_h$. We have

$$\begin{aligned}
\|A_h u_h\| &= \sup_{\psi \in V_h} \frac{(A_h u_h,\ \psi)}{\|\psi\|} \\
&= \sup_{\psi \in V_h} \frac{a\Big(u_h,\ \sum_i \frac{(\psi_i,\ \psi)}{(1,\ \psi_i)}\ \psi_i\Big)}{\|\psi\|} \\
&= \sup_{\psi \in V_h} \frac{\mathbf{b}^T A\mathbf{x}}{\|\psi\|}, \qquad \mathbf{b} = (b_i),\ b_i = \frac{(\psi_i,\ \psi)}{(1,\ \psi_i)}, \\
&\le \|A\mathbf{x}\| \sup_{\psi \in V_h} \frac{\Big(\sum_i \Big(\frac{(\psi_i,\ \psi)}{(1,\ \psi_i)}\Big)^2\Big)^{1/2}}{\|\psi\|} \\
&\le \|A\mathbf{x}\| \sup_{\psi \in V_h} \frac{\Big(\sum_i \Big\|\frac{\psi_i}{(1,\ \psi_i)}\Big\|^2 \|\psi\|^2_{\Omega_i}\Big)^{1/2}}{\|\psi\|} \\
&\le \|A\mathbf{x}\| \sup_{x_i \in \mathcal{N}_h} \Big\|\frac{\psi_i}{(1,\ \psi_i)}\Big\| \sup_{\psi \in V_h} \frac{\big(\sum_i \|\psi\|^2_{\Omega_i}\big)^{1/2}}{\|\psi\|} \\
&\le \|A\mathbf{x}\|\ Ch^{-(d/2)}.
\end{aligned}$$

Here, $\Omega_i = \cup\{\tau \in \mathcal{T}_h\ :\ x_i \in \tau\}$ stands for the support of ψ_i. Note that they have bounded overlap, hence $\big(\sum_i \|\psi\|^2_{\Omega_i}\big)^{1/2} \le C\|\psi\|$. Also, by assumption on $\mathcal{T}_h$, $|\tau| \simeq h^d$, and $|\Omega_i| = \sum_{\tau \subset \Omega_i} |\tau| \simeq h^d$. Finally, note that

$$\frac{\|\psi_i\|^2}{(1,\ \psi_i)^2} = \mathcal{O}\left(\frac{|\Omega_i|}{|\Omega_i|^2}\right) = \mathcal{O}(h^{-d}).$$

We use next the fact that for any $\varphi \in H_0^1(\Omega)$ we can choose a finite element function φ_h such that $\|\varphi_h\|_1 \le C_0\|\varphi\|_1$ and $\|\varphi - \varphi_h\|_0 \le C_0 h\ \|\varphi\|_1$ (cf., [Br93]). Then, based on the L_2-norm bound of $A_h u_h$ the coercivity of $\widetilde{Q}_h$, and Proposition 1.3 (i.e., that $\|A\| \simeq h^{d-2}$), the following estimate is readily seen.

$$\begin{aligned}
\|\widetilde{Q}_h^{-1} A_h u_h\|_{-1} \equiv &\sup_{\varphi \in H_0^1(\Omega)} \frac{(\widetilde{Q}_h^{-1} A_h u_h,\ \varphi)}{\|\varphi\|_1} \\
&\le \sup_{\varphi_h} \frac{(\widetilde{Q}_h^{-1} A_h u_h,\ \varphi_h)}{\|\varphi_h\|_1} \frac{\|\varphi_h\|_1}{\|\varphi\|_1} \\
&\quad + \|\widetilde{Q}_h^{-1} A_h u_h\|_0 \sup_{\varphi \in H_0^1(\Omega)} \frac{\|\varphi - \varphi_h\|_0}{\|\varphi\|_1} \\
&\le C_0 \left(\sup_{\varphi_h} \frac{a(u_h,\ \varphi_h)}{\|\varphi_h\|_1} + h\ \|\widetilde{Q}_h^{-1} A_h u_h\|_0 \right) \\
&\le C \left(\sqrt{a(u_h,\ u_h)} + h^{1-(d/2)}\ \|A\mathbf{x}\| \right) \\
&\le C\big(1 + \|A\|^{\frac{1}{2}} h^{1-(d/2)}\big)\ \|\mathbf{x}\|_A \\
&\le C\ \|\mathbf{x}\|_A.
\end{aligned}$$

Now solve the Poisson equation

$$-\Delta u = f \equiv \widetilde{Q}_h^{-1} A_h u_h \in V_h \subset L_2(\Omega). \tag{1.11}$$

Because $(\widetilde{Q}_h \psi,\ \psi) \geq \delta\ \|\psi\|^2$ for any $\psi \in V_h$ (i.e., $\|\widetilde{Q}_h^{-1}\| \leq (1/\delta)$, based on the assumed a priori estimate

$$\|u\|_2 \leq C\|f\|, \tag{1.12}$$

we get that

$$\|u\|_2 \leq Ch^{-(d/2)}\ \|A\mathbf{x}\|.$$

Note now that u_h and u_H with corresponding coefficient vectors $\mathbf{x}$ and $P\mathbf{x}_c$ are the f.e. solutions to the Poisson problem (1.11). The latter is seen because for any $\psi \in V_h$,

$$a(u,\ \psi) = (f,\ \psi) = (\widetilde{Q}_h^{-1} A_h u_h,\ \psi) = (A_h u_h,\ \widetilde{Q}_h^{-1}\psi) = a(u_h,\ \psi).$$

Then, because $\|u_h - u_H\|_1 \leq \|u - u_h\|_1 + \|u - u_H\|_1 \leq C\ (h + H)\ \|u\|_2 \leq C\ H\ h^{-(d/2)}\|A\mathbf{x}\|$, the following strong approximation property holds,

$$\|\mathbf{x} - P\mathbf{x}_c\|_A \leq C\ H\ h^{-(d/2)}\|A\mathbf{x}\|. \tag{1.13}$$

In the less regular case, we have for an $\alpha \in (0, 1]$, the following estimate,

$$\|u\|_{1+\alpha} \leq C\|f\|_{-1+\alpha}. \tag{1.14}$$

Recall, that $f = \widetilde{Q}_h^{-1} A_h u_h$. We showed that $\|f\|_{-1} \leq C\ \|A^{1/2}\mathbf{x}\|$ and $\|f\|_0 \leq Ch^{-(d/2)}\ \|A\mathbf{x}\|$. Now using a major estimate for the space $H^{-1+\alpha}$, which is an interpolation space between $L_2(\Omega)$ and $H^{-1}(\Omega)$ (the dual of $H_0^1(\Omega)$), the following estimate is seen (cf., e.g., Theorem B.4 in Bramble [Br93])

$$\|f\|_{-1+\alpha} \leq C\ (h^{-(d/2)})^\alpha \|A^{(1-\alpha)(1/2)+\alpha}\mathbf{x}\| = Ch^{-\alpha(d/2)}\ \|A^{(1+\alpha)/2}\mathbf{x}\|.$$

The latter two estimates, combined with a standard error estimate $\|u - u_H\|_1 \leq CH^\alpha\ \|u\|_{1+\alpha}$, lead to the following approximation property, in the less regular case,

$$\|\mathbf{x} - P\mathbf{x}_c\|_A \leq C\left(\frac{H}{h}\right)^\alpha \left(\frac{1}{h^{d-2}}\right)^\alpha \|A^{(1+\alpha)/2}\mathbf{x}\|. \tag{1.15}$$

In what follows, we find a simple relation between $\mathbf{x}_c$ and $\mathbf{x}$, namely, that $\mathbf{x}_c = A_c^{-1} P^T A\mathbf{x}$. Indeed, from the definition of $u_H \in V_H$, we have that it satisfies the Galerkin equations

$$a(u_H,\ \psi) = (f,\ \psi) = a(u_h,\ \psi) \quad \text{for all } \psi \in V_H.$$

Because ψ has a coarse coefficient vector $\mathbf{g}_c$ and hence $P\mathbf{g}_c$ is its fine-grid coefficient vector, we then have

$$\mathbf{g}_c^T A_c \mathbf{x}_c = a(u_H,\ \psi) = a(u_h,\ \psi) = (P\mathbf{g}_c)^T A\mathbf{x}.$$

That is,

$$A_c \mathbf{x}_c = P^T A\mathbf{x}.$$

Hence,

$$P\mathbf{x}_c = \pi_A \mathbf{x} = P A_c^{-1} P^T A\mathbf{x}.$$

The strong approximation property then takes the following matrix–vector form,

$$\|\mathbf{x} - \pi_A \mathbf{x}\|_A \leq C \;\; H\, h^{-\frac{d}{2}} \|A\mathbf{x}\|.$$

Now use Proposition 1.3, that is, that $\|A\| \simeq h^{d-2}$, to conclude with the desired estimate,

$$\|\mathbf{x} - \pi_A \mathbf{x}\|_A^2 \leq \frac{\eta_a}{\|A\|} \|A\mathbf{x}\|^2, \tag{1.16}$$

where $\eta_a \simeq (H/h)^2 = \mathcal{O}(1)$, if h is a constant fraction of H.

In the less regular case, based on estimate (1.15) and Proposition 1.3 (i.e., that $\|A\| \simeq h^{d-2}$) we arrive at the following weaker approximation property,

$$\|\mathbf{x} - \pi_A \mathbf{x}\|_A^2 \leq \frac{\eta_a}{\|A\|^{\alpha}} \|A^{(1+\alpha)/2}\mathbf{x}\|^2, \tag{1.17}$$

where $\eta_a \simeq (H/h)^{2\alpha} = \mathcal{O}(1)$, if h is a constant fraction of H. □

1.6 The coarse-grid correction

Let $\mathbf{x}$ be a current approximation for solving $A\mathbf{x} = \mathbf{b}$. Note that $\mathbf{x}$ is a coefficient vector of some finite element function $u \in V_h$. To look for a coarse-grid correction, in terms of finite elements means, we seek a $u_c \in V_H$ (with coarse coefficient vector $\mathbf{x}_c$), which solves the coarse finite element problem with an r.h.s. r computed on the basis of the current approximation u, for any $\psi_{i_c}^{(H)} \in V_H$, as follows.

$$\begin{aligned} a(u_c, \psi_{i_c}^{(H)}) &= (r, \; \psi_{i_c}^{(H)}) \\ &\equiv (f, \; \psi_{i_c}^{(H)}) - a(u, \; \psi_{i_c}^{(H)}) \\ &= \sum_i \psi_{i_c}^{(H)}(x_i)(f, \; \psi_i^{(h)}) - a(u, \; \psi_{i_c}^{(H)}). \end{aligned}$$

The latter system, in terms of vectors, reads

$$A_c \mathbf{x}_c = \mathbf{r}_c.$$

Recalling that $\psi_{i_c}^{(H)} = \sum_{i=1}^n \psi_{i_c}^{(H)}(x_i)\psi_i^{(h)}$, and $\mathbf{b} = (b_i)$ with $b_i = (f, \; \psi_i^{(h)})$, we get

$$(A_c \mathbf{x}_c)_{i_c} = \sum_i \psi_{i_c}^{(H)}(x_i) b_i - \boldsymbol{\psi}_{i_c}^T A\mathbf{x} = \boldsymbol{\psi}_{i_c}^T (\mathbf{b} - A\mathbf{x}).$$

That is,

$$A_c \mathbf{x}_c = \mathbf{r}_c \equiv \left(\boldsymbol{\psi}_{i_c}^T (\mathbf{b} - A\mathbf{x})\right)_{i_c=1}^{n_c} = P^T (\mathbf{b} - A\mathbf{x}) = P^T \mathbf{r}.$$

Hence,

$$\mathbf{x}_c = A_c^{-1} P^T \mathbf{r}.$$

Then, the new iterate is $u := u + u_c$. In terms of vectors we have

$$\mathbf{x} := \mathbf{x} + P\mathbf{x}_c = \mathbf{x} + PA_c^{-1} P^T \mathbf{r}.$$

We have the following relation between the errors (initial and final),

$$\mathbf{e} = A^{-1}\mathbf{b} - \mathbf{x},$$

noting that $\mathbf{r} = \mathbf{b} - A\mathbf{x} = A\mathbf{e}$,

$$\mathbf{e} := (A^{-1}\mathbf{b} - \mathbf{x}) - PA_c^{-1} P^T A(A^{-1}\mathbf{b} - \mathbf{x}) = (I - PA_c^{-1} P^T A)\mathbf{e}.$$

The matrix $\pi_A = PA_c^{-1} P^T A$ is a projection; that is (recall that $A_c = P^T AP$),

$$\pi_A^2 = PA_c^{-1}(P^T AP)A_c^{-1} P^T A = PA_c^{-1} P^T A = \pi_A.$$

Thus, we have

Proposition 1.8. *The error matrix corresponding to a coarse-grid correction is given by $I - \pi_A$, where $\pi_A = PA_c^{-1} P^T A$ is the coarse-grid projection matrix.*

1.7 A f.e. (geometric) two-grid method

The coarse-grid correction, combined with a few steps of a stationary iterative method, defines the classical two-grid method. In the present f.e. setting, we explore the natural (defined from two nested f.e. spaces $V_H \subset V_h$) interpolation matrix P, the fine-grid matrix A and the coarse one $A_c = P^T AP$. In addition, we need the iteration matrix M that defines a stationary iterative procedure. A typical case is a matrix M that satisfies the following conditions.

(i) M provides a convergent method in the A-norm; namely $\|I - A^{1/2} M^{-1} A^{1/2}\| < 1$.
(ii) M gives rise to an s.p.d. matrix $\overline{M} = M(M^T + M - A)^{-1} M^T$ which is assumed spectrally equivalent to the diagonal D of A (see (1.18) for a motivation).

The conditions (i) and (ii) define the notion of *smoother*. The definition comes from the fact that the lower part of the spectrum of $D^{-1}A$ corresponds to eigenvectors that are geometrically smooth. Recall, that here we consider matrices A that come from f.e., discretization of second-order Laplacian-like PDEs. Therefore, an iterative method with $\omega \simeq \|D^{-(1/2)} AD^{-(1/2)}\|$,

$$\mathbf{x}_k = \mathbf{x}_{k-1} + (\omega D)^{-1} (\mathbf{b} - A\mathbf{x}_{k-1}),$$

rewritten in terms of the errors $\mathbf{e}_k = A^{-1}\mathbf{b} - \mathbf{x}_k$ takes the form

$$\mathbf{e}_k = (I - (\omega D)^{-1} A)\mathbf{e}_{k-1}.$$

Consider the generalized eigenvalue problem

$$A\mathbf{q}_i = \lambda_i D\mathbf{q}_i,$$

with $\lambda_1 \le \lambda_2 \le \cdots \le \lambda_n$.

Note that if we expand $\mathbf{e}_k = \sum_i \beta_i^{(k)} \mathbf{q}_i$, we see that

$$\beta_i^{(k)} = \left(1 - \frac{\lambda_i}{\omega}\right)^k \beta_i^{(0)} \simeq \left(1 - \frac{\lambda_i}{\lambda_n}\right)^k \beta_i^{(0)}$$

for large i are reduced very quickly, whereas the entries corresponding to the lower part of the spectrum hardly change. This effect is referred to as *smoothing*. The combined effect of reducing the highly oscillatory components of the error by the smoother and approximating the smooth components of the error on a related coarse grid gives an intuitive explanation of the potential for the mesh-independent rate of convergence of the two-grid, and by recursion, of the (geometric) multigrid methods. The combined effect of smoothing and coarse-grid approximation lies at the heart of the two-grid (and multigrid) methods as originally observed by R. P. Fedorenko [Fe64, Fe64] and led A. Brandt, originally in [AB77], to generalize and promote it as a general methodology for solving a wide range of problems in the natural sciences. More on the history of MG is found in [TOS, pp. 23–24].

To explain item (ii) above consider the composite iteration

$$\begin{aligned} \mathbf{x}_{k-(1/2)} &= \mathbf{x}_{k-1} + M^{-1}(\mathbf{b} - A\mathbf{x}_{k-1}), \\ \mathbf{x}_k &= \mathbf{x}_{k-(1/2)} + M^{-T}(\mathbf{b} - A\mathbf{x}_{k-(1/2)}). \end{aligned} \tag{1.18}$$

Rewritten in terms of the errors $\mathbf{e}_s = A^{-1}\mathbf{b} - \mathbf{x}_s$, the above composite iteration reads,

$$\begin{aligned} \mathbf{e}_{k-(1/2)} &= (I - M^{-1}A)\mathbf{e}_{k-1}, \\ \mathbf{e}_k &= (I - M^{-T}A)\mathbf{e}_{k-(1/2)}. \end{aligned}$$

The composite iteration matrix E, relating $\mathbf{e}_k$ and $\mathbf{e}_{k-1}$ as $\mathbf{e}_k = E\mathbf{e}_{k-1}$, has then the following product form $E = (I - M^{-T}A)(I - M^{-1}A) = I - (M^{-T} + M^{-1} - M^{-T} AM^{-1})A = I - \overline{M}^{-1} A$ with $\overline{M} = M(M^T + M - A)^{-1} M^T$. It is easily seen then that $\|I - A^{1/2} M^{-1} A^{1/2}\| < 1$ based on the identity $(I - M^{-T}A)(I - M^{-1}A) = I - \overline{M}^{-1} A$ is equivalent to $M^T + M - A$ being s.p.d. Also, we have then $\|I - A^{1/2} M^{-1} A^{1/2}\|^2 = \|I - A^{1/2}\overline{M}^{-1} A^{1/2}\| = 1 - \lambda_{\min}(\overline{M}^{-1} A)$.

In conclusion, we can formulate the following proposition.

Proposition 1.9. *To have M be a convergent smoother for A (in the A-norm) it is equivalent to say that $M^T + M - A$ is s.p.d., and hence, $\overline{M} - A$ is symmetric positive semidefinite where $\overline{M} = M(M + M^T - A)^{-1} M^T$.*

A main example of M is the scaled Jacobi smoother ωD with $\omega \simeq \|D^{-(1/2)} AD^{-(1/2)}\|$ or the Gauss–Seidel smoother, defined from $A = D - L - L^T$, where D is the diagonal of A, and $-L$ is the strictly lower triangular part of A. Then $M = D - L$ is the forward Gauss–Seidel iteration matrix and $\overline{M} = M(M + M^T - A)^{-1}M^T = (D - L)D^{-1}(D - L^T)$ is the symmetric Gauss–Seidel matrix.

To illustrate the smoothing process, we start with $\mathbf{e}_0$ as a linear combination of a smooth and an oscillatory component, and then one, two, and three symmetric Gauss–Seidel iterations applied to $A\mathbf{e} = 0$ are run in succession. That is, we run the iteration $\mathbf{e}_k = (I - \overline{M}^{-1}A)\mathbf{e}_{k-1}$ for $k = 1, 2, 3$. The resulting smoothing phenomenon is illustrated in Figures 1.11 to 1.14.

At the end, we formulate an algorithm implementing the classical two-grid method.

Algorithm 1.7.1 *[Two-grid algorithm]*

Let $A\mathbf{x} = \mathbf{b}$ *be the fine-grid problem.*

Given is A_c *the coarse-grid matrix, related to* A *via the interpolation matrix* P *as* $A_c = P^TAP$, *and let* M *and* M^T *be the given smoother and its transpose.*

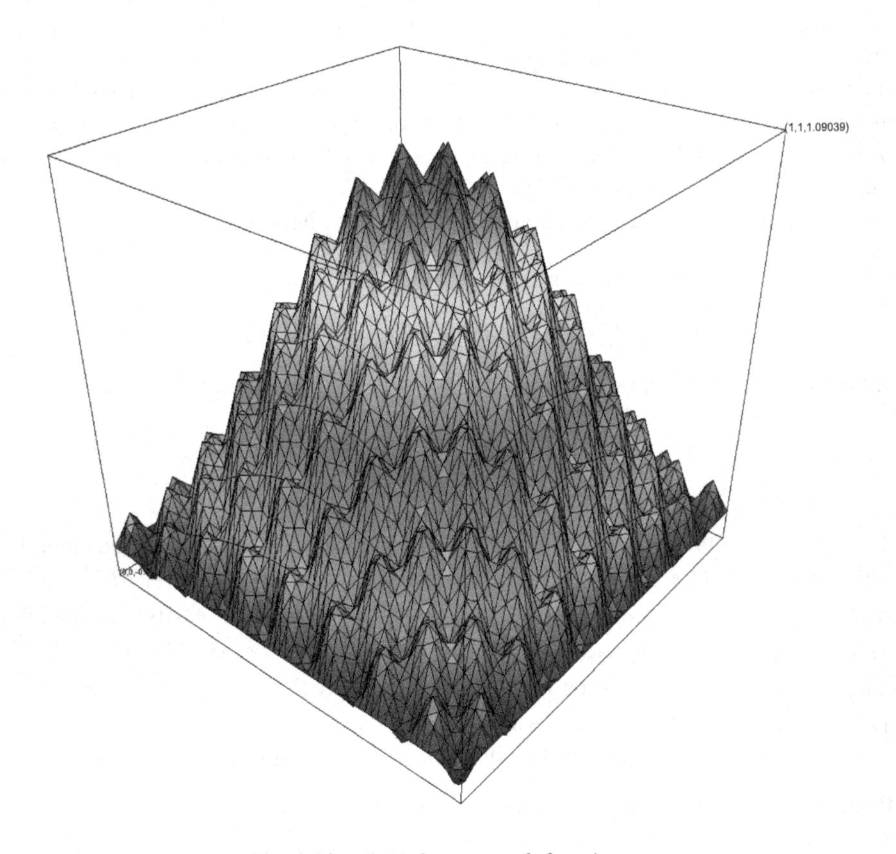

Fig. 1.11. *Initial nonsmooth function.*

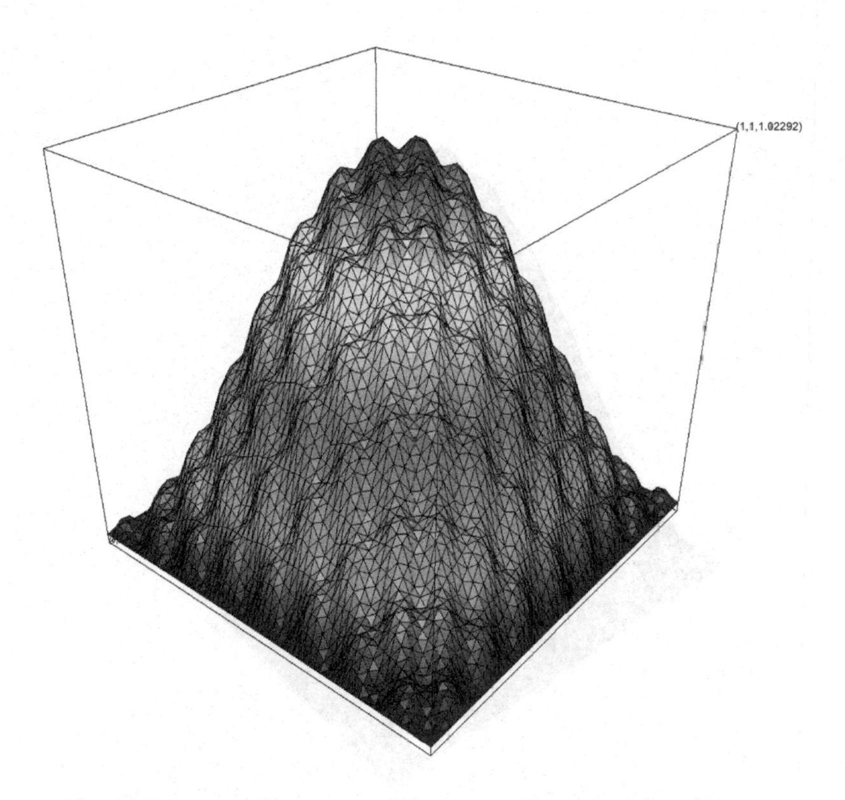

Fig. 1.12. *Result after one step of symmetric Gauss–Seidel smoothing.*

For a current fine-grid iterate $\mathbf{x}_0$, *the (symmetric) two-grid algorithm computes a next fine-grid iterate* $\mathbf{x}_{next}$, *in the following steps.*

(i) "Presmooth", that is, compute $\mathbf{x}_1 = \mathbf{x}_0 + M^{-1}(\mathbf{b} - A\mathbf{x}_0)$.
(ii) "Coarse-grid correction", that is, compute $\mathbf{x}_c$ *from*

$$A_c\mathbf{x}_c = P^T(\mathbf{b} - A\mathbf{x}_1).$$

(iii) "Interpolate" coarse-grid approximation, that is, compute $\mathbf{x}_2 = \mathbf{x}_1 + P\mathbf{x}_c$.
(iv) (An optional) "Postsmoothing" step, that is, compute

$$\mathbf{x}_3 = \mathbf{x}_2 + M^{-T}(\mathbf{b} - A\mathbf{x}_2).$$

(v) The next two-grid iterate is $\mathbf{x}_{next} = \mathbf{x}_3$. □

1.8 Element matrices and matrix orderings

We next introduce the notion of "element matrix". We recall that the matrix A was computed from a bilinear form $a(u, \psi) = \int_\Omega \nabla u \cdot \nabla\psi \, dx$. We can define element

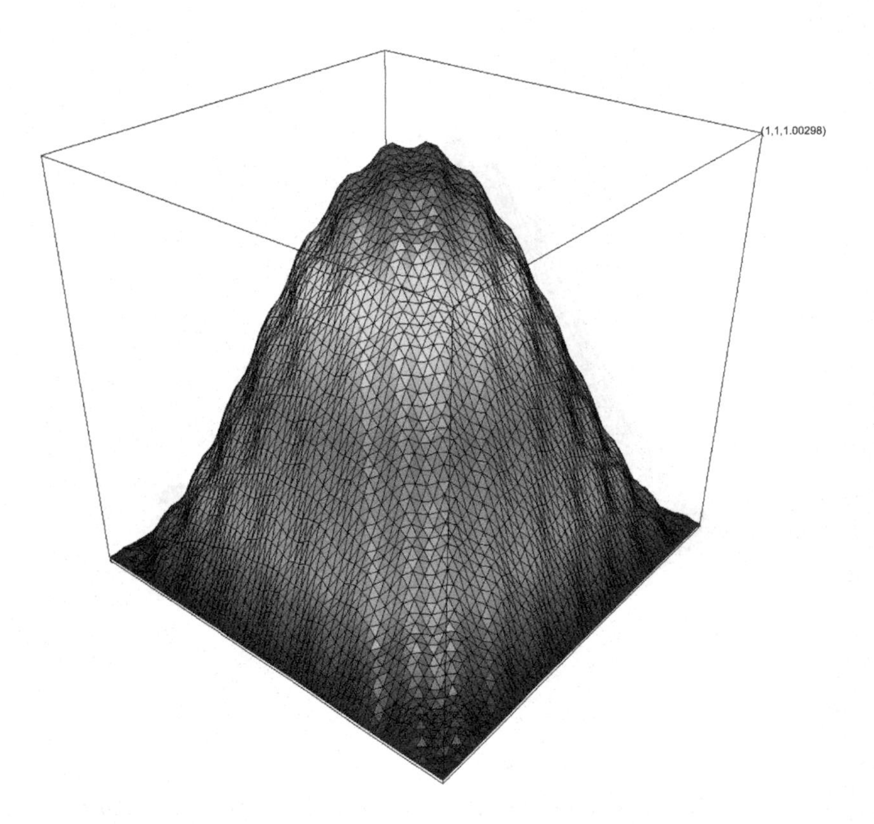

Fig. 1.13. *Result after two steps of symmetric Gauss–Seidel smoothing.*

bilinear forms $a_\tau(\cdot,\ \cdot)$, by restricting the integration over individual elements $\tau \in \mathcal{T}_h$. We then trivially have $a(u,\ \psi) = \sum_{\tau \in \mathcal{T}_h} a_\tau(u,\ \psi)$.

Assembly

Let x_{i_k}, $k = 1, 2, 3$ be the vertices of triangle τ. We can compute the 3×3 matrix $A_\tau = (a_\tau(\psi_{i_l},\ \psi_{i_k}))_{k,l=1}^3$. Define now for any vector $\mathbf{v}$ its restriction $\mathbf{v}_\tau$ to τ; that is, let $\mathbf{v}_\tau = (v(x_{i_k}))_{k=1}^3$. The following identity follows from the definition of element matrices,

$$\mathbf{w}^T A \mathbf{v} = \sum_\tau \mathbf{w}_\tau^T A_\tau \mathbf{v}_\tau.$$

Using this identity for basis vectors $\boldsymbol{\psi}_i$ and $\boldsymbol{\psi}_j$ representing the basis functions ψ_i and ψ_j refers to the popular procedure in the finite element method called *assembly*. The latter means that every nonzero entry $a_{i,j}$ of A is obtained by proper summation of the corresponding entries of the element matrices A_τ (for all elements τ that have

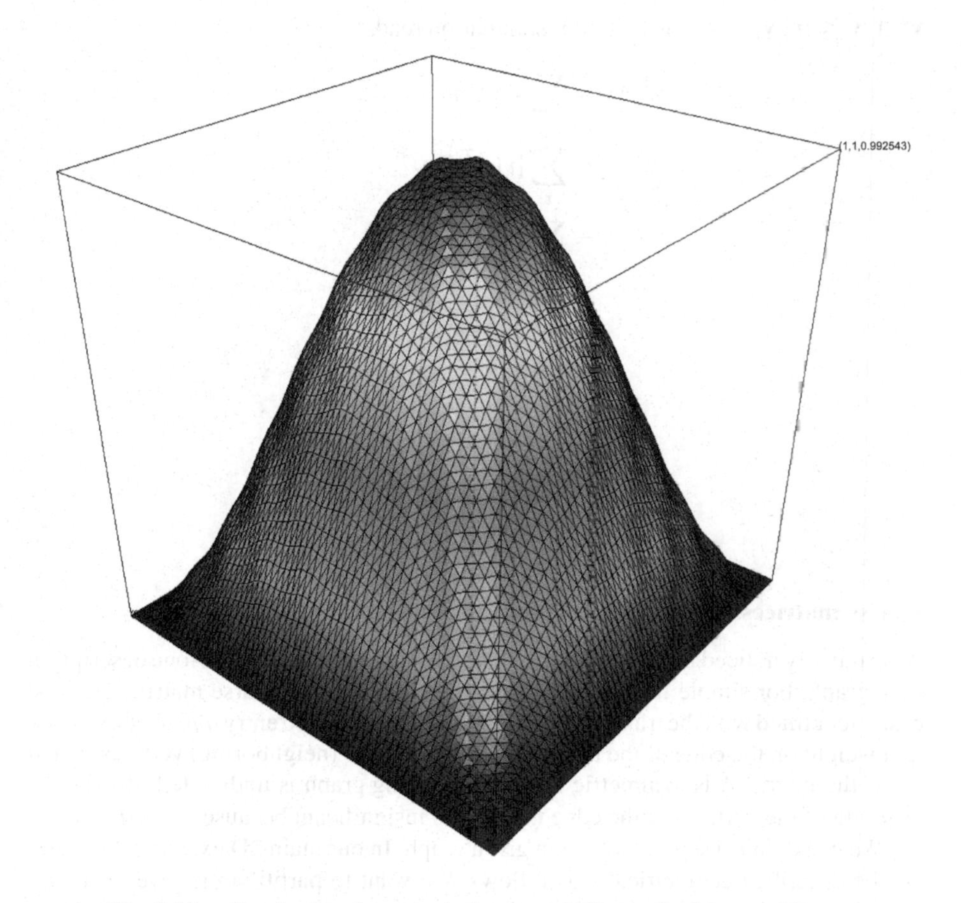

Fig. 1.14. *Result after three steps of symmetric Gauss–Seidel smoothing.*

an edge (x_i, x_j), or vertex if $x_i = x_j$). This in particular implies that the diagonal entries of A_τ, defining D_τ, contribute exactly to the diagonal entries of A; that is, D (the diagonal of A) is assembled from D_τ (the diagonal of the element matrix A_τ).

By applying estimate (1.5) to A_τ and $\mathcal{D}_\tau$, we get the following improved version of (1.5) for A and D.

Proposition 1.10. *The following estimate holds,*

$$\mathbf{v}^T A \mathbf{v} \leq \mathbf{v}^T \chi D \mathbf{v},$$

where either $\chi = \max_i \chi_i$ *or* $\chi = \operatorname{diag}(\chi_i)$, *with* $\chi_i = \max_{\tau \in \mathcal{T}_h:\, i \in \tau} |\tau|$, *where* $|\cdot|$ *stands for cardinality; that is,* $|\tau|$ *equals the number of dofs that belong to* τ. *Thus, in particular, for triangular elements,* $\chi_i = 3$.

Proof. Let $A_\tau = (a_{i,j}^{(\tau)})_{i,j \in \tau}$ and $A = (a_{ij})$. We have $a_{ij} = \sum_{\tau:\, i,j \in \tau} a_{ij}^{(\tau)}$. In particular, $a_{ii} = \sum_{\tau:\, i \in \tau} a_{ii}^{(\tau)}$. From (1.5) applied to A_τ and $\mathcal{D}_\tau$, we get

$\mathbf{v}_\tau^T A_\tau \mathbf{v} \le |\tau|\, \mathbf{v}_\tau^T D_\tau \mathbf{v}_\tau$, which after summation reads,

$$
\begin{aligned}
\mathbf{v}^T A \mathbf{v} &= \sum_\tau \mathbf{v}_\tau^T A_\tau \mathbf{v}_\tau \\
&\le \sum_\tau |\tau|\, \mathbf{v}_\tau^T D \mathbf{v}_\tau \\
&= \sum_\tau |\tau| \sum_{i\in\tau} a_{ii}^{(\tau)}\, v_i^2 \\
&= \sum_i v_i^2 \sum_{\tau:\, i\in\tau} |\tau| a_{ii}^{(\tau)} \\
&\le \sum_i v_i^2 \max_{\tau:\, i\in\tau} |\tau| \sum_{\tau:\, i\in\tau} a_{ii}^{(\tau)} \\
&= \sum_i v_i^2 \chi_i\, a_{ii} \\
&= \mathbf{v}^T \chi D \mathbf{v}.
\end{aligned}
$$

□

Sparse matrices, graphs, separators, and respective block-orderings

As originally noticed by S. Parter [Pa61], a sparse matrix has a one-to-one description by a graph. For simple linear triangular finite elements, the sparse matrix A can be easily identified with the triangular mesh, where each nonzero entry a_{ij} can be assigned as a weight on the edge of the triangle(s) that share the (neighboring) vertices x_i and x_j. If the matrix A is symmetric the corresponding graph is undirected; that is, the ordering of the vertices of the edge $(x_i,\ x_j)$ is insignificant because $a_{ij} = a_{ji}$.

We can define a separator Γ for a given graph. In our main 2D example, the latter can be described geometrically as follows. We want to partition the given geometric domain Ω into two pieces Ω_1 and Ω_2, by drawing a connected path of edges $[x_{i_k},\ x_{i_{k+1}}]$ (the latter connected path defines the separator Γ). The separators are useful for generating special ordering, sometimes referred to as a domain decomposition (or DD) block ordering of the corresponding matrix A. It is easily seen that the entries $a_{r,s}$ of A, for vertices $x_r \in \Omega_1$ and $x_s \in \Omega_2$ satisfy $a_{r,s} = 0$. That is, the following block structure of A (see Figure 1.15) by grouping the vertices first in Ω_1, then in Ω_2 and finally those on Γ, is then very natural;

$$
A = \begin{bmatrix} A_1 & 0 & A_{1,\Gamma} \\ 0 & A_2 & A_{2,\Gamma} \\ A_{\Gamma,1} & A_{\Gamma,2} & A_\Gamma \end{bmatrix}. \tag{1.19}
$$

We notice that the block A_Γ has much a smaller size than the subdomain blocks A_1 and A_2.

Also the blocks $A_{\Gamma,\, i} = A_{i,\, \Gamma}^T = \{a_{j,i_k} :\ j \in \Omega_i,\ i_k \in \Gamma\}$ have nonzero entries only for indices j corresponding to vertices adjacent to Γ.

There is one more block partition of A. We can group the element matrices A_τ into two groups: elements that have vertices in Ω_1 and Γ and elements that have

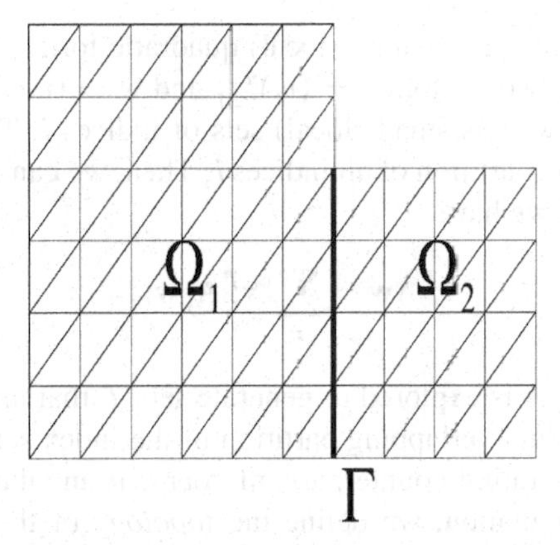

Fig. 1.15. An L-shaped domain partitioned into two squares Ω_1 and Ω_2 by an interface Γ.

vertices from Ω_2 and Γ. Splitting the quadratic form $\mathbf{v}^T A\mathbf{v} = \sum_{\tau\cap\Omega_2\neq\emptyset} \mathbf{v}_\tau^T A_\tau \mathbf{v}_\tau + \sum_{\tau\cap\Omega_1\neq\emptyset} \mathbf{v}_\tau^T A_\tau \mathbf{v}_\tau$, we can get the following block form for A.

$$A = A_1^{(N)} + A_2^{(N)}. \tag{1.20}$$

The superscript N stands for the fact that "natural" (or "no essential") boundary conditions are imposed on Γ. The matrices $A_k^{(N)}$, $k = 1, 2$, have the following block form.

$$A_1^{(N)} = \begin{bmatrix} A_1 & 0 & A_{1,\Gamma} \\ 0 & 0 & 0 \\ A_{\Gamma,1} & 0 & A_{1;\Gamma}^{(N)} \end{bmatrix},$$

and

$$A_2^{(N)} = \begin{bmatrix} 0 & 0 & 0 \\ 0 & A_2 & A_{2,\Gamma} \\ 0 & A_{\Gamma,2} & A_{2;\Gamma}^{(N)} \end{bmatrix}.$$

We notice that A_1 and A_2, and $A_{\Gamma,1}$ and $A_{\Gamma,2}$ are the same as in (1.19). The other important observation is that

$$A_{1;\Gamma}^{(N)} + A_{2;\Gamma}^{(N)} = A_\Gamma.$$

Also, both matrices $A_k^{(N)}$, $k = 1, 2$ are positive semidefinite (because the local matrices A_τ are symmetric positive semidefinite). The latter implies that $A_{k,\Gamma}^{(N)}$ are also symmetric positive semidefinite.

1.9 Element topology

On several occasions throughout the book, we use the fundamental property of the finite element matrices A, namely, that the corresponding quadratic forms $\mathbf{v}^T A\mathbf{w}$

can be represented as a sum of small (local) quadratic forms, resulting from small matrices A_τ. Consider two vectors $\mathbf{v} = (v_i)_{i=1}^n$ and $\mathbf{w} = (w_i)_{i=1}^n$. In what follows, the elements τ are viewed as small (local) sets of indices i. Then the set of all τs provides an overlapping partition of all indices i. Then, we can define $\mathbf{v}_\tau = \mathbf{v}|_\tau$; that is, $\mathbf{v}_\tau = (v_i)_{i \in \tau}$. Then, we have

$$\mathbf{v}^T A \mathbf{w} = \sum_\tau \mathbf{v}_\tau^T A_\tau \mathbf{w}_\tau.$$

The above property of A is explored to generate sets T that are a union of τs such that $\{T\}$ also provides an overlapping partition of the index set $\{1, 2, \ldots, n\}$. Also, for the purpose of generating counterparts of coarse triangulations on the basis of a given fine-grid triangulation, we define the *topology* of the sets T (referred to as agglomerated elements or agglomerates). In particular, we define faces of the agglomerated elements. One application of the "element topology" is to generate a special, so-called *nested dissection* ordering of the given matrix A. Further application of the element matrix topology is to construct element agglomeration algorithms used in element agglomeration AMG (algebraic multigrid) methods in a later chapter.

1.9.1 Main definitions and constructions

By definition, in what follows, an *element* is a list of degrees of freedom (or list of nodes), $e = \{d_1, \ldots, d_{n_e}\}$, and we are given an overlapping partition $\{e\}$ of $\mathcal{D}$ (the set of degrees of freedom or nodes).

In practice, each element e is associated with an element matrix A_e, an $n_e \times n_e$ matrix; then the given sparse matrix A is assembled from the individual element matrices A_e in the usual way. That is,

$$\mathbf{w}^T A \mathbf{v} = \sum_e \mathbf{w}_e^T A_e \mathbf{v}_e.$$

Here, $\mathbf{v}_e = \mathbf{v}|_e$, that is, restricted to subset ($e \subset \mathcal{D}$).

In what follows, we do not assume explicit knowledge of the element matrices A_e; more precisely, the element matrices are needed only in one of the applications but not in the construction of the element topology.

As an illustration, seen in Figure 1.16, we have the following elements as lists (or sets) of nodes.

$$\begin{aligned}
e_1 &= \{1, 2, 6, 7\},\\
e_2 &= \{2, 3, 7, 8\},\\
e_3 &= \{3, 4, 8, 9\},\\
e_4 &= \{4, 5, 9, 10\},\\
e_5 &= \{6, 7, 11, 12\},\\
e_6 &= \{7, 8, 12, 13\},\\
e_7 &= \{8, 9, 13, 14\},
\end{aligned}$$

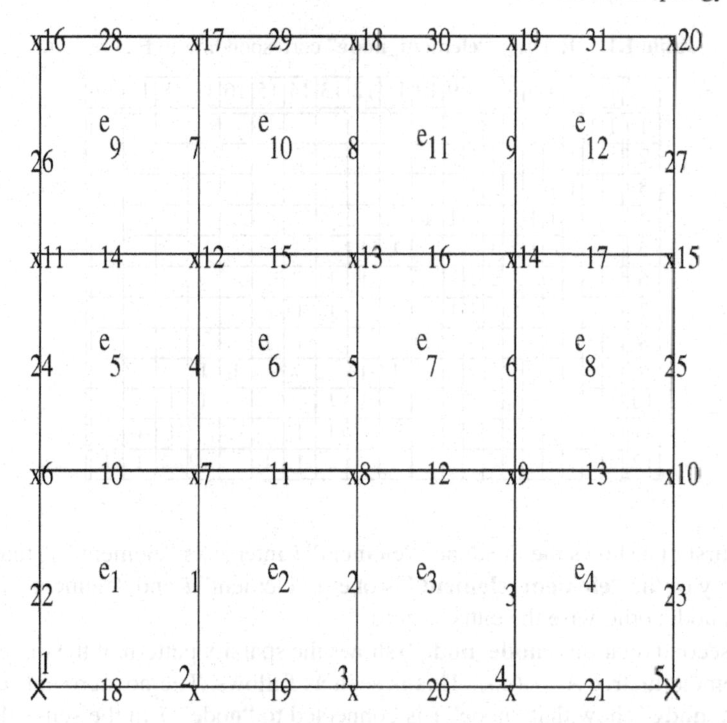

Fig. 1.16. Sample grid: 12 elements, 31 faces, and 20 nodes.

$$
\begin{aligned}
e_8 &= \{9, 10, 14, 15\},\\
e_9 &= \{11, 12, 16, 17\},\\
e_{10} &= \{12, 13, 17, 18\},\\
e_{11} &= \{13, 14, 18, 19\},\\
e_{12} &= \{14, 15, 19, 20\}.
\end{aligned}
$$

Assume that the following relation (in the sense of [Co99]) "**element_node**" is given; that is, the incidence "element" i (rows) contains "node" j (columns), that is, "**element_node**" can be viewed as the rectangular (Boolean) sparse matrix of ones in the (i, j)-position if element i contains node j and zeros elsewhere. The size of the matrix is (number of elements) $\times$ (number of nodes).

The relation "**element_node**" corresponding to Figure 1.16 is shown in Table 1.1. The incidence "node" i belongs to "element" j is simply given by the transpose of the above rectangular sparse matrix; that is, **node_element** = (**element_node**)^T.

We can consider a number of useful relations (easily computable as operations between sparse matrices).

"**element_element**" = "**element_node**" $\times$ "**node_element**",

"**node_node**" = "**node_element**" $\times$ "**element_node**".

Table 1.1. Relation "**element_node**" corresponding to Figure 1.16

	1	2	3	4	5	6	7	8	9	10	11	12	13	14	15	16	17	18	19	20
1	1	1				1	1													
2		1	1				1	1												
3			1	1				1	1											
4				1	1				1	1										
5						1	1				1	1								
6							1	1				1	1							
7								1	1				1	1						
8									1	1				1	1					
9											1	1				1	1			
10												1	1				1	1		
11													1	1				1	1	
12														1	1				1	1

The first one shows the incidence "element" i intersects "element" j; that is, the (i, j) entry of the "**element_element**" is one if "element" i and "element" j have a common node; otherwise the entry is zero.

The second relation ("**node_node**") shows the sparsity pattern of the (assembled) finite element matrix $A = (a_{ij})$. This is seen as follows. The nonzero entries (i, j) of "**node_node**" show that "node" i is connected to "node" j in the sense that they belong to a common element. Hence the corresponding entry $a_{i,j}$ of A is possibly nonzero. This is exactly the case because $a_{i,j}$ can be nonzero only if the nodes i and j belong to the same element. Here, we assume that each node represents a degree of freedom; in other words, it is associated with a finite element basis function whose support is contained in the union of elements sharing that node.

The relation "**node_node**" corresponding to Figure 1.16 is illustrated in Table 1.2.

In practice, we can implement these relations using any available sparse matrix format, such as the popular CSR (compressed sparse row) format (cf., [Sa03]). For parallel implementation, we have to use the appropriate parallel sparse matrix format.

1.9.2 Element faces

In practice, it is typical that a finite element mesh generator can provide the fine-grid element topology, namely the relations

"**element_face**", "**face_element**", "**face_node**", "**face_face**", and so on.

If the initial set of element faces is not given, we can define a "**face**" (as a list of nodes) as a maximal intersection set. Recall that every element is a list (set) of nodes. Consider all pairwise intersections of elements such as $e \cap e_1$, $e_1 \neq e$. Then all maximal sets form the faces of e. Here "maximal" stands for a set that is not a proper subset of any other intersection set. The above definition only gives the set of interior faces. We may assume that additional information about the domain boundary is given in terms of lists of nodes called boundary surfaces. Then, a face is

Table 1.2. Relation "**node_node**" corresponding to Figure 1.16

	1	2	3	4	5	6	7	8	9	10	11	12	13	14	15	16	17	18	19	20
1	1	1				1	1													
2	1	1	1			1	1	1												
3		1	1	1			1	1	1											
4			1	1	1			1	1	1										
5				1	1				1	1										
6	1	1				1	1				1	1								
7	1	1	1			1	1	1			1	1	1							
8		1	1	1			1	1	1			1	1	1						
9			1	1	1			1	1	1			1	1	1					
10				1	1				1	1				1	1					
11						1	1				1	1				1	1			
12						1	1	1			1	1	1			1	1	1		
13							1	1	1			1	1	1			1	1	1	
14								1	1	1			1	1	1			1	1	1
15									1	1				1	1				1	1
16											1	1				1	1			
17											1	1	1			1	1	1		
18												1	1	1			1	1	1	
19													1	1	1			1	1	1
20														1	1				1	1

Table 1.3. Relation "**boundarysurface_node**" corresponding to Figure 1.16. Boundary surface 1 is the left vertical, boundary surface 2 is the bottom horizontal, boundary surface 3 is the right vertical, and boundary surface 4 is the top horizontal

	1	2	3	4	5	6	7	8	9	10	11	12	13	14	15	16	17	18	19	20
1	1					1					1					1				
2	1	1	1	1	1															
3					1					1					1					1
4																1	1	1	1	1

a maximal intersection set of the previous type, or a maximal intersection set of the type $e \cap$ "boundary surface".

In Figure 1.16, we can define four boundary surfaces and can construct the relation "**boundarysurface_node**" shown in Table 1.3.

At any rate, we assume that the faces of the initial set of elements are given either by a mesh generator or they can be computed as the maximal intersection sets. That is, we assume that the relations "**element_face**" and "**face_node**" are given.

We can then construct, based on sparse matrix manipulations, the following relations.

"**face_element**" = ("**element_face**")^T, "**node_face**" = ("**face_node**")^T and "**face_face**" = "**face_node**" × "**node_face**".

1.9.3 Faces of AEs

The purpose of constructing AEs is to define similar topological relations for them and perform further agglomeration steps by recursion. For this reason, we have to be able to define faces of AEs, which we call "AEfaces". Assume that the relation "**AE_element**" has been constructed; then we can build the relation (as a Boolean sparse matrix) "**AE_face**" = "**AE_element**" × "**element_face**". This represents the AEs in terms of the faces of the original elements. The idea is that every two AEs that share a face of the original elements should also share an "AEface". That is, we can define faces of agglomerated elements, "AEfaces", based on "**AE_face**" by simply intersecting the lists (sets) of every two AEs that share a common face, or if the relation "**boundarysurface_face**" is given, by intersecting every AE with a boundary surface if they share a common face of the original elements. By doing so (intersecting two different AEs in terms of faces or intersecting an AE in terms of faces and a boundary surface also in terms of faces), we get the "AEfaces" of the "AE"s in terms of the faces of the original elements. In this way we construct the new relations "**AEface_face**" and "**AE_AEface**". The above definition of the (interior) AEfaces, can be formalized in the following algorithm.

Algorithm 1.9.1 (Creating interior AEfaces) *Given are the relations,*

$$\text{“}\mathbf{AE\_element}\text{”},\ \ \text{“}\mathbf{element\_face}\text{”},$$

implemented as Boolean sparse matrices. In order to produce as an output the new relations

$$\text{“}\mathbf{AEface\_AE}\text{”},\ \textit{and}\ \text{“}\mathbf{AEface\_face}\text{”},$$

we perform the following steps.

- *Form the relations:*
  1.
  $$\text{“}\mathbf{AE\_face}\text{”} = \text{“}\mathbf{AE\_element}\text{”} \times \text{“}\mathbf{element\_face}\text{”};$$
  2.
  $$\text{“}\mathbf{AE\_AE}\text{”} = \text{“}\mathbf{AE\_face}\text{”} \times (\text{“}\mathbf{AE\_face}\text{”})^T.$$
- *Assign an "**AEface**" to each (undirected) pair* $(\mathbf{AE}_1,\ \mathbf{AE}_2)$ *of different AEs from the relation "**AE_AE**". The new relation "**AEface_AE**" is also stored as a Boolean rectangular sparse matrix.*
- *Form the product (including the numerical part of the sparse matrix–matrix multiply):*
  $$\text{“}\mathbf{AEface\_AE\_face}\text{”} \equiv \text{“}\mathbf{AEface\_AE}\text{”} \times \text{“}\mathbf{AE\_face}\text{”}.$$
- *Finally, the required relation*
  $$\text{“}\mathbf{AEface\_face}\text{”}$$
  *is obtained by deleting all entries of "**AEface_AE_face**" with numerical value* 1.

The last step of the above algorithm is motivated as follows. The nonzero entries of the sparse matrix "**AEface_AE_face**" are either 1 or 2 (because a face can belong to at most two AEs). An entry a_{ij} of "**AEface_AE_face**" with value 2 indicates that the "**AEface**" corresponding to the row index "i" of a_{ij} has a face corresponding to the column index "j" with a weight 2. This means that the face "j" is common to the two AEs that define the AEface "i". Therefore the face "j" "belongs" to the AEface "i" (because it is a shared face by the two neighboring AEs which form the AEface "i"). The entries a_{ij} of "**AEface_AE_face**" with value one correspond to a face "j", which is interior to one of the AEs (from the undirected pair of AEs that forms the AEface "i') and hence is of no interest here.

Remark 1.11. If the relation "**boundarysurface_face**" is given we can use it to define the boundary AEfaces. We first form the relation

$$\text{"}\mathbf{AE\_boundarysurface}\text{"} = \text{"}\mathbf{AE\_face}\text{"} \times (\text{"}\mathbf{boundarysurface\_face}\text{"})^T,$$

and then to each AE that is connected to a boundary surface (i.e., to each pair (AE, boundarysurface) from the relation "**AE_boundarysurface**") we assign (a boundary) AEface. Thus the relation "**AE_AEface**" obtained from Algorithm 1.9.1 is augmented with the boundary AEfaces. The list "**AEface_face**" is augmented with the intersection sets

$$(\text{"}\mathbf{AE\_face}\text{"}) \cap (\text{"}\mathbf{boundarysurface\_face}\text{"})$$

for every related pair (AE, boundarysurface) from the relation "**AE_boundarysurface**". This means that we intersect every row of "**AE_face**" with any (related to it) row of "**boundarysurface_face**".

1.9.4 Edges of AEs

We may define edges of AEs. A suitable topological relation for this is the "**AEface_edge**" defined as the product of the relations "**AEface_face**" and "**face_edge**". Thus, we assume that at the fine grid, we have access to the relation faces of elements in terms of the edges of the elements. After we have created the faces of the agglomerates in terms of the faces of the fine-grid elements, we can then generate edges of the agglomerates. The algorithm is based on pairwise intersecting lists for any given AEface F, viewed as a set of fine-grid edges, with its neighboring AEfaces $\overline{F}$, again viewed as sets of fine-grid edges. Any intersection $\mathcal{E}_{F,\,\overline{F}} = F \cap \overline{F}$ is a set of fine-grid edges. The set $\mathcal{E}_{F,\,\overline{F}}$ is a likely candidate for an edge of the agglomerates that share all these edges. The actual definition is as follows.

Definition 1.12 (Definition of AEedges). *For any fine-grid edge e consider the intersection of all AEfaces F (viewed as a set of fine-grid edges) that contain e. It may happen that a fine-grid edge belongs to several such intersection sets. A minimal one defines an AEedge.*

1.9.5 Vertices of AEs

We can further refine the definition of AEedges by splitting the minimal intersection sets into connected components. To do this we need additional information; namely, we need the fine-grid relation "**edge_vertex**". A connected component then is a set of edges that can be ordered into a connected path of 1D fine edges, where two neighboring edges have a single fine-grid vertex in common. Then each connected component of a minimal intersection set from Definition 1.12 is now called a (connected) AEedge. The endpoints (vertices) of the AEedges are referred to as coarse vertices.

1.9.6 Nested dissection ordering

We now adopt a dual notation. First, we consider any given relation "**obj1_obj2**" as a rectangular Boolean sparse matrix, and second, each row of this matrix gives a set of "**obj2**"s; that is, the rows "**obj1**" can be considered as sets consisting of "**obj2**" entries. Hence, we can operate with these rows as sets and in particular we can find their intersection and union. We in particular view a relation "**obj1_obj2**" as the set obtained by the union of its rows.

Assume now that we have generated a sequence of agglomerated elements and their topology. In particular, we need $\{(\text{"}\mathbf{face\_node}\text{"})_k\}$, and $\{(\text{"}\mathbf{AEface\_face}\text{"})_k\}$, $k \geq 0$. (For convenience, we let ("**AEface_face**")$_0$ be the identity Boolean matrix; that is, at the initial fine level $k = 0$ "AEface" equals "face". Similarly, for other purposes, it is also convenient to let ("**AE_element**")$_0$ be the identity relation; that is "AE" equals "element" on the initial level.)

Having the topological information at fine-level $k = 0$, in addition to the nodal information ("**face_node**")$_0$, we first create the topological information recursively; in particular, we create $\{(\text{"}\mathbf{AEface\_face}\text{"})_k\}$, $k \geq 0$. Then, by definition, we set ("**face_node**")$_k$ = ("**AEface_face**")$_k$ $\times$ ("**face_node**")$_{k-1}$ for $k > 0$.

Note that, by construction, ("**face_node**")$_k$ $\subset$ ("**face_node**")$_{k-1}$. This means that each coarse face (i.e., a face at the coarse level k) contains nodes only from the fine-level $k-1$ faces.

Definition 1.13. *The splitting,*

- $\mathcal{S}_0 \equiv \mathcal{D} \setminus (\text{"}\mathbf{face\_node}\text{"})_0$*;*
- *And for* $k > 0$, $\mathcal{S}_k \equiv (\text{"}\mathbf{face\_node}\text{"})_{k-1} \setminus (\text{"}\mathbf{face\_node}\text{"})_k$,

provides a direct decomposition of the original set of nodes $\mathcal{D}$.

In the case of regular refinement (elements of fine-level $k-1$ are obtained by geometrical refinement of coarse-level k elements) the above splitting gives rise to the so-called nested dissection ordering (cf., e.g., Chapter 8 of [GL81]). Thus in a general unstructured grid case, our sparse matrix element topology leads to the following natural extension.

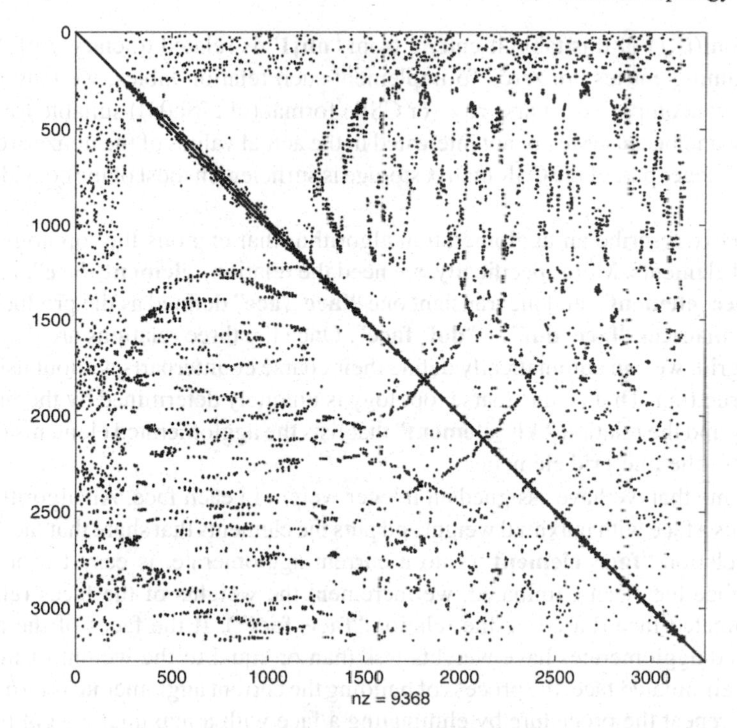

Fig. 1.17. Typical sparsity pattern in the nested dissection ordering.

Definition 1.14 (Nested dissection ordering). *Consider the sets $\mathcal{S}_k$ defined in Definition 1.13. The splitting*

$$\mathcal{D} = \bigcup_{k \geq 0} \mathcal{S}_k, \tag{1.21}$$

gives rise to a block ordering of the assembled sparse matrix A (or of the relation "**node_node**"*) called nested dissection ordering.*

Two examples of a sparsity pattern of the fine-grid assembled matrix in the nested dissection ordering are shown in Figure 1.17.

Nested dissection ordering is useful in direct sparse factorization of A because it tends to minimize the fill-in throughout the factorization (cf. [GL81]). It is also useful in approximate factorization algorithms, due to the same reason.

1.9.7 Element agglomeration algorithms

In what follows, we need some relation tables implemented as Boolean sparse matrices that reflect the topology of the fine-grid elements. Given two sets of objects, "**obj1**" and "**obj2**" indexed from 1 to n_1 and from 1 to n_2, respectively, we construct a Boolean matrix, denoted "**obj1_obj2**". The rows of this matrix represent the entries of "**obj1**" and the columns represent the entries of "**obj2**". We place a nonzero entry at position (i, j) of this table if entry i from "**obj1**" is related to entry j of "**obj2**". All remaining entries are zero. To implement such relation tables, we may use the well-known compressed sparse row (or CSR) format (cf., [Sa03]) suitable for sparse matrices, and because we are not interested in the actual values of the nonzero entries, only the integer part of the CSR matrix storage is sufficient in most of the consideration below.

We next describe an agglomeration algorithm that exploits the topology of the fine-grid elements. More specifically, we need the relation "**element_face**", its transpose "**face_element**", and the transient one "**face_face**" defined as the product of the Boolean matrices "**face_dof**" $\times$ "**dof_face**". Once the three relations are defined on the fine grid, we can automatically define their coarse counterparts without using any dof information. That is, the coarse topology is uniquely determined by the fine-grid topology and the relation "**AE_element**" that lists the agglomerated elements (or AE) in terms of the fine-grid elements.

Assume that we have assigned an integer weight to each face. An algorithm that eliminates a face with maximal weight and puts the elements that share that face (based on the relation "**face_element**") into a current agglomerate, is easy to formulate. After a face has been eliminated, we increment the weights of the faces related to the eliminated face (based on the relation "**face_face**"). If the faces of the already eliminated agglomerates have weights less than or equal to the weight of the most recently eliminated face, the process of building the current agglomerate is terminated. We then repeat the procedure by eliminating a face with a maximal weight (outside the set of faces of already agglomerated elements). To use the algorithm recursively, we have to create the coarse counterparts of the used relations. In particular, we have to define faces of agglomerated elements. Those are easily defined from the relations "**AE_element**" and "**element_face**". We first compute the transient relation "**AE_face**" as the product "**AE_element**" $\times$ "**element_face**". Then, we compute numerically the product "**AE_AE**"= "**AE_face**" $\times$ "**(AE_face)**^{T}". The sparse matrix "**AE_AE**" has nonzero entries equal to one or two. For every entry with a value two, we define an AEface (face of an agglomerated element) by the pair of AEs coming from the row and column indices of the selected entry with value 2. All faces that are shared by the specific pair of AEs (or equivalently by their AEface) define the row of the relation table "**AEface_face**". We define the transient relations, "**AEface_AE**", as the Boolean product "**AEface_face**" $\times$ "**(AE_face)**^{T}". The latter product defines the coarse relation "coarse face"– "coarse element". Finally, computing the triple product "**AEface_face**" $\times$ "**face_face**" $\times$ "**face_AEface**" defines the coarse relation "coarse face"– "coarse face". Thus, the three coarse counterparts of the needed relation to apply the agglomeration recursively have been defined.

We can define some more sophisticated agglomeration algorithms by labeling some faces as unacceptable to eliminate, thus preventing some elements from being agglomerated. In this way, we can generate coarse elements that get coarsened away from certain domains, boundaries, or any given set of given topological entities (i.e., faces of elements).

Other agglomeration algorithms are also possible, for example, based on graph-partitioners. We can use for this purpose the transient relation "**element_element**"

Fig. 1.18. *Initial fine-grid unstructured mesh.*

defined at the initial fine-grid level as the product of the Boolean matrices "**element_face**" × "**face_element**". Once an agglomeration step has been performed (i.e., the relation "**AE_element**" constructed), we define the coarse relation "coarse element"_"coarse element" by"**AE_AE**" which equals the triple product:

"**AE_element**" × "**element_element**" × "**(AE_element)**^{T}".

Another approach is taken in [Wab03], where a bisection algorithm is recursively applied as follows. First, partition the set of elements into two groups. Then each newly created set of elements is further partitioned into two subgroups and so on. At the end, we have ℓ levels of partitioned element sets, which serve as agglomerates in a multilevel hierarchy.

In Figure 1.18, we show a model unstructured mesh, and in Figures 1.19–1.21, one, two, and three levels of agglomerated meshes are shown.

1.10 Finite element matrices on many processors

With the current development of parallel computers having many (sometimes thousands) of processors, the actual generation of the finite element problem (matrix and

Fig. 1.19. *One level of agglomerated coarse elements.*

r.h.s.) may become a nontrivial task. The common practice to generate the problem on a single processor and after proper partitioning to distribute it over the remaining processors has limited applicability due to memory constraints (that single processor has limited memory).

A feasible approach is to derive the pieces of the global problem in parallel, one piece per processor. Thus, we end up with matrices $A_p^{(N)}$ similar to the decomposition (1.20) (with $p = 2$). We need a Boolean mapping P that identifies dofs on a given processor p with their copies in the neighboring processors q. These multiple copies of dofs are identified with a single (master) one, which is sometimes called truedof. Then, the actual matrix A corresponding to the truedofs only is obtained by performing the triple-matrix product

$$P^T \begin{bmatrix} A_1^{(N)} & 0 & \dots & 0 \\ 0 & A_2^{(N)} & 0 & \\ & & \ddots & \\ 0 & \dots & 0 & A_p^{(N)} \end{bmatrix} P.$$

The matrix P^T has for every row (corresponding to a truedof) a number of unit entries identifying it with its copy in the neighboring processors. The latter procedure

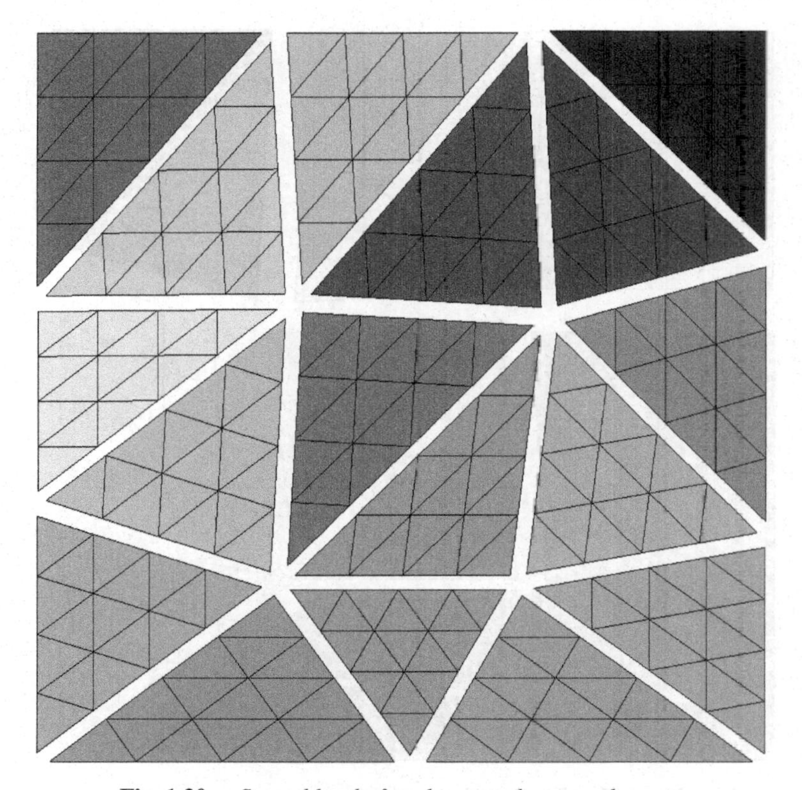

Fig. 1.20. *Second level of agglomerated coarse elements.*

represents the so-called parallel assembly of the global matrix. Note that we can use different matrices P (not necessarily the above Boolean one) to specify how the remaining degrees of freedom are defined from a set of master dofs (or truedofs).

The procedure for assembling the problem r.h.s. $\mathbf{b}$ is similar. We have locally computed r.h.s. $\mathbf{b}_p$; then the global one equals

$$P^T \begin{bmatrix} \mathbf{b}_1 \\ \mathbf{b}_2 \\ \vdots \\ \mathbf{b}_p \end{bmatrix}.$$

1.11 The mortar method

A more sophisticated way to generate the finite element problem in parallel that, in general, can also utilize nonmatching meshes, is based on the so-called mortar method (cf., [Wo00]). Here, we handle two neighboring processors i and j at a time. Let $\Gamma = \Gamma_{ij}$ be the geometrical interface between the subdomains Ω_1 and Ω_2 triangulated by elements $\{\tau\}$ on processor i and elements $\{\tilde{\tau}\}$ on processor j. There

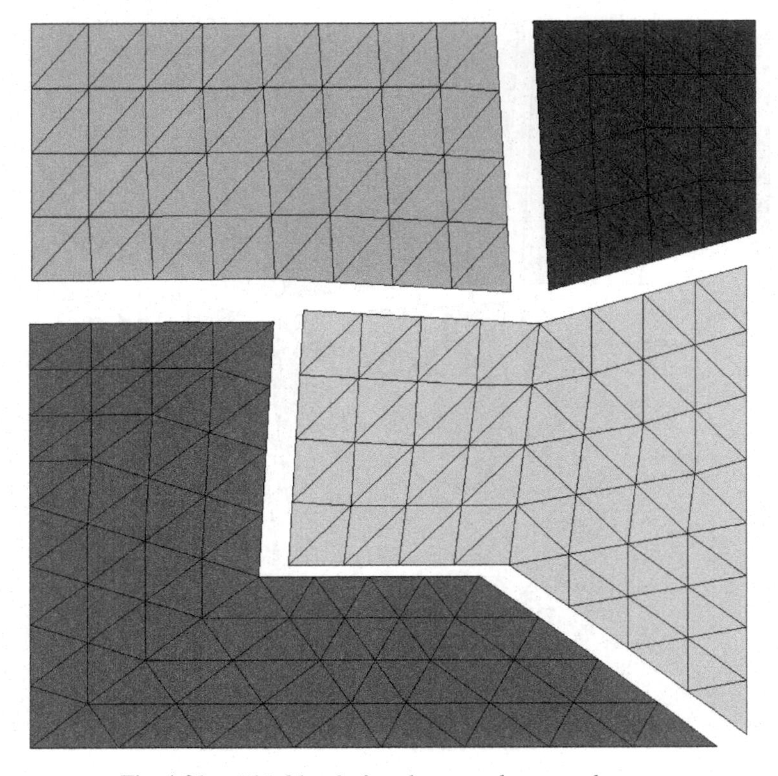

Fig. 1.21. *Third level of agglomerated coarse elements.*

are degrees of freedom associated with τ elements and degrees of freedom associated with $\tilde{\tau}$ elements. The dofs on one side of Γ are chosen as "master" ones, whereas dofs on the other side of Γ are "slave". In the mortar method, the degrees of freedom from both sides of Γ that happen to belong to other interfaces $\Gamma_{r,s}$ are considered to be master dofs. That is, even on a slave side of Γ, the dofs on $\partial\Gamma$ (i.e., belonging to other interfaces as well) are considered as master dofs. Next, we select the dofs in the interior to $\Gamma = \Gamma \setminus \partial\Gamma$ of one of its sides (say, on processor i, assuming $i < j$) to be slave. Then, we have to define a mapping P_Γ that interpolates at these slave dofs from the master (mortar) side (j) and the boundary dofs (on $\partial\Gamma$) from the same side i so that

$$\mathbf{v}_{i;\, \Gamma \setminus \partial\Gamma} = P_\Gamma \begin{bmatrix} \mathbf{v}_{i;\, \partial\Gamma} \\ \mathbf{v}_{j;\, \Gamma} \end{bmatrix}. \tag{1.22}$$

The global problem is formulated as a constrained minimization one as follows. Find $\{\mathbf{v}_i\}$ that solves

$$\sum_i \left[\frac{1}{2} \mathbf{v}_i^T A_i^{(N)} \mathbf{v}_i - \mathbf{v}_i^T \mathbf{b}_i \right] \mapsto \min, \tag{1.23}$$

subject to the constraints (1.22).

That is, we have a quadratic form, a sum of subdomain quadratic forms $a_i(v_i, v_i) - \int_{\Omega_i} v_i f_i \, dx = \frac{1}{2}\mathbf{v}_i^T A_i^{(N)} \mathbf{v}_i - \mathbf{v}_i^T \mathbf{b}_i$ (as in (1.23)), and we impose continuity to define the slave variables in terms of the master ones. In the present finite element setting, the constraints are imposed on every interface $\Gamma = \Gamma_{ij}$ via another quadratic form $(\cdot, \cdot)_{0,\Gamma}$, which is assembled from local (small) quadratic forms. In practice, $(\cdot, \cdot)_{0,\Gamma}$ corresponds to an integral L_2-inner product on any interface boundary Γ between two subdomains.

1.11.1 Algebraic construction of mortar spaces

In this section, we present an algebraic element-based construction of dual mortar multiplier spaces (originated in [KPVb]). This construction generalizes the dual basis approach from [KLPV, Wo00, Wo00] to any type of meshes and (Lagrangian) finite element spaces that are generated independently on each subdomain.

Note that any interface Γ_{ij} is the union of faces from the mortar or the nonmortar side. Thus, we can define finite element spaces on both sides of the interface by taking the trace of the corresponding subdomain finite element spaces V_i and V_j.

A node on Γ_{ij} is called a *boundary* if it also belongs to another interface. The nodes on Γ_{ij} that are not boundaries are called *interior* (to the interface). A face on Γ_{ij} is called an *interior* if it does not contain any boundary nodes, it is called a *boundary* face if it contains interior and boundary nodes, and finally, it is called a *corner* face if it does not contain any interior nodes. This is illustrated in Figure 1.22.

For any given interface Γ_{ij}, the space M^0_{ij} is defined to be the set of functions on the nonmortar (slave) mesh on Γ_{ij} that vanish on the boundary nodes of Γ_{ij}. Let T be a nonmortar face of Γ_{ij} and define $M_{ij}(T)$ to be the restriction of the nonmortar functions to T. We define $\widetilde{M}_{ij} = \oplus M_{ij}(T)$ where the sum is taken over all faces of the nonmortar mesh on Γ_{ij}. The space M_{ij} is a subset of $\widetilde{M}_{ij}$.

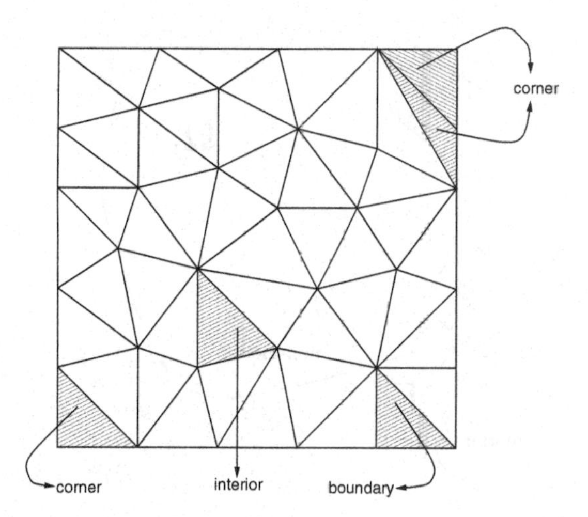

Fig. 1.22. Different types of faces on a nonmortar interface.

The dual basis functions in the finite element case have two important properties:

1. The dual basis functions are constructed locally.
2. The dual basis functions can reproduce constants locally.

These properties are fundamental in the analysis of the approximation properties of the mortar method used as the discretization method (for details cf., [Wo00] or [KLPV]).

To be specific, we consider 3D subdomains Ω_i that are polytopes, triangulated by tetrahedral meshes. Then each interface Γ_{ij} between two subdomains Ω_i and Ω_j is triangulated by two sets of triangular meshes (the restriction of the tetrahedral meshes to Γ_{ij}). The mortar constraints are imposed on the basis of the $L_2(\Gamma_{ij})$ form, denoted by $(.,.)$ and by $(.,.)_T$ if the integration is restricted to an element T. Assuming we have constructed a mortar f.e. space M_{ij}, we impose "weak" continuity conditions as follows,

$$([v],\ \mu_l) = (v_i - v_j,\ \mu_l) = 0.$$

Here, $\mu_l \in M_{ij}$ runs over a basis of M_{ij} and $[v] = v_i - v_j$ stands for the jump of v on Γ_{ij}. In what follows, we also use the notation $v_m = v_i$ for the mortar side of Γ_{ij} and $v_{nm} = v_j$ for the nonmortar side of Γ_{ij}.

Also, let θ_l define a Lagrangian basis on the nonmortar side of Γ_{ij} and similarly let $\{\tilde{\theta}_k\}$ define a basis on the mortar (master) side of Γ_{ij}.

This relation forces the interior nodes on the nonmortar interface to be slaves of those on the boundary and those on the mortar side. This is illustrated in Figure 1.23.

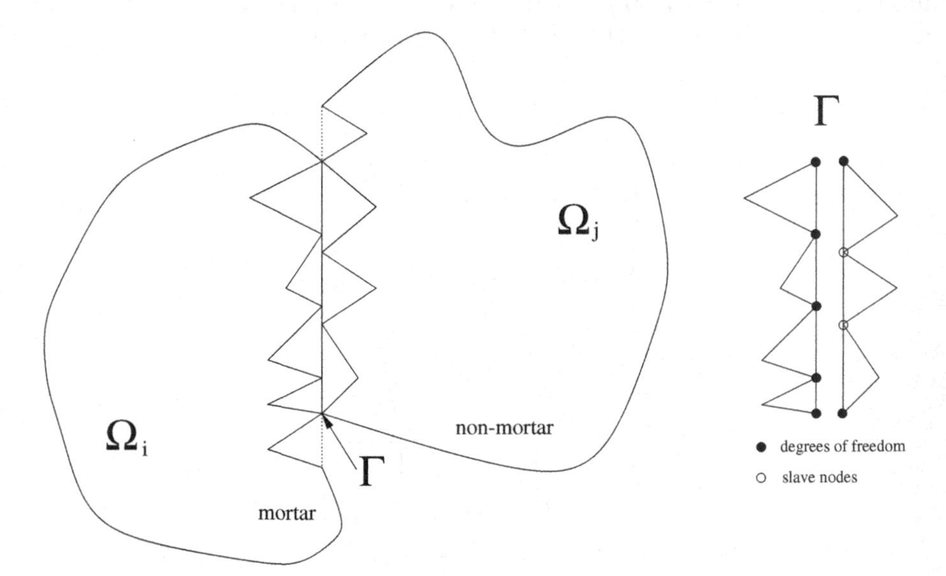

Fig. 1.23. Mortar interface and degrees of freedom.

A local biorthogonal basis

There is a unique function $\hat{\mu}_l \in M_{ij}(T)$ satisfying

$$(\hat{\mu}_l,\ \theta_k)_T = \delta_{ij} = \begin{cases} 1, & l = k, \\ 0, & l \neq k. \end{cases}$$

Here $\{\theta_k\}$ are the basis functions for $M_{ij}(T)$ (these basis functions are restrictions of the basis functions of the nonmortar subdomain to T). In fact,

$$\hat{\mu}_l = \sum_k c_{lk}\theta_k$$

where the coefficients $\underline{c}_l = (c_{lk})$ solve the system

$$G_T \underline{c}_l = \mathbf{e}_l = (\delta_{lk})$$

where $G_T = ((\theta_k,\ \theta_l)_T)$ denotes the local mass matrix for the element T. This system has a unique solution because G_T is invertible.

Using the biorthogonality property $(\hat{\mu}_l,\ \theta_k)_T = \delta_{lk}$, it follows that

$$\alpha_l \equiv \alpha_l^{(T)} = (1,\ \theta_l)_T$$

satisfies

$$\sum_l \alpha_l \hat{\mu}_l = 1 \quad \text{on } T.$$

The biorthogonal mortar basis

The mortar multiplier basis functions $\{\mu_l\}$ are defined only for nodes x_l that are interior to Γ_{ij}. We first assign corner faces T to nearby interior vertices; For example, T is assigned to the nearest interior vertex. For each interior node x_l and face T, we define μ_l on T as follows.

1. If T is a corner face assigned to x_l then $\mu_l = 1$ on T.
2. If T is a face that does not contain x_l (excluding the case of (1) above) then $\mu_l = 0$ on T.
3. If T is a boundary face containing x_l then

$$\mu_l = \alpha_l \hat{\mu}_l + m^{-1} \sum_{k:\ x_k \in \partial\Gamma \cap T} \alpha_k \hat{\mu}_k \quad \text{on } T$$

   where m is the number of interior nodes in T.
4. If T is an interior face containing x_l then $\mu_l = \alpha_l \hat{\mu}_l$ on T.

We then have

$$1 = \sum \mu_l \quad \text{on } T,$$

where the sum is taken over l such that $\mu_l \neq 0$ on T.

The dual basis mortar formulation defines M_{ij} to be a subspace of discontinuous piecewise linear functions on Γ_{ij} (with respect to the nonmortar mesh) that are generated by a dual basis, $\{\chi_l\}$, $l = 1, \ldots, n_{ij}$ satisfying

$$\int_{\Gamma_{ij}} \theta_l \chi_k \, ds = \begin{cases} 1 & \text{if } l = k, \\ 0 & \text{otherwise.} \end{cases} \tag{1.24}$$

Here $\{\theta_l\}$, $l = 1, \ldots, n_{ij}$ is the usual nodal finite element basis for the space of functions M^0_{ij} which are piecewise linear (with respect to the nonmortar mesh) and vanish on $\partial\Gamma_{ij}$.

The construction of the dual basis functions only requires the use of the local mass matrices G_T and local geometric information, such as relations between nodes, edges, and triangles, and whether a node and a face are on the boundary of an interface.

Implementation of the mortar interpolation

The condition (1.24) implies that the nodal value of a finite element function v on the interior node x_l (corresponding to the basis function θ_l) is given by

$$c_l = \int_{\Gamma_{ij}} v_m(x)\chi_l \, dx - \int_{\Gamma_{ij}} v_{nm,0}(x)\chi_l \, dx \tag{1.25}$$

where $v_m(x)$ denotes the trace of v to Γ_{ij} from the mortar subdomain and $v_{nm,0}$ denotes v on the nonmortar side, cut down to zero on the interior nodes. Equation (1.25) defines a row of the desired mortar interpolation (cf., (1.22)). The computation of the right-hand side of (1.25) requires information about how the elements on the subdomain are connected to those on the boundary, as well as the geometric relation between the triangles (faces) on the mortar and those on the nonmortar side and an "interaction mass matrix"

$$\mathbf{M}_{rksm} = \int_{\tau_r \cap \tilde{\tau}_s} \theta_r^k \tilde{\theta}_s^m \, dx, \qquad k, \ m = 1, 2, 3. \tag{1.26}$$

Here τ_r and $\tilde{\tau}_s$ are triangles on Γ_{ij} on the mortar and nonmortar side, respectively, and θ_r^k and $\tilde{\theta}_s^m$ run over the nodal basis functions on τ_r and $\tilde{\tau}_s$. In our particular case, there are three linear basis functions θ_r^k per triangle τ_r and also three basis functions $\tilde{\theta}_s^m$ per triangle $\tilde{\tau}_s$.

The space of mortar multipliers M_{ij} is defined to be the span of $\{\mu_l\}$. Note that the dimension of M_{ij} equals the number of interior nodes x_l on Γ. The dual basis functions $\{\chi_l\}$ satisfying (1.24) are obtained from $\{\mu_l\}$ by obvious scaling.

Note that in general, $\{\mu_l\}$ are discontinuous across the element boundaries. Two examples with piecewise linear finite elements and a nonmortar interface with uniform triangulation in one and two dimensions are presented in Figures 1.24 and 1.25.

Note that the construction here is quite general in that it extends to any element-by-element defined interface functions as long as the element mass matrices are available. Thus, it extends to the types of interface functions resulting from element agglomeration-based (AMGe) procedures described in Section 1.9. For more details, cf. [KPVb].

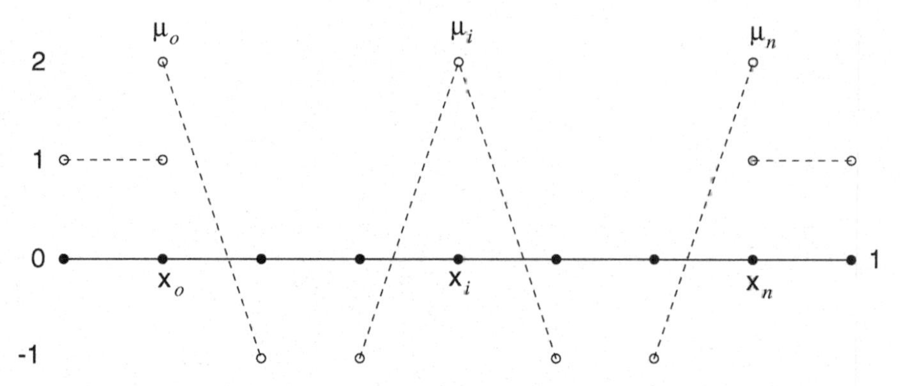

Fig. 1.24. Mortar basis functions for one-dimensional nonmortar interface discretized with a uniform grid.

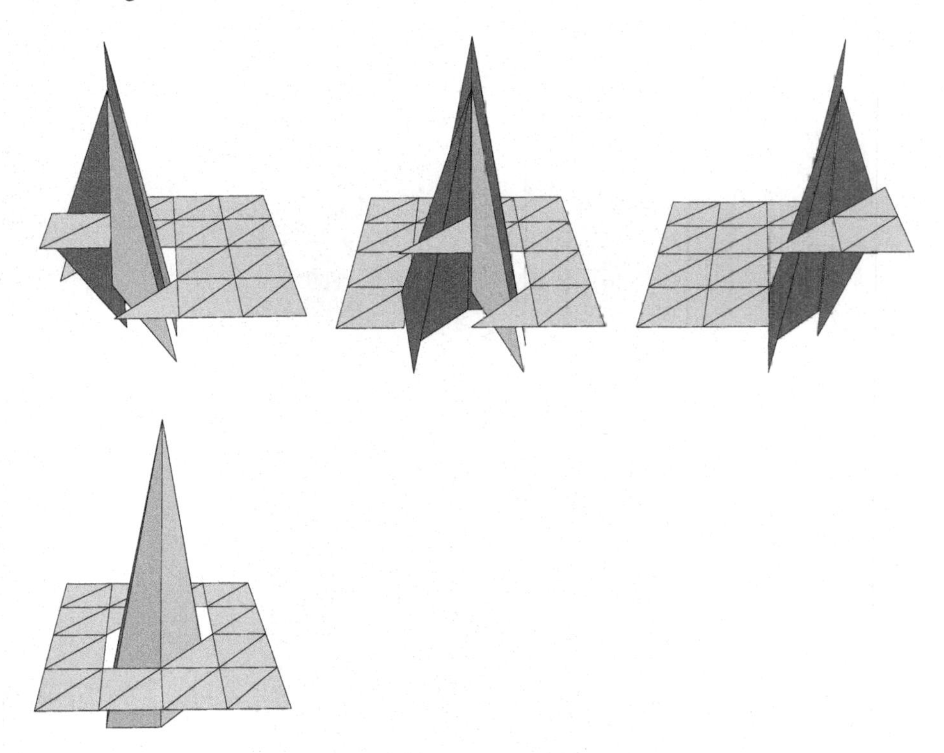

Fig. 1.25. Mortar basis functions for two-dimensional nonmortar interface discretized with a uniform grid.

In conclusion, because the dual basis for M_{ij} and the basis for M_{ij}^0 are related by (1.24), the values of the slave nodes on the interior of the nonmortar interface are given by (1.25). The implementation of this requires the corresponding interaction matrices (1.26).

2

A Main Goal

Given a sparse matrix, for example, symmetric positive definite (or s.p.d. for short) $A = (a_{i,j})_{i,j=1}^n$, our main goal is to devise efficient algorithms for solving the system of linear equations,

$$A\mathbf{x} = \mathbf{b}.$$

In practice, n can be very large (e.g., in the range of millions). A first comment then is that in a massively parallel environment direct solvers are out of the question due to their prohibitive memory requirements.

The fact that A is sparse means that an inexpensive operation is "matrix times vector." Therefore, iterative methods for solving $A\mathbf{x} = \mathbf{b}$ are appealing because they involve computing at every iteration the residual

$$\mathbf{r} = \mathbf{b} - A\mathbf{x},$$

corresponding to a current iterate $\mathbf{x}$. The next iterate is obtained by computing a correction based on $\mathbf{r}$. A stationary iterative method exploits a mapping (sometimes explicitly represented by a matrix) B, which has an easily computable inverse action B^{-1} (in some cases implicitly represented only by a procedure). Then, the new iterate $\mathbf{x}_{\text{new}}$ equals,

$$\mathbf{x}_{\text{new}} = \mathbf{x} + B^{-1}\mathbf{r}.$$

The corresponding new residual $\mathbf{r}_{\text{new}} = \mathbf{b} - A\mathbf{x}_{\text{new}}$ equals

$$\mathbf{r}_{\text{new}} = (I - B^{-1}A)\mathbf{r}.$$

By making the successive residuals $\mathbf{r} \mapsto 0$ (in some norm), we get more accurate approximations $\mathbf{x}$ to the exact solution $A^{-1}\mathbf{b}$.

If B is s.p.d. (symmetric positive definite), the method of choice is the (preconditioned) CG, which initially computes $\mathbf{p} = B^{-1}\mathbf{r}$ and at every successive step updates it based on the preconditioned residual $\overline{\mathbf{r}} = B^{-1}\mathbf{r}$ as $\mathbf{p} := B^{-1}\mathbf{r} + \beta\,\mathbf{p}$ for a proper scalar β. At any rate, major operations here are again (as in a stationary iteration) the

P.S. Vassilevski, *Multilevel Block Factorization Preconditioners*,
doi: 10.1007/978-0-387-71564-3_2,

actions of B^{-1} and matrix vector products with A in addition to some vector inner products. The actual preconditioned CG iteration takes the form:

(0) Initiate: Given a tolerance ϵ and an integer $n_{\text{iter}}^{\max}$ that gives the maximal number of iterations allowed, for an initial iterate $\mathbf{x}$, which we choose $\mathbf{x} = B^{-1}\mathbf{b}$, we
1. Compute $\delta_0 = \mathbf{x}^T\mathbf{b}$.
2. Compute $\mathbf{r} = \mathbf{b} - A\mathbf{x}$,
3. Compute $\overline{\mathbf{r}} = B^{-1}\mathbf{r}$,
4. Let $\mathbf{p} = \overline{\mathbf{r}}$.
5. Compute $\delta = \overline{\mathbf{r}}^T\mathbf{r}$.
6. Test for convergence, if $\delta \le \epsilon^2\,\delta_0$ stop.
7. Set $n_{\text{iter}} = 0$;

(i) Loop: until convergence or max number of iterations reached
1. Compute $\mathbf{h} = A\mathbf{p}$.
2. Compute $\alpha = \delta/\mathbf{p}^T\mathbf{h}$.
3. Compute $\mathbf{x} := \mathbf{x} + \alpha\mathbf{p}$.
4. Compute $\mathbf{r} := \mathbf{r} - \alpha\mathbf{h}$.
5. Compute $\overline{\mathbf{r}} = B^{-1}\mathbf{r}$.
6. Set $\delta_{\text{old}} = \delta$,
7. Compute $\delta = \overline{\mathbf{r}}^T\mathbf{r}$.
8. Set $n_{\text{iter}} := n_{\text{iter}} + 1$;
9. Check for convergence: if either $\delta \le \epsilon^2\,\delta_{\text{old}}$ or $n_{\text{iter}} > n_{\text{iter}}^{\max}$, stop.
10. Compute $\beta = \delta/\delta_{\text{old}}$.
11. Compute $\mathbf{p} := \overline{\mathbf{r}} + \beta\,\mathbf{p}$ and go to (i).

We have the following popular convergence rate result after n_{iter} steps,

$$\delta \le \kappa \left(\frac{2q^{n_{\text{iter}}}}{1 + q^{2n_{\text{iter}}}}\right)^2 \delta_0 \le 4\kappa\, q^{2n_{\text{iter}}}\,\delta_0,$$

where $q = \sqrt{\kappa} - 1/\sqrt{\kappa} + 1$ and $\kappa = \text{Cond}(B^{-1}A)$.

In practice, we typically prove (spectral equivalence) estimates

$$c_1\,\mathbf{v}^T A\mathbf{v} \le \mathbf{v}^T B\mathbf{v} \le c_2\,\mathbf{v}^T A\mathbf{v} \quad \text{for all } \mathbf{v}.$$

Then, because the eigenvalues of $B^{-1}A$ are in $[1/c_2,\ 1/c_1]$, we clearly have $\kappa \le c_2/c_1$.

A simple candidate for an iteration matrix B is based on the diagonal D of A. For example, we can use the diagonal matrix χD, where χ is either a diagonal matrix with entries $\chi_i = \sum_{j:\,a_{i,j}\neq 0} 1$ or just the scalar $\max_i \chi_i$. We have

$$c_1\,\mathbf{v}^T A\mathbf{v} \le \mathbf{v}^T B\mathbf{v} \le c_2\,\mathbf{v}^T A\mathbf{v}$$

with $c_1 = 1$. Unfortunately, for f.e. matrices A, such as the discretized Poisson equation, the estimate from below is mesh-dependent; that is, we typically have

$$c_2^{-1} = \min_{\mathbf{v}} \frac{\mathbf{v}^T A\mathbf{v}}{\mathbf{v}^T\chi D\mathbf{v}} \simeq h^2 \mapsto 0.$$

Hence, Cond $(B^{-1}A) \le c_2/c_1 \simeq h^{-2} \mapsto \infty$ with $h \mapsto 0$.

The goal then is to construct a B such that:

(i) The action of B^{-1} costs as little as possible, the best being $\mathcal{O}(n)$ flops.
(ii) The constants c_1 and c_2 in the spectral equivalence estimate are such that c_2/c_1 is as close as possible to one, for example, being independent of various problem parameters, in particular being independent of n.
(iii)* In a massively parallel computer environment, we also want B^{-1} to be composed of local actions, essentially based on a "hierarchy" of sparse matrix vector products. The latter is achieved by the multilevel preconditioners that are a main topic of the present book.

Based on the convergence estimate, it is clear then that to get an approximate solution to $A\mathbf{x} = \mathbf{b}$ within tolerance ϵ, it is sufficient to perform as many as n_{iter} iterations such that $q^{n_{\text{iter}}} < \frac{1}{2}\kappa^{-1}\ \epsilon$ or $n_{\text{iter}} = \mathcal{O}(\log 1/\epsilon)$. The constant in the $\mathcal{O}$ symbol is reasonable as long as κ is kept under control (not too far away from one).

For large sparse matrices A that come from finite element discretization of elliptic PDEs (partial differential equations), like the Poisson equation, we can achieve both (i) and (ii) (and to a certain extent (iii)*) based on the multilevel preconditioning methods that are the main topic of the present book. Such methods are often referred to as *scalable* iterative methods.

Part II

Block Factorization Preconditioners

3

Two-by-Two Block Matrices and Their Factorization

The topics that are covered in this chapter are as follows. We first study some fundamental properties of block matrices and their Schur complements. We next consider a popular product iteration method, and then the concept of approximate block-factorization is introduced. A main relation between a familiar product iteration algorithm and a basic block-factorization preconditioner is then established. This relation is a cornerstone in proving the spectral equivalence estimates. Next, a sharp spectral equivalence result is proved in a general setting. It provides necessary and sufficient conditions in an abstract form for a preconditioner to be spectrally equivalent to the given matrix. Then two major examples, a two-level and a two-grid preconditioner, are considered and analyzed. Finally the more classical two-by-two (two-level) block-factorization preconditioners are introduced and analyzed. The chapter concludes with a procedure to generate a stable block form of matrices and with an analysis of a respective block-factorization preconditioner.

3.1 Matrices of two-by-two block form

3.1.1 Exact block-factorization. Schur complements

In this section, we consider general s.p.d. matrices of the form

$$A = \begin{bmatrix} \mathcal{A} & \mathcal{R} \\ \mathcal{L} & \mathcal{B} \end{bmatrix}.$$

Assuming that $\mathcal{A}$ is square and invertible, introduce the Schur complement

$$\mathcal{S} = \mathcal{B} - \mathcal{L}\mathcal{A}^{-1}\mathcal{R}.$$

A standard block-factorization of A reads

$$A = \begin{bmatrix} I & 0 \\ \mathcal{L}\mathcal{A}^{-1} & I \end{bmatrix} \begin{bmatrix} \mathcal{A} & 0 \\ 0 & \mathcal{S} \end{bmatrix} \begin{bmatrix} I & \mathcal{A}^{-1}\mathcal{R} \\ 0 & I \end{bmatrix}.$$

P.S. Vassilevski, *Multilevel Block Factorization Preconditioners*,
doi: 10.1007/978-0-387-71564-3_3,

This shows that $\mathcal{S}$ is invertible if A is invertible. Moreover $\mathcal{S}$ is s.p.d. if A is s.p.d. A straightforward computation shows the identity

$$A^{-1} = \begin{bmatrix} I & -\mathcal{A}^{-1}\mathcal{R} \\ 0 & I \end{bmatrix} \begin{bmatrix} \mathcal{A}^{-1} & 0 \\ 0 & \mathcal{S}^{-1} \end{bmatrix} \begin{bmatrix} I & 0 \\ -\mathcal{L}\mathcal{A}^{-1} & I \end{bmatrix}$$

$$= \begin{bmatrix} \mathcal{A}^{-1} + \mathcal{A}^{-1}\mathcal{R}\mathcal{S}^{-1}\mathcal{L}\mathcal{A}^{-1} & -\mathcal{A}^{-1}\mathcal{R}\mathcal{S}^{-1} \\ -\mathcal{S}^{-1}\mathcal{L}\mathcal{A}^{-1} & \mathcal{S}^{-1} \end{bmatrix}.$$

This, in particular, implies the identities for the blocks of

$$A^{-1} = \begin{bmatrix} \mathcal{V} & \mathcal{F} \\ \mathcal{G} & \mathcal{W} \end{bmatrix}.$$

We have $\mathcal{W} = \mathcal{S}^{-1}$, $\mathcal{F} = -\mathcal{A}^{-1}\mathcal{R}\mathcal{S}^{-1}$, $\mathcal{G} = -\mathcal{S}^{-1}\mathcal{L}\mathcal{A}^{-1}$, and $\mathcal{A}^{-1} = \mathcal{V} - \mathcal{F}\mathcal{S}\mathcal{G}$. We may introduce the following Schur complement of A,

$$\mathcal{S}' = \mathcal{A} - \mathcal{R}\mathcal{B}^{-1}\mathcal{L}.$$

It is clear then that $\mathcal{V} = \left(\mathcal{S}'\right)^{-1}$. That is, $\mathcal{A}^{-1}$ and $\mathcal{B}^{-1}$ are Schur complements of A^{-1}. Alternatively, we may say the following.

Proposition 3.1. *The inverse of a principal submatrix of a given matrix is a respective Schur complement of the inverse of the given matrix.*

Another form of the explicit inverse of A is as follows.

$$A^{-1} = \begin{bmatrix} \mathcal{A}^{-1} & 0 \\ 0 & 0 \end{bmatrix} + \begin{bmatrix} -\mathcal{A}^{-1}\mathcal{R} \\ I \end{bmatrix} \mathcal{S}^{-1} [-\mathcal{L}\mathcal{A}^{-1},\ I].$$

The latter shows that A^{-1} can be computed as a "low–rank" update to

$$\begin{bmatrix} \mathcal{A}^{-1} & 0 \\ 0 & 0 \end{bmatrix}.$$

Theorem 3.2. *The Schur complement $\mathcal{S}$ of a s.p.d. matrix A is better conditioned than the matrix. Moreover, we have*

$$\lambda_{\min}(A) \le \lambda_{\min}(\mathcal{S}) \le \lambda_{\max}(\mathcal{S}) \le \lambda_{\max}(\mathcal{B}) \le \lambda_{\max}(A).$$

Proof. The last two inequalities are straightforward because

$$\frac{\mathbf{x}^T\mathcal{S}\mathbf{x}}{\mathbf{x}^T\mathbf{x}} = \frac{\mathbf{x}^T(\mathcal{B} - \mathcal{L}\mathcal{A}^{-1}\mathcal{L}^T)\mathbf{x}}{\mathbf{x}^T\mathbf{x}} \le \frac{\mathbf{x}^T\mathcal{B}\mathbf{x}}{\mathbf{x}^T\mathbf{x}} = \frac{\begin{bmatrix} 0 \\ \mathbf{x} \end{bmatrix}^T A \begin{bmatrix} 0 \\ \mathbf{x} \end{bmatrix}}{0^T 0 + \mathbf{x}^T\mathbf{x}} \le \sup_{\mathbf{v},\,\mathbf{x}} \frac{\begin{bmatrix} \mathbf{v} \\ \mathbf{x} \end{bmatrix}^T A \begin{bmatrix} \mathbf{v} \\ \mathbf{x} \end{bmatrix}}{\mathbf{v}^T\mathbf{v} + \mathbf{x}^T\mathbf{x}}$$

implies

$$\lambda_{\max}(\mathcal{S}) \le \lambda_{\max}(\mathcal{B}) \le \lambda_{\max}(A).$$

Now using the fact that $\mathcal{S}^{-1}$ is a principal submatrix of A^{-1}, we can apply the proved inequality $\lambda_{\max}(B) \le \lambda_{\max}(A)$ for $A := A^{-1}$ and $B := \mathcal{S}^{-1}$, which gives

$$\frac{1}{\lambda_{\min}(\mathcal{S})} = \lambda_{\max}(\mathcal{S}^{-1}) \le \lambda_{\max}(A^{-1}) = \frac{1}{\lambda_{\min}(A)}.$$

That is, the remaining inequality $\lambda_{\min}(A) \le \lambda_{\min}(\mathcal{S})$ is proved. □

In what follows, we often use the following properties of A.

Lemma 3.3. *Given are two rectangular matrices (less columns than rows) J and P, such that $[J,\ P]$ is square and invertible. In other words, any vector $\mathbf{v}$ can be uniquely decomposed as $\mathbf{v} = J\mathbf{w} + P\mathbf{x}$. This shows that the vector spaces Range(J) and Range(P) have a nontrivial angle in a given inner product generated by any s.p.d. matrix A. Let $\gamma \in [0, 1)$ measure that abstract angle; that is, the following strengthened Cauchy–Schwarz inequality holds.*

$$\mathbf{w}^T JAP\mathbf{x} \le \gamma(\mathbf{w}^T J^T AJ\mathbf{w})^{1/2}(\mathbf{x}^T P^T AP\mathbf{x})^{1/2}, \quad \textit{for all } \mathbf{w}, \mathbf{x}. \tag{3.1}$$

The following inequality is an equivalent statement of (3.1)

$$\mathbf{w}^T J^T AJ\mathbf{w} \le \frac{1}{1-\gamma^2} \inf_{\mathbf{x}:\ \mathbf{v}=J\mathbf{w}+P\mathbf{x}} \mathbf{v}^T A\mathbf{v}. \tag{3.2}$$

Due to symmetry, we also have that the following statement is equivalent to (3.1),

$$\mathbf{x}^T P^T AP\mathbf{x} \le \frac{1}{1-\gamma^2} \inf_{\mathbf{w}:\ \mathbf{v}=J\mathbf{w}+P\mathbf{x}} \mathbf{v}^T A\mathbf{v}. \tag{3.3}$$

Finally, consider the special case $[J,\ P] = I$, that is,

$$J = \begin{bmatrix} I \\ 0 \end{bmatrix}, \quad P = \begin{bmatrix} 0 \\ I \end{bmatrix}, \tag{3.4}$$

and let $\mathcal{A} = J^T AJ$, $\mathcal{R} = J^T AP$, $\mathcal{L} = P^T AJ$, and $\mathcal{B} = P^T AP$. Consider also the Schur complement $\mathcal{S} = \mathcal{B} - \mathcal{L}\mathcal{A}^{-1}\mathcal{R}$. Then

$$\inf_{\mathbf{w}:\ \mathbf{v}=J\mathbf{w}+P\mathbf{x}} \mathbf{v}^T A\mathbf{v} = \inf_{\mathbf{w}} \begin{bmatrix} \mathbf{w} \\ \mathbf{x} \end{bmatrix}^T \begin{bmatrix} \mathcal{A} & \mathcal{R} \\ \mathcal{L} & \mathcal{B} \end{bmatrix} \begin{bmatrix} \mathbf{w} \\ \mathbf{x} \end{bmatrix} = \mathbf{x}^T(\mathcal{B} - \mathcal{L}\mathcal{A}^{-1}\mathcal{R})\mathbf{x} = \mathbf{x}^T \mathcal{S}\mathbf{x}. \tag{3.5}$$

Proof. We look at the quadratic form, for any $t \in \mathcal{R}$,

$$\begin{aligned} Q(t) &\equiv (J\mathbf{w} + tP\mathbf{x})^T A(J\mathbf{w} + tP\mathbf{x}) - (1-\gamma^2)\mathbf{w}^T J^T AJ\mathbf{w} \\ &= t^2\mathbf{x}^T P^T AP\mathbf{x} + 2t\mathbf{w}^T J^T AP\mathbf{x} + \gamma^2\mathbf{w}^T J^T AJ\mathbf{w}. \end{aligned}$$

An equivalent statement $Q(t)$ to be nonnegative is its discriminant

$$(\mathbf{w}^T J^T A P \mathbf{x})^2 - \gamma^2 \mathbf{w}^T J^T A J \mathbf{w} \mathbf{x}^T P^T A P \mathbf{x},$$

to be nonpositive. This shows that (3.1) and (3.2) are equivalent.

Finally, in the case (3.4), the "energy" minimization property (3.5) of the Schur complement $\mathcal{S}$ of A follows from the identity

$$(J\mathbf{w} + P\mathbf{x})^T A (J\mathbf{w} + P\mathbf{x}) = \mathbf{x}^T \mathcal{S} \mathbf{x} + (\mathcal{A}\mathbf{w} + \mathcal{R}\mathbf{x})^T \mathcal{A}^{-1} (\mathcal{A}\mathbf{w} + \mathcal{R}\mathbf{x}).$$

It is clear then that $\mathbf{x}^T \mathcal{S} \mathbf{x} = \min_{\mathbf{w}} (J\mathbf{w} + P\mathbf{x})^T A (J\mathbf{w} + P\mathbf{x})$ with the minimum attained at $\mathbf{w} = -\mathcal{A}^{-1}\mathcal{R}\mathbf{x}$. □

The following result was proved in [FVZ05].

Lemma 3.4. *Let $T + N^T N$ be invertible, with T being symmetric positive semidefinite and N a rectangular matrix such that for any vector $\mathbf{v}$ (of proper dimension) the equation $N\mathbf{w} = \mathbf{v}$ has a solution. Introduce the matrix $Z = N(T + N^T N)^{-1} N^T$. We have that Z is s.p.d., and the following identity holds.*

$$\frac{\mathbf{v}^T Z^{-1} \mathbf{v}}{\mathbf{v}^T \mathbf{v}} = 1 + \inf_{\mathbf{w}:\, N\mathbf{w} = \mathbf{v}} \frac{\mathbf{w}^T T \mathbf{w}}{\mathbf{v}^T \mathbf{v}}.$$

Proof. We first mention that N being onto implies that N^T has a full column rank. Indeed, assume that for some vector $\mathbf{v}_0$ we have that $N^T \mathbf{v}_0 = 0$. Because by assumption, the equation $N\mathbf{w}_0 = \mathbf{v}_0$ has a solution $\mathbf{w}_0$, it follows then that $0 = (N^T \mathbf{v}_0)^T \mathbf{w}_0 = \mathbf{v}_0^T (N\mathbf{w}_0) = \mathbf{v}_0^T \mathbf{v}_0$. That is, $\|\mathbf{v}_0\| = 0$, which implies $\mathbf{v}_0 = 0$. Thus, the columns of N^T must be linearly independent, and hence, NN^T is invertible. It is clear also that the matrix $Z = N(T + N^T N)^{-1} N^T$ is s.p.d., hence invertible. This follows from the simple inequalities,

$$\begin{aligned} \mathbf{w}^T N (T + N^T N)^{-1} N^T \mathbf{w} &= (N^T \mathbf{w})^T (T + N^T N)^{-1} (N^T \mathbf{w}) \\ &\geq \frac{1}{\lambda_{\max}(T + N^T N)} \|N^T \mathbf{w}\|^2 \geq 0. \end{aligned}$$

The last expression can be zero only if $N^T \mathbf{w} = 0$, that is, only if $\mathbf{w} = 0$ due to the fact that N^T has a full column rank.

Consider now the following constrained minimization problem.

For a given $\mathbf{v}$ find the solution of

$$J(\mathbf{w}) = \frac{1}{2} \mathbf{w}^T T \mathbf{w} \mapsto \min,$$

subject to the constraint $N\mathbf{w} = \mathbf{v}$. Let $\mathbf{w}_*$ be a solution of this quadratic constrained minimization problem. Introducing the Lagrange multiplier $\boldsymbol{\lambda}$ and taking derivatives of the respective Lagrangian $\mathcal{L}(\mathbf{w}, \boldsymbol{\lambda}) \equiv J(\mathbf{w}) + \boldsymbol{\lambda}^T(N\mathbf{w} - \mathbf{v})$, we end up with the following saddle-point problem for $(\mathbf{w}_*, \boldsymbol{\lambda})$,

$$\begin{bmatrix} T & N^T \\ N & 0 \end{bmatrix} \begin{bmatrix} \mathbf{w}_* \\ \boldsymbol{\lambda} \end{bmatrix} = \begin{bmatrix} 0 \\ \mathbf{v} \end{bmatrix}.$$

The latter system is equivalent to the following transformed one (obtained by multiplying the second equation with N^T and adding it to the first equation),

$$\begin{bmatrix} T + N^T N & N^T \\ N & 0 \end{bmatrix} \begin{bmatrix} \mathbf{w}_* \\ \boldsymbol{\lambda} \end{bmatrix} = \begin{bmatrix} N^T \mathbf{v} \\ \mathbf{v} \end{bmatrix}.$$

The transformed matrix has (by assumption) an invertible (1,1)-block, $T + N^T N$, and also it has a (negative) Schur complement $Z = N(T + N^T N)^{-1} N^T$ which we showed (above) is invertible.

Thus, the solution of the constrained minimization problem $\mathbf{w}_\star$ coincides with the $\mathbf{w}$-component of the solution of the last saddle-point problem. Because $\mathbf{w}_* = (T + N^T N)^{-1} N^T(\mathbf{v} - \boldsymbol{\lambda})$, we have $\mathbf{v} = N\mathbf{w}_* = Z(\mathbf{v} - \boldsymbol{\lambda})$, which shows that $\boldsymbol{\lambda} = \mathbf{v} - Z^{-1}\mathbf{v}$. Thus, using again that $\mathbf{v} = N\mathbf{w}_\star$ and $N^T\boldsymbol{\lambda} = -T\mathbf{w}_\star$, we end up with the equalities,

$$\begin{aligned} \mathbf{v}^T Z^{-1} \mathbf{v} &= \mathbf{v}^T \mathbf{v} - \mathbf{v}^T \boldsymbol{\lambda} \\ &= \mathbf{v}^T \mathbf{v} - (N\mathbf{w}_\star)^T \boldsymbol{\lambda} \\ &= \mathbf{v}^T \mathbf{v} - \mathbf{w}_\star^T (N^T \boldsymbol{\lambda}) \\ &= \mathbf{v}^T \mathbf{v} + \mathbf{w}_\star^T T \mathbf{w}_\star. \end{aligned}$$

Using now the characterization of $\mathbf{w}_\star$, namely, that

$$\mathbf{w}_\star^T T \mathbf{w}_\star = \min_{\mathbf{w}:\ N\mathbf{w} = \mathbf{v}} \mathbf{w}^T T \mathbf{w},$$

we end up with the desired identity,

$$\frac{\mathbf{v}^T Z^{-1} \mathbf{v}}{\mathbf{v}^T \mathbf{v}} = 1 + \min_{\mathbf{w}:\ N\mathbf{w} = \mathbf{v}} \frac{\mathbf{w}^T T \mathbf{w}}{\mathbf{v}^T \mathbf{v}}.$$

□

Finally, we derive a symmetric version of the popular Sherman–Morrison formula.

Proposition 3.5. *Given are three matrices, X, T, and F, such that X and T are invertible; then the following formula holds,*

$$(X + F^T T^{-1} F)^{-1} = X^{-1} - X^{-1} F^T (T + F X^{-1} F^T)^{-1} F X^{-1},$$

provided $T + F X^{-1} F^T$ is invertible.

Proof. Consider the following factored matrix

$$A \equiv \begin{bmatrix} I & 0 \\ F & I \end{bmatrix} \begin{bmatrix} X^{-1} & 0 \\ 0 & T \end{bmatrix} \begin{bmatrix} I & F^T \\ 0 & I \end{bmatrix}.$$

We have the following explicit expression for A

$$A = \begin{bmatrix} X^{-1} & 0 \\ FX^{-1} & T \end{bmatrix} \begin{bmatrix} I & F^T \\ 0 & I \end{bmatrix}$$
$$= \begin{bmatrix} X^{-1} & X^{-1}F^T \\ FX^{-1} & T + FX^{-1}F^T \end{bmatrix}.$$

Form the Schur complement of A,

$$S = X^{-1} - X^{-1}F^T(T + FX^{-1}F^T)^{-1}FX^{-1}.$$

Now, let us derive an explicit expression for A^{-1}. Based on the factored form of A, we get

$$A^{-1} = \begin{bmatrix} I & F^T \\ 0 & I \end{bmatrix}^{-1} \begin{bmatrix} X^{-1} & 0 \\ 0 & T \end{bmatrix}^{-1} \begin{bmatrix} I & 0 \\ F & I \end{bmatrix}^{-1}$$
$$= \begin{bmatrix} I & -F^T \\ 0 & I \end{bmatrix} \begin{bmatrix} X & 0 \\ 0 & T^{-1} \end{bmatrix} \begin{bmatrix} I & 0 \\ -F & I \end{bmatrix}$$
$$= \begin{bmatrix} X & -F^T T^{-1} \\ 0 & T^{-1} \end{bmatrix} \begin{bmatrix} I & 0 \\ -F & I \end{bmatrix}$$
$$= \begin{bmatrix} X + F^T T^{-1} F & -F^T T^{-1} \\ -T^{-1}F & T^{-1} \end{bmatrix}.$$

Then, Proposition 3.1 tells us that $(X + F^T T^{-1} F)^{-1} = S$ which is the desired Sherman–Morrison formula. □

3.1.2 Kato's Lemma

We often use the following classical result of Kato (see Lemma 4 in the appendix of [Kato]). For a survey on this topic we refer to [Sz06].

Lemma 3.6. *Let π be a projection; that is, $\pi^2 = \pi$ and $\pi \neq I,\ 0$. Then for any inner-product norm $\|.\| = \sqrt{(\cdot,\ \cdot)}$, we have $\|\pi\| = \|I - \pi\|$.*

Proof. For any vector $\mathbf{v}$ and any $t \in \mathbb{R}$, consider $\mathbf{v}_t \equiv \pi\mathbf{v} + t(I - \pi)\mathbf{v}$. Notice that $\pi\mathbf{v}_t = \pi\mathbf{v}$. Then,

$$\|\pi\mathbf{v}\|^2 = \|\pi\mathbf{v}_t\|^2 \leq \|\pi\|^2\|\mathbf{v}_t\|^2 = \|\pi\|^2\|\pi\mathbf{v} + t(I - \pi)\mathbf{v}\|^2.$$

The latter expression shows that

$$Q(t) \equiv \left(1 - \frac{1}{\|\pi\|^2}\right)\|\pi\mathbf{v}\|^2 + 2t\,(\pi\mathbf{v},\ (I - \pi)\mathbf{v}) + t^2\|(I - \pi)\mathbf{v}\|^2 \geq 0.$$

This implies that the discriminant of the quadratic form $Q(t)$ must be nonnegative; that is, the following strengthened Cauchy–Schwarz inequality holds,

$$(\pi \mathbf{v},\ (I-\pi)\mathbf{v})^2 \le \left(1 - \frac{1}{\|\pi\|^2}\right) \|\pi \mathbf{v}\|^2 \|(I-\pi)\mathbf{v}\|^2.$$

We prove in the same way

$$(\pi \mathbf{v},\ (I-\pi)\mathbf{v})^2 \le \left(1 - \frac{1}{\|I-\pi\|^2}\right) \|\pi \mathbf{v}\|^2 \|(I-\pi)\mathbf{v}\|^2.$$

Thus, $\|\pi\| = \|I-\pi\|$, because the strengthened Cauchy–Schwarz inequality

$$(\pi \mathbf{v},\ (I-\pi)\mathbf{v})^2 \le \gamma^2\ \|\pi \mathbf{v}\|^2 \|(I-\pi)\mathbf{v}\|^2 \tag{3.6}$$

(i.e., with the best constant $\gamma \in [0,\ 1)$) would imply, by following the above steps in a reverse order, that $\|\pi\|^2 = \|I-\pi\|^2 = 1/(1-\gamma^2)$. □

The above proof also shows the following corollary.

Corollary 3.7. *Consider a vector space equipped with a norm $\|.\|$ generated by an inner product $(\cdot,\ \cdot)$. The norm $\|\pi\|$ of a nontrivial projection π in that vector space, is related to the cosine $\gamma \in [0,\ 1)$ of the angle between the complementary spaces Range(π) and Range$(I-\pi)$ in the inner product $(\cdot,\ \cdot)$ (defined in* (3.6)*), as*

$$\|\pi\| = \frac{1}{\sqrt{1-\gamma^2}}.$$

The latter implies $\|\pi\| = \|I-\pi\|$ (for $\pi \ne I,\ 0$).

3.1.3 Convergent iteration in A-norm

In this section we formulate an auxiliary result that is used many times throughout the book.

Proposition 3.8. *Let A be a s.p.d. matrix. Assume, that for a given nonsingular matrix M the iteration matrix $I-M^{-1}A$ has an A-norm less than one. Equivalently, (because A is s.p.d.) assume that*

$$\|I - A^{1/2}M^{-1}A^{1/2}\| < 1.$$

Consider the matrix $\overline{M} = M(M+M^T-A)^{-1}M^T$ (sometimes referred to as a symmetrization of M).

The following properties hold.

(i) $I - \overline{M}^{-1}A = (I - M^{-T}A)(I-M^{-1}A)$.
(ii) $\overline{M} - A$ *is symmetric positive semidefinite.*
(iii) $\|I - A^{1/2}\overline{M}^{-1}A^{1/2}\| = \|I - A^{1/2}M^{-1}A^{1/2}\|^2$.
(iv) $\|I - A^{1/2}M^{-1}A^{1/2}\| < 1$ *is equivalent to* $M + M^T - A$ *being s.p.d.*

In particular, a stationary iteration for solving systems with A based on M is convergent in the A-norm, if and only if a stationary iteration for solving systems with A based on $\overline{M}$ is convergent in the A-norm. The convergence factor of the latter equals the square of the convergence factor of the former.

Similar results hold for $\widetilde{M} = M^T(M + M^T - A)^{-1}M$ which is in general different from $\overline{M}$ (if $M \neq M^T$) based on the fact $\|X\| = \|X^T\|$ (used here for $X = I - A^{1/2}M^{-1}A^{1/2}$).

Proof. From the explicit expression $\overline{M}^{-1} = M^{-1} + M^{-T} - M^{-1}AM^{-T}$, letting $E = I - M^{-1}A$, it follows that $I - \overline{M}^{-1}A = E - M^{-T}AE = (I - M^{-T}A)E$, which is the first desired identity. This identity admits the following more symmetric form

$$I - A^{1/2}\overline{M}^{-1}A^{1/2} = (I - A^{1/2}M^{-T}A^{1/2})(I - A^{1/2}M^{-1}A^{1/2}).$$

This is exactly the third item of the proposition. It actually shows that $I - A^{1/2}\overline{M}^{-1}A^{1/2} = X^TX$ (for $X = I - A^{1/2}M^{-1}A^{1/2}$). From the identity $I - X^TX = A^{1/2}\overline{M}^{-1}A^{1/2}$ and the fact that $\|X\| < 1$ it follows that $A^{1/2}\overline{M}^{-1}A^{1/2} = I - X^TX$ is s.p.d. This implies that $\overline{M}$ is s.p.d. Equivalently, we have that $M + M^T - A = M^{-1}\overline{M}M^{-T}$ is s.p.d. The identity $X^TX = I - A^{1/2}\overline{M}^{-1}A^{1/2}$ implies that $A^{-1/2}X^TXA^{-1/2} = A^{-1} - \overline{M}^{-1}$ is symmetric positive semidefinite which is equivalent to $\overline{M} - A$ is symmetric positive semidefinite. This shows the second item of the proposition.

We already showed that $\|X\| < 1$ implies $M + M^T - A$ being s.p.d. The converse statement follows from the fact that $M + M^T - A$ being s.p.d. implies that $\overline{M}$ and hence $A^{1/2}\overline{M}^{-1}A^{1/2} = I - X^TX$ are s.p.d. Therefore, $\|X\| < 1$, which is the converse statement. □

At the end we prove one more auxiliary result.

Proposition 3.9. *Let A be a given s.p.d. matrix and M provide a convergent iteration method for A in the A-norm; that is, let $\|I - M^{-1}A\|_A < 1$. Let*

$$I_F = \begin{bmatrix} I \\ 0 \end{bmatrix}$$

specify a principal submatrix $A_F = (I_F)^TAI_F$ of A. Finally, let M_F be such that $\|I - M_F^{-1}A_F\|_{A_F} < 1$. Then, the product iteration matrix $E = (I - M^{-1}A)(I - I_FM_F^{-1}(I_F)^TA)$ satisfies

$$\|E\|_A \leq \|I - M^{-1}A\|_A < 1.$$

That is, the product iteration method based on M^{-1} and $I_FM_F^{-1}(I_F)^T$ is convergent in the A-norm. In particular, the matrix $\widehat{M}$ defined implicitly $I - \widehat{M}^{-1}A = E$ is well defined; that is, $I - E$ is invertible.

Proof. If we show that $\|E\|_A < 1$ then the convergence of the series $(I - E)^{-1} = \sum_{k=1}^{\infty} E^k < \infty$ would imply that $\widehat{M} = A^{-1}(I - E)^{-1}$ is well defined.

We first use the fact that $\|I - M_F^{-1} A_F\|_{A_F} < 1$ is equivalent to the statement that $M_F^{-1} + M_F^{-T} - A_F$ is s.p.d. (see Proposition 3.8).

If we show that $\|I - I_F M_F^{-1} (I_F)^T A\|_A \leq 1$ then $\|E\|_A \leq \|I - M^{-1} A\|_A < 1$ the result is proven. Consider the expression

$$\begin{aligned}
&\|(I - I_F M_F^{-1} (I_F)^T A)\mathbf{v}\|_A^2 \\
&= \mathbf{v}^T \big(I - I_F M_F^{-1} (I_F)^T A\big)^T A \big(I - I_F M_F^{-1} (I_F)^T A\big)\mathbf{v} \\
&= \mathbf{v}^T \big(A - A I_F M_F^{-T} I_F^T A - A I_F M_F^{-1} I_F^T A + A I_F M_F^{-T} (I_F^T A I_F) M_F^{-1} I_F^T A\big)\mathbf{v} \\
&= \mathbf{v}^T \big(A - A I_F M_F^{-T} I_F^T A - A I_F M_F^{-1} I_F^T A + A I_F M_F^{-T} A_F M_F^{-1} I_F^T A\big)\mathbf{v} \\
&= \mathbf{v}^T A \mathbf{v} - (A\mathbf{v})^T I_F \big(M_F^{-T} + M_F^{-1} - M_F^{-T} A_F M_F^{-1}\big) I_F A \mathbf{v} \\
&\leq \mathbf{v}^T A \mathbf{v}.
\end{aligned}$$

That is, $\|I - I_F M_F^{-1} (I_F)^T A\|_A \leq 1$ which was the desired result. □

3.2 Approximate block-factorization

Given the s.p.d. matrix A operating on vectors in $\mathbb{R}^n$, let J and P be two rectangular matrices such that their number of rows equals n. Simple examples are the rectangular matrices

$$J = \begin{bmatrix} I \\ 0 \end{bmatrix}, \quad P = \begin{bmatrix} 0 \\ I \end{bmatrix}.$$

In what follows, we consider general rectangular matrices J and P. Form the subspace matrices $\mathcal{A} = J^T A J$ and $\mathcal{B} = P^T A P$ and let $\mathcal{M}$ and $\mathcal{D}$ be their respective approximations (sometimes called preconditioners).

In the following sections, we show that there is a close relation between certain block-factorization preconditioners that exploit solvers based on $\mathcal{M}$, $\mathcal{D}$, and $\mathcal{M}^T$ and subspace product iteration algorithms of inexact block Gauss–Seidel form.

3.2.1 Product iteration matrix formula

More specifically, consider the problem

$$A\mathbf{u} = \mathbf{b},$$

with a given initial approximation $\mathbf{u}_0$. The procedure to generate a new approximation $\mathbf{u}_{\text{new}}$ exploits updates from two subspaces Range (J) and Range (P), which require approximate solutions of problems with the respective subspace matrices $\mathcal{A} = J^T A J$ and $\mathcal{B} = P^T A P$.

Assume that $\mathcal{M}$ is an approximation to $\mathcal{A}$ and that $\mathcal{D}$ is a s.p.d. approximation to $\mathcal{B}$. Note that $\mathcal{M}$ does not have to be symmetric.

For practical purposes we often assume that

$$\|I - \mathcal{A}^{1/2}\mathcal{M}^{-1}\mathcal{A}^{1/2}\| < 1, \tag{3.7}$$

and

$$\|I - \mathcal{B}^{1/2}\mathcal{D}^{-1}\mathcal{B}^{1/2}\| < 1; \tag{3.8}$$

that is, both $\mathcal{M}$ and $\mathcal{D}$ provide convergent splittings for $\mathcal{A}$ and $\mathcal{B}$ in the "energy" norms $\|.\|_{\mathcal{A}}$ and $\|.\|_{\mathcal{B}}$, respectively. The latter is motivated by the fact that the best approximation to the solution $\mathbf{u}$ from the subspace Range (P), defined as $\|\mathbf{u} - P\mathbf{x}\|_A \mapsto \min$ is given by the Ritz–Galerkin projection $\pi_A\mathbf{u} = P(P^TAP)^{-1}P^TA\mathbf{u} = P\mathcal{B}^{-1}P^T\mathbf{b}$. In the case of approximate solutions, that is, using $\mathcal{D}^{-1}$ instead of $\mathcal{B}^{-1}$, it is natural to assume that $\mathcal{D}^{-1}$ is close to $\mathcal{B}^{-1}$ in the $\mathcal{B}$-norm. The same argument applies to $\mathcal{M}^{-1}$ and $\mathcal{M}^{-T}$ used as approximations to $\mathcal{A}^{-1}$.

The inexact symmetric block Gauss–Seidel procedure of interest takes the following familiar form.

Algorithm 3.2.1 (Product iteration method)

(0) *Let* $\mathbf{u}_0$ *be a current iterate and* $\mathbf{r}_0 = \mathbf{b} - A\mathbf{u}_0$ *be the corresponding residual. We perform steps* (m), (w), *and* (m') *(corresponding to iterations in subspaces Range*(J), *Range*(P), *and Range*(J)*) to define the new approximation* $\mathbf{u}_{new}$.

(m) *Solve approximately for a correction* $J\mathbf{x}_m$ *from the subspace residual equation,*

$$(J^TAJ)\mathbf{x} = J^T\mathbf{r}_0;$$

that is, compute $\mathbf{x}_m = \mathcal{M}^{-1}J^T\mathbf{r}_0$ *and let* $\mathbf{u}_m = \mathbf{u}_0 + J\mathbf{x}_m$. *Compute the new residual* $\mathbf{r}_m = \mathbf{b} - A\mathbf{u}_m = \mathbf{b} - A\mathbf{u}_0 - AJ\mathbf{x}_m = (I - AJ\mathcal{M}^{-1}J^T)\mathbf{r}_0$.

(w) *Solve approximately for a correction* $P\mathbf{w}$ *from the subspace residual equation*

$$(P^TAP)\mathbf{w} = P^T\mathbf{r}_m;$$

that is, let $\mathbf{w} = \mathcal{D}^{-1}P^T\mathbf{r}_m$ *and let* $\mathbf{u}_w = \mathbf{u}_m + P\mathbf{w}$. *Compute the next residual* $\mathbf{r}_w = \mathbf{b} - A\mathbf{u}_w = \mathbf{b} - A\mathbf{u}_m - AP\mathbf{w} = (I - AP\mathcal{D}^{-1}P^T)(I - AJ\mathcal{M}^{-1}J^T)\mathbf{r}_0$.

(m') *Solve approximately for a correction* $J\mathbf{x}_{m'}$ *from the subspace residual equation,*

$$(J^TAJ)\mathbf{x} = J^T\mathbf{r}_w;$$

that is, let $\mathbf{x}_{m'} = \mathcal{M}^{-T}J^T\mathbf{r}_w$ *and the final new iterate equal* $\mathbf{u}_{new} = \mathbf{u}_w + J\mathbf{x}_{m'}$. *Compute the new residual* $\mathbf{r}_{new} = \mathbf{b} - A\mathbf{u}_{new} = \mathbf{b} - A\mathbf{u}_w - AJ\mathbf{x}_{m'} = (I - AJ\mathcal{M}^{-T}J^T)(I - AP\mathcal{D}^{-1}P^T)(I - AJ\mathcal{M}^{-1}J^T)\mathbf{r}_0$.

Thus the residual iteration matrix E_r of the above composite iteration $\mathbf{u}_0 \mapsto \mathbf{u}_{\text{new}}$, which maps $\mathbf{r}_0 \mapsto \mathbf{r}_{\text{new}}$, has the product form $E_r = (I - AJ\mathcal{M}^{-T}J^T)$

$(I - AP\mathcal{D}^{-1}P^T)(I - AJ\mathcal{M}^{-1}J^T)$. For the error iteration matrix E defined from $\mathbf{u} - \mathbf{u}_0 = A^{-1}\mathbf{r}_0 \mapsto \mathbf{u} - \mathbf{u}_{\text{new}} = A^{-1}\mathbf{r}_{\text{new}}$, noticing that $AE = E_r A$, we obtain

$$E = (I - J\mathcal{M}^{-T}J^T A)(I - P\mathcal{D}^{-1}P^T A)(I - J\mathcal{M}^{-1}J^T A) = A^{-1}E_r A.$$

We can easily show the above algorithm does not diverge if (3.7) and (3.8) hold. We have the following.

Lemma 3.10. *Assume that $\mathcal{M}$ and $\mathcal{D}$ satisfy* (3.7) *and* (3.8). *Then* $\|E\mathbf{e}\|_A \le \|\mathbf{e}\|_A$.

Proof. Due to the product form of E, it is sufficient to prove that $\|(I - J\mathcal{M}^{-1}J^T A)\mathbf{e}\|_A \le \|\mathbf{e}\|_A$, $\|(I - J\mathcal{M}^{-T}J^T A)\mathbf{e}\|_A \le \|\mathbf{e}\|_A$, and $\|(I - P\mathcal{D}^{-1}P^T A)\mathbf{e}\|_A \le \|\mathbf{e}\|_A$. Equivalently, it is sufficient to prove that

$$\begin{aligned}\|I - A^{1/2}J\mathcal{M}^{-1}J^T A^{1/2}\| = \|I - A^{1/2}J\mathcal{M}^{-T}J^T A^{1/2}\| \le 1, \quad \text{and}\\ \|I - A^{1/2}P\mathcal{D}^{-1}P^T A^{1/2}\| \le 1.\end{aligned}$$

We prove the inequality involving $\mathcal{M}$ (the result for $\mathcal{D}$ is analogous). From the identity

$$\begin{aligned}&(I - \mathcal{A}^{1/2}\mathcal{M}^{-T}\mathcal{A}^{1/2})(I - \mathcal{A}^{1/2}\mathcal{M}^{-1}\mathcal{A}^{1/2})\\ &\quad = I - \mathcal{A}^{1/2}(\mathcal{M}^{-T} + \mathcal{M}^{-1} - \mathcal{M}^{-T}\mathcal{A}\mathcal{M}^{-1})\mathcal{A}^{1/2},\end{aligned}$$

we see that the assumption $\|I - \mathcal{A}^{1/2}\mathcal{M}^{-1}\mathcal{A}^{1/2}\| \le 1$ is equivalent to $\mathcal{M}^{-1} + \mathcal{M}^{-T} - \mathcal{M}^{-1}\mathcal{A}\mathcal{M}^{-1}$ being symmetric positive semidefinite.

Consider then the expression

$$\begin{aligned}&\|(I - J\mathcal{M}^{-1}J^T A)\mathbf{e}\|_A^2\\ &\quad = \mathbf{e}^T(I - J\mathcal{M}^{-1}J^T A)^T A(I - J\mathcal{M}^{-1}J^T A)\mathbf{e}\\ &\quad = \mathbf{e}^T(A - AJ\mathcal{M}^{-T}J^T A - AJ\mathcal{M}^{-1}J^T A + AJ\mathcal{M}^{-T}(J^T AJ)\mathcal{M}^{-1}J^T A)\mathbf{e}\\ &\quad = \mathbf{e}^T(A - AJ\mathcal{M}^{-T}J^T A - AJ\mathcal{M}^{-1}J^T A + AJ\mathcal{M}^{-T}\mathcal{A}\mathcal{M}^{-1}J^T A)\mathbf{e}\\ &\quad = \mathbf{e}^T A\mathbf{e} - (A\mathbf{e})^T J(\mathcal{M}^{-T} + \mathcal{M}^{-1} - \mathcal{M}^{-T}\mathcal{A}\mathcal{M}^{-1})J^T A\mathbf{e}\\ &\quad \le \mathbf{e}^T A\mathbf{e}.\end{aligned}$$

That is, $\|I - J\mathcal{M}^{-1}J^T A\|_A \le 1$, which was the desired result. □

3.2.2 Block-factorizations and product iteration methods

Define implicitly a matrix B^{-1} from the equation

$$I - B^{-1}A = E = (I - J\mathcal{M}^{-T}J^T A)(I - P\mathcal{D}^{-1}P^T A)(I - J\mathcal{M}^{-1}J^T A). \quad (3.9)$$

We show next that B^{-1} can be constructed as a certain approximate block-factorization matrix given below.

Theorem 3.11. *Let* $\overline{\mathcal{M}} = \mathcal{M}(\mathcal{M}+\mathcal{M}^T - \mathcal{A})^{-1}\mathcal{M}^T$. *Consider the following block-factored matrix.*

$$\widehat{B} = \begin{bmatrix} \mathcal{M} & 0 \\ P^T AJ & I \end{bmatrix} \begin{bmatrix} (\mathcal{M}+\mathcal{M}^T - \mathcal{A})^{-1} & 0 \\ 0 & \mathcal{D} \end{bmatrix} \begin{bmatrix} \mathcal{M}^T & J^T AP \\ 0 & I \end{bmatrix}. \tag{3.10}$$

Then, the matrix $B^{-1} = [J,\ P]\widehat{B}^{-1}[J,\ P]^T$ *solves the equation* (3.9). *Also,* B^{-1} *admits the following more explicit form,*

$$B^{-1} = J\overline{\mathcal{M}}^{-1}J^T + (I - J\mathcal{M}^{-T}J^T A)P\mathcal{D}^{-1}P^T(I - AJ\mathcal{M}^{-1}J^T). \tag{3.11}$$

We remark that if the matrix $[J,\ P]^T$ does not have full column rank, then B^{-1} is just a notation for the product $[J,\ P]\widehat{B}^{-1}[J,\ P]^T$, which is only symmetric positive semidefinite, and not invertible.

Proof. The following explicit expression of $\widehat{B}^{-1}$ is easily derived,

$$\widehat{B}^{-1} = \begin{bmatrix} I & -\mathcal{M}^{-T}J^T AP \\ 0 & I \end{bmatrix} \begin{bmatrix} \overline{\mathcal{M}}^{-1} & 0 \\ 0 & \mathcal{D}^{-1} \end{bmatrix} \begin{bmatrix} I & 0 \\ -P^T AJ\mathcal{M}^{-1} & I \end{bmatrix}.$$

Next, form the product $[J,\ P]\widehat{B}^{-1}[J,\ P]^T$. We have

$$\begin{aligned}
&[J,\ P]\widehat{B}^{-1}[J,\ P]^T \\
&= [J,\ P]\begin{bmatrix} I & -\mathcal{M}^{-T}J^T AP \\ 0 & I \end{bmatrix} \begin{bmatrix} \overline{\mathcal{M}}^{-1}J^T \\ \mathcal{D}^{-1}P^T(I - AJ\mathcal{M}^{-1}J^T) \end{bmatrix} \\
&= [J,\ (I - J\mathcal{M}^{-T}J^T A)P] \begin{bmatrix} \overline{\mathcal{M}}^{-1}J^T \\ \mathcal{D}^{-1}P^T(I - AJ\mathcal{M}^{-1}J^T) \end{bmatrix} \\
&= J\overline{\mathcal{M}}^{-1}J^T + (I - J\mathcal{M}^{-T}J^T A)P\mathcal{D}^{-1}P^T(I - AJ\mathcal{M}^{-1}J^T),
\end{aligned}$$

which proves (3.11).

Notice next, that

$$\overline{\mathcal{M}}^{-1} = \mathcal{M}^{-T}(\mathcal{M}+\mathcal{M}^T - \mathcal{A})\mathcal{M}^{-1} = \mathcal{M}^{-T} + \mathcal{M}^{-1} - \mathcal{M}^{-T}\mathcal{A}\mathcal{M}^{-1},$$

which shows,

$$\begin{aligned}
I - J\overline{\mathcal{M}}^{-1}J^T A &= I - J(\mathcal{M}^{-T} + \mathcal{M}^{-1} - \mathcal{M}^{-T}\mathcal{A}\mathcal{M}^{-1})J^T A \\
&= I - J(\mathcal{M}^{-T} + \mathcal{M}^{-1} - \mathcal{M}^{-T}J^T AJ\mathcal{M}^{-1})J^T A \\
&= (I - J\mathcal{M}^{-T}J^T A)(I - J\mathcal{M}^{-1}J^T A).
\end{aligned}$$

Therefore,

$$
\begin{aligned}
I - [J,\ P]\widehat{B}^{-1}[J,\ P]^T A
&= I - J\overline{\mathcal{M}}^{-1}J^T A - (I - J\mathcal{M}^{-T}J^T A)P\mathcal{D}^{-1}P^T(I - AJ\mathcal{M}^{-1}J^T)A \\
&= I - J\overline{\mathcal{M}}^{-1}J^T A - (I - J\mathcal{M}^{-T}J^T A)P\mathcal{D}^{-1}P^T A(I - J\mathcal{M}^{-1}J^T A) \\
&= I - J\overline{\mathcal{M}}^{-1}J^T A - (I - J\mathcal{M}^{-T}J^T A)(I - J\mathcal{M}^{-1}J^T A) \\
&\quad + (I - J\mathcal{M}^{-T}J^T A)(I - P\mathcal{D}^{-1}P^T A)(I - J\mathcal{M}^{-1}J^T A) \\
&= (I - J\mathcal{M}^{-T}J^T A)(I - P\mathcal{D}^{-1}P^T A)(I - J\mathcal{M}^{-1}J^T A) \\
&= E.
\end{aligned}
$$

Thus, $B^{-1} \equiv [J,\ P]\widehat{B}^{-1}[J,\ P]^T$ solves the equation $I - B^{-1}A = E$, which concludes the proof. □

In conclusion, we proved that the block-factorization preconditioner B defined as $B^{-1} = [J,\ P]\widehat{B}^{-1}[J,\ P]^T$, where $\widehat{B}$ comes from the approximate block-factorization of the two-by-two block matrix $\widehat{A} = [J,\ P]^T A[J,\ P]$, as defined in (3.10), leads to an iteration matrix $I - B^{-1}A$ that admits the product form $(I - J\mathcal{M}^{-1}J^T A)(I - P\mathcal{D}^{-1}P^T A)(I - J\mathcal{M}^{-T}J^T A)$, composed of three simpler iteration matrices coming from $\mathcal{M}$, $\mathcal{D}$, and $\mathcal{M}^T$, acting on the subspaces Range(J), Range(P), and Range(J), respectively.

3.2.3 Definitions of two-level B_{TL} and two-grid B_{TG} preconditioners

Next, we consider two special cases of full-column rank matrices J and P. The first one corresponds to $[J,\ P]$ being a square invertible matrix. The corresponding preconditioner $B = B_{TL}$ is referred to (cf., [BDY88]) as the "two-level" or "hierarchical basis multigrid" (or HBMG).

Definition 3.12 (Two-level preconditioner). *Given an approximation $\mathcal{M}$ for $\mathcal{A} = J^T AJ$ and an "interpolation" matrix P, such that $[J,\ P]$ is square and invertible, and let $\mathcal{D}$ be an s.p.d. approximation to $\mathcal{B} = P^T AP$, such that*

- *$\mathcal{M} + \mathcal{M}^T - \mathcal{A}$ is s.p.d., or equivalently $\|I - \mathcal{A}^{1/2}\mathcal{M}^{-1}\mathcal{A}^{1/2}\| < 1$.*
- *$\mathcal{D} - \mathcal{B}$ is symmetric positive semidefinite.*

We first form

$$
\widehat{B}_{TL} = \begin{bmatrix} \mathcal{M} & 0 \\ P^T AJ & I \end{bmatrix} \begin{bmatrix} (\mathcal{M}^T + \mathcal{M} - A)^{-1} & 0 \\ 0 & \mathcal{D} \end{bmatrix} \begin{bmatrix} \mathcal{M}^T & J^T AP \\ 0 & I \end{bmatrix}.
$$

and then define $B_{TL}^{-1} = [J,\ P]\widehat{B}_{TL}^{-1}[J,\ P]^T$. Or, more explicitly (based on (3.11)*), letting $\overline{\mathcal{M}} = \mathcal{M}(\mathcal{M} + \mathcal{M}^T - \mathcal{A})^{-1}\mathcal{M}^T$,*

$$
B_{TL}^{-1} = J\overline{\mathcal{M}}^{-1}J^T + (I - J\mathcal{M}^{-T}J^T A)P\mathcal{D}^{-1}P^T(I - AJ\mathcal{M}^{-1}J^T),
$$

The case $J = I$ leads to $[J,\ P]^T$ with full column rank, because $[I,\ P][I,\ P]^T = I + PP^T$, which is s.p.d., hence invertible. The corresponding preconditioner $B = B_{TG}$ is referred to as the "two-grid" preconditioner.

Definition 3.13 (Two-grid preconditioner). *Given a "smoother" M for A and an interpolation matrix P, and let $\mathcal{D}$ be a s.p.d. approximation to $\mathcal{B} = P^TAP$, such that*

- *$M + M^T - A$ is s.p.d., or equivalently $\|I - A^{1/2}M^{-1}A^{1/2}\| < 1$.*
- *$\mathcal{D} - \mathcal{B}$ is symmetric positive semidefinite.*

We first form

$$\widehat{B}_{TG} = \begin{bmatrix} M & 0 \\ P^TA & I \end{bmatrix} \begin{bmatrix} (M^T + M - A)^{-1} & 0 \\ 0 & \mathcal{D} \end{bmatrix} \begin{bmatrix} M^T & AP \\ 0 & I \end{bmatrix}.$$

and then define $B_{TG}^{-1} = [I,\ P]\widehat{B}_{TG}^{-1}[I,\ P]^T$. Or, more explicitly (based on (3.11)*) letting $\overline{M} = M(M + M^T - A)^{-1}M^T$,*

$$B_{TG}^{-1} = \overline{M}^{-1} + (I - AM^{-T})P\mathcal{D}^{-1}P^T(I - M^{-1}A).$$

Proposition 3.14. *To implement $B^{-1}\mathbf{b}$, for both $B = B_{TL}$ and $B = B_{TG}$, we can use Algorithm 3.2.1 starting with $\mathbf{u}_0 = 0$. We have then $B^{-1}\mathbf{b} = \mathbf{u}_{new}$. Alternatively, we may use the explicit expression given by* (3.11)*.*

Proof. This is seen from the fact that $E_r = I - AB^{-1}$ relates the initial and final residuals via $\mathbf{r}_{\text{new}} = E_r\mathbf{r}_0$. Noting then that $\mathbf{r}_0 = \mathbf{b} - A\mathbf{u}_0 = \mathbf{b}$ gives $\mathbf{r}_{\text{new}} = \mathbf{b} - A\mathbf{u}_{\text{new}} = (I - AB^{-1})\mathbf{b}$ (i.e., $A\mathbf{u}_{\text{new}} = AB^{-1}\mathbf{b}$), and therefore $\mathbf{u}_{\text{new}} = B^{-1}\mathbf{b}$, which is the desired result. □

3.2.4 A main identity

Consider the block-factorization preconditioner B leading to the product iteration matrix $E = I - B^{-1}A = (I - J\mathcal{M}^{-T}J^TA)(I - P\mathcal{D}^{-1}P^TA)(I - J\mathcal{M}^{-1}J^TA)$.

We assume here that $[J,\ P]^T$ has a full column rank, hence B is well defined from $B^{-1} = [J,\ P]\widehat{B}^{-1}[J,\ P]^T$, which is now s.p.d.

The present section is devoted to the proof of the following main characterization result for B.

Theorem 3.15. *The following main identity holds.*

$$\mathbf{v}^TB\mathbf{v} = \min_{\mathbf{v}=J\mathbf{v}_s+P\mathbf{v}_c} \left[\mathbf{v}_c^T\mathcal{D}\mathbf{v}_c + (\mathcal{M}^T\mathbf{v}_s + J^TAP\mathbf{v}_c)^T(\mathcal{M} + \mathcal{M}^T - \mathcal{A})^{-1} \times (\mathcal{M}^T\mathbf{v}_s + J^TAP\mathbf{v}_c)\right].$$

Recall that $\mathcal{D}$ was a s.p.d. approximation to P^TAP.

Proof. We use the equivalent block-factorization definition of $B = B_{TL}$, Definition 3.12; that is,

$$B^{-1} = [J, \ P]\widehat{B}^{-1}[J, \ P]^T.$$

Then, because $\|X\| = \|X^T\| = 1$ is used for $X = B^{1/2}[J, \ P]\widehat{B}^{-1/2}$, we get the inequality

$$\widehat{\mathbf{v}}^T \widehat{B}^{-1/2}[J, \ P]^T B[J, \ P]\widehat{B}^{-1/2}\widehat{\mathbf{v}} \le \widehat{\mathbf{v}}^T\widehat{\mathbf{v}},$$

or equivalently,

$$\widehat{\mathbf{v}}^T [J, \ P]^T B[J, \ P]\widehat{\mathbf{v}} \le \widehat{\mathbf{v}}^T \widehat{B}\widehat{\mathbf{v}}. \tag{3.12}$$

Given $\mathbf{v}$ decomposed as

$$\mathbf{v} = J\mathbf{v}_s + P\mathbf{v}_c = [J, \ P]\begin{bmatrix}\mathbf{v}_s\\ \mathbf{v}_c\end{bmatrix},$$

based on inequality (3.12) used for

$$\widehat{\mathbf{v}} = \begin{bmatrix}\mathbf{v}_s\\ \mathbf{v}_c\end{bmatrix},$$

we obtain,

$$\mathbf{v}^T B\mathbf{v} = \widehat{\mathbf{v}}^T [J, \ P]^T B[J, \ P]\widehat{\mathbf{v}} \le \begin{bmatrix}\mathbf{v}_s\\ \mathbf{v}_c\end{bmatrix}^T \widehat{B}\begin{bmatrix}\mathbf{v}_s\\ \mathbf{v}_c\end{bmatrix}.$$

That is, we showed that (see (3.10)),

$$\begin{aligned}\mathbf{v}^T B\mathbf{v} &\le \min_{\mathbf{v}=J\mathbf{v}_s+P\mathbf{v}_c} \begin{bmatrix}\mathbf{v}_s\\ \mathbf{v}_c\end{bmatrix}^T \widehat{B}\begin{bmatrix}\mathbf{v}_s\\ \mathbf{v}_c\end{bmatrix}\\ &= \min_{\mathbf{v}=J\mathbf{v}_s+P\mathbf{v}_c} \left[\mathbf{v}_c^T \mathcal{D}\mathbf{v}_c + (\mathcal{M}^T\mathbf{v}_s + J^T AP\mathbf{v}_c)^T(\mathcal{M}+\mathcal{M}^T - \mathcal{A})^{-1}\right.\\ &\qquad\qquad \left.\times (\mathcal{M}^T\mathbf{v}_s + J^T AP\mathbf{v}_c)\right].\end{aligned} \tag{3.13}$$

It remains to show that this upper bound is sharp. For this, consider the problem for $\widehat{\mathbf{v}}$,

$$\widehat{B}\widehat{\mathbf{v}} = [J, \ P]^T B\mathbf{v}.$$

We have $[J, \ P]\widehat{\mathbf{v}} = [J, \ P]\widehat{B}^{-1}[J, \ P]^T B\mathbf{v} = \mathbf{v}$. That is, $[J, \ P]\widehat{\mathbf{v}}$ provides a decomposition for $\mathbf{v}$. For that particular decomposition, we have $\mathbf{v}^T B\mathbf{v} = \widehat{\mathbf{v}}^T [J, \ P]^T B\mathbf{v} = \widehat{\mathbf{v}}^T \widehat{B}\widehat{\mathbf{v}}$. The latter shows that the upper bound in (3.13) is sharp, which is the desired result. □

3.2.5 A simple lower-bound estimate

Recall the product formula $E = (I - J\mathcal{M}^{-T}J^TA)(I - P\mathcal{D}^{-1}P^TA)(I - J\mathcal{M}^{-1}J^TA)$. We also note that when $J = I$, $\mathcal{A} = J^TAJ = A$, then we let $\mathcal{M} = M$ for a given M.

We assume (as in Definitions 3.12–3.13), that

$$\mathbf{w}^T\mathcal{D}\mathbf{w} \geq \mathbf{w}^T\mathcal{B}\mathbf{w}. \tag{3.14}$$

The following simple lower-bound result holds.

Theorem 3.16. *If $\mathcal{D} - \mathcal{B}$ is symmetric positive semidefinite, then AE and $B - A$ are symmetric positive semidefinite.*

Proof. We have the identity,

$$AE = X^T(I - A^{1/2}P\mathcal{D}^{-1}P^TA^{1/2})X, \qquad X = A^{1/2}(I - J\mathcal{M}^{-1}J^TA).$$

If we show that

$$\mathbf{v}^TA^{-1}\mathbf{v} \geq (P^T\mathbf{v})^T\mathcal{D}^{-1}(P^T\mathbf{v}), \tag{3.15}$$

then the middle term $I - A^{1/2}P\mathcal{D}^{-1}P^TA^{1/2}$ in $AE = X^T(*)X$ will be symmetric positive semidefinite, which would imply then that AE itself is symmetric positive semidefinite.

We next prove (3.15). From the definition $\mathcal{B} = P^TAP$, we have $I = Y^TY$, for $Y^T = \mathcal{B}^{-1/2}P^TA^{1/2}$. Thus, $\|Y^T\| = \|Y\| = 1$, implies

$$(P^TA^{1/2}\mathbf{v})^T\mathcal{B}^{-1}(P^TA^{1/2}\mathbf{v}) \leq \mathbf{v}^T\mathbf{v},$$

or equivalently (letting $\mathbf{v} := A^{-1/2}\mathbf{v}$),

$$(P^T\mathbf{v})^T\mathcal{B}^{-1}(P^T\mathbf{v}) \leq \mathbf{v}^TA^{-1}\mathbf{v}.$$

Now use the corollary from (3.14)

$$(P^T\mathbf{v})^T\mathcal{B}^{-1}(P^T\mathbf{v}) \geq (P^T\mathbf{v})^T\mathcal{D}^{-1}(P^T\mathbf{v}),$$

to obtain

$$\mathbf{v}^TA^{-1}\mathbf{v} \geq (P^T\mathbf{v})^T\mathcal{D}^{-1}(P^T\mathbf{v}),$$

which is inequality (3.15) The final result follows then from the fact that $AE = A - AB^{-1}A = A(A^{-1} - B^{-1})A$ is symmetric positive semidefinite is equivalent to $A^{-1} - B^{-1}$, or for that matter, to $B - A$ being symmetric positive semidefinite. The latter concludes the proof. □

3.2.6 Sharp upper bound

Here we assume that $\mathcal{D} = \mathcal{B} = P^TAP$. To establish an estimate from above for B in terms of A, we first assume (as in Definitions 3.12) that $\mathcal{M} + \mathcal{M}^T - \mathcal{A}$ is s.p.d. This is equivalent to saying that the iteration matrix $I - \mathcal{M}^{-T}\mathcal{A}$ has a norm ϱ less than one, in the $\mathcal{A}$-inner product, n; that is,

$$((I - \mathcal{M}^{-T}\mathcal{A})\mathbf{w})^T\mathcal{A}((I - \mathcal{M}^{-T}\mathcal{A})\mathbf{w}) \leq \varrho^2\, \mathbf{w}^T\mathcal{A}\mathbf{w}.$$

Because $\mathcal{D} = \mathcal{B} = \mathcal{P}^T A P$ and $\|I - \mathcal{A}^{1/2}\mathcal{M}^{-1}\mathcal{A}^{1/2}\| < 1$, it is clear then (from Lemma 3.10) that E will have a spectral radius no greater than one. More specifically (recalling that AE is symmetric positive semidefinite shown in Theorem 3.16), we have

$$\mathbf{v}^T AE\mathbf{v} \leq \mathbf{v}^T A\mathbf{v}.$$

We want to find a sharp bound for the spectral radius of E. The largest eigenvalue of

$$\begin{aligned} A^{1/2}EA^{-1/2} &= (I - A^{1/2}J\mathcal{M}^{-T}J^TA^{1/2})(I - A^{1/2}P\mathcal{B}^{-1}P^TA^{1/2}) \\ &\quad \times (I - A^{1/2}J\mathcal{M}^{-1}J^TA^{1/2}) \end{aligned}$$

equals the largest eigenvalue of

$$\begin{aligned} \Theta \equiv\; &(I - A^{1/2}P\mathcal{B}^{-1}P^TA^{1/2})^{1/2}(I - A^{1/2}J\mathcal{M}^{-1}J^TA^{1/2}) \\ &\times (I - A^{1/2}J\mathcal{M}^{-T}J^TA^{1/2})(I - A^{1/2}P\mathcal{B}^{-1}P^TA^{1/2})^{1/2}, \end{aligned}$$

therefore we estimate the last expression.

We notice then that $\overline{\pi}_A = A^{1/2}P\mathcal{B}^{-1}P^TA^{1/2}$ is a projection. Hence, we can remove the square root, such that $I - \overline{\pi}_A = (I - A^{1/2}P\mathcal{D}^{-1}P^TA^{1/2})^{1/2}$. Introduce

$$\widetilde{\mathcal{M}} = \mathcal{M}^T(\mathcal{M} + \mathcal{M}^T - \mathcal{A})^{-1}\mathcal{M}.$$

Note that $\widetilde{\mathcal{M}}$ is in general different from $\overline{\mathcal{M}} = \mathcal{M}(\mathcal{M} + \mathcal{M}^T - \mathcal{A})^{-1}\mathcal{M}^T$.

Consider now the expression

$$\begin{aligned} \mathbf{v}^T\Theta\mathbf{v} &= \mathbf{v}^T(I - \overline{\pi}_A)^2\mathbf{v} - \mathbf{v}^T(I - \overline{\pi}_A)A^{1/2}J\widetilde{\mathcal{M}}^{-1}J^TA^{1/2}(I - \overline{\pi}_A)\mathbf{v} \\ &= \mathbf{v}^T(I - \overline{\pi}_A - (I - \overline{\pi}_A)A^{1/2}J\widetilde{\mathcal{M}}^{-1}J^TA^{1/2}(I - \overline{\pi}_A))\mathbf{v} \\ &\leq \left(1 - \frac{1}{K}\right)\mathbf{v}^T(I - \overline{\pi}_A)\mathbf{v}. \end{aligned}$$

Here,

$$K = \max_{\mathbf{v}} \frac{\mathbf{v}^T(I - \overline{\pi}_A)\mathbf{v}}{\mathbf{v}^T\left((I - \overline{\pi}_A)A^{1/2}J\widetilde{\mathcal{M}}^{-1}J^TA^{1/2}(I - \overline{\pi}_A)\right)\mathbf{v}}, \tag{3.16}$$

is the best (minimal) possible constant. Letting $\mathbf{v} := A^{1/2}\mathbf{v}$ and introducing the new projection $\pi_A = A^{-1/2}\overline{\pi}_A A^{1/2} = P\mathcal{B}^{-1}P^TA$ the above formula takes the form

$$K = \max_{\mathbf{v}} \frac{((I - \pi_A)\mathbf{v})^TA(I - \pi_A)\mathbf{v}}{\mathbf{v}^T((I - \pi_A)AJ\widetilde{\mathcal{M}}^{-1}J^TA(I - \pi_A))\mathbf{v}}. \tag{3.17}$$

3.2.7 The sharp spectral equivalence result

Here we simplify the expression for K (3.17), obtained in the previous section, where K is the best constant taking part in the spectral equivalence relations between A and B,

$$\mathbf{v}^T A\mathbf{v} \leq \mathbf{v}^T B\mathbf{v} \leq K\ \mathbf{v}^T A\mathbf{v}. \tag{3.18}$$

We can avoid the inverses in (3.17) and show the following equivalent result.

Theorem 3.17. *Assume that J and P are such that any vector $\mathbf{v}$ can be decomposed as $\mathbf{v} = J\mathbf{w} + P\mathbf{x}$ and this does not have to be a direct decomposition. Introduce the projections $\pi_A = P\mathcal{B}^{-1}P^TA$, ($\mathcal{B} = P^TAP$), and $\overline{\pi}_A = A^{1/2}P\mathcal{B}^{-1}P^TA^{1/2}$, and let $\widetilde{\mathcal{M}} = \mathcal{M}^T(\mathcal{M} + \mathcal{M}^T - \mathcal{A})^{-1}\mathcal{M}$.*

The best constant K in (3.18) is given by the expression,

$$\begin{aligned} K &= \sup_{\mathbf{v}\in Range(I-\pi_A)}\ \inf_{\mathbf{w}:\ \mathbf{v}=(I-\pi_A)J\mathbf{w}} \frac{\mathbf{w}^T\widetilde{\mathcal{M}}\mathbf{w}}{\mathbf{v}^TA\mathbf{v}} \\ &= \sup_{\mathbf{v}}\ \inf_{\mathbf{w}:\ \mathbf{v}=(I-\pi_A)J\mathbf{w}} \frac{\mathbf{w}^T\widetilde{\mathcal{M}}\mathbf{w}}{\mathbf{v}^TA(I-\pi_A)\mathbf{v}}. \end{aligned} \tag{3.19}$$

In the case of J and P providing unique decomposition, that is $[J,\ P]$ being an invertible square matrix, the following simplified expression for K holds.

$$K = \sup_{\mathbf{w}} \frac{\mathbf{w}^T\widetilde{\mathcal{M}}\mathbf{w}}{\mathbf{w}^TJ^TA(I-\pi_A)J\mathbf{w}}. \tag{3.20}$$

Proof. Let $N = (I - \overline{\pi}_A)A^{1/2}J$. Then

$$N^TN = J^T(A^{1/2}(I-\overline{\pi}_A)^2A^{1/2})J = J^TA(I-\pi_A)J.$$

Define

$$T = \widetilde{\mathcal{M}} - J^TA(I-\pi_A)J = \widetilde{\mathcal{M}} - \mathcal{A} + J^TA\pi_{AJ}.$$

It is clear that T is symmetric positive semidefinite. We also have

$$\widetilde{\mathcal{M}} = T + N^TN.$$

The key point in what follows is to notice that we consider $Z = N(T+N^TN)^{-1}N^T$ as a mapping from Range(N) = Range($(I-\overline{\pi}_A)A^{1/2}J$) into the same space.

Then the "saddle-point" lemma 3.4 gives us the identity for $Z = N(T + N^TN)^{-1}N^T$ (for any $\mathbf{v}$ in the range of N),

$$\frac{\mathbf{v}^TZ^{-1}\mathbf{v}}{\mathbf{v}^T\mathbf{v}} = 1 + \inf_{\mathbf{w}:\ N\mathbf{w}=\mathbf{v}} \frac{\mathbf{w}^TT\mathbf{w}}{\mathbf{v}^T\mathbf{v}}.$$

This implies

$$\max_{\mathbf{v}\in \text{Range}(N)} \frac{\mathbf{v}^T\mathbf{v}}{\mathbf{v}^T Z\mathbf{v}} = \max_{\mathbf{v}\in \text{Range}(N)} \frac{\mathbf{v}^T Z^{-1}\mathbf{v}}{\mathbf{v}^T\mathbf{v}}$$
$$= 1 + \sup_{\mathbf{v}\in \text{Range}(N)} \inf_{\mathbf{w}:\, N\mathbf{w}=\mathbf{v}} \frac{\mathbf{w}^T T\mathbf{w}}{\mathbf{v}^T\mathbf{v}}. \tag{3.21}$$

Note that $Z = (I - \overline{\pi}_A)A^{1/2}J\widetilde{\mathcal{M}}^{-1}J^T A^{1/2}(I - \overline{\pi}_A)$. We also have $\mathbf{v} = N\mathbf{w} = (I - \overline{\pi}_A)A^{1/2}J\mathbf{w}$. Replace now $\mathbf{v} := A^{1/2}\mathbf{v}$. This implies, $A^{-(1/2)}N\mathbf{w} = \mathbf{v}$, or $\mathbf{v} = (I - \pi_A)J\mathbf{w}$. That is, now $\mathbf{v}$ belongs to the space $\text{Range}((I - \pi_A)J)$. It is clear that because any vector $\mathbf{z}$ admits the decomposition $\mathbf{z} = J\mathbf{x} + P\mathbf{y}$ and because $(I - \pi_A)P = 0$, that $(I - \pi_A)\mathbf{z} = (I - \pi_A)J\mathbf{x}$. Therefore, $\text{Range}(I - \pi_A) = \text{Range}((I - \pi_A)J)$. Similarly, $\text{Range}(I - \overline{\pi}_A) = \text{Range}((I - \overline{\pi}_A)^{1/2}J) = \text{Range}(N)$, seen from the fact that any vector $\mathbf{z}$ admits the decomposition $\mathbf{z} = A^{1/2}J\mathbf{x} + A^{1/2}P\mathbf{y}$ which implies $(I - \overline{\pi}_A)\mathbf{z} = (I - \overline{\pi}_A)A^{1/2}J\mathbf{x}$ using the fact that $(I - \overline{\pi}_A)A^{1/2}P = 0$. Identity (3.16), combined with (3.21), takes the form,

$$K = \sup_{\mathbf{v}\in \text{Range}(I-\overline{\pi}_A)} \frac{\mathbf{v}^T\mathbf{v}}{\mathbf{v}^T (I - \overline{\pi}_A)^T A^{\frac{1}{2}} J\widetilde{\mathcal{M}}^{-1}J^T A^{\frac{1}{2}}(I - \overline{\pi}_A)\mathbf{v}}$$
$$= \sup_{\mathbf{v}\in \text{Range}(N)} \frac{\mathbf{v}^T\mathbf{v}}{\mathbf{v}^T Z\mathbf{v}}$$
$$= 1 + \sup_{\mathbf{v}\in \text{Range}(I-\pi_A)} \inf_{\mathbf{w}:\, \mathbf{v}=(I-\pi_A)J\mathbf{w}} \frac{\mathbf{w}^T T\mathbf{w}}{\mathbf{v}^T A\mathbf{v}}.$$

We have

$$\mathbf{w}^T T\mathbf{w} = \mathbf{w}^T\widetilde{\mathcal{M}}\mathbf{w} - \mathbf{w}^T N^T N\mathbf{w} = \mathbf{w}^T\widetilde{\mathcal{M}}\mathbf{w} - \mathbf{v}^T A\mathbf{v}.$$

That is,

$$K = \sup_{\mathbf{v}\in \text{Range}(I-\pi_A)} \inf_{\mathbf{w}:\, \mathbf{v}=(I-\pi_A)J\mathbf{w}} \frac{\mathbf{w}^T\widetilde{\mathcal{M}}\mathbf{w}}{\mathbf{v}^T A\mathbf{v}}.$$

This shows the first desired identity (3.19).

To prove the second one (3.20), note that $\mathbf{v} = (I - \pi_A)J\mathbf{w} = J\mathbf{w} + P(-\mathcal{B}^{-1}P^T A)(J\mathbf{w})$. Then, in the case when J and P provide unique decomposition of $\mathbf{v} = J\mathbf{w} + P\mathbf{x}$, the latter shows that the second component of $\mathbf{v}$, $P\mathbf{x}$, satisfies $\mathbf{x} = -\mathcal{B}^{-1}(P^T A)(J\mathbf{w})$ and it is unique. That is,

$$K = \sup_{\mathbf{w}} \frac{\mathbf{w}^T\widetilde{\mathcal{M}}\mathbf{w}}{\mathbf{w}^T J^T A(I - \pi_A)J\mathbf{w}},$$

which is (3.20). □

3.2.8 Analysis of B_{TL}

We now derive an upper bound of K in the case of J and P providing unique decomposition; that is, we bound $K = K_{TL}$ corresponding to the two-level preconditioner B_{TL} defined in Definition 3.12.

Let $\gamma^2 \in [0, 1)$ be the best constant in the following strengthened Cauchy–Schwarz inequality

$$(\mathbf{w}^T J^T A P \mathbf{x})^2 \leq \gamma^2 \, \mathbf{w}^T J^T A J \mathbf{w} \; \mathbf{x}^T P^T A P \mathbf{x}. \tag{3.22}$$

An equivalent form of this inequality reads (see Lemma 3.3, inequality (3.2)),

$$\mathbf{w}^T \mathcal{A} \mathbf{w} \leq \frac{1}{1-\gamma^2} \inf_{\mathbf{x}} (J\mathbf{w} + P\mathbf{x})^T A \, (J\mathbf{w} + P\mathbf{x}) \, .$$

The latter minimum is attained at $P\mathbf{x} = -\pi_A J \mathbf{w}$. Therefore,

$$\begin{aligned} \mathbf{w}^T \mathcal{A} \mathbf{w} &\leq \frac{1}{1-\gamma^2} \, \mathbf{w}^T (J^T (I - \pi_A)^T A (I - \pi_A) J) \mathbf{w} \\ &= \frac{1}{1-\gamma^2} \, \mathbf{w}^T (J^T A (I - \pi_A) J) \mathbf{w}. \end{aligned}$$

Using the latter estimate in (3.20), for $\mathcal{D} = \mathcal{B}$, we arrive at the following upper bound,

$$K \leq \sup_{\mathbf{w}} \frac{\mathbf{w}^T \widetilde{\mathcal{M}} \mathbf{w}}{\mathbf{w}^T \mathcal{A} \mathbf{w}} \, \frac{1}{1-\gamma^2}.$$

In general, we have the following result.

Theorem 3.18. *Assume that $\mathcal{M}$ provides a convergent splitting for $\mathcal{A}$ in the $\mathcal{A}$-inner product (i.e., that $(\mathcal{M} + \mathcal{M}^T - \mathcal{A})$ is s.p.d.). Let also $\gamma \in [0, 1)$ be the constant in the strengthened Cauchy–Schwarz inequality* (3.22)*. Then B, with $\mathcal{D} = \mathcal{B}$, and A are spectrally equivalent and the following bounds hold,*

$$\mathbf{v}^T A \mathbf{v} \leq \mathbf{v}^T B \mathbf{v} \leq K \mathbf{v}^T A \mathbf{v},$$

where

$$K \leq \frac{1}{1-\gamma^2} \sup_{\mathbf{w}} \frac{\mathbf{w}^T \widetilde{\mathcal{M}} \mathbf{w}}{\mathbf{w}^T \mathcal{A} \mathbf{w}}.$$

We recall that $\widetilde{\mathcal{M}} = \mathcal{M}^T (\mathcal{M} + \mathcal{M}^T - \mathcal{A})^{-1} \mathcal{M}$. In the case of the inexact second block $\mathcal{D}$ which satisfies

$$0 \leq \mathbf{x}^T (\mathcal{D} - \mathcal{B}) \mathbf{x} \leq \delta \, \mathbf{x}^T \mathcal{B} \mathbf{x},$$

the following perturbation result holds,

$$\mathbf{v}^T A \mathbf{v} \leq \mathbf{v}^T B \mathbf{v} \leq \left(K + \frac{\delta}{1-\gamma^2} \right) \mathbf{v}^T A \mathbf{v}.$$

Proof. Denote with B_{exact} the preconditioner with $\mathcal{D} = \mathcal{B}$. We have for $\mathbf{v} = J\mathbf{w} + P\mathbf{x}$,

$$\begin{aligned}
0 \le \mathbf{v}^T(B - A)\mathbf{v} &= \mathbf{v}^T(B_{\text{exact}} - A)\mathbf{v} + \mathbf{x}^T(\mathcal{D} - \mathcal{B})\mathbf{x},\\
&\le \mathbf{v}^T(B_{\text{exact}} - A)\mathbf{v} + \delta\mathbf{x}^T\mathcal{B}\mathbf{x}\\
&\le \mathbf{v}^T(B_{\text{exact}} - A)\mathbf{v} + \frac{\delta}{1-\gamma^2}\,\mathbf{x}^T\mathcal{S}\mathbf{x}\\
&\le \mathbf{v}^T(B_{\text{exact}} - A)\mathbf{v} + \frac{\delta}{1-\gamma^2}\,\mathbf{v}^T A\mathbf{v}\\
&\le \left(K - 1 + \frac{\delta}{1-\gamma^2}\right)\mathbf{v}^T A\mathbf{v}.
\end{aligned}$$

We used the fact that the Schur complement $\mathcal{S}$ of A is spectrally equivalent to the principal submatrix $\mathcal{B}$ of A which is an equivalent statement of the strengthened Cauchy–Schwarz inequality (3.22). □

3.2.9 Analysis of B_{TG}

The analysis for $B = B_{TG}$ defined in Definition 3.13, follows from estimate (3.19) proved in Theorem 3.17. We have now $J = I$, $\mathcal{M} = M$ is a given smoother for A. We assume here that $\mathcal{D} = \mathcal{B} = P^T A P$.

We are estimating the best constant $K = K_{TG}$ such that

$$\mathbf{v}^T A\mathbf{v} \le \mathbf{v}^T B_{TG}\mathbf{v} \le K_{TG}\,\mathbf{v}^T A\mathbf{v}.$$

Based on estimate (3.19), the best constant $K = K_{TG}$ in the present case (with $J = I$ and $\mathcal{M} = M$), is given by the identity

$$K = \sup_{\mathbf{v}} \inf_{\mathbf{w}:\ \mathbf{v}=(I-\pi_A)\mathbf{w}} \frac{\mathbf{w}^T\widetilde{M}\mathbf{w}}{\mathbf{v}^T A\mathbf{v}}.$$

That is,

$$K = \sup_{\mathbf{v}} \frac{\inf_{\mathbf{w}} (\pi_A\mathbf{w} + (I-\pi_A)\mathbf{v})^T\widetilde{M}(\pi_A\mathbf{w} + (I-\pi_A)\mathbf{v})}{((I-\pi_A)\mathbf{v})^T A((I-\pi_A)\mathbf{v})}.$$

Introduce the projection $\pi_{\widetilde{M}} = P(P^T\widetilde{M}P)^{-1}P^T\widetilde{M}$. Let $\widetilde{M}_c = P^T\widetilde{M}P$. The inf over $\mathbf{w}$ is attained at $\mathbf{w}:\ \pi_A(\mathbf{v}-\mathbf{w}) = \pi_{\widetilde{M}}\mathbf{v}$; that is,

$$A_c^{-1}P^T A(\mathbf{v}-\mathbf{w}) = \widetilde{M}^{-1}P^T\widetilde{M}\mathbf{v}.$$

Then $\pi_{\widetilde{M}} = P\widetilde{M}_c^{-1}P^T\widetilde{M}$. Let $\mathbf{w} = P\mathbf{w}_c$, where

$$\mathbf{w}_c = A_c^{-1}P^T A\mathbf{v} - \widetilde{M}_c^{-1}P^T\widetilde{M}\,\mathbf{v}.$$

We then have $\pi_A\mathbf{w} = \mathbf{w} = P\mathbf{w}_c = (\pi_A - \pi_{\widetilde{M}})\mathbf{v}$. Therefore, $\pi_A\mathbf{w} + (I-\pi_A)\mathbf{v} = (I-\pi_{\widetilde{M}})\mathbf{v}$. Thus we arrived at the final estimate which is formulated in the next theorem.

Theorem 3.19. *The two-grid preconditioner B_{TG} defined from the iteration matrix $I - B_{TG}^{-1}A = (I - M^{-1}A)(I - P\mathcal{B}^{-1}P^TA)(I - M^{-T}A)$ with $\mathcal{B} = P^TAP$ or as in Definition 3.13, is spectrally equivalent to A and the following sharp estimate holds,*

$$\mathbf{v}^TA\mathbf{v} \le \mathbf{v}^TB_{TG}\mathbf{v} \le K_{TG}\,\mathbf{v}^TA\mathbf{v},$$

where

$$K_{TG} = \sup_{\mathbf{v}} \frac{((I - \pi_{\widetilde{M}})\mathbf{v})^T\widetilde{M}(I - \pi_{\widetilde{M}})\mathbf{v}}{((I - \pi_A)\mathbf{v})^TA(I - \pi_A)\mathbf{v}} = \sup_{\mathbf{v}} \frac{\mathbf{v}^T\widetilde{M}(I - \pi_{\widetilde{M}})\mathbf{v}}{\mathbf{v}^TA\mathbf{v}}, \qquad (3.23)$$

is the best possible constant. We recall, that $\widetilde{M} = M(M + M^T - A)^{-1}M^T$ is the symmetrized "smoother" and $\pi_{\widetilde{M}} = P(P^T\widetilde{M}P)^{-1}P^T\widetilde{M}$ is the $\widetilde{M}$-based projection.

Proof. It remains to show that the two formulas are the same. Note first that $(I - \pi_{\widetilde{M}})P(*) = 0$, hence $(I - \pi_{\widetilde{M}})(I - \pi_A) = I - \pi_{\widetilde{M}}$. Thus

$$K_{TG} = \sup_{\mathbf{v}=(I-\pi_A)\mathbf{v}} \frac{((I - \pi_{\widetilde{M}})\mathbf{v})^T\widetilde{M}(I - \pi_{\widetilde{M}})\mathbf{v}}{\mathbf{v}^TA\mathbf{v}} \le \sup_{\mathbf{v}} \frac{\mathbf{v}^T\widetilde{M}(I - \pi_{\widetilde{M}})\mathbf{v}}{\mathbf{v}^TA\mathbf{v}}.$$

On the other hand, $\mathbf{v}^TA\mathbf{v} \ge \mathbf{v}^TA(I - \pi_A)\mathbf{v}$, hence

$$\sup_{\mathbf{v}} \frac{\mathbf{v}^T\widetilde{M}(I - \pi_{\widetilde{M}})\mathbf{v}}{\mathbf{v}^TA\mathbf{v}} \le \sup_{\mathbf{v}} \frac{\mathbf{v}^T\widetilde{M}(I - \pi_{\widetilde{M}})\mathbf{v}}{\mathbf{v}^TA(I - \pi_A)\mathbf{v}} = K_{TG}. \qquad \square$$

Corollary 3.20. *Let $\widetilde{M}$ be spectrally equivalent to a s.p.d. matrix D, such that*

$$c_1\,\mathbf{v}^TD\mathbf{v} \le \mathbf{v}^T\widetilde{M}\mathbf{v} \le c_2\,\mathbf{v}^TD\mathbf{v} \quad \text{for all } \mathbf{v}.$$

Then, with $\pi_D = P(P^TDP)^{-1}P^TD$ being the D-based projection on the coarse space Range (P), the following two-sided estimates hold for K_{TG},

$$c_1 \sup_{\mathbf{v}} \frac{\mathbf{v}^TD(I - \pi_D)\mathbf{v}}{\mathbf{v}^TA\mathbf{v}} \le K_{TG} \le c_2 \sup_{\mathbf{v}} \frac{\mathbf{v}^TD(I - \pi_D)\mathbf{v}}{\mathbf{v}^TA\mathbf{v}}.$$

Proof. The proof readily follows from identity (3.23), the property of the projection $\pi_{\widetilde{M}}$,

$$\|(I - \pi_{\widetilde{M}})\mathbf{v}\|_{\widetilde{M}}^2 = \min_{\mathbf{v}_c} \|\mathbf{v} - P\mathbf{v}_c\|_{\widetilde{M}}^2,$$

the spectral equivalence relations between D and $\widetilde{M}$, and similar property of the projection π_D; that is,

$$\|(I - \pi_D)\mathbf{v}\|_D^2 = \min_{\mathbf{v}_c} \|\mathbf{v} - P\mathbf{v}_c\|_D^2. \qquad \square$$

The following two examples often appear in practice.

Example 3.21. Let M be s.p.d. such that $M - A$ is positive semidefinite. Then $\widetilde{M} = M\,(2M - A)^{-1}\,M$ is spectrally equivalent to M such that $\frac{1}{2}\;\mathbf{v}^T M\mathbf{v} \le \mathbf{v}^T \widetilde{M}\mathbf{v} \le \mathbf{v}^T M\mathbf{v}$. Note that with the scaling $M - A$ being positive semidefinite, we guarantee that M is an A-convergent smoother for A. From Corollary 3.20 we have then the estimates

$$\frac{1}{2}\sup_{\mathbf{v}} \frac{\mathbf{v}^T M(I - \pi_M)\mathbf{v}}{\mathbf{v}^T A\mathbf{v}} \le K_{TG} \le \sup_{\mathbf{v}} \frac{\mathbf{v}^T M(I - \pi_M)\mathbf{v}}{\mathbf{v}^T A\mathbf{v}}.$$

The second example deals with the Gauss–Seidel smoother.

Example 3.22. Consider $M = D - L$ coming from the splitting of A, $A = D - L - U$ where D is the diagonal of A and $-L$ is the strictly lower triangular part of A. Then $\widetilde{M} = (D - U)D^{-1}(D - L)$ is spectrally equivalent to D. More specifically, as shown in Proposition 6.12, we have

$$\frac{1}{4}\,\mathbf{v}^T D\mathbf{v} \le \mathbf{v}^T \widetilde{M}\mathbf{v} \le \kappa^2\;\mathbf{v}^T D\mathbf{v},$$

where κ is bounded by the maximum number of nonzero entries of A per row. Then, to estimate K_{TG} it is sufficient to estimate

$$\kappa^2 \sup_{\mathbf{v}} \frac{\mathbf{v}^T D(I - \pi_D)\mathbf{v}}{\mathbf{v}^T A\mathbf{v}}.$$

Necessary conditions for two-grid convergence

Using the inequalities

$$\|(I - \pi_A)\mathbf{w}\|_A = \inf_{\mathbf{v}\in \mathrm{Range}(P)} \|\mathbf{v} - \mathbf{w}\|_A \le \|(I - \pi_{\widetilde{M}})\mathbf{w}\|_A,$$

and (see Proposition 3.8)

$$\mathbf{v}^T A\mathbf{v} \le \mathbf{v}^T \widetilde{M}\mathbf{v},$$

we obtain the following main corollaries from estimate (3.23) of Theorem 3.19, which are hence necessary conditions for two-grid convergence.

Assume that for an A-convergent smoother M and interpolation matrix P the resulting two-grid preconditioner B_{TG} is spectrally equivalent to A and let $K_{TG} \ge 1$ be an upper bound of the spectral equivalence relations $\mathbf{v}^T A\mathbf{v} \le \mathbf{v}^T B_{TG}\mathbf{v} \le K_{TG}\,\mathbf{v}^T A\mathbf{v}$. Then the following two corollaries are necessary conditions for this spectral equivalence to hold.

Corollary 3.23. *For any* $\mathbf{v}$ *in the space Range*$(I - \pi_{\widetilde{M}})$*, we have the spectral equivalence relations,*

$$\mathbf{v}^T A\mathbf{v} \le \mathbf{v}^T \widetilde{M}\mathbf{v} \le K_{TG}\,\mathbf{v}^T A\mathbf{v}.$$

That is, in a space complementary to the coarse space Range(P) the symmetrized smoother $\widetilde{M}$ is an efficient preconditioner for A. If we introduce the matrix J such that Range(J) = Range($I - \pi_{\widetilde{M}}$), we have the following spectral equivalence relations between $J^T AJ$ and $J^T \widetilde{M} J$,

$$\mathbf{v}_s^T J^T AJ\mathbf{v}_s \leq \mathbf{v}_s^T J^T \widetilde{M} J\mathbf{v}_s \leq K_{TG}\, \mathbf{v}_s^T J^T AJ\mathbf{v}_s.$$

Corollary 3.24. *The operator $I - \pi_{\widetilde{M}}$ is bounded in the A-norm; that is, the following estimate holds.*

$$((I - \pi_{\widetilde{M}})\mathbf{v})^T A(I - \pi_{\widetilde{M}})\mathbf{v} \leq K_{TG}\, \mathbf{v}^T A\mathbf{v}.$$

It is equivalent also to say (due to Kato's Lemma 3.6) that $\pi_{\widetilde{M}}$ is bounded in energy norm (with the same constant K_{TG}); that is,

$$(\pi_{\widetilde{M}}\mathbf{v})^T A\pi_{\widetilde{M}}\mathbf{v} \leq K_{TG}\, \mathbf{v}^T A\mathbf{v}.$$

Finally, another equivalent statement is that the spaces Range(J) $\equiv$ Range($I - \pi_{\widetilde{M}}$) and Range(P) have a nontrivial angle in the A inner product; that is,

$$\left(\mathbf{v}_s^T J^T AP\mathbf{x}\right)^2 \leq \left(1 - \frac{1}{K_{TG}}\right) \mathbf{v}_s^T J^T AJ\mathbf{v}_s\, \mathbf{x}^T P^T AP\mathbf{x}, \quad \textit{for any } \mathbf{v}_s \textit{ and } \mathbf{x}.$$

Proof. The last equivalence statements are proved in the same way as Lemma 3.3, by considering the quadratic form $Q(t) = (\pi_{\widetilde{M}}\mathbf{v} + tJ\mathbf{v}_s)^T A(\pi_{\widetilde{M}}\mathbf{v} + tJ\mathbf{v}_s) - (1/K_{TG})(\pi_{\widetilde{M}}\mathbf{v})^T A\pi_{\widetilde{M}}\mathbf{v}$. Note that $\pi_{\widetilde{M}} J\mathbf{v}_s = 0$, hence $\pi_{\widetilde{M}}(\mathbf{v} + tJ\mathbf{v}_s) = \pi_{\widetilde{M}}\mathbf{v}$. This shows that $Q(t) \geq 0$ for any real t if $\pi_{\widetilde{M}}$ is A-bounded. Then, the fact that its discriminant is nonpositive shows the strengthened Cauchy–Schwarz inequality because Range($\pi_{\widetilde{M}}$) = Range(P). The argument goes both ways. Namely, the strengthened Cauchy–Schwarz inequality implies that the discriminant is nonpositive, hence Q is nonnegative; that is, $\pi_{\widetilde{M}}$ is bounded in energy. Due to the symmetry of the strengthened Cauchy–Schwarz inequality, we see that $I - \pi_{\widetilde{M}}$ has the same energy norm as $\pi_{\widetilde{M}}$ if $K_{TG} > 1$. □

3.3 Algebraic two-grid methods and preconditioners; sufficient conditions for spectral equivalence

The last two corollaries 3.23 and 3.24 represent the main foundation of constructing efficient two-grid preconditioners. They motivate us to formulate conditions for two-grid convergence. We show in the remainder of this section that the conditions below are sufficient for two-grid convergence.

Motivated by Corollaries 3.23–3.24, we need to construct a coarse space Range(P) such that there is a complementary one, Range(J), with the properties:

(i) The symmetrized smoother restricted to the subspace Range(J), that is, $J^T \widetilde{M} J$, is spectrally equivalent to the subspace matrix $J^T AJ$.

(ii) The complementary spaces Range(J) and Range(P) have nontrivial angle in the A-inner product; that is, they are almost A-orthogonal.

In practice, we need a sparse matrix P so that the coarse matrix $P^T AP$ is also sparse, whereas the explicit knowledge of the best J is not really needed. If P and J are constructed based solely on A and the smoother M (or the symmetrized one, $\widetilde{M}$), the resulting method (or preconditioner) belongs to the class of "algebraic" two-grid methods (or preconditioners).

In order to guarantee the efficiency of the method, we only need a J (not necessarily the best one defined as Range($I - \pi_{\widetilde{M}}$)) in order to test if the subspace smoother $J^T \widetilde{M} J$ is efficient on the subspace matrix $J^T AJ$. That is, we need an estimate (for the particular J)

$$\mathbf{v}_s^T J^T AJ\mathbf{v}_s \leq \mathbf{v}_s^T J^T \widetilde{M} J\mathbf{v}_s \leq \kappa \ \mathbf{v}_s^T J^T AJ\mathbf{v}_s, \tag{3.24}$$

with a reasonable constant κ. The efficiency of the smoother on a complementary space Range(J) is sometimes referred to as efficient compatible relaxation. The latter notion is due to Achi Brandt (2000), [B00].

The second main ingredient is the energy boundedness of P in the sense that for a small constant η, we want the bound,

$$\mathbf{x}^T P^T AP\mathbf{x} \leq \eta \inf_{\mathbf{v}_s:\ \mathbf{v}=J\mathbf{v}_s+P\mathbf{x}} \mathbf{v}^T A\mathbf{v}. \tag{3.25}$$

Then, we can actually prove the following main result (cf., [FV04]).

Theorem 3.25. *Assume properties (i) and (ii), that is, estimates* (3.24) *and* (3.25). *Then two-grid preconditioner* B_{TG} *is spectrally equivalent to* A *with a constant* $K = K_{TG} \leq \eta\kappa$.

Proof. We have to estimate K defined in (3.23). Because Range(J) is complementary to Range(P) (by assumption), then any $\mathbf{v}$ can be uniquely decomposed as $\mathbf{v} = J\mathbf{v}_s + P\mathbf{x}$. Then the term in the numerator of (3.23) can be estimated as follows.

$$\begin{aligned}((I - \pi_{\widetilde{M}})\mathbf{v})^T \widetilde{M}((I - \pi_{\widetilde{M}})\mathbf{v} &= \inf_{\mathbf{y}} (\mathbf{v} - P\mathbf{y})^T \widetilde{M}(\mathbf{v} - P\mathbf{y}) \\ &\leq (\mathbf{v} - P\mathbf{x})^T \widetilde{M}(\mathbf{v} - P\mathbf{x}) \\ &= \mathbf{v}_s^T J^T \widetilde{M} J\mathbf{v}_s \\ &\leq \kappa \ \mathbf{v}_s^T J^T AJ\mathbf{v}_s.\end{aligned}$$

In the last line we used (3.24).

The energy boundedness of (3.25) implies a strengthened Cauchy–Schwarz inequality for Range(J) and Range(P). That inequality implies (see Lemma 3.3) the following energy boundedness of J,

$$\mathbf{v}_s^T J^T AJ\mathbf{v}_s \leq \eta \inf_{\mathbf{x}:\ \mathbf{v}=J\mathbf{v}_s+P\mathbf{x}} \mathbf{v}^T A\mathbf{v}.$$

Using the projection π_A, we get

$$\mathbf{v}_s^T J^T A J \mathbf{v}_s \leq \eta \, ((I - \pi_A) J \mathbf{v}_s)^T A (I - \pi_A) J \mathbf{v}_s.$$

Finally, because $(I - \pi_A) P\mathbf{x} = 0$, we arrive at the following bound for the denominator of (3.23),

$$\begin{aligned} \mathbf{v}_s^T J^T A J \mathbf{v}_s &\leq \eta \, ((I - \pi_A)(J\mathbf{v}_s + P\mathbf{x}))^T A (I - \pi_A)(J\mathbf{v}_s + P\mathbf{x}) \\ &= \eta \, ((I - \pi_A)\mathbf{v})^T A (I - \pi_A)\mathbf{v}. \end{aligned}$$

Thus, (3.23) is finally estimated as follows.

$$K_{TG} = \sup_{\mathbf{v}} \frac{((I - \pi_{\widetilde{M}})\mathbf{v})^T \widetilde{M} (I - \pi_{\widetilde{M}})\mathbf{v}}{((I - \pi_A)\mathbf{v})^T A (I - \pi_A)\mathbf{v}} \leq \sup_{\mathbf{v}_s} \frac{\kappa \, \mathbf{v}_s^T J^T A J \mathbf{v}_s}{\frac{1}{\eta} \mathbf{v}_s^T J^T A J \mathbf{v}_s} = \kappa \, \eta. \qquad \square$$

A two-grid convergence measure

For a given P and smoother M, let R be a computable restriction matrix such that $RP = I$. This implies that $Q = PR$ is a projection (onto the coarse space Range (P)). Then, Range $(I - Q)$ is a complementary space to the coarse space Range $(P) =$ Range (Q). A typical example is

$$P = \begin{bmatrix} W \\ I \end{bmatrix} \quad \text{and} \quad R = [0, \; I]$$

so that $RP = I$.

Based on a computable projection Q, the following quantity (cf., [FV04]) is sometimes referred to as a measure

$$\mu_{\widetilde{M}} \, (Q, \, \mathbf{e}) = \frac{(\mathbf{e} - Q\mathbf{e})^T \widetilde{M} (I - Q)\mathbf{e}}{\mathbf{e}^T A \mathbf{e}}.$$

Using the minimal distance property of the projection $\pi_{\widetilde{M}}$ in the $\widetilde{M}$-norm, we have the estimate

$$\|(I - \pi_{\widetilde{M}})\mathbf{e}\|_{\widetilde{M}}^2 = \min_{\mathbf{e}_c} \|\mathbf{e} - P\mathbf{e}_c\|_{\widetilde{M}}^2 \leq \|(I - Q)\mathbf{e}\|_{\widetilde{M}}^2.$$

Theorem 3.19 then implies the upper bound $K_{TG} \leq \sup_{\mathbf{e}} \, \mu_{\widetilde{M}} \, (Q, \, \mathbf{e})$. That is, the quantity

$$\sup_{\mathbf{e}} \mu_{\widetilde{M}} \, (Q, \, \mathbf{e}), \tag{3.26}$$

can be used to measure the convergence of the respective two-grid method.

We conclude this section with the comment that we have not so far assumed any particular structure of P. The above example of

$$P = \begin{bmatrix} W \\ I \end{bmatrix}$$

is typical in the case of the algebraic multigrid method (or AMG). Here, the second (identity) block corresponds to rows of A, referred to as "c", or coarse dofs, and the remaining ones to "f", or fine dofs. The latter structure of P is further exploited in the setting of the algebraic multigrid in Chapter 6, especially when specific smoothers of type "c"–"f" relaxation are considered (as in Section 6.8), as well as some other specific topics.

3.4 Classical two-level block-factorization preconditioners

If the matrix admits a stable two-by-two block, in the sense that the off-diagonal block $\mathcal{L} = \mathcal{R}^T$ of

$$A = \begin{bmatrix} \mathcal{A} & \mathcal{R} \\ \mathcal{L} & \mathcal{B} \end{bmatrix}$$

is dominated by its main diagonal such that for a constant $\gamma \in [0, 1)$ we have

$$(\mathbf{w}^T \mathcal{R}\mathbf{x})^2 \leq \gamma \; \mathbf{w}^T \mathcal{A}\mathbf{w} \; \mathbf{x}^T \mathcal{B}\mathbf{x}, \tag{3.27}$$

we can approximate $\mathcal{A}$ and the Schur complement $\mathcal{S} = \mathcal{B} - \mathcal{L}\mathcal{A}^{-1}\mathcal{R}$ with s.p.d. matrices $\mathcal{M}$ and $\mathcal{D}$ and the resulting approximate block-factorization matrix

$$B = \begin{bmatrix} \mathcal{M} & 0 \\ \mathcal{L} & \mathcal{D} \end{bmatrix} \begin{bmatrix} I & \mathcal{M}^{-1}\mathcal{R} \\ 0 & I \end{bmatrix}, \tag{3.28}$$

is spectrally equivalent to A.

We note that B is different from B_{TL} (or B_{TG}) because it does not correspond to a product iteration method (if $\mathcal{M} \neq \mathcal{A}$). Recall that a corresponding B_{TG} takes the following explicit form,

$$B_{TG} = \begin{bmatrix} \mathcal{M} & 0 \\ \mathcal{L} & \mathcal{D} \end{bmatrix} \begin{bmatrix} (2\mathcal{M} - \mathcal{A})^{-1}\mathcal{M} & 0 \\ 0 & I \end{bmatrix} \begin{bmatrix} I & \mathcal{M}^{-1}\mathcal{R} \\ 0 & I \end{bmatrix};$$

that is, the (minor) difference is the extra factor involving $(2\mathcal{M} - \mathcal{A})^{-1}\mathcal{M}$. In the case of $\mathcal{M}$ being spectrally equivalent to $\mathcal{A}$ and also $\mathcal{M}$ giving an $\mathcal{A}$-convergent iteration for solving systems with $\mathcal{A}$ so that the resulting B_{TG} is spectrally equivalent to A, then the middle factor in question can be dropped out without losing overall spectral equivalence. The definition (3.28) of B does not require that $\mathcal{M}$ be scaled so that $2\mathcal{M} - \mathcal{A}$ is s.p.d. (which is equivalent to $\|I - \mathcal{M}^{-1}\mathcal{A}\|_{\mathcal{A}} < 1$).

To implement B^{-1}, we use the standard forward and backward elimination sweeps.

Algorithm 3.4.1 (Computing actions of B^{-1}) *Consider*

$$B \begin{bmatrix} \mathbf{w} \\ \mathbf{x} \end{bmatrix} = \begin{bmatrix} \mathbf{f} \\ \mathbf{g} \end{bmatrix}.$$

To compute $\mathbf{w}$ *and* $\mathbf{x}$*, we perform the following steps.*

- *Compute* $\mathbf{w} = \mathcal{M}^{-1}\mathbf{f}$.
- *Compute* $\mathbf{x} = \mathcal{D}^{-1}\,(\mathbf{g} - \mathcal{L}\mathbf{w})$.
- *Compute* $\mathbf{w} := \mathbf{w} - \mathcal{M}^{-1}\mathcal{R}\mathbf{x}$.

The following is a classical result originated by Axelsson and Gustafsson [AG83].

Theorem 3.26. *Let* $\mathcal{M}$ *and* $\mathcal{D}$ *be s.p.d. spectrally equivalent preconditioners to* $\mathcal{A}$ *and* $\mathcal{B}$*, respectively,*

$$\alpha\ \mathbf{w}^T\mathcal{M}\mathbf{w} \le \mathbf{w}^T\mathcal{A}\mathbf{w} \le \beta\ \mathbf{w}^T\mathcal{M}\mathbf{w},$$

and

$$\sigma\ \mathbf{x}^T\mathcal{D}\mathbf{x} \le \mathbf{x}^T\mathcal{B}\mathbf{x} \le \eta\ \mathbf{x}^T\mathcal{D}\mathbf{x}.$$

Then, if A *is "stable" in the sense of inequality* (3.27)*, then* B *defined in* (3.28) *is spectrally equivalent to* A*, and the following spectral equivalence estimates hold.*

$$b_1\ \mathbf{v}^T B\mathbf{v} \le \mathbf{v}^T A\mathbf{v} \le b_2\ \mathbf{v}^T B\mathbf{v}$$

for positive constants $b_1,\ b_2$ *depending on* α*,* β*,* σ*, and* η *(see the proof). We can also consider the block-diagonal preconditioner*

$$D = \begin{bmatrix} \mathcal{M} & 0 \\ 0 & \mathcal{D} \end{bmatrix}. \tag{3.29}$$

We similarly have the estimates, for two positive constants d_1 *and* d_2 *depending on* α*,* β*,* σ*, and* η *(see the proof),*

$$d_1\ \mathbf{v}^T D\mathbf{v} \le \mathbf{v}^T A\mathbf{v} \le d_2\ \mathbf{v}^T D\mathbf{v}.$$

Proof. We have, for any $\zeta > 0$,

$$\begin{aligned}
\mathbf{v}^T A\mathbf{v} &= \mathbf{w}^T\mathcal{A}\mathbf{w} + 2\mathbf{w}^T\mathcal{R}\mathbf{x} + \mathbf{x}^T\mathcal{B}\mathbf{x} \\
&\le \mathbf{w}^T\mathcal{A}\mathbf{w} + 2\frac{\gamma}{\sqrt{\zeta}}\ \sqrt{\mathbf{w}^T\mathcal{A}\mathbf{w}}\left(\sqrt{\zeta}\ \sqrt{\mathbf{x}^T\mathcal{B}\mathbf{x}}\right) + \mathbf{x}^T\mathcal{B}\mathbf{x} \\
&\le \left(1 + \frac{\gamma^2}{\zeta}\right)\ \mathbf{w}^T\mathcal{A}\mathbf{w} + (1+\zeta)\ \mathbf{x}^T\mathcal{B}\mathbf{x} \\
&\le \left(1 + \frac{\gamma^2}{\zeta}\right)\beta\ \mathbf{w}^T\mathcal{M}\mathbf{w} + (1+\zeta)\eta\ \mathbf{x}^T\mathcal{D}\mathbf{x} \\
&\le d_2\ \mathbf{v}^T D\mathbf{v}.
\end{aligned}$$

The upper bound d_2 is minimal for

$$\zeta = \frac{\beta - \eta + \sqrt{(\beta-\eta)^2 + 4\gamma^2\eta\beta}}{2\eta}$$

and then

$$d_2 = \left(1 + \frac{\gamma^2}{\zeta}\right) \beta = (1+\zeta)\eta = \frac{\beta + \eta + \sqrt{(\beta-\eta)^2 + 4\gamma^2 \eta\beta}}{2\eta}.$$

Similarly, we have the estimate from below:

$$\begin{aligned}
\mathbf{v}^T A \mathbf{v} &= \mathbf{w}^T \mathcal{A} \mathbf{w} + 2\mathbf{w}^T \mathcal{R} \mathbf{x} + \mathbf{x}^T \mathcal{B} \mathbf{x} \\
&\geq \mathbf{w}^T \mathcal{A} \mathbf{w} - 2\frac{\gamma}{\sqrt{\zeta}} \sqrt{\mathbf{w}^T \mathcal{A} \mathbf{w}} \left(\sqrt{\zeta}\ \sqrt{\mathbf{x}^T \mathcal{B} \mathbf{x}}\right) + \mathbf{x}^T \mathcal{B} \mathbf{x} \\
&\geq \left(1 - \frac{\gamma^2}{\zeta}\right) \mathbf{w}^T \mathcal{A} \mathbf{w} + (1-\zeta)\ \mathbf{x}^T \mathcal{B} \mathbf{x} \\
&\geq \left(1 - \frac{\gamma^2}{\zeta}\right) \alpha\ \mathbf{w}^T \mathcal{M} \mathbf{w} + (1-\zeta)\sigma\ \mathbf{x}^T \mathcal{D} \mathbf{x} \\
&\geq d_1\ \mathbf{v}^T D \mathbf{v}.
\end{aligned}$$

The lower bound is maximal for

$$\zeta = \frac{\sigma - \alpha + \sqrt{(\sigma-\alpha)^2 + 4\gamma^2\sigma\alpha}}{2\sigma}$$

and then

$$d_1 = \left(1 - \frac{\gamma^2}{\zeta}\right) \alpha = (1-\zeta)\sigma = \frac{2(1-\gamma^2)\sigma\alpha}{\sigma + \alpha + \sqrt{(\sigma-\alpha)^2 + 4\gamma^2\sigma\alpha}}.$$

To analyze B, we proceed similarly. We have $\mathbf{v}^T B \mathbf{v} = \mathbf{w}^T \mathcal{M} \mathbf{w} + 2\ \mathbf{w}^T \mathcal{R} \mathbf{x} + \mathbf{x}^T \mathcal{D} \mathbf{x} + \mathbf{x}^T \mathcal{L} \mathcal{M}^{-1} \mathcal{R} \mathbf{x}$. Because $\mathbf{x}^T \mathcal{L} \mathcal{M}^{-1} \mathcal{R} \mathbf{x} \leq \beta\ \mathbf{x}^T \mathcal{L} \mathcal{A}^{-1} \mathcal{R} \mathbf{x} \leq \beta\gamma^2\ \mathbf{x}^T \mathcal{B} \mathbf{x} \leq \beta\gamma^2\eta\ \mathbf{x}^T \mathcal{D} \mathbf{x}$, we can easily estimate $\mathbf{v}^T B \mathbf{v}$ from above in terms of $\mathbf{v}^T D \mathbf{v} = \mathbf{w}^T \mathcal{M} \mathbf{w} + \mathbf{x}^T \mathcal{D} \mathbf{x}$. More specifically,

$$\begin{aligned}
\mathbf{v}^T B \mathbf{v} &\leq \mathbf{w}^T \mathcal{M} \mathbf{w} + 2\mathbf{w}^T \mathcal{R} \mathbf{x} + (1 + \beta\eta\gamma^2)\ \mathbf{x}^T \mathcal{D} \mathbf{x} \\
&\leq \left(1 + \frac{\gamma^2}{\zeta}\beta\right) \mathbf{w}^T \mathcal{M} \mathbf{w} + (1 + \beta\eta\gamma^2 + \zeta\eta)\ \mathbf{x}^T \mathcal{D} \mathbf{x} \\
&\leq d_2'\ \mathbf{v}^T D \mathbf{v}.
\end{aligned}$$

Here, $d_2' = 1 + (\gamma^2/\zeta)\beta = 1 + \beta\eta\gamma^2 + \zeta\eta$ which gives

$$\zeta = \frac{-\beta\eta\gamma^2 + \sqrt{(\beta\eta\gamma^2)^2 + 4\eta\beta\gamma^2}}{2\eta}$$

and $d_2' = 1 + \frac{1}{2}\left(\beta\eta\gamma^2 + \sqrt{(\beta\eta\gamma^2)^2 + 4\eta\beta\gamma^2}\right)$.

Finally, in the other direction, we first notice that $\mathbf{v}^T B\mathbf{v} \geq \mathbf{x}^T \mathcal{D}\mathbf{x}$ inasmuch as $\mathcal{D}$ is its Schur complement. Then, proceeding as before, we arrive at the inequalities,

$$\begin{aligned}
\mathbf{v}^T B\mathbf{v} &= \mathbf{w}^T \mathcal{M}\mathbf{w} + 2\,\mathbf{w}^T \mathcal{R}\mathbf{x} + \mathbf{x}^T \mathcal{D}\mathbf{x} + \mathbf{x}^T \mathcal{L}\mathcal{M}^{-1}\mathcal{R}\mathbf{x} \\
&\geq \mathbf{w}^T \mathcal{M}\mathbf{w} - \frac{\gamma^2}{\zeta}\,\mathbf{w}^T \mathcal{A}\mathbf{w} - \zeta\,\mathbf{x}^T \mathcal{B}\mathbf{x} + \mathbf{x}^T \mathcal{D}\mathbf{x} + \mathbf{x}^T \mathcal{L}\mathcal{M}^{-1}\mathcal{R}\mathbf{x} \\
&\geq \left(1 - \frac{\gamma^2}{\zeta}\,\beta\right)\,\mathbf{w}^T \mathcal{M}\mathbf{w} + (-\zeta\eta + 1)\,\mathbf{x}^T \mathcal{D}\mathbf{x} \\
&\geq \left(1 - \frac{\gamma^2}{\zeta}\,\beta\right)\,\mathbf{w}^T \mathcal{M}\mathbf{w} + (-\zeta\eta + 1)\,\mathbf{v}^T B\mathbf{v}.
\end{aligned}$$

Here, we assume that $\zeta\eta \geq 1$. Thus,

$$\mathbf{v}^T B\mathbf{v} \geq \frac{1 - \frac{\gamma^2}{\zeta}\,\beta}{\zeta\eta}\,\mathbf{w}^T \mathcal{M}\mathbf{w}.$$

Letting $\zeta = (1/\eta) + \gamma^2\beta > (1/\eta)$, we get

$$\mathbf{v}^T B\mathbf{v} \geq \frac{1}{(1+\gamma^2\beta\eta)^2}\,\mathbf{w}^T \mathcal{M}\mathbf{w},$$

which together with $\mathbf{v}^T B\mathbf{v} \geq \mathbf{x}^T \mathcal{D}\mathbf{x}$ shows

$$\mathbf{v}^T B\mathbf{v} \geq \frac{1-\theta}{(1+\gamma^2\beta\eta)^2}\,\mathbf{w}^T \mathcal{M}\mathbf{w} + \theta\,\mathbf{v}^T \mathcal{D}\mathbf{x}.$$

The latter estimate for

$$\theta = \frac{1-\theta}{(1+\gamma^2\beta\eta)^2} = \frac{1}{1+(1+\gamma^2\beta\eta)^2} \in (0,\,1]$$

gives

$$\mathbf{v}^T B\mathbf{v} \geq d_1'\,\mathbf{v}^T D\mathbf{v},$$

with $d_1' = 1/(1 + (1+\gamma^2\beta\eta)^2)$.

To bound $\mathbf{v}^T B\mathbf{v}$ in terms of $\mathbf{v}^T A\mathbf{v}$, we combine the proven estimates $d_1'\,\mathbf{v}^T D\mathbf{v} \leq \mathbf{v}^T B\mathbf{v} \leq d_2'\mathbf{v}^T D\mathbf{v}$ and $d_1\,\mathbf{v}^T D\mathbf{v} \leq \mathbf{v}^T A\mathbf{v} \leq d_2\,\mathbf{v}^T D\mathbf{v}$. That is, we can let $b_2 = d_2/d_1'$ and $b_1 = d_1/d_2'$, to demonstrate the final desired estimates $b_1\,\mathbf{v}^T B\mathbf{v} \leq \mathbf{v}^T A\mathbf{v} \leq b_2\,\mathbf{v}^T B\mathbf{v}$. □

3.4.1 A general procedure of generating stable block-matrix partitioning

In practice, to construct good-quality block-factorization preconditioners a given block partitioning of a given matrix may not be suitable. In particular, we may not be able to establish a strengthened Cauchy–Schwarz inequality (3.27) with a good constant γ. A general way to achieve a stable form of A is to use "change of variables" in the following sense (cf., [EV91], [VA94], or [ChV03]). Let P be a rectangular matrix (fewer columns than rows) with bounded A-norm. We assume that P has full-column rank. We may then assume that

$$P = \begin{bmatrix} \mathcal{W} \\ I \end{bmatrix}.$$

Then, consider the square, invertible transformation matrix $Y = [J,\ P]$, where

$$J = \begin{bmatrix} I \\ 0 \end{bmatrix}.$$

The A-boundedness of P then can be stated as follows,

$$\mathbf{x}^T P^T A P \mathbf{x} \le \eta \min_{\mathbf{w}:\ \mathbf{v}=J\mathbf{w}+P\mathbf{x}} \mathbf{v}^T A \mathbf{v}.$$

Finally, let $\widehat{A} = Y^T A Y$ be the transformed matrix. We have

$$\widehat{A} = \begin{bmatrix} \widehat{\mathcal{A}} & \widehat{\mathcal{R}} \\ \widehat{\mathcal{L}} & \widehat{\mathcal{B}} \end{bmatrix}.$$

More explicitly,

$$\widehat{\mathcal{A}} = \mathcal{A}, \qquad \widehat{\mathcal{B}} = P^T A P, \qquad \widehat{\mathcal{L}} = \mathcal{L} + \mathcal{W}^T \mathcal{A}, \qquad \widehat{\mathcal{R}} = \mathcal{R} + \mathcal{A}\mathcal{W}.$$

We notice that $\mathcal{S} = \widehat{\mathcal{B}} - \widehat{\mathcal{L}}\widehat{\mathcal{A}}^{-1}\widehat{\mathcal{R}}$; that is, A and the transformed matrix $\widehat{A}$ have the same first principal blocks and the same Schur complements.

Sometimes $\widehat{A}$ is called the HB ("hierarchical basis") matrix. We can then prove (see Lemma 3.3) that $\widehat{A}$ has a stable block form, in the sense that

$$(\mathbf{w}^T \widehat{\mathcal{R}} \mathbf{x})^2 \le \left(1 - \frac{1}{\eta}\right) \mathbf{w}^T \widehat{\mathcal{A}} \mathbf{w}\ \mathbf{x}^T \widehat{\mathcal{B}} \mathbf{x},$$

or equivalently,

$$((J\mathbf{w})^T A (P\mathbf{x}))^2 \le \left(1 - \frac{1}{\eta}\right) (J\mathbf{w})^T A (J\mathbf{w})\ (P\mathbf{x})^T A (P\mathbf{x}).$$

Next, we can first construct a spectrally equivalent preconditioner $\widehat{B}$ to $\widehat{A}$ based on spectrally equivalent preconditioners $\mathcal{M}$ to $\mathcal{A}$ and $\mathcal{D}$ to $\widehat{\mathcal{B}} = P^T A P$. Then $B = Y^{-T} \widehat{B}\, Y^{-1}$ is a spectrally equivalent block-factorization preconditioner to the original matrix A. To summarize, let

$$P = \begin{bmatrix} \mathcal{W} \\ I \end{bmatrix}$$

satisfy

$$\mathbf{x}^T P^T A P \mathbf{x} \le \eta \min_{\mathbf{w}} \begin{bmatrix} \mathbf{w} \\ \mathbf{x} \end{bmatrix}^T A \begin{bmatrix} \mathbf{w} \\ \mathbf{x} \end{bmatrix},$$

and $\mathcal{M}$ and $\mathcal{D}$ are based on $\mathcal{A}$ and $P^T AP$. Then the transformed inexact block-factorization preconditioner $B = Y^{-T}\widehat{B}Y^{-1}$ is spectrally equivalent to A with the same constants established in Theorem 3.26 applied to $\widehat{B}$ and $\widehat{A}$. More specifically, consider

$$\widehat{B} = \begin{bmatrix} \mathcal{M} & 0 \\ \widehat{\mathcal{L}} & \mathcal{D} \end{bmatrix} \begin{bmatrix} I & \mathcal{M}^{-1}\widehat{R} \\ 0 & I \end{bmatrix}.$$

Then we have the following explicit form for the transformed preconditioner

$$\begin{aligned} B &= Y^{-T}\widehat{B}Y^{-1} \\ &= \begin{bmatrix} I & 0 \\ -\mathcal{W}^T & I \end{bmatrix} \begin{bmatrix} \mathcal{M} & 0 \\ \widehat{\mathcal{L}} & \mathcal{D} \end{bmatrix} \begin{bmatrix} I & \mathcal{M}^{-1}\widehat{R} \\ 0 & I \end{bmatrix} \begin{bmatrix} I & -\mathcal{W} \\ 0 & I \end{bmatrix} \\ &= \begin{bmatrix} \mathcal{M} & 0 \\ \mathcal{L} + \mathcal{W}^T(\mathcal{A} - \mathcal{M}) & \mathcal{D} \end{bmatrix} \begin{bmatrix} I & \mathcal{M}^{-1}\mathcal{R} - (I - \mathcal{M}^{-1}\mathcal{A})\mathcal{W} \\ 0 & I \end{bmatrix}. \end{aligned}$$

The following algorithm can be used to implement

$$\begin{bmatrix} \mathbf{w} \\ \mathbf{x} \end{bmatrix} = B^{-1} \begin{bmatrix} \mathbf{f} \\ \mathbf{g} \end{bmatrix}.$$

Algorithm 3.4.2 (Transformed two-level block-factorization preconditioner)

- *Compute* $\mathbf{w} = \mathcal{M}^{-1}\mathbf{f}$.
- *Compute* $\mathbf{x} = \mathcal{D}^{-1}(\mathbf{g} - \mathcal{L}\mathbf{w} + \mathcal{W}^T(\mathbf{f} - \mathcal{A}\mathbf{w}))$.
- *Compute* $\mathbf{u} = \mathcal{W}\mathbf{x}$.
- *Compute* $\mathbf{w} = \mathbf{w} + \mathbf{u} - \mathcal{M}^{-1}(\mathcal{R}\mathbf{x} + \mathcal{A}\mathbf{u})$.

It is clear that B^{-1} exploits solutions with $\mathcal{M}$ and $\mathcal{D}$ in addition to matrix–vector products based on $\mathcal{L}$, $\mathcal{R}$, $\mathcal{A}$ (the original blocks of A), and $\mathcal{W}$.

There is one special case in practice (originally noted by Y. Notay [Not98], and independently used in [Mc01]) when we can avoid the explicit use of the transformation matrix Y, namely, if we can find a s.p.d. approximation $\mathcal{M}$ to the first block $\mathcal{A}$ of A, which is spectrally equivalent to $\mathcal{A}$ such that

(i) $\mathcal{A} - \mathcal{M}$ is symmetric positive semidefinite, so that
(ii) The perturbed matrix $\begin{bmatrix} \mathcal{M} & \mathcal{R} \\ \mathcal{L} & \mathcal{B} \end{bmatrix}$ is still s.p.d.

In that case, we can show that P with $\mathcal{W} = -\mathcal{M}^{-1}\mathcal{R}$ leads to a $\widehat{\mathcal{B}}$ that is spectrally equivalent to the exact Schur complement $S = \mathcal{B} - \mathcal{L}\mathcal{A}^{-1}\mathcal{R}$ of A, and the following block-factored matrix

$$B = \begin{bmatrix} \mathcal{M} & \mathcal{R} \\ \mathcal{L} & \widehat{\mathcal{B}} + \mathcal{L}\mathcal{M}^{-1}\mathcal{R} \end{bmatrix} = \begin{bmatrix} \mathcal{M} & 0 \\ \mathcal{L} & \widehat{\mathcal{B}} \end{bmatrix} \begin{bmatrix} I & \mathcal{M}^{-1}\mathcal{R} \\ 0 & I \end{bmatrix} \tag{3.30}$$

can be used as a spectrally equivalent approximation to A. Note, that the latter matrix is a perturbation to A and the perturbations occur only on the main diagonal of A.

Theorem 3.27. *Under the assumptions (i) and (ii) above, the block-factorization preconditioner B, (3.30) is spectrally equivalent to A and the following bounds hold,*

$$c_1 \mathbf{v}^T A\mathbf{v} \leq \mathbf{v}^T B\mathbf{v} \leq c_2 \mathbf{v}^T A\mathbf{v}.$$

We can further replace $\widehat{\mathcal{B}}$ (or the exact Schur complement $\mathcal{S} = \mathcal{B} - \mathcal{L}\mathcal{A}^{-1}\mathcal{R}$) with a spectrally equivalent s.p.d. matrix $\mathcal{D}$ and still end up with a spectrally equivalent preconditioner

$$B = \begin{bmatrix} \mathcal{M} & 0 \\ \mathcal{L} & \mathcal{D} \end{bmatrix} \begin{bmatrix} I & \mathcal{M}^{-1}\mathcal{R} \\ 0 & I \end{bmatrix}.$$

Proof. A main observation is that $\widehat{A} = Y^T AY$ with

$$Y = \begin{bmatrix} I & -\mathcal{M}^{-1}\mathcal{R} \\ 0 & I \end{bmatrix}$$

has a stable 2-by-2 block form because we have for its second entry on the diagonal the representation $\widehat{\mathcal{B}} = P^T AP = \mathcal{B} - \mathcal{L}\mathcal{A}^{-1}\mathcal{R} + \mathcal{L}(\mathcal{A}^{-1} - \mathcal{M}^{-1})\mathcal{A}(\mathcal{A}^{-1} - \mathcal{M}^{-1})\mathcal{R}$, which can be viewed as a perturbation of the exact Schur complement $\mathcal{S} = \mathcal{B} - \mathcal{L}\mathcal{A}^{-1}\mathcal{R}$ of A. The following estimates then hold, letting $X = \mathcal{A}^{1/2}\mathcal{M}^{-1}\mathcal{A}^{1/2}$.

$$\begin{aligned}
\mathbf{x}^T(\widehat{\mathcal{B}} - \mathcal{S})\mathbf{x} &= \mathbf{x}^T \mathcal{L}(\mathcal{A}^{-1} - \mathcal{M}^{-1})\mathcal{A}(\mathcal{A}^{-1} - \mathcal{M}^{-1})\mathcal{R}\mathbf{x} \\
&= \mathbf{x}^T \mathcal{R}^T \mathcal{A}^{-(1/2)}(X - I)^2 \mathcal{A}^{-(1/2)}\mathcal{R}\mathbf{x} \\
&\leq \|X - I\|\ \mathbf{x}^T \mathcal{R}^T \mathcal{A}^{-(1/2)}(X - I)\mathcal{A}^{-(1/2)}\mathcal{R}\mathbf{x} \\
&= \|X - I\|\ \mathbf{x}^T(\mathcal{R}^T \mathcal{M}^{-1}\mathcal{R} - \mathcal{R}^T \mathcal{A}^{-1}\mathcal{R})\mathbf{x} \\
&\leq \|X - I\|\ \mathbf{x}^T(\mathcal{B} - \mathcal{R}^T \mathcal{A}^{-1}\mathcal{R})\mathbf{x} \\
&= \|X - I\|\ \mathbf{x}^T \mathcal{S}\mathbf{x} \\
&= \|X - I\|\ \mathbf{x}^T \widehat{\mathcal{S}}\mathbf{x}.
\end{aligned}$$

Above, we first used assumption (i), that is, that $X - I$ is symmetric positive semidefinite, and second, assumption (ii), which implies that the Schur complement $\mathcal{B} - \mathcal{L}\mathcal{M}^{-1}\mathcal{R}$ of the perturbed symmetric positive semidefinite matrix

$$\begin{bmatrix} \mathcal{M} & \mathcal{R} \\ \mathcal{L} & \mathcal{B} \end{bmatrix}$$

is positive semidefinite, that is, that $\mathbf{x}^T \mathcal{R}^T \mathcal{M}^{-1}\mathcal{R}\mathbf{x} \leq \mathbf{x}^T \mathcal{B}\mathbf{x}$. Thus we showed that $\mathbf{x}^T \widehat{\mathcal{B}}\mathbf{x} \leq (1 + \|X - I\|)\ \mathbf{x}^T \widehat{\mathcal{S}}\mathbf{x}$. Equivalently,

$$\mathbf{x}^T P^T AP\mathbf{x} \leq \eta \inf_{\mathbf{v}=J\mathbf{w}+P\mathbf{x}} \mathbf{v}^T A\mathbf{v}, \quad \text{with } \eta = 1 + \|X - I\|.$$

The latter is true, because the Schur complements $\mathcal{S}$ of A and $\widehat{\mathcal{S}}$ of $\widehat{A}$ are the same. Thus the matrix $\widehat{A} = Y^T AY$ can be preconditioned by the block-diagonal matrix

$$\begin{bmatrix} \mathcal{M} & 0 \\ 0 & \widehat{\mathcal{B}} \end{bmatrix} \quad \text{or equivalently by} \quad \begin{bmatrix} \mathcal{M} & 0 \\ 0 & \mathcal{D} \end{bmatrix}.$$

Therefore, A can be preconditioned by

$$B = Y^{-T} \begin{bmatrix} \mathcal{M} & 0 \\ 0 & \widehat{\mathcal{B}} \end{bmatrix} Y^{-1} = \begin{bmatrix} I & 0 \\ \mathcal{L}\mathcal{M}^{-1} & I \end{bmatrix} \begin{bmatrix} \mathcal{M} & 0 \\ 0 & \widehat{\mathcal{B}} \end{bmatrix} \begin{bmatrix} I & \mathcal{M}^{-1}\mathcal{R} \\ 0 & I \end{bmatrix},$$

or by

$$B = \begin{bmatrix} \mathcal{M} & 0 \\ \mathcal{L} & \mathcal{D} \end{bmatrix} \begin{bmatrix} I & \mathcal{M}^{-1}\mathcal{R} \\ 0 & I \end{bmatrix},$$

which is the desired result. □

We comment at the end that the construction of $\mathcal{M}$ that satisfies both (i) and (ii) is a bit tricky in practice. Some possibilities are outlined in Section 4.7 of Chapter 4, where other types of approximate block-factorization matrices are considered as well.

4

Classical Examples of Block-Factorizations

4.1 Block-ILU factorizations

Consider a block form of

$$A = \begin{bmatrix} \mathcal{A} & \mathcal{R} \\ \mathcal{L} & \mathcal{B} \end{bmatrix}$$

in which $\mathcal{A}$ is sparse and well conditioned. Then, as is well-known, $\mathcal{A}^{-1}$ has a certain decay rate (cf., Appendix A.2.4). The latter, in short, means that it admits a good polynomial approximation in terms of $\mathcal{A}$. Alternatively, we may say that $\mathcal{A}^{-1}$ can be well approximated with a sparse matrix $\mathcal{M}^{-1}$. Thus, the approximate Schur complement $\overline{S} \equiv \mathcal{B} - \mathcal{L}\mathcal{M}^{-1}\mathcal{R}$ will also be sparse. This procedure is attractive, if $\mathcal{L}$ (and $\mathcal{R}$) have a single nonzero diagonal. Thus the sparsity pattern of $\overline{S}$ depends on $\mathcal{B}$ and how accurately we want $\mathcal{M}^{-1}$ to approximate $\mathcal{A}^{-1}$. If we keep the sparsity pattern of $\mathcal{L}\mathcal{M}^{-1}\mathcal{R}$ the same as that of $\mathcal{B}$, the procedure can be recursively applied to $\overline{S}$, which leads to the classical block-ILU factorization preconditioners. Those are well defined for M-matrices, which naturally arise from finite difference approximations of second-order elliptic PDEs. It seems that the block-ILU methods were first introduced in Kettler [K82] but have become most popular after the papers [ABI], [CGM85], [AP86], and others have appeared. These methods are very robust and perhaps the most efficient (and parameter-(to estimate) free) preconditioners for matrices coming from 2D second-order elliptic PDEs. By expanding the sparsity pattern (or half-bandwidth) of the approximate Schur complements, we improve the quality of the preconditioner, which in the limit case becomes exact factorization.

We point out that any finite element discretization matrix coming, for example, from elliptic PDEs, can always be reordered so that it admits a block-tridiagonal form with sparse blocks. The blocks are actually banded matrices for 2D meshes. A typical situation is illustrated in Figure 4.1. In summary, the block-tridiagonal case covers the general situation.

P.S. Vassilevski, *Multilevel Block Factorization Preconditioners*,
doi: 10.1007/978-0-387-71564-3_4,

Fig. 4.1. Block-tridiagonal ordering of finite element matrix on unstructured triangular mesh. The blocks correspond to degrees of freedom associated with nodes on each interface boundary obtained by intersecting any two neighboring contiguous slabs of elements (triangles) of two different colors.

Consider the block-tridiagonal matrix

$$A = \begin{bmatrix} A_{11} & A_{12} & 0 & \dots & 0 \\ A_{21} & A_{22} & A_{23} & \dots & 0 \\ & \ddots & \ddots & \ddots & \\ & & A_{n-1,n-2} & A_{n-1,n-1} & A_{n-1,n} \\ 0 & \dots & 0 & A_{n,n-1} & A_{nn} \end{bmatrix}.$$

For a five-point, finite difference discretization of 2D second-order elliptic PDEs on a rectangular mesh, the matrices on the diagonal of A are scalar tridiagonal matrices, and the upper and lower diagonals of A are scalar diagonal matrices. In a similar situation in 3D (7-point stencil), the off-diagonal blocks of A are scalar diagonal whereas the blocks on the diagonal of A have now the sparsity pattern of a 2D block-tridiagonal matrix. In either case the A_{ii} are well conditioned, because they are strictly diagonally dominant. To be specific, we concentrate now on the 2D case.

The approximate block-factorization of A can be written in the form $(X - L)X^{-1}(X - U)$, where $X = \text{diag}(X_i)$ and X_i are the approximate Schur complements computed throughout the factorization, $-U$ is the strictly upper triangular part of A, and $-L$ is correspondingly the strictly lower triangular part of A. Note that L and U have only one nonzero block diagonal. The recursion for X_i reads

(0) For $i = 1$ set $X_i = A_{i,i}$.
(i) For $i = 1, \dots, n-1$ compute a banded approximation Y_i to X_i^{-1} and compute the product

$$A_{i+1,i} Y_i A_{i,i+1}.$$

(*ii*) In order to keep the sparsity under control, we may need to further approximate the above product by a sparser matrix H_{i+1} (e.g., by dropping the nonzero entries of $A_{i+1,i} Y_i A_{i,i+1}$ outside a prescribed sparsity pattern). Finally, define

$$X_{i+1} = A_{i+1,i+1} - H_{i+1}.$$

Then the actual block-factorization matrix $M = (X - L)X^{-1}(X - U)$ has the following more explicit block-tridiagonal form,

$$\begin{bmatrix} X_1 & A_{12} & 0 & \dots & & 0 \\ A_{21} & X_2 + A_{21}X_1^{-1}A_{12} & A_{23} & \dots & & 0 \\ & \ddots & \ddots & \ddots & & \\ & & A_{n-1,n-2} & X_{n-1} + A_{n-1,n-2}X_{n-2}^{-1}A_{n-2,n-1} & & A_{n-1,n} \\ 0 & \dots & 0 & A_{n,n-1} & & X_n + A_{n,n-1}X_{n-1}^{-1}A_{n-1,n} \end{bmatrix}. \tag{4.1}$$

In particular, it is clear that M has the same off-diagonal blocks as A. The difference $M - A$ is block diagonal with blocks

$$\begin{aligned} X_i - A_{ii} + A_{i,i-1}X_{i-1}^{-1}A_{i-1,i} &= A_{i,i-1}X_{i-1}^{-1}A_{i-1,i} - H_i \\ &= A_{i,i-1}\left(X_{i-1}^{-1} - Y_{i-1}\right)A_{i-1,i} \\ &\quad + A_{i,i-1}Y_{i-1}A_{i-1,i} - H_i. \end{aligned} \tag{4.2}$$

It is clear then, if we keep the differences $X_{i-1}^{-1} - Y_{i-1}$ and $A_{i,i-1}Y_{i-1}A_{i-1,i} - H_i$ symmetric positive semidefinite, the block-ILU matrix M will provide a convergent splitting for A; that is, $M - A$ will be symmetric positive semidefinite. This is in general difficult to ensure, however, for the case of A being a s.p.d. M-matrix; we can ensure that $2M - A$ is positive definite, hence $\|I - M^{-1}A\|_A < 1$; that is, M provides a convergent iterative method in the A-norm (see Corollary 4.6). Alternatively, we can use low rank approximations Y_{i-1} to X_{i-1}^{-1} (and $H_i = A_{i,i-1}Y_{i-1}A_{i-1,i}$) as in Section 4.6, thus leading to a matrix M such that $M - A$ is indeed symmetric positive semidefinite.

The purpose of constructing the block-factorization matrix M is so that it can be used as a preconditioner in an iterative method. At every step of the iterative method, we have to solve a system

$$M\mathbf{v} = \mathbf{w}$$

for some (residual) vector $\mathbf{w}$. Because M is factored, the above system is solved in the usual forward and backward recurrences.

(i) **Forward.** Solve,

$$\begin{bmatrix} X_1 & & & 0 \\ A_{21} & X_2 & & \\ & \ddots & \ddots & \\ 0 & & A_{n,n-1} & X_n \end{bmatrix} \begin{bmatrix} \mathbf{u}_1 \\ \mathbf{u}_2 \\ \vdots \\ \mathbf{u}_n \end{bmatrix} = \begin{bmatrix} \mathbf{w}_1 \\ \mathbf{w}_2 \\ \vdots \\ \mathbf{w}_n \end{bmatrix},$$

in the following steps,

$$\mathbf{u}_1 = X_1^{-1}\mathbf{w}_1,$$
$$\mathbf{u}_i = X_i^{-1}(\mathbf{w}_i - A_{i,i-1}\mathbf{u}_{i-1}), \qquad i > 1.$$

- **Backward.** Solve

$$\begin{bmatrix} I & X_1^{-1}A_{12} & & 0 \\ & I & X_2^{-1}A_{23} & \\ & & \ddots & \ddots \\ 0 & & & I \end{bmatrix} \begin{bmatrix} \mathbf{v}_1 \\ \mathbf{v}_2 \\ \vdots \\ \mathbf{v}_n \end{bmatrix} = \begin{bmatrix} \mathbf{u}_1 \\ \mathbf{u}_2 \\ \vdots \\ \mathbf{u}_n \end{bmatrix},$$

in the following steps,

$$\mathbf{v}_n = \mathbf{u}_n,$$
$$\mathbf{v}_i = \mathbf{u}_i - X_i^{-1}A_{i,i+1}\mathbf{v}_{i+1}, \quad \text{for } i = n-1 \text{ down to } 1.$$

For a $(2p+1)$-banded matrix X, we can construct various $(2p+1)$-banded approximations of its inverse based, for example, on the standard $LD^{-1}U$ factorization of X. We can actually compute the exact innermost $2p+1$ banded part of X^{-1} without computing the full inverse.

Details about implementation of algorithms that compute approximate band inverses are given in Section 4.4.

4.2 The M-matrix case

We begin with the definition of an M-matrix.

Definition 4.1 (M-matrix). *A matrix $A = (a_{ij})_{i,j=1}^n$ is called an M-matrix if*

(o) A has nonpositive off-diagonal entries.
(i) A is nonsingular.
(ii) A^{-1} has nonnegative entries.

If A is s.p.d., and M-matrix, A is sometimes called the Stieltjes matrix.

In what follows in the next few sections by $A \geq 0$ or $\mathbf{v} \geq 0$ we mean componentwise inequalities.

Theorem 4.2. *A main property of an M-matrix A is that there exists a positive vector $\mathbf{c} = (c_i)_{i=1}^n$ (i.e., $c_i > 0$) such that $\mathbf{b} = A\mathbf{c} = (b_i)$ is also positive (i.e., $b_i > 0$ for all i). Conversely, if $A = (a_{ij})$ with $a_{ij} \leq 0$ for $i \neq j$ and $\mathbf{b} = A\mathbf{c}$ is a positive vector for a given positive vector $\mathbf{c}$ then A is an M-matrix.*

Proof. The fact that for an M-matrix A there is a positive vector $\mathbf{c}$ such that $\mathbf{b} = A\mathbf{c}$ is also positive follows from the following simple observation. Because A^{-1} exists and has nonnegative entries, it is clear that A^{-1} has at least one strictly positive entry per row (otherwise A^{-1} would have a zero row, which is not possible for a nonsingular matrix). Then for the constant vector $\mathbf{b} = (1)$, we have $\mathbf{c} = A^{-1}\mathbf{b} > 0$ (the row-sums of A^{-1} are strictly positive). The latter shows that for the positive vector $\mathbf{c}$, $A\mathbf{c} = \mathbf{b}$ is also positive.

The converse statement is seen by first forming the diagonal matrix $C = \text{diag}(c_i)$ and looking at the diagonally scaled matrix $AC = (a_{i,j}c_j)$. It is easily seen that AC is strictly diagonally dominant. Indeed, because

$$b_i = a_{i,i}c_i + \sum_{j \neq i} a_{i,j}c_j > 0,$$

we get (using the fact that $-a_{i,j}c_j = |a_{i,j}c_j|$ for $j \neq i$)

$$a_{i,i}c_i > \sum_{j \neq i} |a_{i,j}c_j|.$$

The latter implies that AC is invertible, hence A is invertible. Moreover, because AC is strictly diagonally dominant, it admits an LDU factorization, or more specifically, the following product expansion holds, $AC = L_1 \cdots L_{n-1} D U_{n-1} \cdots U_1$, where each

$$L_i = \begin{bmatrix} I & 0 & 0 \\ 0 & 1 & 0 \\ 0 & \underline{\ell}_i & I \end{bmatrix}, \quad \text{and} \quad U_i = \begin{bmatrix} I & 0 & 0 \\ 0 & 1 & \mathbf{u}_i^T \\ 0 & 0 & I \end{bmatrix}.$$

We can easily see that $\underline{\ell}_i \leq 0$ and $\mathbf{u}_i \leq 0$. Also the diagonal matrix D has positive entries. For a proof of the last fact, see the next lemma, 4.3. Then,

$$A^{-1} = C U_1^{-1} \cdots U_{n-1}^{-1} D^{-1} L_{n-1}^{-1} \cdots L_1^{-1} \geq 0,$$

as a product of nonnegative matrices. We notice that

$$L_i^{-1} = \begin{bmatrix} I & 0 & 0 \\ 0 & 1 & 0 \\ 0 & -\underline{\ell}_i & I \end{bmatrix} \geq 0,$$

and similarly,

$$U_i^{-1} = \begin{bmatrix} I & 0 & 0 \\ 0 & 1 & -\mathbf{u}_i^T \\ 0 & 0 & I \end{bmatrix} \geq 0. \qquad \square$$

In the proof above, we used the following well-known result.

Lemma 4.3. *Given a strictly diagonally dominant matrix*

$$B = \begin{bmatrix} \beta & \mathbf{u}^T \\ \underline{\ell} & G \end{bmatrix}.$$

Then its Schur complement $S = G - \underline{\ell}\beta^{-1}\mathbf{u}^T$ is also strictly diagonally dominant. Also B admits the LDU factorization

$$B = \begin{bmatrix} 1 & 0 \\ \frac{1}{\beta}\underline{\ell} & I \end{bmatrix} \begin{bmatrix} \beta & 0 \\ 0 & S \end{bmatrix} \begin{bmatrix} 1 & \frac{1}{\beta}\mathbf{u}^T \\ 0 & I \end{bmatrix},$$

from which it is clear that $(1/\beta)\underline{\ell} < 0$ and $(1/\beta)\mathbf{u}^T < 0$ if the off-diagonal entries of B are nonpositive, and $\beta > 0$.

Proof. Let $S = (s_{i,j})$, $G = (g_{i,j})$, $\underline{\ell} = (l_i)$, and $\mathbf{u} = (u_i)$. Here, β is a scalar. We have

$$s_{i,j} = g_{i,j} - l_i\beta^{-1}u_j.$$

We would like to show that

$$|s_{i,i}| > \sum_{j\neq i} |g_{i,j} - l_i\beta^{-1}u_j|.$$

Because B is strictly diagonally dominant, using this property for its first row, we get

$$\frac{\sum_j |u_j|}{|\beta|} < 1.$$

Using the strict diagonal dominance for the $(i+1)$st row of B, we get

$$|g_{i,i}| > |l_i| + \sum_{j\neq i} |g_{i,j}|.$$

Combining the last two inequalities, we end up with

$$|g_{i,i}| > \sum_{j\neq i} |g_{i,j}| + |l_i|\frac{\sum_j |u_j|}{|\beta|}.$$

The result then follows from the triangle inequality,

$$\begin{aligned} |s_{i,i}| &= |g_{i,i} - l_i\beta^{-1}u_i| \\ &\geq |g_{i,i}| - |l_i|\frac{|u_i|}{|\beta|} \\ &> \sum_{j\neq i} |g_{i,j}| + |l_i|\frac{\sum_j |u_j|}{|\beta|} - |l_i|\frac{|u_i|}{|\beta|} \\ &= \sum_{j\neq i} \left(|g_{i,j}| + \frac{|l_i||u_j|}{|\beta|}\right) \\ &\geq \sum_{j\neq i} |g_{i,j} - l_i\beta^{-1}u_j| \\ &= \sum_{j\neq i} |s_{i,j}|. \end{aligned}$$

That is, we showed the strict inequality $|s_{i,i}| > \sum_{j\neq i} |s_{i,j}|$. □.

Let us return to the M-matrix case. The following result is immediate.

Lemma 4.4. *Let*

$$A = \begin{bmatrix} \mathcal{A} & \mathcal{R} \\ \mathcal{L} & \mathcal{B} \end{bmatrix}$$

be an M-matrix. Then, both $\mathcal{A}$ and the Schur complement $\mathcal{S} = \mathcal{B} - \mathcal{L}\mathcal{A}^{-1}\mathcal{R}$ are M-matrices.

Proof. We have for a positive vector $\mathbf{c}$, $\mathbf{b} = A\mathbf{c} > 0$. Let

$$\mathbf{c} = \begin{bmatrix} \mathbf{c}_1 \\ \mathbf{c}_2 \end{bmatrix} \quad \text{and} \quad \mathbf{b} = \begin{bmatrix} \mathbf{b}_1 \\ \mathbf{b}_2 \end{bmatrix}.$$

Then, because $\mathcal{R} \leq 0$, the inequality

$$\mathcal{A}\mathbf{c}_1 = \mathbf{b}_1 - \mathcal{R}\mathbf{c}_2 \geq \mathbf{b}_1 > 0$$

shows that $\mathcal{A}$ is an M-matrix. (Note that its off-diagonal entries are nonpositive.)

From $\mathcal{S} = \mathcal{B} - \mathcal{L}\mathcal{A}^{-1}\mathcal{R}$, we get that $\mathcal{S} \leq \mathcal{B}$ because we already proved that $\mathcal{A}$ is an M-matrix; that is, $\mathcal{A}^{-1} \geq 0$, and also $\mathcal{L} \leq 0$ and $\mathcal{R} \leq 0$. Therefore, because the off-diagonal entries of $\mathcal{B}$ are nonpositive from $\mathcal{S} \leq \mathcal{B}$, it follows that the off-diagonal entries of $\mathcal{S}$ are also nonpositive. Finally, from the fact that $\mathcal{S}^{-1}$ is a principal submatrix of A^{-1} (Proposition 3.1) it is clear that $\mathcal{S}^{-1}$ has nonnegative entries (because $A^{-1} \geq 0$). □

Armed with the above main properties of M-matrices, it is not hard to show the existence of block-ILU factorization of block-tridiagonal matrices, a result originally proven in [AP86], and earlier in [CGM85] for diagonally dominant M-matrices.

The following main result holds.

Theorem 4.5. *Let $A = [A_{i,i-1},\ A_{i,i},\ A_{i,i+1}]$ be a block-tridiagonal M-matrix. Consider the following algorithm.*

Algorithm 4.2.1 (Block-ILU factorization).

(0) Let $X_1 = A_{1,1}$. For $i \geq 1$, consider an approximation Y_i of X_i^{-1} that satisfies

$$0 \leq Y_i \leq X_i^{-1}. \tag{4.3}$$

(i) Also, choose an approximation H_{i+1} of the product $A_{i+1,i}Y_iA_{i,i+1}$ that satisfies

$$0 \leq H_{i+1} \leq A_{i+1,i}Y_iA_{i,i+1}. \tag{4.4}$$

The role of H_{i+1} is to control the possible fill-in in the product $A_{i+1,i}Y_iA_{i,i+1}$. That is, in practice, we compute only the entries of $A_{i+1,i}Y_iA_{i,i+1}$ within a prescribed sparsity pattern of X_{i+1}.

(ii) Finally, define the $(i+1)$th approximate Schur complement as

$$X_{i+1} = A_{i+1,i+1} - H_{i+1}.$$

The above algorithm is well defined; that is, X_i are M-matrices for all $i \geq 1$, and hence, a matrix Y_i, which satisfies condition (4.3), *always exists and therefore the choice of H_{i+i} as in* (4.4) *is also feasible.*

Proof. Because A is an M-matrix, its principal submatrix

$$A_i \equiv \begin{bmatrix} A_{1,1} & A_{1,2} & 0 & \dots & 0 \\ A_{2,1} & A_{2,2} & A_{2,3} & \dots & 0 \\ 0 & \ddots & \ddots & \ddots & 0 \\ 0 & \dots & 0 & A_{i,i-1} & A_{i,i} \end{bmatrix}$$

is also an M-matrix. Let Z_i be the exact Schur complements obtained by exact block-factorization of A; that is, $Z_1 = A_{1,1}$ and

$$\begin{aligned} Z_{i+1} &= A_{i+1,i+1} - A_{i+1,i} Z_i^{-1} A_{i,i+1} \\ &= A_{i+1,i+1} - [0, \dots, 0, A_{i+1,i}] A_i^{-1} \begin{bmatrix} 0 \\ \vdots \\ 0 \\ A_{i,i+1} \end{bmatrix}. \end{aligned}$$

Assuming (by induction) that Z_i is a Schur complement of A_i, then Z_i^{-1} is a principal submatrix of A_i^{-1}; that is,

$$A_i^{-1} = \begin{bmatrix} * & * & * & * \\ * & * & * & * \\ \vdots & \ddots & \ddots & \vdots \\ * & * & * & Z_i^{-1} \end{bmatrix}.$$

It is clear then that the product

$$[0, \dots, 0, A_{i+1,i}] A_i^{-1} \begin{bmatrix} 0 \\ \vdots \\ 0 \\ A_{i,i+1} \end{bmatrix}$$

equals $A_{i+1,i} Z_i^{-1} A_{i,i+1}$. That is, $Z_{i+1} = A_{i+1,i+1} - A_{i+1,i} Z_i^{-1} A_{i,i+1}$ is indeed a Schur complement of

$$A_{i+1} = \begin{bmatrix} A_i & \begin{bmatrix} 0 \\ \vdots \\ 0 \\ A_{i,i+1} \end{bmatrix} \\ [0, \dots, 0, A_{i+1,i}] & A_{i+1,i+1} \end{bmatrix}$$

(which confirms the induction assumption).

As a Schur complement of the M-matrix A_i (a principal submatrix of the M-matrix A), Z_i itself is an M-matrix, and hence there is a positive vector $\mathbf{c}_i$ such that $Z_i\mathbf{c}_i > 0$.

The remainder of the proof proceeds by induction. Assume, now that for some $i \geq 1$, $X_i \geq Z_i$. Note that $X_1 = Z_1 = A_{11}$.

It is clear that for any choice of Y_i such that $Y_i \geq 0$, we have $A_{i+1,i}Y_iA_{i,i+1} \geq 0$ and hence a nonnegative choice of H_{i+1} is feasible. With such a choice of H_{i+1} we have $X_{i+1} \leq A_{i+1,i+1}$, hence the off-diagonal entries of X_{i+1} are nonpositive.

Because X_i has nonpositive off-diagonal entries (by construction) and for a positive vector $\mathbf{c}_i$ we have $X_i\mathbf{c}_i \geq Z_i\mathbf{c}_i > 0$, it follows then that X_i is an M-matrix. From $X_i \geq Z_i$, because both $Z_i^{-1} \geq 0$ and $X_i^{-1} \geq 0$, it follows that $Z_i^{-1} - X_i^{-1} = X_i^{-1}(X_i - Z_i)Z_i^{-1} \geq 0$; that is, $Z_i^{-1} \geq X_i^{-1}$. Now choose Y_i as in (4.3). Then, because $-Y_i \geq -X_i^{-1} \geq -Z_i^{-1}$, we have

$$\begin{aligned} X_{i+1} &= A_{i+1,i+1} - H_{i+1} \\ &\geq A_{i+1,i+1} - A_{i+1,i}Y_iA_{i,i+1} \\ &\geq A_{i+1,i+1} - A_{i+1,i}X_i^{-1}A_{i,i+1} \\ &\geq A_{i+1,i+1} - A_{i+1,i}Z_i^{-1}A_{i,i+1} \\ &= Z_{i+1}. \end{aligned} \tag{4.5}$$

That is, the induction assumption is confirmed for $i := i + 1$ and thus the proof is complete. □

Corollary 4.6. *Assume now that A is a symmetric M-matrix. Then, the block-ILU factorization matrix $M = (X - L)X^{-1}(X - U)$ provided by Algorithm 4.2.1 is such that $2M - A$ is symmetric positive definite; that is, M provides a convergent splitting for A in the A-inner product. Equivalently, we have $\|I - M^{-1}A\|_A < 1$.*

Proof. We first notice that $M = (X-L)X^{-1}(X-U)$ is s.p.d. because the symmetric M-matrices X_i are s.p.d. This holds, because any symmetric M-matrix V allows for an LDL^T factorization with a positive diagonal matrix D. Indeed, V being an M-matrix implies that there is a positive vector $\mathbf{c}$ such that $V\mathbf{c}$ is also positive. Let $\mathbf{c} = (c_i)$ and form the diagonal matrix $C = \text{diag}(c_i)$. Then C^TVC is symmetrical and a strictly diagonally dominant matrix (see the proof of Theorem 4.2). Then Lemma 4.3, modified accordingly, implies the existence of the desired factorization of C^TVC and hence of V (because C is diagonal).

The desired result follows from a main result of Varga [Var62] (for a proof, see Theorem 10.3.1 in [Gr97]). In what follows by ϱ, we denote spectral radius. Namely, because $A = M - R$ with $R \geq 0$ (see (4.2)) and $A^{-1} \geq 0$, Varga's result states that

$$\varrho(I - M^{-1}A) = \varrho(M^{-1}R) = \frac{\varrho(A^{-1}R)}{1 + \varrho(A^{-1}R)} < 1.$$

Therefore, in our symmetric case, $\lambda_{\min}(I - M^{-1}A) > -1$, which is equivalent to $\lambda_{\max}(M^{-1}A) < 2$ or $2M - A$ being symmetric positive definite. □

4.3 Decay rates of inverses of band matrices

A main motivation for the block-ILU methods is based on the observation that the inverse of a band matrix can be well approximated by a band matrix. The latter can be more rigorously justified by the decay rate estimate provided at the end of the present section.

Illustration of decay rates

We first demonstrate by graphical representation the decay behavior of the inverse of a number of tridiagonal matrices. Consider first,

$$T_n = \begin{bmatrix} 4 & -1 & & & & 0 \\ -1 & 4 & -1 & & & \\ & -1 & 4 & -1 & & \\ & & \ddots & \ddots & \ddots & \\ & & & -1 & 4 & -1 \\ 0 & & & & -1 & 4 \end{bmatrix}.$$

The decay behavior of T_n^{-1} for $n = 32$ is shown in Figure 4.2.

Consider now the tridiagonal matrix, which is only weakly diagonally dominant,

$$\tau_n = \begin{bmatrix} 2 & -1 & & & & 0 \\ -1 & 2 & -1 & & & \\ & -1 & 2 & -1 & & \\ & & \ddots & \ddots & \ddots & \\ & & & -1 & 2 & -1 \\ 0 & & & & -1 & 2 \end{bmatrix}.$$

Its decay behavior for $n = 32$ is shown in Figure 4.3.

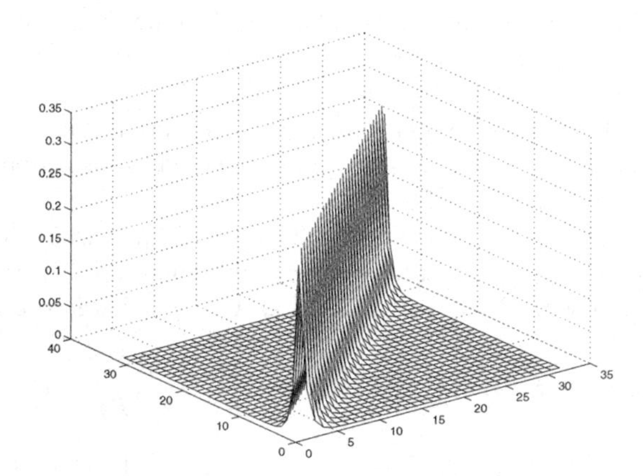

Fig. 4.2. Decay behavior of the inverse of T_{32}.

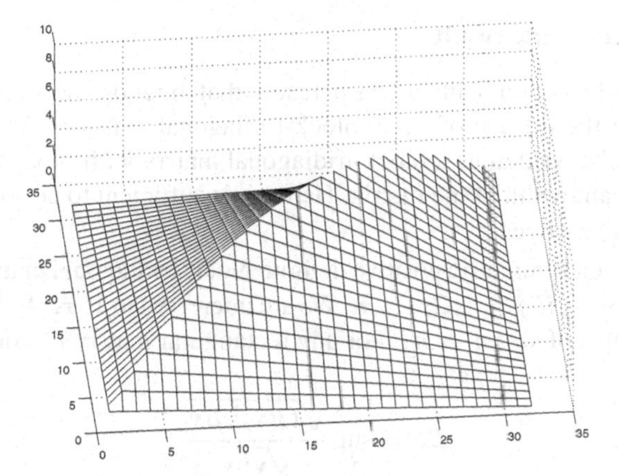

Fig. 4.3. Decay behavior of the inverse of τ_{32}.

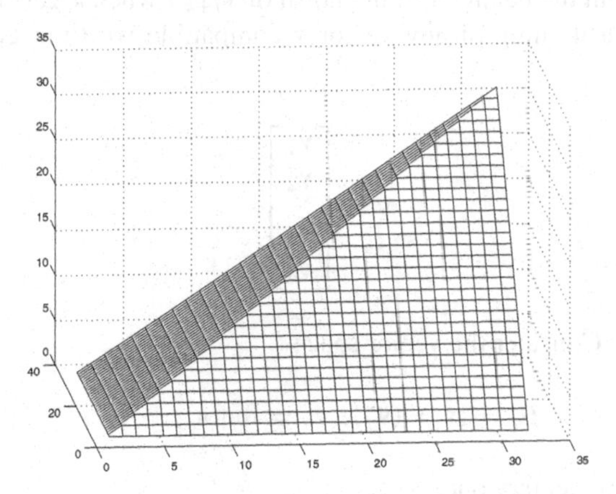

Fig. 4.4. Decay behavior of the inverse of θ_{32}.

Finally, consider the following tridiagonal matrix,

$$\theta_n = \begin{bmatrix} 2 & -1 & & & & 0 \\ -1 & 2 & -1 & & & \\ & -1 & 2 & -1 & & \\ & & \ddots & \ddots & \ddots & \\ & & & -1 & 2 & -1 \\ 0 & & & & -1 & 1 \end{bmatrix}.$$

Its decay behavior for $n = 32$ is shown in Figure 4.4.

Based on the above examples, we may draw the conclusion, that for strictly diagonally dominant matrices, we should expect a very fast (e.g., exponential) decay rate, whereas for weakly dominant matrices, the decay rate may not be as fast, as in the case shown in Figure 4.4 where we only see a linear decay rate.

An algebraic decay rate result

We complete this section with a sharp result that bounds the decay rate of the (block)entries of the inverse of s.p.d. block-tridiagonal matrices. We note that any band matrix can be written as a block-tridiagonal matrix with blocks that are equal size to the half-bandwidth of the matrix. Hence it is sufficient to consider the case of block-tridiagonal matrices A.

Let A be a given symmetric positive definite block-tridiagonal matrix with entries $\{A_{i,j}\}$ of size $n_i \times n_j$, $i, j = 1, 2, \ldots, n$. We are interested in $V \equiv A^{-1} = \{V_{i,j}\}$. The entries $V_{i,j}$ are also of size $n_i \times n_j$. For any rectangular matrix B, consider its norm

$$\|B\| = \sup_{\mathbf{v}} \frac{\sqrt{(B\mathbf{v})^T B\mathbf{v}}}{\sqrt{\mathbf{v}^T\mathbf{v}}}.$$

We are interested in the behavior of the norm of $V_{i+k,i}$ when k gets large. Introduce now the block partitioning of any vector $\mathbf{v}$ compatible with the given tridiagonal matrix A,

$$\mathbf{v} = \begin{bmatrix} \mathbf{v}_1 \\ \mathbf{v}_2 \\ \vdots \\ \mathbf{v}_n \end{bmatrix}.$$

That is, $\mathbf{v}_i \in \mathbb{R}^{m_i}$. Consider the vector space

$$H_i = \{\mathbf{v} = (\mathbf{v}_j),\ \mathbf{v}_j = 0 \text{ for } j > i\},$$

and also its complementary one,

$$H_i^{'} = \{\mathbf{v} = (\mathbf{v}_j),\ \mathbf{v}_j = 0 \text{ for } j \leq i\}.$$

Because A is assumed symmetric positive definite, A can define an inner product. It is then clear that the linearly independent spaces H_i and $H_i^{'}$ will have a nontrivial angle in the A-inner product. That is, there exists a constant $\gamma_i \in [0, 1)$ (i.e., strictly less than one) such that the following strengthened Cauchy–Schwarz inequality holds.

$$\mathbf{v}^T A\mathbf{w} \leq \gamma_i (\mathbf{v}^T A\mathbf{v})^{1/2} (\mathbf{w}^T A\mathbf{w})^{1/2}, \quad \text{for all } \mathbf{v} \in H_i,\ \mathbf{w} \in H_i^{'}.$$

Then the following main result holds ([V90]).

Theorem 4.7.

$$\frac{\|V_{i+k,i}\|}{\|V_{i,i}\|^{1/2}\|V_{i+k,i+k}\|^{1/2}} \leq \prod_{j=1}^{k} \gamma_{i+j}.$$

Proof. Because the matrix A is block-tridiagonal, the strengthened Cauchy–Schwarz inequality actually reads,

$$\mathbf{v}_{i+1}^T A_{i+1,i}\mathbf{v}_i \leq \gamma_i \min_{\mathbf{v}=\begin{bmatrix}\mathbf{v}_1\\ \vdots\\ \mathbf{v}_{i-1}\\ \mathbf{v}_i\\ 0\\ \vdots\\ 0\end{bmatrix}} (\mathbf{v}^T A\mathbf{v})^{1/2} \min_{\mathbf{w}=\begin{bmatrix}0\\ \vdots\\ 0\\ \mathbf{v}_{i+1}\\ \mathbf{v}_{i+2}\\ \vdots\\ \mathbf{v}_n\end{bmatrix}} (\mathbf{w}^T A\mathbf{w})^{1/2}.$$

The respective minimums are first taken with respect to the first $i-1$ components and in the second term with respect to the last $n-i-1$ components.

Now using the Schur complements,

$$S_i = A_{i,i} - [0,\dots,0,A_{i,i-1}] \times \begin{bmatrix} A_{11} & A_{12} & & \dots & 0\\ A_{21} & A_{22} & A_{23} & & \\ & & & & \\ 0 & & & A_{i-1,i-2} & A_{i-1,i-1}\end{bmatrix}^{-1}\begin{bmatrix}0\\ \vdots\\ A_{i-1,i}\end{bmatrix},$$

and

$$S_i^{'} = A_{i+1,i+1} - [A_{i+1,i+2},0,\dots,0] \times \begin{bmatrix} A_{i+2,i+2} & A_{i+2,i+3} & & \dots & 0\\ A_{i+3,i+2} & A_{i+3,i+3} & A_{i+3,i+4} & & \\ & & & & \\ 0 & & & A_{n,n-1} & A_{n,n}\end{bmatrix}^{-1}\begin{bmatrix}A_{i+2,i+1}\\ 0\\ \vdots\\ 0\end{bmatrix},$$

the strengthened Cauchy inequality above takes the simpler form,

$$\mathbf{v}_{i+1}^T A_{i+1,i}\mathbf{v}_i \leq \gamma_i \left(\mathbf{v}_i^T S_i \mathbf{v}_i\right)^{1/2}\left(\mathbf{v}_{i+1}^T S_i^{'}\mathbf{v}_{i+1}\right)^{1/2}.$$

The next useful observation is the identity,

$$\begin{bmatrix} S_i & A_{i,i+1}\\ A_{i+1,i} & S_i^{'}\end{bmatrix}^{-1} = \begin{bmatrix} V_{i,i} & V_{i,i+1}\\ V_{i+1,i} & V_{i+1,i+1}\end{bmatrix}.$$

The latter is seen from the fact that the matrix on the left is a Schur complement of A, and the matrix on the right is a corresponding principal matrix of A^{-1}. The inverse of any Schur complement of a matrix is a corresponding principal matrix of the inverse matrix (cf. Proposition 3.1), therefore the above identity follows. As a corollary of the strengthened Cauchy–Schwarz inequality valid for the Schur complement

$$\begin{bmatrix} S_i & A_{i,i+1}\\ A_{i+1,i} & S_i^{'}\end{bmatrix},$$

we get the inequality,

$$\inf_{\mathbf{v}_i} \frac{\mathbf{v}_i^T V_{i,i}^{-1} \mathbf{v}_i}{\mathbf{v}_i^T S_i \mathbf{v}_i} \geq 1 - \gamma_i^2. \tag{4.6}$$

The latter inequality is seen from the fact that $V_{i,i}^{-1}$ equals the Schur complement $S_i - A_{i,i+1}(S_i^{'})^{-1} A_{i+1,i}$.

Next, we use the explicit formulas for computing the entries of A^{-1}, based on the recursively computed successive Schur complements S_i, $S_1 = A_{1,1}$, and for $i > 1$, $S_i = A_{i,i} - A_{i,i-1} S_{i-1}^{-1} A_{i-1,i}$. Then, the entries of A^{-1} are computed as $V_{n,n} = S_n^{-1}$ and for $i = n-1, \ldots, 1$ based on the recurrence,

$$\begin{aligned} V_{i,i} &= S_i^{-1} + S_i^{-1} A_{i,i+1} V_{i+1,i+1} A_{i+1,i} S_i^{-1}, \\ V_{i,i+k} &= -S_i A_{i,i+1} V_{i+1,i+k}, \qquad k = 1, 2, \ldots, n-i \\ V_{i+k,i} &= -V_{i+k,i+1} A_{i+1,i} S_i^{-1}, \qquad k = 1, 2, \ldots, n-i. \end{aligned}$$

Therefore, by recursion, we get

$$\begin{aligned} V_{i+k,i} &= V_{i+k,i+2}\big(-A_{i+2,i+1} S_{i+1}^{-1}\big)\big(-A_{i+1,i} S_i^{-1}\big) \\ &= \cdots = V_{i+k,i+k} \prod_{j=1}^{k} \big(-A_{i+j,i+j-1} S_{i+j-1}^{-1}\big). \end{aligned}$$

We can then get the identity

$$V_{i+k,i} = V_{i+k,i+k}^{1/2} \left(\prod_{j=i}^{i+k-1} R_j S_j^{-1/2} V_{j,j}^{-1/2} \right) V_{i,i}^{1/2}.$$

Here $R_j = -V_{j+1,j+1}^{1/2} A_{j+1,j} S_j^{-(1/2)}$. Note that $R_j^T R_j = S_j^{1/2}(V_{j,j} - S_j^{-1}) S_j^{1/2} = -I + S_j^{1/2} V_{j,j} S_j^{1/2}$. Therefore

$$\| R_j S_j^{-(1/2)} V_{j,j}^{-(1/2)} \| = \sqrt{1 - \inf_{\mathbf{v}_j} \frac{\mathbf{v}_j^T V_{j,j}^{-1} \mathbf{v}_j}{\mathbf{v}_j^T S_j \mathbf{v}_j}} \leq \gamma_j,$$

where we have used inequality (4.6). This completes the proof of the theorem. □

Let us apply the above theorem to the strictly diagonally dominant matrix T_n. We can show that all $\gamma_i \leq \frac{1}{3}$; that is, they are uniformly bounded away from unity. This is seen from the fact that the Schur complement

$$\begin{bmatrix} S_i & A_{i,i+1} \\ A_{i+1,i} & S_i^{'} \end{bmatrix}$$

in the present case equals

$$\begin{bmatrix} 3+s_i & -1 \\ -1 & 3+s_i^{'} \end{bmatrix}$$

for some nonnegative numbers s_i, $s_i^{'}$. This shows that

$$\gamma_i = \frac{1}{\sqrt{3+s_i}} \frac{1}{\sqrt{3+s_i^{'}}} \leq \frac{1}{3}.$$

Therefore, we indeed get a uniform exponential decay rate of the entries $v_{i+k,i}$ of T_n^{-1}. Actually, we can show that $s_i, s_i^{'} \geq \lim \sigma_i$, where $\sigma_1 = 1, \sigma_i + 3 = 4 - (1/(\sigma_{i-1}+3))$. The latter comes from the two-by-two matrix

$$\begin{bmatrix} 4 & -1 \\ -1 & \sigma_{i-1}+3 \end{bmatrix},$$

and σ_i is defined so that $\sigma_i + 3$ is the Schur complement $4 - (-1)(\sigma_{i-1}+3)^{-1}(-1)$ of the above matrix. In other words, we have the recursion $\sigma_i + 3 = 4 - (1/(\sigma_{i-1}+3))$, $\sigma_1 = 1$. By induction, we get that $\sigma_i \geq -1 + \sqrt{3} = \lim \sigma_i$. Hence,

$$\gamma_i \leq \frac{1}{3 + \lim s_i} = \frac{1}{3 + (\sqrt{3}-1)}.$$

This shows that

$$\gamma_i \leq \frac{1}{2+\sqrt{3}} = 2 - \sqrt{3} \simeq 0.2679.$$

The latter bound incidentally coincides with the quotient $\sqrt{\kappa} - 1/\sqrt{\kappa} + 1$, where $\kappa = \text{cond}(T_n) \simeq 3$. Decay rates of the entries of A^{-1}, based on the square root of the condition number of A, were originally developed in [DMS84].

4.4 Algorithms for approximate band inverses

In this section, we present a number of algorithms that provide banded approximate inverses to banded matrices. Such approximations can be useful in case where the blocks $A_{i,i}$ of a block-tridiagonal matrix A are banded. Then, if we have Algorithm 4.2.1 in mind, we need banded approximations Y_i to the inverses of the successive approximate Schur complements X_i. The bandwidth of X_{i+1} can be kept under control by choosing appropriate banded approximation H_{i+1} to the product $A_{i+1,i}Y_iA_{i,i+1}$. The latter is typically obtained by dropping the nonzero entries of $A_{i+1,i}Y_iA_{i,i+1}$ outside a prescribed sparsity pattern (or bandwidth of certain size $2p+1$).

Given a $(2p+1)$-banded matrix $n \times n$ matrix T, consider its

$$L_1L_2 \cdots L_{n-1}D^{-1}U_{n-1} \cdots U_2U_1, \tag{4.7}$$

factorization. Here

$$L_i = \begin{bmatrix} I & 0 & 0 \\ 0 & 1 & 0 \\ 0 & \ell_i & I \end{bmatrix}$$

is an elementary unit lower triangular matrix;

$$U_i = \begin{bmatrix} I & 0 & 0 \\ 0 & 1 & \mathbf{u}_i^T \\ 0 & 0 & I \end{bmatrix}$$

is an elementary unit upper triangular matrix; and $D = \operatorname{diag}(d_i)$ is a diagonal matrix. We also have

$$\ell_i = \begin{bmatrix} \ell_{i+1,i} \\ \ell_{i+2,i} \\ \vdots \\ \ell_{i+p,i} \\ 0 \\ \vdots \\ 0 \end{bmatrix}, \qquad \mathbf{u}_i = \begin{bmatrix} u_{i,i+1} \\ u_{i,i+2} \\ \vdots \\ u_{i,i+p} \\ 0 \\ \vdots \\ 0 \end{bmatrix}. \tag{4.8}$$

The entries $\ell_{i+k,i}$ and $u_{i,i+k}$ are precisely the entries of the unit triangular factors L and U coming from the $LD^{-1}U$ (Cholesky) factorization of T. The main observation here is that if the matrix is banded, then its L and U factors are also banded. Assuming that the factorization (4.7) has been computed, we can then derive a number of banded approximations to T^{-1} based on the identity

$$T^{-1} = U_1^{-1}U_2^{-1}\cdots U_{n-1}^{-1}DL_{n-1}^{-1}\cdots L_2^{-1}L_1^{-1}.$$

We use the fact that

$$L_i^{-1} = \begin{bmatrix} I & 0 & 0 \\ 0 & 1 & 0 \\ 0 & -\ell_i & I \end{bmatrix} \quad \text{and} \quad U_i^{-1} = \begin{bmatrix} I & 0 & 0 \\ 0 & 1 & -\mathbf{u}_i^T \\ 0 & 0 & I \end{bmatrix}.$$

We begin with an algorithm found in [ABI].

Algorithm 4.4.1 (ABI, the exact $2p+1$-banded innermost part of T^{-1}). *The algorithm* ABI *computes the entries $v_{i+k,i}$ of T^{-1} for $|k| \leq p$ without complete inversion of T.*

Let

$$V_i = \begin{bmatrix} v_{i,i} & v_{i,i+1} & \dots & v_{i,n} \\ v_{i+1,i} & v_{i+1,i+1} & \dots & v_{i+1,n} \\ \vdots & \vdots & \ddots & \vdots \\ v_{n,i} & v_{n,i+1} & \dots & v_{n,n} \end{bmatrix}.$$

Then starting with $V_n = d_n$ *for* $i = n-1, \ldots, 1$ *we have the identity,*

$$V_i = \begin{bmatrix} 1 & -\mathbf{u}_i^T \\ 0 & I \end{bmatrix} \begin{bmatrix} d_i & 0 \\ 0 & V_{i+1} \end{bmatrix} \begin{bmatrix} 1 & 0 \\ -\ell_i & I \end{bmatrix},$$

or

$$V_i = \begin{bmatrix} d_i + \mathbf{u}_i^T V_{i+1}\ell_i & -\mathbf{u}_i^T V_{i+1} \\ -V_{i+1}\ell_i & V_{i+1} \end{bmatrix}.$$

It is clear then that to compute $\tilde{V}_i$*, the* $2p+1$ *banded part of* V_i*, we need only the* $2p+1$ *banded part of* V_{i+1} *(i.e.,* $\tilde{V}_{i+1}$*) because* $\mathbf{u}_i$ *and* ℓ_i *have the special sparse form* (4.8)*. Therefore the following algorithm is applicable, denoting by* I_p *the matrix that zeros the* j*th,* $j > p$*, entries of a vector.*

- $V_n = d_n$.
- *For* $i = n-1$ *down to* 1 *compute the product* $I_p V_{i+1}\ell_i = I_p V_{i+1} I_p \ell_i = I_p \tilde{V}_{i+1}\ell_i$ *and similarly* $\mathbf{u}_i^T V_{i+1} I_p = \mathbf{u}_i^T I_p V_{i+1} I_p = \mathbf{u}_i^T \tilde{V}_{i+1} I_p$*. Then*
- $$\tilde{V}_i = \begin{bmatrix} d_i + \mathbf{u}_i^T \tilde{V}_{i+1}\ell_i & -\mathbf{u}_i^T \tilde{V}_{i+1} I_p \\ -I_p \tilde{V}_{i+1}\ell_i & \tilde{V}_{i+1} \end{bmatrix}.$$
- $\tilde{T}^{-1} = \tilde{V}_1$.

It is clear that the cost of the algorithm is $\mathcal{O}(np^2)$*.*

A possible disadvantage of the above algorithm ABI is that if the decay rate of T^{-1} is not that strong, the approximation $[T^{-1}]^{(p)} \equiv \tilde{T}^{-1}$ (as defined above) may fail to be positive definite. Alternatives to the algorithm ABI are the CHOL and the INVFAC algorithms. The CHOL algorithm (considered in [CGM85]) provides the exact p lower diagonals of L^{-1} (i.e., $\tilde{L}^{-1}$), and the exact p upper diagonals of U^{-1} (i.e., $\tilde{U}^{-1}$) from the factorization $T = LD^{-1}U$, or $T^{-1} = U^{-1}DL^{-1}$. Hence $[T^{-1}]_{\text{CHOL}_p} \equiv \tilde{U}^{-1}D\tilde{L}^{-1}$, which is seen to be symmetric positive definite if T is symmetric positive definite.

Algorithm 4.4.2 (CHOL, a factored banded approximate inverse of T). *Let* $\mathcal{L}_n = 1$*, and for* $i < n$ *set*

$$\mathcal{L}_i = \begin{bmatrix} 1 & 0 \\ -\mathcal{L}_{i+1}\ell_i & \mathcal{L}_{i+1} \end{bmatrix}.$$

Introduce again the projection matrix I_p *(the matrix that zeros all* j*th,* $j > p$*, entries of a vector). We then have* $I_p\mathcal{L}_{i+1}\ell_i = I_p\mathcal{L}_{i+1}I_p\ell_i = I_p\tilde{\mathcal{L}}_{i+1}\ell_i$*, which immediately implies that*

$$\tilde{\mathcal{L}}_i = \begin{bmatrix} 1 & 0 \\ -I_p\tilde{\mathcal{L}}_{i+1}\ell_i & \tilde{\mathcal{L}}_{i+1} \end{bmatrix}.$$

In other words, to compute $\tilde{\mathcal{L}}_i$, the $2p+1$ banded part of $\mathcal{L}_i$, we only need $\tilde{\mathcal{L}}_{i+1}$, the $2p+1$ banded part of $\mathcal{L}_{i+1}$. The same holds for

$$\mathcal{U}_i = \begin{bmatrix} 1 & -\mathbf{u}_i^T \mathcal{U}_{i+i} \\ 0 & \mathcal{U}_{i+1} \end{bmatrix},$$

where $\mathcal{U}_n = 1$. Then, because $\mathbf{u}_i^T \mathcal{U}_i I_p = \mathbf{u}_i^T I_p \mathcal{U}_i I_p = \mathbf{u}_i^T \tilde{\mathcal{U}}_i I_p$, we get the identity

$$\tilde{\mathcal{U}}_i = \begin{bmatrix} 1 & -\mathbf{u}_i^T \tilde{\mathcal{U}}_{i+i} I_p \\ 0 & \tilde{\mathcal{U}}_{i+1} \end{bmatrix}.$$

The CHOL$_p$ *approximation to T^{-1} is then defined by $\tilde{\mathcal{U}}_1 D \tilde{\mathcal{L}}_1$. It is clear that the latter product gives a $2p+1$ banded matrix. We summarize:*

- *Set $\tilde{\mathcal{L}}_n = 1$.*
- *For $i = n-1$ down to 1, compute*

$$\tilde{\mathcal{L}}_i = \begin{bmatrix} 1 & 0 \\ -I_p \tilde{\mathcal{L}}_{i+1} \boldsymbol{\ell}_i & \tilde{\mathcal{L}}_{i+1} \end{bmatrix}.$$

- $\tilde{L}^{-1} = \mathcal{L}_1$.

Similarly,

- *Set $\tilde{\mathcal{U}}_n = 1$.*
- *For $i = n-1$ down to 1, compute*

$$\tilde{\mathcal{U}}_i = \begin{bmatrix} 1 & -\mathbf{u}_i^T \tilde{\mathcal{U}}_{i+i} I_p \\ 0 & \tilde{\mathcal{U}}_{i+1} \end{bmatrix}.$$

- *Set $\tilde{U}^{-1} = \tilde{\mathcal{U}}_1$.*

Finally

$$[T^{-1}]_{\text{CHOL}_p} = \tilde{U}^{-1} D \tilde{L}^{-1}.$$

The cost of the algorithm is readily seen to be $\mathcal{O}(np^2)$

Because in both algorithms we drop certain quantities, in the M-matrix case the approximations $[T^{-1}]$ (both ABI and CHOL) will satisfy the (entrywise) inequality,

$$[T^{-1}] \le T^{-1}. \tag{4.9}$$

The last algorithm that we consider is based on the $2p+1$ banded approximation to the U and L factors from the factorization of T^{-1}; that is,

$$T^{-1} = L_1 L_2 \cdots L_{n-1} D U_{n-1} \cdots U_2 U_1 = LDU.$$

In this latter case, the elementary matrices

$$L_i = \begin{bmatrix} I & 0 & 0 \\ 0 & 1 & 0 \\ 0 & \ell_i & I \end{bmatrix} \quad \text{and} \quad U_i = \begin{bmatrix} I & 0 & 0 \\ 0 & 1 & \mathbf{u}_i^T \\ 0 & 0 & I \end{bmatrix}$$

are not generally banded. One algorithm to compute the vectors ℓ_i and $\mathbf{u}_i$, as well as the entries d_i of the diagonal matrix D, reads as follows.

Let $T_n = t_{n,n}$ be the (n,n)th entry of $T = (t_{i,j})_{i,j=1}^n$. Let T_i be the lower principal submatrix of T; that is,

$$T_i = \begin{bmatrix} t_{i,i} & \mathbf{b}_i^T \\ \mathbf{a}_i & T_{i+1} \end{bmatrix}.$$

Here,

$$\mathbf{a}_i = \begin{bmatrix} t_{i+1,i} \\ \vdots \\ t_{i+p,i} \\ 0 \\ \vdots \\ 0 \end{bmatrix} \quad \text{and similarly} \quad \mathbf{b}_i = \begin{bmatrix} t_{i,i+1} \\ \vdots \\ t_{i,i+p} \\ 0 \\ \vdots \\ 0 \end{bmatrix}.$$

The vectors ℓ_i and $\mathbf{u}_i$ are determined from the identity

$$\begin{bmatrix} 1 & -\mathbf{u}_i^T \\ 0 & I \end{bmatrix} T_i \begin{bmatrix} 1 & 0 \\ -\ell_i & I \end{bmatrix} = \begin{bmatrix} d_i^{-1} & 0 \\ 0 & \mathcal{U}_{i+1}^{-1} D_{i+1}^{-1} \mathcal{L}_{i+1}^{-1} \end{bmatrix}.$$

The latter implies

$$\begin{bmatrix} t_{i,i} - \mathbf{u}_i^T \mathbf{a}_i - \mathbf{b}_i^T \ell_i + \mathbf{u}_i^T T_{i+1} \ell_i & \mathbf{b}_i^T - \mathbf{u}_i^T T_{i+1} \\ \mathbf{a}_i - T_{i+1} \ell_i & T_{i+1} \end{bmatrix} = \begin{bmatrix} d_i^{-1} & 0 \\ 0 & \mathcal{U}_{i+1}^{-1} D_{i+1}^{-1} \mathcal{L}_{i+1}^{-1} \end{bmatrix}.$$

This gives us the relations

$$\begin{aligned} \mathbf{b}_i^T - \mathbf{u}_i^T T_{i+1} &= 0, \\ \mathbf{a}_i - T_{i+1} \ell_i &= 0, \\ T_{i+1} &= \mathcal{U}_{i+1}^{-1} D_{i+1}^{-1} \mathcal{L}_{i+1}^{-1}, \\ d_i^{-1} &= t_{i,i} - \mathbf{u}_i^T \mathbf{a}_i - \mathbf{b}_i^T \ell_i + \mathbf{u}_i^T T_{i+1} \ell_i. \end{aligned}$$

Or equivalently,

$$\begin{aligned} T_{i+1}^{-1} &= \mathcal{L}_{i+1} D_{i+1} \mathcal{U}_{i+1}, \\ \mathbf{u}_i^T &= \mathbf{b}_i^T \mathcal{L}_{i+1} D_{i+1} \mathcal{U}_{i+1}, \\ \ell_i &= \mathcal{L}_{i+1} D_{i+1} \mathcal{U}_{i+1} \mathbf{a}_i, \\ d_i &= \left(t_{i,i} - \mathbf{b}_i^T \mathcal{L}_{i+1} D_{i+1} \mathcal{U}_{i+1} \mathbf{a}_i\right)^{-1}. \end{aligned} \tag{4.10}$$

Finally,

$$T_i^{-1} = \begin{bmatrix} 1 & 0 \\ \ell_i & I \end{bmatrix} \begin{bmatrix} d_i & 0 \\ 0 & \mathcal{L}_{i+1} D_{i+1} \mathcal{U}_{i+1} \end{bmatrix} \begin{bmatrix} 1 & \mathbf{u}_i^T \\ 0 & I \end{bmatrix}.$$

Therefore,

$$T_i^{-1} = \begin{bmatrix} 1 & 0 \\ \ell_i & \mathcal{L}_{i+1} \end{bmatrix} \begin{bmatrix} d_i & 0 \\ 0 & D_{i+1} \end{bmatrix} \begin{bmatrix} 1 & \mathbf{u}_i^T \\ 0 & \mathcal{U}_{i+1} \end{bmatrix} = \mathcal{L}_i D_i \mathcal{U}_i. \tag{4.11}$$

Now assume that we store at each step the $2p+1$ banded parts of $\mathcal{L}_{i+1}$ and $\mathcal{U}_{i+1}$. Then, we have the equations $I_p \ell_i = I_p \mathcal{L}_{i+1} D_{i+1} \mathcal{U}_{i+1} \mathbf{a}_i = I_p \mathcal{L}_{i+1} D_{i+1} \mathcal{U}_{i+1} I_p \mathbf{a}_i = I_p \tilde{\mathcal{L}}_{i+1} D_{i+1} \tilde{\mathcal{U}}_{i+1} \mathbf{a}_i$. Similarly, $\mathbf{u}_i^T I_p = \mathbf{b}_i^T \mathcal{L}_{i+1} D_{i+1} \mathcal{U}_{i+1} I_p = \mathbf{b}_i^T I_p \mathcal{L}_{i+1} D_{i+1} \mathcal{U}_{i+1} I_p = \mathbf{b}_i^T \tilde{\mathcal{L}}_{i+1} D_{i+1} \tilde{\mathcal{U}}_{i+1} I_p$. In other words, the exact first p entries of ℓ_i and $\mathbf{u}_i^T$ are computable from the exact $2p+1$ banded parts of $\mathcal{L}_{i+1}$ and $\mathcal{U}_{i+1}$. The latter is sufficient to compute the exact $2p+1$ banded part of $\mathcal{L}_i$ and of $\mathcal{U}_i$ based on the identity (4.11). The so-called INVFAC approximation to T^{-1} is then defined as

$$[T^{-1}]_{\text{INVFAC}_p} = \mathcal{L}_1 D_1 \mathcal{U}_1. \tag{4.12}$$

Identity (4.11) provides an alternative way to compute the exact $2p+1$ banded part $\tilde{T}_i^{-1}$ of T_i^{-1}. Indeed, we have

$$T_i^{-1} = \begin{bmatrix} d_i & d_i \mathbf{u}_i^T \\ \ell_i d_i & \mathcal{L}_i D_i \mathcal{U}_i + \ell_i d_i \mathbf{u}_i^T \end{bmatrix}. \tag{4.13}$$

Therefore,

$$\tilde{T}_i^{-1} = \begin{bmatrix} d_i & d_i \mathbf{u}_i^T I_p \\ I_p \ell_i d_i & I_p \mathcal{L}_i D_i \mathcal{U}_i I_p + I_p \ell_i d_i \mathbf{u}_i^T I_p \end{bmatrix} = \begin{bmatrix} d_i & d_i \mathbf{u}_i^T I_p \\ I_p \ell_i d_i & \tilde{\mathcal{L}}_i D_i \tilde{\mathcal{U}}_i + I_p \ell_i d_i \mathbf{u}_i^T I_p \end{bmatrix}.$$

Then,

$$[T^{-1}]^{(p)} = \tilde{T}_1^{-1}.$$

Therefore, the following algorithm provides enough information to compute both the INVFAC and the ABI approximations to T^{-1}.

Algorithm 4.4.3 (Banded approximate inverse factorization of T). *Given is the $2p+1$ banded matrix $T = (t_{i,j})_{i,j=1}^n$. Introduce,*

$$\mathbf{a}_i = \begin{bmatrix} t_{i+1,i} \\ \vdots \\ t_{i+p,i} \\ 0 \\ \vdots \\ 0 \end{bmatrix}, \quad \mathbf{b}_i = \begin{bmatrix} t_{i,i+1} \\ \vdots \\ t_{i,i+p} \\ 0 \\ \vdots \\ 0 \end{bmatrix}.$$

- *Initiate:*

$$d_n = \frac{1}{t_{n,n}}, \quad \tilde{\mathcal{L}}_n = 1, \quad \tilde{\mathcal{U}}_n = 1.$$

- *For $i = n - 1$ down to 1, compute*

$$\mathbf{u}_i^T I_p = \mathbf{b}_i^T \tilde{\mathcal{L}}_{i+1} D_{i+1} \tilde{\mathcal{U}}_{i+1};$$
$$I_p \boldsymbol{\ell}_i = \tilde{\mathcal{L}}_{i+1} D_{i+1} \tilde{\mathcal{U}}_{i+1} \, \mathbf{a}_i;$$
$$d_i = \left(t_{i,i} - \mathbf{b}_i^T \tilde{\mathcal{L}}_{i+1} D_{i+1} \tilde{\mathcal{U}}_{i+1} \mathbf{a}_i\right)^{-1}.$$

- *Then set*

$$\tilde{\mathcal{L}}_i = \begin{bmatrix} 1 & 0 \\ I_p \boldsymbol{\ell}_i & \tilde{\mathcal{L}}_{i+1} \end{bmatrix};$$
$$\tilde{\mathcal{U}}_i = \begin{bmatrix} 1 & \mathbf{u}_i^T I_p \\ 0 & \tilde{\mathcal{U}}_{i+1} \end{bmatrix};$$
$$D_i = \begin{bmatrix} d_i & 0 \\ 0 & D_{i+1} \end{bmatrix}.$$

It is clear, because in the above algorithm we drop certain quantities, that in the M-matrix case the INVFAC approximation, $[T^{-1}]_{\text{INVFAC}_p}$, of T^{-1} satisfies the entrywise inequality

$$[T^{-1}]_{\text{INVFAC}_p} \leq T^{-1}. \tag{4.14}$$

4.5 Wittum's frequency filtering decomposition

A general scheme

Here we briefly describe an approximate block-factorization algorithm that utilizes certain vectors throughout the factorization process. With a certain choice of the vectors, the method was originally proposed by Wittum in [Wi92], and further developed in [Wag97] and [WW97].

Consider a two-by-two block matrix A,

$$A = \begin{bmatrix} \mathcal{A} & \mathcal{R} \\ \mathcal{L} & \mathcal{B} \end{bmatrix}.$$

We assume that A is s.p.d.. Given is a block-vector

$$\mathbf{1} = \begin{bmatrix} \mathbf{x} \\ \mathbf{y} \end{bmatrix}.$$

We begin with the case where the first component $\mathbf{x}$ is chosen to satisfy the equation $\mathcal{A}\mathbf{x} + \mathcal{R}\mathbf{y} = 0$ for a given $\mathbf{y}$.

Let $\mathcal{M}$ be an s.p.d. approximation to $\mathcal{A}$ such that

$$\mathcal{M}\mathbf{x} = \mathcal{A}\mathbf{x}.$$

Finally, let $\mathcal{Y}$ be an s.p.d. matrix (specified later on). The matrix $\mathcal{Y}$ is in general an approximation to the exact Schur complement $\mathcal{Z} = \mathcal{B} - \mathcal{L}\mathcal{A}^{-1}\mathcal{R}$ of A. To this end, define the block-factored approximation matrix M to A, assuming in addition that $2\mathcal{M} - \mathcal{A}$ is s.p.d.,

$$M = \begin{bmatrix} \mathcal{M} & 0 \\ \mathcal{L} & I \end{bmatrix} \begin{bmatrix} (2\mathcal{M} - \mathcal{A})^{-1} & 0 \\ 0 & \mathcal{Y} \end{bmatrix} \begin{bmatrix} \mathcal{M} & \mathcal{R} \\ 0 & I \end{bmatrix}.$$

Below, we formulate conditions on $\mathcal{M}$ and $\mathcal{Y}$ which guarantee that $M\mathbf{1} = A\mathbf{1}$.

More specifically, the following result holds.

Proposition 4.8. *Let* $\mathbf{x}$ *satisfy* $\mathcal{A}\mathbf{x} + \mathcal{R}\mathbf{y} = 0$ *and* $\mathcal{M}$ *be constructed such that* $\mathcal{M}\mathbf{x} = \mathcal{A}\mathbf{x}$. *Also, let* $\mathcal{Y}$ *satisfy* $\mathcal{Y}\mathbf{y} = \mathcal{Z}\mathbf{y}$. *Then,* $M\mathbf{1} = A\mathbf{1}$.

Proof. We have

$$M\begin{bmatrix} \mathbf{x} \\ \mathbf{y} \end{bmatrix} = \begin{bmatrix} \mathcal{M}(2\mathcal{M} - \mathcal{A})^{-1}(\mathcal{M}\mathbf{x} + \mathcal{R}\mathbf{y}) \\ \mathcal{L}(2\mathcal{M} - \mathcal{A})^{-1}(\mathcal{M}\mathbf{x} + \mathcal{R}\mathbf{y}) + \mathcal{Y}\mathbf{y} \end{bmatrix}.$$

Because

$$A\begin{bmatrix} \mathbf{x} \\ \mathbf{y} \end{bmatrix} = \begin{bmatrix} \mathcal{A}\mathbf{x} + \mathcal{R}\mathbf{y} \\ \mathcal{L}\mathbf{x} + \mathcal{B}\mathbf{y} \end{bmatrix},$$

we see that if $\mathcal{A}\mathbf{x} = \mathcal{M}\mathbf{x}$ and $\mathbf{x}: \ \mathcal{A}\mathbf{x} + \mathcal{R}\mathbf{y} = 0$ that $\mathcal{M}\mathbf{x} + \mathcal{R}\mathbf{y} = 0$. Hence, to guarantee that $M\mathbf{1} = A\mathbf{1}$, we need to satisfy $\mathcal{Y}\mathbf{y} = \mathcal{L}\mathbf{x} + \mathcal{B}\mathbf{y} = (\mathcal{B} - \mathcal{L}\mathcal{A}^{-1}\mathcal{R})\mathbf{y} = \mathcal{Z}\mathbf{y}$, which we have assumed. □

We next construct an approximation $\mathcal{Y}$ to the Schur complement $\mathcal{Z} = \mathcal{B} - \mathcal{L}\mathcal{A}^{-1}\mathcal{U}$ of A, which has the property

$$\mathcal{Y}\mathbf{y} = \mathcal{Z}\mathbf{y}.$$

The approximate Schur complement $\mathcal{Y}$ can be constructed as follows. Let

$$P = \begin{bmatrix} \mathcal{W} \\ I \end{bmatrix}$$

for a block-matrix $\mathcal{W}$ to be determined. Then $\mathcal{Y} = P^T A P$, that is, obtained variationally from A. It is clear that $\mathbf{z}^T \mathcal{Y} \mathbf{z} \geq \mathbf{z}^T \mathcal{Z} \mathbf{z} \geq 0$ for any $\mathbf{z}$ (due to the minimization property of the s.p.d. Schur complement $\mathcal{Z}$).

We select $\mathcal{W}$ such that $P\mathbf{y} = \mathbf{1}$. Then, by construction

$$\mathcal{Y}\mathbf{y} = P^T A(P\mathbf{y}) = P^T A\mathbf{1} = P^T \begin{bmatrix} \mathcal{A}\mathbf{x} + \mathcal{R}\mathbf{y} \\ \mathcal{L}\mathbf{x} + \mathcal{B}\mathbf{y} \end{bmatrix} = [\mathcal{W}^T,\ I] \begin{bmatrix} 0 \\ \mathcal{Z}\mathbf{y} \end{bmatrix} = \mathcal{Z}\mathbf{y}.$$

To summarize:

(i) Construct a $\mathcal{W}$ such that $P\mathbf{y} = \mathbf{1}$. One algorithm for computing $P = (\boldsymbol{\psi}_i)$, where $\boldsymbol{\psi}_i$ stands for the ith column of P, can be based on constrained minimization (studied in detail in Section 6.3),

$$\sum_i \boldsymbol{\psi}_i^T A \boldsymbol{\psi}_i \mapsto \min,$$

subject to the constraint $P\mathbf{y} = \mathbf{1}$.

(ii) The approximate Schur complement $\mathcal{Y}$ is then computed as

$$\mathcal{Y} = P^T A P = \mathcal{B} + \mathcal{W}^T \mathcal{R} + \mathcal{L}\mathcal{W} + \mathcal{W}^T \mathcal{A}\mathcal{W}.$$

Now, let us consider the more general case for vectors $\mathbf{x}$ and $\mathbf{y}$. The following result holds.

Proposition 4.9. *Assume that $\mathcal{M}$ has been constructed such that*

$$\mathcal{M}\mathbf{x} = \mathcal{A}\mathbf{x},$$

and let $\mathcal{A}\mathbf{e} = \mathbf{r} \equiv \mathcal{A}\mathbf{x} + \mathcal{R}\mathbf{y}$. Assume in addition that

$$\mathcal{M}\mathbf{e} = \mathcal{A}\mathbf{e}.$$

That is, now $\mathcal{M}$ and $\mathcal{A}$ have the same actions on two vectors, $\mathbf{x}$ and $\mathbf{e} = \mathbf{x} + \mathcal{A}^{-1}\mathcal{R}\mathbf{y}$. Then, assuming in addition that $\mathcal{Y}\mathbf{y} = \mathcal{Z}\mathbf{y}$, we have that $A\mathbf{1} = M\mathbf{1}$.

Proof. We first show that $\mathcal{M}(2\mathcal{M} - \mathcal{A})^{-1}\mathbf{r} = \mathbf{r}$ or equivalently, $\mathbf{r} = (2\mathcal{M} - \mathcal{A})\mathcal{M}^{-1}\mathbf{r} = 2\mathbf{r} - \mathcal{A}\mathcal{M}^{-1}\mathbf{r}$. That is, $\mathcal{M}^{-1}\mathbf{r} = \mathcal{A}^{-1}\mathbf{r}$. The latter is true, because $\mathcal{M}\mathbf{e} = \mathcal{A}\mathbf{e} = \mathbf{r}$. Because $\mathcal{M}\mathbf{x} = \mathcal{A}\mathbf{x}$, it is clear that $\mathbf{r} = \mathcal{M}\mathbf{x} + \mathcal{R}\mathbf{y}$. Thus, we showed that

$$\mathcal{M}(2\mathcal{M} - \mathcal{A})^{-1}(\mathcal{M}\mathbf{x} + \mathcal{R}\mathbf{y}) = \mathcal{M}(2\mathcal{M} - \mathcal{A})^{-1}\mathbf{r} = \mathbf{r} = \mathcal{A}\mathbf{x} + \mathcal{R}\mathbf{y}.$$

That is, $M\mathbf{1}$ and $A\mathbf{1}$ have the same first block component (equal to $\mathbf{r}$). The second block component of $M\mathbf{1}$ equals

$$\begin{aligned} \mathcal{L}(2\mathcal{M} - \mathcal{A})^{-1}(\mathcal{M}\mathbf{x} + \mathcal{R}\mathbf{y}) + \mathcal{Y}\mathbf{y} &= \mathcal{L}(2\mathcal{M} - \mathcal{A})^{-1}\mathbf{r} + \mathcal{Y}\mathbf{y} \\ &= \mathcal{L}\mathcal{A}^{-1}\mathbf{r} + \mathcal{Y}\mathbf{y} \\ &= \mathcal{L}\mathcal{A}^{-1}(\mathcal{A}\mathbf{x} + \mathcal{R}\mathbf{y}) + \mathcal{Y}\mathbf{y} \\ &= \mathcal{L}\mathbf{x} + \mathcal{L}\mathcal{A}^{-1}\mathcal{R}\mathbf{y} + \mathcal{Y}\mathbf{y}. \end{aligned}$$

The second block component of $A\mathbf{1}$ equals $\mathcal{B}\mathbf{y} + \mathcal{L}\mathbf{x}$. Thus if $\mathcal{Y}\mathbf{y} + \mathcal{L}\mathcal{A}^{-1}\mathcal{R}\mathbf{y} + \mathcal{L}\mathbf{x} = \mathcal{B}\mathbf{y} + \mathcal{L}\mathbf{x}$, that is, $\mathcal{Y}\mathbf{y} = \mathcal{Z}\mathbf{y}$, we finally get that $A\mathbf{1} = M\mathbf{1}$. □

Some applications

The factorization process from the preceding section can be applied to a two-by-two block structure of A where $\mathcal{A}$ is a sparse well-conditioned matrix. Then, we need an easily invertible matrix $\mathcal{M}$ that has the same actions as $\mathcal{A}$ on two vectors, $\mathbf{x}$ and $\mathbf{e}: \ \mathcal{A}\mathbf{e} = \mathbf{r} = \mathcal{A}\mathbf{x} + \mathcal{R}\mathbf{y}$. This imposes two conditions on $\mathcal{M}$. In [Wi92], Wittum proposed an algorithm that constructs a (scalar) symmetric tridiagonal matrix $\mathcal{M}$ that has the same action as a given s.p.d. matrix $\mathcal{A}$ on two prescribed vectors. Applying the same algorithm recursively (now to $\mathcal{Y}$), we can end up with a multilevel block-ILU factorization matrix. For various possible venues in this direction we refer to [BW99] and [BaS02].

We observe that there is one vector ($\mathbf{1}$) that drives the construction of the block-ILU factored matrix M. Namely, after partitioning $\mathbf{1}$ gives rise to two smaller vectors $\mathbf{x}$ and $\mathbf{y}$ that are then used to define $\mathbf{r} = \mathcal{A}\mathbf{x} + \mathcal{R}\mathbf{y}$. The pair $\mathbf{x}, \ \mathbf{r}$ is used to construct a suitable $\mathcal{M}$, whereas the block $\mathbf{y}$, extended to

$$\mathbf{1}_0 \equiv \begin{bmatrix} \mathbf{x}_0 \\ \mathbf{y} \end{bmatrix}$$

where $\mathcal{A}\mathbf{x}_0 + \mathcal{R}\mathbf{y} = 0$, is used to construct

$$P = \begin{bmatrix} \mathcal{W} \\ I \end{bmatrix}$$

such that $P\mathbf{y} = \mathbf{1}_0$. To get the vector $\mathbf{1}_0$, we need to solve a system with $\mathcal{A}$. This can be practical if, for example, $\mathcal{A}$ is well conditioned. With a P in hand, the approximate Schur complement $\mathcal{Y}$ is computed as P^TAP. The latter choice of $\mathbf{1}_0$ and the construction of P such that $P\mathbf{y} = \mathbf{1}_0$ guarantees that $\mathcal{Y}\mathbf{y} = \mathcal{Z}\mathbf{y}$. As proved in the preceding section, we have ensured at the end that $M\mathbf{1} = A\mathbf{1}$. By setting $A := \mathcal{Y}$ and $\mathbf{1} := \mathbf{y}$ after a successive two-by-two partitioning of A and respective partitioning of $\mathbf{1}$, we can apply the same construction by recursion.

Another application is considered in the following section where we formally have $\mathcal{M} = \mathcal{A}$; hence, we only need to construct an approximation $\mathcal{Y}$ to the exact Schur complement $\mathcal{Z}$ that has the same action on a prescribed vector $\mathbf{y}$.

Application of the "filtering" approximate block-factorization to block-tridiagonal matrices

Consider the block-tridiagonal matrix

$$A = \begin{bmatrix} A_{11} & A_{12} & 0 & \dots & 0 \\ A_{21} & A_{22} & A_{23} & \dots & 0 \\ & \ddots & \ddots & \ddots & \\ & & A_{n-1,n-2} & A_{n-1,n-1} & A_{n-1,n} \\ 0 & \dots & 0 & A_{n,n-1} & A_{nn} \end{bmatrix}.$$

Let A_i be the principal submatrix of A containing its first $i \times i$ blocks, $\mathcal{B} = A_{i+1,i+1}$. Finally, let

$$\mathcal{R} = \begin{bmatrix} 0 \\ \vdots \\ 0 \\ A_{i,i+1} \end{bmatrix}$$

and $\mathcal{L} = \mathcal{R}^T$ be the resulting off-diagonal blocks of A_{i+1}, which is the principal submatrix of A consisting of its first $(i+1) \times (i+1)$ blocks. That is, we have

$$A_{i+1} = \begin{bmatrix} A_i & \mathcal{R} \\ \mathcal{L} & \mathcal{B} \end{bmatrix}.$$

Let $\mathbf{1} = (\mathbf{1}_k)_{k=1}^n$ be a given block vector.

Assume that we have constructed a block-factored matrix M_i such that $M_i \mathbf{x}_i = A_i \mathbf{x}_i$, where $\mathbf{x}_i = (\mathbf{1}_k)_{k=1}^i$. Consider then the partially factored matrix

$$\widehat{A}_{i+1} = \begin{bmatrix} M_i & \mathcal{R} \\ \mathcal{L} & \mathcal{B} \end{bmatrix}.$$

We obviously have $\widehat{A}_{i+1}\mathbf{x}_{i+1} = A_{i+1}\mathbf{x}_{i+1}$. Based on the two-by-two block structure of $\widehat{A}_{i+1}$, we construct a block-factored matrix M_{i+1} to approximate $\widehat{A}_{i+1}$ and hence A_{i+1} such that $M_{i+1}\mathbf{x}_{i+1} = \widehat{A}_{i+1}\mathbf{x}_{i+1} = A_{i+1}\mathbf{x}_{i+1}$. As demonstrated in the previous section, for this to hold, we needed to construct an approximate Schur complement $\mathcal{Y}$ such that $\mathcal{Y}\mathbf{y} = \mathcal{Z}\mathbf{y}$ where $\mathcal{Z} = \mathcal{B} - \mathcal{L}M_i^{-1}\mathcal{R}$ is the exact Schur complement of $\widehat{A}_{i+1}$. Because we deal with block-tridiagonal matrices, we have $\mathcal{Z} = A_{i+1,i+1} - A_{i+1,i}Y_i^{-1}A_{i,i+1}$. We denote $Y_{i+1} = \mathcal{Y}$ to be that approximation, which hence satisfies

$$Y_{i+1}\mathbf{1}_{i+1} = \left(A_{i+1,i+1} - A_{i+1,i}Y_i^{-1}A_{i,i+1}\right)\mathbf{1}_{i+1}.$$

The latter is referred to as a "filter condition" (cf., e.g., [WW97]) and the resulting block-factorization, sometimes referred to as "tangential frequency filtering" decomposition if the vectors $\mathbf{1}_i$ come from a lower part of the spectrum of the Schur complements $\mathcal{Z}$. For vectors $\mathbf{1}_i$ chosen adaptively, we end up with the so-called "adaptive filtering" method as proposed in [WW97]. An adaptive vector is constructed by looking at the error $\mathbf{1}_{\text{new}} = (I - M^{-1}A)\mathbf{1}$. Then, based on the new vector $\mathbf{1}_{\text{new}}$, a new tangential frequency filtering decomposition matrix M_{new} is constructed. The procedure is repeated with $M^{-1} := M^{-1} + M_{\text{new}}^{-1} - M_{\text{new}}^{-1}AM^{-1}$, and if needed, a new adaptive vector is computed based on the thus modified M^{-1}.

4.6 Block-ILU factorizations with block-size reduction

Another way to keep the complexity of the block-ILU factorization algorithms as described in Section 4.1, is to construct a low-rank matrix Y_i that approximates the inverse of the previously computed approximate Schur complement X_i. In this case, we do not need to further approximate the product $A_{i+1,i} Y_i A_{i,i+1}$ by another matrix H_{i+1}. That is, formally, we can let $H_{i+1} = A_{i+1,i} Y_i A_{i,i+1}$.

The way we construct low-rank approximations Y_i to X_i^{-1} exploits the concept of block-size reduction ([ChV95]). Let $\{R_i\}$ be a set of reduction matrices. These are rectangular matrices that have a relatively small number of rows. A typical choice is a block-diagonal rectangular matrix

$$R_i = \begin{bmatrix} \mathbf{1}_1^{(i)^T} & 0 & 0 & \dots & 0 \\ 0 & \mathbf{1}_2^{(i)^T} & 0 & \dots & 0 \\ & & \ddots & \ddots & \\ & & & \ddots & 0 \\ 0 & 0 & \dots & 0 & \mathbf{1}_m^{(i)^T} \end{bmatrix},$$

for a small integer $m \geq 1$. The vectors $\mathbf{1}_k^{(i)}$, $k = 1, \dots, m$, can come from a nonoverlapping partition of a single vector $\mathbf{1}^{(i)} = (\mathbf{1}_k^{(i)})$. A simple choice is $\mathbf{1}^{(i)} = (1)$, and hence, all $\mathbf{1}_k^{(i)} = (1)$ being constant vectors. That is, R_i^T corresponds to a piecewise constant interpolation.

Having the (full-rank) matrix R_i in hand and X_i being computed at the preceding step ($X_1 = A_{1,1}$), we define

$$Y_i = R_i^T \left(R_i X_i R_i^T\right)^{-1} R_i. \tag{4.15}$$

Then, as before, we let

$$X_{i+1} = A_{i+1,i+1} - A_{i+1,i} Y_i A_{i,i+1} = A_{i+1,i+1} - A_{i+1,i} R_i^T \left(R_i X_i R_i^T\right)^{-1} R_i A_{i,i+1}.$$

Note that $R_i X_i R_i^T$ is a dense matrix but of small size, hence its explicit inverse is computationally feasible. The low-rank approximation Y_i to X_i^{-1} has the following key property. For any vector $\mathbf{v}_i$, we have

$$0 \leq \mathbf{v}_i^T Y_i \mathbf{v}_i \leq \mathbf{v}_i^T X_i^{-1} \mathbf{v}_i. \tag{4.16}$$

This is seen from the fact that $\|C_i\| = \|C_i^T\|$ applied to $C_i = \widetilde{X}_i^{-1/2} R_i X_i^{1/2}$, where

$$\widetilde{X}_i \equiv R_i X_i R_i^T.$$

Note now that $C_i C_i^T = I$, hence $\|C_i^T\| = 1 = \|C_i\|$. Using $\|C_i\| = 1$ implies,

$$\begin{aligned} \mathbf{v}_i^T \mathbf{v}_i &\geq \mathbf{v}_i^T C_i^T C_i \mathbf{v}_i \\ &= \mathbf{v}_i^T X_i^{1/2} R_i^T \widetilde{X}_i^{-1} R_i X_i^{1/2} \mathbf{v}_i. \\ &= \mathbf{v}_i^T X_i^{1/2} Y_i X_i^{1/2} \mathbf{v}_i. \end{aligned}$$

The latter is equivalent to (4.16).

Assuming by induction that $X_i - Z_i$ are symmetric positive semidefinite, based on property (4.16), we easily show that $X_{i+1} - Z_{i+1}$ is also symmetric positive semidefinite. We recall that Z_i is the successive Schur complement computed during an exact factorization of A; that is, $Z_1 = A_{1,1}$ and $Z_{i+1} = A_{i+1,i+1} - A_{i+1,i} Z_i^{-1} A_{i,i+1}$. More specifically, we have the following main result.

Theorem 4.10. *The block-size reduction algorithm based on Y_i defined as in* (4.15), *and $X_{i+1} = A_{i+1,i+1} - A_{i+1,i} Y_i A_{i,i+1}$ is well defined. It provides a block-ILU factorization matrix $M = (X - L)X^{-1}(X - U)$ that is s.p.d., and also the difference $M - A$ is symmetric positive semidefinite.*

Proof. To demonstrate the existence of the factorization means to show that X_i are s.p.d., hence invertible. Following the proof of Theorem 4.5, we show by induction that $X_i - Z_i$ are symmetric positive semidefinite. We have to interpret the inequalities in (4.5) not entrywise but in terms of inner products.

Looking at the expression (4.2) (recall that now $H_i = A_{i,i-1} Y_{i-1} A_{i-1,i}$), we immediately see (due to (4.16)) that $M - A$ is symmetric positive semidefinite. □

Because M is meant to be used as a preconditioner, we need efficient algorithms to solve systems with M, which is based on solving systems with the blocks X_i; that is, we need algorithms to evaluate the actions of X_i^{-1}. Here, we can take advantage of the fact that X_i is a low-rank update of the original block $A_{i,i}$. Assuming that the inverse actions of $A_{i,i}$ are easy to compute, the Sherman–Morrison formula can be handy here to compute the inverse actions of X_i. In general though, we may need to further approximate $A_{i,i}$ with some s.p.d. matrices $B_{i,i}$ such that $B_{i,i}^{-1}$ have readily available actions.

A computational version of the block-size reduction ILU algorithm is as follows. Compute the coarse block-tridiagonal matrix $\widetilde{A} = (R_i A_{ij} R_j^T)$. It is reasonable to assume that its exact block-factorization $\widetilde{A} = (\widetilde{X} - \widetilde{L})\widetilde{X}^{-1}(\widetilde{X} - \widetilde{U})$ is inexpensive because the blocks $\widetilde{A}_{ij} = R_i A_{ij} R_j^T$ have a relatively small size. Here, $\widetilde{X} = \text{diag}(\widetilde{X}_i)$ and $\widetilde{L} =$ lower triangular with blocks on the first lower diagonal $-\widetilde{A}_{i,i-1}$, and similarly, the upper triangular part $\widetilde{U}$ is defined from the blocks $-\widetilde{A}_{i-1,i}$. Then the following approximations of the Schur complements of A are feasible,

$$X_i = A_{ii} - A_{i,i-1} R_{i-1}^T \widetilde{X}_{i-1}^{-1} R_{i-1} A_{i-1,i}.$$

Note that based on the symmetric version of the Sherman–Morrison formula (see Proposition 3.5), we have

$$X_i^{-1} = A_{ii}^{-1} + A_{ii}^{-1} A_{i,i-1} R_{i-1}^T \widetilde{V}_{i-1} R_{i-1} A_{i-1,i} A_{ii}^{-1}.$$

Here,

$$\widetilde{V}_{i-1}^{-1} = \widetilde{X}_{i-1} - R_{i-1} A_{i-1,i} A_{ii}^{-1} A_{i,i-1} R_{i-1}^T.$$

That is, the inverse actions of X_i are based on the inverse actions of A_{ii} and on the inverse of the small matrix $\widetilde{V}_{i-1}^{-1}$. Let $m \times m$ be the size of $\widetilde{V}_{i-1}^{-1}$; its entries then can be computed by m actions of A_{ii}^{-1}. Thus, the resulting block-factorization preconditioner is practical if m is really small. A second step would be to approximate A_{ii}^{-1} with computationally feasible matrices B_{ii}^{-1}. We assume that

$$\widehat{B}_i \equiv \begin{bmatrix} \widetilde{X}_{i-1} & R_{i-1}A_{i-1,i} \\ A_{i,i-1}R_{i-1}^T & B_{ii} \end{bmatrix}$$

is s.p.d. The latter can be ensured if

$$\mathbf{v}_i^T A_{ii}\mathbf{v}_i \leq \mathbf{v}_i^T B_{ii}\mathbf{v}_i, \tag{4.17}$$

which is the case, for example, for B_{ii} coming from a block-factored form of A_{ii} described in the previous sections or simply being a symmetric Gauss–Seidel preconditioner for A_{ii}. To see that $\widehat{B}_i$ is s.p.d., with the above choice (4.17) of B_{ii}, it is sufficient to show that the matrix

$$\widehat{A}_i \equiv \begin{bmatrix} \widetilde{X}_{i-1} & R_{i-1}A_{i-1,i} \\ A_{i,i-1}R_{i-1}^T & A_{ii} \end{bmatrix}$$

is s.p.d. Letting $A_i = (A_{r,s})_{r,s=1}^i$ be the ith principal submatrix of A, then the main observation is that $\widehat{A}_i$ is a Schur complement of the following partially coarse s.p.d. matrix,

$$\begin{bmatrix} R_1 & 0 & & 0 \\ 0 & \ddots & 0 & \\ & 0 & R_{i-1} & 0 \\ & & 0 & I \end{bmatrix} A_i \begin{bmatrix} R_1^T & 0 & & 0 \\ 0 & \ddots & 0 & \\ & 0 & R_{i-1}^T & 0 \\ & & 0 & I \end{bmatrix}.$$

With the choice (4.17) of B_{ii}, the following more practical version of the block-size reduction ILU algorithm is of interest.

Algorithm 4.6.1 (Block-size reduction ILU).

(i) *Compute the coarse matrix* $\widetilde{A} = (R_i A_{i,j} R_j^T)$ *and its factorization,* $\widetilde{X}_1 = \widetilde{A}_{1,1}$, $\widetilde{X}_{i+1} = \widetilde{A}_{i+1,i+1} - \widetilde{A}_{i+1,i}\widetilde{X}_i^{-1}\widetilde{A}_{i,i+1}$.

(ii) *Compute the inverse* $\widetilde{V}_{i-1}$ *of the Schur complement of the partially coarse matrix* $\widehat{B}_i$; *that is, compute,*

$$\widetilde{V}_{i-1} = \left(\widetilde{X}_{i-1} - R_{i-1}A_{i-1,i}B_{ii}^{-1}A_{i,i-1}R_{i-1}^T\right)^{-1}.$$

(iii) *Define*

$$X_i = B_{ii} - A_{i,i-1}R_{i-1}^T\widetilde{X}_{i-1}^{-1}R_{i-1}A_{i-1,i},$$

which is not really needed. What we need is the expression for the inverse action of X_i, *based on the Sherman–Morrison formula:*

$$X_i^{-1} = B_{ii}^{-1} + B_{ii}^{-1}A_{i,i-1}R_{i-1}^T\widetilde{V}_{i-1}R_{i-1}A_{i-1,i}B_{ii}^{-1}. \tag{4.18}$$

Note that we do not carry out the matrix multiplications in (4.18), *which can be costly in terms of storage and flops. We use the expression to compute* $X_i^{-1}\mathbf{v}_i$ *for a given vector* $\mathbf{v}_i$. *The latter involves two solutions with* $B_{i,i}$ *and matrix–vector* *products with $A_{i,i-1}$, R_{i-1}^T, $\widetilde{V}_{i-1}$, R_{i-1}, and $A_{i-1,i}$. The latter matrices are either sparse, or dense but small.*

This resulting block-ILU factorization matrix is well defined as long as $\widehat{A}_i$ is s.p.d., which we can ensure. Because $X_i = B_{ii} - A_{i,i-1}Y_{i-1}A_{i-1,i}$, we have that the difference $M - A$ is block-diagonal with blocks on the diagonal equal to $X_i + A_{i,i-1}X_{i-1}^{-1}A_{i-1,i} - A_{i,i} = B_{i,i} - A_{i,i} + A_{i,i-1}(X_{i-1}^{-1} - Y_{i-1})A_{i-1,i}$, which shows that $M - A$ will be symmetric positive semidefinite as long as $B_{i,i} - A_{i,i}$ are symmetric positive semidefinite (recalling (4.16)). We summarize as follows.

Theorem 4.11. *The block-size reduction ILU algorithm based on inexact blocks $\{B_{i,i}\}$ leading to $\widetilde{V}_{i-1}^{-1} = \widetilde{X}_{i-1} - R_{i-1}A_{i-1,i}B_{i,i}^{-1}A_{i,i-1}R_{i-1}^T$ and corresponding X_i as in* (4.18) *is well defined. Moreover, the difference $M - A$ is symmetric positive semidefinite. This holds for any full-rank restriction matrices R_i and matrices $B_{i,i}$ such that $B_{i,i} - A_{i,i}$ are symmetric positive semidefinite.*

4.7 An alternative approximate block-LU factorization

Let

$$A = \begin{bmatrix} \mathcal{A} & \mathcal{R} \\ \mathcal{L} & \mathcal{B} \end{bmatrix}$$

be s.p.d. and assume that we can derive s.p.d. approximations $\mathcal{M}$ to $\mathcal{A}^{-1}$ and $\mathcal{A}_c$ to the exact Schur complement $\mathcal{S} = \mathcal{B} - \mathcal{L}\mathcal{A}^{-1}\mathcal{R}$. With a special choice of $\mathcal{M}$, the following approximate block-factorization matrix,

$$M = \begin{bmatrix} \mathcal{M}^{-1} & 0 \\ \mathcal{L} & \mathcal{A}_c \end{bmatrix} \begin{bmatrix} I & \mathcal{M}\mathcal{R} \\ 0 & I \end{bmatrix},$$

leads to a spectrally equivalent preconditioner to A. This type of preconditioner was introduced and analyzed in Section 3.4.1. Here, we specify some ways to construct $\mathcal{M}$ so that the conditions (assumed in Section 3.4.1) both $\mathcal{A} - \mathcal{M}^{-1}$, and

$$\begin{bmatrix} \mathcal{M}^{-1} & \mathcal{R} \\ \mathcal{L} & B \end{bmatrix}$$

are symmetric positive semidefinite, are met.

A local procedure for choosing the first block

In the setting of two-grid methods, we typically define $\mathcal{A}_c = P^T AP$ to be a coarse matrix obtained from A by an interpolation matrix P. We then have to guarantee that $\mathbf{x}^T P^T AP\mathbf{x} \leq \eta \inf_{\mathbf{v}:=J\mathbf{w}+P\mathbf{x}} \mathbf{v}^T A\mathbf{v}$, where

$$J = \begin{bmatrix} I \\ 0 \end{bmatrix}.$$

It is clear that this energy boundedness of

$$P = \begin{bmatrix} W \\ I \end{bmatrix},$$

implies the estimate $\mathbf{x}^T \mathcal{A}_c \mathbf{x} \leq \eta \ \mathbf{x}^T \mathcal{S} \mathbf{x}$. Another more specific choice is given later on in this section.

The construction of $\mathcal{M}$ is a bit tricky because it has to satisfy two types of inequalities, namely, $\mathbf{v}^T \mathcal{A}^{-1} \mathbf{v} \leq \mathbf{v}^T \mathcal{M} \mathbf{v}$, and

$$\alpha \ \mathbf{x}^T \mathcal{S} \mathbf{x} \leq \mathbf{x}^T (\mathcal{B} - \mathcal{L} \mathcal{M} \mathcal{R}) \mathbf{x}.$$

These two estimates, in particular, mean that the approximate Schur complement $\mathcal{B} - \mathcal{L}\mathcal{M}\mathcal{R}$ should be spectrally equivalent to the exact Schur complement $\mathcal{S}$; that is,

$$\alpha \ \mathbf{x}^T \mathcal{S} \mathbf{x} \leq \mathbf{x}^T (\mathcal{B} - \mathcal{L} \mathcal{M} \mathcal{R}) \mathbf{x} \leq \mathbf{x}^T \mathcal{S} \mathbf{x}.$$

Such a construction is possible if A is assembled from local matrices $\{\mathcal{A}_\tau\}$ in the sense

$$\mathbf{v}^T A \mathbf{v} = \sum_\tau \mathbf{v}_\tau^T A_\tau \mathbf{v}_\tau, \tag{4.19}$$

where $\mathbf{v}_\tau = \mathbf{v}|_\tau$; that is, $\mathbf{v}_\tau$ is the restriction of $\mathbf{v}$ to the subset of indices τ. We assume here that $\{\tau\}$ provides an overlapping partition of the indices of the vectors $\mathbf{v}$. Next, partition the local matrices A_τ as A accordingly,

$$A_\tau = \begin{bmatrix} \mathcal{A}_\tau & \mathcal{R}_\tau \\ \mathcal{L}_\tau & \mathcal{B}_\tau \end{bmatrix}.$$

Let $\mathcal{D}_\tau$ be a s.p.d. local matrix such that

$$\mathbf{w}_\tau^T \mathcal{A}_\tau \mathbf{w}_\tau \geq \mathbf{w}_\tau^T \mathcal{D}_\tau \mathbf{w}_\tau \geq \mathbf{w}_\tau^T \mathcal{R}_\tau (\mathcal{B}_\tau)^{-1} \mathcal{L}_\tau \mathbf{w}_\tau. \tag{4.20}$$

For example, it may be possible to choose $\mathcal{D}_\tau$ being diagonal. The latter estimate implies the spectral equivalence

$$\begin{aligned} \begin{bmatrix} \mathbf{w}_\tau \\ \mathbf{x}_\tau \end{bmatrix}^T \begin{bmatrix} \mathcal{A}_\tau & \mathcal{R}_\tau \\ \mathcal{L}_\tau & \mathcal{B}_\tau \end{bmatrix} \begin{bmatrix} \mathbf{w}_\tau \\ \mathbf{x}_\tau \end{bmatrix} &\geq \begin{bmatrix} \mathbf{w}_\tau \\ \mathbf{x}_\tau \end{bmatrix}^T \begin{bmatrix} \mathcal{D}_\tau & \mathcal{R}_\tau \\ \mathcal{L}_\tau & \mathcal{B}_\tau \end{bmatrix} \begin{bmatrix} \mathbf{w}_\tau \\ \mathbf{x}_\tau \end{bmatrix} \\ &\geq \delta \begin{bmatrix} \mathbf{w}_\tau \\ \mathbf{x}_\tau \end{bmatrix}^T \begin{bmatrix} \mathcal{A}_\tau & \mathcal{R}_\tau \\ \mathcal{L}_\tau & \mathcal{B}_\tau \end{bmatrix} \begin{bmatrix} \mathbf{w}_\tau \\ \mathbf{x}_\tau \end{bmatrix}, \end{aligned}$$

inasmuch as both matrices are symmetric positive semidefinite and (4.20) imply that they have a common potential null space. We assume that the constant δ is uniform with respect to τ.

Define now the global matrix $\mathcal{D}$ by assembling $\{\mathcal{D}_\tau\}$

$$\mathbf{w}^T \mathcal{D} \mathbf{w} = \sum_\tau \mathbf{w}_\tau^T \mathcal{D}_\tau \mathbf{w}_\tau.$$

Then let $\mathcal{M} = \mathcal{D}^{-1}$. Note that if $\{\mathcal{D}_\tau\}$ are diagonal, then $\mathcal{D}$, and hence $\mathcal{M}$, is also diagonal. It is clear that

$$\mathbf{w}^T \mathcal{A}\mathbf{v} \geq \mathbf{w}^T \mathcal{M}^{-1}\mathbf{w},$$

and that

$$\begin{aligned}
\mathbf{x}^T(\mathcal{B} - \mathcal{L}\mathcal{M}\mathcal{R})\mathbf{x} &= \inf_{\mathbf{w}} \sum_\tau \begin{bmatrix} \mathbf{w}_\tau \\ \mathbf{x}_\tau \end{bmatrix}^T \begin{bmatrix} \mathcal{D}_\tau & \mathcal{R}_\tau \\ \mathcal{L}_\tau & \mathcal{B}_\tau \end{bmatrix} \begin{bmatrix} \mathbf{w}_\tau \\ \mathbf{x}_\tau \end{bmatrix} \\
&\geq \sum_\tau \min_{\mathbf{w}_\tau} \begin{bmatrix} \mathbf{w}_\tau \\ \mathbf{x}_\tau \end{bmatrix}^T \begin{bmatrix} \mathcal{D}_\tau & \mathcal{R}_\tau \\ \mathcal{L}_\tau & \mathcal{B}_\tau \end{bmatrix} \begin{bmatrix} \mathbf{w}_\tau \\ \mathbf{x}_\tau \end{bmatrix} \\
&\geq \sum_\tau \min_{\mathbf{w}_\tau} \delta \begin{bmatrix} \mathbf{w}_\tau \\ \mathbf{x}_\tau \end{bmatrix}^T \begin{bmatrix} \mathcal{A}_\tau & \mathcal{R}_\tau \\ \mathcal{L}_\tau & \mathcal{B}_\tau \end{bmatrix} \begin{bmatrix} \mathbf{w}_\tau \\ \mathbf{x}_\tau \end{bmatrix} \\
&= \delta \sum_\tau \mathbf{x}_\tau^T \mathcal{S}_\tau \mathbf{x}_\tau \\
&\simeq \mathbf{x}^T \mathcal{S}\mathbf{x}.
\end{aligned}$$

We have also assumed that the matrix assembled from the local Schur complements $\{\mathcal{S}_\tau\}$ (which is another good choice for $\mathcal{A}_c$) is spectrally equivalent to the global Schur complement. This can be rigorously proved if we can construct a bounded in A-norm, element-based, interpolation matrix P. The latter means that $P\mathbf{x}|_\tau = P_\tau \mathbf{x}_\tau$, where $\mathbf{x}_\tau = \mathbf{x}|_\tau$, for some local interpolation matrices P_τ. (For more details, cf., Section 6.9.)

For the purpose of the present analysis, assume that we can construct an element-based P such that its restrictions to every τ,

$$P_\tau = \begin{bmatrix} W_\tau \\ I \end{bmatrix},$$

for a τ–independent constant η, satisfy

$$\mathbf{x}_\tau^T P_\tau^T A_\tau P_\tau \mathbf{x}_\tau \leq \eta \inf_{\mathbf{w}_\tau} \begin{bmatrix} \mathbf{w}_\tau + W_\tau \mathbf{x}_\tau \\ \mathbf{x}_\tau \end{bmatrix}^T A_\tau \begin{bmatrix} \mathbf{w}_\tau + W_\tau \mathbf{x}_\tau \\ \mathbf{x}_\tau \end{bmatrix} = \eta\, \mathbf{x}_\tau^T \mathcal{S}_\tau \mathbf{x}_\tau.$$

Then, because $\mathcal{S}_\tau$ is a Schur complement of the symmetric positive semidefinite matrix A_τ, and

$$P_\tau = \begin{bmatrix} W_\tau \\ I \end{bmatrix},$$

we easily get

$$\mathbf{x}_\tau^T \mathcal{S}_\tau \mathbf{x}_\tau = \inf_{\mathbf{w}_\tau} \begin{bmatrix} \mathbf{w}_\tau \\ \mathbf{x}_\tau \end{bmatrix}^T A_\tau \begin{bmatrix} \mathbf{w}_\tau \\ \mathbf{x}_\tau \end{bmatrix} \leq \mathbf{x}_\tau^T P_\tau^T A_\tau P_\tau \mathbf{x}_\tau.$$

That is, we see that the matrix obtained by assembling the element-based Schur complements is spectrally equivalent to $\mathcal{A}_c$ because

$$\sum_\tau \mathbf{x}_\tau^T \mathcal{S}_\tau \mathbf{x}_\tau \leq \sum_\tau \mathbf{x}_\tau^T P_\tau^T A_\tau P_\tau \mathbf{x}_\tau = \mathbf{x}^T \mathcal{A}_c \mathbf{x} \leq \eta \sum_\tau \mathbf{x}_\tau^T \mathcal{S}_\tau \mathbf{x}_\tau.$$

On the other hand, it is also straightforward to prove that $\mathcal{A}_c$ is spectrally equivalent to the exact Schur complement $\mathcal{S}$. We have

$$\begin{aligned}
\mathbf{x}^T \mathcal{S} \mathbf{x} &= \inf_{\mathbf{w}} \begin{bmatrix} \mathbf{w} \\ \mathbf{x} \end{bmatrix}^T A \begin{bmatrix} \mathbf{w} \\ \mathbf{x} \end{bmatrix} \\
&\leq \mathbf{x}^T P^T A P \mathbf{x} \\
&= \sum_\tau \mathbf{x}_\tau^T P_\tau^T A_\tau P_\tau \mathbf{x}_\tau \\
&\leq \sum_\tau \eta\, \mathbf{x}_\tau^T \mathcal{S}_\tau \mathbf{x}_\tau \\
&\leq \eta \sum_\tau \begin{bmatrix} \mathbf{w}_\tau \\ \mathbf{x}_\tau \end{bmatrix}^T A_\tau \begin{bmatrix} \mathbf{w}_\tau \\ \mathbf{x}_\tau \end{bmatrix} \\
&= \eta \begin{bmatrix} \mathbf{w} \\ \mathbf{x} \end{bmatrix}^T A \begin{bmatrix} \mathbf{w} \\ \mathbf{x} \end{bmatrix}.
\end{aligned}$$

Because $\mathbf{w}$ can be arbitrary, by taking inf over it, we obtain that $\mathbf{x}^T \mathcal{S} \mathbf{x} \leq \mathbf{x}^T \mathcal{A}_c \mathbf{x} \leq \eta\, \mathbf{x}^T \mathcal{S} \mathbf{x}$. That is, the global Schur complement $\mathcal{S}$ is spectrally equivalent to $\mathcal{A}_c$, and hence, is also spectrally equivalent to the matrix obtained by assembling the local Schur complements $\mathcal{S}_\tau$.

Thus, we showed that if $\mathcal{M}$ is spectrally equivalent to $\mathcal{A}$ and based on $\mathcal{D}_\tau$ such that (4.20) holds, the two conditions on $\mathcal{M}$ are met, which implies that the corresponding two-grid preconditioner M will be spectrally equivalent to A. We comment that to implement M^{-1}, we do not need the interpolation matrix P once $\mathcal{A}_c$ has been constructed (either by $P^T A P$ or by assembling the local element Schur complements $\mathcal{S}_\tau$), which makes the method somewhat different from the two-grid AMG methods.

We conclude this section with the following simple example of local matrices that satisfy the local condition (4.20).

Consider,

$$A_\tau = \begin{bmatrix} \mathcal{A}_\tau & \mathcal{R}_\tau \\ \mathcal{L}_\tau & \mathcal{B}_\tau \end{bmatrix},$$

where

$$\mathcal{A}_\tau = \begin{bmatrix} 4 & -1 & -1 \\ -1 & 4 & -1 \\ -1 & -1 & 4 \end{bmatrix}, \quad \mathcal{R}_\tau = \begin{bmatrix} -1 & -1 & 0 \\ 0 & -1 & -1 \\ -1 & 0 & -1 \end{bmatrix}, \quad \mathcal{B}_\tau = \begin{bmatrix} 2 & 0 & 0 \\ 0 & 2 & 0 \\ 0 & 0 & 2 \end{bmatrix},$$

and $\mathcal{L}_\tau = \mathcal{R}_\tau^T$. Define then

$$\mathcal{D}_\tau = \begin{bmatrix} 2 & 0 & 0 \\ 0 & 2 & 0 \\ 0 & 0 & 2 \end{bmatrix}.$$

We first see that

$$\mathcal{A}_\tau - \mathcal{D}_\tau = \begin{bmatrix} 2 & -1 & -1 \\ -1 & 2 & -1 \\ -1 & -1 & 2 \end{bmatrix}$$

is positive semidefinite, and it is readily seen that the approximate Schur complement $\mathcal{B}_\tau - \mathcal{L}_\tau \mathcal{D}_\tau^{-1} \mathcal{L}_\tau^T$ is also positive semidefinite. Indeed,

$$\begin{aligned}
\mathcal{B}_\tau - \mathcal{L}_\tau \mathcal{D}_\tau^{-1} \mathcal{L}^T &= \mathcal{B}_\tau - \frac{1}{2}\mathcal{L}_\tau \mathcal{L}_\tau^T \\
&= \begin{bmatrix} 2 & 0 & 0 \\ 0 & 2 & 0 \\ 0 & 0 & 2 \end{bmatrix} - \begin{bmatrix} 1 & \frac{1}{2} & \frac{1}{2} \\ \frac{1}{2} & 1 & \frac{1}{2} \\ \frac{1}{2} & \frac{1}{2} & 1 \end{bmatrix} \\
&= \begin{bmatrix} 1 & -\frac{1}{2} & -\frac{1}{2} \\ -\frac{1}{2} & 1 & -\frac{1}{2} \\ -\frac{1}{2} & -\frac{1}{2} & 1 \end{bmatrix},
\end{aligned}$$

which is positive semidefinite. This shows that the block-matrix

$$M_\tau = \begin{bmatrix} \mathcal{D}_\tau & \mathcal{R}_\tau \\ \mathcal{L}_\tau & \mathcal{B}_\tau \end{bmatrix}$$

is positive semidefinite. Finally notice that A_τ and M_τ have the same null space (the constant vectors in $\mathbb{R}^6$). Alternatively, we have

$$\begin{aligned}
\mathcal{D}_\tau - \mathcal{R}_\tau \mathcal{B}_\tau^{-1} \mathcal{L}_\tau &= \mathcal{D}_\tau - \frac{1}{2}\mathcal{R}_\tau \mathcal{L}_\tau \\
&= \begin{bmatrix} 2 & 0 & 0 \\ 0 & 2 & 0 \\ 0 & 0 & 2 \end{bmatrix} - \begin{bmatrix} 1 & \frac{1}{2} & \frac{1}{2} \\ \frac{1}{2} & 1 & \frac{1}{2} \\ \frac{1}{2} & \frac{1}{2} & 1 \end{bmatrix} \\
&= \begin{bmatrix} 1 & -\frac{1}{2} & -\frac{1}{2} \\ -\frac{1}{2} & 1 & -\frac{1}{2} \\ -\frac{1}{2} & -\frac{1}{2} & 1 \end{bmatrix},
\end{aligned}$$

which is positive semidefinite. This shows the r.h.s. inequality in (4.20).

A reduction to an M-matrix

As demonstrated by the example at the end of the preceding section, the local construction of M^{-1} seemed feasible, in general, for (Stieltjes) symmetric M-matrices. This gives a motivation to replace A with a spectrally equivalent M-matrix $\overline{A}$ and apply the method studied in the present section to $\overline{A}$ instead. The M-matrix approximation can be achieved as follows. By adding semidefinite matrices of the type

$$\begin{bmatrix} 0 & 0 & 0 & 0 & 0 \\ 0 & d & 0 & -d & 0 \\ 0 & 0 & 0 & 0 & 0 \\ 0 & -d & 0 & d & 0 \\ 0 & 0 & 0 & 0 & 0 \end{bmatrix} \begin{matrix} \\ \} \; i''\text{th position} \\ \\ \} \; j''\text{th position} \\ \\ \end{matrix}$$

to A, we can make all positive off-diagonal entries $a_{ij} = d$ of A zero. The resulting matrix $\overline{A}$ is a s.p.d. M-matrix that satisfies $\mathbf{v}^T\overline{A}\mathbf{v} \geq \mathbf{v}^T A\mathbf{v}$, and it is likely spectrally equivalent to A. This is a typical case for finite element matrices A coming from second-order elliptic PDEs. This fact is easily seen if the above positive-entry diagonal compensation is performed on an element matrix level, that is, on A_τ (see (4.19)). Thus, at least in theory, we may reduce the problem of constructing preconditioners for A to a problem of constructing preconditioners to the s.p.d. M (or Stieltjes) matrix $\overline{A}$. We also notice that in the M-matrix case (i.e., when the local matrices A_τ are semidefinite M-matrices), their Schur complements $\mathcal{S}_\tau$ are semidefinite M-matrices as well. Thus, if we define $\mathcal{A}_c$ based on $\{\mathcal{S}_\tau\}$, it will be an M-matrix, and in principle, a recursion involving only M-matrices (i.e., factorizing $\mathcal{A}_c$ in the same way as A and ending up with a new $\mathcal{A}_c$ that is an M-matrix) is feasible. Such a procedure will lead to a multilevel approximate block-factorization of A. We have not specified how the two-by-two block structure of A (and later on $\mathcal{A}_c$) can be chosen. One viable choice of block structure of A can be based, for example, as described in Section 6.9.

4.8 Odd–even modified block-ILU methods

One approach that can be applied to 3D discretization matrices on tensor product meshes is using the unknowns as blocks within planes parallel to each other. Each plane block can be accurately approximated by a 2D block-ILU factorization matrix. The approximate Schur complements can then be computed in a Galerkin way, that is, in the form $P^T AP$, for appropriate matrix P. More specifically, partition the block-tridiagonal matrix A as follows.

$$A = \begin{bmatrix} A_{\text{odd}} & A_{o,e} \\ A_{e,o} & A_{\text{even}} \end{bmatrix}.$$

The matrices A_{odd} and A_{even} are block-diagonal and equal to $\text{diag}(A_{2i-1,2i-1})$ and $\text{diag}(A_{2i,2i})$, respectively. For a positive vector

$$\mathbf{1} = \begin{bmatrix} \mathbf{1}_o \\ \mathbf{1}_e \end{bmatrix},$$

typically chosen such that $A_{\text{odd}}\mathbf{1}_o + A_{\text{even}}\mathbf{1}_e = 0$, we construct an interpolation matrix

$$P = \begin{bmatrix} W \\ I \end{bmatrix}$$

such that $P\mathbf{1}_e = \mathbf{1}$. Then the approximate Schur complement $A_c = P^T AP = W^T A_{\text{odd}} W + A_{e,o} W + W^T A_{o,e} + A_{\text{even}}$, is positive definite and again easily seen to be block-tridiagonal if W couples only two neighboring odd-planes. We also have that $A_c\mathbf{1}_e = (A_{\text{even}} - A_{e,o}A_{\text{odd}}^{-1}A_{o,e})\mathbf{1}_e$.

In our particular case of block–tridiagonal matrices, a suitable way to construct P (or rather W) in order to keep the sparsity pattern of A_c under control relies on the following observation. Let

$$P = \begin{bmatrix} & 0 & & \\ \ddots & W_{2i-3,2i-2} & 0 & \\ \ddots & W_{2i-1,2i-2} & W_{2i-1,2i} & 0 \\ & 0 & W_{2i+1,2i} & W_{2i+1,2i+2} \\ & & 0 & W_{2i+3,2i+2} \\ & & & \\ & 0 & & \\ & I & 0 & \ddots \\ & 0 & I & 0 \\ & \ddots & 0 & I \\ & \ddots & \ddots & 0 \end{bmatrix}.$$

That is, for any column $(2i)$ of P, there are only two nonzero block entries of W, denoted by $W_{2i-1,2i}$ and $W_{2i+1,\,2i}$. Then, a direct computation of the product $A_c = P^T AP$ shows that the resulting matrix is also block-tridiagonal with block entries $[A^c_{2i,2i-2}, A^c_{2i,2i}, A^c_{2i,2i+2}]$ and is defined as follows.

$$\begin{aligned} A^c_{2i,2i-2} &= W^T_{2i-1,2i}(A_{2i-1,2i-1}W_{2i-1,2i-2} + A_{2i-1,2i-2}) + A_{2i,2i-1}W_{2i-1,2i-2}, \\ A^c_{2i,2i} &= W^T_{2i-1,2i}(A_{2i-1,2i-1}W_{2i-1,2i} + A_{2i-1,2i}) \\ &\quad + W^T_{2i+1,2i}(A_{2i+1,2i+1}W_{2i+1,2i} + A_{2i+1,2i}) \\ &\quad + A_{2i,2i-1}W_{2i-1,2i} + A_{2i,2i+1}W_{2i+1,2i} + A_{2i,2i}, \\ A^c_{2i,2i+2} &= W^T_{2i+1,2i}(A_{2i+1,2i+1}W_{2i+1,2i+2} + A_{2i+1,2i+2}) + A_{2i,2i+1}W_{2i+1,2i+2}. \end{aligned}$$

The simplest choice for $W_{2i-1,2i}$ and $W_{2i+1,2i}$ is to be diagonal.

It is clear then that the sparsity pattern of $A^c_{2i,2i-2}$ is determined by the sparsity pattern of $A_{2i-1,2i-1}$, $A_{2i-1,2i-2}$, and $A_{2i,2i-1}$. The sparsity pattern of $A^c_{2i,2i}$ is determined by the one of $A_{2i-1,2i-1}$, $A_{2i-1,2i}$, $A_{2i+1,2i+1}$, $A_{2i+1,2i}$, $A_{2i,2i-1}$, $A_{2i,2i+1}$, and $A_{2i+1,2i+2}$. In conclusion, if the original matrix A has blocks with a regular sparsity pattern, the choice of W with diagonal blocks $W_{2i-1,2i}$ and $W_{2i+1,2i}$ will lead to a matrix A_c again with blocks having the same regular sparsity pattern.

The entries of $W_{2i-1,2i}$ and $W_{2i+1,2i}$ can be determined from the conditions (cf., [SS98])

$$\begin{aligned} W_{2i-1,2i}\mathbf{1}_{2i} &= -A^{-1}_{2i-1,2i-1}A_{2i-1,2i}\mathbf{1}_{2i}, \\ W_{2i+1,2i}\mathbf{1}_{2i} &= -A^{-1}_{2i+1,2i+1}A_{2i+1,2i}\mathbf{1}_{2i}. \end{aligned}$$

It is clear then that

$$\begin{aligned} (P\mathbf{1})|_{2i-1} &= W_{2i-1,2i-2}\mathbf{1}_{2i-2} + W_{2i-1,2i}\mathbf{1}_{2i} \\ &= -A^{-1}_{2i-1,2i-1}(A_{2i-1,2i-2}\mathbf{1}_{2i-2} + A_{2i-1,2i}\mathbf{1}_{2i}) \\ &= \mathbf{1}_{2i-1}; \end{aligned}$$

that is, $P\mathbf{1}_e = \mathbf{1}$.

The two-level scheme is then applied to define the actual two-level modified block-ILU preconditioner.

As mentioned above, we may derive a very accurate block-ILU factorization matrix $(X-L)X^{-1}(X-U)$ for the block-diagonal matrix

$$\text{diag}(A_{i,\,i}) = \begin{bmatrix} A_{\text{odd}} & 0 \\ 0 & A_{\text{even}} \end{bmatrix}.$$

Let

$$(X-L)X^{-1}(X-U) = \begin{bmatrix} M_{\text{odd}} & 0 \\ 0 & M_{\text{even}} \end{bmatrix}.$$

Consider then the following block upper-triangular matrix

$$M = \begin{bmatrix} M_{\text{odd}} & A_{e,o} \\ 0 & M_{\text{even}} \end{bmatrix}.$$

Such an M is referred to as a "c–f" plane relaxation. Then, following the general definition of two-level methods, we first define

$$\widehat{B} = \begin{bmatrix} M & 0 \\ P^T A & I \end{bmatrix} \begin{bmatrix} (M+M^T-A)^{-1} & 0 \\ 0 & A_c \end{bmatrix} \begin{bmatrix} M^T & AP \\ 0 & I \end{bmatrix},$$

and then $B^{-1} = [I,\ P]\widehat{B}^{-1}[I,\ P]^T$. The above method is well defined as long as $M + M^T - A$ is positive definite. In our particular case, we have

$$M + M^T - A = \begin{bmatrix} 2M_{\text{odd}} - A_{\text{odd}} & 0 \\ 0 & 2M_{\text{even}} - A_{\text{even}} \end{bmatrix}.$$

Because the block-ILU methods provide convergent splittings (Corollary 4.6) the differences $2M_{\text{odd}} - A_{\text{odd}}$ and $2M_{\text{even}} - A_{\text{even}}$ are symmetric positive definite, hence $M + M^T - A$ is s.p.d.

As already mentioned, A_c is also block-tridiagonal, and in principle, we can apply recursion to it if its sparsity pattern is being kept under control.

Alternatively, we may simply define a block-ILU factorization of A by using a block-ILU factorization of A_{odd} only. In other words, let $(X_{\text{odd}} - L_{\text{odd}})X_{\text{odd}}^{-1}(X_{\text{odd}} - U_{\text{odd}})$ be an accurate block-ILU factorization of the block-diagonal matrix A_{odd}. Let $M_{\text{odd}} = X_{\text{odd}} - L_{\text{odd}}$. The following more traditional way of defining a block-ILU factorization matrix then reads,

$$B = \begin{bmatrix} I & 0 \\ A_{e,o}M_{\text{odd}}^{-1} & I \end{bmatrix} \begin{bmatrix} M_{\text{odd}}X_{\text{odd}}^{-1}M_{\text{odd}}^T & 0 \\ 0 & A_c \end{bmatrix} \times \begin{bmatrix} I & M_{\text{odd}}^{-T}A_{o,e} \\ 0 & I \end{bmatrix}.$$

The difference between the two-grid definition of B and the more classical block-ILU definition is that the latter does not use the interpolation matrix P after the approximate Schur complement A_c has been constructed. Another difference is that in the more classical block-ILU definition, we use "smoothing" only on the odd blocks.

For more details, regarding 3D finite difference matrices A on tensor product meshes, we refer here to [SS98].

4.9 A nested dissection (approximate) inverse

A nested dissection solver

Given an s.p.d. matrix A partitioned by two separators "b" and "B", as follows

$$A = \begin{bmatrix} A_I & A_{IB} \\ A_{BI} & A_B \end{bmatrix},$$

where A_I is typically block-diagonal, which by itself is partitioned by the separator "b" in a similar fashion,

$$A_I = \begin{bmatrix} A_i & A_{ib} \\ A_{bi} & A_b \end{bmatrix}.$$

We now compute (approximately) A^{-1} based on some partial knowledge of A_I^{-1}. We only need specific entries of A^{-1}, namely, the entries $(A^{-1})_{ij}$ for $i, j \in$ "B" $\cup$ "b". The entries that we need from A_I^{-1} correspond to the set "b" and the set $\partial I \equiv \{j \in I;\ a_{i,j} \neq 0 \text{ for some } i \in \text{"B"}\}$. The latter is motivated by the fact that in order to compute the Schur complement $S_B = A_B - A_{BI}A_I^{-1}A_{IB}$, we only need the entries of A_I^{-1} from exactly the set ∂I. Also, in order to compute the required entries of A^{-1} based on the formula

$$A^{-1} = \begin{bmatrix} A_I^{-1} + A_I^{-1}A_{IB}S_B^{-1}A_{BI}A_I^{-1} & -A_I^{-1}A_{IB}S_B^{-1} \\ -S_B^{-1}A_{BI}A_I^{-1} & S_B^{-1} \end{bmatrix},$$

it is clear that we need $(A_I^{-1})_{ij}$ for $i, j \in$ "b" $\cup\, \partial I$ and all the entries of S_B^{-1}, in order to be able to compute the entries of A^{-1} corresponding to $B \cup \Gamma$ for any set $\Gamma \subset$ ("b" $\cup\, \partial I$).

We now comment on the fact that having the above-mentioned entries of A^{-1} and A_I^{-1} available is sufficient to solve the following problem with sparse r.h.s.,

$$A\mathbf{x} = \mathbf{b} = \begin{bmatrix} 0 \\ \mathbf{b}_b \\ \mathbf{b}_B \end{bmatrix}$$

for $\mathbf{x}$ on "B" and "b". We can proceed as follows.

1. Solve

$$A_I \mathbf{x}_I = \mathbf{b}_I = \begin{bmatrix} 0 \\ \mathbf{b}_b \end{bmatrix}$$

for $\mathbf{x}_I$ on ∂I. Because we have (by assumption) the entries $(A_I^{-1})_{i,j}$ for $i \in \partial I$ and $j \in$ "b", the latter equals $(\mathbf{x}_I)_i = \sum_{j \in \text{"b"}} (A_I^{-1})_{ij} (\mathbf{b}_b)_j$.
2. Compute the residual

$$\mathbf{r} = \begin{bmatrix} 0 \\ \mathbf{b}_b \\ \mathbf{b}_B \end{bmatrix} - A \begin{bmatrix} \mathbf{x}_I \\ 0 \end{bmatrix}.$$

It is clear that $\mathbf{r}$ is nonzero only on the separator set B and its entries on B are equal to

$$\mathbf{b}_B - A_{BI}\mathbf{x}_I = \mathbf{b}_B - A_{BI} \begin{bmatrix} * \\ \mathbf{x}_I|_{\partial I} \end{bmatrix}.$$

These entries are computable inasmuch as A_{BI} is nonzero only on ∂I.
3. Solve

$$A\mathbf{y} = \mathbf{r},$$

for $\mathbf{y}_B = \mathbf{y}|_B$ (and perhaps for some other entries). We notice, because

$$A\left(\mathbf{y} + \begin{bmatrix} \mathbf{x}_I \\ 0 \end{bmatrix}\right) = \mathbf{b},$$

that $\mathbf{y}_B = \mathbf{x}_B$ is the true solution on the separator set B.
4. After $\mathbf{x}_B$ has been computed, we solve

$$A_I \mathbf{x}_I = \mathbf{b}_I - A_{IB}\mathbf{x}_B.$$

The latter r.h.s. is nonzero on $b \cup \partial I$. Therefore, we can compute $\mathbf{x}_b$ because A_I^{-1} is available for entries (i, j), $i \in b$ and $j \in b \cup \partial I$.

The complete solution algorithm is then obtained by applying the same process recursively, now to A_I based on the separator b and A_i replacing A_I. If the separators are used as in the nested dissection ordering (cf., Section 1.9.6), we end up with a nested dissection-based (approximate) inverse of A, which can also be used to solve the system $A\mathbf{x} = \mathbf{b}$ with a general r.h.s. The assumed sparsity of $\mathbf{b}$ is automatically obtained (seen by induction; e.g., the sparsity of $\mathbf{b}$ implied that the residual $\mathbf{r}$ is also sparse; $\mathbf{r}$ is nonzero only on the separator B).

Approximate inverses based on low-rank matrices

The idea is to compute both the Schur complement S_B and its inverse S_B^{-1} only approximately by saving memory and operations.

Assume, for example, that the needed principal submatrix of A_I^{-1} is approximated by a low-rank matrix $Q_I \Lambda_I^{-1} Q_I^T$ where Q_I has $m \geq$ columns, for a small m. Then, the Schur complement S_B can be approximated by the expression

$$X_B \equiv A_{BB} - A_{BI} Q_{\partial I} \Lambda_I^{-1} Q_{\partial I}^T A_{IB}.$$

We recall that low-rank updates to compute approximate Schur complements were used in Section 4.6. In general, given an approximate inverse X_I^{-1} to A_I and letting X_B^{-1} be another approximate inverse to $S_B \approx A_{BB} - A_{BI} \Lambda_I^{-1} A_{IB}$, the required and now only approximate entries of A^{-1} can be computed from

$$\begin{bmatrix} X_I^{-1} + X_I^{-1} A_{IB} X_B^{-1} A_{BI} X_I^{-1} & -X_I^{-1} A_{IB} X_B^{-1} \\ -X_B^{-1} A_{BI} X_I^{-1} & X_B^{-1} \end{bmatrix},$$

where the multiplications are carried out only approximately.

For more details on how to operate on a certain class of matrices exploiting "low-rankness", referred to as *hierarchical* matrices, see, for example, [H05], or referred to as "semiseparable" matrices, see, for example, [ChG05].

5

Multigrid (MG)

5.1 From two-grid to multigrid

We recall now the classical two-grid method from a matrix point of view introduced in Section 3.2.3. We are given a $n \times n$ s.p.d. matrix A. The two-grid method exploits an interpolation matrix $P : \mathbb{R}^{n_c} \mapsto \mathbb{R}^n$ and a smoother M. Then, we define the coarse matrix $A_c = P^T AP$. The smoother M is assumed to provide convergent iteration in the A-norm. As we well know (cf. Proposition 3.8), this is equivalent to having the symmetrized smoother $\overline{M} = M(M + M^T - A)^{-1} M^T$ satisfy the inequality,

$$\mathbf{v}^T A\mathbf{v} \le \mathbf{v}^T \overline{M}\mathbf{v}. \tag{5.1}$$

The classical two-grid method is defined as a stationary iterative procedure, which is based on composite iterations; a presmoothing step with M, coarse-grid correction, and a postsmoothing step with M^T. This leads to an iteration matrix E that admits the following product form

$$E = (I - M^{-T} A)(I - PA_c^{-1} P^T A)(I - M^{-1} A).$$

A corresponding two-grid matrix B can be defined from the equation $E = I - B^{-1} A$ which leads to the expression

$$B^{-1} = M^{-T} (M + M^T - A) M^{-1} + (I - M^{-T} A) PA_c^{-1} P^T (I - AM^{-1}).$$

We may equivalently define B as a block-factorization of A. Namely, introduce

$$\overline{B} = \begin{bmatrix} I & 0 \\ P^T AM^{-1} & I \end{bmatrix} \begin{bmatrix} M(M + M^T - A)^{-1} M^T & 0 \\ 0 & A_c \end{bmatrix} \begin{bmatrix} I & M^{-T} AP \\ 0 & I \end{bmatrix}.$$

Note that $\overline{B}$ has bigger size than B and A; namely, its size equals the fine-grid vector size plus the coarse-grid vector size. Then, a straightforward computation shows the following identity

$$B^{-1} = [I,\ P]\, \overline{B}^{-1} [I,\ P]^T .$$

P.S. Vassilevski, *Multilevel Block Factorization Preconditioners*,
doi: 10.1007/978-0-387-71564-3_5,

This definition shows that B is s.p.d. if $M+M^T-A$ is positive definite, or equivalently if M is a convergent smoother in the A inner product (which we have assumed).

The above two-grid (or TG) definitions are the basis for extending the TG method to multiple levels by simply replacing the exact inverse A_c^{-1} with an approximate one exploiting the same procedure defined by recursion on the preceding coarse levels, which leads to various definitions of multigrid (or MG).

Assume that we have $\ell \geq 1$ levels (or grids). To be specific, we define next a symmetric $V(1,1)$-cycle MG. Let $A_0 = A$ and P_k be the interpolation matrices from coarse-grid $k+1$ to fine-grid k and $A_{k+1} = P_k^T A_k P_k$ be the coarse-grid $k+1$ matrix. We assume that $P_k : \mathbf{V}_{k+1} \mapsto \mathbf{V}_k$; that is, $P_k \mathbf{V}_{k+1} \subset \mathbf{V}_k$. Here, each vector space $\mathbf{V}_k$ is identified with $\mathbb{R}^{n_k}$, and $n_{k+1} < n_k$ are their respective dimensions. Finally, for any k, let M_k be a convergent smoother for A_k so that $\|I - A_k^{1/2} M_k^{-1} A_k^{1/2}\| < 1$. The symmetric $V(1,1)$-cycle MG that we define below exploits smoothing iterations based on the inverse actions of both M_k and M_k^T.

The traditional definition of MG is based on an algorithm that provides actions of an approximate inverse B_k^{-1} to A_k.

Definition 5.1 (MG algorithm). *At the coarsest level, we set $B_\ell = A_\ell$. Then for $k = \ell - 1, \ldots, 0$, assuming that B_{k+1}^{-1} has been defined, we perform the following steps to define the actions $B_k^{-1}\mathbf{r}$ for any given vector $\mathbf{r}$.*

(i) "Presmooth;" that is, solve

$$M_k \mathbf{x}_k = \mathbf{r}.$$

(ii) Compute the residual $\mathbf{d} = \mathbf{r} - A_k \mathbf{x}_k = (I - A_k M_k^{-1})\mathbf{r}$.

(iii) Compute a coarse-grid correction by applying B_{k+1}^{-1} to the restricted residual $P_k^T \mathbf{d}$; that is, compute

$$\mathbf{x}_{k+1} = B_{k+1}^{-1} P_k^T \left(I - A_k M_k^{-1}\right)\mathbf{r}.$$

(iv) Interpolate the coarse-grid correction and update the current approximation; that is, compute

$$\mathbf{x}_k := \mathbf{x}_k + P_k \mathbf{x}_{k+1} = M_k^{-1}\mathbf{r} + P_k B_{k+1}^{-1} P_k^T \left(I - A_k M_k^{-1}\right)\mathbf{r}.$$

(v) "Postsmooth;" that is, solve

$$M_k^T \mathbf{y} = \mathbf{r} - A_k \mathbf{x}_k,$$

and finally set

$$B_k^{-1}\mathbf{r} = \mathbf{x}_k + \mathbf{y}.$$

□

The following more explicit form of B_k^{-1} is readily seen from

$$\mathbf{x}_k = M_k^{-1}\mathbf{r} + P_k B_{k+1}^{-1} P_k^T \left(I - A_k M_k^{-1}\right)\mathbf{r},$$

and

$$\mathbf{y} = M_k^{-T}(\mathbf{r} - A_k\mathbf{x}_k),$$

based on the definition $B_k^{-1}\mathbf{r} = \mathbf{x}_k + \mathbf{y}$. We have

$$\begin{aligned}
B_k^{-1}\mathbf{r} &= \mathbf{x}_k + \mathbf{y} \\
&= \mathbf{x}_k + M_k^{-T}(\mathbf{r} - A_k\mathbf{x}_k) \\
&= M_k^{-1}\mathbf{r} + P_k B_{k+1}^{-1} P_k^T \left(I - A_k M_k^{-1}\right)\mathbf{r} \\
&\quad + M_k^{-T}\left(\mathbf{r} - A_k\left(M_k^{-1}\mathbf{r} + P_k B_{k+1}^{-1} P_k^T \left(I - A_k M_k^{-1}\right)\mathbf{r}\right)\right) \\
&= \left(M_k^{-1} + M_k^{-T} - M_k^{-1} A_k M_k^{-T}\right. \\
&\quad \left. + \left(I - M_k^{-T} A_k\right) P_k B_{k+1}^{-1} P_k^T \left(I - A_k M_k^{-1}\right)\right)\mathbf{r}.
\end{aligned}$$

Thus, the following recursive definition can be used instead.

Definition 5.2 (Recursive definition of MG). *Set $B_\ell = A_\ell$. For $k = \ell - 1, \ldots, 0$, introduce the symmetrized smoothers $\overline{M}_k = M_k(M_k^T + M_k - A_k)^{-1} M_k^T$, and then let,*

$$B_k^{-1} = \overline{M}_k^{-1} + \left(I - M_k^{-T} A_k\right) P_k B_{k+1}^{-1} P_k^T \left(I - A_k M_k^{-1}\right). \qquad \square$$

Definition 5.2 and Theorem 3.11 give us one more equivalent definition of the symmetric $V(1,1)$-cycle MG as a recursive two-by-two block-factorization preconditioner, which is a direct generalization of the TG one.

Definition 5.3 (MG as block-factorization preconditioner). *Starting with $B_\ell = A_\ell$ and assuming that B_{k+1} for $k \leq \ell - 1$, has already been defined, we first form*

$$\overline{B}_k = \begin{bmatrix} I & 0 \\ P_k^T A_k M_k^{-1} & I \end{bmatrix} \begin{bmatrix} M_k \left(M_k + M_k^T - A_k\right)^{-1} M_k^T & 0 \\ 0 & B_{k+1} \end{bmatrix} \begin{bmatrix} I & M_k^{-T} A_k P_k \\ 0 & I \end{bmatrix}, \tag{5.2}$$

and then B_k is defined from,

$$B_k^{-1} = [I,\ P_k]\, \overline{B}_k^{-1}\, [I,\ P_k]^T. \qquad \square$$

Introduce next the composite interpolation matrices $\overline{P}_k = P_0 \cdots P_{k-1}$ from kth-level coarse vector space $\mathbf{V}_k$ all the way up to the finest-level vector space $\mathbf{V} = \mathbf{V}_0$. The following result allows us to view the symmetric $V(1,1)$-cycle MG as a product iterative method performed on the finest-level. The iterations exploit corrections from the subspaces $\overline{P}_k \mathbf{V}_k$ of the original vector space $\mathbf{V} = \mathbf{V}_0$. Such methods are sometimes called *subspace correction* methods (cf. [Xu92a]).

Proposition 5.4. *The following recursive relation between the subspace iteration matrices $I - \overline{P}_k B_k^{-1} \overline{P}_k^T A$ and $I - \overline{P}_{k+1} B_{k+1}^{-1} \overline{P}_{k+1}^T A$ holds,*

$$I - \overline{P}_k B_k^{-1} \overline{P}_k^T A = \left(I - \overline{P}_k M_k^{-T} \overline{P}_k^T A\right)\left(I - \overline{P}_{k+1} B_{k+1}^{-1} \overline{P}_{k+1}^T A\right)\left(I - \overline{P}_k M_k^{-1} \overline{P}_k^T A\right).$$

Proof. We have, from Definition 5.2,

$$\overline{P}_k B_k^{-1} \overline{P}_k^T = \overline{P}_k \overline{M}_k^{-1} \overline{P}_k^T + \overline{P}_k (I - M_k^{-T} A_k) P_k B_{k+1}^{-1} P_k^T (I - A_k M_k^{-1}) \overline{P}_k^T.$$

Now use the fact that $A_k = \overline{P}_k^T A \overline{P}_k$ and $\overline{P}_{k+1} = \overline{P}_k P_k$ to arrive at the expression

$$\overline{P}_k B_k^{-1} \overline{P}_k^T = \overline{P}_k \overline{M}_k^{-1} \overline{P}_k^T + (I - \overline{P}_k M_k^{-T} \overline{P}_k^T A) \overline{P}_{k+1} B_{k+1}^{-1} \overline{P}_{k+1}^T (I - A \overline{P}_k M_k^{-1} \overline{P}_k^T).$$

Then forming $I - \overline{P}_k B_k^{-1} \overline{P}_k^T A$ gives

$$\begin{aligned} I - \overline{P}_k B_k^{-1} \overline{P}_k^T A = I &- \overline{P}_k \overline{M}_k^{-1} \overline{P}_k^T A \\ &- (I - \overline{P}_k M_k^{-T} \overline{P}_k^T A) \overline{P}_{k+1} B_{k+1}^{-1} \overline{P}_{k+1}^T A (I - \overline{P}_k M_k^{-1} \overline{P}_k^T A). \end{aligned}$$

It remains to notice that $\overline{M}_k^{-1} = M_k^{-1} + M_k^{-T} - M_k^{-T} A_k M_k^{-1} = M_k^{-1} + M_k^{-T} - M_k^{-T} \overline{P}_k^T A \overline{P}_k M_k^{-1}$ implies

$$I - \overline{P}_k \overline{M}_k^{-1} \overline{P}_k^T A = (I - \overline{P}_k M_k^{-T} \overline{P}_k^T A)(I - \overline{P}_k M_k^{-1} \overline{P}_k^T A),$$

which combined with the previous identity gives the desired result. □

The following proposition shows that the symmetric $V(1,1)$-cycle MG preconditioners B_k provide convergent splittings for A_k. More specifically, the following result holds.

Proposition 5.5. *Under the assumption that the smoothers M_k are convergent in the A_k-norm (i.e., $\|I - A_k^{1/2} M_k^{-1} A_k^{1/2}\| < 1$), the symmetric $V(1,1)$-cycle preconditioner B_k is such that $B_k - A_k$ is symmetric positive semidefinite.*

Proof. From Proposition 5.4, letting $E_k = I - \overline{P}_k M_k^{-1} \overline{P}_k^T A$, we have that

$$A - A \overline{P}_k B_k^{-1} \overline{P}_k^T A = E_k^T (A - A \overline{P}_{k+1} B_{k+1}^{-1} \overline{P}_{k+1}^T A) E_k,$$

which shows by induction that $A - A \overline{P}_k B_k^{-1} \overline{P}_k^T A$ is symmetric positive semidefinite. For $k = \ell$, we have $B_\ell = A_\ell = \overline{P}_\ell^T A \overline{P}_\ell$. Letting $G = A^{1/2} \overline{P}_\ell A_\ell^{-1/2}$, we have $G^T G = I$. Because $\|G\| = \|G^T\| = 1$, we also have

$$\mathbf{v}^T A^{1/2} \overline{P}_\ell A_\ell^{-1} \overline{P}_\ell^T A^{1/2} \mathbf{v} = \mathbf{v}^T G G^T \mathbf{v} \leq \mathbf{v}^T \mathbf{v}.$$

Letting $\mathbf{v} := A^{1/2} \mathbf{v}$ above, we arrive at

$$\mathbf{v}^T A \overline{P}_\ell A_\ell^{-1} \overline{P}_\ell^T A \mathbf{v} \leq \mathbf{v}^T A \mathbf{v}.$$

That is, $A - A \overline{P}_k B_k^{-1} \overline{P}_k^T A$ for $k = \ell$ is symmetric positive semidefinite. The fact that $A - A \overline{P}_k B_k^{-1} \overline{P}_k^T A$ is symmetric positive semidefinite (for all k) implies that $\overline{P}_k^T (A - A \overline{P}_k B_k^{-1} \overline{P}_k^T A) \overline{P}_k = A_k - A_k B_k^{-1} A_k$, or equivalently, that $A_k^{-1} - B_k^{-1}$ is symmetric positive semidefinite. Therefore, we have that $B_k - A_k$ is symmetric positive semidefinite (because we showed that B_k is s.p.d.). Thus the proof is complete. □

5.2 MG as block Gauss–Seidel

The following relation between MG and a certain inexact block Gauss–Seidel factorization of an extended matrix was first explored by M. Griebel in [Gr94]. Namely, consider the composite interpolation matrices $\overline{P}_k = P_0, \ldots, P_{k-1}$ from coarse-level k vector space $\mathbf{V}_k$ all the way up to the finest-level vector space $\mathbf{V} = \mathbf{V}_0$. Let also $\overline{P}_0 = I$. Then, form the following extended block-matrix $T = (T_{ij})_{i,j=0}^{\ell}$ with blocks $T_{ij} = \overline{P}_i^T A \overline{P}_j$. Note that the diagonal blocks $T_{ii} = A_i$ are simply the ith coarse-level matrices. Let M_i be the smoother at level $i < \ell$ and at the coarsest-level ℓ let $M_\ell = A_\ell$. Then form the block-lower triangular matrix,

$$L_B = \begin{bmatrix} M_0 & 0 & \ldots & 0 \\ T_{10} & M_1 & \ldots & 0 \\ \vdots & \ldots & \ddots & 0 \\ T_{\ell,0} & \ldots & T_{\ell,\ell-1} & M_\ell \end{bmatrix}.$$

The inverse of the MG preconditioner B_{MG}^{-1} satisfies the identity,

$$B_{MG}^{-1} = [\overline{P}_0, \ldots, \overline{P}_\ell] L_B^{-T} (\mathrm{diag}(M_i^T + M_i - A_i)_{i=0}^{\ell}) L_B^{-1} [\overline{P}_0, \ldots, \overline{P}_\ell]^T. \quad (5.3)$$

The fact that B_{MG} defined in (5.3) actually coincides with the one in (5.2) follows by induction from Theorem 3.11. We also have then the familiar product representation of the iteration matrix $I - B_{MG}^{-1} A$,

$$\begin{aligned} I - B_{MG}^{-1} A &= \left(I - \overline{P}_0 M_0^{-T} \overline{P}_0^T A\right) \cdots \left(I - \overline{P}_\ell M_\ell^{-T} \overline{P}_\ell^T A\right) \\ &\quad \times \left(I - \overline{P}_{\ell-1} M_{\ell-1}^{-1} \overline{P}_{\ell-1}^T A\right) \cdots \left(I - \overline{P}_0 M_0^{-1} \overline{P}_0^T A\right). \end{aligned}$$

Because $M_\ell = A_\ell = \overline{P}_\ell^T A \overline{P}_\ell$, we notice that $I - \overline{P}_{\ell-1} M_\ell^{-T} \overline{P}_{\ell-1}^T A$ is a projection. Hence the following more symmetric expression for B_{MG}^{-1} holds.

$$\begin{aligned} I - B_{MG}^{-1} A &= \left(I - \overline{P}_0 M_0^{-T} \overline{P}_0^T A\right) \cdots \left(I - \overline{P}_{\ell-1} M_\ell^{-T} \overline{P}_{\ell-1}^T A\right) \\ &\quad \times \left(I - \overline{P}_{\ell-1} M_\ell^{-1} \overline{P}_{\ell-1}^T A\right) \cdots \left(I - \overline{P}_0 M_0^{-1} \overline{P}_0^T A\right). \end{aligned}$$

In [Gr94], it was actually proposed to transform a given system $A\mathbf{x} = \mathbf{b}$ based on the fact that any $\mathbf{x}$ allows for a (nonunique) decomposition $\mathbf{x} = \sum_{k=0}^{\ell} \overline{P}_k \mathbf{x}_k^f$ and then after forming $\overline{P}_k^T A\mathbf{x} = \sum_{l=0}^{\ell} \overline{P}_k^T A \overline{P}_l \mathbf{x}_l^f = \overline{P}_k^T \mathbf{b}$ to end up with the following consistent extended system,

$$T \begin{bmatrix} \mathbf{x}_0^f \\ \vdots \\ \mathbf{x}_\ell^f \end{bmatrix} = \begin{bmatrix} \overline{P}_0^T \mathbf{b} \\ \vdots \\ \overline{P}_\ell^T \mathbf{b} \end{bmatrix}.$$

Note that the matrix of this system T is symmetric and only positive semidefinite. The latter consistent semidefinite system is solved then by the CG method using the symmetric Gauss–Seidel matrix

$$L_B\left(\text{diag}\left(M_i^T + M_i - A_i\right)_{i=0}^{\ell}\right)^{-1} L_B^T,$$

as preconditioner. The original solution is recovered then as

$$\mathbf{x} = \left[\overline{P}_0, \ldots, \overline{P}_\ell\right] \begin{bmatrix} \mathbf{x}_0^f \\ \vdots \\ \mathbf{x}_\ell^f \end{bmatrix} = \sum_{k=0}^{\ell} \overline{P}_k \mathbf{x}_k^f.$$

5.3 A MG analysis in general terms

The multilevel convergence analysis relies on stable multilevel decompositions of the form

$$\mathbf{v} = \sum_k \overline{\mathbf{v}}_k^f,$$

where $\overline{\mathbf{v}}_k^f \in \overline{\mathbf{V}}_k$, the kth-level coarse space viewed as a subspace of the fine-grid vector space $\mathbf{V} = \overline{\mathbf{V}}_0$. That is, $\overline{\mathbf{V}}_k = \text{Range}(P_0, \ldots, P_{k-1})$. The stability means that for a desirably level independent constant $\sigma > 0$, we have

$$\sum_k \left(\overline{\mathbf{v}}_k^f\right)^T A \overline{\mathbf{v}}_k^f \le \sigma\ \mathbf{v}^T A \mathbf{v}.$$

Equivalently, because $\overline{\mathbf{v}}_k^f = (P_0, \ldots, P_{k-1})\mathbf{v}_k^f$, with $\mathbf{v}_k^f \in \mathbf{V}_k$ (the actual coarse vector space) the same estimate reads

$$\sum_k \left(\mathbf{v}_k^f\right)^T A_k \mathbf{v}_k^f \le \sigma\ \mathbf{v}^T A \mathbf{v}. \tag{5.4}$$

Introduce, for the purpose of the following analysis, the subspace $\mathbf{V}_j^f \subset \mathbf{V}_j$, which is complementary to the coarse space $P_j \mathbf{V}_{j+1}$. The space $\mathbf{V}_j^f$ is chosen so that the symmetrized smoother $\overline{M}_j = M_j(M_j^T + M_j - A_j)^{-1} M_j^T$ when restricted to $\mathbf{V}_j^f$ is efficient. The latter means that A_j and $\overline{M}_j$ are spectrally equivalent uniformly w.r.t. j on the subspace $\mathbf{V}_j^f$. Then, we can replace (5.4) with an estimate that involves the symmetrized smoothers

$$\sum_k \left(\mathbf{v}_k^f\right)^T \overline{M}_k \mathbf{v}_k^f \le \sigma\ \mathbf{v}^T A \mathbf{v}. \tag{5.5}$$

Note that the subspaces $\mathbf{V}_j^f$ are not needed in the actual MG algorithm.

Decompose $\mathbf{v}_j = \mathbf{v}_j^f + P_j \mathbf{v}_{j+1}$, with $\mathbf{v}_j^f \in \mathbf{V}_j$ and $\mathbf{v}_{j+1} \in \mathbf{V}_{j+1}$, for $j = k, k+1, \ldots, \ell - 1$. From the definition of $B_k^{-1} = [I,\ P_k]\, \overline{B}_k^{-1}\, [I,\ P_k]^T$, we have that $I = GG^T$, where $G = B_k^{1/2}\, [I,\ P_k]\, \overline{B}_k^{-(1/2)}$. This shows that G has a spectral norm not greater than 1. Therefore, the following inequality holds,

$$\begin{bmatrix} \mathbf{v}_k^f \\ \mathbf{v}_{k+1} \end{bmatrix}^T [I,\ P_k]\, B_k\, [I,\ P_k]^T \begin{bmatrix} \mathbf{v}_k^f \\ \mathbf{v}_{k+1} \end{bmatrix} \le \begin{bmatrix} \mathbf{v}_k^f \\ \mathbf{v}_{k+1} \end{bmatrix}^T \overline{B}_k \begin{bmatrix} \mathbf{v}_k^f \\ \mathbf{v}_{k+1} \end{bmatrix}. \tag{5.6}$$

Because

$$\mathbf{v}_k = [I,\ P_k] \begin{bmatrix} \mathbf{v}_k^f \\ \mathbf{v}_{k+1} \end{bmatrix},$$

based on the above inequality and the explicit form of $\overline{B}_k$, one arrives at the estimates

$$\begin{aligned} 0 &\le \mathbf{v}_k^T (B_k - A_k) \mathbf{v}_k \\ &\le \left((M_k)^T \mathbf{v}_k^f + A_k P_k \mathbf{v}_{k+1}\right)^T \left(M_k + M_k^T - A_k\right)^{-1} \left((M_k)^T \mathbf{v}_k^f + A_k P_k \mathbf{v}_{k+1}\right) \\ &\quad + \mathbf{v}_{k+1}^T (B_{k+1} - A_{k+1}) \mathbf{v}_{k+1} + \left(\mathbf{v}_{k+1}^T A_{k+1} \mathbf{v}_{k+1} - \mathbf{v}_k^T A_k \mathbf{v}_k\right) \\ &\le \sum_{j=k}^{\ell-1} \left[\left((M_j)^T \mathbf{v}_j^f + A_j P_j \mathbf{v}_{j+1}\right)^T \left(M_j + M_j^T - A_j\right)^{-1} \left((M_j)^T \mathbf{v}_j^f + A_j P_j \mathbf{v}_{j+1}\right)\right] \\ &\quad + \mathbf{v}_\ell^T A_\ell \mathbf{v}_\ell - \mathbf{v}_k^T A_k \mathbf{v}_k. \end{aligned} \tag{5.7}$$

Note that we have the freedom to choose the decomposition $\mathbf{v}_j = \mathbf{v}_j^f + P_j \mathbf{v}_{j+1}$. In particular, we can choose $\mathbf{v}_j^f \in \mathbf{V}_j^f \subset \mathbf{V}_j$ so that we have (by assumption) the estimate

$$\sum_{j \ge k} \left(\mathbf{v}_j^f\right)^T \overline{M}_j \mathbf{v}_j^f \le \sigma\ \mathbf{v}_k^T A_k \mathbf{v}_k. \tag{5.8}$$

If it happens also that

$$\sum_{j \ge k} \mathbf{v}_{j+1}^T P_j^T A_j \left(M_j + M_j^T - A_j\right)^{-1} A_j P_j \mathbf{v}_{j+1} \le \mu\ \mathbf{v}_k^T A_k \mathbf{v}_k. \tag{5.9}$$

and

$$\mathbf{v}_\ell^T A_\ell \mathbf{v}_\ell \le \sigma_c\ \mathbf{v}_k^T A_k \mathbf{v}_k, \tag{5.10}$$

we would then have the following spectral equivalence result,

$$0 \le \mathbf{v}_k^T (B_k - A_k) \mathbf{v}_k \le (\sigma_c + 2(\sigma + \mu) - 1)\ \mathbf{v}_k^T A_k \mathbf{v}_k. \tag{5.11}$$

Note that typically, for a constant $\delta > 0$, we can ensure that

$$\delta\ \mathbf{v}_j^T A_j \mathbf{v}_j \le \mathbf{v}_j^T \left(M_j + M_j^T - A_j\right) \mathbf{v}_j. \tag{5.12}$$

Equivalently, we can ensure $(1+\delta)\mathbf{v}_j^T A_j \mathbf{v}_j \le 2\ \mathbf{v}_j^T M_j \mathbf{v}_j$ generally achievable by scaling M_j. If M_j is s.p.d. and $M_j - A_j$ is symmetric positive semidefinite, we can simply let $\delta = 1$ in (5.12). Estimate (5.12) together with the following strong stability estimate

$$\sum_{j>k} \mathbf{v}_j^T A_j \mathbf{v}_j \le \eta\ \mathbf{v}_k^T A_k \mathbf{v}_k, \tag{5.13}$$

imply estimate (5.9) with $\mu = \eta/\delta$.

We remark that (5.13) follows from (5.8) with level-dependent constant (of order $\mathcal{O}((\ell - k)^2)$. Because $\overline{M}_j$ comes from an A_j-convergent smoother, we first have $\mathbf{v}_j^T A_j \mathbf{v}_j \le \mathbf{v}_j^T \overline{M}_j \mathbf{v}_j$. Then, because by construction $\mathbf{v}_j = \mathbf{v}_j^f + P_j \mathbf{v}_{j+1}$, we easily get the estimate $\|\mathbf{v}_k\|_{A_k} \le \sum_{j\ge k} \|\mathbf{v}_j^f\|_{A_j} + \|\mathbf{v}_\ell\|_{A_\ell}$. Hence,

$$(\|\mathbf{v}_k\|_{A_k} - \|\mathbf{v}_\ell\|_{A_\ell})^2 \le (\ell - k) \sum_{j\ge k} \left(\mathbf{v}_j^f\right)^T A_j \mathbf{v}_j^f \le (\ell - k) \sum_{j\ge k} \left(\mathbf{v}_j^f\right)^T \overline{M}_j \mathbf{v}_j^f.$$

Therefore,

$$\begin{aligned}
\sum_{j>k} \mathbf{v}_j^T A_j \mathbf{v}_j &\le 2 \sum_{j>k} (\|\mathbf{v}_j\|_{A_j} - \|\mathbf{v}_\ell\|_{A_\ell})^2 + (2(\ell - k) - 1)\mathbf{v}_\ell^T A_\ell \mathbf{v}_\ell \\
&\le (2(\ell - k) - 1)\mathbf{v}_\ell^T A_\ell \mathbf{v}_\ell + 2 \sum_{j>k} (\ell - j) \sum_{s\ge j} \left(\mathbf{v}_s^f\right)^T \overline{M}_s \mathbf{v}_s^f \\
&\le (2(\ell - k) - 1)\mathbf{v}_\ell^T A_\ell \mathbf{v}_\ell + (\ell - k - 1)(\ell - k) \sum_{j\ge k} \left(\mathbf{v}_j^f\right)^T \overline{M}_j \mathbf{v}_j^f.
\end{aligned}$$

Thus, if only estimate (5.12) holds (provided the smoothers are also properly scaled as in (5.12) and the coarse component is "energy" stable as in (5.10)), we still have MG convergence, however, with weakly level-dependent bounds.

Remark 5.6. In practice, the most difficult estimate with a level-independent bound is (5.9). In the case of matrices A_j coming from second-order elliptic bilinear form $a(\cdot,\ \cdot)$ and respective finite element space V_j, with M_j simply being the diagonal of A_j (properly scaled), estimate (5.9) reads

$$\sum_j h_j^2 \|\widehat{A}_j P_j \mathbf{v}_{j+1}\|_0^2 \le C \|\mathbf{v}\|_A^2.$$

Here, $\|.\|_0$ comes from the inner product based on the L_2-mass matrix G_j, h_j is the jth-level mesh-size, and $\widehat{A}_j = G_j^{-1} A_j$ are the operators typically used in the finite element analysis of MG. If $P_j \mathbf{v}_{j+1}$ stands for the vector representation of $Q_{j+1} v$ where Q_{j+1} is the L_2-projection onto the $(j+1)$st-level finite element space V_{j+1}, the above estimate in terms of finite element functions v, finite element operators $\widehat{A}_j : V_j \mapsto V_j$ defined via the relation $(\widehat{A}_j v_j, w_j)_0 = a(v_j, w_j) = (\nabla v_j, \nabla w_j)_0$

for all $v_j, w_j \in V_j$, and the L_2-projections $Q_j : L_2 \mapsto V_j$ defined via $(Q_j v, v_j)_0 = (v, v_j)_0$ for all $v_j \in V_j$, reads

$$\sum_j h_j^2 \|\widehat{A}_j Q_j v\|_0^2 \leq C \|\nabla v\|_0^2.$$

Such an estimate with uniform bound C was proven in [VW97] based on a well-known strengthened Cauchy–Schwarz inequality (cf., [Y93], or see Proposition F.1 in the appendix).

We conclude with the following main MG convergence result formulated in general terms.

Theorem 5.7. *Consider A_j-convergent smoothers M_j, $j = 0, \ldots, \ell - 1$ used in the definition of the symmetric $V(1,1)$-cycle MG preconditioner $B = B_0$ for $A = A_0$. If any fine-grid vector $\mathbf{v} = \mathbf{v}_0$ allows for a decomposition based on vector components $\mathbf{v}_j^f = \mathbf{v}_j - P_j \mathbf{v}_{j+1}$, $j = 0, 1, \ldots, \ell - 1$, such that:*

- *The smoothers M_j are efficient on the components $\mathbf{v}_j^f$ in the sense that the estimate*

$$\sum_j \left(\mathbf{v}_j^f\right)^T \overline{M}_j \mathbf{v}_j^f \leq \sigma \, \mathbf{v}^T A \mathbf{v},$$

  holds.
- *The smoothers M_j are scaled as follows,*

$$(1 + \delta) \mathbf{v}_j^T A_j \mathbf{v}_j \leq \mathbf{v}_j^T \left(M_j^T + M_j\right) \mathbf{v}_j = 2 \mathbf{v}_j^T M_j \mathbf{v}_j.$$

- *The coarse component $\mathbf{v}_\ell$ is stable in energy; that is,*

$$\mathbf{v}_\ell^T A_\ell \mathbf{v}_\ell \leq \sigma_c \, \mathbf{v}^T A \mathbf{v}.$$

Then, the symmetric $V(1,1)$-cycle MG preconditioner $B = B_0$ is spectrally equivalent to $A = A_0$ with the following suboptimal bound,

$$\mathbf{v}^T A \mathbf{v} \leq \mathbf{v}^T B \mathbf{v} \leq \left(\sigma_c \left[1 + \frac{2(2\ell - 1)}{\delta} \right] + \sigma \left[\frac{2(\ell - 1)\ell}{\delta} + 1 \right] \right) \mathbf{v}^T A \mathbf{v}.$$

If, in addition, the smoothers M_j are efficient on the components $A_j P_j \mathbf{v}_{j+1}$ so that there holds

$$\sum_j \mathbf{v}_{j+1}^T P_j^T A_j \left(M_j + M_j^T - A_j\right)^{-1} A_j P_j \mathbf{v}_{j+1} \leq \mu \, \mathbf{v}^T A \mathbf{v},$$

then the MG preconditioner B is uniformly spectrally equivalent to A; that is, we have

$$\mathbf{v}^T A \mathbf{v} \leq \mathbf{v}^T B \mathbf{v} \leq (\sigma_c + 2(\sigma + \mu)) \mathbf{v}^T A \mathbf{v}.$$

Stable decomposition of vectors can generally be derived based on the finite element functions from which they come. The latter was a topic of intensive research in the last 20 years. Some details about computable stable decomposition of functions in various Sobolev norms are found in Appendix C, and specific applications of Theorem 5.7 to a number of finite element bilinear forms and spaces are found in Appendix F. Stable vector decompositions are further investigated in Section 5.7.

A different analysis of MG exploiting its relation with the product iteration method was originally developed in [BPWXii] and [BPWXi]. Those were breakthrough results that led to the understanding of the importance of providing stable multilevel decompositions of finite element spaces.

On the sharpness of (5.7)

We show that for some special decompositions $\mathbf{v}_k = [I,\ P_k]\,\overline{\mathbf{v}}_k$, where

$$\overline{\mathbf{v}}_k = \begin{bmatrix} \mathbf{v}_k^f \\ \mathbf{v}_{k+1} \end{bmatrix}$$

inequalities (5.6) and hence (5.7) hold as equalities.

Consider B_k and $\overline{B}_k$. We drop the subscript k whenever appropriate. We also need the following useful lemma.

Lemma 5.8. *Consider* $B^{-1} = [I,\ P]\,\overline{B}^{-1}\,[I,\ P]^T$. *For any given vector* $\mathbf{v}$ *solve*

$$\overline{B}\overline{\mathbf{w}} = [I,\ P]^T\, B\mathbf{v},$$

for $\overline{\mathbf{w}}$. *Then,*

(i) $[I,\ P]\,\overline{\mathbf{w}} = \mathbf{v}$. *That is,*

$$\overline{\mathbf{w}} = \begin{bmatrix} \mathbf{v}^f \\ \mathbf{v}^c \end{bmatrix}$$

represents a decomposition of $\mathbf{v} = \mathbf{v}^f + P\mathbf{v}^c$.

(ii) The decomposition from (i) has some minimal norm property; namely, we have

$$\overline{\mathbf{w}}^T\overline{B}\overline{\mathbf{w}} = \mathbf{v}^T B\mathbf{v} = \min_{\overline{\mathbf{v}}:\mathbf{v}=[I,\ P]\overline{\mathbf{v}}} \overline{\mathbf{v}}^T\overline{B}\overline{\mathbf{v}}.$$

Proof. We have

$$[I,\ P]\,\overline{\mathbf{w}} = [I,\ P]\,\overline{B}^{-1}\,[I,\ P]^T\, B\mathbf{v} = \mathbf{v},$$

which is (i). Also, from the definition of $\overline{\mathbf{w}}$ and (i), we get

$$\overline{\mathbf{w}}^T\overline{B}\overline{\mathbf{w}} = \overline{\mathbf{w}}^T\,[I,\ P]^T\, B\mathbf{v} = ([I,\ P]\,\overline{\mathbf{w}})^T\, B\mathbf{v} = \mathbf{v}^T B\mathbf{v}.$$

Finally, we already showed in (5.6), that for any decomposition $\mathbf{v} = [I,\ P]\,\overline{\mathbf{v}}$, we have $\mathbf{v}^T B\mathbf{v} \le \overline{\mathbf{v}}^T\overline{B}\overline{\mathbf{v}}$. The latter two facts represent the proof of (ii). □

In conclusion, for vectors

$$\begin{bmatrix} \mathbf{v}_k^f \\ \mathbf{v}_{k+1} \end{bmatrix} \equiv \overline{\mathbf{w}}_k = \overline{B}_k^{-1}\,[I,\ P_k]^T\, B_k\mathbf{v}_k$$

(i.e., defined as in Lemma 5.8 (used recursively)), the estimates (5.7) hold as equalities. Thus the following main result holds.

Theorem 5.9. *Consider for any* $\mathbf{v}$ *decompositions of the form:*

(o) $\mathbf{v}_0 = \mathbf{v}$.

(i) For $k = 0, \ldots, \ell - 1$ *let* $\mathbf{v}_k = [I, \; P_k] \begin{bmatrix} \mathbf{v}_k^f \\ \mathbf{v}_{k+1} \end{bmatrix}$.

Then the following main identity holds, for any $k \geq 0$ *and* $\ell \geq k$,

$$\mathbf{v}_k^T B_k \mathbf{v}_k = \inf_{(\mathbf{v}_j = \mathbf{v}_j^f + P_j \mathbf{v}_{j+1})_{j=k}^{\ell-1}} \left[\mathbf{v}_\ell^T B_\ell \mathbf{v}_\ell + \sum_{j=k}^{\ell-1} \left(M_j^T \mathbf{v}_j^f + A_j P_j \mathbf{v}_{j+1} \right)^T \times \left(M_j + M_j^T - A_j \right)^{-1} \left(M_j^T \mathbf{v}_j^f + A_j P_j \mathbf{v}_{j+1} \right) \right].$$

Note that at the coarsest-level ℓ, *we typically set* $B_\ell = A_\ell$.

If we use the representation (5.3) for $B_{MG} = B_0$, we can reformulate Theorem 5.9 as follows.

Theorem 5.10. *Let* $\overline{P}_k = P_0, \ldots, P_{k-1}$ *be the composite interpolation matrices from coarse-level* k *all the way up to the finest-level* 0. *Consider the extended matrix*

$$T = (T_{i,j}) = \left(\overline{P}_i^T A \overline{P}_j \right)_{i,j=0}^{\ell}$$

and form the following block-lower triangular matrix

$$L_B = \begin{bmatrix} M_0 & 0 & \ldots & 0 \\ T_{10} & M_1 & \ldots & 0 \\ \vdots & \ldots & \ddots & 0 \\ T_{\ell,0} & \ldots & T_{\ell,\ell-1} & M_\ell \end{bmatrix}.$$

Then, the following identity holds

$$\mathbf{v}^T B_{MG} \mathbf{v} = \inf_{\mathbf{v} = \sum_{k=0}^{\ell} \overline{P}_k \mathbf{v}_k^f} \begin{bmatrix} \mathbf{v}_0^f \\ \vdots \\ \mathbf{v}_\ell^f \end{bmatrix}^T L_B \left(\operatorname{diag}\left(M_k^T + M_k - A_k \right)_{k=0}^{\ell} \right)^{-1} L_B^T \begin{bmatrix} \mathbf{v}_0^f \\ \vdots \\ \mathbf{v}_\ell^f \end{bmatrix}. \tag{5.14}$$

The following corollary is needed later on.

Corollary 5.11. *For any* $\ell \geq k$, *let* $K_{MG}^{\ell \mapsto k}$ *bound the condition number of the MG V-cycle with exact coarse solution at level* ℓ. *This V-cycle exploits the same smoothers as the original MG V-cycle at levels* $\ell - 1, \ldots, k$. *Assume that* K_ℓ *is the bound of the condition number of* B_ℓ *in terms of* A_ℓ; *that is,* $\mathbf{v}_\ell^T B_\ell \mathbf{v}_\ell \leq K_\ell \; \mathbf{v}_\ell^T A_\ell \mathbf{v}_\ell$. *Then*

Theorem 5.9 applied to any vector decompositions starting with $\mathbf{v}_k = \mathbf{v}$*, and for* $j = k, k+1, \ldots, \ell-1$, $\mathbf{v}_j = \mathbf{v}_j^f + P_j \mathbf{v}_{j+1}$*, gives (noting that* $K_\ell \geq 1$*)*

$$\begin{aligned}
\mathbf{v}^T B_k \mathbf{v} &\leq \inf_{(\mathbf{v}_j)} \Bigg[\mathbf{v}_\ell^T B_\ell \mathbf{v}_\ell + \sum_{j=k}^{\ell-1} \left(M_j^T \mathbf{v}_j^f + (A_j - M_j^T) P_j \mathbf{v}_{j+1} \right)^T \\
&\qquad \times \left(M_j + M_j^T - A_j \right)^{-1} \left(M_j^T \mathbf{v}_j^f + (A_j - M_j^T) P_j \mathbf{v}_{j+1} \right) \Bigg] \\
&\leq K_\ell \inf_{(\mathbf{v}_j)} \Bigg[\mathbf{v}_\ell^T A_l \mathbf{v}_\ell + \sum_{j=k}^{\ell-1} \left(M_j^T \mathbf{v}_j^f + (A_j - M_j^T) P_j \mathbf{v}_{j+1} \right)^T \\
&\qquad \times \left(M_j + M_j^T - A_j \right)^{-1} \left(M_j^T \mathbf{v}_j^f + (A_j - M_j^T) P_j \mathbf{v}_{j+1} \right) \Bigg] \\
&= K_\ell \, K_{MG}^{\ell \mapsto k} \, \mathbf{v}^T A_k \mathbf{v}.
\end{aligned}$$

That is,

$$K_k \leq K_\ell K_{MG}^{\ell \mapsto k}.$$

Here, $K_{TG}^{\ell \mapsto k}$ *is the relative condition number of the exact V-cycle MG method corresponding to fine matrix* A_k*, smoother* M_k*, and coarse-level ones* A_j*,* M_j*, and interpolation matrices* P_j *from level* $j+1$ *to level* j*, and exact coarse grid solution with* A_ℓ *at level* ℓ*.*

5.4 The XZ identity

In this section, we relate the identity proven in [XZ02] in its simplified equivalent form found in [LWXZ] with a subspace correction block-factorization preconditioner in the form defined in [V98] now in a somewhat more general setting.

Let A be a given $n \times n$ s.p.d. matrix. For $k = 1, 2, \ldots, \ell$, let $\overline{P}_k : \mathbb{R}^{n_k} \mapsto \mathbb{R}^n$ be given full column rank interpolation matrices, where $n_k \leq n$. Introduce the $n_k \times n_k$ s.p.d. matrices $A_k = \overline{P}_k^T A \overline{P}_k$ and let M_k be given matrices that provide A_k-convergent iteration for solving systems with A_k. Consider also $\overline{M}_k = M_k(M_k^T + M_k - A_k)^{-1} M_k^T$, the symmetrized versions of M_k.

Let $\mathbf{V}_k \equiv \mathbb{R}^{n_k}$ and $\mathbf{V} \equiv \mathbb{R}^n$. Define the vector spaces $\overline{\mathbf{V}}_k = \text{Range } \overline{P}_k \subset \mathbf{V}$. As an example, in the setting of the MG method from the preceding sections we can define $\overline{P}_k = P_0, \ldots, P_{k-1}$. With this definition the resulting spaces $\overline{\mathbf{V}}_k$ are nested; that is, $\overline{\mathbf{V}}_{k+1} \subset \overline{\mathbf{V}}_k$. In what follows, to derive the XZ identity the spaces $\overline{\mathbf{V}}_k$ need not be nested. Another example is given in Chapter 7.

We introduce the following auxiliary spaces $\widehat{\mathbf{V}}_k = \overline{\mathbf{V}}_k + \overline{\mathbf{V}}_{k+1} + \cdots + \overline{\mathbf{V}}_\ell$. They are not needed in the implementation of the resulting product iteration method, but are useful in its analysis.

The inner product $\widehat{\mathbf{v}}_k^T \widehat{A}_k \widehat{\mathbf{w}}_k \equiv \widehat{\mathbf{v}}_k^T A \widehat{\mathbf{w}}_k$ for any $\widehat{\mathbf{v}}_k,\ \widehat{\mathbf{w}}_k \in \widehat{\mathbf{V}}_k$ defines an operator $\widehat{A}_k : \widehat{\mathbf{V}}_k \mapsto \widehat{\mathbf{V}}_k$. Let $\widehat{Q}_k : \mathbf{V} \mapsto \widehat{\mathbf{V}}_k$ be the ℓ_2-projection onto $\widehat{\mathbf{V}}_k$; that is, $\widehat{Q}_k \mathbf{v} \in \widehat{\mathbf{V}}_k$ is defined via the identity $\widehat{\mathbf{w}}_k^T \widehat{Q}_k \mathbf{v} = \widehat{\mathbf{w}}_k^T \mathbf{v}$ for any $\widehat{\mathbf{w}}_k \in \widehat{\mathbf{V}}_k$. We have $\widehat{Q}_k = \widehat{Q}_k^T$, $\widehat{Q}_k^2 = \widehat{Q}_k$, and $\widehat{Q}_k \overline{P}_k = \overline{P}_k$. Define $A_{k,k+1} = \overline{P}_k^T A \widehat{Q}_{k+1}$ and $A_{k+1,k} = \widehat{Q}_{k+1} A \overline{P}_k$.

Using the decomposition $\widehat{\mathbf{V}}_k = \overline{\mathbf{V}}_k + \widehat{\mathbf{V}}_{k+1}$, the actions of the operator $\widehat{A}_k$ can be computed based on the following two-by-two block form,

$$\left(\overline{P}_k \mathbf{w}_k + \widehat{\mathbf{w}}_{k+1}\right)^T \widehat{A}_k \left(\overline{P}_k \mathbf{v}_k + \widehat{\mathbf{v}}_{k+1}\right) = \begin{bmatrix} \mathbf{w}_k \\ \widehat{\mathbf{w}}_{k+1} \end{bmatrix}^T \begin{bmatrix} A_k & A_{k,k+1} \\ A_{k+1,k} & \widehat{A}_{k+1} \end{bmatrix} \begin{bmatrix} \mathbf{v}_k \\ \widehat{\mathbf{v}}_{k+1} \end{bmatrix}.$$

The above two-by-two block form of $\widehat{A}_k$ serves as a motivation for the next definition of the preconditioner of $\widehat{A}_k$ as an approximate block-factorization.

Definition 5.12 (Subspace correction preconditioner). *Let $\widehat{B}_\ell : \widehat{\mathbf{V}}_\ell \mapsto \widehat{\mathbf{V}}_\ell$ be defined from the identity*

$$\left(\overline{P}_\ell \mathbf{w}_\ell\right)^T \widehat{B}_\ell \left(\overline{P}_\ell \mathbf{v}_\ell\right) = \mathbf{w}_\ell^T \overline{M}_\ell \mathbf{v}_\ell, \quad \textit{for all } \mathbf{v}_\ell,\ \mathbf{w}_\ell \in \mathbf{V}_\ell.$$

Recall that $\widehat{\mathbf{V}}_\ell = \overline{\mathbf{V}}_\ell = \text{Range}\ (\overline{P}_\ell)$.
For $k < \ell$, assuming (by induction) that $\widehat{B}_{k+1} : \widehat{\mathbf{V}}_{k+1} \mapsto \widehat{\mathbf{V}}_{k+1}$ has been defined, we first define a mapping $\widetilde{B}_k : [\mathbf{V}_k,\ \widehat{\mathbf{V}}_{k+1}] \mapsto [\mathbf{V}_k,\ \widehat{\mathbf{V}}_{k+1}]$ in the following factored form,

$$\widetilde{B}_k = \begin{bmatrix} I & 0 \\ A_{k+1,k} M_k^{-1} & I \end{bmatrix} \begin{bmatrix} \overline{M}_k & 0 \\ 0 & \widehat{B}_{k+1} \end{bmatrix} \begin{bmatrix} I & M_k^{-T} A_{k,k+1} \\ 0 & I \end{bmatrix}$$

and then let

$$\widehat{B}_k^{-1} = \left[\overline{P}_k,\ \widehat{Q}_{k+1}\right] \widetilde{B}_k^{-1} \left[\overline{P}_k,\ \widehat{Q}_{k+1}\right]^T.$$

More explicitly, because

$$\widetilde{B}_k^{-1} = \begin{bmatrix} I & -M_k^{-T} A_{k,k+1} \\ 0 & I \end{bmatrix} \begin{bmatrix} \overline{M}_k^{-1} & 0 \\ 0 & \widehat{B}_{k+1}^{-1} \end{bmatrix} \begin{bmatrix} I & 0 \\ -A_{k+1,k} M_k^{-1} & I \end{bmatrix},$$

where we have assumed (by induction) that $\widehat{B}_{k+1}$ is invertible on $\widehat{\mathbf{V}}_{k+1}$, we have

$$\widehat{B}_k^{-1} = \overline{P}_k \overline{M}_k^{-1} \overline{P}_k^T + \left(\widehat{Q}_{k+1} - \overline{P}_k M_k^{-T} A_{k,k+1}\right) \widehat{B}_{k+1}^{-1} \left(\widehat{Q}_{k+1} - A_{k+1,k} M_k^{-1} \overline{P}_k^T\right).$$

This expression shows that in fact $\widehat{B}_k^{-1}$ is s.p.d. (and invertible) on $\widehat{B}_k$. Based on the properties $\overline{P}_k^T \widehat{Q}_k = \overline{P}_k^T$ and $\widehat{Q}_{k+1} \widehat{Q}_k = \widehat{Q}_{k+1}$, and the identity

$$I - \overline{P}_k \overline{M}_k^{-1} \overline{P}_k^T A = \left(I - \overline{P}_k M_k^{-T} \overline{P}_k^T A\right)\left(I - \overline{P}_k M_k^{-1} \overline{P}_k^T A\right),$$

it is straightforward to show the following product iteration formula.

$$I - \widehat{B}_k^{-1}\widehat{Q}_k A = \left(I - \overline{P}_k M_k^{-T}\overline{P}_k^T A\right)\left(I - \widehat{B}_{k+1}^{-1}\widehat{Q}_{k+1}A\right)\left(I - \overline{P}_k M_k^{-1}\overline{P}_k^T A\right). \tag{5.15}$$

We show next that $\widehat{B}_\ell^{-1}\widehat{Q}_\ell A$ can be computed without the knowledge of $\widehat{Q}_\ell$. More specifically, we derive the expression

$$I - \widehat{B}_\ell^{-1}\widehat{Q}_\ell A = \left(I - \overline{P}_\ell M_\ell^{-T}\overline{P}_\ell^T A\right)\left(I - \overline{P}_\ell M_\ell^{-1}\overline{P}_\ell^T A\right). \tag{5.16}$$

Clearly, the above expression holds if we show that

$$\widehat{B}_\ell^{-1}\widehat{Q}_\ell = \overline{P}_\ell M_\ell^{-T}\overline{P}_\ell^T + \overline{P}_\ell M_\ell^{-1}\overline{P}_\ell^T - \overline{P}_\ell M_\ell^{-T} A_\ell M_\ell^{-1}\overline{P}_\ell^T,$$

or equivalently,

$$\widehat{Q}_\ell = \widehat{B}_\ell \overline{P}_\ell \left(M_\ell^{-T} + M_\ell^{-1} - M_\ell^{-T} A_\ell M_\ell^{-1}\right)\overline{P}_\ell^T = \widehat{B}_\ell \overline{P}_\ell \overline{M}_\ell^{-1}\overline{P}_\ell^T.$$

Based on the definition of $\widehat{B}_\ell$

$$\left(\overline{P}_\ell \mathbf{w}_\ell\right)^T \widehat{B}_\ell \left(\overline{P}_\ell \mathbf{v}_\ell\right) = \mathbf{w}_\ell^T \overline{M}_\ell \mathbf{v}_\ell,$$

we will have then, for any $\mathbf{w}_\ell \in \mathbf{V}_\ell$ and $\mathbf{v} \in \mathbf{V}$,

$$\begin{aligned}\left(\overline{P}_\ell \mathbf{w}_\ell\right)^T \widehat{Q}_\ell \mathbf{v} &= \left(\overline{P}_\ell \mathbf{w}_\ell\right)^T \widehat{B}_\ell \overline{P}_\ell \overline{M}_\ell^{-1}\overline{P}_\ell^T \mathbf{v} \\ &= \mathbf{w}_\ell^T \overline{M}_\ell \overline{M}_\ell^{-1}\overline{P}_\ell^T \mathbf{v} \\ &= \mathbf{w}_\ell^T \overline{P}_\ell^T \mathbf{v}.\end{aligned}$$

That is, we have to show that $\overline{P}_\ell^T \widehat{Q}_\ell = \overline{P}_\ell^T$ or its transpose $\widehat{Q}_\ell \overline{P}_\ell = \overline{P}_\ell$ which is the case. It is clear that we can repeat the above steps in reverse order thus ending up with the desired expression (5.16). In conclusion, combining (5.15) and (5.16), we end up with the following result.

Theorem 5.13. *The subspace correction preconditioner $\widehat{B}_k$, defined in Definition 5.12, for any $k \leq \ell$, can be implemented as a subspace iteration algorithm for solving systems with A giving rise to the product iteration formula*

$$\begin{aligned}I - \widehat{B}_k^{-1}\widehat{Q}_k A = &\left(I - \overline{P}_k \overline{M}_k^{-T}\overline{P}_k^T A\right)\cdots\left(I - \overline{P}_\ell \overline{M}_\ell^{-T}\overline{P}_\ell^T A\right) \\ &\times \left(I - \overline{P}_\ell \overline{M}_\ell^{-1}\overline{P}_\ell^T A\right)\cdots\left(I - \overline{P}_k \overline{M}_k^{-1}\overline{P}_k^T A\right).\end{aligned}$$

The definition of $\widehat{B}_k$ implies $I = \widehat{B}_k^{1/2}[\overline{P}_k,\ \widehat{Q}_{k+1}]\widetilde{B}_k^{-1}[\overline{P}_k,\ \widehat{Q}_{k+1}]^T \widehat{B}_k^{1/2}$. From this equality, based on the fact that $\|G\| = \|G^T\|$ used for $G = \widetilde{B}_k^{-(1/2)}$ $[\overline{P}_k, \widehat{Q}_{k+1}]^T\ \widehat{B}_k^{1/2}$ it follows that the difference

$$\left[\overline{P}_k,\ \widehat{Q}_{k+1}\right]^T \widehat{B}_k \left[\overline{P}_k,\ \widehat{Q}_{k+1}\right] - \widetilde{B}_k$$

is symmetric negative semidefinite on $[\mathbf{V}_k, \widehat{\mathbf{V}}_{k+1}]$. Therefore, we have for any $\widehat{\mathbf{v}}_k \in \widehat{\mathbf{V}}_k$ decomposed as

$$\widehat{\mathbf{v}}_k = \overline{P}_k \mathbf{v}_k + \widehat{\mathbf{v}}_{k+1} = \left[\overline{P}_k, \ \widehat{Q}_{k+1}\right] \begin{bmatrix} \mathbf{v}_k \\ \widehat{\mathbf{v}}_{k+1} \end{bmatrix}, \quad \text{for } \widehat{\mathbf{v}}_{k+1} \in \widehat{\mathbf{V}}_{k+1},$$

$$\begin{aligned} \widehat{\mathbf{v}}_k^T \widehat{B}_k \widehat{\mathbf{v}}_k &= \begin{bmatrix} \mathbf{v}_k \\ \widehat{\mathbf{v}}_{k+1} \end{bmatrix}^T \left[\overline{P}_k, \ \widehat{Q}_{k+1}\right]^T \widehat{B}_k \left[\overline{P}_k, \ \widehat{Q}_{k+1}\right] \begin{bmatrix} \mathbf{v}_k \\ \widehat{\mathbf{v}}_{k+1} \end{bmatrix} \\ &\le \begin{bmatrix} \mathbf{v}_k \\ \widehat{\mathbf{v}}_{k+1} \end{bmatrix}^T \widetilde{B}_k \begin{bmatrix} \mathbf{v}_k \\ \widehat{\mathbf{v}}_{k+1} \end{bmatrix}. \end{aligned}$$

Using the explicit form of $\widetilde{B}_k$, we obtain

$$\widehat{\mathbf{v}}_k^T \widehat{B}_k \widehat{\mathbf{v}}_k \le \widehat{\mathbf{v}}_{k+1}^T \widehat{B}_{k+1} \widehat{\mathbf{v}}_{k+1} + \left(\mathbf{v}_k + M_k^{-T} \overline{P}_k^T A \widehat{\mathbf{v}}_{k+1}\right)^T \overline{M}_k \left(\mathbf{v}_k + M_k^{-T} \overline{P}_k^T A \widehat{\mathbf{v}}_{k+1}\right).$$

Using recursion on k, for any decomposition $\widehat{\mathbf{v}}_k = \overline{P}_k \mathbf{v}_k + \cdots + \overline{P}_\ell \mathbf{v}_\ell$ setting $\widehat{\mathbf{v}}_j = \overline{P}_j \mathbf{v}_j + \cdots + \overline{P}_\ell \mathbf{v}_\ell$ for $j \ge k$, we arrive at the inequality

$$\widehat{\mathbf{v}}_k^T \widehat{B}_k \widehat{\mathbf{v}}_k \le \mathbf{v}_\ell^T \overline{M}_\ell \mathbf{v}_\ell + \sum_{j=k}^{\ell-1} \left(\mathbf{v}_j + M_j^{-T} \overline{P}_j^T A \widehat{\mathbf{v}}_{j+1}\right)^T \overline{M}_j \left(\mathbf{v}_j + M_j^{-T} \overline{P}_j^T A \widehat{\mathbf{v}}_{j+1}\right).$$

Equivalently, because the decomposition $\widehat{\mathbf{v}}_k = \overline{P}_k \mathbf{v}_k + \cdots + \overline{P}_\ell \mathbf{v}_\ell$ was arbitrary, we have

$$\begin{aligned} &\widehat{\mathbf{v}}_k^T \widehat{B}_k \widehat{\mathbf{v}}_k \\ &\le \min_{\widehat{\mathbf{v}}_k = \overline{P}_k \mathbf{v}_k + \cdots + \overline{P}_\ell \mathbf{v}_\ell} \left[\mathbf{v}_\ell^T \overline{M}_\ell \mathbf{v}_\ell + \sum_{j=k}^{\ell-1} \| M_j^T \mathbf{v}_j + \overline{P}_j^T A \widehat{\mathbf{v}}_{j+1} \|^2_{(M_j^T + M_j - A_j)^{-1}} \right]. \end{aligned}$$

The fact that this is actually an equality is proven similarly as in Lemma 5.8 (or Theorem 3.15). That is, we have the following main identity which is sometimes referred to as the XZ identity.

Theorem 5.14. *The subspace correction preconditioner $\widehat{B}_k$ defined in Definition 5.12 satisfies, for any $k \le \ell$, the identity:*

$$\begin{aligned} \widehat{\mathbf{v}}_k^T \widehat{B}_k \widehat{\mathbf{v}}_k = &\min_{\widehat{\mathbf{v}}_k = \overline{P}_k \mathbf{v}_k + \cdots + \overline{P}_\ell \mathbf{v}_\ell} \\ &\times \left[\mathbf{v}_\ell^T \overline{M}_\ell \mathbf{v}_\ell + \sum_{j=k}^{\ell-1} \| M_j^T \mathbf{v}_j + \overline{P}_j^T A \widehat{\mathbf{v}}_{j+1} \|^2_{(M_j^T + M_j - A_j)^{-1}} \right]. \end{aligned} \tag{5.17}$$

The XZ identity is traditionally formulated in terms of the operators $T_k = \overline{P}_k M_k^{-1} \overline{P}_k^T A$, $T_k^* = \overline{P}_k M_k^{-T} \overline{P}_k^T A$ and $\overline{T}_k = \overline{P}_k \overline{M}_k^{-1} \overline{P}_k^T A$. The operators $\overline{T}_k$ are invertible on $\overline{\mathbf{V}}_k$. By definition, $\overline{T}_k^{-1} \overline{P}_k \mathbf{v}_k = \overline{P}_k \mathbf{x}_k$ where $\mathbf{x}_k$ is determined from the equation

$\overline{P}_k \mathbf{v}_k = \overline{T}_k \overline{P}_k \mathbf{x}_k$. Equivalently, we have $\overline{P}_k \mathbf{v}_k = \overline{P}_k \overline{M}_k^{-1} \overline{P}_k^T A \overline{P}_k \mathbf{x}_k$. That is, because $\overline{P}_k$ has full column rank, we obtain $\mathbf{x}_k = A_k^{-1} \overline{M}_k \mathbf{v}_k$. Hence, we have the following explicit expression for $\overline{T}_k^{-1} \overline{P}_k$,

$$\overline{T}_k^{-1} \overline{P}_k = \overline{P}_k A_k^{-1} \overline{M}_k.$$

The operators T_j and T_j^* give rise to the following product iteration formula for any $k \leq \ell$,

$$(I - T_k^*) \cdots (I - T_\ell^*)(I - T_\ell) \cdots (I - T_k).$$

We proved in Theorem 5.13 that this formula defines a preconditioner $\widehat{B}_k$ for the operator $\widehat{A}_k$. The preconditioner $\widehat{B}_k$ is s.p.d. on $\widehat{\mathbf{V}}_k$ and can be defined either (implicitly) via the relation

$$I - \widehat{B}_k^{-1} \widehat{Q}_k A = (I - T_k^*) \cdots (I - T_\ell^*)(I - T_\ell) \cdots (I - T_k),$$

or more explicitly as in Definition 5.12.

We are now in a position to formulate the XZ identity in the form found in [LWXZ].

Theorem 5.15. *The subspace correction preconditioner $\widehat{B}_k$ defined in Definition 5.12 satisfies, for any $k \leq \ell$, the identity:*

$$\widehat{\mathbf{v}}_k^T \widehat{B}_k \widehat{\mathbf{v}}_k = \min_{\widehat{\mathbf{v}}_k = \overline{P}_k \mathbf{v}_k + \cdots + \overline{P}_\ell \mathbf{v}_\ell} \Bigg[\left(\overline{T}_\ell^{-1} \overline{P}_\ell \mathbf{v}_\ell, \overline{P}_\ell \mathbf{v}_\ell\right)_A + \sum_{j=k}^{\ell-1} \left(\overline{T}_j^{-1} \left(\overline{P}_j \mathbf{v}_j + T_j^* \widehat{\mathbf{v}}_{j+1}\right), \left(\overline{P}_j \mathbf{v}_j + T_j^* \widehat{\mathbf{v}}_{j+1}\right)\right)_A \Bigg]. \tag{5.18}$$

Here, $T_j = \overline{P}_j M_j^{-1} \overline{P}_j^T A$, $T_j^* = \overline{P}_j M_j^{-T} \overline{P}_j^T A$, and $\overline{T}_j^{-1} : \overline{\mathbf{V}}_j \mapsto \overline{\mathbf{V}}_j$ is such that $\overline{T}_j^{-1} \overline{P}_j = \overline{P}_j A_j^{-1} \overline{M}_j$. We also used the notation $(\mathbf{u}, \mathbf{w})_A = \mathbf{w}^T A \mathbf{u}$.

5.5 Some classical upper bounds

We next prove an upper bound that is useful in the analysis of the V-cycle with several smoothing steps. We consider for the time being two consecutive levels k and $k+1$. For this reason, we omit the subscript k, and for the coarse quantities, the subscript $k+1$ is replaced with "c".

We recall two matrices that combine smoothing with M and M^T,

$$\overline{M} = M(M + M^T - A)^{-1} M^T,$$

and

$$\widetilde{M} = M^T (M + M^T - A)^{-1} M.$$

The following identities for the corresponding iteration matrices hold,

$$\begin{aligned} I - \overline{M}^{-1} A &= (I - M^{-T} A)(I - M^{-1} A) \quad \text{and} \\ I - \widetilde{M}^{-1} A &= (I - M^{-1} A)(I - M^{-T} A). \end{aligned} \tag{5.19}$$

Introducing $\overline{E} = I - A^{1/2} M^{-1} A^{1/2}$, we then have

$$A^{1/2}(I - \overline{M}^{-1} A) A^{-(1/2)} = \overline{E}^T \overline{E} \text{ and } A^{1/2}(I - \widetilde{M}^{-1} A) A^{-(1/2)} = \overline{E}\,\overline{E}^T. \tag{5.20}$$

By definition, the following explicit relation between B^{-1} and the coarse one B_c^{-1} holds.

$$B^{-1} = \overline{M}^{-1} + (I - M^{-T} A) P B_c^{-1} P^T (I - A M^{-1}).$$

Using the identity $A^{1/2} \overline{M}^{-1} A^{1/2} = I - \overline{E}^T \overline{E}$, we end up with the following relation

$$A^{1/2} B^{-1} A^{1/2} = I - \overline{E}^T \overline{E} + \overline{E}^T A^{1/2} P B_c^{-1} P^T A^{1/2} \overline{E}.$$

Assume now, by induction, that

$$0 \leq \mathbf{v}_c^T (B_c - A_c) \mathbf{v}_c \leq \eta_c \, \mathbf{v}_c^T A_c \mathbf{v}_c.$$

Then, the following upper bound holds, introducing the projection $\overline{\pi}_A = A^{1/2} P A_c^{-1} P^T A^{1/2}$,

$$\begin{aligned} \frac{\mathbf{v}^T B \mathbf{v}}{\mathbf{v}^T A \mathbf{v}} &\leq \sup_{\mathbf{v}} \frac{\mathbf{v}^T A^{-1/2} B A^{-1/2} \mathbf{v}}{\mathbf{v}^T \mathbf{v}} \\ &\leq \sup_{\mathbf{v}} \frac{\mathbf{v}^T \left(I - \overline{E}^T \overline{E} + \frac{1}{1+\eta_c} \overline{E}^T \overline{\pi}_A \overline{E} \right)^{-1} \mathbf{v}}{\mathbf{v}^T \mathbf{v}} \\ &= \sup_{\mathbf{v}} \frac{\mathbf{v}^T \mathbf{v}}{\mathbf{v}^T \left(I - \overline{E}^T \overline{E} + \frac{1}{1+\eta_c} \overline{E}^T \overline{\pi}_A \overline{E} \right) \mathbf{v}} \\ &= (1 + \eta_c) \sup_{\mathbf{v}} \frac{\mathbf{v}^T \mathbf{v}}{\mathbf{v}^T \left[\eta_c \left(I - \overline{E}^T \overline{E} \right) + I - \overline{E}^T (I - \overline{\pi}_A) \overline{E} \right] \mathbf{v}}. \end{aligned} \tag{5.21}$$

The assumption (A) below provides perhaps the shortest convergence proof for the V-cycle MG. We show next that (A) is equivalent to assumption (A*) originally used in [Mc84, Mc85]. Assumption (A) is found as inequality (4.82) in [Sh95].

(A) There is a constant $\eta_s > 0$ such that,

$$\begin{aligned} &\mathbf{v}^T A (I - M^{-T} A)(I - \pi_A)(I - M^{-1} A) \mathbf{v} \\ &\quad \leq \eta_s [\mathbf{v}^T A \mathbf{v} - \mathbf{v}^T A (I - M^{-T} A)(I - M^{-1} A) \mathbf{v}]. \end{aligned}$$

This assumption can equivalently be stated as

$$\mathbf{v}^T\overline{E}^T(I - \overline{\pi}_A)\overline{E}\mathbf{v} \leq \eta_s \left[\mathbf{v}^T\mathbf{v} - \mathbf{v}^T\overline{E}^T\overline{E}\mathbf{v}\right]. \tag{5.22}$$

Using the latter inequality in (5.21), we get

$$\begin{aligned}\frac{1}{1+\eta_c}\,\frac{\mathbf{v}^T B\mathbf{v}}{\mathbf{v}^T A\mathbf{v}} &\leq \sup_{\mathbf{v}} \frac{\mathbf{v}^T\mathbf{v}}{\mathbf{v}^T\left(\eta_c\left(I - \overline{E}^T\overline{E}\right) + I - \eta_s\, I + \eta_s\overline{E}^T\overline{E}\right)\mathbf{v}} \\ &= \sup_{\mathbf{v}} \frac{\mathbf{v}^T\mathbf{v}}{\mathbf{v}^T\left(I + (\eta_c - \eta_s)\left(I - \overline{E}^T\overline{E}\right)\right)\mathbf{v}}.\end{aligned}$$

Assuming (by induction) that $\eta_c \geq \eta_s$, the induction assumption $\mathbf{v}_c^T B_c\mathbf{v}_c \leq (1+\eta_c)\ \mathbf{v}_c^T A_c\mathbf{v}_c$ is confirmed at the next level, because with $\eta = \eta_c$, we get from the last estimate above, $\mathbf{v}^T B\mathbf{v} \leq (1+\eta)\ \mathbf{v}^T A\mathbf{v}$.

Thus we proved the following main theorem.

Theorem 5.16. *Under the assumption (A), valid for $A = A_k$ at levels $k \leq \ell$, the V-cycle preconditioner $B := B_k$ is uniformly (in $k \leq \ell$) spectrally equivalent to $A := A_k$, and the following estimate holds*

$$\mathbf{v}^T A\mathbf{v} \leq \mathbf{v}^T B\mathbf{v} \leq (1+\eta_s)\ \mathbf{v}^T A\mathbf{v},$$

where $\eta_s > 0$ is from the main assumption (A).

The following is a sufficient condition for (A) to hold.

Lemma 5.17. *If the smoother M is efficient on the A-orthogonal complement to the coarse space Range $(I - \pi_A)$, in the sense that*

$$\mathbf{v}_s^T\overline{M}\mathbf{v}_s \leq \eta_s\ \mathbf{v}_s^T A\mathbf{v}_s \quad \textit{for any } \mathbf{v}_s = (I - \pi_A)\mathbf{v}, \tag{5.23}$$

then condition (A) holds. If M is symmetric and properly scaled so that

$$\mathbf{v}^T A\mathbf{v} \leq \mathbf{v}^T M\mathbf{v},$$

then (5.23) *can equivalently be formulated in terms of M instead.*

Proof. We have, for any $\mathbf{w}$,

$$\begin{aligned}\mathbf{w}^T A(I - \pi_A)\mathbf{w} &\leq \left(\overline{M}^{-1/2}A\mathbf{w}\right)^T\overline{M}^{1/2}(I - \pi_A)\mathbf{w} \\ &\leq \|A\mathbf{w}\|_{\overline{M}^{-1}}((I - \pi_A)\mathbf{w})^T\overline{M}(I - \pi_A)\mathbf{w})^{1/2} \\ &\leq \sqrt{\eta_s}\ \|A\mathbf{w}\|_{\overline{M}^{-1}}(((I - \pi_A)\mathbf{w})^T A(I - \pi_A)\mathbf{w})^{1/2} \\ &= \sqrt{\eta_s}\ \|A\mathbf{w}\|_{\overline{M}^{-1}}(\mathbf{w}^T A(I - \pi_A)\mathbf{w})^{1/2}.\end{aligned}$$

That is, we have

$$\mathbf{w}^T A(I - \pi_A)\mathbf{w} \leq \eta_s\ \mathbf{w}^T A\overline{M}^{-1}A\mathbf{w}. \tag{5.24}$$

Choose now $\mathbf{w} = (I - M^{-1}A)\mathbf{v}$. The left-hand side above becomes then the left-hand side of (A). For the r.h.s. of (5.24) use the identity (see (5.19)) $A\overline{M}^{-1}A = AM^{-T}(M + M^T - A)M^{-1}A = A - A(I - M^{-T}A)(I - M^{-1}A)$ which equals exactly the r.h.s. of (A).

In the case $M = M^T$ and $M - A$ being positive semidefinite, we have that M and $\widetilde{M} = \overline{M} = M(2M - A)^{-1}M$ are spectrally equivalent because then

$$\frac{1}{2}\mathbf{v}^T M\mathbf{v} \leq \mathbf{v}^T \overline{M}\mathbf{v} \leq \mathbf{v}^T M\mathbf{v}.$$

Also, $M - A$ being positive semidefinite implies that M is an A-convergent smoother (for A). Finally, if $\mathbf{v}_s^T M\mathbf{v}_s \leq \eta_s\ \mathbf{v}_s^T A\mathbf{v}_s$ for any $\mathbf{v}_s = (I - \pi_A)\mathbf{v}$, we also have $\mathbf{v}_s^T \overline{M}\mathbf{v}_s \leq \mathbf{v}_s^T M\mathbf{v}_s \leq \eta_s\ \mathbf{v}_s^T A\mathbf{v}_s$ and the proof proceeds as before. □

Remark 5.18. We comment here that the assumption (5.23) is much stronger than one of the necessary conditions for two-grid convergence formulated in Corollary 3.23 in the case $M = M^T$, hence $\widetilde{M} = \overline{M}$. This is seen from the estimates

$$\begin{aligned}\|(I - \pi_{\widetilde{M}})\mathbf{v}\|_{\widetilde{M}}^2 &= \min_{\mathbf{v}_c} \|\mathbf{v} - P\mathbf{v}_c\|_{\widetilde{M}}^2 \leq \|(I - \pi_A)\mathbf{v}\|_{\widetilde{M}}^2 \\ &\leq \eta_s\ \|(I - \pi_A)\mathbf{v}\|_A^2 \leq \eta_s\ \|\mathbf{v}\|_A^2.\end{aligned}$$

That is, we have then

$$T_{TG} = \sup_{\mathbf{v}} \frac{\mathbf{v}^T\widetilde{M}(I - \pi_{\widetilde{M}})\mathbf{v}}{\mathbf{v}^T A\mathbf{v}} \leq \eta_s.$$

Therefore, the two-grid convergence factor satisfies $\varrho_{TG} = 1-(1/K_{TG}) \leq 1-(1/\eta_s)$. The condition (5.23), however, implies much more than a two-grid convergence because it also implies condition (A) and hence, we have a uniform V-cycle convergence (due to Theorem 5.16).

Consider now the following assumption.

(A*) There is a constant $\delta_s \in [0, 1)$ such that,

$$\|(I - M^{-T}A)\mathbf{v}\|_A^2 \leq \delta_s\ \|(I - \pi_A)\mathbf{v}\|_A^2 + \|\pi_A\mathbf{v}\|_A^2.$$

Assumption (A*) has the following interpretation. The smoother M^T reduces (in energy norm) by a factor of $\delta_s^{1/2}$ the "oscillatory" error components (referring to the space Range$(I - \pi_A)$), whereas at the same time it does not amplify the "smooth" error components (referring to the coarse space Range(P) = Range(π_A)).

We note that (A*) implies (5.24) for $\mathbf{w} = \mathbf{v} = (I - \pi_A)\mathbf{v}$ and $\eta_s = 1/(1 - \delta_s)$, and therefore condition (A) holds. Moreover, the following equivalence result actually holds.

Proposition 5.19. *Assumptions (A) and (A*) are equivalent with* $\delta_s = \eta_s/(1 + \eta_s)$.

Proof. Consider assumption (A) in the form (5.22). By rearranging terms, we arrive at

$$\mathbf{v}^T \overline{E}^T \left(I - \frac{1}{1+\eta_s} \overline{\pi}_A \right) \overline{E}\mathbf{v} \leq \frac{\eta_s}{1+\eta_s} \mathbf{v}^T \mathbf{v}. \tag{5.25}$$

Using the fact that $\overline{\pi}_A$ is a projection, that is, that $\overline{\pi}_A(I - \overline{\pi}_A) = 0$, we also have

$$I - \frac{1}{1+\eta_s} \overline{\pi}_A = (I - \overline{\pi}_A) + \delta_s \overline{\pi}_A = (I - \overline{\pi}_A)^2 + \delta_s \overline{\pi}_A^2 = (I - \overline{\pi}_A + \sqrt{\delta_s} \overline{\pi}_A)^2.$$

Therefore (because $\overline{\pi}_A$ is symmetric), we can rewrite (5.25) as follows.

$$\|((I - \overline{\pi}_A) + \sqrt{\delta_s} \overline{\pi}_A)\overline{E}\mathbf{v}\|^2 \leq \delta_s \mathbf{v}^T \mathbf{v}.$$

The fact $\|X\| = \|X^T\|$ used for $X = \overline{E}^T ((I - \overline{\pi}_A) + \sqrt{\delta_s} \overline{\pi}_A)$ shows then the estimate

$$\|\overline{E}^T ((I - \overline{\pi}_A) + \sqrt{\delta_s} \overline{\pi}_A)\mathbf{w}\|^2 \leq \delta_s \mathbf{w}^T \mathbf{w}. \tag{5.26}$$

Finally, using again the orthogonality of $I - \overline{\pi}_A$ and $\overline{\pi}_A$, we first see that $((I - \overline{\pi}_A) + \sqrt{\delta_s}\overline{\pi}_A)^{-1} = (I - \overline{\pi}_A) + (1/\sqrt{\delta_s})\overline{\pi}_A$, which together with (5.26) then shows

$$\begin{aligned} \|\overline{E}^T \mathbf{v}\|^2 &\leq \delta_s \|((I - \overline{\pi}_A) + \sqrt{\delta_s}\overline{\pi}_A)^{-1}\mathbf{v}\|^2 \\ &= \delta_s \left\|((I - \overline{\pi}_A) + \frac{1}{\sqrt{\delta_s}}\overline{\pi}_A)\mathbf{v}\right\|^2 \\ &= \delta_s \|(I - \overline{\pi}_A)\mathbf{v}\|^2 + \|\overline{\pi}_A \mathbf{v}\|^2. \end{aligned}$$

Letting $\mathbf{v} := A^{1/2}\mathbf{v}$ the estimate (A*) is finally obtained.

The converse statement follows by repeating the above argument in a reverse order. □

Some auxiliary estimates

Assumption (A) is commonly verified (see Lemma 5.21) based on a boundedness assumption of the projection π_A, namely,

(B) "ℓ_2-Boundedness" of π_A:

$$\|A\| \, \|(I - \pi_A)\mathbf{v}\| \leq \eta_b \, \|A\mathbf{v}\|.$$

We can prove an estimate such as (B) if the following strong approximation property holds.

(C) "Strong approximation property":
For every $\mathbf{v}$, there is a coarse interpolant $P\mathbf{v}_c$ such that

$$(\mathbf{v} - P\mathbf{v}_c)^T A(\mathbf{v} - P\mathbf{v}_c) \leq \frac{\eta_a}{\|A\|} \|A\mathbf{v}\|^2.$$

We verified such an estimate for f.e. matrices coming from the Poisson equation $-\Delta u = f$ in Ω and $u = 0$ on $\partial\Omega$, which admits full regularity; that is, $\|u\|_2 \le C\ \|f\|$. Such regularity estimates are available for convex polygonal domains Ω (cf., e.g., [TW05]). We proved estimate (1.16) in that case.

Estimate (C) is also proved for a purely algebraic two-grid method described in Section 6.11. More precisely, we show there (see the second inequality from the bottom in (6.47)) that for any $\mathbf{e}$, there is an $\epsilon \in \text{Range}\ (P)$ such that,

$$(\mathbf{e}-\epsilon)^T A(\mathbf{e}-\epsilon) \le \|A\|\|\mathbf{e}-\epsilon\|^2 \le \frac{\delta}{\eta\|A\|}\ \|A\mathbf{e}\|^2.$$

Lemma 5.20. *Assumption (C) implies (B) with $\eta_b = \eta_a$.*

Proof. The proof is based on the so-called Aubin–Nitsche trick. Consider $\mathbf{e} = (I - \pi_A)\mathbf{v}$ and let $\mathbf{u} : A\mathbf{u} = \mathbf{e}$. We have, noting that $\mathbf{e}$ is A-orthogonal to the coarse space, letting $\overline{\eta}_a = \eta_a/\|A\|$,

$$\begin{aligned}
\|\mathbf{e}\|^2 &= \mathbf{e}^T A\mathbf{u} \\
&= \mathbf{e}^T A(\mathbf{u} - P\mathbf{u}_c) \\
&\le \|\mathbf{e}\|_A \|\mathbf{u} - P\mathbf{u}_c\|_A \\
&\le \|\mathbf{e}\|_A \sqrt{\overline{\eta}_a}\|A\mathbf{u}\| \\
&= \|\mathbf{e}\|_A \sqrt{\overline{\eta}_a}\|\mathbf{e}\|.
\end{aligned}$$

That is,

$$\|\mathbf{e}\|^2 \le \overline{\eta}_a\ \mathbf{e}^T A\mathbf{e} = \overline{\eta}_a\ \mathbf{e}^T A\mathbf{v} \le \overline{\eta}_a\ \|A\mathbf{v}\|\,\|\mathbf{e}\|.$$

This implies the required boundedness estimate (B) of the projection π_A,

$$\|(I-\pi_A)\mathbf{v}\| \le \overline{\eta}_a\ \|A\mathbf{v}\| = \frac{\eta_a}{\|A\|}\ \|A\mathbf{v}\|. \qquad \square$$

At the end, we prove an estimate of the form (A).

Lemma 5.21. *Assumption (B) implies (A) with $\eta_s = \eta_b\ \|\widetilde{M}\|/\|A\|$.*

Proof. We have, with $\widetilde{\mathbf{v}} = E\mathbf{v}$, $E = I - M^{-1}A$,

$$\widetilde{\mathbf{v}}^T A(I-\pi_A)\widetilde{\mathbf{v}} \le \|A\widetilde{\mathbf{v}}\|\|(I-\pi_A)\widetilde{\mathbf{v}}\| \le \frac{\eta_b}{\|A\|}\ \|A\widetilde{\mathbf{v}}\|^2.$$

Also, recalling that $A^{1/2}\widetilde{M}^{-1}A^{1/2} = I - \overline{E}\overline{E}^T$ ((5.20))

$$\begin{aligned}
\|A\widetilde{\mathbf{v}}\|^2 &\le \|\widetilde{M}\|\ \mathbf{v}^T E^T A\widetilde{M}^{-1}AE\mathbf{v} \\
&= \|\widetilde{M}\|\ (A^{1/2}\mathbf{v})^T A^{-1/2}E^T A\widetilde{M}^{-1}AEA^{-(1/2)}(A^{1/2}\mathbf{v}) \\
&= \|\widetilde{M}\|\ (A^{1/2}\mathbf{v})^T\overline{E}^T\left(I - \overline{E}\overline{E}^T\right)\overline{E}(A^{1/2}\mathbf{v}) \\
&= \|\widetilde{M}\|\ (A^{1/2}\mathbf{v})^T\left(\overline{E}^T\overline{E} - \left(\overline{E}^T\overline{E}\right)^2\right)(A^{1/2}\mathbf{v}) \\
&\le \|\widetilde{M}\|\ (A^{1/2}\mathbf{v})^T\left(I - \overline{E}^T\overline{E}\right)(A^{1/2}\mathbf{v}) \\
&= \|\widetilde{M}\|\ \mathbf{v}^T\left(A - A(I - M^{-T}A)(I - M^{-1}A)\right)\mathbf{v}.
\end{aligned}$$

We used the elementary inequality $t - t^2 \leq 1 - t$ for the symmetric matrix $\overline{E}^T\overline{E}$ (because its eigenvalues are between zero and one). Thus, we proved (A) with $\eta_s = (\|\widetilde{M}\|/\|A\|)\ \eta_b$. □

More smoothing steps

Here, we consider a smoother M_k that can be a combined one; that is, M_k is implicitly defined from $m \geq 1$ steps of a given (not necessarily symmetric) smoother $M_k^{(0)}$, as follows.

$$I - M_k^{-1}A_k = \begin{cases} \left(I - \overline{M}_k^{(0)^{-1}} A_k\right)^{m_0}, & m = 2m_0, \\ \left(I - M_k^{(0)^{-1}} A_k\right)\left(I - \overline{M}_k^{(0)^{-1}} A_k\right)^{m_0 - 1}, & m = 2m_0 - 1. \end{cases} \tag{5.27}$$

Recall that

$$\overline{M}_k^{(0)} = M_k^{(0)}\left(M_k^{(0)^T} + M_k^{(0)} - A_k\right)^{-1} M_k^{(0)^T}$$

and

$$\widetilde{M}_k^{(0)} = M_k^{(0)^T}\left(M_k^{(0)^T} + M_k^{(0)} - A_k\right)^{-1} M_k^{(0)}.$$

Also, in the above formula for $I - M_k^{-1}A_k$, we have $I - \overline{M}_k^{(0)^{-1}} A_k = (I - M_k^{(0)^{-T}} A_k)(I - M_k^{(0)^{-1}} A_k)$; that is, we use both $M_k^{(0)}$ and $M_k^{(0)^T}$ in an alternating fashion. We notice that (in both cases),

$$A_k(I - M_k^{-T} A_k)(I - M_k^{-1}A_k) = A_k(I - \overline{M}_k^{(0)^{-1}} A_k)^m.$$

That is, the resulting symmetrized smoother $\overline{M}_k = M_k(M_k + M_k^T - A_k)^{-1}M_k^T$ satisfies the identity

$$A_k\left(I - \overline{M}_k^{-1} A_k\right) = A_k\left(I - \overline{M}_k^{(0)^{-1}} A_k\right)^m.$$

We omit in what follows the level index k.

Introduce the smoothing iteration matrices

$$\overline{E}^{(0)} = I - A^{1/2} M^{(0)^{-1}} A^{1/2} \quad \text{and} \quad E = I - M^{-1}A.$$

We then have,

$$\overline{E} \equiv A^{1/2} E A^{-(1/2)} = \begin{cases} \left(\overline{E}^{(0)^T} \overline{E}^{(0)}\right)^{m_0}, & m = 2m_0, \\ \overline{E}^{(0)} \left(\overline{E}^{(0)^T} \overline{E}^{(0)}\right)^{m_0 - 1}, & m = 2m_0 - 1. \end{cases}$$

For a given $m \geq 1$, the resulting MG preconditioner is referred to as a $V(m,\ m)$-cycle one.

We observe that in both cases (m odd or even) $\overline{E}^T\overline{E} = (\overline{E}^{(0)^T}\overline{E}^{(0)})^m$. Also, because $I - \widetilde{M}^{(0)^{-1}} A = (I - M^{(0)^{-1}} A)(I - M^{(0)^{-T}} A)$, we get

$$A^{1/2}\left(\widetilde{M}^{(0)^{-1}} A\right) A^{-(1/2)} = I - \overline{E}^{(0)} \overline{E}^{(0)^T}.$$

Similarly,

$$A^{1/2}(\overline{M}^{(0)^{-1}} A)A^{-(1/2)} = I - \overline{E}^{(0)^T}\overline{E}^{(0)}.$$

With this combined smoother, the following strong smoothing property can be proved.

First, consider the case m-odd; that is, $m = 2m_0 - 1$. We have, with $\widetilde{\mathbf{v}} = E\mathbf{v}$, recalling that $A^{1/2}\widetilde{M}^{(0)^{-1}}A^{1/2} = I - \overline{E}^{(0)}\overline{E}^{(0)^T}$,

$$\begin{aligned}
\|A\widetilde{\mathbf{v}}\|^2 &\le \|\widetilde{M}^{(0)}\|(AE\mathbf{v})^T\widetilde{M}^{(0)^{-1}}(AE\mathbf{v}) \\
&= \|\widetilde{M}^{(0)}\|(A^{1/2}\mathbf{v})^T A^{-(1/2)}E^T A\widetilde{M}^{(0)^{-1}}AEA^{-(1/2)}(A^{1/2}\mathbf{v}) \\
&= \|\widetilde{M}^{(0)}\|\,(A^{1/2}\mathbf{v})^T\big(\overline{E}^T\big(I - \overline{E}^{(0)}\overline{E}^{(0)^T}\big)\big)\overline{E}(A^{1/2}\mathbf{v}) \\
&= \|\widetilde{M}^{(0)}\|\,(A^{1/2}\mathbf{v})^T\big(\big(\overline{E}^{(0)^T}\overline{E}^{(0)}\big)^m - \big(\overline{E}^{(0)^T}\overline{E}^{(0)}\big)^{m+1}\big)(A^{1/2}\mathbf{v}) \\
&\le \frac{1}{m}\,\|\widetilde{M}^{(0)}\|\,(A^{1/2}\mathbf{v})^T\big(I - \big(\overline{E}^{(0)^T}\overline{E}^{(0)}\big)^m\big)(A^{1/2}\mathbf{v}) \\
&= \frac{1}{m}\,\|\widetilde{M}^{(0)}\|\,(A^{1/2}\mathbf{v})^T\big(I - \overline{E}^T\overline{E}\big)(A^{1/2}\mathbf{v}) \\
&= \frac{1}{m}\,\|\widetilde{M}^{(0)}\|\,\mathbf{v}^T(A - A(I - M^{-T}A)(I - M^{-1}A))\mathbf{v}.
\end{aligned}$$

We used above the elementary inequality (as in [Br93] or [BS96]) for any $t \in [0, 1]$, $t^m \le t^k, 0 \le k \le m-1$, which implies

$$(1-t)t^m \le (1-t)\,\frac{1}{m}\sum_{k=0}^{m-1} t^k = \frac{1}{m}(1 - t^m),$$

noticing that the spectrum of the symmetric matrix $\overline{E}^{(0)^T}\overline{E}^{(0)}$ is contained in $[0, 1]$. The case m-even is handled analogously. We start then with the inequality

$$\|A\widetilde{\mathbf{v}}\|^2 \le \|\overline{M}^{(0)}\|\,(AE\mathbf{v})^T\overline{M}^{(0)^{-1}}(AE\mathbf{v}).$$

Using the fact that $A^{1/2}(\overline{M}^{(0)^{-1}}A)A^{-(1/2)} = I - \overline{E}^{(0)^T}\overline{E}^{(0)}$, we end up with the same type of inequality as before;

$$\begin{aligned}
\|A\widetilde{\mathbf{v}}\|^2 &\le \|\overline{M}^{(0)}\|\,(A^{1/2}\mathbf{v})^T\big(\overline{E}^T\big(I - \overline{E}^{(0)^T}\overline{E}^{(0)}\big)\big)\overline{E}(A^{1/2}\mathbf{v}) \\
&= \|\overline{M}^{(0)}\|\,(A^{1/2}\mathbf{v})^T\big(\big(\overline{E}^{(0)^T}\overline{E}^{(0)}\big)^m - \big(\overline{E}^{(0)^T}\overline{E}^{(0)}\big)^{m+1}\big)(A^{1/2}\mathbf{v}) \\
&\le \frac{1}{m}\,\|\overline{M}^{(0)}\|\,(A^{1/2}\mathbf{v})^T\big(I - \big(\overline{E}^{(0)^T}\overline{E}^{(0)}\big)^m\big)(A^{1/2}\mathbf{v}) \\
&= \frac{1}{m}\,\|\overline{M}^{(0)}\|\,(A^{1/2}\mathbf{v})^T(I - \overline{E}^T\overline{E})(A^{1/2}\mathbf{v}) \\
&= \frac{1}{m}\,\|\overline{M}^{(0)}\|\,\mathbf{v}^T(A - A(I - M^{-T}A)(I - M^{-1}A))\mathbf{v}.
\end{aligned}$$

Thus, we proved a smoothing property (A) (assuming (B)) with

$$\eta_s = \eta_b\,\frac{\|\widetilde{M}^{(0)}\|}{\|A\|}\,\frac{1}{m} \quad \text{or} \quad \eta_s = \eta_b\,\frac{\|\overline{M}^{(0)}\|}{\|A\|}\,\frac{1}{m}.$$

This shows uniform convergence of the resulting V–cycle MG with a rate that improves with increasing m, the number of smoothing steps. That is, we have the following result, originating in Braess and Hackbusch [BH83].

Theorem 5.22. *Under the assumption (B), which holds if the strong approximation property (C) holds, using m combined pre- and postsmoothing steps as defined in* (5.27)*, we have the following uniform estimate for the resulting V-cycle preconditioner,*

$$\mathbf{v}^T A\mathbf{v} \le \mathbf{v}^T B\mathbf{v} \le \begin{cases} \left(1+\eta_b \dfrac{\|\widetilde{M}^{(0)}\|}{\|A\|}\dfrac{1}{m}\right)\mathbf{v}^T A\mathbf{v}, & m = 2m_0 - 1,\\ \left(1+\eta_b \dfrac{\|\overline{M}^{(0)}\|}{\|A\|}\dfrac{1}{m}\right)\mathbf{v}^T A\mathbf{v}, & m = 2m_0. \end{cases}$$

In particular, the following corollary holds for the window spectral AMG method, because a strong approximation property (C) holds for it (proved in Theorem 6.19).

Corollary 5.23. *The two-level window-based spectral AMG method from Section 6.11 improves its convergence factor ϱ_{TG} linearly with increasing m, the number of smoothing steps; that is, we have*

$$\varrho_{TG} \le \frac{c_0}{c_0+m}, \qquad c_0 = \frac{\delta}{\eta}\frac{\|\widetilde{M}^{(0)}\|}{\|A\|} \le \frac{\delta}{\eta}\frac{1}{\omega(2-\omega)}.$$

The constants δ and η are defined in Theorem 6.19. The last inequality for c_0 holds if the Richardson smoother $M^{(0)} = \|A\|/\omega\ I$, $\omega \in (0,2)$ is used.

5.5.1 Variable V-cycle

In this section, we present a first attempt to stabilize the V-cycle by increasing the number of smoothing steps at coarse levels. The latter is referred to as a variable V-cycle originating in Bramble and Pasciak [BP87]. We first analyze the complexity of a V-cycle with a variable number of smoothing steps.

Let n_k be the number of degrees of freedom at level k and the smoothing and interpolation procedures take $\mathcal{O}(n_k)$ operations. Assume also, a geometric ratio of coarsening; that is, $n_{k+\ell} \simeq q^\ell\ n_k$, for some $q \in (0,1)$. Then the asymptotic work w_0 (at the finest-level $k = 0$) of the resulting variable V-cycle preconditioner with $m_k \ge 1$ (level-dependent) number of smoothing steps, can be readily estimated as

$$w_0 \simeq \sum_{k\ge 0} m_k\, n_k.$$

Assume now that for an $\alpha \in (0,1]$ and a given $\sigma > 0$, m_k, for a fixed $m_0 \ge 0$, grows as,

$$(m_k + 1 - m_0)^{-\alpha} \simeq (1+k)^{-(1+\sigma)}; \tag{5.28}$$

that is,

$$m_k + 1 \simeq m_0 + (1+k)^{(1+\sigma)/\alpha}. \tag{5.29}$$

Then, the work estimate takes the form

$$w_0 \simeq n_0 \sum_{k \geq 0} q^k (k+1)^{(1+\sigma)/\alpha} \simeq n_0.$$

That is, at the finest-level $k = 0$, the total work w_0 is of optimal order.

Theorem 5.24. *Let K_k be the relative condition number of the V-cycle preconditioner B_k with respect to A_k at level k, and let $K_{TG,\,k}$ be the one of the respective two-grid preconditioners at level k (i.e., with exact coarse solution at level $k+1$). Both exploit the same smoother at level k with the same number of smoothing steps $m_k \geq 1$. Assume that at coarser levels, the two-grid methods get more accurate, so that, for a constant $\eta > 0$ and a fixed $\sigma > 0$,*

$$K_{TG,\,k} \leq 1 + \frac{\eta}{(1+k)^{1+\sigma}}, \quad \text{at all levels } k \geq 0,$$

The latter can be guaranteed (as shown later on, depending on certain approximation properties of the coarse spaces, cf., Theorem 5.27), if we perform $m_k \geq 1$ (i.e., level-dependent) number of smoothing steps. More specifically, we assume that the following asymptotic TG convergence behavior holds,

$$K_{TG,\,k} \simeq 1 + \eta\, m_k^{-\alpha}, \tag{5.30}$$

for a fixed $\alpha \in (0, 1]$. Then, if we select m_k, for a fixed $m_0 \geq 0$, as in (5.28) or (5.29), then the resulting variable V-cycle is both of optimal complexity and its spectral relative condition number is bounded independently of the number of levels. More specifically, the following bound holds,

$$K_0 \simeq K_{TG,\,0} = 1 + \eta\, m_0^{-\alpha}.$$

The latter can be made sufficiently close to one by choosing m_0 sufficiently large.

Proof. For any two levels, a fine-level k and a coarse-level $\ell \geq k$, we have (see, e.g., Corollary 5.11),

$$K_k \leq K_{TG,\,k} K_{TG,\,k+1} \cdots K_{TG,\,\ell-1}\, K_\ell.$$

Then from the assumption (5.30) and the choice of m_k in (5.28)–(5.29), we immediately get

$$K_0 \leq K_{TG,\,0} \prod_{k \geq 1} \left(1 + \frac{\eta}{(1+k)^{1+\sigma}}\right) \simeq K_{TG,\,0} = 1 + \eta\, m_0^{-\alpha}.$$

The choice of m_k as in (5.29) as already shown guarantees that the variable V-cycle is of optimal complexity. □

Less regular problems

In this section, we consider a little more sophisticated case of less regular problems; namely, we assume the following.

(D) "Weaker approximation property":
For an $\alpha \in (0, 1]$, there is a constant η_a such that the following weaker approximation property holds,

$$\|(I - \pi_A)\mathbf{v}\|_A^2 \leq (\mathbf{v} - P\mathbf{v}_c)^T A(\mathbf{v} - P\mathbf{v}_c) \leq \frac{\eta_a}{\|A\|^\alpha} \|A^{(1+\alpha)/2}\mathbf{v}\|^2.$$

Here, for a given $\mathbf{v}$, $P\mathbf{v}_c$ is some coarse interpolant that satisfies the above approximation property.

We show next that the following boundedness estimate holds.

Corollary 5.25. *Estimate (D) implies the following corollary,*

$$\|A^{(1-\alpha)/2}(I - \pi_A)\mathbf{v}\| \leq \frac{\eta_a}{\|A\|^\alpha} \|A^{(1+\alpha)/2}\mathbf{v}\|.$$

Proof. Let $\mathbf{e} = (I - \pi_A)\mathbf{v}$ and consider the problem $A\mathbf{u} = A^{1-\alpha}\mathbf{e}$. With $\eta = \eta_a/\|A\|^\alpha$, noticing the $\mathbf{e}$ is A-orthogonal to the coarse space Range (P), we have

$$\begin{aligned}
\mathbf{e}^T A^{1-\alpha}\mathbf{e} &= \mathbf{e}^T A\mathbf{u} \\
&= \mathbf{e}^T A(\mathbf{u} - P\mathbf{u}_c) \\
&\leq \|\mathbf{e}\|_A \|\mathbf{u} - P\mathbf{u}_c\|_A \\
&\leq \|\mathbf{e}\|_A \sqrt{\eta}\, \|A^{(1+\alpha)/2}\mathbf{u}\| \\
&= \sqrt{\eta}\, \|\mathbf{e}\|_A \|A^{(1-\alpha)/2}\mathbf{e}\|.
\end{aligned}$$

That is, we have,

$$\mathbf{e}^T A^{1-\alpha}\mathbf{e} \leq \eta\, \mathbf{e}^T A\mathbf{e} = \eta\, \mathbf{e}^T A\mathbf{v} \leq \eta\, \|A^{(1-\alpha)/2}\mathbf{e}\| \|A^{(1+\alpha)/2}\mathbf{v}\|,$$

which implies the desired result. □

Lemma 5.26. *The less strong approximation property (D) implies the following weaker version of (A); namely,*

(A_w)

$$\begin{aligned}
&\mathbf{v}^T A(I - M^{-T}A)(I - \pi_A)(I - M^{-1}A)\mathbf{v} \\
&\quad \leq \eta_s\big[\mathbf{v}^T A\mathbf{v} - \alpha\, \mathbf{v}^T A(I - M^{-T}A)(I - M^{-1}A)\mathbf{v}\big].
\end{aligned}$$

where $\eta_s = \eta_a\, \|\widetilde{M}\|^\alpha/\|A\|^\alpha$.

In the case of a combined smoother as defined in (5.27)*, we have the following improving with $m \mapsto \infty$ upper bound $\eta_s = \eta_a\, (\|\widetilde{M}\|^\alpha/\|A\|^\alpha)\, (1/m^\alpha)$ for $m = 2m_0 - 1$ and $\eta_s = \eta_a\, (\|\overline{M}^{(0)}\|^\alpha/\|A\|^\alpha)(1/m^\alpha)$ for $m = 2m_0$.*

Proof. Consider the series $(1-t)^\alpha = 1 - \sum_{k\ge 1} \alpha_k t^k$, noticing that $\alpha_k > 0$, and $\alpha_1 = \alpha$. Then the following estimate for any $t \in [0,1]$ is immediate

$$t\left(1 - \sum_{k\ge 1} \alpha_k t^k\right) \le 1 - \sum_{k\ge 1} \alpha_k t^k \le 1 - \alpha_1 t = 1 - \alpha t. \tag{5.31}$$

Use now estimate (D). It gives, for $\widetilde{\mathbf{v}} = E\mathbf{v}$, $E = I - M^{-1}A$,

$$\mathbf{v}^T A E_{TG} \mathbf{v} = (E\mathbf{v})^T A (I - \pi_A)(E\mathbf{v}) \le \frac{\eta_a}{\|A\|^\alpha} \, \|A^{(1+\alpha)/2} \widetilde{\mathbf{v}}\|^2.$$

We apply next Lemma G.3, that is, the fact that for any two symmetric positive definite matrices U and V, the inequality $\mathbf{v}^T U \mathbf{v} \le \mathbf{v}^T V \mathbf{v}$ implies $\mathbf{v}^T U^\alpha \mathbf{v} \le \mathbf{v}^T V^\alpha \mathbf{v}$, for any $\alpha \in [0,1]$. Choosing $U = A$ and $V = \|\widetilde{M}\| A^{1/2} \widetilde{M}^{-1} A^{1/2}$, we then have

$$\mathbf{w}^T A^\alpha \mathbf{w} \le \|\widetilde{M}\|^\alpha \; \mathbf{w}^T \left(A^{1/2} \widetilde{M}^{-1} A^{1/2}\right)^\alpha \mathbf{w} = \|\widetilde{M}\|^\alpha \; \mathbf{w}^T \left(I - \overline{E}\,\overline{E}^T\right)^\alpha \mathbf{w}. \tag{5.32}$$

We let now $\mathbf{w} = A^{1/2} \widetilde{\mathbf{v}} = A^{1/2} E \mathbf{v} = \overline{E} A^{1/2} \mathbf{v}$. The desired result then follows using inequality (5.31) for the eigenvalues of the matrix $\overline{E}^T (I - \overline{E}\,\overline{E}^T)^\alpha \overline{E} = \overline{E}^T \overline{E} - \sum_{k\ge 1} \alpha_k (\overline{E}^T \overline{E})^{k+1}$, which have the form $t(1 - \sum_{k \ge 1} \alpha_k t^k)$ for $t \in [0,1]$ being an eigenvalue of the symmetric positive semidefinite matrix $\overline{E}^T \overline{E}$.

Next, we analyze the case of the combined smoother defined in (5.27). The term $\|A^{(1+\alpha)/2} \widetilde{\mathbf{v}}\|^2 = \|A^{\alpha/2} \overline{E} (A^{1/2}\mathbf{v})\|^2$ is estimated below. Consider the case $m = 2m_0 - 1$ (the case $m = 2m_0$ is analyzed analogously). Use as before (see (5.32)) the inequality

$$\|A^{(1+\alpha)/2} \widetilde{\mathbf{v}}\|^2 \le \|\widetilde{M}^{(0)}\|^\alpha (A^{1/2}\mathbf{v})^T \overline{E}^T \left(I - \overline{E}^{(0)} \overline{E}^{(0)^T}\right)^\alpha \overline{E} A^{1/2} \mathbf{v}.$$

We have $\overline{E} = \overline{E}^{(0)} (\overline{E}^{(0)^T} \overline{E}^{(0)})^{m_0 - 1}$. Therefore, because $\|\overline{E}^{(0)}\| = \|\overline{E}^{(0)^T}\| < 1$, we also have

$$\begin{aligned}
&\overline{E}^T \left(I - \overline{E}^{(0)} \overline{E}^{(0)^T}\right)^\alpha \overline{E} \\
&= \left(\overline{E}^{(0)^T} \overline{E}^{(0)}\right)^{m_0 - 1} \overline{E}^{(0)^T} \left(I - \overline{E}^{(0)} \overline{E}^{(0)^T}\right)^\alpha \overline{E}^{(0)} \left(\overline{E}^{(0)^T} \overline{E}^{(0)}\right)^{m_0 - 1} \\
&= \left(\overline{E}^{(0)^T} \overline{E}^{(0)}\right)^{m_0 - 1} \overline{E}^{(0)^T} \left(I - \sum_{k \ge 1} \alpha_k \left(\overline{E}^{(0)} \overline{E}^{(0)^T}\right)^k\right) \overline{E}^{(0)} \left(\overline{E}^{(0)^T} \overline{E}^{(0)}\right)^{m_0 - 1} \\
&= \left(\overline{E}^{(0)^T} \overline{E}^{(0)}\right)^{m_0} \left(I - \overline{E}^{(0)^T} \overline{E}^{(0)}\right)^\alpha \left(\overline{E}^{(0)^T} \overline{E}^{(0)}\right)^{m_0 - 1} \\
&= \left(\overline{E}^{(0)^T} \overline{E}^{(0)}\right)^m \left(I - \overline{E}^{(0)^T} \overline{E}^{(0)}\right)^\alpha.
\end{aligned} \tag{5.33}$$

Next, use the elementary inequalities for $t \in [0,1]$ (and $\alpha \in (0,1]$)

$$t^{m/\alpha}(1-t) \le t^m (1-t) \le \frac{1}{m} \sum_{k=0}^{m-1} t^k (1-t) \le \frac{1}{m} (1 - t^m).$$

It implies,

$$t^m (1-t)^\alpha \le \frac{1}{m^\alpha} (1-t^m)^\alpha \le \frac{1}{m^\alpha} (1-\alpha t^m).$$

Applying the last inequality for the symmetric positive semidefinite matrix $\overline{E}^{(0)^T} \overline{E}^{(0)}$ in (5.33) proves estimate (A_w) with $\eta_s = \eta_a \ (\|\widetilde{M}^{(0)}\|^\alpha / \|A\|^\alpha) \ (1/m^\alpha)$. In the case $m = 2m_0$, we have similar estimate with $\eta_s = \eta_a \ (\|\overline{M}^{(0)}\|^\alpha / \|A\|^\alpha) \ (1/m^\alpha)$. □

Next, we show that (A_w) implies two-grid convergence if η_s is sufficiently small, such that

$$\eta_s < \frac{1}{1-\alpha}. \tag{5.34}$$

Recalling estimate (5.21) for $\eta_s > 0$,

$$\begin{aligned}
&\frac{\mathbf{v}^T B \mathbf{v}}{\mathbf{v}^T A \mathbf{v}} \\
&\quad \le \sup_{\mathbf{v}} \frac{\mathbf{v}^T A^{-(1/2)} B A^{-(1/2)} \mathbf{v}}{\mathbf{v}^T \mathbf{v}} \\
&\quad \le \sup_{\mathbf{v}} \frac{\mathbf{v}^T \left(I - \overline{E}^T \overline{E} + \frac{1}{1+\eta_s} \overline{E}^T \overline{\pi}_A \overline{E} \right)^{-1} \mathbf{v}}{\mathbf{v}^T \mathbf{v}} \\
&\quad = \sup_{\mathbf{v}} \frac{\mathbf{v}^T \mathbf{v}}{\mathbf{v}^T \left(I - \overline{E}^T \overline{E} + \frac{1}{1+\eta_s} \overline{E}^T \overline{\pi}_A \overline{E} \right) \mathbf{v}} \\
&\quad = (1+\eta_s) \sup_{\mathbf{v}} \frac{\mathbf{v}^T \mathbf{v}}{\mathbf{v}^T [\eta_s (I - \overline{E}^T \overline{E}) + I - \overline{E}^T (I - \overline{\pi}_A) \overline{E}] \mathbf{v}} \\
&\quad = (1+\eta_s) \sup_{\mathbf{v}} \frac{\mathbf{v}^T \mathbf{v}}{\mathbf{v}^T [\eta_s (I - \alpha \, \overline{E}^T \overline{E}) - \eta_s (1-\alpha) \, \overline{E}^T \overline{E} + I - \overline{E}^T (I - \overline{\pi}_A) \overline{E}] \mathbf{v}} \\
&\quad \le (1+\eta_s) \sup_{\mathbf{v}} \frac{\mathbf{v}^T \mathbf{v}}{\mathbf{v}^T [(1 - (1-\alpha)\eta_s)) I + \eta_s (I - \alpha \, \overline{E}^T \overline{E}) - \overline{E}^T (I - \overline{\pi}_A) \overline{E}] \mathbf{v}}.
\end{aligned}$$

Using the fact that $\eta_s (I - \alpha \ \overline{E}^T \overline{E}) - \overline{E}^T (I - \overline{\pi}_A) \overline{E}$ is symmetric positive semidefinite (due to (A_w)) the following TG convergence result follows (with η_s as in (5.34))

$$\mathbf{v}^T B_{TG} \mathbf{v} \le \frac{(1+\eta_s)}{1-(1-\alpha)\eta_s} \ \mathbf{v}^T A \mathbf{v}.$$

To ensure inequality (5.34), we use the combined m–step smoother as defined in (5.27). We showed in Lemma 5.26 that $\eta_s = \eta_a \ (\|\widetilde{M}\|^\alpha / \|A|^\alpha)(1/m^\alpha)$ for $m = 2m_0 - 1$ and $\eta_s = \eta_a \ (\|\overline{M}\|^\alpha / \|A\|^\alpha)(1/m^\alpha)$ for $m = 2m_0$. Therefore, in conclusion we proved that assumption (A_w) implies uniform convergence of the resulting TG method with a rate that improves when increasing the number of smoothing steps m.

Theorem 5.27. *Under the assumption (A_w), using combined $m+1$ pre–and post-smoothing steps as defined in (5.27), we have the following uniform estimate for the resulting TG preconditioner, for $m = 2m_0 - 1$,*

$$\mathbf{v}^T A\mathbf{v} \le \mathbf{v}^T B_{TG}\mathbf{v} \le \frac{1+\eta_a \frac{\|\widetilde{M}^{(0)}\|^\alpha}{\|A\|^\alpha}\frac{1}{m^\alpha}}{1-(1-\alpha)\eta_a \frac{\|\widetilde{M}^{(0)}\|^\alpha}{\|A\|^\alpha}\frac{1}{m^\alpha}}\, \mathbf{v}^T A\mathbf{v}.$$

If $m = 2m_0$, the same estimate holds with $\|\widetilde{M}^{(0)}\|$ replaced by $\|\overline{M}^{(0)}\|$. It is clear that when $m \mapsto \infty$ the upper bound tends to unity.

5.6 MG with more recursive cycles; W-cycle

5.6.1 Definition of a ν-fold MG-cycle; complexity

We can generalize the definition of the MG preconditioner by replacing the Schur complement B_{k+1} of $\overline{B}_k$ with a more accurate approximation to A_{k+1} thus ending up with a multilevel preconditioner that is much closer to the respective two-grid one (at a given level k). A simple choice is to use, for a given integer $\nu \ge 1$, the following polynomial approximation to A_{k+1}^{-1},

$$B_{k+1}^{(\nu)^{-1}} = (I-(I-B_{k+1}^{-1}A_{k+1})^\nu)A_{k+1}^{-1} = \sum_{l=0}^{\nu-1}(I-B_{k+1}^{-1}A_{k+1})^l\, B_{k+1}^{-1}. \tag{5.35}$$

It is clear that with $\nu \mapsto \infty$, we get $B_{k+1}^{(\nu)} \mapsto A_{k+1}$.

The modified $\overline{B}_k$ reads

$$\overline{B}_k = \begin{bmatrix} I & 0 \\ P_k^T A_k M_k^{-1} & I \end{bmatrix} \begin{bmatrix} M_k\left(M_k + M_k^T - A_k\right)^{-1} M_k^T & 0 \\ 0 & B_{k+1}^{(\nu)} \end{bmatrix} \begin{bmatrix} I & M_k^{-T}A_k P_k \\ 0 & I \end{bmatrix},$$

and then B_k is defined as before,

$$B_k^{-1} = [I,\ P_k]\,\overline{B}_k^{-1}\,[I,\ P_k]^T.$$

The case of $\nu = 1$ gives the original, called a V-cycle MG preconditioner, whereas the multilevel preconditioner corresponding to $\nu = 2$ is referred to as a W-cycle MG.

It is clear that we cannot choose ν too large due to the increasing cost to implement the resulting ν-fold multilevel preconditioner. The latter cost can be estimated as follows.

Let w_k stand for the cost in terms of number of flops to implement one action of B_k^{-1}. The following assumptions are met for uniformly refined triangulations $\mathcal{T}_k$, the resulting matrices A_k, and for reasonably chosen smoothers M_k (such as Gauss–Seidel or scaled Jacobi).

- Let $n_k \sim \mu n_{k+1}$ be the number of nodes $\mathcal{N}_k$ (or degrees of freedom) corresponding to the triangulation $\mathcal{T}_k$. In the case of 2D uniformly refined triangular

meshes, we can let $\mu = 4$. In three dimensions ($d = 3$) a typical behavior is $\mu = 8$.
- Let one action of the smoother M_k^{-1} (and M_k^{-T}) take $\mathcal{O}(n_k)$ flops.
- Let the restriction P_k^T and interpolation P_k require $\mathcal{O}(n_k)$ flops.
- Finally, one action of A_k is $\mathcal{O}(n_k)$ flops.

From the formula

$$B_k^{-1} = \overline{M}_k^{-1} + \left(I - M_k^{-T} A_k\right) P_k B_{k+1}^{(\nu)^{-1}} P_k^T \left(I - A_k M_k^{-1}\right)$$

and the definition of $B_{k+1}^{(\nu)^{-1}}$ (which is based on ν actions of B_{k+1}^{-1}, ν residual computations on the basis of A_{k+1}, and respective ν vector updates at a cost n_{k+1} each), we easily get the recursion

$$w_k \leq C_\nu n_k + \nu\, w_{k+1}.$$

The latter implies

$$w_k \leq C_\nu n_k \sum_{j=k}^{\ell} \left(\frac{\nu}{\mu}\right)^{j-k}.$$

Thus if $\nu < \mu$, (i.e., $\nu < 4$ for 2D problems, and $\nu < 8$ for 3D problems) the resulting ν-fold preconditioner can be implemented with optimal cost; that is, $w_k = \mathcal{O}(n_k)$.

5.6.2 AMLI-cycle multigrid

Other types of cycles, in general varying with the level index, that is, $\nu = \nu_k$ (see the definition in the preceding Section 5.6.1), are also possible. Also, if we are willing to estimate the spectrum of the preconditioner at a given level k, we can then use the best (appropriately scaled and shifted Chebyshev) polynomials instead of the simpler one $(I - t)^\nu$ (used in (5.35)). The resulting technique, leading to a multilevel cycle, sometimes referred to as the algebraic multilevel iteration (or AMLI), described in the present section, was first applied to the hierarchical basis MG method; see [V92b], and for a special case, earlier in [AV89] and [AV90]. The word "algebraic" in AMLI stands for the fact that certain inner polynomial iterations are used in the definition of the multilevel cycle. It is not be confused with the "algebraic" in AMG, which stands for the way of constructing the coarse spaces (or interpolation matrices).

The main assumption that leads to AMLI-cycle MG methods of optimal condition number is that all V-cycles based on exact solutions at their coarse-level ℓ up to finer-level $k < \ell$ with bounded-level difference $\ell - k \leq k_0$, have bounded condition number $K_{MG}^{\ell \mapsto k}$. The latter may in general grow with $k_0 \geq 1$ but for a fixed k_0 is assumed bounded. Such estimates are feasible for finite element discretizations of second-order elliptic PDEs, as well as, for some less standard forms such as $H(\mathrm{div})$ (cf., [CGP]), without assuming any regularity of the underlined PDE. Moreover, the constants involved in the estimates can typically be estimated locally (on an element-by-element basis) and can be shown to be independent of possible jumps in the PDE coefficients (provided that those occur only across element boundaries on the coarsest mesh).

In geometric multigrid, applied to second-order finite element elliptic problems, denoting $H = h_\ell$ and $h = h_k$ the mesh-sizes at respective levels, and using the linear interpolation inherited by the f.e.m. between two consecutive levels, and for example, using a Gauss–Seidel smoother, the following asymptotic estimate holds,

$$K_{MG}^{\ell \mapsto k} \simeq \begin{cases} \log^2\left(\dfrac{H}{h}\right), & d = 2, \\ \dfrac{H}{h}, & d = 3. \end{cases} \tag{5.36}$$

Details about the last two estimates ($d = 2$ and $d = 3$) are found in Appendix B. We note that typically (for uniform refinement) $H/h \simeq 2^{\ell - k}$.

We are now in position to define the AMLI-cycle.

Definition 5.28 (AMLI-cycle MG). *For a given $\nu \geq 1$ and a fixed-level difference $k_0 \geq 1$, let $p_\nu = p_\nu(t)$ be a given polynomial of degree ν, which is nonnegative in $[0, 1]$ and scaled so that $p_\nu(0) = 1$. We also assume the standard components of a MG, that is, kth-level matrices A_k, smoothers M_k and M_k^T, and coarse-to-fine interpolation matrices P_k, such that $A_{k+1} = P_k^T A_k P_k$.*

The AMLI-cycle is defined as a ν–fold MG cycle with a variable $\nu = \nu_k$. More specifically, for a given integer $\nu \geq 1$, and another fixed integer $k_0 \geq 1$, we set $\nu_{sk_0} = \nu > 1$ for $s = 1, 2, \ldots,$ and $\nu_k = 1$ otherwise.

Let $B_\ell = A_\ell$ and assume that for $k + 1 \leq \ell$, B_{k+1} has already been defined. If $k + 1 = (s + 1)k_0$, based on B_{k+1} we let

$$\left(B_{(s+1)k_0}^{(\nu)}\right)^{-1} = \left(I - p_\nu\left(B_{(s+1)k_0}^{-1} A_{(s+1)k_0}\right)\right)\left(A_{(s+1)k_0}\right)^{-1},$$

For all other indices $k + 1$, $\nu = 1$, and hence $B_{k+1}^{(\nu)} = B_{k+1}$. Then, at the kth level, we set

$$B_k^{-1} = \overline{M}_k^{-1} + (I - M_k^{-T} A_k) P_k B_{k+1}^{(\nu)^{-1}} P_k^T (I - A_k M_k^{-1}).$$

We recall that $\overline{M}_k = M_k(M_k^T + M_k - A_k)^{-1} M_k^T$ is the symmetrized smoother.

5.6.3 Analysis of AMLI

Theorem 5.29. *With proper choice of the parameters k_0 and ν, all fixed for the AMLI-cycle, but sufficiently large in general, and for a proper choice of the polynomial $p_\nu(t)$, the condition number of $B_k^{-1} A_k$ can be uniformly bounded provided the V-cycle preconditioners with bounded-level difference $\ell - k \leq k_0$ have uniformly bounded* *condition numbers* $K_{MG}^{\ell\mapsto k}$. *More specifically, let for a fixed* k_0, $\nu > \sqrt{K_{MG}^{\ell\mapsto k}}$, *and choose* $\alpha > 0$, *such that*

$$\alpha K_{MG}^{\ell\mapsto k} + K_{MG}^{\ell\mapsto k} \frac{(1-\alpha)^\nu}{\left[\sum_{j=1}^{\nu}(1+\sqrt{\alpha})^{\nu-j}(1-\sqrt{\alpha})^{j-1}\right]^2} \leq 1. \tag{5.37}$$

Then, consider the polynomial

$$p_\nu(t) = \frac{1 + T_\nu\left(\frac{1+\alpha-2t}{1-\alpha}\right)}{1 + T_\nu\left(\frac{1+\alpha}{1-\alpha}\right)}. \tag{5.38}$$

Here, T_ν *is the Chebyshev polynomial of the first kind of degree* ν. *If* $p_\nu(t) = (1-t)^\nu$, *for* $\nu > K_{MG}^{\ell\mapsto k}$, *we can choose* $\alpha \in (0, 1)$ *such that*

$$\alpha K_{MG}^{\ell\mapsto k} + K_{MG}^{\ell\mapsto k} \frac{(1-\alpha)^\nu}{\sum_{j=1}^{\nu}(1-\alpha)^{j-1}} \leq 1. \tag{5.39}$$

The resulting AMLI-cycle preconditioner $B = B_0$, *as defined in Definition 5.28 for both choices of polynomial* p_ν, *is spectrally equivalent to* $A = A_0$ *and the following estimate holds,*

$$\mathbf{v}^T A\mathbf{v} \leq \mathbf{v}^T B\mathbf{v} \leq \frac{1}{\alpha}\,\mathbf{v}^T A\mathbf{v},$$

with the respective $\alpha \in (0, 1]$ *depending on the choice of the polynomial.*

Proof. First, it is clear that inequality (5.37) has a solution $\alpha > 0$. This is seen because for $\alpha \mapsto 0$ the left-hand side of (5.37) tends to $K_{MG}^{\ell\mapsto k}/\nu^2$, which is less than one due to the choice of ν. It is also clear that $\alpha K_{MG}^{\ell\mapsto k} < 1$; that is, $\alpha < 1/K_{MG}^{\ell\mapsto k} \leq 1$.

Choose $s \geq 0$ and assume that for some $\delta_{s+1} \geq 0$ the eigenvalues of $A_{(s+1)k_0}^{-1} B_{(s+1)k_0}$ are in the interval $[1,\ 1+\delta_{s+1}]$. Next, we estimate the spectrum of $A_{sk_0}^{-1} B_{sk_0}$. Let $\ell = (s+1)k_0$ and $k = sk_0$. Assume, by induction that

$$\alpha \leq \frac{1}{1+\delta_{s+1}}.$$

The latter holds for the V-cycle $K_{MG}^{\ell\mapsto k}$ starting from the coarsest level ℓ because $\alpha K_{MG}^{\ell\mapsto k} < 1$.

The eigenvalues of $A_k^{-1} B_k$ are contained in an interval $[1,\ 1+\delta_s]$ which we want to estimate. First, we have that the eigenvalues of $A_\ell^{-1} B_\ell^{(\nu)}$ are contained in the interval $[1,\ 1+\widetilde{\delta}_{s+1}^{(\nu)}]$, where

$$\begin{aligned}
\widetilde{\delta}_{s+1}^{(\nu)} &= \sup\left\{\frac{1}{1-p_\nu(t)} - 1, \quad t \in \left[\frac{1}{1+\delta_{s+1}},\ 1\right]\right\} \\
&\leq \sup\left\{\frac{p_\nu(t)}{1-p_\nu(t)}, \quad t \in [\alpha,\ 1]\right\},
\end{aligned}$$

where we have used $\left[1/(1+\delta_{s+1}),\ 1\right] \subset [\alpha,\ 1]$.

Because

$$\sup_{t\in[\alpha,\,1]} \left| T_\nu\left(\frac{1+\alpha-2t}{1-\alpha}\right)\right| = 1,$$

we obtain

$$\begin{aligned}\sup_{t\in[\alpha,\,1]} p_\nu(t) &= \frac{2}{1+T_\nu\left(\frac{1+\alpha}{1-\alpha}\right)}\\ &= \frac{2}{1+\frac{1+q^{2\nu}}{2q^\nu}}, \quad q = \frac{1-\sqrt{\alpha}}{1+\sqrt{\alpha}}\\ &= \frac{4q^\nu}{(1+q^\nu)^2}.\end{aligned}$$

Hence, because $p/(1-p)$ is an increasing function of $p \in [0, 1)$, we have

$$\begin{aligned}\widetilde{\delta}_{s+1}^{(\nu)} &\le \frac{\sup\,\{p_\nu(t),\ t\in[\alpha,\,1]\}}{1-\sup\,\{p_\nu(t),\ t\in[\alpha,\,1]\}}\\ &= \frac{4q^\nu}{(1+q^\nu)^2-4q^\nu}\\ &= \frac{4q^\nu}{(q^\nu-1)^2}\\ &= \frac{(1-\alpha)^\nu}{\alpha\left[\sum_{j=1}^{\nu}(1+\sqrt{\alpha})^{\nu-j}(1-\sqrt{\alpha})^{j-1}\right]^2}.\end{aligned}$$

Now, use the estimate based on Corollary 5.11, to bound the multilevel cycle by an inexact with fixed level–difference cycle, which gives

$$1+\delta_s \le (1+\widetilde{\delta}_{s+1}^{(\nu)})K_{MG}^{\ell\mapsto k} \tag{5.40}$$

and therefore,

$$1+\delta_s \le K_{MG}^{\ell\mapsto k}\left(1+\frac{1}{\alpha}\frac{(1-\alpha)^\nu}{\left[\sum_{j=1}^{\nu}(1+\sqrt{\alpha})^{\nu-j}(1-\sqrt{\alpha})^{j-1}\right]^2}\right).$$

In order to confirm the induction assumption, we need to choose ν and α such that

$$1+\delta_s \le K_{MG}^{\ell\mapsto k}\left(1+\frac{1}{\alpha}\frac{(1-\alpha)^\nu}{\left[\sum_{j=1}^{\nu}(1+\sqrt{\alpha})^{\nu-j}(1-\sqrt{\alpha})^{j-1}\right]^2}\right) \le \frac{1}{\alpha},$$

which is equivalent to inequality (5.37).

Consider now the simpler polynomial $p_\nu(t) = (1-t)^\nu$. Inequality (5.39) has a solution $\alpha > 0$ (sufficiently small) for $\nu > K_{MG}^{\ell \mapsto k}$. That solution satisfies $\alpha < 1/K_{MG}^{\ell \mapsto k}$. The remainder of the proof is the same as before with the only difference being

$$\delta_{s+1}^{(\nu)} \leq \sup \left\{ \frac{p_\nu(t)}{1 - p_\nu(t)},\ t \in [\alpha, 1] \right\} = \frac{(1-\alpha)^\nu}{\alpha \sum_{j=1}^{\nu} (1-\alpha)^{j-1}}. \qquad \square$$

5.6.4 Complexity of the AMLI-cycle

The complexity of the AMLI-cycle MG is readily estimated as follows. Let n_k be the number of degrees of freedom (dofs) at level k and assume uniform refinement; that is, $n_k = \mu^d\ n_{k-1}$, $d = 2$, or $d = 3$, and typically $\mu = 2$.

Assume that the V-cycle from coarse-level ℓ and fine-level k, with bounded-level difference can be implemented for $w_{MG}^{\ell \mapsto k_0} \simeq n_k$ flops. The latter cost does not involve coarse-grid solution at level ℓ. At that level, we use ν inner iterations based on $\mathcal{B}_\ell$ in the AMLI method and their cost can be estimated by $\nu w_{s+1} + C\nu n_\ell$ flops. Here w_j stands for the cost of implementing $B_{jk_0}^{-1}$.

Thus, letting $\ell = (s+1)k_0$ and $k = sk_0$, the recursive work estimate holds:

$$w_s \leq \nu w_{s+1} + C\nu n_\ell + w_{MG}^{\ell \mapsto k_0} \leq \nu w_{s+1} + C\nu n_\ell.$$

Then,

$$w_s \leq \nu w_{s+1} + C\nu n_{sk_0} \leq C\ n_{sk_0} \sum_{j=0}^{s} \left(\frac{\nu}{\mu^{dk_0}} \right)^j.$$

Thus, to have a method of optimal complexity, we have to balance ν and k_0 as follows,

$$\nu < \mu^{dk_0} = 2^{dk_0}. \tag{5.41}$$

On the other hand, for optimal condition number (based on Theorem 5.29), we have to choose ν sufficiently large such that

$$\nu > \sqrt{K_{MG}^{\ell \mapsto k}} \simeq \begin{cases} k_0, & d = 2, \\ 2^{\frac{k_0}{2}}, & d = 3. \end{cases} \tag{5.42}$$

The last expression comes from (5.36) in the case of second-order finite element elliptic problems.

It is clear then that for sufficiently large but fixed k_0, we can choose ν such that both (5.41) and (5.42) hold, which implies that the AMLI-cycle preconditioner is optimal for second-order finite element elliptic problems in the case of uniform refinement.

If we use the simple polynomial $p_\nu(t) = (1-t)^\nu$ to define the AMLI-cycle, the condition for ν, which implies uniform bound $1/\alpha$ on the condition number of the respective AMLI-cycle MG, reads

$$\nu > K_{MG}^{\ell \mapsto k} \simeq \begin{cases} k_0^2, & d = 2, \\ 2^{k_0}, & d = 3. \end{cases} \tag{5.43}$$

It is again clear that for sufficiently large but fixed k_0, we can choose ν such that both (5.43) and (5.41) hold, which implies that the AMLI-cycle preconditioner is optimal for second-order finite element elliptic problems in the case of uniform refinement and simple polynomial $(1-t)^\nu$. The latter choice of polynomial does not need the explicit knowledge of α in order to construct the polynomial.

The AMLI-cycle MG with the optimal choice of polynomial p_ν has more or less mostly theoretical value. In practice, we should use either the simple polynomial $(1-t)^\nu$ or the variable-step multilevel preconditioner presented in Section 10.3. The latter one is more practical because in its implementation no estimation of α is needed and no polynomial is explicitly constructed.

5.6.5 Optimal W-cycle methods

For $k_0 = 1$ and $\nu = 2$ the AMLI-cycle MG has the complexity of a W-cycle MG and for the simple polynomial $p_\nu = (1-t)^\nu$ (and $\nu = 2$), it is actually identical with a W-cycle MG. Applying Theorem 5.29 with $k_0 = 1$, hence $K_{TG} = K_{MG}^{k+1 \mapsto k}$ stands for uniform bound of the TG method (exact solution at coarse-level $k+1$), tells us that if $K_{TG} < 2$ then the inequality (5.39) has a solution $\alpha > 0$ (sufficiently small) for $\nu = 2 > K_{TG} \geq K_{MG}^{\ell \mapsto k}$ and the respective W-cycle preconditioner B satisfies the spectral equivalence relations

$$\mathbf{v}^T A\mathbf{v} \leq \mathbf{v}^T B\mathbf{v} \leq \frac{1}{\alpha}\, \mathbf{v}^T A\mathbf{v}.$$

Inequality (5.39) with the best α reduces to

$$\alpha + \frac{(1-\alpha)^2}{2-\alpha} = \frac{1}{K_{TG}},$$

or $(1/K_{TG})(2-\alpha) = \alpha(2-\alpha) + (1-\alpha)^2 = 1$. That is,

$$\alpha = 2 - K_{TG}.$$

In terms of convergence factors $\varrho_{W\text{-cycle}} = \varrho(I - B^{-1}A) \leq 1-\alpha$ and $\varrho_{TG} = 1 - (1/K_{TG})$, we have

$$\varrho_{W\text{-cycle}} \leq 1 - \alpha = K_{TG} - 1 = \frac{\varrho_{TG}}{1 - \varrho_{TG}}.$$

In conclusion, we have the following result.

Corollary 5.30. *If the two-grid method at any level k (with exact solution at coarse level $k+1$) has a uniformly bounded convergence factor $\varrho_{TG} < \frac{1}{2}$, then the respective W-cycle MG has a uniformly bounded convergence factor $\varrho_{W\text{-cycle}}$ that satisfies the estimate*

$$\varrho_{W\text{-cycle}} \leq \frac{\varrho_{TG}}{1-\varrho_{TG}} < 1.$$

An alternative, more qualitative analysis of the W-cycle methods (or AMLI-cycle with $k_0 = 1$) is based on the following approach. Assume that B_c is a s.p.d. approximation to A_c such that

$$\mathbf{v}_c^T B_c \mathbf{v}_c \leq (1+\delta_c)\, \mathbf{v}_c^T A_c \mathbf{v}_c.$$

Consider the inexact two-grid preconditioner with A_c approximated by B_c. The following characterization of B holds (use Theorem 5.9 in the case of two levels).

$$\begin{aligned}\mathbf{v}^T B\mathbf{v} = \inf_{\mathbf{v}_c} \big[&\mathbf{v}_c^T B_c \mathbf{v}_c + (M^T(\mathbf{v} - P\mathbf{v}_c) + AP\mathbf{v}_c)^T (M^T + M - A)^{-1} \\ &\times (M^T(\mathbf{v} - P\mathbf{v}_c) + AP\mathbf{v}_c)\big].\end{aligned}$$

Let $\mathbf{v}_c$ be a vector that is constructed on the basis of any given $\mathbf{v}$, such that the following two estimates hold,

(i) Stability,

$$\mathbf{v}_c^T A_c \mathbf{v}_c \leq \eta^* \, \mathbf{v}^T A\mathbf{v},$$

and

(ii) Approximation property, with $\overline{M} = M(M^T + M - A)^{-1} M^T$ being the symmetrized smoother, we have

$$(\mathbf{v} - P\mathbf{v}_c)^T \overline{M} (\mathbf{v} - P\mathbf{v}_c) \leq \delta^* \, \mathbf{v}^T A\mathbf{v}.$$

In other words, $P\mathbf{v}_c$ is a stable and accurate interpolant of $\mathbf{v}$. Assume in addition to (i)–(ii), the following estimate for the smoother M and A,

$$\delta_0 \, \mathbf{v}^T A\mathbf{v} \leq \mathbf{v}^T (M^T + M - A)\mathbf{v}.$$

The latter estimate, can always be guaranteed by proper scaling of the smoother M (such that $(1+\delta_0)\mathbf{v}^T A\mathbf{v} \leq \mathbf{v}^T (M^T + M)\mathbf{v} = 2\mathbf{v}^T M\mathbf{v}$).

Then B can be estimated in terms of A as follows, for any choice of $\mathbf{v}_c$,

$$\mathbf{v}^T B\mathbf{v} \leq (1+\delta_c)\, \mathbf{v}_c^T A_c \mathbf{v}_c + 2\left[(\mathbf{v} - P\mathbf{v}_c)^T \overline{M} (\mathbf{v} - P\mathbf{v}_c) + \frac{1}{\delta_0}\, \mathbf{v}_c^T A_c \mathbf{v}_c\right].$$

Based on our assumptions for the particular choice of $\mathbf{v}_c$, the following additive estimate then holds,

$$\mathbf{v}^T B\mathbf{v} \le \left[\eta^*\delta_c + \eta^* + 2\left(\delta^* + \frac{\eta^*}{\delta_0}\right)\right] \mathbf{v}^T A\mathbf{v}. \qquad (5.44)$$

In other words, the error δ_c that we commit with inexact coarse-grid solvers affects only the term involving the particular coarse interpolant $P\mathbf{v}_c$ of $\mathbf{v}$ and does not involve the quality of the smoother. Then, an optimal AMLI-cycle MG will be feasible with the simple polynomial $p_\nu(t) = (1-t)^\nu$ if $\eta^*/\nu < 1$.

This is seen by using (5.44) in place of (5.40) and δ_c in place of $\delta_{s+1}^{(\nu)}$ in the proof of Theorem 5.29. This observation leads to the following inequality for $\alpha > 0$,

$$\eta^* \frac{(1-\alpha)^\nu}{\alpha \sum_{j=1}^{\nu}(1-\alpha)^{j-1}} + \eta^* + 2\left(\delta^* + \frac{\eta^*}{\delta_0}\right) \le \frac{1}{\alpha},$$

which indeed has a solution for $\nu > \eta^*$; just multiply the above inequality by $\alpha > 0$ and then let $\alpha \mapsto 0$, which leads to $(\eta^*/\nu) < 1$. In particular, for $\nu = 2$ we can have an optimal W-cycle MG if $\eta^* < 2$. With the optimal choice of p_ν the estimate then becomes $\nu = 2 > \sqrt{\eta^*}$; that is, $\eta^* < 4$.

It is unclear that we can always find a coarse-grid interpolant such that (i)–(ii) hold with the stability constant η^* independent of the quality of the exact two-grid preconditioner B_{TG}, that is, to have $\eta^* < 4$. However, based essentially on the main result in [AV89]–[AV90], reformulated now in terms of the AMLI-cycle MG, the following optimal convergence result is available.

Theorem 5.31. *Consider matrices coming from triangular piecewise linear elements and second-order elliptic finite element problems. Estimates (i) and (ii) are feasible with $P\mathbf{v}_c$ being the standard nodal interpolant. More specifically, we have then $\eta^* = 1/(1-\gamma^2)$, where $\gamma < (\sqrt{3}/2)$ is the constant in the strengthened Cauchy–Schwarz inequality between the coarse f.e. space V_{2h} and its hierarchical complement in V_h (see* (B.29)–(B.30) *in the appendix). Also, (ii) holds with $\overline{M}$ being any smoother spectrally equivalent to the diagonal of A, for example, the Gauss–Seidel smoother (see Proposition 6.12 for such conditions). The choice $\nu = 2 > \sqrt{\eta^*} = \sqrt{1/(1-\gamma^2)}$ gives an optimal Chebyshev polynomials-based (see* (5.38)*) AMLI-cycle MG that has the complexity of the W-cycle.*

5.7 MG and additive MG

We present here an additive version of MG and an additive representation of the traditional MG. We use the recursive matrix factorization definition of MG found in Section 5.1. The same notation introduced there is used here.

5.7.1 The BPX-preconditioner

Based on a sequence of smoothers M_k, we can define the following, somewhat simpler than the traditional, multilevel preconditioner, originally proposed in [BPX].

Definition 5.32 (The BPX method). *Let $B_\ell = A_\ell$ and assume that B_{k+1}, for $k \leq \ell - 1$, has already been defined. We first define*

$$\overline{B}_k = \begin{bmatrix} M_k\left(M_k + M_k^T - A_k\right)^{-1} M_k^T & 0 \\ 0 & B_{k+1} \end{bmatrix},$$

and then B_k, the kth-level BPX preconditioner, is defined from

$$B_k^{-1} = [I, \; P_k]\overline{B}_k^{-1} \begin{bmatrix} I \\ P_k^T \end{bmatrix} = \left(M_k\left(M_k + M_k^T - A_k\right)^{-1} M_k^T\right)^{-1} + P_k B_{k+1}^{-1} P_k^T.$$

It is clear that the BPX preconditioner is obtained from the MG one by simply removing the off-diagonal blocks $M_k^{-T} A_k P_k$ and $P_k^T A_k M_k^{-1}$ of $\overline{B}_k$ defined in (5.2).

More explicitly, we have for ℓ levels, (with $B_\ell = A_\ell$),

$$\begin{aligned}\left(B_{MG}^{\text{add}}\right)^{-1} &= P_0 \cdots P_{\ell-1} A_\ell^{-1} P_{\ell-1}^T \cdots P_0^T \\ &\quad + \sum_{k=1}^{\ell-1} P_0 \cdots P_{k-1} M_k^{-T} (M_k + M_k^T - A_k) M_k^{-1} P_{k-1}^T \cdots P_0^T \\ &\quad + M_0^{-T} (M_0 + M_0^T - A_0) M_0^{-1}.\end{aligned}$$

The symmetrized smoother is not really needed here, because $(B_{MG}^{\text{add}})^{-1}$ (unless properly scaled) does not provide convergent splitting for $A = A_0$. Thus, we need a s.p.d. smoother Λ_k for A_k and the resulting additive MG takes then the form,

$$\begin{aligned}\left(B_{MG}^{\text{add}}\right)^{-1} &= P_0 \cdots P_{\ell-1} A_\ell^{-1} P_{\ell-1}^T \cdots P_0^T \\ &\quad + \sum_{k=1}^{\ell-1} P_0 \cdots P_{k-1} \Lambda_k^{-1} P_{k-1}^T \cdots P_0^T + \Lambda_0^{-1}.\end{aligned} \tag{5.45}$$

If Λ_k is such that Λ_k^{-1} is sparse, the simplest choice being diagonal, for example, $\Lambda_k = \text{diag}(A_k)$, it is clear then that $(B^{\text{add}})^{-1}$ is a linear combination of products of sparse matrices plus the term involving the inverse of A_ℓ, which is dense but typically has very small size.

5.7.2 Additive representation of MG

If we introduce the so-called "smoothed" interpolant $\overline{P}_k = (I - M_k^{-T} A_k) P_k$, and let $\Lambda_k = M_k (M_k^T + M_k - A_k)^{-1} M_k^T$ be the symmetrized smoother, formula (5.2) reduces to one of the additive MGs (the difference is only in the interpolation matrices used); that is,

$$B_k^{-1} = \Lambda_k^{-1} + \overline{P}_k B_{k+1}^{-1} \overline{P}_k^T.$$

Therefore, if both M_k^{-1} and M_k are explicitly available and sparse, then both $\overline{P}_k$ and Λ_k^{-1} will be explicitly available and sparse. This is a rare case in general and essentially means that M_k is diagonal. If only M_k^{-1} is explicitly available and sparse, we can explicitly form $\overline{P}_k = (I - M_k^{-T} A_k) P_k$, which then will be sparse (as a product of sparse matrices), and use a different smoother Λ_k coming from A_k, such that Λ_k^{-1} is s.p.d. and sparse. That is, we can view the traditional MG as an additive one because the following explicit formula holds for $B^{-1} = B_0^{-1}$,

$$B^{-1} = \overline{P}_0 \cdots \overline{P}_{\ell-1} A_\ell^{-1} \overline{P}_{\ell-1}^T \cdots \overline{P}_0^T + \sum_{k=1}^{\ell-1} \overline{P}_0 \cdots \overline{P}_{k-1} \Lambda_k^{-1} \overline{P}_{k-1}^T \cdots \overline{P}_0^T + \Lambda_0^{-1}.$$

This representation of the MG preconditioner offers the flexibility to utilize one smoother M_k in the construction of $\overline{P}_k = (I - M_k^{-T} A_k) P_k$ and another one (Λ_k) that does not have to be necessarily related to M_k.

5.7.3 Additive MG; convergence properties

Similarly to the traditional MG, the following main result holds.

Theorem 5.33. *Consider for any* $\mathbf{v}$ *decompositions of the form:*

(o) $\mathbf{v}_0 = \mathbf{v}$.
(i) For $k = 0, \ldots, \ell - 1$ *let*

$$\mathbf{v}_k = [I, \; P_k] \begin{bmatrix} \mathbf{v}_k^f \\ \mathbf{v}_{k+1} \end{bmatrix}.$$

Then for the additive MG based on s.p.d. smoothers Λ_k *for* A_k *(e.g.,* $\Lambda_k = M_k(M_k^T + M_k - A_k)^{-1} M_k^T$*) the following identity holds. For any* $k \geq 0$ *and* $\ell \geq k$,

$$\mathbf{v}_k^T B_k^{add} \mathbf{v}_k = \inf_{(\mathbf{v}_j = \mathbf{v}_j^f + P_j \mathbf{v}_{j+1})_{j=k}^{\ell-1}} \left[\mathbf{v}_\ell^T B_\ell \mathbf{v}_\ell + \sum_{j=k}^{\ell-1} (\mathbf{v}_j^f)^T \Lambda_j \mathbf{v}_j^f \right].$$

Note that at the coarsest-level ℓ*, we typically set* $B_\ell = A_\ell$.

Proof. We have to note that because the additive MG is also defined via a relation $B_k^{-1} = [I, \; P_k] \overline{B}_k^{-1} [I, \; P_k]^T$ the same proof as for the standard MG applies in this case, as well. □

Based on the last theorem the following estimate of A in terms of B for the additive MG holds.

Corollary 5.34. *Let* Λ_k *be s.p.d. smoothers for* A_k *scaled such that* $\mathbf{v}_k^T A_k \mathbf{v}_k \leq \mathbf{v}_k^T \Lambda_k \mathbf{v}_k$. *The following estimate then holds,*

$$\mathbf{v}_k^T A_k \mathbf{v}_k \leq (\ell + 1 - k) \; \mathbf{v}_k^T B_k^{add} \mathbf{v}_k.$$

Proof. Consider any decomposition sequence $\mathbf{v}_j = \mathbf{v}_j^f + P_j \mathbf{v}_{j+1}$ for $j \geq 0$ starting with a given $\mathbf{v}_0$. Assuming by induction that $\mathbf{v}_j^T A_j \mathbf{v}_j \leq \; (\ell + 1 - j) \; \mathbf{v}_j^T B_j^{\text{add}} \mathbf{v}_j$

(which trivially holds for $j = \ell$), we get, based on Cauchy–Schwarz inequalities,

$$
\begin{aligned}
\mathbf{v}_{j-1}^T A_{j-1} \mathbf{v}_{j-1} &= \left(\mathbf{v}_{j-1}^f + P_{j-1}\mathbf{v}_j\right)^T A_{j-1} \left(\mathbf{v}_{j-1}^f + P_{j-1}\mathbf{v}_j\right) \\
&= \left\|\mathbf{v}_{j-1}^f + P_{j-1}\mathbf{v}_j\right\|_{A_{j-1}}^2 \\
&\le \left(\left\|\mathbf{v}_{j-1}^f\right\|_{A_{j-1}} + \|P_{j-1}\mathbf{v}_j\|_{A_{j-1}}\right)^2 \\
&= \left(\left\|\mathbf{v}_{j-1}^f\right\|_{A_{j-1}} + \|\mathbf{v}_j\|_{A_j}\right)^2 \\
&\le \left(\left\|\mathbf{v}_{j-1}^f\right\|_{\Lambda_{j-1}} + (\ell + 1 - j)^{\frac{1}{2}} \|\mathbf{v}_j\|_{B_j^{\text{add}}}\right)^2 \\
&\le (1 + (\ell + 1 - j)) \left(\left\|\mathbf{v}_{j-1}^f\right\|_{\Lambda_{j-1}}^2 + \|\mathbf{v}_j\|_{B_j^{\text{add}}}^2\right) \\
&= (\ell + 1 - (j-1)) \left(\left(\mathbf{v}_{j-1}^f\right)^T \Lambda_{j-1} \mathbf{v}_{j-1}^f + \mathbf{v}_j^T B_j^{\text{add}} \mathbf{v}_j\right).
\end{aligned}
$$

Because we can take the minimum over the decomposition of $\mathbf{v}_{j-1} = \mathbf{v}_{j-1}^f + P_{j-1}\mathbf{v}_j$ based on Theorem 5.33, we get

$$
\mathbf{v}_{j-1}^T A_{j-1} \mathbf{v}_{j-1} \le (\ell + 1 - (j-1))\, \mathbf{v}_{j-1}^T B_{j-1}^{\text{add}} \mathbf{v}_{j-1}.
$$

The latter confirms the induction assumption for $j := j - 1$ and the proof is complete. □

Based on the above result, the following suboptimal relation between the conventional (multiplicative) MG preconditioner B and its additive counterpart B^{add} is easily seen.

Theorem 5.35. *Consider the multiplicative MG preconditioner B and let B^{add} be the additive one that exploits the symmetrized smoothers*

$$
\Lambda_k = M_k \left(M_k^T + M_k - A_k\right)^{-1} M_k^T = \overline{M}_k.
$$

Assume also that M_k is properly scaled such that $\mathbf{v}_k^T (M_k + M_k^T - A_k)\mathbf{v}_k \ge \delta\, \mathbf{v}_k^T A_k \mathbf{v}_k$. Then, the following upper bound of B in terms of B^{add} holds,

$$
\mathbf{v}^T B \mathbf{v} \le \left(1 + \frac{\ell(\ell+1)}{4\delta} \left(3 + \sqrt{1 + \frac{4\delta}{\ell}}\right)\right) \mathbf{v}^T B^{add} \mathbf{v}.
$$

Proof. For any decomposition $\mathbf{v}_k = \mathbf{v}_k^f + P_k \mathbf{v}_{k+1}$, starting with $\mathbf{v}_0 = \mathbf{v}$ for any given $\mathbf{v}$, we have

$$
\begin{aligned}
\mathbf{v}^T B \mathbf{v} \le \mathbf{v}_\ell^T A_\ell \mathbf{v}_\ell + \sum_{k=0}^{\ell-1} \left(M_k^T \mathbf{v}_k^f + A_k P_k \mathbf{v}_{k+1}\right)^T \left(M_k + M_k^T - A_k\right)^{-1} \\
\times \left(M_k^T \mathbf{v}_k^f + A_k P_k \mathbf{v}_{k+1}\right).
\end{aligned} \tag{5.46}
$$

By assumption, $\mathbf{v}_k^T(M_k + M_k^T - A_k)\mathbf{v}_k \geq \delta\, \mathbf{v}_k^T A_k \mathbf{v}_k$. Then, use the Cauchy–Schwarz inequality and Corollary 5.34 for A_k and B_k^{add} to get for any $\tau > 0$,

$$\begin{aligned}
\mathbf{v}^T B\mathbf{v} &\leq \mathbf{v}_\ell^T A_\ell \mathbf{v}_\ell + (1+\tau)\sum_{k=0}^{\ell-1} (\mathbf{v}_k^f)^T \overline{M}_k \mathbf{v}_k^f \\
&\quad + \left(1 + \frac{1}{\tau}\right)\frac{1}{\delta}\sum_{k=0}^{\ell-1} \mathbf{v}_{k+1}^T A_{k+1}\mathbf{v}_{k+1} \\
&\leq \mathbf{v}_\ell^T A_\ell \mathbf{v}_\ell + (1+\tau)\sum_{k=0}^{\ell-1} (\mathbf{v}_k^f)^T \overline{M}_k \mathbf{v}_k^f \\
&\quad + \left(1 + \frac{1}{\tau}\right)\frac{1}{\delta}\sum_{k=0}^{\ell-1} (\ell - k)\, \mathbf{v}_{k+1}^T B_{k+1}^{\text{add}}\mathbf{v}_{k+1}.
\end{aligned}$$

Therefore,

$$\begin{aligned}
\mathbf{v}^T B\mathbf{v} &\leq \mathbf{v}_\ell^T A_\ell \mathbf{v}_\ell + (1+\tau)\sum_{k=0}^{\ell-1} (\mathbf{v}_k^f)^T \overline{M}_k \mathbf{v}_k^f \\
&\quad + \left(1 + \frac{1}{\tau}\right)\frac{1}{\delta}\sum_{k=0}^{\ell-1} (\ell - k)\left(\mathbf{v}_\ell^T A_\ell \mathbf{v}_\ell + \sum_{j=k+1}^{\ell-1} (\mathbf{v}_j^f)^T \overline{M}_j \mathbf{v}_j^f\right) \\
&= \left(1 + \frac{\ell(\ell+1)}{2\delta}\left(1 + \frac{1}{\tau}\right)\right)\mathbf{v}_\ell^T A_\ell \mathbf{v}_\ell + (1+\tau)\sum_{k=0}^{\ell-1} (\mathbf{v}_k^f)^T \overline{M}_k \mathbf{v}_k^f \\
&\quad + \left(1 + \frac{1}{\tau}\right)\frac{1}{\delta}\sum_{j=1}^{\ell-1} (\mathbf{v}_j^f)^T \overline{M}_j \mathbf{v}_j^f \sum_{k=j-1}^{\ell-1} (\ell - k).
\end{aligned}$$

That is,

$$\begin{aligned}
\mathbf{v}^T B\mathbf{v} &\leq \left(1 + \frac{\ell(\ell+1)}{2\delta}\left(1 + \frac{1}{\tau}\right)\right)\mathbf{v}_\ell^T A_\ell \mathbf{v}_\ell + (1+\tau)\sum_{k=0}^{\ell-1} (\mathbf{v}_k^f)^T \overline{M}_k \mathbf{v}_k^f \\
&\quad + \frac{\ell(\ell-1)}{2\delta}\left(1 + \frac{1}{\tau}\right)\sum_{j=1}^{\ell-1} (\mathbf{v}_j^f)^T \overline{M}_j \mathbf{v}_j^f \\
&\leq \left(1 + \frac{\ell(\ell+1)}{2\delta}\left(1 + \frac{1}{\tau}\right)\right)\left(\mathbf{v}_\ell^T A_\ell \mathbf{v}_\ell + \sum_{k=0}^{\ell-1} (\mathbf{v}_k^f)^T \overline{M}_k \mathbf{v}_k^f\right).
\end{aligned}$$

Here $\tau > 0$ is such that $1 + (\ell(\ell+1)/2\delta)(1 + (1/\tau)) = 1 + \tau + (\ell(\ell-1)/2\delta)(1 + (1/\tau))$, which gives $\tau = (\ell/\delta)(1 + (1/\tau))$ and after solving the quadratic equation for $\tau > 0$, we get $\tau = 2\ell/(\ell + \sqrt{\ell^2 + 4\ell\delta})$.

Because the decomposition of $\mathbf{v}$ based on $\{\mathbf{v}_k^f\}$ was arbitrary, by taking minimum (based on Theorem 5.33) we arrive at the desired result. □

Thus, based on the above result, if we are able to bound the additive preconditioner B^{add} in terms of A, we can get a bound for the multiplicative preconditioner B in terms of A with an extra factor of order ℓ^2.

It is also clear, from the above proof, that if $\mathbf{v}_k^T A_k \mathbf{v}_k \leq C\ \mathbf{v}_k^T B_k^{\text{add}} \mathbf{v}_k$, we can bound B in terms of B^{add} only with a factor of order ℓ. The latter can be further (substantially) improved to $\mathcal{O}(1 + (\log \ell/2))$, based on the following lemma by Griebel and Oswald ([GO95]).

Lemma 5.36. *Consider a symmetric, positive semidefinite block-matrix* $T = (T_{ij})_{i,j=1}^{\ell+1}$ *with square diagonal blocks* T_{ii}. *Let* $L = (L_{ij})_{i,j=1}^{\ell+1}$ *be its strictly lower-triangular part, that is,* $L_{ij} = T_{ij}$ *for* $i > j$ *and* $L_{ij} = 0$ *otherwise. Then, the following estimate holds,*

$$\|L\| \leq \frac{1}{2} \log \ell\ \|T\|.$$

Here, for any matrix B *we define* $\|B\| = \sup_{\mathbf{v},\ \mathbf{w}}\ \mathbf{w}^T B \mathbf{v}/\|\mathbf{v}\|\,\|\mathbf{w}\|$, *and* $\|\mathbf{v}\|^2 = \mathbf{v}^T \mathbf{v}$.

Proof. Partition L into a two-by-two block structure as follows.

$$L = \begin{bmatrix} L_1 & 0 \\ L_{21} & L_2 \end{bmatrix}.$$

The proof proceeds by induction with respect to the block–size of L and L_1 and L_2, and is based on the following two observations. First, consider the diagonal of L,

$$D_L = \begin{bmatrix} L_1 & 0 \\ 0 & L_2 \end{bmatrix}.$$

We have, using the definition of norm and the Cauchy–Schwarz inequality,

$$\begin{aligned} \|D_L\| &= \sup_{\mathbf{v}_1,\ \mathbf{v}_2,\ \mathbf{w}_1,\ \mathbf{w}_2} \frac{\mathbf{w}_1^T L_1 \mathbf{v}_1 + \mathbf{w}_2^T L_2 \mathbf{v}_2}{\sqrt{\|\mathbf{v}_1\|^2 + \|\mathbf{v}_2\|^2}\sqrt{\|\mathbf{w}_1\|^2 + \|\mathbf{w}_2\|^2}} \\ &\leq \max\{\|L_1\|,\ \|L_2\|\} \sup_{\mathbf{v}_1,\ \mathbf{v}_2,\ \mathbf{w}_1,\ \mathbf{w}_2} \frac{\|\mathbf{v}_1\|\,\|\mathbf{w}_1\| + \|\mathbf{v}_2\|\,\|\mathbf{w}_2\|}{\sqrt{\|\mathbf{v}_1\|^2 + \|\mathbf{v}_2\|^2}\sqrt{\|\mathbf{w}_1\|^2 + \|\mathbf{w}_2\|^2}} \\ &\leq \max\{\|L_1\|,\ \|L_2\|\}. \end{aligned}$$

The second observation concerns the norm of the strictly lower triangular part L_{21} of L, which is also a strictly lower triangular part of T. We have

$$\begin{bmatrix} \mathbf{w}_1 \\ \mathbf{w}_2 \end{bmatrix}^T \begin{bmatrix} 0 & 0 \\ L_{21} & 0 \end{bmatrix} \begin{bmatrix} \mathbf{v}_1 \\ \mathbf{v}_2 \end{bmatrix} = \mathbf{w}_2^T L_{21} \mathbf{v}_1.$$

Now, use the identity $L_{21} = [0,\ I] T \left[\begin{smallmatrix} I \\ 0 \end{smallmatrix}\right]$ and hence, due to the symmetry of T,

$$4\mathbf{w}_2^T L_{21} \mathbf{v}_1 = \begin{bmatrix} \mathbf{v}_1 \\ \mathbf{w}_2 \end{bmatrix}^T T \begin{bmatrix} \mathbf{v}_1 \\ \mathbf{w}_2 \end{bmatrix} - \begin{bmatrix} \mathbf{v}_1 \\ -\mathbf{w}_2 \end{bmatrix}^T T \begin{bmatrix} \mathbf{v}_1 \\ -\mathbf{w}_2 \end{bmatrix}.$$

The fact that T is positive semidefinite then shows the inequality

$$\mathbf{w}_2^T L_{21}\mathbf{v}_1 \le \frac{\|T\|}{4}\left(\|\mathbf{v}_1\|^2 + \|\mathbf{w}_2\|^2\right) \le \frac{\|T\|}{4}\left(\|\mathbf{v}_1\|^2 + \|\mathbf{v}_2\|^2 + \|\mathbf{w}_1\|^2 + \|\mathbf{w}_2\|^2\right).$$

That is, based on the last inequality for $\|\mathbf{v}_1\|^2 + \|\mathbf{v}_2\|^2 = \|\mathbf{w}_1\|^2 + \|\mathbf{w}_2\|^2 = 1$, the following estimate is obtained,

$$\|L_{21}\| \le \frac{\|T\|}{2}.$$

In conclusion, we proved the following estimate for $\|L\| \le \|L_{21}\| + \|L_D\|$,

$$\|L\| \le \max\{\|L_1\|,\ \|L_2\|\} + \frac{\|T\|}{2}.$$

Assume now by induction that $\|L_1\|,\ \|L_2\| \le ((k-1)/2)\ \|T\|$ where the block-size of L_1 and L_2 is not greater than 2^{k-1}, then we get for any L of block-size two times bigger than that of L_1 and L_2 the estimate $\|L\| \le \|T\|(((k-1)/2) + 1/2) = \|T\|\ (k/2)$, which confirms the induction assumption and hence the proof is complete. □

Introduce now the composite interpolants $\overline{P}_k = P_0, \ldots, P_{k-1}$ that map the kth-level coarse vector space into the finest vector space and let $\overline{P}_0 = I$. We can then consider the block-matrix $T = (T_{ij})$

$$T_{ij} = \overline{P}_i^T A \overline{P}_j.$$

Note that the block-diagonal part D_T of T has entries $T_{ii} = A_i$ (the ith-level matrices).

Recall the block Gauss–Seidel matrix appearing in the characterization of $B = B_{MG}$ in (5.14). The quadratic form Q corresponding to this block Gauss–Seidel matrix can be given in terms of $T = D_T + L_T + U_T$, where L_T has nonzero blocks equal to T_{ij} for $i > j$. More specifically, we have

$$Q(\mathbf{v}) = Q\left(\mathbf{v}_0^f, \ldots, \mathbf{v}_\ell^f\right) = \begin{bmatrix} \mathbf{v}_0^f \\ \vdots \\ \mathbf{v}_\ell^f \end{bmatrix}^T (D_T + L_T) D_T^{-1} (D_T + U_T) \begin{bmatrix} \mathbf{v}_0^f \\ \vdots \\ \mathbf{v}_\ell^f \end{bmatrix}. \tag{5.47}$$

Here, $\mathbf{v} = \sum_{k=0}^{\ell} \overline{P}_k \mathbf{v}_k^f$.

Define now $\overline{T}_{ij} = A_i^{-(1/2)} T_{ij} A_j^{-(1/2)}$ and let $\overline{T} = (\overline{T}_{ij})$. Based on Lemma 5.36 applied to $\overline{L}_T = D_T^{-(1/2)} L_T D_T^{-(1/2)}$, the following bound holds,

$$\|\overline{L}_T\| \le \frac{\log \ell}{2}\ \|\overline{T}\|. \tag{5.48}$$

The latter results in the following estimate for the quadratic form Q; that is,

$$Q(\mathbf{v}) = Q(\mathbf{v}_0^f, \ldots, \mathbf{v}_\ell^f) \le \left(1 + \frac{\log \ell}{2}\ \|\overline{T}\|\right)^2 \sum_{k=0}^{\ell} (\mathbf{v}_k^f)^T A_k \mathbf{v}_k^f. \tag{5.49}$$

It is easily seen that $\|\overline{T}\|$ is the best upper bound in the estimate,

$$\mathbf{v}^T A\mathbf{v} \leq \|\overline{T}\| \inf_{(\mathbf{v}_k^f):\mathbf{v}=\sum_k \overline{P}_k \mathbf{v}_k^f} \sum_k (\mathbf{v}_k^f)^T A_k \mathbf{v}_k^f.$$

Now, we are in position to improve Theorem 5.35 as follows ([GO95]).

Theorem 5.37. *Consider the multiplicative MG preconditioner B and let B^{add} be the additive one that exploits the symmetrized smoothers*

$$\Lambda_k = \overline{M}_k = M_k\left(M_k^T + M_k - A_k\right)^{-1} M_k^T.$$

Assume also that M_k is properly scaled such that $\mathbf{v}_k^T(M_k+M_k^T-A_k)\mathbf{v}_k \geq \delta\ \mathbf{v}_k^T A_k \mathbf{v}_k$. Then, the following upper bound on B in terms of B^{add} holds,

$$\mathbf{v}^T B\mathbf{v} \leq \left(1 + \frac{1}{\sqrt{\delta}} \frac{\log \ell}{2} \|\overline{T}\|\right)^2 \mathbf{v}^T B^{add}\mathbf{v}.$$

Here, $\|\overline{T}\|$ can be characterized as the (best) upper bound in the estimate

$$\mathbf{v}^T A\mathbf{v} \leq \|\overline{T}\| \inf_{(\mathbf{v}_k^f):\mathbf{v}=\sum_{k=0}^{\ell} \overline{P}_k \mathbf{v}_k^f} \sum_{k=0}^{\ell} \left(\mathbf{v}_k^f\right)^T A_k \mathbf{v}_k^f. \tag{5.50}$$

A trivial estimate is $\|\overline{T}\| \leq \ell+1$. That is, the following suboptimal result holds then,

$$\mathbf{v}^T B\mathbf{v} \leq \left(1 + \frac{1}{\sqrt{\delta}} (\ell+1) \frac{\log \ell}{2}\right)^2 \mathbf{v}^T B^{add}\mathbf{v}. \tag{5.51}$$

Proof. Use the main identity of Theorem 5.10 (which is equivalent to (5.46))

$$\mathbf{v}^T B\mathbf{v} \leq \inf_{(\mathbf{v}_k^f):\mathbf{v}=\sum_{k=0}^{\ell} \overline{P}_k \mathbf{v}_k^f} \begin{bmatrix} \mathbf{v}_0^f \\ \vdots \\ \mathbf{v}_\ell^f \end{bmatrix}^T L_B\left(\text{diag}\left(M_k^T + M_k - A_k\right)\right)^{-1} L_B^T \begin{bmatrix} \mathbf{v}_0^f \\ \vdots \\ \mathbf{v}_\ell^f \end{bmatrix}. \tag{5.52}$$

Here,

$$L_B = \begin{bmatrix} M_0 & 0 & \dots & 0 \\ T_{10} & M_1 & \dots & 0 \\ \vdots & \dots & \ddots & 0 \\ T_{\ell,0} & \dots & T_{\ell,\ell-1} & M_\ell \end{bmatrix}.$$

Let $\widehat{M} = \text{diag}(M_k)_{k=0}^{\ell}$, $\widehat{A} = \text{diag}(A_k)_{k=0}^{\ell}$, and $\widehat{\Delta} = \text{diag}(M_k^T + M_k - A_k)_{k=0}^{\ell}$. Then the following estimate is readily seen based on the coercivity of $M_k + M_k^T - A_k$ in terms of A_k,

$$\|\widehat{\Delta}^{-(1/2)} \widehat{A}^{1/2}\|^2 \leq \frac{1}{\delta}. \tag{5.53}$$

Also, let

$$\widehat{\mathbf{v}} = \begin{bmatrix} \mathbf{v}_0^f \\ \vdots \\ \mathbf{v}_\ell^f \end{bmatrix}.$$

Then the desired estimate would follow if we bound the norm of the symmetric Gauss–Seidel matrix $L_B \Delta^{-1} L_B^T = (\widehat{M} + L_T)\widehat{\Delta}^{-1}(\widehat{M} + L_T)^T$ where L_T is exactly the strictly lower triangular part of T, which was the matrix representation of the quadratic form Q (see (5.47)). Based on the triangle inequality, estimate (5.53), and the fact that $\overline{M}_k - A_k$ is symmetric positive semidefinite, and the proven estimate for the diagonally scaled strictly off-diagonal part of T, (5.48) (i.e., $\|\widehat{A}^{-(1/2)} L_T^T \widehat{A}^{-(1/2)}\| = \|\widehat{A}^{-(1/2)} L_T \widehat{A}^{-(1/2)}\| \le \log \ell/2 \; \|\overline{T}\|$), we get

$$\begin{aligned}
&\left\|\widehat{\Delta}^{-(1/2)}(\widehat{M}^T + L_T^T)\widehat{\mathbf{v}}\right\| \\
&\quad \le \left\|\widehat{\Delta}^{-(1/2)}\widehat{M}^T\widehat{\mathbf{v}}\right\| + \frac{1}{\sqrt{\delta}} \left\|\widehat{A}^{-(1/2)} L_T \widehat{A}^{-(1/2)}\right\| \left\|\widehat{A}^{1/2}\widehat{\mathbf{v}}\right\| \\
&\quad \le \left[\sum_{k=0}^{\ell} (\mathbf{v}_k^f)^T \overline{M}_k \mathbf{v}_k^f\right]^{1/2} + \frac{1}{\sqrt{\delta}} \frac{\log \ell}{2} \|\overline{T}\| \left[\sum_{k=0}^{\ell} (\mathbf{v}_k^f)^T A_k \mathbf{v}_k^f\right]^{1/2} \\
&\quad \le \left(1 + \frac{1}{\sqrt{\delta}} \frac{\log \ell}{2} \|\overline{T}\|\right) \left(\sum_{k=0}^{\ell} (\mathbf{v}_k^f)^T \overline{M}_k \mathbf{v}_k^f\right)^{1/2}.
\end{aligned}$$

Now, using Theorem 5.33, that is, the identity

$$\inf_{(\mathbf{v}_k^f):\mathbf{v}=\sum_k \overline{P}_k \mathbf{v}_k^f} \sum_k (\mathbf{v}_k^f)^T \overline{M}_k \mathbf{v}_k^f = \mathbf{v}^T B^{\mathrm{add}} \mathbf{v}$$

in the previous estimate to bound (5.52), gives the desired one. □

The level-independent boundedness of $\|\overline{T}\|$ in (5.50) can be proved (see Appendix E) for matrices A_k corresponding to the discrete Laplacian which is a fundamental result due to Oswald [0s94]; see also [DK92].

We conclude with the comment that for geometric MG applied to second-order elliptic finite element problems, all extra factors containing weak dependence on the number of levels in estimating B in terms of B^{add} can be removed due to the following strengthened Cauchy–Schwarz inequality for a $\delta \in (0, 1)$,

$$\left((\mathbf{v}_k^f)^T T_{kl} \mathbf{v}_l^f\right)^2 \le C\, \delta^{|k-l|}\, (\mathbf{v}_k^f)^T A_k \mathbf{v}_k^f\, (\mathbf{v}_l^f)^T A_l \mathbf{v}_l^f.$$

That is, the block entries of T have certain decay away from its main diagonal for certain particular decomposition $\mathbf{v} = \sum_k \overline{P}_{k-1} \mathbf{v}_k^f$. To prove such a strengthened Cauchy–Schwarz inequality some additional properties of the finite element spaces are needed. For the respective details, we refer to Proposition F.1 in the appendix, or to the survey papers by Yserentant [Y93] and Xu [Xu92a]; see also [Zh92] and [BP93].

5.7.4 MG convergence based on results for matrix subblocks

Let A be a given s.p.d. matrix and let I_1 and I_2 be extensions by zero of vector of smaller dimension to the dimension of A. Consider the principal subblocks of $A_i^{(0)} \equiv I_i^T A I_i$, $i = 1, 2$ of A.

We assume that every vector $\mathbf{v}$ can be decomposed as

$$\mathbf{v} = I_1 \mathbf{v}_1 + I_2 \mathbf{v}_2, \tag{5.54}$$

such that for a constant $\eta_s > 0$, we have

$$\sum_i \mathbf{v}_i^T A_i^{(0)} \mathbf{v}_i = \sum_i \mathbf{v}_i^T I_i^T A I_i \mathbf{v}_i \leq \eta_s \; \mathbf{v}^T A \mathbf{v}. \tag{5.55}$$

Assume that we have constructed a MG method for A such that the interpolation matrices P_k respect the local vector spaces Range (I_i), $i = 1, 2$. That is, there is a sequence of zero extension matrices $I_i^{(k)}$, $i = 1, 2$, for every level $k \geq 0$, such that the restriction $P_i^{(k)}$, $i = 1, 2$ of the interpolation matrices P_k to the local vector spaces satisfies the property,

$$I_i^{(k-1)} P_i^{(k)} \mathbf{v}_i = P_k I_i^{(k)} \mathbf{v}_i.$$

That is, if we first interpolate locally a vector $\mathbf{v}_i$ obtaining a local fine-grid vector $P_i^{(k)} \mathbf{v}_i$ and then extend it by zero giving rise to $I_i^{(k-1)} P_i^{(k)} \mathbf{v}_i$, it is the same as if we first extend by zero the local coarse vector $\mathbf{v}_i$ to a global coarse vector $I_i^{(k)} \mathbf{v}_i$ and then interpolate it to end up with a (global) fine-grid vector $P_k I_i^{(k)} \mathbf{v}_i$.

A typical case is that the set of dofs $\mathcal{N}_k$ at every level $k \geq 0$ is partitioned into two overlapping groups $\mathcal{N}_k = \mathcal{N}_1^{(k)} \cup \mathcal{N}_2^{(k)}$, and we then have

$$I_i^{(k)} = \begin{bmatrix} 0 \\ I \end{bmatrix} \begin{matrix} \} \, \mathcal{N}_k \setminus \mathcal{N}_i^{(k)} \\ \} \, \mathcal{N}_i^{(k)} \end{matrix},$$

and

$$P_k = \begin{bmatrix} * & 0 \\ * & P_i^{(k)} \end{bmatrix} \begin{matrix} \} \, \mathcal{N}_{k-1} \setminus \mathcal{N}_i^{(k-1)} \\ \} \, \mathcal{N}_i^{(k-1)} \end{matrix}, \quad P_k^T = \begin{bmatrix} * & * \\ 0 & P_i^{(k)T} \end{bmatrix} \begin{matrix} \} \, \mathcal{N}_k \setminus \mathcal{N}_i^{(k)} \\ \} \, \mathcal{N}_i^{(k)} \end{matrix}.$$

In other words P_k interpolates local coarse-grid vectors

$$\mathbf{v}_c = \begin{bmatrix} 0 \\ \mathbf{v}_i^c \end{bmatrix} \begin{matrix} \} \, \mathcal{N}_k \setminus \mathcal{N}_i^{(k)} \\ \} \, \mathcal{N}_i^{(k)} \end{matrix},$$

that is, that vanish outside $\mathcal{N}_i^{(k)}$ based only on $P_i^{(k)}$ keeping the result zero outside $\mathcal{N}_i^{(k-1)}$,

$$P_k \begin{bmatrix} 0 \\ \mathbf{v}_i^c \end{bmatrix} = \begin{bmatrix} 0 \\ P_i^{(k)} \mathbf{v}_i^c \end{bmatrix} \begin{matrix} \} \, \mathcal{N}_{k-1} \setminus \mathcal{N}_i^{(k-1)} \\ \} \, \mathcal{N}_i^{(k-1)} \end{matrix}.$$

Based on $P_i^{(k)}$ and the smoothers $M_i^{(k)} = I_i^{(k)^T} M_k I_i^{(k)}$, we can define V-cycle MG preconditioners B_i for the local matrices $A_i^{(0)}$.

We assume that B_i are spectrally equivalent to $A_i^{(0)}$. Then our goal is to show that B is spectrally equivalent to A.

One application of the above result would be if A corresponds to a discretization of the Poisson equation on an L-shaped domain Ω. Note that Ω can be decomposed as $\Omega = \Omega_1 \cup \Omega_2$, where Ω_i are rectangles (i.e., convex polygonal domains). For each of Ω_i (because the Dirichlet problem for the Poisson equation on Ω_i allows for solutions with two derivatives (as in (1.11)–(1.12)), a MG V-cycle preconditioner B_i will be spectrally equivalent to $A_i^{(0)}$ (the local submatrices of A corresponding to the convex polygonal subdomains Ω_i). Then the result we prove gives that the V-cycle preconditioner B is spectrally equivalent to A, as long as we can prove the assumed estimate (5.54)–(5.55). For the latter, see Example E.1 in the appendix due to Lions (cf., [Li87], pp. 8–9) and the related Section E.1.1.

In what follows, we consider the additive MG (or BPX) only and prove a uniform upper bound for B in terms of A.

Theorem 5.38. *Let P_k and Λ_k and A_k for $k = 0, 1, \ldots, \ell$; define an additive MG preconditioner B for $A = A_0$. Let I_i, $i = 1, 2$ induce submatrices $A_i^{(0)} = I_i^T A I_i$ and the P_k induce local interpolation matrices $P_i^{(k)}$, $i = 1, 2$ such that for zero extension matrices*

$$I_i^{(k)} = \begin{bmatrix} 0 \\ I \end{bmatrix} \begin{matrix} \} \mathcal{N}_k \setminus \mathcal{N}_i^{(k)} \\ \} \mathcal{N}_i^{(k)} \end{matrix}$$

we have $P_k I_i^{(k)} = I_i^{(k-1)} P_i^{(k)}$. This implies $A_i^{(k)} \equiv I_i^{(k)^T} A_k I_i^{(k)} = P_i^{(k)^T} A_i^{(k+1)} P_i^{(k)}$. Let $\Lambda_i^{(k)} = I_i^{(k)^T} \Lambda_k I_i^{(k)}$ be the respective principal submatrices of the s.p.d. smoother Λ_k. Then $\Lambda_i^{(k)}$, $P_i^{(k)}$, and $A_i^{(k)}$ define spectrally equivalent additive MG preconditioners B_i for the principal submatrices $\overline{A}_i^{(0)} = I_i^T A I_i$, $i = 1,\ 2$. Let K_i be bounds on the maximal eigenvalue of $B_i A_i^{(0)^{-1}}$. (Recall that the minimal eigenvalue can be estimated at the worst as $1/(\ell + 1)$; cf., Corollary 5.34.)

Then under the stability estimates (5.54)–(5.55) *valid for any vector $\mathbf{v} = \mathbf{v}_0$, the global additive MG preconditioner B is bounded from above in terms of A. More specifically, the following upper bound holds,*

$$\mathbf{v}^T B \mathbf{v} \leq 2\, \eta_s \max_{i=1,\,2} K_i\, \mathbf{v}^T A \mathbf{v}.$$

Proof. To prove the stated result, we use our main identity from Theorem 5.9:

$$\mathbf{v}^T B \mathbf{v} = \inf_{(\mathbf{v}_j = \mathbf{v}_j^f + P_j \mathbf{v}_{j+1})_{j=0}^{\ell-1}} \left[\mathbf{v}_\ell^T A_\ell \mathbf{v}_\ell + \sum_{j=0}^{\ell-1} \left(\mathbf{v}_j^f\right)^T \Lambda_j \mathbf{v}_j^f \right]. \tag{5.56}$$

By assumption we have $\mathbf{v}_0 = I_1\mathbf{v}_1^{(0)} + I_2\mathbf{v}_2^{(0)}$ with stable components $\mathbf{v}_i^{(0)}$, $i = 1, 2$. Now use Theorem 5.9 applied to all of the blocks $A_i^{(0)}$ and their respective additive MG preconditioners B_i. It implies that for $i = 1, 2$, there is a multilevel decomposition $\mathbf{v}_i^{(j)} = \mathbf{v}_{f,\,i}^{(j)} + P_i^{(j)}\mathbf{v}_i^{(j+1)}$, $j = 0, \ldots, \ell - 1$, such that

$$K_i \; \mathbf{v}_i^{(0)^T} A_i^{(0)}\mathbf{v}_i^{(0)} \geq \mathbf{v}_i^{(0)^T} B_i\mathbf{v}_i^{(0)} = \left[\mathbf{v}_i^{(\ell)^T} A_i^{(\ell)}\mathbf{v}_i^{(\ell)} + \sum_{j=0}^{\ell-1} \left(\mathbf{v}_{f,\,i}^{(j)}\right)^T \Lambda_i^{(j)}\mathbf{v}_{f,\,i}^{(j)}\right].$$

Consider now the vectors

$$\mathbf{v}_j = I_1^{(j)}\mathbf{v}_1^{(j)} + I_2^{(j)}\mathbf{v}_2^{(j)}, \quad \mathbf{v}_j^{(f)} = I_1^{(j)}\mathbf{v}_{f,\,1}^{(j)} + I_2^{(j)}\mathbf{v}_{f,\,2}^{(j)}.$$

We have

$$\begin{aligned} P_j\mathbf{v}_{j+1} &= P_j\left(I_1^{(j+1)}\mathbf{v}_1^{(j+1)} + I_2^{(j+1)}\mathbf{v}_2^{(j+1)}\right) \\ &= I_1^{(j)} P_i^{(j)}\mathbf{v}_1^{(j+1)} + I_2^{(j)} P_2^{(j)}\mathbf{v}_2^{(j+1)}. \end{aligned}$$

Therefore, we have the decompositions,

$$\begin{aligned} \mathbf{v}_j &= I_1^{(j)}\mathbf{v}_1^{(j)} + I_2^{(j)}\mathbf{v}_2^{(j)} \\ &= I_1^{(j)}\left(\mathbf{v}_{f,\,1}^{(j)} + P_1^{(j)}\mathbf{v}_1^{(j+1)}\right) + I_2^{(j)}\left(\mathbf{v}_{f,\,2}^{(j)} + P_2^{(j)}\mathbf{v}_2^{(j+1)}\right) \\ &= \mathbf{v}_j^{(f)} + P_j\mathbf{v}_{j+1}. \end{aligned}$$

Now use this particular decomposition in (5.56). We have, ($\mathbf{v} = \mathbf{v}_0$),

$$\begin{aligned} \mathbf{v}^T B\mathbf{v} &\leq \left[\mathbf{v}_\ell^T A_\ell\mathbf{v}_\ell + \sum_{j=0}^{\ell-1} \left(\mathbf{v}_j^f\right)^T \Lambda_j\mathbf{v}_j^f\right] \\ &\leq 2\left[\sum_{i=1}^{2} \mathbf{v}_i^{(\ell)^T} A_i^{(\ell)}\mathbf{v}_i^{(\ell)} + \sum_{j=0}^{\ell-1}\sum_{i=1}^{2} \left(I_i^{(j)}\mathbf{v}_{f,\,i}^{(j)}\right)^T \Lambda_j\left(I_i^{(j)}\mathbf{v}_{f,\,i}^{(j)}\right)\right] \\ &= 2\sum_{i=1}^{2}\left[\mathbf{v}_i^{(\ell)^T} A_i^{(\ell)}\mathbf{v}_i^{(\ell)} + \sum_{j=0}^{\ell-1} \left(\mathbf{v}_{f,\,i}^{(j)}\right)^T \Lambda_i^{(j)}\mathbf{v}_{f,\,i}^{(j)}\right] \\ &\leq 2\sum_{i=1}^{2} K_i \; \mathbf{v}_i^{(0)^T} A_i^{(0)}\mathbf{v}_i^{(0)} \\ &\leq 2\left(\max_{i=1,\,2} \; K_i\right) \sum_{i=1}^{2} \mathbf{v}_i^{(0)^T} A_i^{(0)}\mathbf{v}_i^{(0)} \\ &\leq 2\left(\max_{i=1,\,2} \; K_i\right) \eta_s \; \mathbf{v}_0^T A_0\mathbf{v}_0 \\ &= 2\left(\max_{i=1,\,2} \; K_i\right) \eta_s \; \mathbf{v}^T A\mathbf{v}. \end{aligned}$$

□

5.8 Cascadic multigrid

The cascadic multigrid has been proposed by Deuflhard et al. [DLY89] (see also [Dfl94]), and analyzed in Shaidurov [Sh94], and [BD96].

The main ingredient is the following smoothing property of the CG method. The CG method applied to $A\mathbf{v} = \mathbf{b}$ with zero initial iterate leads to an approximation $\mathbf{v}_m$ after $m \geq 1$ iterations, that satisfies the estimate

$$\|\mathbf{v} - \mathbf{v}_m\|_A \leq \frac{\|A\|^{1/2}}{2m+1} \|\mathbf{v}\|.$$

(A) Assume now that $\mathbf{v} = (I - \pi_A)\mathbf{v}$ is A-orthogonal to the coarse space Range(P). Here, $\pi_A = P(P^TAP)^{-1}P^TA$. Then, assume that

(B) $\pi_A = PA_c^{-1}P^TA$ is ℓ_2-bounded,

$$\|A\| \|(I - \pi_A)\mathbf{v}\|^2 \leq \eta_a \, \mathbf{v}^T A\mathbf{v}.$$

Based on Lemma 5.20, we can ensure (B) if the following strong approximation property holds.

(C) Strong approximation property:
for every $\mathbf{v}$, there is a coarse interpolant $P\mathbf{v}_c$ such that

$$(\mathbf{v} - P\mathbf{v}_c)^T A(\mathbf{v} - P\mathbf{v}_c) \leq \frac{\eta_a}{\|A\|} \|A\mathbf{v}\|^2.$$

We verified such an estimate (cf. (1.16)) for f.e. matrices coming from the Poisson equation $-\Delta u = f$ in Ω and $u = 0$ on $\partial\Omega$, which admits full regularity; that is, $\|u\|_2 \leq C \; \|f\|$. Such regularity estimates are available for convex polygonal domains Ω (cf., e.g., [TW05]).

The following "cascadic" two-grid (or CTG) algorithm is of interest.

Algorithm 5.8.1 (Two-grid cascadic method)*Consider* $A\mathbf{x} = \mathbf{b}$. *Let* P *be an interpolation matrix, and* $A_c = P^TAP$ *the respective coarse matrix. Let also* Λ *be a s.p.d. preconditioner to* A *(such as symmetric Gauss–Seidel, or simply Jacobi). The two-grid cascadic algorithm computes an approximation* $\mathbf{x}_{CTG}$ *to the exact solution* $\mathbf{x} = A^{-1}\mathbf{b}$ *in the following steps.*

(i) Solve the coarse-grid problem

$$A_c\mathbf{x}_c = P^T\mathbf{b}.$$

(ii) Interpolate and compute the residual, $\mathbf{r} = \mathbf{b} - AP\mathbf{x}_c = (I - \pi_A)^T\mathbf{b}$, *where* $\pi_A = PA_c^{-1}P^TA$ *is the coarse-grid projection.*

(iii) Apply $m \geq 1$ *PCG iterations to* $A\mathbf{v} = \mathbf{r}$ *with initial iterate* $\mathbf{v}_0 = 0$. *Let* $\mathbf{v}_m$ *be the resulting* m*th iterate.*

(iv) Compute the cascadic TG approximation $\mathbf{x}_{CTG} = \mathbf{v}_m + P\mathbf{x}_c$ *to the exact solution* $\mathbf{x} = A^{-1}\mathbf{b}$.

The following error estimate is immediate, letting $\overline{A} = \Lambda^{-(1/2)} A \Lambda^{-(1/2)}$,

$$\begin{aligned}
\|\mathbf{v} - \mathbf{v}_m\|_A &\leq \frac{\|\overline{A}\|^{1/2}}{2m+1} \|\mathbf{v}\|_\Lambda \\
&= \frac{\|\overline{A}\|^{1/2}}{2m+1} \|A^{-1}\mathbf{r}\|_\Lambda \\
&= \frac{\|\overline{A}\|^{1/2}}{2m+1} \|(I - \pi_A)A^{-1}\mathbf{b}\|_\Lambda \\
&= \frac{\|\overline{A}\|^{1/2}}{2m+1} \|(I - \pi_A)\mathbf{x}\|_\Lambda \\
&\leq \frac{\|\overline{A}\|^{1/2}}{2m+1} \|\Lambda\|^{1/2} \|(I - \pi_A)\mathbf{x}\| \\
&\leq \frac{\sqrt{\eta_a}}{2m+1} \frac{\left\|\Lambda^{-(1/2)} A \Lambda^{-(1/2)}\right\|^{1/2} \|\Lambda\|^{1/2}}{\|A\|^{1/2}} \|(I - \pi_A)\mathbf{x}\|_A.
\end{aligned}$$

In the last step, we used the ℓ_2-boundedness (B) of π_A.

Letting

$$\sqrt{\overline{\eta}_a} = \sqrt{\eta_a} \frac{\left\|\Lambda^{-(1/2)} A \Lambda^{-(1/2)}\right\|^{1/2} \|\Lambda\|^{1/2}}{\|A\|^{1/2}},$$

the overall error then can be estimated as follows.

$$\begin{aligned}
\|\mathbf{x} - \mathbf{x}_{CTG}\|_A &= \|A(\mathbf{x} - P\mathbf{x}_c) - A\mathbf{v}_m\|_{A^{-1}} \\
&= \|A(\mathbf{v} - \mathbf{v}_m)\|_{A^{-1}} \\
&= \|\mathbf{v} - \mathbf{v}_m\|_A \\
&\leq \frac{\sqrt{\overline{\eta}_a}}{2m+1} \|(I - \pi_A)\mathbf{x}\|_A.
\end{aligned}$$

The multilevel version of the cascadic MG (or CMG) replaces the exact solution at Step (i) above with a $P\mathbf{x}_c$, which is the coarse-grid approximation at hand (at the initial coarse-level $\ell \geq 1$ we use the exact solution).

The analysis then is similar as before. We have for $\mathbf{r} = \mathbf{b} - AP\mathbf{x}_c = \mathbf{b} - A\pi_A\mathbf{x} + A(\pi_A\mathbf{x} - P\mathbf{x}_c)$. Note that here $\pi_A\mathbf{x}$ is the exact coarse-grid solution. Then using the best polynomial approximation property of the PCG method, we have the estimate

$$\begin{aligned}
\|\mathbf{v} - \mathbf{v}_m\|_A &\leq \inf_{p_m : p_m(0)=1} \left\| p_m\left(\Lambda^{-(1/2)} A \Lambda^{-(1/2)}\right)\mathbf{v}\right\|_A \\
&\leq \frac{\sqrt{\overline{\eta}_a}}{2m+1} \|\mathbf{x} - \pi_A\mathbf{x}\|_A + \|\pi_A\mathbf{x} - P\mathbf{x}_c\|_A.
\end{aligned} \tag{5.57}$$

Here, we use $p_m(t)$ coming from the Chebyshev polynomial T_{2m+1} defined as (cf., section 6.13.2),

$$T_{2m+1}(t) = (-1)^\nu \, (2m+1)t \; p_m\left(\|\overline{A}\|t^2\right).$$

Above, we used the following optimal property of p_m,

$$\min_{p_m: p_m(0)=1} \max_{t \in [0,\, \|\overline{A}\|]} |\sqrt{t}\, p_m(t)| = \max_{t \in [0,\, \|\overline{A}\|]} |\sqrt{t}\, p_m(t)| = \frac{\sqrt{\|\overline{A}\|}}{2m+1}.$$

In a multilevel setting, we let $A_0 = A$ be the finest-grid matrix, P_k be the interpolation matrix from coarse-level $k+1$ to fine-level k and let $A_{k+1} = P_k^T A_k P_k$ be the coarse $k+1$-level matrix. Finally, let Λ_k be the s.p.d. preconditioner that will be used for the kth-level PCG iterations. Simple examples of Λ_k are the symmetric Gauss–Seidel or Jacobi preconditioners for A_k. The resulting cascadic MG algorithm takes the following form.

Algorithm 5.8.2 (Cascadic MG) *Consider $A\mathbf{x} = \mathbf{b}$. The cascadic MG algorithm computes $\mathbf{x}_{CMG} = \overline{\mathbf{x}}_0$ in the following steps.*

- *Let $\mathbf{b}_0 = \mathbf{b}$. For $k = 1, \ldots, \ell$ compute $\mathbf{b}_k = P_{k-1}^T \mathbf{b}_{k-1}$.*

(o) Solve the coarse-grid problem

$$A_\ell \overline{\mathbf{x}}_\ell = \mathbf{b}_\ell.$$

For $k = \ell - 1, \ldots, 0$ perform the following steps:

(i) Interpolate $\mathbf{x}^{(0)} = P_k \overline{\mathbf{x}}_{k+1}$.
(ii) Compute residual $\mathbf{r} = \mathbf{b}_k - A_k \mathbf{x}^{(0)}$.
(iii) Apply $m = m_k \geq 1$ PCG iterations to $A_k \mathbf{v} = \mathbf{r}$ with initial iterate $\mathbf{v}_0 = 0$. Let $\mathbf{v}_m$ be the resulting mth iterate.
(iv) Compute the kth-level cascadic MG approximation $\overline{\mathbf{x}}_k = \mathbf{v}_m + \mathbf{x}^{(0)}$.

- *Finally, set $\mathbf{x}_{CMG} = \overline{\mathbf{x}}_0$.*

Introducing the composite level k-to-0 interpolation matrices $\overline{P}_{k-1} = P_0 \ldots P_{k-1}$ and respective projections $\pi_k = \overline{P}_{k-1} A_k^{-1} \overline{P}_{k-1}^T A$, noticing then that $\pi_j \mathbf{x}$ ($j = k-1,\ k$) represent the jth-level exact solutions $\overline{P}_{j-1} A_j^{-1} \mathbf{b}_j$ interpolated to the finest-level, the last two-level estimate (5.57) translates to (with $m = m_k$),

$$\|\pi_{k-1}\mathbf{x} - \mathbf{x}_{k-1}\|_A \leq \frac{\sqrt{\overline{\eta}_a}}{2m_k + 1} \|\pi_{k-1}\mathbf{x} - \pi_k \mathbf{x}\|_A + \|\pi_k \mathbf{x} - \mathbf{x}_k\|_A,$$

where $\mathbf{x}_j = \overline{P}_{j-1} \overline{\mathbf{x}}_j$ stands for the approximation computed at level j, interpolated to the finest-level 0 (for $\overline{\mathbf{x}}_j$ see Step (iv) of Algorithm 5.8.2). Note also, that

$$\sqrt{\overline{\eta}_a} = \sqrt{\eta_a} \frac{\|\Lambda_{k-1}^{-(1/2)} A_{k-1} \Lambda_{k-1}^{-(1/2)}\|^{1/2} \|\Lambda_{k-1}\|^{1/2}}{\|A_{k-1}\|^{1/2}},$$

and η_a is a uniform constant that relates two consecutive levels, k and $k-1$,

$$\|A_{k-1}\| \|(I - P_{k-1} A_k^{-1} P_{k-1}^T A_{k-1})\mathbf{v}\|^2 \leq \eta_a \|\mathbf{v}\|_{A_{k-1}}^2. \tag{5.58}$$

Also,

$$\overline{\eta}_a = \eta_a \frac{\|\Lambda_{k-1}^{-(1/2)} A_{k-1} \Lambda_{k-1}^{-(1/2)}\| \|\Lambda_{k-1}\|}{\|A_{k-1}\|}$$

can be assumed bounded independently of k (recall that Λ_{k-1} was the preconditioner for A_{k-1} at level $k-1$, such as the diagonal of A_{k-1}, e.g.). Therefore, we have the following bound on the multilevel error $\mathbf{x} - \mathbf{x}_{CMG}$, assuming that at level k we have performed $m_k \geq 1$ PCG smoothing iterations, and because $\pi_\ell \mathbf{x} = \mathbf{x}_\ell$ (i.e., we use exact solve at the coarsest level ℓ),

$$\|\mathbf{x} - \mathbf{x}_{CMG}\|_A \leq \sqrt{\overline{\eta}_a} \sum_{k=1}^{\ell} \frac{1}{2m_k + 1} \|(\pi_{k-1} - \pi_k)\mathbf{x}\|_A. \tag{5.59}$$

Using the Cauchy–Schwarz inequality, we have the following estimate as a corollary,

$$\|\mathbf{x} - \mathbf{x}_{CMG}\|_A \leq \sqrt{\overline{\eta}_a} \left(\sum_{k=1}^{\ell} \frac{1}{(2m_k + 1)^2} \right)^{1/2} \|\mathbf{x}\|_A.$$

Here, we used the fact that $\sum_k \|(\pi_{k-1} - \pi_k)\mathbf{x}\|_A^2 = \sum_k (\|\pi_{k-1}\mathbf{x}\|_A^2 - \|\pi_k \mathbf{x}\|_A^2) \leq \|\mathbf{x}\|_A^2$.

The latter estimate will provide a uniform bound (less than one) if the number of smoothing steps increases geometrically with k; that is, we have the following main result.

Theorem 5.39. *Let m_k be the number of smoothing PCG iterations at level k that satisfy $2m_k + 1 = (2m_0 + 1)\mu^k$, for a $\mu > 1$. Then, under the uniform assumption (B) (valid, at every two levels $k-1$ and k as in (5.58)), the cascadic MG method provides an approximation $\mathbf{x}_{CMG}$ to the exact solution $\mathbf{x}$ of $A\mathbf{x} = \mathbf{b}$ ($A = A_0$), such that*

$$\|\mathbf{x} - \mathbf{x}_{CMG}\|_A \leq \frac{\mu \sqrt{\overline{\eta}_a}}{\mu - 1} \frac{1}{2m_0 + 1} \|\mathbf{x}\|_A.$$

Thus, if m_0 is sufficiently large, the CMG method provides an approximate inverse to A defined as $\mathbf{b} \mapsto \mathbf{x}_{CMG} \approx \mathbf{x} = A^{-1}\mathbf{b}$. Assuming that n_k, the number of unknowns at level k, satisfy $n_k = \beta\, n_{k+1}$ for a $\beta > 1$, we have the restriction $\mu < \beta$ in order to have optimal complexity $\mathcal{O}(n_0)$ of the resulting CMG.

Proof. The complexity of the CMG is readily seen to be of order $\sum_k m_k n_k = n_0 \sum_k m_k/\beta^k \leq n_0(m_0 + \frac{1}{2}) \sum_k (\mu/\beta)^k$, which is of order $\mathcal{O}(n_0)$ if $\mu < \beta$. □

Cascadic MG with stationary smoothing

The PCG smoothing steps in Algorithm 5.8.2 can be replaced by a more standard stationary method. The "smoothing" rate of $\sim 1/(2m + 1)$ will then be generally reduced. More specifically, the following result holds.

Let M be a matrix (not necessarily symmetric), which provides a convergent iteration for $A\mathbf{x} = \mathbf{b}$ in the A-norm. That is, let $\|I - A^{1/2}M^{-1}A^{1/2}\| < 1$. Equivalently, let $M^T + M - A$ be s.p.d. Let

$$\overline{L} = M(M + M^T - A)^{-(1/2)}, \quad \overline{A} = \overline{L}^{-1} A \overline{L}^{-T}, \quad \text{and} \quad \overline{\mathbf{b}} = \overline{L}^{-1}\mathbf{b}.$$

Consider the following iteration, starting with $\mathbf{x}_0 = 0$, for $k \geq 1$,

$$\left(\overline{L}^T \mathbf{x}_k\right) = \left(\overline{L}^T \mathbf{x}_{k-1}\right) + \left(\overline{\mathbf{b}} - \overline{A}(\overline{L}^T \mathbf{x}_{k-1})\right).$$

Its computationally feasible equivalent version reads,

$$\begin{aligned}
\mathbf{x}_k &= \mathbf{x}_{k-1} + \overline{L}^{-T}\overline{L}^{-1}(\mathbf{b} - A\mathbf{x}_{k-1}) \\
&= \mathbf{x}_{k-1} + M^{-T}(M + M^T - A)M^{-1}(\mathbf{b} - A\mathbf{x}_{k-1}) \\
&= \mathbf{x}_{k-1} + \left(M^{-T} + M^{-1} - M^{-T}AM^{-1}\right)(\mathbf{b} - A\mathbf{x}_{k-1}).
\end{aligned}$$

Another, more familiar form of the above iteration reads,

$$\begin{aligned}
\mathbf{x}_{k-(1/2)} &= \mathbf{x}_{k-1} + M^{-1}(\mathbf{b} - A\mathbf{x}_{k-1}) \\
\mathbf{x}_k &= \mathbf{x}_{k-(1/2)} + M^{-T}(\mathbf{b} - A\mathbf{x}_{k-(1/2)}).
\end{aligned} \tag{5.60}$$

Introducing $\overline{E} = I - \overline{A}$, $\mathbf{e}_k = \mathbf{x} - \mathbf{x}_k$, $\mathbf{e}_0 = \mathbf{x} = A^{-1}\mathbf{b}$, noticing that $\overline{L}^T \mathbf{e}_k = \overline{E}(\overline{L}^T \mathbf{e}_{k-1})$, after m iterations, we have the following error estimate,

$$\begin{aligned}
\|\overline{L}^T \mathbf{e}_m\|_{\overline{A}} &= \|\overline{E}^m \, \overline{L}^T \mathbf{e}_0\|_{\overline{A}} \\
&= \|\overline{A}^{1/2}(I - \overline{A})^m \, \overline{L}^T \mathbf{e}_0\| \\
&\leq \max_{t \in [0,1]} t^{1/2}(1 - t)^m \, \|\overline{L}^T \mathbf{e}_0\| \\
&= \frac{1}{\sqrt{m+1}}\left(1 - \frac{1}{m+1}\right)^m \|\overline{L}^T \mathbf{e}_0\|.
\end{aligned}$$

That is, with the symmetrized smoother $\overline{M} = M(M + M^T - A)^{-1}M^T$, the following smoothing rate holds.

Lemma 5.40. *Consider M that provides a convergent iteration in the A-norm, for $A\mathbf{x} = \mathbf{b}$; that is, $\|I - A^{1/2}M^{-1}A^{1/2}\| < 1$ (or equivalently, let $M + M^T - A$ be s.p.d.). Perform $m \geq 1$ combined smoothing steps as in (5.60), that is, effectively based on the symmetrized smoother $\overline{M} = M(M + M^T - A)^{-1}M^T$, starting with $\mathbf{x}_0 = 0$. The error $\mathbf{e}_m = \mathbf{x} - \mathbf{x}_m$ satisfies the estimate,*

$$\|\mathbf{e}_m\|_A \leq \frac{1}{\sqrt{m+1}} \|\mathbf{e}_0\|_{\overline{M}}.$$

Assuming now the same estimate (5.58) letting $\eta_a \; (\|\overline{M}_{k-1}\|/\|A_{k-1}\|) \leq \overline{\eta}_a$, which we assume bounded independently of k, estimate (5.59) takes the following form,

$$\|\mathbf{x} - \mathbf{x}_{CMG}\|_A \leq \sqrt{\overline{\eta}_a} \sum_{k=1}^{\ell} \frac{1}{\sqrt{m_k + 1}} \|(\pi_{k-1} - \pi_k)\mathbf{x}\|_A. \tag{5.61}$$

Using again the Cauchy–Schwarz inequality as a corollary, we have the following estimate,

$$\|\mathbf{x} - \mathbf{x}_{CMG}\|_A \leq \sqrt{\eta_a} \left(\sum_{k=1}^{\ell} \frac{1}{m_k + 1} \right)^{1/2} \|\mathbf{x}\|_A.$$

Finally, the following analogue of Theorem 5.39 holds.

Theorem 5.41. *Let m_k be the number of stationary smoothing iterations based on M_k and M_k^T at level k. Let m_k satisfy $m_k + 1 = (m_0 + 1)\mu^k$, for a $\mu > 1$. Then, under the uniform assumption (B) (valid, at every two levels $k-1$ and k as in (5.58)), the cascadic MG method (with stationary smoothing) provides an approximation $\mathbf{x}_{CMG}$ to the exact solution $\mathbf{x}$ of $A\mathbf{x} = \mathbf{b}$ ($A = A_0$), such that*

$$\|\mathbf{x} - \mathbf{x}_{CMG}\|_A \leq \sqrt{\frac{\eta_a \mu}{\mu - 1}} \frac{1}{\sqrt{m_0 + 1}} \|\mathbf{x}\|_A.$$

Thus, for m_0 sufficiently large, the CMG method provides an approximate inverse to A defined as $\mathbf{b} \mapsto \mathbf{x}_{CMG} \approx \mathbf{x} = A^{-1}\mathbf{b}$. Assuming that n_k, the number of unknowns at level k, satisfy $n_k = \beta\, n_{k+1}$ for a $\beta > 1$, we have the restriction $\mu < \beta$ in order to have optimal complexity $\mathcal{O}(n_0)$ of the resulting CMG.

5.8.1 Convergence in a stronger norm

We can also prove convergence of CMG in a stronger norm.

Based on the strong approximation property (C) as stated in Theorem 1.7, we have that

$$\|A\| \|(\pi_{k-1} - \pi_k)\mathbf{x}\|_A^2 \leq \eta_a \|A\mathbf{x}\|^2.$$

Here η_a depends on the ratio of the fine-grid mesh-size $h = h_0$ and the kth-level mesh-size $h_k = 2^k h_0$. That is, $\eta_a \simeq (h_k/h_0)^2 = 2^{2k}$. Next, because $\|A(\mathbf{x} - \mathbf{x}_{CMG})\| \leq \|A\|^{1/2} \|\mathbf{x} - \mathbf{x}_{CMG}\|_A$, based on estimate (5.61), for example, and the above strong approximation property, we arrive at

$$\|A(\mathbf{x} - \mathbf{x}_{CMG})\| \leq \sqrt{\eta_a} \left(\sum_{k=1}^{\ell} \frac{2^k}{\sqrt{m_k + 1}} \right) \|A\mathbf{x}\|.$$

If we use PCG as a smoother based on estimate (5.59), the following estimate is similarly derived,

$$\|A(\mathbf{x} - \mathbf{x}_{CMG})\| \leq \sqrt{\eta_a} \left(\sum_{k=1}^{\ell} \frac{2^k}{2m_k + 1} \right) \|A\mathbf{x}\|.$$

Thus the following results hold.

Theorem 5.42. *Consider the CMG algorithm with either PCG or stationary iteration as smoother. Assume that degrees of freedom n_{k-1} and n_k, at level k and level $k-1$, respectively, satisfy $n_{k-1} \simeq 2^d n_k$, $d = 2, 3$, and let the constants η_a in the level k to finest-level 0, $\|(I - \pi_k)\mathbf{x}\|_A^2 \leq (\eta_a/\|A\|) \;\|A\mathbf{x}\|^2$ grows like 2^{2k} (based on Theorem 1.7). Then, choose $2m_k + 1 = (2m_0 + 1)\mu^k$ in the case of PCG as the smoother or $m_k + 1 = (m_0 + 1)\mu^k$ in the stationary smoother case. For optimal complexity (i.e., the total work to be of order n_0), we need*

$$\sum_k m_k n_k \simeq n_0 \sum_k \left(\frac{\mu}{2^d}\right)^k \simeq n_0.$$

That is, $\mu < 2^d$. To have the CMG method convergent in the $\|A(.)\|$-norm (i.e.,

$$\|A(\mathbf{x} - \mathbf{x}_{CMG})\| \leq \frac{c}{1 + 2m_0} \;\|A\mathbf{x}\|,$$

with PCG as smoother), we need $\mu > 2$, and in the case of a stationary smoother, in order to have an estimate of the form

$$\|A(\mathbf{x} - \mathbf{x}_{CMG})\| \leq \frac{c}{\sqrt{1 + m_0}} \;\|A\mathbf{x}\|,$$

we need $\sqrt{\mu} > 2$. In summary, for the PCG smoother the conditions $2^d > \mu > 2$ are possible for $d = 2,\; 3$, whereas in the case of a stationary smoother both inequalities $2^d > \mu > 4$ are possible only for $d = 3$.

Relation to the variable V-cycle

We should note that the cascadic MG with stationary smoothing can be seen to give the second (coarse-to-fine) half of a variable V-cycle method as described in Theorem 5.24. The latter is seen because both are product iteration methods with subspaces Range $(\overline{P}_{k-1})$, $k = \ell, \ldots, 1$ and the original vector space itself. More specifically, the iteration matrix E_{CMG} of the CMG with stationary smoothers M_k admits the product form (cf. Section 3.2.1),

$$(I - M_0^{-1}A)^{m_0}(I - P_0 M_1^{-1} P_0^T A)^{m_1} \cdots (I - \overline{P}_{\ell-2} M_{\ell-1}^{-1} \overline{P}_{\ell-2}^T A)^{m_{\ell-1}}(I - \pi_\ell).$$

Whereas, the iteration matrix E_{MG} of the variable V-cycle MG admits the following product form,

$$E_{MG} = E_{CMG} A^{-1} E_{CMG}^T A.$$

The (minor) difference is that the cascadic MG solves at the coarsest-level problem with a particular r.h.s., namely, $\overline{P}_{\ell-1}^T \mathbf{b}$, whereas the variable V-cycle coarse-level ℓ problem has generally a different r.h.s. It is given by $\overline{P}_{\ell-1}^T(\mathbf{b} - A\mathbf{x}_0)$, where $\mathbf{x}_0$ is the approximation provided by the first (fine-to-coarse) half of the variable V-cycle. Note, that $\mathbf{x}_0$ will generally be nonzero.

Let $E_{MG} = I - B_{MG}^{-1}A$ be the error propagation matrix of the variable V-cycle. Note that the relation $E_{MG} = E_{CMG}A^{-1}(E_{CGM})^T A$ allows us to analyze nonsymmetric smoothers in cascadic MG because we have the following identity,

$$\mathbf{e}^T A E_{MG}\mathbf{e} = \|A^{-(1/2)}E_{CMG}^T A\mathbf{e}\|^2.$$

Thus, with $X = A^{1/2}E_{CMG}A^{-(1/2)}$ based on fact that $\|X\| = \|X^T\|$, we have $\varrho_{CMG}^2 = \|A^{1/2}E_{CMG}A^{-(1/2)}\|^2 = \|A^{-(1/2)}E_{CMG}^T A^{1/2}\|^2 = \|A^{1/2}E_{MG}A^{-(1/2)}\| = \varrho_{MG}$. That is, the convergence factor of the cascadic MG equals the square root of the convergence factor of the corresponding variable V-cycle MG. The latter was estimated in Theorem 5.24.

Cascadic MG as discretization method

Another feature of the CMG is that it can be used as a discretization procedure. If we solve the Poisson equation $-\Delta u = f$ on a sequence of uniformly refined meshes of size $h_{k+1} = 2h_k$, we typically have the following error behavior, $|u - u_h|_1 \leq Ch\ \|f\|_0$. Here $\|.\|_0$ stands for the integral L_2-norm, and $|v|_1 = \|\nabla v\|_0$ is the L_2-norm of the gradient of v. The latter error estimate, translates to the following matrix–vector analogue (at discretization level k), assuming that the initial coarse mesh size $h_\ell = \mathcal{O}(1)$,

$$\|(I - \pi_k)\mathbf{v}\|_A \leq C\,\frac{1}{2^{\ell-k}}\,\|f\|_0.$$

Then, it is clear, we can get an estimate (based on (5.59))

$$\|\mathbf{x} - \mathbf{x}_{CMG}\|_A \leq C\,\frac{1}{2^\ell}\,\|f\|_0,$$

if $(2m_k + 1) \simeq (2m_0 + 1)\mu^k$ such that $\sum_k 2^k/\mu^k$ is finite, that is, if $\mu > 2$. Let n_k be the number of unknowns at level k. We typically have $n_k = 2^d\ n_k$, where $d = 2$ or $d = 3$ is the dimension of the domain where the Poisson equation is posed.

The complexity of the cascadic MG is then readily seen to be of order $\sum_k m_k n_k = n_0 \sum_k \mu^k 2^{-dk}$ ($d = 2$ or 3). Thus, if we can satisfy the following inequalities,

$$2 < \mu < 2^d$$

(possible for $d > 1$; e.g., $\mu = 3$) the CMG provides a discretization method of optimal complexity.

If we use a stationary smoother, the inequalities read

$$4 < \mu < 2^d,$$

which is possible for $d = 3$. The case $d = 2$ leads to suboptimal estimates. The latter can be avoided if a different coarsening factor is used, for example, $h_k = \frac{1}{4}h_{k+1}$. Then, the conditions are $4 < \mu < 4^d$, which is possible for $d > 1$.

5.9 The hierarchical basis (HB) method

5.9.1 The additive multilevel HB

The multilevel counterpart of B_{TL} defined in Definition 3.12 and analyzed in Section 3.2.8 leads to the classical hierarchical basis (or HB) method, originally considered by Yserentant in [Yhb] and its multiplicative version (referred to as the HBMG) in [BDY88].

Assume a fine-grid vector space $\mathbf{V}$ and a coarse one $\mathbf{V}_c$ such that for a given interpolation matrix P, $P\mathbf{V}_c \subset \mathbf{V}$. We assume that P admits a natural block form,

$$P = \begin{bmatrix} * \\ I \end{bmatrix}$$

where the identity block reflects the embedding of the coarse dofs (or nodes) $\mathcal{N}_c$ into the fine-grid dofs $\mathcal{N}$. That is, for any $i_c \in \mathcal{N}_c$ there is a unique $i = i(i_c) \in \mathcal{N}$, so that we can write $\mathcal{N}_c \subset \mathcal{N}$ (for details refer to Section 1.2).

We consider a two-level direct decomposition $\mathbf{v} = J\mathbf{v}_f + P\mathbf{v}_c$, where

$$J = \begin{bmatrix} I \\ 0 \end{bmatrix} \quad \text{and} \quad P = \begin{bmatrix} * \\ I \end{bmatrix}.$$

Let $R = [0,\ I]$. Define then $\mathcal{I} = PR$, the so–called nodal interpolation operator. Note that $\mathcal{I}$ is a projection (i.e., $\mathcal{I}^2 = \mathcal{I}$), because $RP = I$. Using the nodal interpolation operator $\mathcal{I}$, we can rewrite the above direct decomposition as follows,

$$\mathbf{v} = (I - \mathcal{I})\mathbf{v} + \mathcal{I}\mathbf{v}.$$

Because

$$I - \mathcal{I} = \begin{bmatrix} * \\ 0 \end{bmatrix}$$

has the same range as J and $\mathcal{I}$ has the same range as P, it is clear that the two direct decompositions are the same.

For vector spaces $\mathbf{V}_c$ and $\mathbf{V}$ corresponding to two nested finite element spaces V_H and V_h, the corresponding nodal interpolation operator $\mathcal{I} = \mathcal{I}_h^H$ is not stable in energy (A-norm) when $(H/h) \simeq 2^k$ grows (for details, see (G.5) and (G.4)). Assume now a sequence of finite element spaces $V_k = V_{h_k}$ and respective kth-level interpolation operators $\mathcal{I}_k = \mathcal{I}_{h_0}^{h_k}$, $(h_k = 2^k h_0)$, which relates the kth coarse-level and the finest-level 0 vector spaces. Denote the vector spaces $\overline{\mathbf{V}}_k$ corresponding to the finite element space V_k. Note that we view here $\overline{\mathbf{V}}_k$ as subspaces of the fine-grid vector space $\mathbf{V} = \overline{\mathbf{V}}_0$.

The direct multilevel decomposition of interest then reads,

$$\mathbf{v} = \sum_{k=0}^{\ell-1} (\mathcal{I}_k - \mathcal{I}_{k+1})\mathbf{v} + \mathcal{I}_\ell \mathbf{v}. \tag{5.62}$$

The following estimate is easily derived (based on (G.5), and (G.4)),

$$\|\mathcal{I}_\ell \mathbf{v}\|_A^2 + \sum_{k=0}^{\ell-1} \|(\mathcal{I}_k - \mathcal{I}_{k+1})\mathbf{v}\|_A^2 \leq C_\ell \ \|\mathbf{v}\|_A^2, \tag{5.63}$$

where C_ℓ grows like ℓ^2 in two dimensions (the main result in [Yhb]) and like 2^ℓ in three dimensions (see [O97]).

To bound the sum on the left is a necessary and sufficient condition for a convergent additive multilevel HB method. The latter is defined similarly to Definition 5.32. More specifically, let A_k be the stiffness matrix coming from the space V_k and the bilinear form $a(\cdot, \ \cdot)$. The matrix A_k admits a natural two-by-two block form (referred to as an "f"–"c" block form),

$$A_k = \begin{bmatrix} A_{k,\,ff} & A_{k,\,fc} \\ A_{k,\,cf} & A_{k,\,cc} \end{bmatrix} \begin{matrix} \} & \text{“}f\text{''-dofs} \equiv \mathcal{N}_k \setminus \mathcal{N}_{k+1} \\ \} & \text{“}c\text{''-dofs} \equiv \mathcal{N}_{k+1} \end{matrix} .$$

In the finite element case, we have that $A_{k,\,ff}$ is spectrally equivalent to its diagonal (proven on an element matrix level). For more details, see Appendix B. Hence, we can easily find a s.p.d. matrix Λ_k that is spectrally equivalent to $A_{k,\,ff}$, for example, the symmetric Gauss–Seidel one. Denote by J_k the extension by zero of vectors defined on $\mathcal{N}_k \setminus \mathcal{N}_{k+1}$ to vectors in $\mathcal{N}_0$; that is, let

$$J_k = \begin{bmatrix} 0 \\ I \\ 0 \end{bmatrix} \begin{matrix} \} & \mathcal{N}_0 \setminus \mathcal{N}_k \\ \} & \mathcal{N}_k \setminus \mathcal{N}_{k+1} \\ \} & \mathcal{N}_{k+1} \end{matrix} .$$

Note that for $k = \ell$, we have

$$J_\ell = \begin{bmatrix} 0 \\ I \end{bmatrix} \begin{matrix} \} & \mathcal{N}_0 \setminus \mathcal{N}_\ell \\ \} & \mathcal{N}_\ell \end{matrix} .$$

Then the additive multilevel HB preconditioner B_{HB} is defined as

$$B_{HB}^{-1} = \mathcal{I}_\ell J_\ell A_\ell^{-1} J_\ell^T (\mathcal{I}_\ell)^T + \sum_{k=0}^{\ell-1} (\mathcal{I}_k - \mathcal{I}_{k+1})\, J_k \Lambda_k^{-1} J_k^T\, (\mathcal{I}_k - \mathcal{I}_{k+1})^T . \tag{5.64}$$

To implement B_{HB}^{-1}, we need to be able to identify the $\mathcal{N}_{k+1}$ as subset of $\mathcal{N}_k$ represented by a mapping I_{k+1}^k and to interpolate from level $k+1$ to the next fine-level k. The latter is typically represented by the interpolation mapping P_k.

The definition of the "nodal interpolation" operator in matrix form translates to

$$\mathcal{I}_k = P_0 P_1 \cdots P_{k-1} (I_1^0 \ldots I_k^{k-1})^T . \tag{5.65}$$

Note that up to a proper reordering, we have

$$I_{k+1}^k = \begin{bmatrix} 0 \\ I \end{bmatrix} \begin{matrix} \} & \mathcal{N}_k \setminus \mathcal{N}_{k+1} \\ \} & \mathcal{N}_{k+1} \end{matrix}$$

and similarly

$$P_k = \begin{bmatrix} * \\ I \end{bmatrix} \begin{matrix} \} & \mathcal{N}_k \setminus \mathcal{N}_{k+1} \\ \} & \mathcal{N}_{k+1} \end{matrix}.$$

The latter is based on the fact P_k is the identity at the coarse nodes $\mathcal{N}_{k+1}$. Therefore, $(I_{k+1}^k)^T P_k = I$. The latter property implies the following result.

Lemma 5.43. *The nodal interpolation operators defined in* (5.65) *are projections (i.e.,* $\mathcal{I}_k^2 = \mathcal{I}_k$*) and satisfy the identity* $\mathcal{I}_{k+1}\mathcal{I}_k = \mathcal{I}_{k+1}$.

Our goal is to derive the following representation of the additive multilevel HB preconditioner.

Theorem 5.44. *Introducing* $\overline{P}_k = P_0 P_1 \cdots P_{k-1}$, *the composite interpolation matrix from level k all the way to the finest-level 0, we have the representation*

$$B_{HB}^{-1} = \overline{P}_\ell A_\ell^{-1} \overline{P}_\ell^T + \sum_{k=0}^{\ell-1} \overline{P}_k J_k^{k+1} \Lambda_k^{-1} (J_k^{k+1})^T \overline{P}_k^T. \tag{5.66}$$

Here,

$$J_k^{k+1} = \begin{bmatrix} I \\ 0 \end{bmatrix} \begin{matrix} \} & \mathcal{N}_k \setminus \mathcal{N}_{k+1} \\ \} & \mathcal{N}_{k+1} \end{matrix},$$

represents the extension by zero of vectors defined on the set of "f"–dofs (i.e., on $\mathcal{N}_k \setminus \mathcal{N}_{k+1}$*), to vectors defined on* $\mathcal{N}_k$.

Proof. The representation (5.65) implies the identity,

$$\mathcal{I}_k - \mathcal{I}_{k+1} = P_0 P_1 \cdots P_{k-1}\big(I - P_k (I_{k+1}^k)^T\big)\big(I_1^0 \cdots I_k^{k-1}\big)^T.$$

Note next that

$$(I_1^0)^T J_k = \begin{bmatrix} 0 \\ I \\ 0 \end{bmatrix} \begin{matrix} \} & \mathcal{N}_1 \setminus \mathcal{N}_k \\ \} & \mathcal{N}_k \setminus \mathcal{N}_{k+1} \\ \} & \mathcal{N}_{k+1} \end{matrix},$$

and similarly

$$\big(I_1^0 \cdots I_k^{k-1}\big)^T J_k = \begin{bmatrix} I \\ 0 \end{bmatrix} \begin{matrix} \} & \mathcal{N}_k \setminus \mathcal{N}_{k+1} \\ \} & \mathcal{N}_{k+1} \end{matrix} = J_k^{k+1}.$$

We also have $\mathcal{I}_\ell J_\ell = P_0 P_1 \cdots P_{\ell-1}$. Recalling that $\overline{P}_k = P_0 P_1 \cdots P_{k-1}$, the expression for B_{HB}^{-1} reduces to

$$B_{HB}^{-1} = \overline{P}_\ell A_\ell^{-1} (\overline{P}_\ell)^T + \sum_{k=0}^{\ell-1} \overline{P}_k (I - P_k (I_{k+1}^k)^T) J_k^{k+1} \Lambda_k^{-1} (J_k^{k+1})^T (I - I_{k+1}^k P_k^T) \overline{P}_k^T.$$

Finally, noticing that

$$\left[J_k^{k+1},\ I_{k+1}^k\right] = \begin{bmatrix} I & 0 \\ 0 & I \end{bmatrix} \begin{matrix} \} & \mathcal{N}_k \setminus \mathcal{N}_{k+1} \\ \} & \mathcal{N}_{k+1} \end{matrix}$$

and hence $(J_k^{k+1})^T I_{k+1}^k = 0$, we obtain the desired simplified expression for B_{HB}^{-1} (5.66). □

If we compare B_{HB}^{-1} and $(B_{MG}^{\text{add}})^{-1}$ from expression (5.45) the seemingly small difference is the term $J_k^{k+1}\Lambda_k^{-1}(J_k^{k+1})^T$ in the former. We have

$$J_k^{k+1}\Lambda_k^{-1}(J_k^{k+1})^T = \begin{bmatrix} \Lambda_k^{-1} & 0 \\ 0 & 0 \end{bmatrix} \begin{matrix} \} & \mathcal{N}_k \setminus \mathcal{N}_{k+1} \\ \} & \mathcal{N}_{k+1} \end{matrix}.$$

In the expression for $(B_{MG}^{\text{add}})^{-1}$ in (5.45) we have a Λ_k^{-1} that is defined for vectors on $\mathcal{N}_k$ and not only on the subset of "f"–dofs (i.e., on the hierarchical complement $\mathcal{N}_k \setminus \mathcal{N}_{k+1}$ of $\mathcal{N}_{k+1}$ in $\mathcal{N}_k$). In summary, in the HB method, we smooth only the "f" matrix block, whereas in the additive MG (or BPX), we smooth all the dofs at a given level.

5.9.2 A stable multilevel hierarchical (direct) decomposition

Assume now that we have a hierarchy of projections $\mathcal{Q}_k$, for $k = 1, \ldots, \ell$ and let $\mathcal{Q}_0 = I$, that provide a decomposition

$$\mathbf{v} = \sum_{k=1}^{\ell-1}(\mathcal{Q}_k - \mathcal{Q}_{k+1})\mathbf{v} + \mathcal{Q}_\ell \mathbf{v},$$

which is more stable than the HB one (5.62), in the sense the constant C_Q in the estimate

$$\|\mathcal{Q}_\ell \mathbf{v}\|_A^2 + \sum_{k=1}^{\ell-1} \|(\mathcal{Q}_k - \mathcal{Q}_{k+1})\mathbf{v}\|_A^2 \le C_Q\ \|\mathbf{v}\|_A^2, \tag{5.67}$$

is much smaller than C_ℓ from estimate (5.63). In practice, the projections $\mathcal{Q}_k$ may give rise to dense matrices (see next section), and therefore they may not be computationally feasible (as $\mathcal{I}_k$, e.g.) to define a multilevel preconditioner that is based only on sparse matrix operations and hence have optimal complexity.

The abstract setting is that we have access to $\{\mathcal{Q}_k^a\}$ which have sparse matrix representation and at the same time approximate the true projections $\mathcal{Q}_k$ well. We assume that in a given norm $\|.\|_0$, the following uniform in $\mathbf{v}$ estimate holds,

$$\|(\mathcal{Q}_k - \mathcal{Q}_k^a)\mathbf{v}\|_0 \le \tau\ \|\mathcal{Q}_k \mathbf{v}\|_0, \tag{5.68}$$

for a sufficiently small constant $\tau \in [0, 1)$.

We assume that

$$\mathcal{Q}_{k+1}\mathcal{Q}_k = \mathcal{Q}_{k+1}. \tag{5.69}$$

The latter property holds if the projections $\mathcal{Q}_k$ are computed from a same inner product $(.,\ .)_0$; that is, $\mathcal{Q}_k\mathbf{v}$ is defined from the Galerkin relation (recalling that $\overline{\mathbf{V}}_k = \text{Range}\ (\overline{P}_k)$),

$$(\mathcal{Q}_k\mathbf{v},\ \overline{\mathbf{w}}_k)_0 = (\mathbf{v},\ \overline{\mathbf{w}}_k)_0, \qquad \text{for all } \overline{\mathbf{w}}_k \in \overline{\mathbf{V}}_k. \tag{5.70}$$

Because $\overline{\mathbf{V}}_{k+1} \subset \overline{\mathbf{V}}_k$, we see then that property (5.69) holds.

Definition 5.45 (Modified nodal projections). *Letting $\overline{\pi}_0 = I$ and for $k = 1, \ldots, \ell$ define*

$$\overline{\pi}_k = (\mathcal{I}_k + \mathcal{Q}_k^a(\mathcal{I}_{k-1} - \mathcal{I}_k))\overline{\pi}_{k-1}.$$

Note that if $\mathcal{Q}_k^a = 0$, we have $\overline{\pi}_k = \mathcal{I}_k\overline{\pi}_{k-1}$, which due to Lemma 5.43 equals $\mathcal{I}_k$. That is, $\overline{\pi}_k = \mathcal{I}_k$ in that case. The other limiting case is when $\mathcal{Q}_k^a = \mathcal{Q}_k$. Then assuming by induction that $\overline{\pi}_{k-1} = \mathcal{Q}_{k-1}$, we have, again due to Lemma 5.43, that $\overline{\pi}_k = (\mathcal{I}_k + \mathcal{Q}_k(\mathcal{I}_{k-1} - \mathcal{I}_k))\mathcal{Q}_{k-1} = (\mathcal{I}_k + \mathcal{Q}_k\mathcal{I}_{k-1} - \mathcal{Q}_k\mathcal{I}_k)\mathcal{Q}_{k-1} = \mathcal{Q}_k\mathcal{I}_{k-1}\mathcal{Q}_{k-1} = \mathcal{Q}_k\mathcal{Q}_{k-1} = \mathcal{Q}_k$. The last identity is by assumption (5.69). That is, the operators $\overline{\pi}_k$ can be viewed as "interpolation" between the two projections $\mathcal{I}_k$ and $\mathcal{Q}_k$. The first one is unstable (see estimate (5.63)), whereas the second one is stable (by assumption (5.67)).

Lemma 5.46. *The operators $\overline{\pi}_k$ are projections that satisfy $\overline{\pi}_k\overline{\pi}_{k-1} = \overline{\pi}_k$.*

Proof. Because for any $\overline{\mathbf{v}}_k \in \overline{\mathbf{V}}_k \subset \overline{\mathbf{V}}_{k-1}$, we have $\mathcal{I}_{k-1}\overline{\mathbf{v}}_k = \mathcal{I}_k\overline{\mathbf{v}}_k = \overline{\mathbf{v}}_k$ due to Lemma 5.43, therefore, by induction $\overline{\pi}_{k-1}\overline{\mathbf{v}}_k = \overline{\mathbf{v}}_k$ (because $\overline{\mathbf{v}}_k \in \overline{\mathbf{V}}_{k-1}$), we have $\overline{\pi}_k\overline{\mathbf{v}}_k = (\mathcal{I}_k + \mathcal{Q}_k^a(\mathcal{I}_{k-1} - \mathcal{I}_k))\overline{\mathbf{v}}_k = \overline{\mathbf{v}}_k$. The latter fact implies that $\overline{\pi}_k$ is a projection (similar to $\mathcal{I}_k$). □

We are interested in the direct multilevel decomposition based on the projections $\overline{\pi}_k$, namely,

$$\mathbf{v} = \sum_{k=0}^{\ell-1}(\overline{\pi}_k - \overline{\pi}_{k+1})\mathbf{v} + \overline{\pi}_\ell\mathbf{v}. \tag{5.71}$$

We want to show a stability estimate of the form

$$\|\overline{\pi}_\ell\mathbf{v}\|_A^2 + \sum_{k=1}^{\ell-1}\|(\overline{\pi}_k - \overline{\pi}_{k+1})\mathbf{v}\|_A^2 \le C\ \|\mathbf{v}\|_A^2. \tag{5.72}$$

Recall that if $\mathcal{Q}_k^a = \mathcal{Q}_k$ then $\overline{\pi}_k = \mathcal{Q}_k$. Thus estimate (5.72) can be viewed as a perturbation of (5.67) for $\mathcal{Q}_k^a \approx \mathcal{Q}_k$.

To analyze the stability of the decomposition (5.71) we need the following additional assumptions.

(e) The following estimate holds in the $\|\cdot\|_0$–norm,

$$\|(\mathcal{Q}_k - \mathcal{Q}_{k-1})\mathbf{v}\|_0 \le C_e\ h_k\ \|\mathbf{v}\|_A,$$

where $h_{k+1} = 2h_k$, hence $h_k = 2^k h_0$. Here, $h_0 = 2^{-\ell}\ H$ is the fine-grid mesh size and $H = h_\ell = \mathcal{O}(1)$ is the coarsest mesh-size.

(b) The operator $\mathcal{I}_k - \mathcal{I}_{k+1}$ restricted to $\overline{\mathbf{V}}_k$ is bounded in $\|\cdot\|_0$ norm; that is, we have the estimate

$$\|(\mathcal{I}_k - \mathcal{I}_{k+1})\overline{\mathbf{v}}_k\|_0 \leq C_R \|\overline{\mathbf{v}}_k\|_0 \quad \text{for all } \overline{\mathbf{v}}_k \in \overline{\mathbf{V}}_k.$$

The analysis relies on the following two lemmas.

Lemma 5.47. *Define the deviation* $\mathbf{e}_k = (\overline{\pi}_k - \mathcal{Q}_k)\mathbf{v}$ *for any given vector* $\mathbf{v}$. *The following recursive relation then holds.*

$$\mathbf{e}_{s+1} = (\mathcal{Q}_{s+1} + \mathcal{R}_{s+1})\mathbf{e}_s + \mathcal{R}_{s+1}(\mathcal{Q}_s - \mathcal{Q}_{s+1})\mathbf{v}, \tag{5.73}$$

where $\mathcal{R}_{s+1} = (\mathcal{Q}_{s+1} - \mathcal{Q}^a_{s+1})(\mathcal{I}_{s+1} - \mathcal{I}_s)$.

Proof. We have the following identities.

$$\begin{aligned}
\mathbf{e}_{s+1} &= \overline{\pi}_{s+1}\mathbf{v} - \mathcal{Q}_{s+1}\mathbf{v} \\
&= \big(\mathcal{I}_{s+1} + \mathcal{Q}^a_{s+1}(\mathcal{I}_s - \mathcal{I}_{s+1})\big)\overline{\pi}_s\mathbf{v} - \mathcal{Q}_{s+1}\mathbf{v} \\
&= \big(\mathcal{Q}_{s+1} - \mathcal{Q}^a_{s+1}\big)\mathcal{I}_{s+1}\overline{\pi}_s\mathbf{v} + \mathcal{Q}^a_{s+1}\overline{\pi}_s\mathbf{v} - \mathcal{Q}_{s+1}\mathbf{v}.
\end{aligned}$$

Thus, we have

$$\begin{aligned}
\mathbf{e}_{s+1} &= \big(\mathcal{Q}_{s+1} - \mathcal{Q}^a_{s+1}\big)\mathcal{I}_{s+1}(\overline{\pi}_s\mathbf{v} - \mathcal{Q}_s\mathbf{v}) + \mathcal{Q}^a_{s+1}(\overline{\pi}_s\mathbf{v} - \mathcal{Q}_s\mathbf{v}) \\
&\quad + \big(\mathcal{Q}_{s+1} - \mathcal{Q}^a_{s+1}\big)\mathcal{I}_{s+1}\mathcal{Q}_s\mathbf{v} + \mathcal{Q}^a_{s+1}\mathcal{Q}_s\mathbf{v} - \mathcal{Q}_{s+1}\mathbf{v} \\
&= \big(\mathcal{Q}_{s+1} - \mathcal{Q}^a_{s+1}\big)\mathcal{I}_{s+1}\mathbf{e}_s + \mathcal{Q}^a_{s+1}\mathbf{e}_s + \big(\mathcal{Q}_{s+1} - \mathcal{Q}^a_{s+1}\big)\big(\mathcal{I}_{s+1}\mathcal{Q}_s\mathbf{v} - \mathcal{Q}_s\mathbf{v}\big) \\
&= \big(\mathcal{Q}_{s+1} - \mathcal{Q}^a_{s+1}\big)(\mathcal{I}_{s+1} - \mathcal{I}_s)\mathbf{e}_s + \big(\mathcal{Q}_{s+1} - \mathcal{Q}^a_{s+1}\big)\mathbf{e}_s + \mathcal{Q}^a_{s+1}\mathbf{e}_s \\
&\quad + \big(\mathcal{Q}_{s+1} - \mathcal{Q}^a_{s+1}\big)(\mathcal{I}_{s+1} - \mathcal{I}_s)\mathcal{Q}_s\mathbf{v} \\
&= \big(\mathcal{Q}_{s+1} - \mathcal{Q}^a_{s+1}\big)(\mathcal{I}_{s+1} - \mathcal{I}_s)\mathbf{e}_s + \mathcal{Q}_{s+1}\mathbf{e}_s + \big(\mathcal{Q}_{s+1} - \mathcal{Q}^a_{s+1}\big)(\mathcal{I}_{s+1} - \mathcal{I}_s)\mathcal{Q}_s\mathbf{v} \\
&= \big[\mathcal{Q}_{s+1} + \big(\mathcal{Q}_{s+1} - \mathcal{Q}^a_{s+1}\big)(\mathcal{I}_{s+1} - \mathcal{I}_s)\big]\mathbf{e}_s + \big(\mathcal{Q}_{s+1} - \mathcal{Q}^a_{s+1}\big)(\mathcal{I}_{s+1} - \mathcal{I}_s)\mathcal{Q}_s\mathbf{v}.
\end{aligned}$$

The latter together with the fact that $(\mathcal{I}_{s+1} - \mathcal{I}_s)\mathcal{Q}_{s+1} = 0$ implies the desired recursive relation (5.73). □

The next lemma estimates a weighted sum of the squared norms of the deviations.

Lemma 5.48. *Under the assumptions (e) and (b) and the uniform estimate* (5.68), *the following bound holds,*

$$\sum_{j=1}^{\ell} h_j^{-2}\|\mathbf{e}_j\|_0^2 \leq \frac{C_R^2}{(1-q)^2}\tau^2 \sum_{j=1}^{\ell-1} h_j^{-2}\|(\mathcal{Q}_j - \mathcal{Q}_{j+1})\mathbf{v}\|_0^2, \quad \forall \mathbf{v} \in \overline{\mathbf{V}}_0 = \mathbf{V}. \tag{5.74}$$

The constant $q \in (0, 1)$ *is chosen such that* $(1 + C_R\tau)/2 \leq q$, *which is possible for sufficiently small* $\tau \in [0, 1)$ *(independently of the level index* k*).*

Proof. Recall that C_R is a level-independent bound of the $\|.\|_0$-norm of $\mathcal{I}_s - \mathcal{I}_{s+1} : \overline{\mathbf{V}}_s \mapsto \overline{\mathbf{V}}_s$. Then,

$$\|\mathcal{R}_{s+1}\overline{\mathbf{v}}_s\|_0 \leq C_R\tau\|\overline{\mathbf{v}}_s\|_0 \qquad \text{for all } \overline{\mathbf{v}}_s \in \overline{\mathbf{V}}_s. \tag{5.75}$$

For sufficiently small $\tau \in [0, 1)$, we have

$$\frac{1 + C_R\tau}{2} \leq q = \text{Const} < 1\cdot \tag{5.76}$$

Next, observe that $\mathbf{e}_0 = 0$. Then a recursive use of (5.73) leads to

$$\begin{aligned}\|\mathbf{e}_{s+1}\|_0 &\leq (1 + C_R\tau)\|\mathbf{e}_s\|_0 + C_R\tau\|(\mathcal{Q}_s - \mathcal{Q}_{s+1})\mathbf{v}\|_0 \\ &\leq C_R\tau\sum_{j=0}^{s}(1 + C_R\tau)^{s-j}\|(\mathcal{Q}_j - \mathcal{Q}_{j+1})\mathbf{v}\|_0\cdot\end{aligned}$$

Therefore, with $h_j = 2^{j-\ell}h_\ell$ and $h_\ell = H = \mathcal{O}(1)$ being the coarsest mesh-size,

$$\begin{aligned}\|e_{s+1}\|_0 &\leq C_R\tau h_{s+1}\sum_{j=0}^{s}(1 + C_R\tau)^{s-j}h_{s+1}^{-1}\|(\mathcal{Q}_j - \mathcal{Q}_{j+1})\mathbf{v}\|_0 \\ &= C_R\tau h_{s+1}\sum_{j=0}^{s}(1 + C_R\tau)^{s-j}\frac{h_{j+1}}{h_{s+1}}h_{j+1}^{-1}\|(\mathcal{Q}_j - \mathcal{Q}_{j+1})v\|_0 \\ &= C_R\tau h_{s+1}\sum_{j=0}^{s}(1 + C_R\tau)^{s-j}\left(\frac{1}{2}\right)^{s-j}h_{j+1}^{-1}\|(\mathcal{Q}_j - \mathcal{Q}_{j+1})\mathbf{v}\|_0 \\ &\leq C_R\tau h_{s+1}\sum_{j=0}^{s}q^{s-j}h_{j+1}^{-1}\|(Q_j - Q_{j+1})\mathbf{v}\|_0 \\ &\leq C_R\tau h_{s+1}\frac{1}{\sqrt{1-q}}\left[\sum_{j=0}^{s}q^{s-j}h_{j+1}^{-2}\|(\mathcal{Q}_j - \mathcal{Q}_{j+1})\mathbf{v}\|_0^2\right]^{1/2}.\end{aligned} \tag{5.77}$$

The latter inequality shows

$$\begin{aligned}\sum_{s=0}^{\ell-1} h_{s+1}^{-2}\|\mathbf{e}_{s+1}\|_0^2 &\leq C_R^2\tau^2\frac{1}{1-q}\sum_{s=0}^{\ell-1}\sum_{j=0}^{s} q^{s-j}h_{j+1}^{-2}\|(\mathcal{Q}_j - \mathcal{Q}_{j+1})\mathbf{v}\|_0^2 \\ &\leq C_R^2\tau^2\frac{1}{(1-q)^2}\sum_{j=0}^{\ell-1} h_{j+1}^{-2}\|(\mathcal{Q}_j - \mathcal{Q}_{j+1})\mathbf{v}\|_0^2\cdot\end{aligned}$$

which proves the lemma. □

To prove the final stability estimate, notice that $\overline{\pi}_k\mathbf{v} = \mathbf{e}_k + \mathcal{Q}_k\mathbf{v}$, which implies $\|(\overline{\pi}_k - \overline{\pi}_{k+1})\mathbf{v}\|_0 \leq \|\mathbf{e}_k\|_0 + \|\mathbf{e}_{k+1}\|_0 + \|(\mathcal{Q}_k - \mathcal{Q}_{k+1})\mathbf{v}\|_0$. Thus the following stability estimate is immediate,

$$\|\overline{\pi}_\ell \mathbf{v}\|_A^2 + \sum_{k=0}^{\ell-1} h_k^{-2} \|(\overline{\pi}_k - \overline{\pi}_{k+1})\mathbf{v}\|_0^2$$
$$\leq C \left(\|\mathcal{Q}_\ell \mathbf{v}\|_A^2 + \|\mathbf{e}_\ell\|_A^2 + \sum_{k=0}^{\ell-1} h_{k+1}^{-2} \|(\mathcal{Q}_k - \mathcal{Q}_{k+1})\mathbf{v}\|_0^2 \right).$$

Define for any $k = 0, \ldots, \ell$,

$$\varrho(A_k) = \sup_{\overline{\mathbf{v}}_k \in \overline{\mathbf{V}}_k} \frac{\overline{\mathbf{v}}_k^T A \overline{\mathbf{v}}_k}{\|\overline{\mathbf{v}}_k\|_0^2}. \tag{5.78}$$

Then, from (5.77), we get

$$\|\mathbf{e}_\ell\|_A^2 \leq C_R^2 \tau^2 \varrho(A_\ell)\ H^2\ \frac{1}{1-q} \sum_{j=0}^{\ell-1} h_{j+1}^{-2} \|(\mathcal{Q}_j - \mathcal{Q}_{j+1})\mathbf{v}\|_0^2.$$

Assuming that the coarsest problem is of fixed size, we then have $\varrho(A_\ell)\ H^2 = \mathcal{O}(1)$. Moreover, the following main stability result holds based on (e) and (5.67).

Theorem 5.49. *Assume that the spectral radius* $\varrho(A_k)$ *(defined in* (5.78)*) of* A_k *(the matrix* A *restricted to the subspace* $\overline{\mathbf{V}}_k$*) satisfies the condition* $\varrho(A_k) \simeq h_{k+1}^{-2}$*; then the following main stability estimate (see* (5.72)*) holds.*

$$\|\overline{\pi}_\ell \mathbf{v}\|_A^2 + \sum_{k=0}^{\ell-1} \|(\overline{\pi}_k - \overline{\pi}_{k+1})\mathbf{v}\|_A^2 \leq C \left(\|\mathcal{Q}_\ell \mathbf{v}\|_A^2 + \sum_{k=0}^{\ell-1} \|(\mathcal{Q}_k - \mathcal{Q}_{k+1})\mathbf{v}\|_A^2 \right)$$
$$\leq C\ \mathbf{v}^T A \mathbf{v}.$$

Remark 5.50. The assumption (e) and the estimate $\varrho(A_k) \simeq h_k^{-2}$ are valid for finite element matrices and $\|.\|_0$ coming from the integral L_2-norm. The projections $\mathcal{Q}_k$ correspond then to the matrix representation of the L_2-based projections $Q_k : L_2 \mapsto V_k$ where V_k is the kth-level finite element space. The latter form of $\mathcal{Q}_k$ is studied in some detail in the following section. Finally, the fact that Q_k and hence $\mathcal{Q}_k$ provide estimates of the form (e) can be found in [Br93]. Assumption (b) follows from the fact that the integral L_2-norm, and the discrete ℓ_2 one, up to a weighting, are equivalent when restricted to V_k (see Theorem 1.6).

5.9.3 Approximation of L_2-projections

In the present section, we consider the case of finite element matrices A in the setting of the introductory Chapter 1. We have a sequence of nested finite element spaces $V_k \subset V_{k-1}$, and V_k is spanned by the standard nodal (Lagrangian) basis functions $\{\phi_i^{(k)}\}_{x_i \in \mathcal{N}_k}$. $\mathcal{N}_k$ stands for the set of vertices x_i of the triangles τ from the kth-level triangulation $\mathcal{T}_k$. Due to the refinement construction of $\mathcal{N}_{k-1}$ from $\mathcal{N}_k$, we have that $\mathcal{N}_k \subset \mathcal{N}_{k-1}$. The vector spaces $\mathbf{V}_k$ correspond to the coefficient vectors of the functions $v \in V_k$ expanded in terms of the kth-level basis $\{\phi_i^{(k)}\}_{x_i \in \mathcal{N}_k}$. The vector space $\overline{\mathbf{V}}_k$ corresponds to the coefficient vectors of functions $v \in V_k$ expanded in terms of the nodal basis of the finest f.e. space $V = V_0$. We have interpolation matrices P_{k-1} that relate the coarse space $\mathbf{V}_k$ and $\mathbf{V}_{k-1}$ in the sense that for any $v \in V_k$, if $\mathbf{v}_k$ is its coefficient vector from $\mathbf{V}_k$, then $P_{k-1}\mathbf{v}_k \in \mathbf{V}_k$ is its coefficient vector as a function from V_{k-1} (in terms of the the $(k-1)$th-level nodal basis).

Consider the well-conditioned Gram (or mass) matrices G_k (as defined in Section 1.4). More specifically G_k is defined based on the L_2-inner product $(\cdot,\ \cdot)$, as follows, $G_k = \{(\phi_j^{(k)}, \phi_i^{(k)})\}_{x_j, x_i \in \mathcal{N}_k}$.

Define the L_2-projection $Q_k : L_2 \mapsto V_k$ as $(Q_k v,\ \phi) = (v,\ \phi)$ for all $\phi \in V_k$.

We show that there is a certain relation between G_k and G_0 coming from the equation $(Q_k v,\ w) = (v,\ w)$ for all $w \in V_k$, for any $v \in V = V_0$. More specifically, the latter problem admits the following matrix–vector form,

$$\mathbf{w}_k^T G_k \mathbf{v}_k = (\overline{P}_k \mathbf{w}_k)^T G_0 \mathbf{v}, \qquad \forall \mathbf{v}_k \in \mathbf{V}_k.$$

Here $\mathbf{v}_k$ and $\mathbf{w}_k$ are the nodal coefficient vectors of $Q_k v$ and $w \in V_k$ at the kth level, respectively. Therefore, we need to solve the following mass matrix problem.

$$G_k \mathbf{v}_k = \overline{P}_k^T G_0 \mathbf{v}. \tag{5.79}$$

In other words, the exact L^2-projection $Q_k v$ has a coefficient vector that is actually given by

$$G_k^{-1} \overline{P}_k^T G_0 \mathbf{v}.$$

In the preceding section, we used the projections $\mathcal{Q}_k : \mathbf{V} \mapsto \overline{\mathbf{V}}_k$. Recall that $\overline{\mathbf{V}}_k \subset \overline{\mathbf{V}}_0 = \mathbf{V}$. Therefore, the matrix $\mathcal{Q}_k$ as a mapping from $\mathbf{V} \mapsto \mathbf{V}$ has the form

$$\overline{P}_k G_k^{-1} \overline{P}_k^T G_0.$$

Then

$$\|\mathcal{Q}_k \mathbf{v}\|_0^2 = \mathbf{v}^T \mathcal{Q}_k^T G_0 \mathcal{Q}_k \mathbf{v} = \mathbf{v}^T G_0^T \overline{P}_k G_k^{-1} \overline{P}_k^T G_0 \mathbf{v}. \tag{5.80}$$

If we define $\mathcal{Q}_k^a = \overline{P}_k \widetilde{G}_k^{-1} \overline{P}_k^T G_0$ where $\widetilde{G}_k^{-1}$ is a (sparse) approximation to G_k^{-1}, the estimate (5.68) takes the following particular matrix–vector form.

$$\begin{aligned}
\|(\mathcal{Q}_k - \mathcal{Q}_k^a)\mathbf{v}\|_0^2 &= \mathbf{v}^T \mathcal{Q}_k^T G_0 \mathcal{Q}_k \mathbf{v} \\
&= \mathbf{v}^T G_0^T \overline{P}_k (G_k^{-1} - \widetilde{G}_k^{-1}) G_k (G_k^{-1} - \widetilde{G}_k^{-1}) \overline{P}_k^T G_0 \mathbf{v} \\
&\le \tau^2\ \|\mathcal{Q}_k \mathbf{v}\|_0^2 \\
&= \tau^2\ \mathbf{v}^T \mathcal{Q}_k^T G_0 \mathcal{Q}_k \mathbf{v} \\
&= \tau^2\ \mathbf{v}^T G_0^T \overline{P}_k G_k^{-1} \overline{P}_k^T G_0 \mathbf{v}.
\end{aligned}$$

Because $\mathbf{v}_k = \overline{P}_k^T G_0 \mathbf{v}$ can be any vector in $\mathbf{V}_k$ the above estimate actually reads

$$\mathbf{v}_k^T (G_k^{-1} - \widetilde{G}_k^{-1}) G_k (G_k^{-1} - \widetilde{G}_k^{-1}) \mathbf{v}_k \le \tau^2\ \mathbf{v}_k^T G_k^{-1} \mathbf{v}_k.$$

To have computationally feasible projections $\overline{\pi}_k$, we approximate $\mathcal{Q}_k$ with a $\mathcal{Q}_k^a$ by replacing G_k^{-1} by some approximations $\widetilde{G}_k^{-1}$ whose actions can be computed by simple iterative methods applied to (5.79). Such iterative methods lead to the following polynomial approximations of G_k^{-1},

$$\widetilde{G}_k^{-1} = [I - p_m(G_k)]\, G_k^{-1}.$$

Here p_m is a polynomial of degree $m \geq 1$ such that p_m satisfies $p_m(0) = 1$ and $0 \leq p_m(t) < 1$ for $t \in [\alpha, \beta]$. The latter interval contains the spectrum of the mass matrix G_k. Because G_k is well conditioned, we can choose the interval $[\alpha, \beta]$ independent of k. Thus, the polynomial degree m can be chosen to be level-independent so that a given prescribed accuracy $\tau > 0$ in (5.68) is guaranteed. More precisely, given a tolerance $\tau > 0$, we can choose $m = m(\tau)$ satisfying

$$\begin{aligned}
\|\mathcal{Q}_k^a \mathbf{v} - \mathcal{Q}_k \mathbf{v}\|_0 &= \left\|G_k^{1/2}(G_k^{-1} - \widetilde{G}_k^{-1})\overline{P}_k^T G_0 \mathbf{v}\right\| \\
&= \left\|G_k^{1/2} p_m(G_k)\, G_k^{-1} \overline{P}_k^T G_0 \mathbf{v}\right\| \\
&\leq \max_{t \in [\alpha,\beta]} p_m(t) \left\|G_k^{-(1/2)} \overline{P}_k^T G_0 \mathbf{v}\right\| \\
&= \max_{t \in [\alpha,\beta]} p_m(t)\ \|\mathcal{Q}_k \mathbf{v}\|_0.
\end{aligned}$$

Here we have used identity (5.80) and the properties of p_m. The last estimate implies the validity of (5.68) with

$$\tau \geq \max_{t \in [\alpha,\beta]} p_m(t).$$

A simple choice of $p_m(t)$ is the truncated series

$$(1 - p_m(t))t^{-1} = p_{m-1}(t) \equiv \beta^{-1} \sum_{k=0}^{m-1} \left(1 - \frac{1}{\beta} t\right)^k, \tag{5.81}$$

which yields $\widetilde{G}_k^{-1} = p_{m-1}(G_k)$. We remark that (5.81) was obtained from the following expansion,

$$1 = t\beta^{-1} \sum_{k=0}^{\infty} (1 - t\beta^{-1})^k, \qquad t \in [\alpha, \beta].$$

With the above choice on the polynomial $p_m(t)$, we have

$$p_m(t) = 1 - t p_{m-1}(t) = t\beta^{-1} \sum_{k \geq m} (1 - \beta^{-1} t)^k = (1 - \beta^{-1} t)^m.$$

It follows that

$$\max_{t \in [\alpha,\beta]} p_m(t) = \left(1 - \frac{\alpha}{\beta}\right)^m.$$

In general, by a careful selection on p_m we have $\max_{t\in[\alpha,\beta]} p_m(t) \leq Cq^m$ for some constants $C > 0$ and $q \in (0, 1)$, both independent of k. Because the restriction on τ was that τ be sufficiently small, then we must have

$$m = O(\log \tau^{-1}). \tag{5.82}$$

The requirement (5.82) obviously imposes a very mild restriction on m. In practice, we expect to use reasonably small m (e.g., $m = 1, 2$). This observation is confirmed by the numerical experiments performed in [VW99].

5.9.4 Construction of bases in the coordinate spaces

Based on the projections $\mathcal{Q}_k$ and their sparse approximations $\mathcal{Q}_k^a$ (as defined in the previous section) and the nodal interpolation mappings $\mathcal{I}_k$ (see (5.65)), we can modify the nodal basis $\{\phi_i^{(k)}\}_{x_i\in\mathcal{N}_k\setminus\mathcal{N}_{k+1}}$, or rather their vector representations as elements of the hierarchical coordinate spaces Range$(\mathcal{I}_k - \mathcal{I}_{k+1})$. The latter is complementary to the coarse space Range$(\mathcal{I}_{k+1})$ = Range$(\overline{P}_{k+1})$. Note that the basis functions $\{\phi_i^{(k)}\}_{x_i\in\mathcal{N}_k}$ span the kth-level finite element space V_k. The procedure described in what follows gives rise to a computable basis of the "coordinate" spaces Range$(\overline{\pi}_k - \overline{\pi}_{k+1})$ from the direct decomposition (5.71). In what follows, we construct a computable basis in the "coordinate" space Range$(\overline{\pi}_k - \overline{\pi}_{k+1})$. First, note the following result.

Lemma 5.51. *The range of $\overline{\pi}_k - \overline{\pi}_{k+1}$ is the same as the range of $(\overline{\pi}_k - \overline{\pi}_{k+1})(\mathcal{I}_k - \mathcal{I}_{k+1})$. More specifically, it coincides with the range of $(I - \mathcal{Q}_{k+1}^a)(\mathcal{I}_k - \mathcal{I}_{k+1})$. Also, any basis $\{\varphi_i\}$ of the space Range$(\mathcal{I}_k - \mathcal{I}_{k+1})$ provides a basis of Range$(\overline{\pi}_k - \overline{\pi}_{k+1})$ defined by $\{(I - \mathcal{Q}_{k+1}^a)\varphi_i\}$.*

Proof. Based on Lemma 5.46, we have for the components in the direct decomposition (5.71),

$$\overline{\pi}_k - \overline{\pi}_{k+1} = \overline{\pi}_k^2 - \overline{\pi}_{k+1}\overline{\pi}_k = (\overline{\pi}_k - \overline{\pi}_{k+1})\overline{\pi}_k.$$

That is, Range $(\overline{\pi}_k - \overline{\pi}_{k+1}) = (\overline{\pi}_k - \overline{\pi}_{k+1})\overline{\mathbf{V}}_k$. Because $\overline{\mathbf{V}}_k$ can be decomposed based on the components Range $(\mathcal{I}_k - \mathcal{I}_{k+1})$ and $\overline{\mathbf{V}}_{k+1}$, and because $(\overline{\pi}_k - \overline{\pi}_{k+1})\overline{\mathbf{V}}_{k+1} = 0$, we see that Range $(\overline{\pi}_k - \overline{\pi}_{k+1})$ = Range $(\overline{\pi}_k - \overline{\pi}_{k+1})(\mathcal{I}_k - \mathcal{I}_{k+1})$, which proves the first statement of the lemma.

Recalling Definition 5.45,

$$\overline{\pi}_{k+1} = (\mathcal{I}_{k+1} + \mathcal{Q}_{k+1}^a(\mathcal{I}_k - \mathcal{I}_{k+1}))\overline{\pi}_k,$$

with $\overline{\pi}_0 = I$, and noting that $\mathcal{I}_k\overline{\pi}_k = \overline{\pi}_k$ (due to Lemma 5.43), the following representation holds,

$$\begin{aligned}\overline{\pi}_k - \overline{\pi}_{k+1} &= \mathcal{I}_k\overline{\pi}_k - (\mathcal{I}_{k+1} + \mathcal{Q}_{k+1}^a(\mathcal{I}_k - \mathcal{I}_{k+1}))\overline{\pi}_k \\ &= (\mathcal{I}_k - \mathcal{I}_{k+1})\overline{\pi}_k - \mathcal{Q}_{k+1}^a(\mathcal{I}_k - \mathcal{I}_{k+1})\overline{\pi}_k \\ &= (I - \mathcal{Q}_{k+1}^a)(\mathcal{I}_k - \mathcal{I}_{k+1})\overline{\pi}_k.\end{aligned}$$

From Lemma 5.46, we have $\overline{\pi}_k \overline{\mathbf{v}}_k = \overline{\mathbf{v}}_k$ if $\overline{\mathbf{v}}_k \in \overline{\mathbf{V}}_k$. This shows that $\overline{\pi}_k(\mathcal{I}_k - \mathcal{I}_{k+1}) = \mathcal{I}_k - \mathcal{I}_{k+1}$. Similarly, Lemma 5.43 implies that $\mathcal{I}_{k+1}\mathcal{I}_k = \mathcal{I}_{k+1}^2 = \mathcal{I}_{k+1}$; that is, $\mathcal{I}_{k+1}(\mathcal{I}_k - \mathcal{I}_{k+1}) = 0$ and $(\mathcal{I}_k - \mathcal{I}_{k+1})^2 = \mathcal{I}_k - \mathcal{I}_{k+1}$. Thus we end up with the following main identity (which is the second statement of the lemma)

$$(\overline{\pi}_k - \overline{\pi}_{k+1})(\mathcal{I}_k - \mathcal{I}_{k+1}) = (I - \mathcal{Q}_{k+1}^a)(\mathcal{I}_k - \mathcal{I}_{k+1}).$$

It is clear then that any basis $\{\varphi_i\}$ of the space Range $(\mathcal{I}_k - \mathcal{I}_{k+1})$ will produce a basis $\{(I - \mathcal{Q}_{k+1}^a)\varphi_i\}$ of Range $(I - \mathcal{Q}_{k+1}^a)(\mathcal{I}_k - \mathcal{I}_{k+1})$ because $(I - \mathcal{Q}_{k+1}^a)\sum_i c_i\varphi_i = 0$ would imply then that $\varphi \equiv \sum_i c_i\varphi_i = \mathcal{Q}_{k+1}^a\varphi \in \overline{\mathbf{V}}_{k+1}$.

Note now that $\varphi \in$ Range $(\mathcal{I}_k - \mathcal{I}_{k+1})$, which is a complementary space to Range $(\mathcal{I}_{k+1}) = \overline{\mathbf{V}}_{k+1}$. Thus $\varphi = 0$, and therefore $c_i = 0$ for all i. That is, the set $\{(I - \mathcal{Q}_{k+1}^a)\varphi_i\}$ is linearly independent, hence provides a basis of the coordinate space Range $(\overline{\pi}_k - \overline{\pi}_{k+1})$. □

5.9.5 The approximate wavelet hierarchical basis (or AWHB)

Let the nodes in $\mathcal{N}_k$ be ordered by first keeping the nodes from $\mathcal{N}_{k+1}$ and then adding the complementary ones from $\mathcal{N}_k \setminus \mathcal{N}_{k+1}$. The latter nodes are labeled $i = n_{k+1} + 1, \ldots, n_k$. Here, n_k stands for the number of nodes $\mathcal{N}_k$ at level k. Because level $k + 1$ (in our notation) is coarser than level k, we have $n_{k+1} < n_k$.

Then, we can consider the following modified multilevel hierarchical basis:

$$\{\phi_i^{(\ell)}, i = 1, \ldots, n_\ell\} \bigcup_{k=\ell-1,\ldots,1,0} \left\{(I - \mathcal{Q}_{k+1}^a)\phi_i^{(k)},\ i = n_{k+1} + 1, \ldots, n_k\right\}. \tag{5.83}$$

Note that here we consider every basis function $\phi_i^{(k)}$ as a vector from $\mathbf{V}$ (i.e., interpolated all the way up to the finest-level 0). That is, as-such, a vector $\phi_i^{(k)}$ actually has the form

$$\overline{P}_k \begin{bmatrix} 0 \\ \vdots \\ 1 \\ \vdots \\ 0 \end{bmatrix},$$

where the vector

$$\begin{bmatrix} 0 \\ \vdots \\ 1 \\ \vdots \\ 0 \end{bmatrix}$$

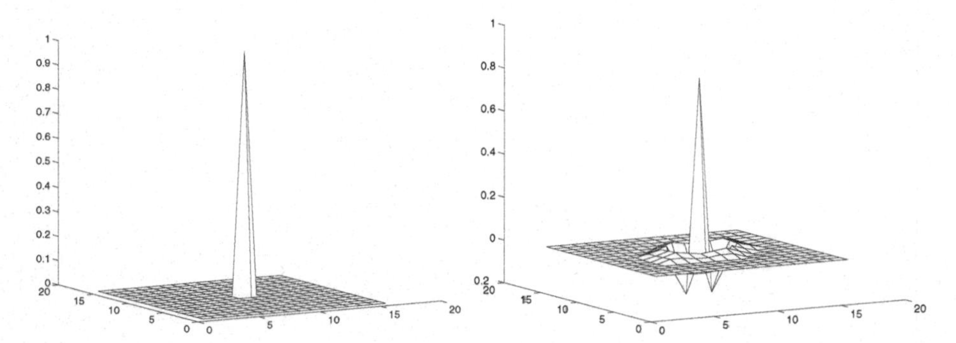

Fig. 5.1. An HB function (no modification) left, and a wavelet-modified HB function ($m = 2$), right.

is the ith coordinate vector in $\mathbb{R}^{n_k}$. The above components $\{(I - \mathcal{Q}^a_{j+1})\phi^{(j)}_i, \ i = n_{j+1}+1, \dots, n_j\}$ can be seen as a modification of the classical hierarchical basis components based on the nodal interpolation operator $\mathcal{I}_k$ because $(I - \mathcal{Q}^a_{j+1})\phi^{(j)}_i = (I - \mathcal{Q}^a_{j+1})(\mathcal{I}_j - \mathcal{I}_{j+1})\phi^{(j)}_i$. The modification of the classical hierarchical basis components $\{(\mathcal{I}_j - \mathcal{I}_{j+1})\phi^{(j)}_i, \ i = n_{j+1}+1, \dots, n_j\}$ comes from the additional term $\mathcal{Q}^a_{j+1}(\mathcal{I}_j - \mathcal{I}_{j+1})\phi^{(j)}_i$. In other words, the modification was made by subtracting from each nodal hierarchical basis function $\phi^{(j)}_i$ its approximate L^2-projection $\mathcal{Q}^a_{j+1}\phi^{(j)}_i$ onto the coarse level $j+1$.

An illustration of an HB function and its modification by an approximate mass matrix inverse provided by $m = 2$ steps of the CG method is shown in Figure 5.1. It can be seen that the modified hierarchical basis functions are close relatives of the known Battle–Lemarié wavelets [D92]. Based on the similarity with the wavelet bases, sometimes the basis (5.83) is referred to as a waveletlike modified HB or approximate wavelet HB (AWHB).

Concluding comments for the chapter

A detailed list of references and additional results on MG, as well as notes on the history of MG, are found, for example, in Hackbusch [H85], Bramble [Br93], Oswald [0s94], Trottenberg et al. [TOS], and Shaidurov [Sh95].

6

Topics on Algebraic Multigrid (AMG)

The algebraic multigrid (or AMG) concept was introduced in [BMcRr, BMcR] (see also [B86]), and gained popularity after the paper [RS87] appeared. Since then, much progress has been made and the present chapter reflects a number of major developments in this area.

There is no single AMG method. Loosely speaking, if the coarse hierarchy defined by respective interpolation matrices $\{P_k\}$ is constructed by the user, the resulting MG method defines a class of AMG. AMG is typically defined as a two-level method, and the construction is used recursively, thus ending up with a multilevel AMG. That is why, for the most part of the presentation, we omit the level subscript k in the present chapter.

6.1 Motivation for the construction of P

Typically, to construct the interpolation matrices P, we utilize some (very often assumed) a priori knowledge of the lower part of the spectrum of $D^{-1}A$ where D is, for example, the diagonal part of A (or the symmetrized smoother to be used, which comes from a convergent splitting of A). The vectors that are spanned by the eigenvectors corresponding to the lower part of the spectrum of $D^{-1}A$ (sometimes called "algebraically smooth" vectors) are attempted to be approximated well by the coarse space Range (P) for a proper choice of P (assuming a two-level setting). The latter condition is more rigorously studied in this section.

We are given a sparse s.p.d. matrix A and a convergent in $\|.\|_A$-norm smoother M. The latter property, as we have very often used it, is equivalent to $M + M^T - A$ being s.p.d. The two-grid algorithm, based on M, its transpose M^T, an interpolation matrix P, and respective coarse matrix $A_c = P^TAP$, gives rise to the product iteration matrix

$$E_{TG} = (I - M^{-T}A)(I - \pi_A)(I - M^{-1}A), \qquad \pi_A = PA_c^{-1}P^TA.$$

P.S. Vassilevski, *Multilevel Block Factorization Preconditioners*,
doi: 10.1007/978-0-387-71564-3_6,

Equivalently, the two-grid preconditioner B_{TG} is defined from the identity $E_{TG} = I - B_{TG}^{-1}A$ or more explicitly,

$$B_{TG}^{-1} = \overline{M}^{-1} + (I - M^{-T}A)PA_c^{-1}P^T(I - AM^{-1}), \tag{6.1}$$

where $\overline{M} = M(M + M^T - A)^{-1}M^T$ is the symmetrized smoother.

Given A and M, to motivate the construction of P, we start from the characterization of the two-grid convergence factor $\varrho(E_{TG})$ studied in Section 3.2.9; namely, Theorem 3.19 states, that $\varrho(E_{TG}) = 1 - (1/K_{TG})$ where

$$K_{TG} = \sup_{\mathbf{v}} \frac{\mathbf{v}^T \widetilde{M}(I - \pi_{\widetilde{M}})\mathbf{v}}{\mathbf{v}^T A\mathbf{v}}, \qquad \pi_{\widetilde{M}} = P(P^T\widetilde{M}P)^{-1}P^T\widetilde{M}.$$

We recall that $\widetilde{M} = M^T(M + M^T - A)^{-1}M$ for a given M, is also a symmetrized smoother. Note that, in general, $\widetilde{M}$ is different from $\overline{M}$ (used in (6.1)). Also, note that $\pi_{\widetilde{M}}$ (similarly to π_A) is a projection onto the coarse space, now based on the $\widetilde{M}$-inner product.

Traditionally, we assume that P has the form

$$P = \begin{bmatrix} W \\ I \end{bmatrix}.$$

Let $\{\boldsymbol{\delta}_{i_c}\}$ be the basis of unit coordinate vectors in $\mathbb{R}^{n_c}$. It is clear that $\boldsymbol{\psi}_{i_c} = P\boldsymbol{\delta}_{i_c}$, $i_c = 1, \ldots, n_c$ form a basis in the space Range $(P(P^T\widetilde{M}P)^{-1}P^T\widetilde{M})$. Indeed, we have $(P^T\widetilde{M}P)^{-1}P^T\widetilde{M}\boldsymbol{\psi}_{i_c} = \boldsymbol{\delta}_{i_c}$, which shows that $\{\boldsymbol{\psi}_i\}$ are linearly independent. Consider now the direct decomposition

$$\mathbf{v} = \begin{bmatrix} \overline{\mathbf{v}}_f \\ 0 \end{bmatrix} + P\mathbf{v}_c.$$

Hence,

$$\begin{aligned}
K_{TG} &= \sup_{\mathbf{v}} \frac{\mathbf{v}^T \widetilde{M}(I - \pi_{\widetilde{M}})\mathbf{v}}{\mathbf{v}^T A\mathbf{v}} \\
&= \sup_{\mathbf{v}} \frac{\begin{bmatrix} \overline{\mathbf{v}}_f \\ 0 \end{bmatrix}^T \widetilde{M}(I - \pi_{\widetilde{M}}) \begin{bmatrix} \overline{\mathbf{v}}_f \\ 0 \end{bmatrix}}{\mathbf{v}^T A\mathbf{v}} \\
&= \sup_{\overline{\mathbf{v}}_f} \sup_{\mathbf{v}_c} \sup_{t\in\mathbb{R}} \frac{\begin{bmatrix} \overline{\mathbf{v}}_f \\ 0 \end{bmatrix}^T \widetilde{M}(I - \pi_{\widetilde{M}}) \begin{bmatrix} \overline{\mathbf{v}}_f \\ 0 \end{bmatrix}}{\left(\begin{bmatrix} \overline{\mathbf{v}}_f \\ 0 \end{bmatrix} + tP\mathbf{v}_c\right)^T A \left(\begin{bmatrix} \overline{\mathbf{v}}_f \\ 0 \end{bmatrix} + tP\mathbf{v}_c\right)} \\
&= \sup_{\overline{\mathbf{v}}_f} \sup_{\mathbf{v}_c} \frac{\begin{bmatrix} \overline{\mathbf{v}}_f \\ 0 \end{bmatrix}^T \widetilde{M}(I - \pi_{\widetilde{M}}) \begin{bmatrix} \overline{\mathbf{v}}_f \\ 0 \end{bmatrix}}{\begin{bmatrix} \overline{\mathbf{v}}_f \\ 0 \end{bmatrix}^T A \begin{bmatrix} \overline{\mathbf{v}}_f \\ 0 \end{bmatrix} - \frac{\left(\begin{bmatrix} \overline{\mathbf{v}}_f \\ 0 \end{bmatrix}^T AP\mathbf{v}_c\right)^2}{\mathbf{v}_c^T P^T AP\mathbf{v}_c}}.
\end{aligned}$$

If we drop the term $\pi_{\widetilde{M}}$ in the numerator, we get an upper bound for K_{TG}. That is, we get then

$$K_{TG} \le \sup_{\overline{\mathbf{v}}_f} \sup_{\mathbf{v}_c} \frac{\begin{bmatrix} \overline{\mathbf{v}}_f \\ 0 \end{bmatrix}^T \widetilde{M} \begin{bmatrix} \overline{\mathbf{v}}_f \\ 0 \end{bmatrix}}{\begin{bmatrix} \overline{\mathbf{v}}_f \\ 0 \end{bmatrix}^T A \begin{bmatrix} \overline{\mathbf{v}}_f \\ 0 \end{bmatrix} - \frac{\left(\begin{bmatrix} \overline{\mathbf{v}}_f \\ 0 \end{bmatrix}^T AP\mathbf{v}_c\right)^2}{\mathbf{v}_c^T P^T AP\mathbf{v}_c}}.$$

Notice now that the numerator no longer depends on P. If we want to minimize the above upper bound for K_{TG} without imposing any restrictions on P, the best bound we can get is to choose P such that

$$\begin{bmatrix} \overline{\mathbf{v}}_f \\ 0 \end{bmatrix}^T AP\mathbf{v}_c = 0 \quad \text{for all } \mathbf{v}_c \text{ and } \overline{\mathbf{v}}_f.$$

The latter shows, if we partition A as

$$A = \begin{bmatrix} A_{ff} & A_{fc} \\ A_{cf} & A_{cc} \end{bmatrix},$$

that W solves the equation

$$A_{ff} W + A_{fc} = 0,$$

or equivalently the "best" P, P_*, is given by

$$P_* = \begin{bmatrix} -A_{ff}^{-1} A_{fc} \\ I \end{bmatrix}.$$

For P_*, we get the following upper bound for K_{TG},

$$K_{TG} \le \sup_{\overline{\mathbf{v}}_f} \frac{\overline{\mathbf{v}}_f^T \widetilde{M}_{ff} \overline{\mathbf{v}}_f}{\overline{\mathbf{v}}_f^T A_{ff} \overline{\mathbf{v}}_f} = \frac{1}{\lambda_{\min}(\widetilde{M}_{ff}^{-1} A_{ff})}.$$

It is clear then that if the symmetrized smoother $\widetilde{M}$ restricted to the "f" set is spectrally equivalent to A_{ff}, then we have a spectrally equivalent two-grid method (based on P_*). We actually know (see Theorem 3.25) that if $\widetilde{M}$ has the above property, plus if for some P, PR (for $R = [0,\ I]$) is bounded in the A-norm, then the respective two-grid method is also spectrally equivalent to A with $K_{TG} \le \|PR\|_A \ (1/\lambda_{\min}(\widetilde{M}_{ff}^{-1} A_{ff}))$.

For now, we can conclude that a reasonable guideline to construct P is to find, for any coarse unit coordinate vector $\boldsymbol{\delta}_{i_c} \in \mathbb{R}^{n_c}$, an approximate solution to

$$A_{ff} \mathbf{w}_{i_c} = -A_{fc} \boldsymbol{\delta}_{i_c}.$$

Then

$$\boldsymbol{\psi}_{i_c} = \begin{bmatrix} \mathbf{w}_{i_c} \\ \boldsymbol{\delta}_{i_c} \end{bmatrix}$$

defines the i_cth column of P.

6.2 On the classical AMG construction of P

Here we assume that the set $\mathcal{N}_c$ of coarse dofs has been selected. Given a dof $i \in \mathcal{N}_f$ (the set of dofs complementary to $\mathcal{N}_c$), consider a given neighborhood $\Omega(i)$ that contains all coarse dofs i_c allowed to interpolate to i (denoted by C_i), the dof i, plus possibly some other dofs. The ith entry of the interpolation $(W\mathbf{v}_c)_i = \sum_{i_c \in C_i} w_{i,i_c}(\mathbf{v}_c)_{i_c}$ will be determined from the ith row of the equation

$$A_{ff}\mathbf{v}_f = -A_{fc}\mathbf{v}_c.$$

The ith row of the above equation takes the form, after introducing $A_{ff}|_i = [a_{ii},\ \mathbf{a}_{i,\,\mathcal{X}}^T]$ and $A_{fc}|_i = \mathbf{a}_{i,c}^T$,

$$a_{ii}v_i = -\mathbf{a}_{i,\,\mathcal{X}}^T\mathbf{v}_{i,\,\mathcal{X}} - \mathbf{a}_{i,\,c}^T\mathbf{v}_c.$$

If we have values assigned to $\mathbf{v}_{i,\,\mathcal{X}}$ in terms of $\mathbf{v}_c$ (cf., [HV01]), then from the above equation, we can compute the needed mapping $\mathbf{v}_c \mapsto (\mathbf{v}_f)_i$, which will give the ith row of P. In other words, if we have an "extension" mapping (cf., [HV01]) $E = [E_{\mathcal{X},\,i},\ E_{\mathcal{X},\,c}]$ that for given v_i, $\mathbf{v}_c$ defines the values

$$\mathbf{v}_{\mathcal{X}} = [E_{\mathcal{X},\,i},\ E_{\mathcal{X},\,c}]\begin{bmatrix} v_i \\ \mathbf{v}_c \end{bmatrix},$$

we then get

$$\left(a_{ii} + \mathbf{a}_{i,\,\mathcal{X}}^T E_{\mathcal{X},\,i}\right)v_i = -\left(\mathbf{a}_{i,\,\mathcal{X}}^T E_{\mathcal{X},\,c} + \mathbf{a}_{i,\,c}^T\right)\mathbf{v}_c.$$

That is, assuming $a_{ii} + \mathbf{a}_{i,\,\mathcal{X}}^T E_{\mathcal{X},\,i} \neq 0$, we obtain the following expression for the ith row of P,

$$-\left(a_{ii} + \mathbf{a}_{i,\,\mathcal{X}}^T E_{\mathcal{X},\,i}\right)^{-1}\left(\mathbf{a}_{i,\,\mathcal{X}}^T E_{\mathcal{X},\,c} + \mathbf{a}_{i,\,c}^T\right).$$

The classical Ruge–Stüben interpolation rule corresponds to the following extension mapping E. We first partition the dofs in set $\mathcal{X}$ into two groups; dofs that are "weakly" connected to i, and dofs that are "strongly" connected to i.

We mention here one possible definition of the notion of "strong dependence" as introduced in [RS87].

Definition 6.1 (Strong dependence). *For a chosen tolerance $\theta \in (0, 1]$, we say that a dof i is strongly influenced by (or depends strongly on) dof $j \neq i$ if*

$$-a_{ij} \geq \theta \max_{k \neq i}(-a_{ik})$$

Equivalently, we say then that dof j is strongly connected to dof i. For all other dofs k, we say that k is weakly connected to i.

The extension mapping $[v_i, \mathbf{v}_c] \mapsto (v_{i_\mathcal{X}}) = E[v_i, \mathbf{v}_c]$ then reads,

$$v_{i_\mathcal{X}} = \begin{cases} v_i, & \text{if } i_\mathcal{X} \text{ is weakly connected to } i, \\ \dfrac{\sum_{i_c \in C_i} a_{i_\mathcal{X}, i_c} v_{i_c}}{\sum_{i_c \in C_i} a_{i_\mathcal{X}, i_c}}, & \text{if } i_\mathcal{X} \text{ is strongly connected to } i. \end{cases}$$

Introducing $W_i = \{j \in \Omega_i : j$ is weakly connected to $i\}$ and $S_i = \{j \in \Omega_i : j$ is strongly connected to $i\}$, the resulting (i, i_c) entry of the Ruge–Stüben interpolation matrix reads,

$$-\left(a_{ii} + \sum_{i_\mathcal{X} \in W_i} a_{i,\, i_\mathcal{X}}\right)^{-1} \left(a_{i,\, i_c} + \sum_{i_\mathcal{X} \in S_i} a_{i,\, i_\mathcal{X}} \frac{a_{i_\mathcal{X},\, i_c}}{\sum_{j_c \in C_i} a_{i_\mathcal{X},\, j_c}}\right).$$

The above formula is well defined if $a_{i,\, i_\mathcal{X}}$ are either positive or small negative for all $i_\mathcal{X}$ weakly connected to i, and if $a_{i_\mathcal{X},\, i_c}$ have the same sign for all $i_\mathcal{X}$ strongly connected to i. These conditions are met in practice for matrices A that are close to M-matrices. The above extension mapping has the property if $[v_i, \mathbf{v}_c]$ is a constant vector, then $E[v_i, \mathbf{v}_c]$ is also (the same) constant.

Given a vector $\mathbf{1}$, we can easily construct extension mapping E such that it has a prescribed value $\mathbf{1}_\mathcal{X} = E[\mathbf{1}_i, \mathbf{1}_c]$. The corresponding formula reads

$$v_{i_\mathcal{X}} = \begin{cases} \dfrac{(\mathbf{1})_{i_\mathcal{X}}}{(\mathbf{1})_i} v_i, & \text{if } i_\mathcal{X} \text{ is weakly connected to } i, \\ \mathbf{1}_{i_\mathcal{X}} \dfrac{\sum_{i_c \in C_i} a_{i_\mathcal{X}, i_c} v_{i_c}}{\sum_{i_c \in C_i} a_{i_\mathcal{X}, i_c} (\mathbf{1})_{i_c}}, & \text{if } i_\mathcal{X} \text{ is strongly connected to } i. \end{cases}$$

The resulting interpolation matrix P (cf., [aAMG]) has its (i, i_c) entry equal to

$$-\left(a_{ii} + \sum_{i_\mathcal{X} \in W_i} a_{i,\, i_\mathcal{X}} \frac{(\mathbf{1})_{i_\mathcal{X}}}{(\mathbf{1})_i}\right)^{-1} \left(a_{i,\, i_c} + \sum_{i_\mathcal{X} \in S_i} a_{i,\, i_\mathcal{X}} \frac{a_{i_\mathcal{X},\, i_c} (\mathbf{1})_{i_\mathcal{X}}}{\sum_{j_c \in C_i} a_{i_\mathcal{X},\, j_c} (\mathbf{1})_{j_c}}\right).$$

The following property holds (cf., [aAMG]).

Proposition 6.2. *The following identity holds,*

$$(P\mathbf{1}_c)_i = (\mathbf{1})_i - \frac{(A\mathbf{1})_i}{a_{ii} + \sum_{i_\mathcal{X} \in W_i} a_{i,\, i_\mathcal{X}} \frac{(\mathbf{1})_{i_\mathcal{X}}}{(\mathbf{1})_i}}.$$

Then if $(A\mathbf{1})_i = 0$, *we have that* $(P\mathbf{1}_c)_i = (\mathbf{1})_i$. *In particular, if* $\mathbf{1}_f = -A_{ff}^{-1} A_{fc} \mathbf{1}_c$ *then*

$$P\mathbf{1}_c = \mathbf{1} = \begin{bmatrix} \mathbf{1}_f \\ \mathbf{1}_c \end{bmatrix}.$$

Proof. We have, letting $\Delta = -1/(a_{ii} + \sum_{i_\mathcal{X} \in W_i} a_{i,\,i_\mathcal{X}}((\mathbf{1})_{i_\mathcal{X}}/(\mathbf{1})_i))$,

$$\begin{aligned}
(P\mathbf{1}_c)_i &= \Delta \left(\sum_{i_c \in C_i} a_{i,\,i_c}(\mathbf{1})_{i_c} + \sum_{i_c \in C_i} \sum_{i_\mathcal{X} \in S_i} a_{i,\,i_\mathcal{X}}(\mathbf{1})_{i_\mathcal{X}} \frac{a_{i_\mathcal{X},\,i_c}(\mathbf{1})_{i_c}}{\sum_{j_c \in C_i} a_{i_\mathcal{X},\,j_c}(\mathbf{1})_{j_c}} \right) \\
&= \Delta \left(\sum_{i_c \in C_i} a_{i,\,i_c}(\mathbf{1})_{i_c} + \sum_{i_\mathcal{X} \in S_i} a_{i,\,i_\mathcal{X}}(\mathbf{1})_{i_\mathcal{X}} \sum_{i_c \in C_i} \frac{a_{i_\mathcal{X},\,i_c}(\mathbf{1})_{i_c}}{\sum_{j_c \in C_i} a_{i_\mathcal{X},\,j_c}(\mathbf{1})_{j_c}} \right) \\
&= \Delta \left(\sum_{i_c \in C_i} a_{i,\,i_c}(\mathbf{1})_{i_c} + \sum_{i_\mathcal{X} \in S_i} a_{i,\,i_\mathcal{X}}(\mathbf{1})_{i_\mathcal{X}} \right) \\
&= \Delta \left((A\mathbf{1})_i - a_{ii}(\mathbf{1})_i - \sum_{i_\mathcal{X} \in W_i} a_{i,\,i_\mathcal{X}}(\mathbf{1})_{i_\mathcal{X}} \right) \\
&= (\mathbf{1})_i - \frac{(A\mathbf{1})_i}{a_{ii} + \sum_{i_\mathcal{X} \in W_i} a_{i,\,i_\mathcal{X}} \frac{(\mathbf{1})_{i_\mathcal{X}}}{(\mathbf{1})_i}}.
\end{aligned}$$

□

Assume that A_{ff} is an M-matrix and let $A_{fc} \le 0$. Then for any positive vector $\mathbf{1}_c$ the vector $\mathbf{1}_f \equiv -A_{ff}^{-1} A_{fc} \mathbf{1}_c$ will be positive.

Note that even if A is an M-matrix, its coarse counterpart $P^T AP$ may not be. We can instead construct an auxiliary M-matrix $\overline{A}$ associated with A for the purpose of constructing P, for example, by adding to A symmetric positive semidefinite matrices $a_{i,j} V_{i,j}$, where

$$V_{i,j} = \begin{bmatrix} \ddots & & & & \\ & \alpha & & -1 & \\ & & \ddots & & \\ & -1 & & \frac{1}{\alpha} & \\ & & & & \ddots \end{bmatrix} \begin{matrix} \\ \} \; i \\ \\ \} \; j \\ \\ \end{matrix}$$

has only four nonzero entries. Here, (i, j) runs over all positive off-diagonal entries $a_{i,j}$ of A. The coefficient α is chosen so that $V_{ij}\mathbf{1} = 0$, hence $\alpha(\mathbf{1})_i - (\mathbf{1})_j = 0$ and $-(\mathbf{1})_i + (1/\alpha)(\mathbf{1})_j = 0$. This gives $\alpha = (\mathbf{1})_j/(\mathbf{1})_i > 0$, for any vector $\mathbf{1}$ with positive entries. Note that the eigenvalues λ of $V_{i,j}$ solve the equation $(\alpha - \lambda)$ $((1/\alpha) - \lambda) - 1 = 0$. That is, $\lambda(\lambda - \alpha - (1/\alpha)) = 0$ which gives either $\lambda = 0$ or $\lambda = \alpha + (1/\alpha) > 0$. This shows that $V_{i,j}$ is positive semidefinite. It is clear then that $\overline{A} := A + a_{i,j} V_{ij}$ is s.p.d. and has zero entry at position (i, j). Moreover, $(A + a_{i,j} V_{ij})\mathbf{1} = A\mathbf{1}$. After running over all pair of indices (i, j), $i < j$, such that $a_{i,j} > 0$, we end up with an M-matrix $\overline{A} = A + \sum_{i<j:\, a_{i,j}>0} a_{i,j} V_{i,j}$ for which $\overline{A}\mathbf{1} = A\mathbf{1}$. In addition, $\overline{A} - A$ is symmetric positive semidefinite. We can then build a P on the basis of $\overline{A}$ and the positive vector $\mathbf{1}$. If we have $(A\mathbf{1})_f = 0$, then the same would apply to $\overline{A}$, and all the formulas for P (based on $\overline{A}$) will then be well defined.

Note that $\overline{A}$ will have exactly the entry $\overline{a}_{ii} = a_{ii} + \sum_{i\chi} a_{i,\, i\chi}(((\mathbf{1})_{i\chi})/((\mathbf{1})_i))$ for all positive off-diagonal entries $a_{i,\, i\chi}$ of A if treated as weakly connected to i.

6.3 On the constrained trace minimization construction of *P*

Consider a given s.p.d. sparse matrix A and a "f"–"c" partitioning of the index set $\mathcal{N}$ into two disjoint groups $\mathcal{N}_f$ and $\mathcal{N}_c$. We want to construct the interpolation matrix P with columns $\boldsymbol{\psi}_i$ for $i \in \mathcal{N}_c$ that have a prescribed sparsity pattern, referred to as the support of $\boldsymbol{\psi}_i$. It is assumed that $\boldsymbol{\psi}_i|_{\mathcal{N}_c}$ equals the ith unit coarse coordinate vector. This gives rise to the form

$$P = \begin{bmatrix} W \\ I \end{bmatrix}$$

of the interpolation matrix. The construction of P under consideration exploits a given vector $\mathbf{1}$, which P is supposed to recover, that is, if

$$\mathbf{1} = \begin{bmatrix} \mathbf{1}_f \\ \mathbf{1}_c \end{bmatrix}$$

then the condition is that $P\mathbf{1}_c = \mathbf{1}$, or equivalently $W\mathbf{1}_c = \mathbf{1}_f$.

After a sparsity pattern of W, or equivalently the support of the columns $\boldsymbol{\psi}_i$ of P is chosen, the actual entries of W are computed from a norm-minimization principle, as originally proposed in [WCS00]; see also [Wag96]. The actual algorithm that we present here was analyzed in [XZ04].

The constrained trace norm minimization problem under consideration reads: Find

$$P = \begin{bmatrix} W \\ I \end{bmatrix} = (\boldsymbol{\psi}_i)_{i \in \mathcal{N}_c},$$

with prescribed sparsity pattern of W, such that

$$\text{trace}\,(P^T A P) \equiv \sum_{i \in \mathcal{N}_c} \boldsymbol{\psi}_i^T A \boldsymbol{\psi}_i \mapsto \min, \tag{6.2}$$

subject to the constraint

$$P\mathbf{1}_c = \mathbf{1}. \tag{6.3}$$

The solution to this problem was given in [XZ04].

Theorem 6.3. *Let I_i be the mapping representing extension by zero outside the (prescribed fixed) support of $\boldsymbol{\psi}_i$. Introduce then $A_i = I_i^T A I_i$, which is the principal submatrix of A corresponding to the support of $\boldsymbol{\psi}_i$. Finally, define the symmetric positive semidefinite matrices $T_i = I_i A_i^{-1} I_i^T$. Then, the solution to the constrained minimization problem* (6.2)–(6.3) *is given by $\boldsymbol{\psi}_i = (\mathbf{1}_c)_i T_i T^{-1} \mathbf{1}$, where $T = \sum_{i \in \mathcal{N}_c} (\mathbf{1}_c)_i^2 T_i$ is the so–called additive Schwarz operator. Here, $(\mathbf{1}_c)_i$ stands for the ith entry of $\mathbf{1}_c$.*

Proof. To solve the problem, we form the Lagrangian

$$\mathcal{L}(P, \underline{\lambda}) = \sum_{i \in \mathcal{N}_c} \boldsymbol{\psi}_i^T A \boldsymbol{\psi}_i + \underline{\lambda}^T (P\mathbf{1}_c - \mathbf{1}).$$

Recall that $P = [\boldsymbol{\psi}_1, \boldsymbol{\psi}_2, \ldots, \boldsymbol{\psi}_{n_c}]$ and hence $\underline{\lambda} \in \mathbb{R}^n$. Then, by varying $\boldsymbol{\psi}_i := \boldsymbol{\psi}_i + I_i \mathbf{g}_i$ for any vector $\mathbf{g}_i$ defined on the support of $\boldsymbol{\psi}_i$, from the necessary conditions for a minimum, we get

$$(I_i \mathbf{g}_i)^T (A\boldsymbol{\psi}_i + (\mathbf{1}_c)_i \underline{\lambda}) = 0,$$

for all $i = 1, \ldots, n_c$. The derivative w.r.t. $\underline{\lambda}$ gives the constraint,

$$P\mathbf{1}_c - \mathbf{1} = \sum_{i=1}^{n_c} (\mathbf{1}_c)_i \boldsymbol{\psi}_i - \mathbf{1} = 0.$$

The first set of equations actually reads, based on the fact that $\boldsymbol{\psi}_i = I_i I_i^T \boldsymbol{\psi}_i$,

$$A_i I_i^T \boldsymbol{\psi}_i = -(\mathbf{1}_c)_i I_i^T \underline{\lambda}.$$

Therefore, multiplying by $I_i A_i^{-1}$, we obtain

$$\boldsymbol{\psi}_i = I_i I_i^T \boldsymbol{\psi}_i = -(\mathbf{1}_c)_i I_i A_i^{-1} I_i^T \underline{\lambda} = -(\mathbf{1}_c)_i T_i \underline{\lambda}. \tag{6.4}$$

Multiplying the latter equation by $(\mathbf{1}_c)_i$ and summing over all $i = 1, \ldots, n_c$ we end up with the expression

$$\sum_{i=1}^{n_c} (\mathbf{1}_c)_i \boldsymbol{\psi}_i = \sum_{i=1}^{n_c} (\mathbf{1}_c)_i I_i I_i^T \boldsymbol{\psi}_i = -\left(\sum_{i=1}^{n_c} (\mathbf{1}_c)_i^2 T_i \right) \underline{\lambda}.$$

That is, using the constraint $\mathbf{1} = P\mathbf{1}_c = \sum_{i=1}^{n_c} (\mathbf{1}_c)_i \boldsymbol{\psi}_i$ and the definition of $T = \sum_{i=1}^{n_c} (\mathbf{1}_c)_i^2 T_i$, the following expression for the Lagrangian multiplier $\underline{\lambda}$ is obtained,

$$\underline{\lambda} = -T^{-1}\mathbf{1}.$$

Using the latter expression in (6.4), the desired result

$$\boldsymbol{\psi}_i = (\mathbf{1}_c)_i T_i T^{-1} \mathbf{1}$$

follows. It is clear that $P\mathbf{1}_c = \sum_{i=1}^{n_c} (\mathbf{1}_c)_i \boldsymbol{\psi}_i = \sum_{i=1}^{n_c} (\mathbf{1}_c)_i^2 T_i T^{-1} \mathbf{1} = \mathbf{1}$; that is, the constraint is satisfied. □

To actually compute $T^{-1}\mathbf{1}$ in practice, we can use the preconditioned CG method with the following diagonal matrix $\sum_{i=1}^{n_c} (\mathbf{1}_c)_i^2 I_i D_i^{-1} I_i^T$ as a preconditioner. Here, D_i stands for the diagonal of A_i. The inverses of A_i can be approximated by a fixed number of symmetric Gauss–Seidel steps. Thus, in practice, P can be based on M_i^{-1} where M_i^{-1} represents a fixed number of symmetric Gauss–Seidel iterations, for example, one or two. That is, the modified additive Schwarz operator reads $T = \sum_{i=1}^{n_c} (\mathbf{1}_c)_i^2 I_i M_i^{-1} I_i^T$ and hence $\boldsymbol{\psi}_i = (\mathbf{1}_c)_i I_i M_i^{-1} I_i^T T^{-1} \mathbf{1}$. The preconditioner for T is again the diagonal matrix $\sum_{i=1}^{n_c} (\mathbf{1}_c)_i^2 I_i D_i^{-1} I_i^T$ with D_i being the diagonal of A_i. Numerical experiments with the inexact additive Schwarz-based construction of P are found in [VZ05]. A final comment is that we have (implicitly) assumed that $\mathbf{1}_c$ has nonzero entries (otherwise T may not be invertible). A common case in practice is $\mathbf{1}_c$ being the constant vector $[1, \ldots, 1]^T \in \mathbb{R}^{n_c}$.

6.4 On the coarse-grid selection

The selection of coarse-grid $\mathcal{N}_c$ is the least rigorous part of AMG. Part of the problem is that many choices of coarse grids can lead to AMG methods with comparable performance. For the element agglomeration AMG (in Section 6.9) and the window-based spectral AMG (in Section 6.11), we provide coarse-grid selection algorithms that guarantee provable two-grid AMG convergence.

For some practical coarse-grid selection algorithms, we refer to [RS87], the chapter on AMG by K. Stüben in [TOS], or the tutorial [MGT00]. State-of-the-art parallel coarse-grid selection algorithms are found in [PMISi], and in combination with scalable ("distance-two") interpolation algorithms, are found in [PMISii].

6.5 On the sparsity pattern of P

After a coarse set $\mathcal{N}_c$ has been selected, letting $\mathcal{N}_f = \mathcal{N} \setminus \mathcal{N}_c$ be the complementary set of f dofs, we want to compute the sparsity pattern of P, that is, the support set of each column $\boldsymbol{\psi}_i$ of P. One possible strategy is to look at the decay behavior of $-A_{ff}^{-1} A_{fc}$, where the blocks A_{ff} and A_{fc} come from the natural "f"–"c" partitioning of A, given by

$$A = \begin{bmatrix} A_{ff} & A_{fc} \\ A_{cf} & A_{cc} \end{bmatrix} \begin{matrix} \} & \mathcal{N}_f \\ \} & \mathcal{N}_c. \end{matrix}$$

The latter strategy has been utilized in [BZ06]. Namely, we perform source PCG iterations

$$A_{ff} \mathbf{x}_f = -A_{fc} \mathbf{e}_{c,\,i}, \qquad \mathbf{e}_{c,\,i} = \begin{bmatrix} 0 \\ \vdots \\ 1 \\ \vdots \\ 0 \end{bmatrix},$$

where the only nonzero entry of $\mathbf{e}_{c,\,i}$ is at position $i \in \mathcal{N}_c$. Then, after computing an approximation $\mathbf{x}_f$ to $-A_{ff}^{-1} A_{fc} \mathbf{e}_{c,\,i}$, we look at the decay of the entries of $\mathbf{x}_f$

around dof i. The entries at positions $j \in \mathcal{N}_f$ that have relatively large values, form the support of ψ_i. More specifically, with D_{ff} being the diagonal of A_{ff}, we look at the magnitude of the entries of $D_{ff}^{1/2}\mathbf{x}_f$. The reason for this is that $D_{ff}^{1/2}\mathbf{x}_f \approx \overline{A}_{ff}^{-1} D_{ff}^{1/2} A_{fc}\mathbf{e}_{c,\,i}$, where $\overline{A}_{ff}$ is A_{ff} scaled symmetrically by $D_{ff}^{-(1/2)}$ so that $\overline{A}_{ff}^{-1}$ has a more uniform decay rate. If we can choose $\mathcal{N}_c$ such that A_{ff} is spectrally equivalent to its diagonal D_{ff}, then $\overline{A}_{ff}$ will be well conditioned. A geometric decay of $\overline{A}_{ff}^{-1}$ with a rate depending only on the condition number of $\overline{A}_{ff}$ can be proved (see Section A.2.4 or estimate (6.56)).

After the support of ψ_i has been determined, the actual entries of ψ_i can then be computed (as in [BZ06]) based on the constrained trace minimization based on a given vector $\mathbf{1}$ that was described in Section 6.3.

6.6 Coarsening by compatible relaxation

The goal of the compatible relaxation (or CR) is to select a set of coarse degrees of freedom, based solely on the smoother, such that after a proper interpolation matrix is constructed later, then the resulting two-grid method exhibits fast convergence. The notion of compatible relaxation was introduced by Achi Brandt in [B00] and studied later on in some detail in [Li04] and [FV04].

We outline the main principles of CR and show some basic estimates.

If we look at the characterization of K_{TG}, we have that for a matrix J_* such that Range $(J_*) = $ Range $(I - PR_*)$, $R_* = (P^T \widetilde{M} P)^{-1} P^T \widetilde{M}$, we have the inequality

$$\mathbf{v}_s^T J_*^T \widetilde{M} J_* \mathbf{v}_s \leq K_{TG}\, \mathbf{v}_s^T J_*^T A J_* \mathbf{v}_s.$$

This shows that A has a principal matrix that is spectrally equivalent to the same principal submatrix of the symmetrized smoother $\widetilde{M}$. CR refers to the process of selecting a J based on a preselected R, typically $R = [0,\ I]$, such that the constant κ_{CR} in the inequality

$$\mathbf{v}_s^T J^T A J \mathbf{v}_s \leq \mathbf{v}_s^T J^T \widetilde{M} J \mathbf{v}_s \leq \kappa_{CR}\, \mathbf{v}_s^T J^T A J \mathbf{v}_s,$$

is close to one. If

$$J = \begin{bmatrix} I \\ 0 \end{bmatrix},$$

we indeed look for a principal submatrix A_{ff} of A that is spectrally equivalent to the corresponding principal submatrix of $\widetilde{M}$. However, we may look for more sophisticated choices of R and J. They can change adaptively throughout an iterative procedure.

In the following few sections, we provide algorithms that test if a given (tentative) coarse space (associated with a given (tentative) interpolation matrix P) provides fast to converge CR.

6.6.1 Smoothing property and compatible relaxation

Consider an $n \times n$ s.p.d. matrix A and an A-convergent smoother M for A (forward Gauss–Seidel, e.g.) in the sense that $\|I - M^{-1}A\|_A < 1$. Let $\widetilde{M} = M^T(M^T + M - A)^{-1}M$ denote the symmetrized smoother that gives rise to the iteration matrix $I - \widetilde{M}^{-1}A = (I - M^{-1}A)(I - M^{-T}A)$.

The following "smoothing" property was derived in Lemma 5.40. For any $m \geq 1$ and any vector $\mathbf{e}$, we have

$$\|(I - \widetilde{M}^{-1}A)^m \mathbf{e}\|_A \leq \frac{1}{\sqrt{m+1}} \|\mathbf{e}\|_{\widetilde{M}}.$$

Note that the left-hand side uses the A-norm, whereas the right-hand side uses the $\widetilde{M}$-norm.

Note also that, for standard smoothers such as Gauss–Seidel, $\|.\|_{\widetilde{M}} \simeq \|.\|_D$ where $D = \text{diag}(A)$ (for more details, see Proposition 6.12). Now, let $\mathbf{e} = (I - Q)\mathbf{e}$, where Q is any projection onto a given coarse space. Assuming that Q provides a standard approximation property given by $\|(I - Q)\mathbf{e}\|_{\widetilde{M}} \leq \delta\ \|\mathbf{e}\|_A$, then the following estimate is obtained.

$$\|(I - \widetilde{M}^{-1}A)^m (I - Q)\mathbf{e}\|_A \leq \frac{1}{\sqrt{m+1}} \|(I - Q)\mathbf{e}\|_{\widetilde{M}} \leq \frac{\delta}{\sqrt{m+1}} \|\mathbf{e}\|_A.$$

For the case that Q is a coarse-grid projection based on the $\widetilde{M}$-inner product, we can derive estimates in the $\widetilde{M}$-norm, which is our focus from now on. We first prove an auxiliary estimate.

Lemma 6.4. *Let Q be an $\widetilde{M}$-orthogonal projection that satisfies the following weak approximation property,*

$$\|(I - Q)\mathbf{e}\|_{\widetilde{M}} \leq \delta\ \|\mathbf{e}\|_A. \tag{6.5}$$

Then the following estimate holds,

$$\|(I - Q)\mathbf{e}\|_{\widetilde{M}A^{-1}\widetilde{M}} \leq \delta\ \|(I - Q)\mathbf{e}\|_{\widetilde{M}}. \tag{6.6}$$

Proof. Consider the problem

$$A\mathbf{u} = \widetilde{M}(I - Q)\mathbf{e}.$$

The following estimates are readily obtained using the fact that $\widetilde{M}(I - Q)\mathbf{e}$ is orthogonal to any vector in Range (Q),

$$\mathbf{u}^T A\mathbf{u} = \mathbf{u}^T \widetilde{M}(I - Q)\mathbf{e} = (\mathbf{u} - Q\mathbf{u})^T \widetilde{M}(I - Q)\mathbf{e} \leq \|(I - Q)\mathbf{u}\|_{\widetilde{M}}\ \|(I - Q)\mathbf{e}\|_{\widetilde{M}}.$$

Using now the approximation property (6.5) leads to

$$\mathbf{u}^T A\mathbf{u} \leq \delta\ \|\mathbf{u}\|_A \|(I - Q)\mathbf{e}\|_{\widetilde{M}}.$$

That is,

$$\|\mathbf{u}\|_A \leq \delta\ \|(I - Q)\mathbf{e}\|_{\widetilde{M}}.$$

Because $\mathbf{u} = A^{-1}\widetilde{M}(I - Q)\mathbf{e}$, the desired estimate (6.6) follows. □

The following result holds for any $\widetilde{M}$-orthogonal projection Q.

Lemma 6.5. *Assume that Q is an $\widetilde{M}$-orthogonal projection onto the coarse space satisfying the weak approximation property* (6.5)*. Then, for any $\mathbf{e} = (I - Q)\mathbf{e}$ and any integer $m \geq 1$, the following estimate holds,*

$$\|(I - \widetilde{M}^{-1}A)^m \mathbf{e}\|_{\widetilde{M}} \leq \varrho \; \|\mathbf{e}\|_{\widetilde{M}},$$

where $\varrho = \delta/\sqrt{m+1}$ and δ is the constant in the weak approximation property (6.5).

Proof. Let $p_m(t) = (1-t)^m$ and $\overline{T} = \widetilde{M}^{-(1/2)} A \widetilde{M}^{-(1/2)}$. Note that the spectrum of $\overline{T}$ is contained in (0, 1]. Then

$$\begin{aligned}
\|(I - \widetilde{M}^{-1}A)^m \mathbf{e}\|_{\widetilde{M}} &= \|\widetilde{M}^{1/2}(I - \widetilde{M}^{-1}A)^m \widetilde{M}^{-(1/2)} \widetilde{M}^{1/2}\mathbf{e}\| \\
&= \|\big(p_m(\overline{T})\overline{T}^{1/2}\big)\overline{T}^{-(1/2)} \widetilde{M}^{1/2}\mathbf{e}\| \\
&\leq \max_{t \in [0,1]} \; t^{1/2}(1-t)^m \; \|(I - Q)\mathbf{e}\|_{\widetilde{M}A^{-1}\widetilde{M}} \\
&\leq \frac{1}{\sqrt{m+1}} \; \delta \; \|(I - Q)\mathbf{e}\|_{\widetilde{M}}.
\end{aligned}$$

In the last line, we used (6.6). □

Consider the following two-grid process.

Algorithm 6.6.1 (Smoothing coarse-grid corrected error). *Consider the homogeneous equation $A\mathbf{x} = 0$. Letting $\mathbf{e}$ be a random initial iterate and $m = 1$, perform the following steps.*

1. *Compute*

$$\mathbf{e}_0 = (I - Q)\mathbf{e}.$$

2. *Smooth:*

$$\mathbf{e}_m = (I - M^{-1}A)(I - M^{-T}A)\mathbf{e}_{m-1}.$$

3. *Monitor convergence in the $\widetilde{M}$-norm; that is, compute $\|\mathbf{e}_m\|_{\widetilde{M}}/\|\mathbf{e}_0\|_{\widetilde{M}}$. If convergence is "slow", use the error $\mathbf{e}_m$ to augment the current coarse space by constructing a new Q, and then increment m and go to Step (2). Otherwise, consider the process to have converged and exit.*

Note that Step (1) is performed only once, outside the inner smoothing loop on m. Also, when $\|.\|_{\widetilde{M}} \simeq \|.\|_D$, we can monitor the convergence in $\|.\|_D$-norm in Step (3) above.

Lemma 6.5 implies that if $m \geq 1$ is sufficiently large, then the above two-grid process must be convergent for the case that Q is an $\widetilde{M}$-orthogonal projection satisfying (6.5).

The following compatible relaxation result holds, where Q now is any projection (not necessarily an $\widetilde{M}$-orthogonal one).

Lemma 6.6. *Suppose for a reasonably small $m \geq 1$ and a given projection Q that Algorithm 6.6.1 provides a convergent process in the $\widetilde{M}$-norm, so that for some $\varrho \in [0, 1)$, the following estimate holds.*

$$\|(I - \widetilde{M}^{-1}A)^m (I - Q)\mathbf{e}\|_{\widetilde{M}} \leq \varrho \ \|\mathbf{e}\|_{\widetilde{M}}.$$

Then the following spectral bound holds for some positive constant δ that depends only on m,

$$\|(I - Q)\mathbf{e}\|_{\widetilde{M}} \leq \delta \ \|(I - Q)\mathbf{e}\|_{A}.$$

A simpler bound, where the roles of A and $\widetilde{M}$ are reversed (and $\delta = 1$), follows from our assumption that M is an A-convergent smoother. Thus, A and the symmetrized smoother $\widetilde{M}$ are spectrally equivalent on the subspace Range $(I - Q)$.

Proof. Let $p_m(t) = (1 - t)^m$ and $T = \widetilde{M}^{-(1/2)} A \widetilde{M}^{-(1/2)}$. Then

$$\begin{aligned}
\|(I - Q)\mathbf{e}\|_{\widetilde{M}} &\leq \|(I - p_m(\widetilde{M}^{-1}A))(I - Q)\mathbf{e}\|_{\widetilde{M}} + \|p_m(\widetilde{M}^{-1}A)(I - Q)\mathbf{e}\|_{\widetilde{M}} \\
&= \|(I - p_m(T))T^{-(1/2)}T^{1/2}\widetilde{M}^{1/2}(I - Q)\mathbf{e}\| + \varrho \ \|(I - Q)\mathbf{e}\|_{\widetilde{M}} \\
&\leq \max_{t \in (0,1]} \frac{1 - p_m(t)}{\sqrt{t}} \ \|T^{1/2}\widetilde{M}^{1/2}(I - Q)\mathbf{e}\| + \varrho \ \|(I - Q)\mathbf{e}\|_{\widetilde{M}} \\
&= \max_{t \in (0,1]} \frac{1 - p_m(t)}{\sqrt{t}} \ \|(I - Q)\mathbf{e}\|_{A} + \varrho \ \|(I - Q)\mathbf{e}\|_{\widetilde{M}}.
\end{aligned}$$

Noting that $p_m(0) = 1$, we have $\max_{t \in (0,1]} ((1 - p_m(t))/\sqrt{t}) \leq \delta_m$ for some positive constant δ_m. Therefore, we finally obtain

$$(1 - \varrho) \ \|(I - Q)\mathbf{e}\|_{\widetilde{M}} \leq \delta_m \ \|(I - Q)\mathbf{e}\|_{A}.$$

Note that δ_m increases with m, but, for a fixed m, it is a fixed constant. □

The following result holds for the desired weak approximation property (6.5).

Corollary 6.7. *In addition to the assumptions of Lemma 6.6, suppose that Q is bounded in energy, so that for some $\eta < \infty$, we have*

$$\|(I - Q)\mathbf{e}\|_{A} \leq \eta \ \|\mathbf{e}\|_{A}.$$

Then Lemma 6.6 implies the weak approximation property

$$\|(I - Q)\mathbf{e}\|_{\widetilde{M}} \leq \frac{\delta_m \eta}{1 - \varrho} \ \|\mathbf{e}\|_{A}.$$

We recall the fact that the weak approximation property implies two-grid convergence. It is in fact a measure (upper bound) for the TG convergence (cf., (3.26)).

6.6.2 Using inexact projections

Here we assume that the projection $\pi = \pi_{\widetilde{M}}$ is approximated by a mapping π^a that is close to π in the sense that, for some given tolerance $\tau \in [0, 1)$ and any $\mathbf{e}$, the following deviation estimate holds,

$$\|(\pi - \pi^a)\mathbf{e}\|_{\widetilde{M}} \leq \tau \; \|\pi \mathbf{e}\|_{\widetilde{M}}.$$

We assume that Range $(\pi^a) \subset$ Range (P); that is, $\pi^a = P(*)$.

Let $Q = PR$ be a simple projection operator for which $RP = I$, such as

$$P = \begin{bmatrix} W \\ I \end{bmatrix} \quad \text{and} \quad R = [0, \, I].$$

The constant δ_Q in the weak approximation property for Q,

$$\|(I - Q)\mathbf{e}\|_{\widetilde{M}} \leq \delta_Q \; \|\mathbf{e}\|_A, \tag{6.7}$$

may not be as good the one for $\pi = \pi_{\widetilde{M}}$,

$$\|(I - \pi)\mathbf{e}\|_{\widetilde{M}} \leq \delta \; \|\mathbf{e}\|_A. \tag{6.8}$$

Consider then the new modified (projection) operator

$$\overline{\pi} = Q + \pi^a(I - Q). \tag{6.9}$$

We show below that $\overline{\pi}$ gives a more stable version of Q. Note that if $\pi^a = \pi$, then $\overline{\pi} = \pi$ and, if $\pi^a = 0$, then $\overline{\pi} = Q$. The construction in (6.9) was introduced (in a geometric MG setting) in [VW97].

We first show that $\overline{\pi}$ is indeed a projection. Because Q is a projection, then $Q(I - Q) = 0$. Also, $QP = P$ implies $(I - Q)\pi^a = (I - Q)P(*) = 0$ and $Q\pi^a = QP(*) = P(*) = \pi^a$. It is clear then that $\overline{\pi}^2 = Q^2 + Q\pi^a(I - Q) = Q + \pi^a(I - Q) = \overline{\pi}$; that is, $\overline{\pi}$ is indeed a projection.

To show that $\overline{\pi}$ satisfies a "weak approximation property" with a better constant than Q, first note that the actions of $\overline{\pi}$ involve actions of Q and π^a, which are assumed to be much less expensive than the actions of the exact projection π itself. The following identity holds,

$$(\overline{\pi} - \pi)\mathbf{e} = (\pi^a - \pi)(I - Q)\mathbf{e}.$$

Thus, the desired result follows from the inequalities

$$\|(I - \overline{\pi})\mathbf{e}\|_{\widetilde{M}} \leq \|(I - \pi)\mathbf{e}\|_{\widetilde{M}} + \|(\pi - \overline{\pi})\mathbf{e}\|_{\widetilde{M}} \leq (\delta + \tau \; \delta_Q)\|\mathbf{e}\|_A.$$

Note that $\overline{\delta} \equiv \delta + \tau\delta_Q \ll \delta_Q$ if $\delta \ll \delta_Q$ for τ sufficiently small.

With the projection $\overline{\pi}$, Algorithm 6.6.1 takes the following modified form.

Algorithm 6.6.2 (Smoothing approximate projection corrected error). *Consider the homogeneous equation* $A\mathbf{x} = 0$, *the simple projection* $Q = PR$, *and the approximation* π^a *to the* $\widetilde{M}$*-based projection* π. *Letting* $\mathbf{e}$ *be a random initial iterate and* $m = 1$, *perform the following steps.*

1. *Compute*

$$\mathbf{e}_0 = (I - \overline{\pi})\mathbf{e} = (I - \pi^a)(I - Q)\mathbf{e}.$$

2. *Smooth:*

$$\mathbf{e}_m = (I - M^{-1}A)(I - M^{-T}A)\mathbf{e}_{m-1}.$$

3. *Monitor convergence in the $\widetilde{M}$-norm; that is, compute $\|\mathbf{e}_m\|_{\widetilde{M}}/\|\mathbf{e}_0\|_{\widetilde{M}}$. If convergence is "slow", use the error $\mathbf{e}_m$ to augment the current coarse space by constructing a new P (which leads to new Q and π), and then increment m and go to Step (2). Otherwise, consider the process to have converged and exit.*

Note that Step (1) is again performed only once, outside the inner smoothing loop on m.

Finally, we comment on a possible choice for π^a. Recall that $\pi = P\widetilde{M}_c^{-1}P^T\widetilde{M}$, where $\widetilde{M}_c = P^T\widetilde{M}P$. Given an approximation $\widetilde{M}_c^a$ to $\widetilde{M}_c$ such that the actions of $(\widetilde{M}_c^a)^{-1}$ are readily available, that is, based on one or a few Gauss–Seidel iterations applied to $\widetilde{M}_c$, then a natural candidate for π^a is

$$P\big(\widetilde{M}_c^a\big)^{-1}P^T\widetilde{M}.$$

6.7 The need for adaptive AMG

Consider the compatible relaxation process (Algorithm 6.6.1 with $m = 1$)

$$\mathbf{x} := (I - \widetilde{M}^{-1}A)(I - \pi_{\widetilde{M}})\mathbf{x}. \tag{6.10}$$

Based on the identity $I - \widetilde{M}^{-1}A = (I - M^{-1}A)(I - M^{-T}A)$, (6.10) can be reformulated (slightly modified) as

$$\mathbf{x} := (I - M^{-T}A)(I - \pi_{\widetilde{M}})(I - M^{-1}A)\mathbf{x}. \tag{6.11}$$

The latter process has the same convergence properties as (6.10).

The iteration (6.11) resembles the exact (symmetric) two-grid cycle

$$\mathbf{x} := (I - M^{-T}A)(I - \pi_A)(I - M^{-1}A)\mathbf{x}.$$

The difference is in the projections used. In general, a V-cycle iteration takes the form

$$\mathbf{x} := (I - M^{-T}A)(I - PB_c^{-1}P^TA)(I - M^{-1}A)\mathbf{x}.$$

Here, B_c^{-1} stands for the next (coarse) level V-cycle. The last iterations make sense if we have already built an initial V-cycle. By testing the current method available, we eventually end up with a component $\mathbf{x}$ that the current level V-cycle cannot handle; that is, the A-norms of two successive iterates $\mathbf{x}$ and $\mathbf{x}_{\text{new}}$ are not too different; that is,

$$\mathbf{x}_{\text{new}}^T A \mathbf{x}_{\text{new}} \simeq \mathbf{x}^T A \mathbf{x}.$$

The reasons for this to happen could be, either,

- The current coarse space cannot approximate well

$$\mathbf{e} = \mathbf{x} - P\mathbf{x}_c = \begin{bmatrix} \mathbf{e}_f \\ 0 \end{bmatrix},$$

and/or

- The coarse V-cycle B_c cannot successfully damp the coarse interpolant $\mathbf{x}_c$ of $\mathbf{x}$.

A possible remedy to the above is to improve the coarse space and/or the coarse solver B_c^{-1} by augmenting the interpolation matrix $\overline{P} = [P,\ P_{\text{new}}]$, where

$$\overline{P} = \begin{bmatrix} W & \overline{P}_{\text{new}} \\ I & 0 \end{bmatrix}.$$

The new columns of $\overline{P}$ are based on additional coarse dofs $\mathcal{N}_{c,\text{ new}} \subset \mathcal{N} \setminus \mathcal{N}_c$. Note that $\mathbf{e}$ vanishes at the current coarse dofs set $\mathcal{N}_c$; that is, $\mathbf{e}|_{\mathcal{N}_c} = 0$.

The additional coarse dofs can be chosen by some independent set algorithm, utilizing a pointwise (or any other locally computable) measure of the interpolation error $\mathbf{e} = \mathbf{x} - P\mathbf{x}_c$. Some details are found in Section 6.10.

We comment at the end that adaptive AMG algorithms originated in [BR02] (the "Bootstrap" AMG) and were developed in [aSA], [aAMG], and [Mc01].

6.8 Smoothing based on "c"–"f" relaxation

Consider the case of interpolation matrix

$$P = \begin{bmatrix} W \\ I \end{bmatrix}$$

and let

$$M = \begin{bmatrix} M_{ff}^T & A_{fc} \\ 0 & M_{cc}^T \end{bmatrix} \tag{6.12}$$

be the so-called "c"–"f" relaxation matrix. It comes from the natural two-by-two block partitioning of

$$A = \begin{bmatrix} A_{ff} & A_{cf} \\ A_{fc} & A_{cc} \end{bmatrix}$$

induced by the interpolation matrix P. Note that the blocks M_{ff} and M_{cc} need not be symmetric. A special case of interest is when $W \approx -M_{ff}^{-T} A_{fc}$. Another limit case is obtained for $M_{cc} = \tau\ D_{cc}$, where D_{cc} is the diagonal of A_{cc} and $\tau > 0$ is sufficiently large. We denote $M = M_\tau$ in this case. Noting that

$$M_\tau^{-1} = \begin{bmatrix} I & -M_{ff}^{-T} A_{fc} \\ 0 & I \end{bmatrix} \begin{bmatrix} M_{ff}^{-T} & 0 \\ 0 & \frac{1}{\tau} D_{cc}^{-1} \end{bmatrix} \mapsto \begin{bmatrix} M_{ff}^{-T} & 0 \\ 0 & 0 \end{bmatrix}, \qquad \tau \mapsto \infty,$$

the results of the present section, as long as they do not depend on τ, also apply to the following two-grid operator

$$B_\tau^{-1} = \overline{M}_\tau^{-1} + (I - M_\tau^{-T} A) P A_c^{-1} P^T (I - A M_\tau^{-1}),$$

and its limit one, as $\tau \mapsto \infty$, referred to as the "hierarchical basis MG" (or HBMG, [BDY88]). We have, letting

$$J = \begin{bmatrix} I \\ 0 \end{bmatrix}, \quad \text{and} \quad \overline{M}_{ff} = M_{ff}^T \left(M_{ff} + M_{ff}^T - A_{ff}\right)^{-1} M_{ff},$$

$$B_{HBMG}^{-1} = J^T \overline{M}_{ff}^{-1} J + (I - J M_{ff}^{-1} J^T A) P A_c^{-1} P^T (I - A J M_{ff}^{-T} J^T). \tag{6.13}$$

We recall the definition of the symmetrized smoother $\widetilde{M} = M^T (M + M^T - A)^{-1} M$, which takes part in the exact convergence factor of the two-grid method based on M and P for a given matrix A (given in Theorem 3.19).

The goal of the analysis in the present section is to compare the exact two-grid convergence factor $\varrho_{TG} = 1 - 1/K_{TG}$, where

$$K_{TG} = \sup_{\mathbf{v}} \frac{\mathbf{v}^T \widetilde{M} (I - \pi_{\widetilde{M}}) \mathbf{v}}{\mathbf{v}^T A \mathbf{v}} = \sup_{\mathbf{v}} \frac{\mathbf{v}^T \widetilde{M} (I - P R_\star) \mathbf{v}}{\mathbf{v}^T A \mathbf{v}},$$

where $R_* = \widetilde{M}_c^{-1} P^T \widetilde{M}$ with $\widetilde{M}_c = P^T \widetilde{M} P$, as characterized in Theorem 3.19, and its upper bound given by the maximum over $\mathbf{e}$ of the measure (in the form introduced in [FV04]):

$$\mu_{\widetilde{M}}(Q, \mathbf{e}) = \frac{((I - Q)\mathbf{e})^T \widetilde{M} (I - Q)\mathbf{e}}{\mathbf{e}^T A \mathbf{e}}, \tag{6.14}$$

where $Q = PR$ with $R = [0, \; I]$ being the trivial injection mapping.

We first derive some useful identities to be needed in the analysis. We have, for the symmetrized smoother,

$$\begin{aligned}
\widetilde{M} &= M^T (M + M^T - A)^{-1} M \\
&= \begin{bmatrix} M_{ff} & 0 \\ A_{cf} & M_{cc} \end{bmatrix} \begin{bmatrix} \left(M_{ff} + M_{ff}^T - A_{ff}\right)^{-1} & 0 \\ 0 & \left(M_{cc} + M_{cc}^T - A_{cc}\right)^{-1} \end{bmatrix} \\
&\quad \times \begin{bmatrix} M_{ff}^T & A_{fc} \\ 0 & M_{cc}^T \end{bmatrix}.
\end{aligned} \tag{6.15}$$

Then,

$$\begin{aligned}
\widetilde{M}_c &= P^T \widetilde{M} P \\
&= \left(M_{ff}^T W + A_{fc}\right)^T \left(M_{ff} + M_{ff}^T - A_{ff}\right)^{-1} \left(M_{ff}^T W + A_{fc}\right) \\
&\quad + M_{cc} \left(M_{cc} + M_{cc}^T - A_{cc}\right)^{-1} M_{cc}^T.
\end{aligned} \tag{6.16}$$

Compute next $P^T \widetilde{M}$. We have, letting $D_f = M_{ff} + M_{ff}^T - A_{ff}$ and $D_c = M_{cc}^T + M_{cc} - A_{cc}$,

$$\begin{aligned}
P^T \widetilde{M} &= [W^T,\ I] \begin{bmatrix} M_{ff} & 0 \\ A_{cf} & M_{cc} \end{bmatrix} \begin{bmatrix} D_f^{-1} & 0 \\ 0 & D_c^{-1} \end{bmatrix} \begin{bmatrix} M_{ff}^T & A_{fc} \\ 0 & M_{cc}^T \end{bmatrix} \\
&= \left[W^T M_{ff} + A_{cf},\ M_{cc} \right] \begin{bmatrix} D_f^{-1} M_{ff}^T & D_f^{-1} A_{fc} \\ 0 & D_c^{-1} M_{cc}^T \end{bmatrix} \\
&= \left[(W^T M_{ff} + A_{cf}) D_f^{-1} M_{ff}^T,\ (W^T M_{ff} + A_{cf}) D_f^{-1} A_{fc} + M_{cc} D_c^{-1} M_{cc}^T \right] \\
&= \left[(W^T M_{ff} + A_{cf}) D_f^{-1} M_{ff}^T,\ \widetilde{M}_c - (W^T M_{ff} + A_{cf}) D_f^{-1} M_{ff}^T W \right].
\end{aligned}$$

We readily see then that $R_* \equiv \widetilde{M}_c^{-1} P^T \widetilde{M} = [X_{cf}, I - X_{cf} W] = [0, I] + X_{cf}[I, -W]$, where

$$\begin{aligned}
X_{cf} &= \widetilde{M}_c^{-1} (A_{cf} + W^T M_{ff})(M_{ff} + M_{ff}^T - A_{ff})^{-1} M_{ff}^T \\
&= \widetilde{M}_c^{-1} \left(A_{cf} M_{ff}^{-1} + W^T \right) \widetilde{M}_{ff}.
\end{aligned} \tag{6.17}$$

Here, $\widetilde{M}_{ff} = M_{ff}^T D_f^{-1} M_{ff}$. Note that X_{cf} is close to zero if $W \approx -M_{ff}^{-T} A_{fc}$. The optimal $J_* \equiv I - \pi_{\widetilde{M}} = I - P \widetilde{M}_c^{-1} P^T \widetilde{M}$ gets the following form,

$$\begin{aligned}
J_* \equiv I - P R_* &= \begin{bmatrix} I & 0 \\ 0 & I \end{bmatrix} - \begin{bmatrix} W \\ I \end{bmatrix} [X_{cf},\ I - X_{cf} W] \\
&= \begin{bmatrix} I - W X_{cf} & -(I - W X_{cf}) W \\ -X_{cf} & X_{cf} W \end{bmatrix} \\
&= \left(\begin{bmatrix} I \\ 0 \end{bmatrix} - \begin{bmatrix} W \\ I \end{bmatrix} X_{cf} \right) [I,\ -W].
\end{aligned}$$

It is clear that

$$J_* P = (*)\, [I,\ -W] \begin{bmatrix} W \\ I \end{bmatrix} = 0.$$

Finally, it is also clear that

$$\text{Range}(J_*) = \text{Range} \left(\begin{bmatrix} I \\ 0 \end{bmatrix} - \begin{bmatrix} W \\ I \end{bmatrix} X_{cf} \right).$$

We formulate the latter results in the following theorem.

Theorem 6.8. *Let*

$$P = \begin{bmatrix} W \\ I \end{bmatrix}$$

be a general interpolation mapping (i.e., $W \neq -M_{ff}^{-T} A_{fc}$*). Define* X_{cf} *as in* (6.17)*. Then the optimal* $R_* = \widetilde{M}_c^{-1} P^T \widetilde{M}$ *and optimal* $J_* = I - PR_*$ *are given by the formulas:*

(i)

$$R_* = \left[X_{fc},\ I - X_{cf}W\right] = [0,\ I] + X_{cf}[I,\ -W],$$

(ii)

$$\begin{aligned} J_* &= \begin{bmatrix} I - WX_{cf} & -(I - WX_{cf})W \\ -X_{cf} & X_{cf}W \end{bmatrix} \\ &= \left(\begin{bmatrix} I \\ 0 \end{bmatrix} - \begin{bmatrix} W \\ I \end{bmatrix} X_{cf}\right)[I,\ -W]. \end{aligned}$$

Formulas (i) and (ii) in the above Theorem 6.8 can be viewed as perturbations of the commonly used mappings

$$R = [0,\ I] \quad \text{and} \quad J_0 = \begin{bmatrix} I \\ 0 \end{bmatrix}[I,\ -W] = I - PR.$$

Estimates of K_{TG}

We estimate below how much we can overestimate K_{TG} when using R and J instead of their optimal values R_* and J_*. To do this, assume that

"PR is bounded in $\widetilde{M}$–norm"; that is,

$$\mathbf{v}^T(PR)^T\widetilde{M}(PR)\mathbf{v} \le \eta_{\widetilde{M}}\ \mathbf{v}^T\widetilde{M}\mathbf{v}.$$

Because PR is a projection, the same norm bound holds for $I - PR$ (due to Kato's lemma 3.6); that is, we have $\mathbf{v}^T(I - PR)^T\widetilde{M}(I - PR)\mathbf{v} \le \eta_{\widetilde{M}}\ \mathbf{v}^T\widetilde{M}\mathbf{v}$.

Letting

$$J = \begin{bmatrix} I \\ 0 \end{bmatrix},$$

the latter norm bound can equivalently be stated as

$$\mathbf{v}_f^T J^T\widetilde{M}J\mathbf{v}_f \le \frac{1}{1 - \gamma_{\widetilde{M}}^2}\ \inf_{\mathbf{v}_c}\ (J\mathbf{v}_f + P\mathbf{v}_c)^T\widetilde{M}(J\mathbf{v}_f + P\mathbf{v}_c),$$

for $\gamma_{\widetilde{M}}^2 = 1 - (1/\eta_{\widetilde{M}}) \in [0,\ 1)$. The left-hand side expression actually simplifies and the estimate reduces to

$$\mathbf{v}_f^T\widetilde{M}_{ff}\mathbf{v}_f \le \frac{1}{1 - \gamma_{\widetilde{M}}^2}\ \mathbf{v}^T\widetilde{M}\mathbf{v}, \quad \text{for any } \mathbf{v} = J\mathbf{v}_f + P\mathbf{v}_c. \tag{6.18}$$

With

$$J = \begin{bmatrix} I \\ 0 \end{bmatrix},$$

we have $J_* = (J - PX_{cf})[I, \ -W]$. Based on inequality (6.18) for $\mathbf{v} := (J - PX_{cf})\widehat{\mathbf{v}}_f$, $\widehat{\mathbf{v}}_f = [I, \ -W]\mathbf{v}$, we get

$$\widehat{\mathbf{v}}_f^T (J - PX_{cf})^T \widetilde{M} (J - PX_{cf}) \widehat{\mathbf{v}}_f \geq \left(1 - \gamma_{\widetilde{M}}^2\right) \widehat{\mathbf{v}}_f^T J^T \widetilde{M} J \widehat{\mathbf{v}}_f$$
$$= \left(1 - \gamma_{\widetilde{M}}^2\right) ((I - PR)\mathbf{v})^T \widetilde{M} (I - PR)\mathbf{v}.$$

We used the fact that $J\widehat{\mathbf{v}}_f = J[I, \ -W]\mathbf{v} = (I - PR)\mathbf{v}$.

The latter implies the following important lower bound for K_{TG} (recall that $Q = PR$),

$$K_{TG} = \sup_{\mathbf{e}} \frac{\mathbf{e}^T J_*^T \widetilde{M} J_* \mathbf{e}}{\mathbf{e}^T A \mathbf{e}} \geq \left(1 - \gamma_{\widetilde{M}}^2\right) \sup_{\mathbf{e}} \mu_{\widetilde{M}}(Q, \ \mathbf{e}).$$

That is, the following result holds.

Theorem 6.9. *Assume that P is bounded in the $\widetilde{M}$-norm as in* (6.18). *Then the following relations hold,*

$$\left(1 - \gamma_{\widetilde{M}}^2\right) \sup_{\mathbf{e}} \mu_{\widetilde{M}}(Q, \ \mathbf{e}) \leq K_{TG} \leq \sup_{\mathbf{e}} \mu_{\widetilde{M}}(Q, \ \mathbf{e}).$$

The latter result shows that the two-grid convergence factor bound predicted by the measure (see (6.14)) can overestimate ϱ_{TG} at the most by

$$1 - \frac{1}{\sup_{\mathbf{e}} \mu_{\widetilde{M}}(Q, \ \mathbf{e})} \leq 1 - \frac{1 - \gamma_{\widetilde{M}}^2}{K_{TG}} = \varrho_{TG} + \gamma_{\widetilde{M}}^2 (1 - \varrho_{TG}) = \varrho_{TG}\left(1 - \gamma_{\widetilde{M}}^2\right) + \gamma_{\widetilde{M}}^2.$$

We show in the next section, that under reasonable assumptions (which guarantee good two-grid convergence) that $\gamma_{\widetilde{M}}$ cannot get too close to one. Thus the overestimation of K_{TG} by $\sup_{\mathbf{e}} \mu_{\widetilde{M}}(Q, \ \mathbf{e})$ cannot be too pessimistic.

Estimating P in $\widetilde{M}$-norm

We show in the present section how bad $\gamma_{\widetilde{M}}^2$, or equivalently, $\eta = 1/(1 - \gamma_{\widetilde{M}}^2)$, could actually get. The following result holds.

Theorem 6.10. *Let*

$$P = \begin{bmatrix} W \\ I \end{bmatrix}, \quad R = [0, \ I].$$

Assume that PR is bounded in the A-norm; that is, for a constant $\gamma \in [0, 1)$, we have

$$\|PR\mathbf{v}\|_A^2 \leq \frac{1}{1 - \gamma^2} \mathbf{v}^T A \mathbf{v},$$

and that the compatible relaxation (or CR for short) for A_{ff} is convergent; that is, for another constant $\varrho_{CR} \in [0, 1)$ we also have $\|I - A_{ff}^{1/2} M_{ff}^{-T} A_{ff}^{1/2}\| =$

$\|I - A_{ff}^{1/2} M_{ff}^{-1} A_{ff}^{1/2}\| \le \varrho_{CR}$, *or equivalently, the symmetrized smoother* $\widetilde{M}_{ff} = M_{ff}(M_{ff}^T + M_{ff} - A_{ff})^{-1} M_{ff}^T$ *satisfies*

$$\mathbf{v}_f^T A_{ff} \mathbf{v}_f \le \mathbf{v}_f^T \widetilde{M}_{ff} \mathbf{v}_f \le \frac{1}{1-\varrho_{CR}^2}\, \mathbf{v}_f^T A_{ff} \mathbf{v}_f.$$

Similarly, assume that M_{cc} *is a convergent smoother for* A_{cc} *in the* A_{cc}*-norm, which implies that*

$$\mathbf{v}_c^T M_{cc} \left(M_{cc}^T + M_{cc} - A_{cc}\right)^{-1} M_{cc}^T \mathbf{v}_c \ge \mathbf{v}^T A_{cc} \mathbf{v}_c. \quad (6.19)$$

Then, the following norm bound holds,

$$\|P R \mathbf{v}\|_{\widetilde{M}}^2 \le \eta\, \mathbf{v}^T \widetilde{M} \mathbf{v},$$

with $\eta \le (1/(1-\gamma^2))(1/(1-\varrho_{CR}^2))$.

Proof. The estimate we are interested in reduces to the following one,

$$\mathbf{v}_c^T P^T \widetilde{M} P \mathbf{v}_c \le \eta\, \mathbf{v}_c^T S_{\widetilde{M}} \mathbf{v}_c = \eta\, \mathbf{v}_c^T M_{cc} \left(M_{cc} + M_{cc}^T - A_{cc}\right)^{-1} M_{cc}^T \mathbf{v}_c,$$

because $S_{\widetilde{M}} = M_{cc}(M_{cc} + M_{cc}^T - A_{cc})^{-1} M_{cc}^T$ is the Schur complement of $\widetilde{M}$. Furthermore, using formula (6.16), we see that the above norm estimate reduces to

$$\mathbf{v}_c^T \left(M_{ff}^T W + A_{fc}\right)^T \left(M_{ff} + M_{ff}^T - A_{ff}\right)^{-1} \left(M_{ff}^T W + A_{fc}\right) \mathbf{v}_c \le (\eta - 1)\, \mathbf{v}_c^T S_{\widetilde{M}} \mathbf{v}_c. \quad (6.20)$$

Alternatively, in terms of $\widetilde{M}_{ff} = M_{ff}(M_{ff}^T + M_{ff} - A_{ff})^{-1} M_{ff}^T$, we have

$$\mathbf{v}_c^T \left(W + M_{ff}^{-T} A_{fc}\right)^T \widetilde{M}_{ff} \left(W + M_{ff}^{-T} A_{fc}\right) \mathbf{v}_c \le (\eta - 1)\, \mathbf{v}_c^T S_{\widetilde{M}} \mathbf{v}_c.$$

Because by assumption,

$$\mathbf{w}_f^T \widetilde{M}_{ff} \mathbf{w}_f \le \frac{1}{1-\varrho_{CR}^2}\, \mathbf{w}_f^T A_{ff} \mathbf{w}_f,$$

it is clear that it is sufficient to prove the bound

$$\frac{1}{1-\varrho_{CR}^2}\, \mathbf{v}_c^T \left(W + M_{ff}^{-T} A_{fc}\right)^T A_{ff} \left(W + M_{ff}^{-T} A_{fc}\right) \mathbf{v}_c \le (\eta - 1)\, \mathbf{v}_c^T S_{\widetilde{M}} \mathbf{v}_c.$$

Also, by assumption, we have that P is bounded in the A-norm, that is, that for a constant $\gamma \in [0,\ 1)$, we have

$$\mathbf{v}_c^T P^T A P \mathbf{v}_c \le \frac{1}{1-\gamma^2}\, \mathbf{v}_c^T S_A \mathbf{v}_c.$$

Here, $S_A = A_{cc} - A_{cf} A_{ff}^{-1} A_{fc}$ is the Schur complement of A. The latter estimate, combined with the identity (recalling that

$$P = \begin{bmatrix} W \\ I \end{bmatrix}),$$

$$\begin{aligned} \mathbf{v}_c^T P^T A P \mathbf{v}_c &= (P\mathbf{v}_c)^T \begin{bmatrix} A_{ff} & 0 \\ A_{cf} & I \end{bmatrix} \begin{bmatrix} A_{ff}^{-1} & 0 \\ 0 & S_A \end{bmatrix} \begin{bmatrix} A_{ff} & A_{fc} \\ 0 & I \end{bmatrix} P\mathbf{v}_c \\ &= \mathbf{v}_c^T S_A \mathbf{v}_c + \mathbf{v}_c^T (A_{cf} + W^T A_{ff}) A_{ff}^{-1} (A_{ff} W + A_{fc}) \mathbf{v}_c, \end{aligned}$$

imply

$$\mathbf{v}_c^T (A_{cf} + W^T A_{ff}) A_{ff}^{-1} (A_{ff} W + A_{fc}) \mathbf{v}_c \le \frac{\gamma^2}{1-\gamma^2}\, \mathbf{v}_c^T S_A \mathbf{v}_c,$$

or equivalently,

$$\mathbf{v}_c^T \left(W + A_{ff}^{-1} A_{fc}\right)^T A_{ff} \left(W + A_{ff}^{-1} A_{fc}\right) \mathbf{v}_c \le \frac{\gamma^2}{1-\gamma^2}\, \mathbf{v}_c^T S_A \mathbf{v}_c. \tag{6.21}$$

Then, because the compatible relaxation is convergent, based on the identity

$$I - \widetilde{M}_{ff}^{-1} A_{ff} = \left(I - M_{ff}^{-T} A_{ff}\right)\left(I - M_{ff}^{-1} A_{ff}\right),$$

the assumption on $\widetilde{M}_{ff}$

$$\mathbf{v}_f^T A_{ff} \mathbf{v}_f \le \mathbf{v}_f^T \widetilde{M}_{ff} \mathbf{v}_f \le \frac{1}{1-\varrho_{CR}^2}\, \mathbf{v}_f^T A_{ff} \mathbf{v}_f,$$

can be reformulated as $\|I - A_{ff}^{1/2} M_{ff}^{-1} A_{ff}^{1/2}\| = \|I - A_{ff}^{1/2} M_{ff}^{-T} A_{ff}^{1/2}\| \le \varrho_{CR}$. Hence, we have the estimate

$$\mathbf{w}_f^T \left(M_{ff}^{-T} - A_{ff}^{-1}\right)^T A_{ff} \left(M_{ff}^{-T} - A_{ff}^{-1}\right) \mathbf{w}_f \le \varrho_{CR}^2\, \mathbf{w}_f^T A_{ff}^{-1} \mathbf{w}_f. \tag{6.22}$$

Using (6.22) for $\mathbf{w}_f = A_{fc}\mathbf{v}_c$, together with (6.21) and the triangle inequality, gives

$$\begin{aligned} &\mathbf{v}_c^T \left(W + M_{ff}^{-T} A_{fc}\right)^T A_{ff} \left(W + M_{ff}^{-T} A_{fc}\right) \mathbf{v}_c \\ &\quad = \mathbf{v}_c^T \left(W + A_{ff}^{-1} A_{fc} + \left(M_{ff}^{-T} - A_{ff}^{-1}\right) A_{fc}\right)^T A_{ff} \\ &\quad\quad \times \left(W + A_{ff}^{-1} A_{fc} + \left(M_{ff}^{-T} - A_{ff}^{-1}\right) A_{fc}\right) \mathbf{v}_c \\ &\quad = \left\| \left(M_{ff}^{-T} - A_{ff}^{-1}\right)(A_{fc}\mathbf{v}_c) + \left(W + A_{ff}^{-1} A_{fc}\right) \mathbf{v}_c \right\|_{A_{ff}}^2 \\ &\quad \le \left(\varrho_{CR} \left(\mathbf{v}_c^T A_{cf} A_{ff}^{-1} A_{fc} \mathbf{v}_c\right)^{1/2} + \frac{\gamma}{\sqrt{1-\gamma^2}} \left(\mathbf{v}_c^T S_A \mathbf{v}_c\right)^{1/2} \right)^2 \end{aligned}$$

$$\leq \left(\varrho_{CR}^2 + \frac{\gamma^2}{1-\gamma^2}\right) \left[\mathbf{v}_c^T A_{cf} A_{ff}^{-1} A_{fc} \mathbf{v}_c + \mathbf{v}_c^T S_A \mathbf{v}_c\right]$$
$$= \left(\varrho_{CR}^2 + \frac{\gamma^2}{1-\gamma^2}\right) \mathbf{v}_c^T A_{cc} \mathbf{v}_c$$
$$\leq (1-\varrho_{CR}^2)(\eta - 1)\, \mathbf{v}_c^T S_{\widetilde{M}} \mathbf{v}_c.$$

Above, we used also the elementary inequality $(ab+cd)^2 \leq (a^2+c^2)(b^2+d^2)$, the fact that $S_A + A_{cf} A_{ff}^{-1} A_{fc} = A_{cc}$, and inequality (6.19). This shows that we can let

$$\eta_{\widetilde{M}} = \eta = 1 + \frac{1}{1-\varrho_{CR}^2} \left(\varrho_{CR}^2 + \frac{\gamma^2}{1-\gamma^2}\right)$$
$$= \frac{1}{1-\gamma^2} \frac{1}{1-\varrho_{CR}^2}. \qquad \square$$

Therefore, for $\gamma_{\widetilde{M}}^2 = 1 - 1/\eta$, we have

$$1 - \gamma_{\widetilde{M}}^2 = \frac{1}{\eta} = (1-\gamma^2)(1-\varrho_{CR}^2);$$

that is, $\gamma_{\widetilde{M}}^2$ cannot get too close to one.

Corollary 6.11. *Assume that we use the "c"–"f" relaxation (as in (6.12)) and that we can construct a*

$$P = \begin{bmatrix} W \\ I \end{bmatrix}$$

such that for $R = [0,\ I]$,

- *PR is bounded in the $\widetilde{M}$ norm.*

Then, a necessary condition for a two-grid convergence is that

- *The compatible relaxation be convergent, or equivalently that $\widetilde{M}_{ff}$ be spectrally equivalent to A_{ff},*

and that

- *PR be bounded in A-norm.*

Proof. We proved (in Theorem 6.9) that if $Q = PR$ is bounded in the $\widetilde{M}$-norm, then

$$K_{TG} \geq \left(1 - \gamma_{\widetilde{M}}^2\right) \sup_{\mathbf{e}}\ \mu_{\widetilde{M}}(Q,\ \mathbf{e}).$$

Thus, the measure $\mu_{\widetilde{M}}(Q,\ \mathbf{e})$ is bounded by $K_{TG}/(1-\gamma_{\widetilde{M}}^2)$. Boundedness of the measure ([FV04]) implies that the compatible relaxation is convergent and that PR is bounded in the A-norm. This is seen as follows. Based on the fact that $(I - Q)\mathbf{e} = (I - PR)\mathbf{e} = J\mathbf{v}_f$, for any $\mathbf{e} = J\mathbf{v}_f$, we get

$$\begin{aligned}\sup_{\mathbf{e}} \mu_{\widetilde{M}}(Q, \mathbf{e}) &= \sup_{\mathbf{e}} \frac{\mathbf{e}^T (I - Q)^T \widetilde{M} (I - Q)\mathbf{e}}{\mathbf{e}^T A\mathbf{e}} \\ &\geq \sup_{\mathbf{e}=J\mathbf{v}_f} \frac{\mathbf{v}_f^T J^T \widetilde{M} J\mathbf{v}_f}{\mathbf{v}_f^T J^T AJ\mathbf{v}_f} = \frac{1}{1 - \varrho_{CR}^2}.\end{aligned}$$

That is, $(1/(1 - \varrho_{CR}^2)) \leq (K_{TG}/(1 - \gamma_{\widetilde{M}}^2))$. This shows that the compatible relaxation convergence factor is bounded as follows.

$$\varrho_{CR} = \|I - A_{ff}^{1/2} M_{ff}^{-1} A_{ff}^{1/2}\| \leq \sqrt{1 - \frac{1 - \gamma_{\widetilde{M}}^2}{K_{TG}}}.$$

Because by assumption M is a convergent smoother in the A-norm for A, hence $\widetilde{M} - A$ is positive semidefinite, we also have,

$$\mathbf{e}^T (I - PR)^T A (I - PR)\mathbf{e} \leq \mathbf{e}^T (I - PR)^T \widetilde{M} (I - PR)\mathbf{e} \leq \eta_{\widetilde{M}} K_{TG}\, \mathbf{e}^T A\mathbf{e}.$$

That is, $I - PR$, hence PR as a projection (due to Kato's lemma 3.6)) is bounded in the A-norm by $\eta_{\widetilde{M}}\ K_{TG}$. □

We also showed in Theorem 6.10, that PR is bounded in the $\widetilde{M}$-norm, if (i) PR is bounded in the A-norm, and (ii) the compatible relaxation is convergent.

We note that very often M is such that $\widetilde{M}$ is spectrally equivalent to D (the diagonal of A). For example, the following conditions on M lead to an $\widetilde{M}$ which is spectrally equivalent to D.

Proposition 6.12. *Let M be such that*

$$\mathbf{v}^T (M + M^T - A)\mathbf{v} \geq \delta_0\, \mathbf{v}^T D\mathbf{v},$$

and

$$\|D^{-(1/2)} M D^{-(1/2)}\| \leq \delta_1.$$

Then $\widetilde{M}$ is spectrally equivalent to D. In particular, for the example $M = D + L$ being the lower-triangular part of $A = D + L + L^T$, giving rise to the forward Gauss-Seidel smoother, then $\delta_0 = 1$ because then $M + M^T - A = D$, and δ_1 is bounded by the maximum number of nonzero entries per row of A.

Proof. We have

$$\begin{aligned}\mathbf{v}^T \widetilde{M}\mathbf{v} &= \mathbf{v}^T M^T (M + M^T - A)^{-1} M\mathbf{v} \\ &\leq \delta_0^{-1} (D^{1/2}\mathbf{v})^T D^{-(1/2)} M^T D^{-(1/2)} D^{-(1/2)} M D^{-(1/2)} (D^{1/2}\mathbf{v}) \\ &\leq \frac{\delta_1^2}{\delta_0} \mathbf{v}^T D\mathbf{v}.\end{aligned}$$

On the other hand, for $X = D^{-(1/2)}MD^{-(1/2)}$, we have

$$2\mathbf{v}^T X\mathbf{v} = \mathbf{v}^T(X^T + X)\mathbf{v} \geq \delta_0 \mathbf{v}^T\mathbf{v}.$$

Therefore, with $\mathbf{v} := X^{-1}\mathbf{v}$, we get

$$\delta_0 \|X^{-1}\mathbf{v}\|^2 \leq 2(X^{-1}\mathbf{v})^T\mathbf{v} \leq 2\|X^{-1}\mathbf{v}\|\|\mathbf{v}\|.$$

That is, $\|X^{-1}\| = \|X^{-T}\| \leq 2/\delta_0$. Then,

$$\begin{aligned}\mathbf{v}^T D^{1/2}\widetilde{M}^{-1}D^{1/2}\mathbf{v} &= \mathbf{v}^T(X^{-1} + X^{-T} - X^{-T}D^{-(1/2)}AD^{-(1/2)}X^{-1})\mathbf{v} \\ &\leq 2\|X^{-1}\|\mathbf{v}^T\mathbf{v} \leq \frac{4}{\delta_0}\mathbf{v}^T\mathbf{v}.\end{aligned}$$

Thus, we proved the spectral equivalence relations,

$$\frac{\delta_0}{4}\mathbf{v}^T D\mathbf{v} \leq \mathbf{v}^T\widetilde{M}\mathbf{v} \leq \frac{\delta_1^2}{\delta_0}\mathbf{v}^T D\mathbf{v}.$$

□

Proposition 6.13. *Assume that*

$$P = \begin{bmatrix} W \\ I \end{bmatrix}$$

is such that

$$\|D_{ff}^{1/2} W D_{cc}^{-(1/2)}\| \leq C,$$

where D_{ff} and D_{cc} are the diagonals of A_{ff} and A_{cc}. Consider a "c"–"f" smoother

$$M = \begin{bmatrix} M_{ff}^T & A_{fc} \\ 0 & M_{cc}^T \end{bmatrix}$$

and assume that M_{ff} and D_{ff}, as well as M_{cc} and D_{cc}, satisfy the conditions of Proposition 6.12. Then, M and

$$D = \begin{bmatrix} D_{ff} & 0 \\ 0 & D_{cc} \end{bmatrix}$$

satisfy the conditions of Proposition 6.12, as well. Moreover, PR with $R = [0,\ I]$ is bounded in the $\widetilde{M}$-norm.

We first comment on the assumption on P (or W). Very often in practice, assuming good CR convergence (note that then A_{ff} is spectrally equivalent to D_{ff}), we can choose a sparse $W \approx -A_{ff}^{-1}A_{fc}$. Thus the estimate

$$\begin{aligned}\|D_{ff}^{1/2} W D_{cc}^{-(1/2)}\| &\leq \|D_{ff}^{1/2} A_{ff}^{-1} D_{ff}^{1/2}\| \|D_{ff}^{-(1/2)} A_{fc} D_{cc}^{-(1/2)}\| \\ &\leq \text{cond}(D^{-(1/2)}A_{ff})\| \|D_{ff}^{-(1/2)} A_{fc} D_{cc}^{-(1/2)}\| \leq \text{const},\end{aligned}$$

reduces to $\|D_{ff}^{-(1/2)} A_{fc} D_{cc}^{-(1/2)}\|$ to be uniformly bounded, which is the case for any sparse s.p.d. matrix A.

Proof. We have

$$\begin{bmatrix} \mathbf{w}_f \\ \mathbf{w}_c \end{bmatrix}^T M \begin{bmatrix} \mathbf{v}_f \\ \mathbf{v}_c \end{bmatrix} = \mathbf{w}_f^T (M_{ff}^T \mathbf{v}_f + A_{fc}\mathbf{v}_c) + \mathbf{w}_c^T M_{cc}^T \mathbf{v}_c$$
$$\leq \delta_1 \left[\mathbf{w}_f^T D_{ff} \mathbf{w}_f\right]^{1/2} \left[\mathbf{v}_f^T D_{ff} \mathbf{v}_f\right]^{1/2} + \mathbf{w}_f^T A_{fc} \mathbf{v}_c$$
$$+ \delta_1 \left[\mathbf{w}_c^T D_{cc} \mathbf{w}_c\right]^{1/2} \left[\mathbf{v}_c^T D_{cc} \mathbf{v}_c\right]^{1/2}.$$

The boundedness of $D^{-(1/2)} M D^{-(1/2)}$ follows then from the boundedness of $D_{ff}^{-(1/2)} A_{fc} D_{cc}^{-(1/2)}$ due to the sparsity of A. The coercivity also holds, because

$$\mathbf{v}^T (M + M^T - A)\mathbf{v} = \mathbf{v}_f^T (M_{ff} + M_{ff}^T - A_{ff})\mathbf{v}_f + \mathbf{v}_c^T (M_{cc} + M_{cc}^T - A_{cc})\mathbf{v}_c$$
$$\geq \delta_0 \left[\mathbf{v}_f^T D_{ff} \mathbf{v}_f + \mathbf{v}_c^T D_{cc} \mathbf{v}_c\right]$$
$$= \delta_0 \, \mathbf{v}^T D \mathbf{v}.$$

In the present case of M leading to an $\widetilde{M}$, which is spectrally equivalent to D (the diagonal of A), the condition on PR to be bounded in the $\widetilde{M}$-norm is equivalent to PR being bounded in the D-norm. The latter simply means

$$\mathbf{v}_c^T (W^T D_{ff} W + D_{cc})\mathbf{v}_c \leq \eta_D \, \mathbf{v}_c^T D_{cc} \mathbf{v}_c.$$

That is, $\|D_{ff}^{1/2} W D_{cc}^{-(1/2)}\| \leq \sqrt{\eta_D - 1}$, which holds by assumption. □

Proposition 6.14. *Consider the smoother*

$$M = M_\tau = \begin{bmatrix} M_{ff}^T & A_{fc} \\ 0 & \tau \, D_{cc} \end{bmatrix}$$

for $\tau > 0$ sufficiently large. Assume that M_{ff} and A_{ff} satisfy the conditions of Proposition 6.12. Then

$$PR = \begin{bmatrix} W \\ I \end{bmatrix} [0, \; I]$$

is bounded in the $\widetilde{M}$-norm as long as τ is sufficiently large and the block W of P is such that $\|D_{ff}^{1/2} W D_{cc}^{-(1/2)}\| \leq const.$

Proof. The estimate (6.20) reads in the present setting

$$\mathbf{v}_c^T \left(M_{ff}^T W + A_{fc}\right)^T \left(M_{ff} + M_{ff}^T - A_{ff}\right)^{-1} \left(M_{ff}^T W + A_{fc}\right)\mathbf{v}_c$$
$$\leq (\eta - 1)\, \tau \, \mathbf{v}_c^T D_{cc} \left(2D_{cc} - \frac{1}{\tau} A_{cc}\right)^{-1} D_{cc} \mathbf{v}_c. \tag{6.23}$$

Note that the left-hand side is independent of τ. Then, if τ is sufficiently large, we have that $D_{cc} - (1/\tau)A_{cc}$ is positive definite and $D_{cc}(2D_{cc} - (1/\tau)A_{cc})^{-1} D_{cc} \simeq D_{cc}$.

It is clear then that it is sufficient to show

$$\sup_{\mathbf{v}_c} \frac{\mathbf{v}_c^T \left(M_{ff}^T W + A_{fc}\right)^T \left(M_{ff} + M_{ff}^T - A_{ff}\right)^{-1} \left(M_{ff}^T W + A_{fc}\right)\mathbf{v}_c}{\mathbf{v}_c^T D_{cc} \mathbf{v}_c} \leq \text{const}.$$

First, from Proposition 6.12, we have $M_{ff}(M_{ff}+M_{ff}^T-A_{ff})^{-1}M_{ff}^T \simeq D_{ff} \leq \text{const}$. Estimate (6.23) reduces then to

$$\left\| D_{ff}^{1/2} \left(W + M_{ff}^{-T} A_{fc}\right) D_{cc}^{-(1/2)} \right\| \leq \text{const}.$$

Then, noticing that $\| D_{ff}^{1/2} M_{ff}^{-T} D_{ff}^{1/2} \| \leq \text{const}$, (see the proof of Proposition 6.12), and by assumption $\| D_{ff}^{(1/2)} W D_{cc}^{-(1/2)} \| \leq \text{const}$, based on the triangle inequality, we arrive at,

$$\begin{aligned}
\left\| D_{ff}^{1/2} \left(W + M_{ff}^{-T} A_{fc}\right) D_{cc}^{-(1/2)} \right\| &\leq \left\| D_{ff}^{1/2} W D_{cc}^{-(1/2)} \right\| + \left\| D_{ff}^{1/2} M_{ff}^{-T} A_{fc} D_{cc}^{-(1/2)} \right\| \\
&\leq C + \left\| D_{ff}^{1/2} M_{ff}^{-T} D_{ff}^{1/2} \right\| \left\| D_{ff}^{-(1/2)} A_{fc} D_{cc}^{-(1/2)} \right\| \\
&\leq C + C \left\| D_{ff}^{-(1/2)} A_{fc} D_{cc}^{-(1/2)} \right\|.
\end{aligned}$$

Finally noticing that $\| D_{ff}^{-(1/2)} A_{fc} D_{cc}^{-(1/2)} \| \leq \text{const}$ (equal to the maximum number of nonzeros per row) for any sparse matrix, the desired result then follows. □

In conclusion, for smoothers M_τ (and $\tau > 0$ sufficiently large) the $\widetilde{M}$-boundedness of PR is straightforward to achieve based on mild restrictions on W (the block of P). Thus, the conditions in Corollary 6.11 on M_{ff} and PR are necessary practical guidelines to guarantee a good two-grid convergence. In particular, this holds for the HBMG (i.e., $\tau = \infty$) two-level method in (6.13).

6.9 AMGe: An element agglomeration AMG

In this section, we consider the practically important case when a s.p.d. matrix A can be assembled from local symmetric positive semidefinite matrices $\{A_\tau\}$, where τ runs over a set of "elements" $\mathcal{T}$. The latter means that each τ is a set of degrees of freedom (indices) and $\{\tau\}$ provide an overlapping partition of the set of degrees of freedom (or the given index set). Let $\mathbf{v}_\tau$ stand for the restriction of a vector $\mathbf{v}$ to the set of indices τ. Then, by "assembly" we mean that the quadratic form $\mathbf{v}^T A \mathbf{v}$ can be computed by simply summing up the local quadratic forms $\mathbf{v}_\tau^T A_\tau \mathbf{v}_\tau$, or more generally,

$$\mathbf{v}^T A \mathbf{w} = \sum_\tau \mathbf{v}_\tau^T A_\tau \mathbf{w}_\tau,$$

for any two vectors $\mathbf{v}$ and $\mathbf{w}$.

AMG methods that exploit element matrices were proposed in [AMGe] and [Ch03]; see also [Br99]. Methods that generate element matrices on coarse levels and hence allow for recursion were proposed in [JV01] and [Ch05]; see also [VZ05].

In what follows, we focus on the second class of methods in as much as they allow for recursion.

Given the set of elements $\{\tau\}$, it is natural to consider a coarsening procedure that first constructs agglomerated elements T by joining together a small number of connected fine-grid elements τ. Here, T is viewed as a set of fine degrees of freedom (or fine dofs). If we can select a subset of the degrees of freedom to define the coarse dofs, denoted by $\mathcal{N}_c$, such that every T has some coarse dofs, then $T \cap \mathcal{N}_c$ defines the actual coarse element. Finally, if we are able to construct a set of local interpolation matrices P_T for every T such that P_T maps a vector defined on the coarse set $T \cap \mathcal{N}_c$ into a vector defined on T, and in addition the collection of $\{P_T\}$ is compatible in the sense that for every shared dof, the values of $P_T(\mathbf{v}^c)_{T \cap \mathcal{N}_c}$ at that dof are the same for any T that shares it, the resulting global interpolation matrix P exhibits the property that

$$\mathbf{v}_c^T P^T A P \mathbf{v}_c = \sum_T ((\mathbf{v}^c)_{T \cap \mathcal{N}_c})^T P_T^T A_T P_T (\mathbf{v}^c)_{T \cap \mathcal{N}_c}.$$

Here, A_T are assembled from $\{A_\tau\}$ for all τs that form T; that is,

$$\mathbf{v}_T^T A_T \mathbf{w}_T = \sum_{\tau \subset T} (\mathbf{v}_\tau)^T A_\tau \mathbf{w}_\tau.$$

The matrices $P_T^T A_T P_T$ naturally define coarse element matrices. The latter property allows for recursive use of the respective algorithm.

6.9.1 Element-based construction of P

We next study ways to construct element-based interpolation matrices P. We adopt a columnwise approach; that is, let $P = (\boldsymbol{\psi}_i)$ for $i = 1, \ldots, n_c$ be the columns of P that have to be computed. In general, we are looking at

$$P = \begin{bmatrix} W \\ I \end{bmatrix}$$

where the identity block I defines the coarse dofs as subset of the fine dofs. Then,

$$\boldsymbol{\psi}_i = \begin{bmatrix} \mathbf{w}_i \\ \mathbf{e}_i \end{bmatrix}$$

where $\mathbf{e}_i$ is the ith unit coordinate vector in $\mathbb{R}^{n_c}$.

We have to first select the sparsity pattern of the component $\mathbf{w}_i$ and second, the actual entries of $\mathbf{w}_i$ have to be eventually computed. Actually, to begin with, first we have to choose the coarse dofs in terms of the fine-grid dofs, which reflects the choice of the identity block in the block structure of P.

The algorithm that we describe in what follows exploits the coarsening process that will be guided by an "algebraically smooth" vector $\mathbf{x}$. For example, $\mathbf{x}$ can be an approximation to the minimal eigenvector of the problem $A\mathbf{x} = \lambda_{\min} \overline{M} \mathbf{x}$, where $\overline{M} = M(M + M^T - A)^{-1} M^T$ is the symmetrized smoother that appears in the resulting multigrid method.

Assume that an element agglomeration step has been performed (see Section 1.9.7) and a set of agglomerated elements $\{T\}$ has been constructed. Based on the overlapping sets $\{T\}$, we can partition the fine-degrees of freedom into nonoverlapping groups $\{\mathcal{I}\}$ such that the dofs in a given group $\mathcal{I}$ belong to the same sets of agglomerated elements T and only to them. Based on the vector $\mathbf{x} = (x_i)$, we look at the numbers $d_i x_i^2$, where d_i is the ith diagonal entry of A. For every group $\mathcal{I}$, we select a dof $i_{\max}$ as a coarse one such that $d_{i_{\max}} x_{i_{\max}}^2 = \max_{i \in \mathcal{I}} d_i x_i^2$. We can select more dofs by allowing a portion of the dofs in a given group $\mathcal{I}$ with values close to $d_{i_{\max}} x_{i_{\max}}^2$ to form the set of coarse dofs $\mathcal{N}_c$.

Assume that a coarse set $\mathcal{N}_c$ has been selected. We next decide on the sparsity pattern of $\boldsymbol{\psi}_i$. Because, we are restricted to choose P_T that are element-compatible in a given sense (explained earlier), the nonzero entries of a column $\boldsymbol{\psi}_i$ should be in $\cup T$ for all Ts that share the group $\mathcal{I}$ where $\mathcal{I}$ is the unique set that contains i. From the set $\cup\{T : T \cap \mathcal{I} \neq \emptyset\}$, we exclude all coarse dofs different from i as well as all other dofs that belong to a different T which does not intersect $\mathcal{I}$. The resulting set defines the row indices of $\boldsymbol{\psi}_i$ where it is allowed to have nonzero entries.

Now, having the sparsity pattern of all $\boldsymbol{\psi}_i$ defined, we can compute the actual entries of $\boldsymbol{\psi}_i$. We adopt the following local procedure. For every agglomerate T and all coarse dofs $i \in T$, we first compute a $\overline{P}_T = (\boldsymbol{\varphi}_i)$ where $\boldsymbol{\varphi}_i$ (defined only on T) have the same sparsity pattern as $\boldsymbol{\psi}_i$ restricted to T. The vectors $\boldsymbol{\varphi}_i$ are computed by solving the following local constrained minimization problem,

$$\sum_{i \in T \cap \mathcal{N}_c} \boldsymbol{\varphi}_i^T A_T \boldsymbol{\varphi}_i \mapsto \min, \tag{6.24}$$

subject to

$$\sum_{i \in T \cap \mathcal{N}_c} x_i \, \boldsymbol{\varphi}_i = \mathbf{x}_T. \tag{6.25}$$

It is clear that x_i are the ith entries of $\mathbf{x}$. This is the case, because we seek $\boldsymbol{\varphi}_i$ to form a Lagrangian basis on T; that is, $\boldsymbol{\varphi}_i$ has zero entry at any other coarse dof different from i and has entry one at the particular coarse dof i. Thus the ith row of the above constraint on the left equals x_i and on the right equals the ith entry of $\mathbf{x}$ (which is x_i). The above constrained minimization problem has a small size and obviously has a computable solution. The solution $\{\boldsymbol{\varphi}_i\}$ depends on T, which we indicate by $\boldsymbol{\varphi}_i = \boldsymbol{\varphi}_i^{(T)}$.

After $\boldsymbol{\varphi}_i = \boldsymbol{\varphi}_i^{(T)}$ are computed, we are not yet done with the construction of P_T. The problem is that we cannot let $\boldsymbol{\psi}_i|_T = \boldsymbol{\varphi}_i^{(T)}$ because $\boldsymbol{\varphi}_i^{(T)}$ for different Ts do not necessarily match at dofs that are shared by two (or more) agglomerates. To fix this problem, we choose weights $d_{T,\,j} = \|A_T\|/(\sum_{T':\, j \in T'} \|A_{T'}\|)$ for every $j \in T$. Other choices are also possible, but the important thing is to ensure the following averaging property,

$$\sum_{T:\, j \in T} d_{T,\,j} = 1.$$

Then, the jth entry of $\boldsymbol{\psi}_i$ equals the averaged value

$$\sum_{T:\, j \in T} d_{T,\,j}\, \left(\boldsymbol{\varphi}_i^{(T)}\right)_j.$$

The above formula defines $\boldsymbol{\psi}_i$ uniquely with nonzero entries only at its prescribed support. Finally, we notice, that $\sum_i x_i \boldsymbol{\psi}_i = \mathbf{x}$ holds because the averaging preserves the constraint. Due to the same reason, $\{\boldsymbol{\psi}_i\}$ form a Lagrangian basis; that is, $\boldsymbol{\psi}_i$ vanishes at all coarse dofs different from i and has value one at coarse dof i. The latter property translates to the fact that $P = (\boldsymbol{\psi}_i)$ admits the block form

$$\begin{bmatrix} W \\ I \end{bmatrix}.$$

6.9.2 On various norm bounds of P

In this section, following [KV06], we derive energy norm bounds for projections $Q = PR$, where P is an interpolation and R is a restriction mapping such that $RP = I$.

A general local to global norm bound

We recall our main assumptions. We are given a set of coarse degrees of freedom (or coarse-grid dofs) $\mathcal{N}_c$. We view $\mathcal{N}_c$, as a subset of $\mathcal{N}$. We have vectors (or grid functions) defined on $\mathcal{N}$ (fine-grid vectors) and vectors defined on $\mathcal{N}_c$ (coarse-grid vectors). Let $n = |\mathcal{N}|$ and $n_c = |\mathcal{N}_c|$, and $n_c < n$ be the respective size (cardinality) of $\mathcal{N}$ and $\mathcal{N}_c$. On each agglomerated element T, let $n_T = |T|$ and $n_{T,c} = |T \cap \mathcal{N}_c|$. The space of coarse-grid vectors is identified with $\mathbb{R}^{n_c}$, and similarly, the fine-grid vectors are identified with $\mathbb{R}^n$. We have global mappings $R \colon \mathbb{R}^n \mapsto \mathbb{R}^{n_c}$ and $P \colon \mathbb{R}^{n_c} \mapsto \mathbb{R}^n$. Respectively, there are local mappings $R_T \colon \mathbb{R}^{n_T} \mapsto \mathbb{R}^{n_{T,c}}$ and $P_T \colon \mathbb{R}^{n_{T,c}} \mapsto \mathbb{R}^{n_T}$ such that R_T restricts a local fine-grid vector defined on T to a local (coarse-grid) vector defined on $T \cap \mathcal{N}_c$, whereas the local interpolation mapping P_T interpolates a local coarse vector defined on $T \cap \mathcal{N}_c$ to vectors defined on T. Let $I_T^c \colon \mathbb{R}^{n_{T,c}} \mapsto \mathbb{R}^{n_c}$ be the extension by zero of coarse vectors defined on $T \cap \mathcal{N}_c$ to a vector defined on $\mathcal{N}_c$. Therefore, $(I_T^c)^T$ restricts a coarse vector defined on $\mathcal{N}_c$ to a coarse vector defined on $T \cap \mathcal{N}_c$. The following matrix definitions are then in place,

$$R^T = \begin{bmatrix} 0 \\ I \end{bmatrix} \begin{matrix} \} \; \mathcal{N} \setminus \mathcal{N}_c \\ \} \; \mathcal{N}_c \end{matrix}, \quad R_T^T = \begin{bmatrix} 0 \\ I \end{bmatrix} \begin{matrix} \} \; (\mathcal{N} \setminus \mathcal{N}_c) \cap T \\ \} \; \mathcal{N}_c \cap T \end{matrix}, \quad \text{and} \quad I_T^c = \begin{bmatrix} 0 \\ I \end{bmatrix} \begin{matrix} \} \; \mathcal{N}_c \setminus T \\ \} \; T \cap \mathcal{N}_c \end{matrix}.$$

We seek local interpolation P_T that has the following form,

$$P_T = \begin{bmatrix} * \\ I \end{bmatrix} \begin{matrix} \} \;\; T \setminus \mathcal{N}_c \\ \} \;\; T \cap \mathcal{N}_c. \end{matrix} \tag{6.26}$$

Note that, in general, P_T may not agree on dofs that are shared by more than one subdomain T. That is why we need a partition of unity nonnegative diagonal weight matrices $D_T = (d_{T,\,i})_{i \in T}$ that are defined for vectors restricted to T. Let also

$$I_T = \begin{bmatrix} 0 \\ I \end{bmatrix} \begin{matrix} \} \ \mathcal{N} \setminus T \\ \} \ T \end{matrix},$$

be the $n \times n_T$ matrix representing extension by zero outside of T. It is clear that $\mathbf{v}_T = I_T^T \mathbf{v}$ and $A = \sum_T I_T A_T I_T^T$. Note that we have also assumed that the row indices of P_T are in T. Partition of unity means that

$$\sum_T I_T D_T I_T^T = I.$$

In other words, $\sum_{T:\, i \in T} d_{T,\,i} = 1$ for every dof i. With the help of the partition of unity diagonal matrices, we are in a position to define a global P as follows,

$$P = \sum_T I_T D_T P_T \left(I_T^c\right)^T. \tag{6.27}$$

The global $Q = PR$ takes the form

$$Q = \sum_T I_T D_T P_T R_T I_T^T.$$

The latter holds because $R_T I_T^T = (I_T^c)^T R$. We can also see that $RP = I$, which implies that Q is a projection.

From local to global estimates

We assume that for another s.p.d. matrix M, which can be assembled from $\{M_T\}$ in the same way as A, there is a stable local procedure that defines P_T such that for a mapping $R_T: \ R_T P_T = I$, we have the bound

$$(\mathbf{v}_T - P_T R_T \mathbf{v}_T)^T M_T \, (\mathbf{v}_T - P_T R_T \mathbf{v}_T) \le \eta_T \ \mathbf{v}_T^T A_T \mathbf{v}_T \,, \tag{6.28}$$

for any $\mathbf{v}_T \in \mathbb{R}^{n_T}$. A simple example of M can be the (scaled) diagonal of A. Then, M_T will be the (scaled) diagonal of A_T. Another choice is $M_T = A_T$. In either case, we assume that $P_T R_T$ recovers exactly any potential null vectors of A_T. For given local matrices $\{M_T\}$, we let

$$d_{T,\,i} = \frac{\|M_T\|}{\sum_{T':\, i \in T'} \|M_{T'}\|}. \tag{6.29}$$

Assuming that P_T, or rather $Q_T = P_T R_T$, are bounded in the M_T-energy norm in terms of A_T as in (6.28), we prove a similar M-energy bound for the global P.

Based on the partition of unity property of $\{D_T\}$, we have $\mathbf{v}-Q\mathbf{v} = \sum_T I_T D_T (I - P_T R_T) I_T^T \mathbf{v}$. Hence, letting $\mathbf{w}_T = (I - P_T R_T) I_T^T \mathbf{v}$, we get

$$\begin{aligned}((I-Q)\mathbf{v})^T M (I-Q)\mathbf{v} &= ((I-Q)\mathbf{v})^T M \sum_T I_T D_T (I - P_T R_T) I_T^T \mathbf{v} \\ &\le (((I-Q)\mathbf{v})^T M (I-Q)\mathbf{v})^{1/2} \\ &\quad \times \left(\left(\sum_T I_T D_T \mathbf{w}_T\right)^T \sum_{T'} I_{T'} M_{T'} I_{T'}^T \sum_T I_T D_T \mathbf{w}_T \right)^{1/2}.\end{aligned}$$

Introduce for a moment the quantity (cf. [Man93] for a different application)

$$K = \sup_{(\mathbf{w}_T:\ P_T R_T \mathbf{w}_T = 0)} \frac{\sum_{T'} \| I_{T'}^T \sum_T I_T D_T \mathbf{w}_T \|_{M_{T'}}^2}{\sum_T \mathbf{w}_T^T M_T \mathbf{w}_T}. \tag{6.30}$$

Then the following bound on P (or $Q = PR$) holds.

$$\begin{aligned}((I-Q)\mathbf{v})^T M (I-Q)\mathbf{v} &\le \sum_{T'} \left(\sum_T I_T D_T \mathbf{w}_T\right)^T I_{T'} M_{T'} I_{T'}^T \sum_T I_T D_T \mathbf{w}_T \\ &\le K \sum_T \left((I - P_T R_T) I_T^T \mathbf{v}\right)^T M_T (I - P_T R_T) I_T^T \mathbf{v} \\ &\le K \sum_T \eta_T\ \mathbf{v}^T I_T A_T I_T^T \mathbf{v} \\ &\le K \left(\max_T\ \eta_T\right) \mathbf{v}^T A \mathbf{v}.\end{aligned} \tag{6.31}$$

Next, we show how to estimate K in (6.30).

Lemma 6.15. *Let $Cond(M_T) = \lambda_{\max}(M_T)/\lambda_{\min}(M_T)$ denote the condition number of M_T, and let $\kappa \ge 1$ be the maximum number of subdomains T that share any given dof. Then, assuming the specific form* (6.29) *of the weight matrices, the quantity K defined in* (6.30) *can be estimated as $K \le \kappa \max_T Cond(M_T)$.*

Proof. Let $w_{T,\,i}$ stand for the ith entry of $\mathbf{w}_T$. We have

$$\begin{aligned}\left\| I_{T'}^T \sum_T I_T D_T \mathbf{w}_T \right\|_{M_{T'}}^2 &\le \|M_{T'}\| \left\| I_{T'}^T \sum_T I_T D_T \mathbf{w}_T \right\|^2 \\ &= \|M_{T'}\| \sum_{i \in T'} \left(\sum_{T:\, i \in T} d_{T,\,i} w_{T,\,i} \right)^2 \\ &= \sum_{i \in T'} \left(\sum_{T:\, i \in T} \|M_{T'}\|^{1/2} d_{T,\,i} w_{T,\,i} \right)^2.\end{aligned}$$

Therefore,

$$\sum_{T'} \|I_{T'}^T \sum_T I_T D_T \mathbf{w}_T\|_{M_{T'}}^2 \le \sum_{T'} \sum_{i \in T'} \left(\sum_{T:\, i \in T} \|M_{T'}\|^{1/2} d_{T,\,i} w_{T,\,i} \right)^2.$$

Because $\|M_{T'}\| d_{T,\,i} \le \|M_T\|$ for $i \in T \cap T'$ and $\sum_{T:\, i \in T} d_{T,\,i} = 1$ using the Cauchy–Schwarz inequality and noting that $\mathbf{w}_T = (I - Q_T)\mathbf{w}_T$, $(Q_T = P_T R_T)$, we get

$$\begin{aligned}
\sum_{T'} \left\| I_{T'}^T \sum_T I_T D_T \mathbf{w}_T \right\|_{M_{T'}}^2 &\le \sum_{T'} \sum_{i \in T'} \left(\sum_{T:\, i \in T} \|M_T\|^{1/2} (d_{T,\,i})^{1/2} |w_{T,\,i}| \right)^2 \\
&\le \sum_{T'} \sum_{i \in T'} \sum_{T:\, i \in T} \|M_T\| w_{T,\,i}^2 \sum_{T:\, i \in T} d_{T,\,i} \\
&= \sum_{T'} \sum_{i \in T'} \sum_{T:\, i \in T} \|M_T\| w_{T,\,i}^2.
\end{aligned}$$

Therefore, with $Q_T \mathbf{w}_T = 0$, we have

$$\begin{aligned}
\sum_{T'} \left\| I_{T'}^T \sum_T I_T D_T \mathbf{w}_T \right\|_{M_{T'}}^2 &\le \sum_{T'} \sum_{i \in T'} \sum_{T:\, i \in T} \mathrm{Cond}(M_T) \lambda_{\min}^+(M_T)\, w_{T,\,i}^2 \\
&\le \max_T \mathrm{Cond}(M_T) \sum_{T'} \sum_{i \in T'} \sum_{T:\, i \in T} \lambda_{\min}^+(M_T)\, w_{T,\,i}^2 \\
&= \max_T \mathrm{Cond}(M_T) \sum_T \sum_{i \in T} \lambda_{\min}^+(M_T)\, w_{T,\,i}^2 \sum_{T':\, i \in T \cap T'} 1 \\
&\le \max_T \mathrm{Cond}(M_T)\, \kappa \sum_T \sum_{i \in T} \lambda_{\min}^+(M_T)\, w_{T,\,i}^2 \\
&= \max_T \mathrm{Cond}(M_T)\, \kappa \sum_T \lambda_{\min}^+(M_T) \|\mathbf{w}_T\|^2 \\
&= \max_T \mathrm{Cond}(M_T)\, \kappa \sum_T \lambda_{\min}^+(M_T) \|(I - Q_T)\mathbf{w}_T\|^2 \\
&\le \max_T \mathrm{Cond}(M_T)\, \kappa \sum_T \|(I - Q_T)\mathbf{w}_T\|_{M_T}^2 \\
&= \max_T \mathrm{Cond}(M_T)\, \kappa \sum_T \mathbf{w}_T^T M_T \mathbf{w}_T.
\end{aligned}$$

Thus, we showed that $K \le \kappa\ \max_T \mathrm{Cond}(M_T)$ where $\kappa \ge 1$ is the maximum number of subdomains that share any given dof, and $\mathrm{Cond}(M_T) = \lambda_{\max}(M_T)/\lambda_{\min}^+(M_T)$, denotes the effective condition number of M_T. More specifically, $\lambda_{\min}^+(M_T) > 0$ is the minimal (nonzero) eigenvalue of M_T in the subspace Range $(P_T R_T)$; that is, we have

$$\lambda_{\min}^+(M_T)\ \|\mathbf{w}_T\|^2 \le \mathbf{w}_T^T M_T \mathbf{w}_T, \quad \text{for all } \mathbf{w}_T = (I - P_T R_T)\mathbf{w}_T.$$

Note that to have $\lambda^+_{\min}(M_T) > 0$, this means that the projection $P_T R_T$ should recover exactly the potential null vectors of M_T that we have assumed. □

In summary, estimate (6.31), based on Lemma 6.15, gives the following main result.

Theorem 6.16. *Let M be an s.p.d. matrix such that the local estimates* (6.28) *hold. Let P be defined by* (6.27) *based on the local interpolation matrices $\{P_T\}$ and the diagonal weight matrices $\{D_T\}$ with coefficients given in* (6.29)*. Then the following global norm bound holds,*

$$\mathbf{v}^T(I-Q)^T M(I-Q)\mathbf{v} \leq \kappa \max_T Cond(M_T) \left(\max_T \eta_T\right) \mathbf{v}^T A\mathbf{v},$$

where $\kappa \geq 1$ is the maximal number of subdomains T that contain any given dof. Here, $Cond(M_T)$ stands for the effective condition number of M_T defined as $\lambda_{\max} M_T/\lambda^+_{\min}(M_T)$, where $\lambda^+_{\min}$ is the minimal eigenvalue of M_T in the subspace Range $(I - Q_T)$. This minimal eigenvalue is positive if $Q_T = P_T R_T$ recovers exactly the potential null vectors of M_T.

In addition, suppose that the local matrices M_T have uniformly bounded effective condition number; that is,

$$\max_T \mathrm{Cond}(M_T) \leq C_M\,. \tag{6.32}$$

This is trivially the case, for example, if M is the diagonal of A, hence M_T is the diagonal of A_T. Furthermore, let

$$\mathbf{v}^T A\mathbf{v} \leq \mathbf{v}^T M\mathbf{v} \quad \text{for all } \mathbf{v} \in \mathbb{R}^n, \tag{6.33}$$

which can be ensured after proper scaling of M. Suppose also that the constants in the local estimates (6.28) are bounded

$$\max_T \eta_T \leq C_\eta\,. \tag{6.34}$$

Then the global estimate in Theorem 6.16 shows that A is spectrally equivalent to M in the subspace Range$(I - Q)$. The latter space is complementary to the coarse space. Therefore, the following corollary holds (based on Theorem 3.25).

Corollary 6.17. *The two-grid method based on P defined by* (6.27) *and a smoother M satisfying* (6.32) *through* (6.34) *is optimally convergent with a convergence factor bounded by $1 - 1/\theta$, where $\theta = \kappa\, C_M\, C_\eta$.*

On a local element level, for finite element matrices A_T, it is very often the case when A_T and its diagonal M_T are spectrally equivalent (independently of the mesh-size) on the subspace complementary to the potential null space of A_T. In such a case, assuming that $P_T R_T$ recovers exactly the potential null space of A_T, we also have that M_T and A_T are spectrally equivalent on the subspace Range $(I - P_T R_T)$. Then, assume that we have established a local energy boundedness of $I - P_T R_T$, that is, an estimate

$$\|(I - P_T R_T)\mathbf{v}_T\|_{A_T}^2 \le \eta_T' \; \mathbf{v}_T^T A_T \mathbf{v}_T,$$

for some mesh-independent constants η_T'. Due to the spectral equivalence of A_T and M_T on the subspace Range $(I - P_T R_T)$, we then have similar estimates for M_T; that is,

$$\|(I - P_T R_T)\mathbf{v}_T\|_{M_T}^2 \le \eta_T \; \mathbf{v}_T^T A_T \mathbf{v}_T.$$

Based on the above Corollary 6.17, we obtain then a uniform two-grid convergence for these finite element matrices A. This covers the case of elliptic-type finite element matrices, either scalar elliptic, or systems such as elasticity. For the scalar elliptic case, the assumption on $P_T R_T$ is that it should recover locally the constant vectors exactly, whereas for the elasticity case, the assumption is that $P_T R_T$ should recover the so-called rigid body modes (or r.b.m.) locally. In 2D, we have three r.b.m., whereas in 3D, there are six such modes.

The rest of this section deals with the practical construction of interpolation, which fits the null vectors of A_T exactly. One way to do this is to first coarsen the null space and then proceed with the rest of the interpolation as described below.

Assume an initial coarse grid $\mathcal{N}_c^0$ such that its restriction to every T is rich enough; that is, there are at least as many coarse dofs per element T as the dimension of the null space of A_T. For example, in the 3D elasticity example, it is sufficient to have six coarse dofs per element T. Let $P_{0,\,T}$ be local interpolation matrices (having the form (6.26)) such that a subset of its columns provides a basis for the nullspace of A_T. Together with the restriction mapping $R_{0,\,T} = [0, I]$ and the projection $Q_{0,\,T} = P_{0,\,T} R_{0,\,T}$ based on $P_{0,T}$ using averaging, we can define a global initial interpolation matrix P_0 via (6.27). We construct interpolation matrices P_T that are defined on a complementary coarse grid $\mathcal{N}_c'$; that is, $\mathcal{N}_c' \cap \mathcal{N}_c^0 = \emptyset$. The resulting composite coarse grid $\mathcal{N}_c$ equals $\mathcal{N}_c^0 \cup \mathcal{N}_c'$. It is natural to assume that for every T, $R_{0,\,T} P_T = 0$. In particular, we assume that P_T have zero rows corresponding to the set $\mathcal{N}_c^0$ (viewed as a subset of the fine-grid dofs).

Define the composite interpolation

$$\overline{P}_T = [P_{0,\,T},\; P_T],$$

and the composite restriction mapping

$$\overline{R}_T = \begin{bmatrix} R_{0,\,T} \\ R_T(I - P_{0,\,T} R_{0,\,T}) \end{bmatrix}.$$

Then the following simple identity holds,

$$I - \overline{P}_T \overline{R}_T = (I - P_T R_T)(I - P_{0,\,T} R_{0,\,T}). \tag{6.35}$$

Therefore, the composite interpolation satisfies $\mathbf{v}_T = \overline{P}_T \overline{R}_T \mathbf{v}_T$ for any $\mathbf{v}_T$ in the null space of A_T as needed in the local estimate (6.28) (with $M_T = A_T$). Furthermore,

$\overline{Q}_T = \overline{P}_T \overline{R}_T$ is a projection. Indeed,

$$\begin{aligned}
\overline{R}_T \overline{P}_T &= \begin{bmatrix} R_{0,\,T} \\ R_T(I - P_{0,\,T} R_{0,\,T}) \end{bmatrix} [P_{0,\,T},\ P_T] \\
&= \begin{bmatrix} R_{0,\,T} P_{0,\,T} & R_{0,\,T} P_T \\ R_T(I - P_{0,\,T} R_{0,\,T}) P_{0,\,T} & R_T(I - P_{0,\,T} R_{0,\,T}) P_T \end{bmatrix} \\
&= \begin{bmatrix} R_{0,\,T} P_{0,\,T} & 0 \\ 0 & R_T P_T \end{bmatrix} \\
&= \begin{bmatrix} I & 0 \\ 0 & I \end{bmatrix}.
\end{aligned}$$

We used the fact that $R_{0,\,T} P_{0,\,T} = I$, $R_T P_T = I$, and $R_{0,\,T} P_T = 0$.

Now, let $\overline{Q}$ be the global mapping defined using the partition of unity diagonal matrices as in (6.27). The conclusion of Theorem 6.16 is that based on local estimates such as

$$\|(I - \overline{Q}_T)\mathbf{v}_T\|_{A_T}^2 \leq \eta_T \ \|\mathbf{v}_T\|_{A_T}^2;$$

the global one,

$$\|(I - \overline{Q})\mathbf{v}\|_A^2 \leq \eta \ \|\mathbf{v}\|_A^2,$$

follows, where η depends only on local quantities (η_T and the nonzero spectrum of A_T).

6.10 Multivector fitting interpolation

In Section 6.2, we described an algorithm that constructs a P that fits (approximately) one given "algebraically" smooth vector in the sense of Proposition 6.2. A procedure that gives a P that fits a given vector exactly, both locally and globally, was described in Section 6.9.

Assume now, that we have constructed a P that contains in its range a number of vectors $\mathbf{x}_1, \ldots, \mathbf{x}_{m-1}$, and we want to construct a modified $\overline{P}$ that keeps in its range the previous vectors $\mathbf{x}_1, \ldots, \mathbf{x}_{m-1}$ and one additional vector $\mathbf{x}_m$. The idea is (cf., [VZ05]) to construct a $P_{\text{new}} = P_m$ and augment P to $\overline{P} = [P,\ P_m]$ such that $\overline{P}$ has a full column rank. To ensure that $\overline{P}$ has a full column rank, let us assume (by induction) that P has the following block form $P = [P_1, \ldots, P_{m-1}]$, and overall:

$$P = \begin{bmatrix} * & * & \ldots & * \\ * & * & \ldots & I \\ \vdots & \vdots & \ddots & 0 \\ * & I & 0 & 0 \\ I & 0 & 0 & 0 \end{bmatrix} \begin{matrix} \}\ \mathcal{N} \setminus (\mathcal{N}_1 \cup \ldots \mathcal{N}_{m-1}) \\ \}\ \mathcal{N}_{m-1} \\ \}\ \vdots \\ \}\ \mathcal{N}_2 \\ \}\ \mathcal{N}_1 \end{matrix}.$$

That is, every block column P_k of P comes with its own coarse set $\mathcal{N}_k$ that is complementary to the previous coarse sets. This hierarchical structure ensures that the resulting P has a full column rank.

The next step is clear then. We compute the interpolation error $\mathbf{e} = \mathbf{e}_{m-1}$ in the following steps.

(i) Let $\mathbf{e}_0 = \mathbf{x}_m$.
(ii) For $k = 1, \ldots, m-1$ compute

$$\mathbf{e}_k = \mathbf{e}_{k-1} - P_k(\mathbf{e}_{k-1}|_{\mathcal{N}_{k-1}}).$$

We notice that $\mathbf{e}_k$ vanishes at $\mathcal{N}_1 \cup \cdots \cup \mathcal{N}_{k-1}$.

At the end, $\mathbf{e} = \mathbf{e}_{m-1}$ vanishes at the current global coarse-grid $\mathcal{N}_1 \cup \cdots \cup \mathcal{N}_{m-1}$. Based on the entries of $D^{1/2}\mathbf{e}$ with maximal absolute values, where D is the diagonal of A, we select a complementary coarse set $\mathcal{N}_m$ (i.e., $\mathcal{N}_m \subset \mathcal{N} \setminus (\mathcal{N}_1 \cup \cdots \cup \mathcal{N}_{m-1})$). The latter can be done, utilizing a certain partitioning of $\mathcal{N}$ into nonoverlapping sets (groups) $\{\mathcal{I}\}$ by selecting a new dof $i_{\mathcal{I}}$ per $\mathcal{I}$ such that $D^{1/2}\mathbf{e}|_{\mathcal{I}}$ has a local (in $\mathcal{I}$) maximal value at $i_{\mathcal{I}}$. Then, we construct a P_m that fits $\mathbf{e}$. For this, we may use any interpolation algorithm that fits a single vector. It is a simple observation to show that $\mathbf{x}_m$ is in the range of $\overline{P}$. This is indeed easily seen, because by construction $\mathbf{e} \in \text{Range}(P_m)$, and the rest follows from the identity

$$\mathbf{x} = \sum_{k=1}^{m-1} P_k \mathbf{e}_{k-1} + \mathbf{e} \in \text{Range}[P_1, \ldots, P_{m-1}, P_m].$$

6.11 Window-based spectral AMG

In the present section, we provide a purely algebraic way of selecting coarse degrees of freedom and a way to construct an energy-bounded interpolation matrix P (cf., [FVZ05]). In the analysis, we use a simple Richardson iteration as a smoother. If local element matrices are available (as in Section 6.9) the following construction and analysis simplifies (cf., [Ch03]). All definitions and constructions below are valid in the case when A is positive and only semidefinite; that is, A may have nonempty null space Null(A).

We consider the problem $A\mathbf{x} = \mathbf{b}$ and reformulate it in the following equivalent least squares minimization,

$$\mathbf{x} = \arg\min_{\mathbf{v}} \sum_{w} \|A_w \mathbf{v} - \mathbf{b}_w\|^2_{D_w}. \tag{6.36}$$

In the least squares formulation, each w is a subset of $\{1, \ldots, n\}$, and we assume that

$$\cup w = \{1, \ldots, n\},$$

where the decomposition can be overlapping. The sets w are called windows, and represent a grouping of the rows of A. The corresponding rectangular matrices we denote by A_w; that is, $A_w = \{A_{ij}\}_{i\in w,\ 1\leq j\leq n}$. Thus, we have that $A_w \in \mathbb{R}^{|w|\times n}$, where $|\cdot|$ (in the present section) stands for cardinality. Accordingly in (6.36), $\mathbf{b}_w = \mathbf{b}|_w = \{\mathbf{b}_i\}_{i\in w}$ denotes a restriction of $\mathbf{b}$ to a subset and $D_w = (D_w(i))_{i\in w}$ are diagonal matrices with nonnegative entries, such that for any i, $\sum_{w:\ i\in w} D_w(i) = 1$; that is, $\{D_w\}_w$ provides a partition of unity. Vanishing the first variation of the least squares functional, we obtain that the solution to the minimization problem (6.36) satisfies

$$\sum_w (A_w)^T D_w A_w \mathbf{x} = \sum_w (A_w)^T D_w \mathbf{b}_w. \tag{6.37}$$

With the specific choice of $\{D_w\}_w$, it is clear that (6.36) is equivalent to the standard least squares problem,

$$\sum_w \|A_w\mathbf{v} - \mathbf{b}_w\|^2_{D_w} = \|A\mathbf{v} - \mathbf{b}\|^2.$$

Therefore, we obtain the identity

$$\mathbf{v}^T\left(\sum_w (A_w)^T D_w A_w\right)\mathbf{v} = \mathbf{v}^T A^T A\mathbf{v}. \tag{6.38}$$

We emphasize that we do not solve the equivalent least squares problem (6.37), and it has only been introduced as a motivation to consider the "local" matrices $(A_w)^T D_w A_w$ as a tool for constructing sparse (and hence local) interpolation mapping P, which we explain below. Of interest are the Schur complements S_w, that are obtained from the matrices $(A_w)^T D_w A_w$ by eliminating the entries outside w. More specifically, let (after possible reordering of the columns of A_w)

$$A_w = [A_{ww} \quad A_{w,\chi}], \tag{6.39}$$

where A_{ww} is the square principal submatrix of A corresponding to the subset w and $A_{w,\chi}$ corresponds to the remaining columns of A_w with indices outside w. As any Schur complement of symmetric positive semidefinite matrices, S_w is characterized by the identity

$$\mathbf{v}_w^T S_w \mathbf{v}_w = \inf_{\mathbf{v}_\chi} \begin{bmatrix}\mathbf{v}_w\\ \mathbf{v}_\chi\end{bmatrix}^T (A_w)^T D_w A_w \begin{bmatrix}\mathbf{v}_w\\ \mathbf{v}_\chi\end{bmatrix}. \tag{6.40}$$

An explicit expression for S_w is readily available. Let $A^T_{w,\chi} D_w A_{w,\chi} = Q^T \Lambda Q$ with $Q^T = Q^{-1}$ and $\Lambda = \text{diag}(\lambda)$ being a diagonal matrix with eigenvalues that are nonnegative. Letting $\Lambda^+ = \text{diag}(\lambda^+)$, where $\lambda^+ = 0$ if $\lambda = 0$, and $\lambda^+ = \lambda^{-1}$ if $\lambda > 0$, we have the expression

$$S_w = (A_{ww})^T D_w A_{ww} - (A_{ww})^T D_w A_{w,\chi} Q^T \Lambda^+ Q A^T_{w,\ \chi} D_w A_{ww}.$$

Note that S_w is symmetric and positive semidefinite by construction (see (6.40)), and we have the inequality

$$(\mathbf{v}_w)^T S_w \mathbf{v}_w \leq \mathbf{v}^T (A_w)^T D_w A_w \mathbf{v}, \qquad \mathbf{v}_w = \mathbf{v}|_w.$$

Hence,

$$\sum_w (\mathbf{v}_w)^T S_w \mathbf{v}_w \leq \mathbf{v}^T \left(\sum_w (A_w)^T D_w A_w \right) \mathbf{v} = \mathbf{v}^T A^T A \mathbf{v} \leq \|A\| \, \mathbf{v}^T A \mathbf{v}. \tag{6.41}$$

This inequality implies (letting $\mathbf{v} = 0$ outside w) that

$$(\mathbf{v}_w)^T S_w \mathbf{v}_w \leq \|A\| \, \mathbf{v}^T A \mathbf{v} \leq \|A\|^2 \, \mathbf{v}^T \mathbf{v} = \|A\|^2 \, \mathbf{v}_w^T \mathbf{v}_w;$$

that is,

$$\|S_w\| \leq \|A\|^2. \tag{6.42}$$

Selecting coarse degrees of freedom

Our goal is to select a coarse space. The way we do that is by fixing a window and associating with it a number $m_w \leq |w|$. Then we construct m_w basis vectors (columns of P) corresponding to this window in the following way. All the eigenvectors and eigenvalues of S_w are computed and the eigenvectors corresponding to the first m_w eigenvalues are chosen. Because generally the windows have overlap, another partition of unity is constructed, with nonnegative diagonal matrices $\{Q_w\}$ where each Q_w is nonzero only on w and the set $\{Q_w\}$ satisfies $\sum_w Q_w = I$. From the first m_w eigenvectors of S_w extended by zero outside w, we form the local interpolation matrix P_w columnwise, which hence has m_w columns. The global interpolation matrix is then defined as

$$P = \sum_w Q_w [0, \; P_w, \; 0].$$

Here, for a global coarse vector $\mathbf{v}^c = (\mathbf{v}_w^c)$, the action of $[0, \; P_w, \; 0]$ is defined such that $[0, \; P_w, \; 0]\mathbf{v}^c = P_w(\mathbf{v}^c|_w) = P_w \mathbf{v}_w^c$.

The first result concerns the null space of A, namely, that it is contained in the range of the interpolation P.

Lemma 6.18. *Suppose that m_w is such that $m_w \geq$ dim Null (S_w) for every window w. Then Null $(A) \subset$ Range (P); that is, if $A\mathbf{v} = 0$, then there exists a $\mathbf{v}^c \in \mathbb{R}^{n_c}$ such that $\mathbf{v} = P\mathbf{v}^c$.*

Proof. Let $A\mathbf{v} = 0$. Then from inequality (6.41) it follows that $S_w \mathbf{v}_w = 0$, where $\mathbf{v}_w = \mathbf{v}|_w$ and we extend $\mathbf{v}_w$ by zero outside w whenever needed. Hence, by our assumption on m_w there exists a local coarse grid vector $\mathbf{v}_w^c$ such that $\mathbf{v}_w = P_w \mathbf{v}_w^c$. Let $\mathbf{v}^c$ be the composite coarse grid vector that agrees with $\mathbf{v}_w^c$ on w, for each w. This is simply the collection $\mathbf{v}^c = (\mathbf{v}_w^c)$. Then,

$$P\mathbf{v}^c = \sum_w Q_w P_w \mathbf{v}_w^c = \sum_w Q_w \mathbf{v}_w = \sum_w Q_w \mathbf{v} = \mathbf{v}. \qquad \square$$

Two-grid convergence

First, we prove a main coarse-grid "weak-approximation property."

Lemma 6.19. *Assume that the windows $\{w\}$ are selected in a "quasiuniform" manner such that for all w, the following uniform estimate holds,*

$$\|S_w\| \geq \eta \|A\|^2. \tag{6.43}$$

Note that $\eta \leq 1$ (see (6.42)). Assume that we have chosen m_w so well that for a constant $\delta > 0$ uniformly in w, we have

$$\|S_w\| \leq \delta \lambda_{m_w+1}(S_w). \tag{6.44}$$

Here $\lambda_{m_w+1}(S_w)$ denotes the $(m_w + 1)$st smallest eigenvalue of S_w. It is clear that $\delta \geq 1$. Then, for any vector $\mathbf{e} \in \mathbb{R}^n$, there exists a global interpolant $\boldsymbol{\epsilon}$ in the range of P such that

$$(\mathbf{e} - \boldsymbol{\epsilon})^T A(\mathbf{e} - \boldsymbol{\epsilon}) \leq \|A\| \|\mathbf{e} - \boldsymbol{\epsilon}\|^2 \leq \frac{\delta}{\eta}\, \mathbf{e}^T A\mathbf{e}. \tag{6.45}$$

Before we present the proof of the lemma, we illustrate how the assumptions (6.43) and (6.44) can be verified. Consider the simple example, when A corresponds to a finite element discretization of the Laplace operator on uniform triangular mesh on the unit square domain Ω with Neumann boundary conditions. We first notice that the entries of A are mesh-independent. Therefore $\|A\|$ is bounded above by a mesh-independent constant ($\|A\| \leq 8$). Let $h = 1/m_0 m$ be the fine-grid mesh-size for a given integer m and a fixed (independently of m) integer $m_0 > 1$. Let $H = 1/m$. This implies that Ω can be covered exactly by m^2 equal coarse rectangles of size $H = m_0 h$. Each coarse rectangle defines a window as the set of indices corresponding to the fine-grid nodes contained in that coarse rectangle. There are $(m_0 + 1)^2$ nodes per rectangle, and all the rectangles form an overlapping partition of the grid. A simple observation is that any such rectangle can have 0, 1, or 2 common sides with the boundary of Ω and therefore, there are only three different types of window matrices A_w and respective Schur complements S_w. It is clear then that inequalities of the type (6.43) and (6.44) are feasible for a mesh-independent constant η and for a mesh-independent choice of m_w. For the simple example in consideration, fix $m_0 > 3$, hence $(m_0+1)^2 > 4(m_0+1)$; that is, let the number of nodes in w be larger than the number of its outside boundary nodes (i.e., nodes outside w, that are connected to w through nonzero entries of $A_{w,\chi}$). From (6.40) it is clear that if $S_w \mathbf{v}_w = 0$, then there is a $\mathbf{v}_\chi$ such that $A_{ww}\mathbf{v}_w + A_{w,\chi}\mathbf{v}_\chi = 0$. Because in our case A_{ww} is invertible, we have then that the dimension of the null space of S_w equals the dimension of Range$(A_{w,\chi})$. The latter is bounded above by $4(m_0 + 1)$ (which is the number of nodes outside w that are connected to w through nonzero entries of $A_{w,\chi}$). Therefore, we may choose any fixed integer $m_w \geq 4(m_0 + 1)$ (and $m_w < (m_0 + 1)^2$) to guarantee estimate (6.44) because then $\lambda_{m_w+1}(S_w) > 0$. The number of coarse degrees of freedom (or dofs) then equals $m_w m^2$. This implies that the coarsening factor, defined below, will satisfy

$$\frac{\#\text{ fine dofs}}{\#\text{ coarse dofs}} = \frac{(mm_0+1)^2}{m_w m^2} = \frac{\left(m_0+\frac{1}{m}\right)^2}{m_w} \simeq \frac{m_0^2}{m_w}.$$

For example, if we choose $m_w = 4(m_0+1)$, the coarsening factor is $\simeq m_0/(4+\frac{4}{m_0})$. It is strictly greater than 1 if $m_0 > 4$, and it can be made as large as needed by increasing m_0. (The latter, of course, reflects the size of the windows.) In conclusion, in this simple example, we can easily see that the bounds $\eta = \min_w(\|S_w\|/\|A\|^2) \le 1$ and $\delta = \max_w(\|S_w\|/\lambda_{m_w+1}(S_w)) \ge 1$ are fixed mesh-independent constants. This is true, because the matrices S_w are a finite number, the number m_w is fixed, and therefore the eigenvalues $\lambda_{m_w+1}(S_w)$ are also a finite number, and all these numbers have nothing to do with m (or the mesh-size $h \mapsto 0$). Similar reasoning can be applied to more general quasiuniform meshes. This is the case if the windows can be chosen such that the matrices $(A_w)^T D_w A_w$ and S_w are spectrally equivalent to a finite number of mesh-independent reference ones. The constants in the spectral equivalence then will only depend on the angles in the mesh.

Proof of Lemma 6.19. Let $\mathbf{e} \in \mathbb{R}^n$ be given. Note that our assumption on m_w is equivalent to the assumption that for any window w, there exists a $\boldsymbol{\epsilon}_w$ in the range of P_w such that

$$\|S_w\|\,\|\mathbf{e}_w - \boldsymbol{\epsilon}_w\|^2 \le \delta\, \mathbf{e}_w^T S_w \mathbf{e}_w, \tag{6.46}$$

where $\mathbf{e}_w = \mathbf{e}|_w$ and whenever needed, we consider $\mathbf{e}_w$ and $\boldsymbol{\epsilon}_w$ extended by zero outside w. We now construct an $\boldsymbol{\epsilon}$ in the range of P that will satisfy (6.45). Namely, we set $\boldsymbol{\epsilon} = \sum_w Q_w \boldsymbol{\epsilon}_w$. We notice that $\sum_w Q_w \boldsymbol{\epsilon} = \boldsymbol{\epsilon} = \sum_w Q_w \boldsymbol{\epsilon}_w$. Hence,

$$\begin{aligned}
\|\mathbf{e} - \boldsymbol{\epsilon}\|^2 &= (\mathbf{e} - \boldsymbol{\epsilon})^T \left(\sum_w Q_w(\mathbf{e} - \boldsymbol{\epsilon})\right) \\
&= (\mathbf{e} - \boldsymbol{\epsilon})^T \left(\sum_w Q_w(\mathbf{e}_w - \boldsymbol{\epsilon}_w)\right) \\
&= \sum_w \left(Q_w^{1/2}(\mathbf{e} - \boldsymbol{\epsilon})\right)^T \left(Q_w^{1/2}(\mathbf{e}_w - \boldsymbol{\epsilon}_w)\right).
\end{aligned}$$

Therefore,

$$\begin{aligned}
\|\mathbf{e} - \boldsymbol{\epsilon}\|^2 &\le \left[\sum_w (\mathbf{e} - \boldsymbol{\epsilon})^T Q_w (\mathbf{e} - \boldsymbol{\epsilon})\right]^{1/2} \left[\sum_w \|Q_w^{1/2}(\mathbf{e}_w - \boldsymbol{\epsilon}_w)\|^2\right]^{1/2} \\
&= \|\mathbf{e} - \boldsymbol{\epsilon}\| \left[\sum_w \|Q_w^{1/2}(\mathbf{e}_w - \boldsymbol{\epsilon}_w)\|^2\right]^{1/2}.
\end{aligned}$$

That is,

$$\|\mathbf{e} - \boldsymbol{\epsilon}\|^2 \le \sum_w \|Q_w^{1/2}(\mathbf{e}_w - \boldsymbol{\epsilon}_w)\|^2.$$

Therefore, based on (6.46), the quasiuniformity of $\{w\}$, and inequality (6.41), we get

$$\begin{aligned}\|\mathbf{e}-\boldsymbol{\epsilon}\|^2 &\le \sum_w \|Q_w^{1/2}(\mathbf{e}_w-\boldsymbol{\epsilon}_w)\|^2 \le \sum_w \|\mathbf{e}_w-\boldsymbol{\epsilon}_w\|^2\\ &\le \delta\sum_w \frac{\mathbf{e}_w^T S_w \mathbf{e}_w}{\|S_w\|} \le \frac{\delta}{\eta\|A\|^2}\sum_w \mathbf{e}_w^T S_w \mathbf{e}_w\\ &\le \frac{\delta}{\eta\|A\|^2}\,\mathbf{e}^T A^T A\mathbf{e} \le \frac{\delta}{\eta\|A\|}\,\mathbf{e}^T A\mathbf{e}.\end{aligned} \tag{6.47}$$

□

We use estimate (6.45) to show that the two-grid method with the Richardson iteration matrix $M = (\|A\|/\omega)I$, $\omega \in (0, 2)$, which leads to $\widetilde{M} = M(2M - A)^{-1}M = (\|A\|^2/\omega^2)(2(\|A\|/\omega)I - A)^{-1}$, is uniformly convergent. More specifically, we have the following main spectral equivalence result.

Theorem 6.20. *The algebraic two-grid preconditioner B, based on the Richardson smoother $M = (\|A\|/\omega)\ I$, $\omega \in (0, 2)$, and the coarse space based on P constructed by the window spectral AMG method, is spectrally equivalent to A and the following estimate holds.*

$$\mathbf{v}^T A\mathbf{v} \le \mathbf{v}^T B\mathbf{v} \le \frac{\delta}{\eta\omega(2-\omega)}\,\mathbf{v}^T A\mathbf{v}.$$

The term δ/η comes from the coarse-grid approximation property (6.45).

Proof. We first notice that

$$\mathbf{w}^T\widetilde{M}\mathbf{w} = \frac{\|A\|^2}{\omega^2}\,\mathbf{w}^T\left(2\frac{\|A\|}{\omega}I - A\right)^{-1}\mathbf{w} \le \frac{\|A\|}{\omega(2-\omega)}\,\mathbf{w}^T\mathbf{w} = \frac{1}{2-\omega}\,\mathbf{w}^T M\mathbf{w}.$$

Then, based on the $\widetilde{M}$-norm minimization property of the projection $\pi_{\widetilde{M}}$, we have

$$\begin{aligned}((I-\pi_{\widetilde{M}})\mathbf{v})^T\widetilde{M}(I-\pi_{\widetilde{M}})\mathbf{v} &= \inf_{\boldsymbol{\epsilon}\in\text{Range}(P)} (\mathbf{v}-\boldsymbol{\epsilon})^T\widetilde{M}(\mathbf{v}-\boldsymbol{\epsilon})\\ &\le \frac{1}{2-\omega}\inf_{\boldsymbol{\epsilon}\in\text{Range}(P)} (\mathbf{v}-\boldsymbol{\epsilon})^T M(\mathbf{v}-\boldsymbol{\epsilon})\\ &= \frac{\|A\|}{\omega(2-\omega)}\inf_{\boldsymbol{\epsilon}\in\text{Range}(P)} \|\mathbf{v}-\boldsymbol{\epsilon}\|^2\\ &\le \frac{1}{\omega(2-\omega)}\,\frac{\delta}{\eta}\,\mathbf{v}^T A\mathbf{v}.\end{aligned}$$

Thus, based on Theorem 3.19, we have that the corresponding two-grid preconditioner B is spectrally equivalent to A with the best constant

$$K_{TG} = \sup_{\mathbf{v}} \frac{((I-\pi_{\widetilde{M}})\mathbf{v})^T\widetilde{M}(I-\pi_{\widetilde{M}})\mathbf{v}}{\mathbf{v}^T A\mathbf{v}} \le \frac{\delta}{\eta\,\omega(2-\omega)}.$$

□

As it has been shown earlier (see Theorem 5.22 in Section 5.6), we know that the two-grid window-based spectral AMG method improves its convergence by increasing the number of smoothing steps. This is due to the "strong approximation property" (see (6.47) one inequality before the last); namely, given $\mathbf{e}$, there is a coarse interpolant $\boldsymbol{\epsilon}$ such that

$$\|\mathbf{e} - \boldsymbol{\epsilon}\|_A^2 \leq \|A\| \|\mathbf{e} - \boldsymbol{\epsilon}\|^2 \leq \frac{\delta}{\eta \|A\|} \|A\mathbf{e}\|^2.$$

6.12 Two-grid convergence of vector-preserving AMG

This section provides a two-grid convergence analysis of a constant vector-preserving AMG with application to matrices coming from second-order elliptic PDEs.

Problem formulation

Here, we first describe the particular AMG method of interest. Consider a s.p.d. sparse matrix A partitioned into the common two-by-two block form,

$$A = \begin{bmatrix} A_{ff} & A_{fc} \\ A_{cf} & A_{cc} \end{bmatrix} \begin{matrix} \} \mathcal{N}_f \\ \} \mathcal{N}_c \end{matrix}. \tag{6.48}$$

As usual, $\mathcal{N}_c$ is the set of coarse degrees of freedom, or coarse dofs. In our application to follow, we assume that A_{ff} is s.p.d. and spectrally equivalent to its diagonal part D_{ff}.

We describe next a two-grid AMG. Let $\mathbf{1}_c$ be a given vector with indices from $\mathcal{N}_c$. In the application to follow, it is assumed that $\mathbf{1}_c$ is a vector with constant entries. Define

$$\mathbf{1}_f = -A_{ff}^{-1} A_{fc} \mathbf{1}_c.$$

Then, by definition

$$\mathbf{1} = \begin{bmatrix} \mathbf{1}_f \\ \mathbf{1}_c \end{bmatrix}.$$

Given is a rectangular matrix W with the same dimension as the off-diagonal block A_{fc} of A. Our next main assumption is that W satisfy

$$A_{fc}\mathbf{1}_c + A_{ff} W \mathbf{1}_c = 0. \tag{6.49}$$

For each coarse dof i, we associate a neighborhood set $\mathcal{A}_i \subset \mathcal{N} \equiv \mathcal{N}_f \cup \mathcal{N}_c$, such that $\cup_{i \in \mathcal{N}_c} \mathcal{A}_i = \mathcal{N}$. The sets $\{\mathcal{A}_i\}_{i \in \mathcal{N}_c}$ can be overlapping. Also, consider neighborhood sets $\Omega_{c,\, i} \subset \mathcal{N}_c$, which have large enough support about the coarse dof i. More specifically, we define $\Omega_{c,i}$ to contain the set

$$\cup_{k \in \mathcal{A}_i} \{j \in \mathcal{N}_c : \; (A_{ff} W)_{k,\, j} \neq 0 \text{ or } (A_{fc})_{k,\, j} \neq 0\}.$$

How large this set is depends on the sparsity of $A_{ff} W$ and A_{fc}.

Define the coarse vector (supported in $\Omega_{c,\,i}$),

$$\mathbf{1}_{c,\,i} = \begin{bmatrix} 0 \\ \mathbf{1}|_{\Omega_{c,\,i}} \\ 0 \end{bmatrix}.$$

Then, due to the choice of $\Omega_{c,\,i}$, and recalling (6.49), we have

$$\begin{aligned}
(A_{ff} W \mathbf{1}_{c,\,i})|_{\mathcal{A}_i} &= \left(A_{ff} W \begin{bmatrix} 0 \\ \mathbf{1}|_{\Omega_{c,\,i}} \\ 0 \end{bmatrix} \right)\Bigg|_{\mathcal{A}_i} \\
&= \left(A_{ff} W \begin{bmatrix} \star \\ \mathbf{1}|_{\Omega_{c,\,i}} \\ \star \end{bmatrix} \right)\Bigg|_{\mathcal{A}_i} \\
&= (A_{ff} W \mathbf{1}_c)|_{\mathcal{A}_i} \\
&= -(A_{fc} \mathbf{1}_c)|_{\mathcal{A}_i} \\
&= -(A_{fc} \mathbf{1}_{c,\,i})|_{\mathcal{A}_i}.
\end{aligned} \tag{6.50}$$

The latter identities use the fact that $\mathbf{1}_{c,\,i}$ coincides with $\mathbf{1}$ on $\Omega_{c,\,i}$, and due to the choice of $\Omega_{c,\,i}$ to be a sufficiently large neighborhood of the coarse dof i. That is, to compute $A_{ff} W \mathbf{1}_c$ restricted to $\mathcal{A}_i$ due to the sparsity of $A_{ff} W$, we need the entries of $\mathbf{1}_c$ only from a subset of $\Omega_{c,\,i}$. A similar argument applies to the vector $A_{fc} \mathbf{1}_{c,\,i}$ restricted to $\mathcal{A}_i$.

In what follows, our goal is to estimate the deviation between $W_* \equiv -A_{ff}^{-1} A_{fc}$ and W. More specifically, we are interested in bounding the inner product $(A_{ff}(W_* - W)\mathbf{e}_c,\ (W_* - W)\mathbf{e}_c)$, for any vector $\mathbf{e}_c$ in terms of $(S\mathbf{e}_c,\ \mathbf{e}_c)$. Here,

$$S = A_{cc} - A_{cf} A_{ff}^{-1} A_{fc} \tag{6.51}$$

is the Schur complement of A.

In the present section, $(\mathbf{u},\ \mathbf{v})$ denotes the standard Euclidean vector inner product $\mathbf{v}^T \mathbf{u}$. Also, by definition $\|\mathbf{w}\|_{\mathcal{A}_i} = \|\mathbf{w}|_{\mathcal{A}_i}\|$.

With D_{ff} being the diagonal of A_{ff}, letting $\alpha = \|D_{ff}^{1/2} A_{ff}^{-1} D_{ff}^{1/2}\|$, we have

$$\begin{aligned}
(A_{ff}(W_* - W)\mathbf{e}_c,\ (W_* - W)\mathbf{e}_c) &\le \alpha\ (A_{ff}(W_* - W)\mathbf{e}_c,\ D_{ff}^{-1} A_{ff}(W_* - W)\mathbf{e}_c) \\
&\le \alpha \sum_{i \in \mathcal{N}_c} \left\| \left(D_{ff}^{-(1/2)} A_{ff}(W_* - W)\mathbf{e}_c\right)\Big|_{\mathcal{A}_i} \right\|^2 \\
&= \alpha \sum_{i \in \mathcal{N}_c} \left\| D_{ff}^{-(1/2)} (A_{fc} + A_{ff} W)\mathbf{e}_c \right\|_{\mathcal{A}_i}^2.
\end{aligned}$$

Let now $\mathbf{e}_{c,\,i} = \text{const}\ \mathbf{1}_{c,\,i}$ be the average value of $\mathbf{e}_c$ restricted to $\Omega_{c,\,i}$. By the definition of average value, we then have

$$\|\mathbf{e}_c - \mathbf{e}_{c,\,i}\|_{\Omega_{c,\,i}} = \min_{t \in \mathbb{R}} \|\mathbf{e}_c - t\ \mathbf{1}_{c,\,i}\|_{\Omega_{c,\,i}}. \tag{6.52}$$

Then, because (see (6.50)) $(A_{ff} W \mathbf{e}_{c,\,i})|_{\mathcal{A}_i} = -(A_{fc}\mathbf{e}_{c,\,i})|_{\mathcal{A}_i}$, we get

$$(A_{ff}(W_* - W)\mathbf{e}_c,\ (W_* - W)\mathbf{e}_c) \leq \alpha \sum_{i \in \mathcal{N}_c} \|D_{ff}^{-(1/2)}(A_{fc} + A_{ff}W)(\mathbf{e}_c - \mathbf{e}_{c,\,i})\|^2_{\mathcal{A}_i}.$$

Let D_{cc} be the diagonal of A_{cc} and D_i be the diagonal of A restricted to any fixed neighborhood that contains $\Omega_{c,\,i}$. Then $\|D_{cc}|_{\Omega_{c,\,i}}\| \leq \|D_i\|$. Let $V(G)$ stand for the vector space of vectors defined on a given index set G. By definition, the notation $\|B\|_{V(G) \mapsto V(D)}$ stands for the norm of the mapping (matrix) B with domain $V(G)$ and range $V(D)$. With this notation and the choice of D_i, we then have

$$\begin{aligned}
&(A_{ff}(W_* - W)\mathbf{e}_c,\ (W_* - W)\mathbf{e}_c) \\
&\leq \alpha \max_{i \in \mathcal{N}_c} \|D_{ff}^{-(1/2)}(A_{fc} + A_{ff}W)D_{cc}^{-(1/2)}\|^2_{V(\Omega_{c,\,i}) \mapsto V(\mathcal{A}_i)} \\
&\quad \times \sum_{i \in \mathcal{N}_c} \|D_i\|\ \|\mathbf{e}_c - \mathbf{e}_{c,\,i}\|^2_{\Omega_{c,\,i}}.
\end{aligned} \tag{6.53}$$

Our goal is to bound the sum $\sum_{i \in \mathcal{N}_c} \|D_i\|\ \|\mathbf{e}_c - \mathbf{e}_{c,\,i}\|^2_{\Omega_{c,\,i}}$, in the case of Laplace-like discretization matrices A, by a constant times $(S\mathbf{e}_c,\ \mathbf{e}_c)$. This seems feasible because $\mathbf{e}_{c,\,i}$ is an average value of $\mathbf{e}_c$ restricted to $\Omega_{c,\,i}$. The constant will generally depend on the overlap of $\Omega_{c,\,i}$, which is assumed bounded. This fact is proven in the following section.

Boundedness of P assuming weak approximation property for Laplacian-like matrices

In this section, we assume that A can be assembled from local element matrices A_τ where the set of elements $\{\tau\}$ provides an overlapping partition of the global set of dofs $\mathcal{N}$. Note that the two-grid method based on a

$$P = \begin{bmatrix} W \\ I \end{bmatrix}$$

with W that satisfies the main equation (6.49) does not require the explicit knowledge of any element matrices A_τ.

Let $A_i^{(N)}$ be a local matrix assembled from element matrices for elements that cover a sufficiently large set $\overline{\Omega}_i$ such that $\Omega_{c,\,i} \subset \overline{\Omega}_i \cap \mathcal{N}_c$. Then, without loss of generality, we can assume that D_i is a principal submatrix of the diagonal of $A_i^{(N)}$ (if $\overline{\Omega}_i$ is sufficiently large).

Assume that the piecewise constant interpolant satisfies the local weak approximation property,

$$\|D_i\| \|\mathbf{e}_i - \overline{\mathbf{e}}_i\|^2_{\overline{\Omega}_i} \leq \delta\ (A_i^{(N)}\ \mathbf{e}_i,\ \mathbf{e}_i), \tag{6.54}$$

for any vector $\mathbf{e}_i$ supported in $\overline{\Omega}_i$. Here (and in what follows), for any vector $\mathbf{v}$ by $\overline{\mathbf{v}}$, we denote its average value over a given set. Here, $A_i^{(N)}$ stands for the local stiffness matrix corresponding to $\overline{\Omega}_i$ (viewed as union of fine-grid elements that cover the set of fine-grid dofs from $\overline{\Omega}_i$).

We choose $\overline{\Omega}_i$ as a set, union of fine-grid elements, such that $\overline{\Omega}_i \cap \mathcal{N}_c$ covers the set $\Omega_{c,\,i}$. Let $\overline{\Omega} = \cup_i \overline{\Omega}_i$ be the domain where the elliptic boundary value problem under consideration is posed. The matrix A then comes from a finite element approximation of an underlined elliptic PDE. Let $\kappa \geq 1$ be the maximal number of overlapping subdomains $\overline{\Omega}_i$, which is assumed bounded.

Given a vector $\mathbf{e}_c$ define $\mathbf{e}_f = -A_{ff}^{-1} A_{fc}\mathbf{e}_c$. Introduce also

$$\mathbf{e} = \begin{bmatrix} \mathbf{e}_f \\ \mathbf{e}_c \end{bmatrix}$$

and let

$$\mathbf{e}_{f,\,i} = -(A_{ff}^{-1} A_{fc}\mathbf{e}_c)|_{\overline{\Omega}_i \cap \mathcal{N}_f}, \quad \text{and} \quad \mathbf{e}_i = \mathbf{e}|_{\overline{\Omega}_i}.$$

For a fixed i define $t_{*,\,i} \in \mathbb{R}$ such that

$$\|\mathbf{e}_i - t_{*,\,i}\mathbf{1}_i\|^2_{\overline{\Omega}_i} = \min_{t\in\mathbb{R}} \left\| \begin{bmatrix} \mathbf{e}_{f,\,i} \\ \mathbf{e}_{c,\,i} \end{bmatrix} - t \begin{bmatrix} \mathbf{1}_{f,\,i} \\ \mathbf{1}_{c,\,i} \end{bmatrix} \right\|^2_{\overline{\Omega}_i} = \|\mathbf{e}_i - \overline{\mathbf{e}}_i\|^2_{\overline{\Omega}_i}.$$

It is clear then, from the definition (6.52) of average value and the assumption $\Omega_{c,\,i} \subset \overline{\Omega}_i \cap \mathcal{N}_c$, that

$$\begin{aligned}
\|\mathbf{e}_{c,\,i} - \overline{\mathbf{e}}_{c,\,i}\|^2_{\Omega_{c,\,i}} &\leq \|\mathbf{e}_{c,\,i} - t_{*,\,i}\mathbf{1}_{c,\,i}\|^2_{\Omega_{c,\,i}} \\
&\leq \|\mathbf{e}_{c,\,i} - t_{*,\,i}\mathbf{1}_{c,\,i}\|^2_{\Omega_{c,\,i}} + \|\mathbf{e}_{f,\,i} - t_{*,\,i}\mathbf{1}_{f,\,i}\|^2_{\overline{\Omega}_i \cap \mathcal{N}_f} \\
&\leq \|\mathbf{e}_{c,\,i} - t_{*,\,i}\mathbf{1}_{c,\,i}\|^2_{\overline{\Omega}_i \cap \mathcal{N}_c} + \|\mathbf{e}_{f,\,i} - t_{*,\,i}\mathbf{1}_{f,\,i}\|^2_{\overline{\Omega}_i \cap \mathcal{N}_f} \\
&= \min_t \left\| \begin{bmatrix} \mathbf{e}_{f,\,i} \\ \mathbf{e}_{c,\,i} \end{bmatrix} - t \begin{bmatrix} \mathbf{1}_{f,\,i} \\ \mathbf{1}_{c,\,i} \end{bmatrix} \right\|^2_{\overline{\Omega}_i} \\
&= \|\mathbf{e}_i - \overline{\mathbf{e}}_i\|^2_{\overline{\Omega}_i}.
\end{aligned} \tag{6.55}$$

Now, given a vector $\mathbf{e}_c$, define $\mathbf{e}_f = -A_{ff}^{-1} A_{fc}\mathbf{e}_c$ and let $\mathbf{e}_i$ be the restriction of

$$\mathbf{e} = \begin{bmatrix} \mathbf{e}_f \\ \mathbf{e}_c \end{bmatrix}$$

to $\overline{\Omega}_i$ for $i \in \mathcal{N}_c$. Then, based on (6.54) and the last estimate (6.55), we have

$$\begin{aligned}
\sum_i \|D_i\|\, \|\mathbf{e}_{c,\,i} - \overline{\mathbf{e}}_{c,\,i}\|^2_{\Omega_{c,\,i}} &\leq \sum_i \|D_i\|\, \|\mathbf{e}_i - \overline{\mathbf{e}}_i\|^2_{\overline{\Omega}_i} \\
&\leq \delta \sum_i \left(A_i^{(N)}\mathbf{e}_i,\ \mathbf{e}_i\right) \\
&\leq \kappa\, \delta\, (A\mathbf{e},\ \mathbf{e}) \\
&= \kappa\, \delta\, (S\mathbf{e}_c,\ \mathbf{e}_c).
\end{aligned}$$

Recall that

$$\mathbf{e}_f : A\mathbf{e} = A \begin{bmatrix} \mathbf{e}_f \\ \mathbf{e}_c \end{bmatrix} = \begin{bmatrix} 0 \\ S\mathbf{e}_c \end{bmatrix}$$

(see (6.51) for S). Also, $\kappa \geq 1$ stands for the maximal number of overlapping subdomains $\overline{\Omega}_j$ that share the coarse dof i.

Combining the last estimate with (6.53), we arrive at

$$\begin{aligned} &(A_{ff}(W_* - W)\mathbf{e}_c, \ (W_* - W)\mathbf{e}_c) \\ &\quad \leq \alpha \max_{i \in \mathcal{N}_c} \|D_{ff}^{-(1/2)}(A_{fc} + A_{ff}W)D_{cc}^{-(1/2)}\|^2_{V(\Omega_{c,\,i}) \mapsto V(\mathcal{A}_i)} \kappa \ \delta \ (S\mathbf{e}_c, \ \mathbf{e}_c). \end{aligned}$$

Now, introduce the "optimal" interpolation matrix

$$P_* = \begin{bmatrix} -A_{ff}^{-1} A_{fc} \\ I \end{bmatrix} \begin{matrix} \} \ \mathcal{N}_f \\ \} \ \mathcal{N}_c \end{matrix},$$

its sparse approximation

$$P = \begin{bmatrix} W \\ I \end{bmatrix} \begin{matrix} \} \ \mathcal{N}_f \\ \} \ \mathcal{N}_c \end{matrix},$$

and the restriction matrix $R = [0, \ I]$. We get the following deviation estimate.

$$\begin{aligned} (A(P_* - P)R\mathbf{e}, \ (P_* - P)R\mathbf{e}) &= (A_{ff}(W_* - W)\mathbf{e}_c, \ (W_* - W)\mathbf{e}_c) \\ &\leq \alpha \max_{i \in \mathcal{N}_c} \left\| D_{ff}^{-(1/2)}(A_{fc} + A_{ff}W) \right. \\ &\quad \left. \times \ D_{cc}^{-(1/2)} \right\|^2_{V(\Omega_{c,\,i}) \mapsto V(\mathcal{A}_i)} \kappa \ \delta \ (S\mathbf{e}_c, \ \mathbf{e}_c) \\ &\leq \alpha \max_{i \in \mathcal{N}_c} \left\| D_{ff}^{-(1/2)}(A_{fc} + A_{ff}W) \right. \\ &\quad \left. \times \ D_{cc}^{-(1/2)} \right\|^2_{V(\Omega_{c,\,i}) \mapsto V(\mathcal{A}_i)} \kappa \ \delta \ (A\mathbf{e}, \ \mathbf{e}). \end{aligned}$$

Thus, the following main result holds.

Theorem 6.21. *The projection PR is bounded in energy and the following estimate holds,*

$$(A(\mathbf{e} - PR\mathbf{e}), \mathbf{e} - PR\mathbf{e}) \leq K \ (A\mathbf{e}, \ \mathbf{e}),$$

where

$$K = 1 + \alpha \ \kappa \ \delta \max_{i \in \mathcal{N}_c} \left\| D_{ff}^{-(1/2)}(A_{fc} + A_{ff}W)D_{cc}^{-(1/2)} \right\|^2_{V(\Omega_{c,\,i}) \mapsto V(\mathcal{A}_i)},$$

and $\alpha = \|D_{ff}^{1/2} A_{ff}^{-1} D_{ff}^{1/2}\|$.

It is clear that the local constants $\|D_{ff}^{-(1/2)}(A_{fc} + A_{ff}W)D_{cc}^{-(1/2)}\|^2_{V(\Omega_{c,\,i})\mapsto V(\mathcal{A}_i)}$ can be bounded (or even made small) if $WI_{\Omega_{c,\,i}}$ approximates $-A_{ff}^{-1}A_{fc}I_{\Omega_{c,\,i}}$ on every local set $\mathcal{A}_i$. Here, $I_{\Omega_{c,\,i}}$ stands for the characteristic matrix (function) of the local coarse set $\Omega_{c,\,i}$.

In conclusion, assuming that $D_{ff}^{-(1/2)}A_{ff}D_{ff}^{-(1/2)}$ is well conditioned based on Theorem 3.25 used for

$$J = \begin{bmatrix} I \\ 0 \end{bmatrix} \begin{matrix} \} & \mathcal{N}_f \\ \} & \mathcal{N}_c \end{matrix},$$

the above Theorem 6.21 then implies the following TG convergence result.

Corollary 6.22. *The two-grid AMG method based on* P *and* ωD *(scaled Jacobi) smoother has a convergence factor* $\varrho_{TG} = 1 - 1/(K_{TG})$*, where* $K_{TG} \leq \omega/(\lambda_{\min}(D_{ff}^{-(1/2)}A_{ff}D_{ff}^{-(1/2)}))\,\|I - PR\|_A^2$*. The term* $\|I - PR\|_A^2 = \|PR\|_A^2$ *was bounded by the constant* K *in Theorem 6.21. Here,* $\omega \geq \|D^{-(1/2)}AD^{-(1/2)}\| \simeq \lambda_{\max}(D_{ff}^{-(1/2)}A_{ff}D_{ff}^{-(1/2)})$.

On the constrained trace minimization construction of P

Here, we study the boundedness of the factors

$$\|D_{ff}^{-(1/2)}(A_{fc} + A_{ff}W)D_{cc}^{-(1/2)}\|^2_{V(\Omega_{c,\,i})\mapsto V(\mathcal{A}_i)},$$

when W is constructed based on trace minimization of P^TAP (described in Section 6.3). In that case, the constraint reads $W\mathbf{1}_c = \mathbf{1}_f \equiv -A_{ff}^{-1}A_{fc}\mathbf{1}_c$, or equivalently $P\mathbf{1}_c = \mathbf{1}$. Note that our main assumption (6.49) holds then.

The constrained trace norm minimization problem (cf. Section 6.3) reads: Find

$$P = \begin{bmatrix} W \\ I \end{bmatrix} = (\boldsymbol{\psi}_i)_{i\in\mathcal{N}_c},$$

with prescribed sparsity pattern of W, such that

$$\text{trace}(P^TAP) \equiv \sum_{i\in\mathcal{N}_c} \boldsymbol{\psi}_i^TA\boldsymbol{\psi}_i \mapsto \min$$

subject to the constraint $P\mathbf{1}_c = \mathbf{1}$. The solution to this problem is given by (see Section 6.3) $\boldsymbol{\psi}_i = T_iT^{-1}\mathbf{1}$, where $T = \sum_{i\in\mathcal{N}_c} T_i$, $T_i = I_iA_i^{-1}I_i^T$, and I_i is the characteristic matrix (function) of the (prescribed fixed) support of $\boldsymbol{\psi}_i$. The matrix $A_i = I_i^TAI_i$ is the principal submatrix of A corresponding to the support of $\boldsymbol{\psi}_i$.

We next study the decay behavior of T^{-1} following a main result in [DMS84] (see also Section A.2.4 in the appendix). We first note that T is sparse. For example, we can partition T into blocks $T_{i,j}$ according to the aggregates $\mathcal{A}_i$. It is clear that for any i there are only a bounded number of indices j such that $T_{i,j} \neq 0$.

Given a matrix B, for any two vectors $\mathbf{v}_i$ and $\mathbf{v}_j$ corresponding to indices from two given sets (row indices) Ω_i and (column indices) $\mathcal{N}_j$ for the corresponding block $B_{i,j}$ of B, we then have

$$\mathbf{v}_i^T B_{i,j} \mathbf{v}_j = \left(\begin{bmatrix} 0 \\ \mathbf{v}_i \\ 0 \end{bmatrix} \begin{matrix} \} \\ \} \; \Omega_i \\ \} \end{matrix} \right)^T B \left(\begin{bmatrix} 0 \\ \mathbf{v}_j \\ 0 \end{bmatrix} \begin{matrix} \} \\ \} \; \mathcal{N}_j \\ \} \end{matrix} \right) \leq \|B\| \|\mathbf{v}_i\| \|\mathbf{v}_j\|.$$

That is, $\|B_{i,j}\| \leq \|B\|$. In the application below, we have $\mathcal{N}_j = \mathcal{A}_j$ (aggregate) whereas the row index set Ω_i is the support of $\boldsymbol{\psi}_i$ which is contained in a union of bounded number of aggregates $\mathcal{A}_k$.

Next, we apply the above-mentioned result from [DMS84] (or see Section A.2.4) in our setting. Let $B = T^{-1}$ and let $[\alpha, \beta]$ be an interval that contains the spectrum of T. For any polynomial p_k of degree $k \geq 0$, consider the matrix $B - p_k(T)$. Let Ω_i and $\mathcal{A}_j$ be at a large enough graph distance apart from each other so that for any pair of indices $i' \in \Omega_i$ and $j' \in \mathcal{A}_j$, the entry $(p_k(T))_{i',j'} = 0$. Note that $p_k(T)$ has the sparsity pattern of T^k. Then, we have

$$\|B_{i,j}\| \leq \|B - p_k(T)\| = \|T^{-1} - p_k(T)\| \leq \sup_{\lambda \in [\alpha, \beta]} |\lambda^{-1} - p_k(\lambda)|.$$

Because p_k is arbitrary, we also have

$$\|B_{i,j}\| \leq \min_{p_k} \sup_{\lambda \in [\alpha, \beta]} |\lambda^{-1} - p_k(\lambda)|.$$

Thus, the following simple upper bound holds,

$$\|B_{ij}\| \leq \frac{1}{\alpha} \inf_{p_k} \sup_{\lambda \in [\alpha, \beta]} |1 - \lambda p_k(\lambda)| = \frac{1}{\alpha} \frac{2q^{k+1}}{1 + q^{2(k+1)}}, \quad q = \frac{\sqrt{\kappa} - 1}{\sqrt{\kappa} + 1}, \quad \kappa = \frac{\beta}{\alpha}. \tag{6.56}$$

In the last estimate, we use

$$p_k : \; 1 - t p_k(t) = \frac{\mathcal{T}_{k+1}\left(\frac{\alpha + \beta - 2t}{\beta - \alpha}\right)}{\mathcal{T}_{k+1}\left(\frac{\alpha + \beta}{\beta - \alpha}\right)},$$

where $\mathcal{T}_{k+1}$ is the well-known Chebyshev polynomial of degree $k + 1$.

Thus, we have the geometric decay $\|B_{i,j}\| \leq (2/\alpha)\, q^{k+1}$, where k: $(T^k)_{i',j'} = 0$ for all $i' \in \Omega_i$ and $j' \in \mathcal{A}_j$. That is, if the graph distance between i and j denoted by

$d(i, j)$ is large, then $\|B_{i,j}\|$ is small. This shows that the estimate for $\boldsymbol{\psi}_i = T_i T^{-1}\mathbf{1}$ is

$$\begin{aligned}
\|\boldsymbol{\psi}_i\| &\leq \|A_i^{-1}\| \|I_i^T (T^{-1}\mathbf{1})\| \\
&\leq \|A_i^{-1}\| \left\| \sum_j B_{i,j}\, \mathbf{1}_j \right\| \\
&\leq \max_j \|\mathbf{1}_j\| \; \|A_i^{-1}\| \sum_j \|B_{i,j}\| \\
&\leq \max_j \|\mathbf{1}_j\| \; \|T\| \frac{2}{\alpha} \sum_j q^{d(i,j)} \\
&\leq \max_j \|\mathbf{1}_j\| \; \frac{2\beta}{\alpha} \sum_j q^{d(i,j)} \\
&\leq C \max_j \|\mathbf{1}_j\|.
\end{aligned} \tag{6.57}$$

It is clear that the constant C depends only on the decay rate of T^{-1}, which based on estimate (6.56) is seen to depend only on the condition number of T. Because for small Schwarz blocks A_i is spectrally equivalent to its diagonal part, it is clear then that T is spectrally equivalent to a diagonal matrix, and hence, T (up to a diagonal scaling) is well conditioned, which we assume.

Consider $W^T = (p_{j,\,i})_{j \in \mathcal{N}_c,\, i \in \mathcal{N}_f}$, where $p_{j,\,i} \neq 0$ for a bounded number of indices $i \in \Omega_j$. We have, for any vector $\mathbf{v} = (v_i)$,

$$\begin{aligned}
\|W^T \mathbf{v}\|^2 &= \sum_j \Big(\sum_{i \in \Omega_j} p_{j,\,i} v_i \Big)^2 \\
&\leq \sum_j \sum_{i \in \Omega_j} p_{j,\,i}^2 \sum_{i \in \Omega_j} v_i^2 \\
&\leq \max_j \Big(\sum_{i \in \Omega_j} p_{j,i}^2 \Big) \sum_j \sum_{i \in \Omega_j} v_i^2 \\
&\leq \max_j \|\boldsymbol{\psi}_j\|^2 \; C \; \|\mathbf{v}\|^2,
\end{aligned}$$

where the constant C depends on the sparsity pattern of W.

Thus, based on the last estimate, $\|W\| = \|W^T\| \leq C \; \max_i \; \|\boldsymbol{\psi}_i\|$ (with a constant C that depends only on the sparsity of W), estimate (6.57), and the triangle inequality, we get the required boundedness

$$\begin{aligned}
&\|D_{ff}^{-(1/2)} (A_{fc} + A_{ff} W) D_{cc}^{-(1/2)}\|_{V(\Omega_{c,\,i}) \mapsto V(\mathcal{A}_i)} \\
&\leq \|D_{ff}^{-(1/2)} A_{fc} D_{cc}^{-(1/2)}\|_{V(\Omega_{c,\,i}) \mapsto V(\mathcal{A}_i)} + \|D_{ff}^{-(1/2)} A_{ff} D_{ff}^{-(1/2)}\|_{V(\overline{\Omega}_i) \mapsto V(\overline{\Omega}_i)} \\
&\quad \times \|D_{ff}^{-(1/2)} W D_{cc}^{-(1/2)}\|_{V(\Omega_{c,\,i}) \mapsto V(\overline{\Omega}_i)} \\
&\leq \kappa + C\|W\|,
\end{aligned}$$

where the constant κ depends only on the sparsity of A (see Proposition 1.1). Therefore, we proved the following corollary.

Corollary 6.23. *The constrained trace minimization construction leads to a P that has bounded block W, which leads to a bounded constant K in Theorem 6.21. This is under the assumption of $D_{ff}^{-(1/2)} A_{ff} D_{ff}^{-(1/2)}$ being well conditioned, the basis functions (columns of P) ψ_i having bounded support with bounded overlap, which hence leads to a well-conditioned additive Schwarz operator T.*

Remark 6.24. Note that based on Proposition 6.2, it is feasible to prove two-grid convergence for M-matrices A coming from second-order diffusion type elliptic equations. Then $A\mathbf{1} = 0$ for the interior dofs i, where $\mathbf{1} = [1,\ 1, \ldots, 1]^T$ is the constant vector. That is, the classical (Ruge–Stüben) AMG interpolation, is vector preserving (in the interior of the domain) and for such problems we can ensure the local weak approximation property (6.54).

6.13 The result of Vaněk, Brezina, and Mandel

The smoothed aggregation (or SA) algebraic MG method was proposed by Petr Vaněk in [VSA] who was motivated by some early work on aggregation-based MG studied by R. Blaheta in [Bl86] and in his dissertation [Bl87]; see also the more recent paper [DB95].

We present here perhaps the only known multilevel convergence result for the algebraic multigrid, namely, the suboptimal convergence of the smoothed aggregation AMG. The original proof is found in [SA].

The construction of coarse bases exploits smoothing of pieces of a null space vector of a given sparse positive semidefinite matrix A. The pieces of the vector correspond to a number of sets, called aggregates. These aggregates form a nonoverlapping partition of the original set of degrees of freedom, $\mathcal{N}$. The aggregates are assumed to satisfy certain properties that later reflect the sparsity of the coarse-level matrices. At every coarsening level $k \geq 0$, in order to construct the coarse $(k+1)$th-level basis, we use certain Chebyshev-like optimal polynomials with argument the matrix A_k (diagonally scaled) applied to every piece of the respectively partitioned vector. Due to the properties of the matrix polynomial, the thus-constructed local basis vectors have a guaranteed energy bound, and the resulting ℓ levels smoothed aggregation $V(1, 1)$-cycle method has a provable convergence factor bounded by $1 - C/\ell^3$.

6.13.1 Null vector-based polynomially smoothed bases

This section illustrates a typical construction of polynomially smoothed bases utilizing null vectors. We note that after this introductory section, the vectors that are used to construct the SA method need not necessarily be null vectors of a given symmetric positive semidefinite matrix. One of the main assumptions will be a "weak approximation property" of certain coarse spaces of piecewise constant vectors. Namely, that a finite element function v can be approximated in L_2 by a piecewise constant interpolant $I_H v$ with order $\mathcal{O}(H)$. The interpolant is defined based on sets $\mathcal{A}_i$ (referred to as aggregates) with diameter $\mathcal{O}(H)$. On each such set $\mathcal{A}_i$, $I_H v$ is constant equal to an average value of v over $\mathcal{A}_i$; for example, we can set $I_H v = 1/|\mathcal{A}_i| \int_{\mathcal{A}_i} v \, dx$. Then, if A comes from a Laplace-like discrete problem, the following is a standard estimate in L_2 in terms of the energy norm $\|.\|_A$,

$$\|v - I_H v\|_0 \leq c_a H \; \|v\|_A.$$

Rewriting this in terms of coefficient vectors leads to the following one

$$h^{d/2} \; \|\mathbf{v} - \underline{I_H v}\| \leq c_a H \; \|\mathbf{v}\|_A,$$

where $d = 2$ or $d = 3$ stands for the dimension of the (geometrical) domain where the corresponding PDE (Laplacian-like) is posed. Then, because $\|A\| \simeq h^{d-2}$ (see Proposition 1.3), we arrive at

$$\|\mathbf{v} - \underline{I_H v}\| \leq c_a \frac{H}{h} \, \frac{1}{\|A\|^{1/2}} \; \|\mathbf{v}\|_A.$$

In the application of the SA, we have $(H/h) \simeq (2\nu + 1)^{k+1}$, where $\nu \geq 1$ is the polynomial degree of the polynomial used to smooth out the piecewise constant interpolants with which we start. Also, $k = 0, 1, \ldots, \ell$ stands for the coarsening level. We summarize this estimate as our main assumption. Given are the nonoverlapping sets $\mathcal{A}_i^{(k)}$ (aggregates) at coarsening level $k \geq 0$ that we view as sets of fine-grid dofs. Let $\overline{Q}_k$ be the block-diagonal ℓ_2-projection that for every vector $\mathbf{v}$ restricted to an aggregate $\mathcal{A}_i^{(k)}$ assigns a scalar value $\overline{\mathbf{v}}_i$, the average of $\mathbf{v}|_{\mathcal{A}_i}$ over $\mathcal{A}_i$. Finally, let I_k interpolate them back all the way up to the finest-level as constants over $\mathcal{A}_i^{(k)}$ (equal to the average value $\overline{\mathbf{v}}_i$). Finally, assume that the diameter of $\mathcal{A}_i^{(k)}$ is of order $(2\nu + 1)^{k+1} h$ where h is the finest mesh-size. Then, the following approximation property is our main assumption

$$\|\mathbf{v} - I_k \overline{Q}_k \mathbf{v}\| \leq c_a \frac{(2\nu + 1)^{k+1}}{\|A\|^{1/2}} \; \|\mathbf{v}\|_A. \tag{6.58}$$

The latter assumption is certainly true if the matrix A comes from elliptic PDEs discretized on a uniformly refined mesh, and the corresponding aggregates at every level k are constructed based on the uniform hierarchy of the geometric meshes. In the applications, when we have access to the fine-grid matrix only (and possibly to the fine-grid mesh) when constructing the hierarchy of aggregates, we have to follow the rule that they are "quasiuniform" in the sense that their graph diameter grows as $(2\nu + 1)^{k+1}$. A common choice in practice is $\nu = 1$.

Construction of locally supported basis by SA

To illustrate the method, we assume in the present section that A is a given symmetric positive semidefinite matrix, and let $\mathbf{1}$ be a given null vector of A; that is, $A\mathbf{1} = 0$.

The method is applied to a matrix A_0 that is obtained from A (after certain boundary conditions are imposed).

For a given integer $\nu \geq 1$, partition the set of degrees of freedom of A, that is, the fine-grid into nonoverlapping sets $\mathcal{A}_i$ such that $\mathcal{A}_i$ contains an index i with the following polynomial property. Namely, for any integer $s \leq \nu$, the entries of $(A^s)_{ij}$ away from i are zero. More specifically, we assume

$$(A^s)_{ij} = 0 \quad \text{for all indices } j \text{ outside } \mathcal{A}_i. \tag{6.59}$$

Let $\mathbf{1}_i = \mathbf{1}|_{\mathcal{A}_i}$ and extend it by zero outside $\mathcal{A}_i$. It is clear that

$$\sum_i \mathbf{1}_i = \mathbf{1}. \tag{6.60}$$

For a given diagonal matrix D (specified later on), let $\overline{A} = D^{-(1/2)}AD^{-(1/2)}$.

The method utilizes a polynomial of degree $\nu \geq 1$ (also, specified later on) $\varphi_\nu(t) = 1 - tq_{\nu-1}(t)$. Note that $\varphi_\nu(0) = 1$.

Sometimes we use the notation $\mathbf{v}(x_i)$ to denote the ith entry of the vector $\mathbf{v}$. The latter notation is motivated by the fact that very often in practice the vectors $\mathbf{v}$ are coefficient vectors of functions v expanded in terms of a given (e.g., Lagrangian finite element) basis.

Define now,

$$\boldsymbol{\psi}_i = (I - D^{-1}Aq_{\nu-1}(D^{-1}A))\mathbf{1}_i. \tag{6.61}$$

We have

$$\sum_i \boldsymbol{\psi}_i = (I - D^{-1}Aq_{\nu-1}D^{-1}A)\sum_i \mathbf{1}_i = (I - D^{-1}Aq_{\nu-1}(D^{-1}A))\mathbf{1} = \mathbf{1}, \tag{6.62}$$

because $A\mathbf{1}=0$. Also, $\mathbf{1}(x_i) = \sum_j (\boldsymbol{\psi}_j)(x_i) = (\mathbf{1}_i)(x_i) - (D^{-1}Aq_{\nu-1}(D^{-1}A)\mathbf{1}_i)(x_i)$, because $(A^s\mathbf{1}_j)_i = 0$, for all $s \leq \nu$ and $j \neq i$. The latter implies

$$(D^{-1}Aq_{k-1}(D^{-1}A)\mathbf{1}_i)(x_i) = 0,$$

and hence

$$\boldsymbol{\psi}_i(x_i) = \mathbf{1}(x_i).$$

The vectors $\boldsymbol{\psi}_i$ form our coarse basis. Note that these have local support and form a partition of unity (in the sense of identity (6.62)) and they also provide a Lagrangian basis. Thus, we can at least expect reasonable two-grid convergence (cf., Theorem 6.21).

To continue the process by recursion, define $\mathbf{1}_c = [1, \ldots, 1]^T \in \mathbb{R}^{n_c}$. We have, $A_c\mathbf{1}_c = P^T A \sum_i \boldsymbol{\psi}_i = P^T A\mathbf{1} = 0$. Here $P = [\boldsymbol{\psi}_1, \ldots, \boldsymbol{\psi}_{n_c}]$ is the interpolation matrix. Due to the Lagrangian property of the basis $\{\boldsymbol{\psi}_i\}$ (i.e., $\boldsymbol{\psi}_i(x_j) = \delta_{ij}$) it follows that P has a full column rank.

We then generate coarse aggregates with the corresponding polynomial property (6.59). Note that we have the flexibility to change ν (i.e., to have $\nu = \nu_k$ depending on the level number). This choice, however, is not considered in what follows. That is, we assume that $\nu \geq 1$ is fixed independently of k. The classical SA method corresponds to the choice $\nu = 1$.

Assume that we have generated $\ell \geq 1$ levels and at every level k, we have constructed the respective interpolation matrices P_k. Then, after a proper choice of smoothers M_k, we end up with a symmetric $V(1,1)$-cycle smoothed aggregation AMG. Our goal is to analyze the method by only assuming that the vector $\mathbf{1}$ ensures the multilevel approximation property (6.58). The fact that it is in the (near)-null space of A is not needed. That is why the resulting coarse bases are not necessarily Lagrangian. Nevertheless, convergence is guaranteed as we show next.

6.13.2 Some properties of Chebyshev-like polynomials

Consider the Chebyshev polynomials $T_k(t)$ defined by recursion as follows, $T_0 = 1$, $T_1(t) = t$, and for $k \geq 1$, $T_{k+1}(t) = 2tT_k(t) - T_{k-1}(t)$. Letting $t = \cos\alpha \in [-1, 1]$, we have the explicit representation $T_k(t) = \cos k\alpha$, which is seen from the trigonometric identity $\cos(k+1)\alpha + \cos(k-1)\alpha = 2\cos\alpha \cos k\alpha$.

We now prove some properties of T_k that are needed in the analysis of the SA method. The polynomial of main interest that we introduce in (6.63) below was used in [BD96] (see also [Br99]). Similar polynomials were used in [Sh95] (p. 133).

Proposition 6.25. *We have the expansion* $T_{2k+1}(t) = c_{2k+1}t + tQ_k(t^2)$, $c_{2k+1} = (-1)^k(2k+1)$, *for* $k \geq 0$, *where* Q_k *is a polynomial of degree k such that* $Q_k(0) = 0$. *Similarly,* $T_{2k}(t) = (-1)^k + P_k(t^2)$, *where* P_k *is a polynomial of degree k such that* $P_k(0) = 0$.

Proof. We have $T_1 = t$, $T_2 = 2tT_1 - T_0 = 2t^2 - 1$, and $T_3 = 2tT_2 - T_1 = 2t(2t^2-1) - t = 4t^3 - 3t$. That is, assume by induction that for $k \geq 1$, $T_{2k-1}(t) = c_{2k-1}t + tQ_{k-1}(t^2)$ and $T_{2k}(t) = (-1)^k + P_k(t^2)$ for some polynomials Q_{k-1} and P_k of respective degrees $k-1$ and k, and such that $Q_{k-1}(0) = 0$ and $P_k(0) = 0$. Then, from $T_{2k+1} = 2tT_{2k} - T_{2k-1}$, we get

$$\begin{aligned} T_{2k+1} &= 2t((-1)^k + P_k(t^2)) - (-1)^{k-1}(2k-1)t - tQ_{k-1}(t^2) \\ &= (-1)^k(2k+1)t + t\big(2P_k(t^2) - Q_{k-1}(t^2)\big). \end{aligned}$$

That is, the induction assumption for T_{2k+1} is confirmed with $Q_k(t) = 2P_k - Q_{k-1}$, and hence, $Q_k(0) = 0$. Similarly, for T_{2k+2}, we have

$$\begin{aligned} T_{2k+2} &= -T_{2k} + 2tT_{2k+1} \\ &= -(-1)^k - P_k(t^2) + 2t((-1)^k(2k+1)t + tQ_k(t^2)) \\ &= (-1)^{k+1} + (2(-1)^k(2k+1)t^2 + 2t^2Q_k(t^2) - P_k(t^2)). \end{aligned}$$

The latter confirms the induction assumption for T_{2k+2} with $P_{k+1}(t^2) = 2(-1)^k(2k+1)t^2 + 2t^2Q_k(t^2) - P_k(t^2)$ and hence $P_{k+1}(0) = 0$. □

Proposition 6.26. *The following estimate holds for any $t \in [0, 1]$,*

$$|T_{2k+1}(t)| \le (2k+1)t.$$

Proof. Note that for $t = \cos\alpha \in [-1, 1]$, $|T_k(t)| = |\cos k\alpha| \le 1$. Therefore, assuming by induction that $|T_{2k-1}(t)| \le (2k-1)t$ for $t \in [0, 1]$, we have

$$|T_{2k+1}(t)| = |2tT_{2k}(t) - T_{2k-1}(t)| \le 2t + (2k-1)t = (2k+1)t,$$

which confirms the induction assumption. □

Proposition 6.27. *For a given $b > 0$, consider for $t \in [0, b]$ the function*

$$\varphi_\nu(t) = (-1)^\nu \frac{1}{2\nu+1} \frac{\sqrt{b}}{\sqrt{t}} T_{2\nu+1}\left(\frac{\sqrt{t}}{\sqrt{b}}\right). \tag{6.63}$$

We have that $\varphi_\nu(t)$ is a polynomial of degree ν such that $\varphi_\nu(0) = 1$; that is, $\varphi_\nu(t) = 1 - tq_{\nu-1}(t)$ for some polynomial $q_{\nu-1}(t)$ of degree $\nu - 1$.

Proof. For $\nu = 0$, $\varphi_\nu = 1$. Consider the case $\nu \ge 1$. Due to Proposition 6.25, we have with $\lambda = \sqrt{t/b} \in [0, 1]$, that $\varphi_\nu(t) = (1/(c_{2\nu+1}))(1/\lambda)\, \lambda(c_{2\nu+1} + Q_\nu(\lambda)) = 1 - \lambda q_{\nu-1}(\lambda)$, because $Q_\nu(0) = 0$ hence $(1/(c_{2\nu+1}))Q_\nu(\lambda) = -\lambda q_{\nu-1}(\lambda)$ for some polynomial $q_{\nu-1}(\lambda)$ of degree $\nu - 1$. That is, we showed that $\varphi_\nu(t)$ as defined in (6.63) is a polynomial of degree ν such that $\varphi_\nu(0) = 1$. □

Proposition 6.28. *The polynomial φ_ν defined in (6.63) has the following optimality property.*

$$\min_{p_\nu:\ p_\nu(0)=1}\ \max_{t\in[0,\ b]} |\sqrt{t}\ p_\nu(t)| = \max_{t\in[0,\ b]} |\sqrt{t}\ \varphi_\nu(t)| = \frac{\sqrt{b}}{2\nu+1}. \tag{6.64}$$

We have $\varphi_\nu(0) = 1$ and also

$$\max_{t\in[0,\ b]} |\varphi_\nu(t)| = 1. \tag{6.65}$$

Proof. The first fact follows from the optimality property of the Chebyshev polynomials, because letting $\lambda = \sqrt{t/b} \in [0, 1]$ $\sqrt{t}\varphi_\nu(t)$ equals $T_{2\nu+1}(\lambda)$ times a constant.

The fact that $|\varphi_\nu(t)| \le 1$ follows from Proposition 6.26. □

Here are some particular cases of the polynomials φ_ν.

Using the definition of the Chebyshev polynomials, $T_0 = 1$, $T_1 = t$, $T_{k+1} = 2tT_k - T_{k-1}$, for $k \ge 1$, we get $T_2 = 2t^2 - 1$ and hence

$$T_3(t) = 4t^3 - 3t.$$

Thus,

$$\varphi_1(t) = -\frac{1}{3}\sqrt{b}\left(4\frac{t}{b^{3/2}} - \frac{3}{\sqrt{b}}\right) = 1 - \frac{4}{3}\frac{t}{b}.$$

This in particular shows that

$$\sup_{t\in(0,\, b]} \frac{|1-\varphi_1(t)|}{\sqrt{t}} = \frac{4}{3}\frac{1}{\sqrt{b}}.$$

The next polynomial is based on $T_5 = 2tT_4 - T_3 = 2t(2tT_3 - T_2) - T_3 = (4t^2 - 1)(4t^3 - 3t) - 4t^3 + 2t = 16t^5 - 20t^3 + 5t$. Therefore,

$$\varphi_2(t) = \frac{1}{5}\sqrt{\frac{b}{t}}\left(16\sqrt{t}t^2\frac{1}{b^{5/2}} - 20\sqrt{t}t\frac{1}{b^{3/2}} + 5\sqrt{t}\frac{1}{\sqrt{b}}\right).$$

This shows,

$$\varphi_2(t) = \frac{16}{5}\frac{t^2}{b^2} - 4\frac{t}{b} + 1.$$

We also have,

$$\sup_{t\in(0,\, b]} \frac{1-\varphi_2(t)}{\sqrt{t}} = \frac{4}{\sqrt{b}} \sup_{x\in(0,1]} \left(x - \frac{4}{5}x^3\right) = \frac{4}{3}\sqrt{\frac{5}{3}}\frac{1}{\sqrt{b}}.$$

In general, it is clear that the following result holds.

Proposition 6.29. *There is a constant C_ν independent of b such that the following estimate holds,*

$$\sup_{t\in(0,\, b]} \frac{|1-\varphi_\nu(t)|}{\sqrt{t}} \leq C_\nu \frac{1}{b^{1/2}}. \tag{6.66}$$

Proof. We have $1 - \varphi_\nu(t) = tq_{\nu-1}(t)$, that is, $(1-\varphi_\nu)/\sqrt{t} = \sqrt{t}\; q_{\nu-1}(t)$ and therefore the quotient in question is bounded for $t \in (0, b]$. More specifically, the following dependence on b is seen,

$$\sup_{t\in(0,\, b]} \frac{|1-\varphi_\nu(t)|}{\sqrt{t}} = \frac{1}{b^{1/2}} \sup_{\lambda\in(0,\, 1]} \frac{\left|1 - \frac{(-1)^\nu}{2\nu+1}\frac{T_{2\nu+1}(\sqrt{\lambda})}{\sqrt{\lambda}}\right|}{\sqrt{\lambda}}.$$

Clearly, the constant

$$C_\nu = \sup_{\lambda\in(0,\, 1]} \frac{\left|1 - \frac{(-1)^\nu}{2\nu+1}\frac{T_{2\nu+1}(\sqrt{\lambda})}{\sqrt{\lambda}}\right|}{\sqrt{\lambda}}$$

is independent of b. $\square$

6.13.3 A general setting for the SA method

In this section, we select the parameters of the smoothed aggregation method.

To simplify the analysis, we assume that $\nu \geq 1$ is independent of k. We assume that we are given a set of block-diagonal matrices $\overline{I}_{k-1} : \mathbb{R}^{n_k} \mapsto \mathbb{R}^{n_{k-1}}$,

$$\overline{I}_{k-1} = \begin{bmatrix} \mathbf{1}_1 & 0 & 0 & \dots & 0 \\ 0 & \mathbf{1}_2 & 0 & \dots & 0 \\ \vdots & \ddots & \ddots & \ddots & \vdots \\ 0 & \dots & 0 & \mathbf{1}_{n_k-1} & 0 \\ 0 & 0 & \dots & 0 & \mathbf{1}_{n_k} \end{bmatrix} \begin{matrix} \} \; \mathcal{A}_1 \\ \} \; \mathcal{A}_2 \\ \} \; \vdots \\ \} \; \mathcal{A}_{n_k-1} \\ \} \; \mathcal{A}_{n_k} \end{matrix}$$

where, for $k > 1$,

$$\mathbf{1}_i = \begin{bmatrix} 1 \\ \vdots \\ 1 \end{bmatrix}.$$

Note that the vector $\mathbf{1}_i \in \mathbb{R}^{|\mathcal{A}_i|}$ has as many entries of ones as the size of the fine-grid set (called aggregate) $\mathcal{A}_i$ to which they interpolate. Once the first piecewise-constant interpolant $\overline{I}_0$ is specified, then the SA method is well defined. We outlined, in the first section, a choice of $\overline{I}_0$ based on a null vector of A. In practice, we can select other initial coarse-level interpolants, for example, ones that can fit several a priori given vectors, such as the rigid body modes in the case of elasticity problem.

Let $\overline{I}_{k-1}$ be the piecewise constant interpolant from level k to level $k-1$, and let $I_{k-1} = \overline{I}_0 \cdots \overline{I}_{k-1}$ be the composite one. We define $D_k = I_{k-1}^T I_{k-1}$. Denote then $\overline{A}_{k-1} = D_{k-1}^{-(1/2)} A_{k-1} D_{k-1}^{-(1/2)}$. Then, the interpolation matrix P_{k-1} is constructed as before on the basis of A_{k-1}, D_{k-1} and the norm of $\overline{A}_{k-1}$ for our fixed ν. More specifically, we have

$$P_{k-1} = S_{k-1} \overline{I}_{k-1},$$

where

$$S_{k-1} = \varphi_\nu \left(D_{k-1}^{-1} A_{k-1} \right),$$

and

$$\varphi_\nu(t) = (-1)^\nu \frac{1}{2\nu+1} \frac{\sqrt{b}}{\sqrt{t}} T_{2\nu+1} \left(\frac{\sqrt{t}}{\sqrt{b}} \right) \quad \text{for } b = b_{k-1} \geq \|\overline{A}_{k-1}\|.$$

We show later (in Lemma 6.30) that $b = b_{k-1} \leq \|A\| / ((2\nu+1)^{2(k-1)})$.

The smoother M_k is chosen such that

$$M_k \simeq \|\overline{A}_k\| \, D_k.$$

More specifically, we assume that, M_k is s.p.d. and spectrally equivalent to the diagonal matrix $\|\overline{A}_k\| \; D_k$, and scaled so that,

$$\mathbf{v}^T A_k \mathbf{v} \leq \|\overline{A}_k\| \mathbf{v}^T D_k \mathbf{v} \leq \mathbf{v}^T M_k \mathbf{v}. \tag{6.67}$$

Based on the above choice of P_k, A_k, and M_k, for $0 \leq k \leq \ell$, starting with $B_\ell = A_\ell$, for $k = \ell - 1, \ldots, 1, \; 0$, we recursively define a V-cycle preconditioner B_k to A_k in the following standard way,

$$I - B_k^{-1} A_k = \left(I - M_k^{-T} A_k\right)\left(I - P_k B_{k+1}^{-1} P_k^T A_k\right)\left(I - M_k^{-1} A_k\right).$$

Letting $B = B_0$, we are concerned in what follows with the (upper) bound K_* in the estimate

$$\mathbf{v}^T A \mathbf{v} \leq \mathbf{v}^T B \mathbf{v} \leq K_\star \, \mathbf{v}^T A \mathbf{v}. \tag{6.68}$$

Preliminary estimates

Our second main assumption is that we can construct at every level $k \geq 1$ aggregates with the polynomial property (6.59). The latter is needed to keep the sparsity pattern of the resulting coarse matrices under control. We also assume that the composite aggregates are quasiuniform, that is, that the size of the composite aggregates coming from level k onto the finest level satisfy the estimate:

$$\max_{i \in \mathcal{N}_k} |\mathcal{A}_i| \simeq \min_{i \in \mathcal{N}_k} |\mathcal{A}_i| \simeq (2\nu + 1)^{dk}. \tag{6.69}$$

Here $d \geq 1$ is the dimension of the grids $\mathcal{N}_k$. As already mentioned, the above assumption is easily met in practice for meshes that are obtained by uniform refinement. For more general unstructured finite element meshes, this assumption is only a rule on how to construct the coarse-level aggregates. Assumption (6.69) implies that the diagonal matrices D_k are uniformly well conditioned; that is,

$$\min_{i \in \mathcal{N}_k} |\mathcal{A}_i| \leq \lambda_{\min}(D_k) \simeq \lambda_{\max}(D_k) \leq \max_{i \in \mathcal{N}_k} |\mathcal{A}_i| \simeq (2\nu + 1)^{dk}, \tag{6.70}$$

and hence

$$\left\| D_k^{-(1/2)} \right\| \left\| D_k^{1/2} \right\| \leq \kappa_D \simeq 1. \tag{6.71}$$

Our analysis closely follows [SA].

Lemma 6.30. *The following main estimate holds,*

$$\|A_k\| \leq \|D_k\| \frac{\|A\|}{(2\nu + 1)^{2k}} \simeq (2\nu + 1)^{(d-2)k} \; \|A\|,$$

assuming that the composite aggregates are quasiuniform (i.e., estimates (6.70)–(6.71)*).*

Proof. Recall that $D_{k+1} = I_k^T I_k$. Then, with $S_k = I - D_k^{-1} A_k q_\nu (D_k^{-1} A_k)$, using the fact that $P_k = S_k \overline{I}_k$ and $D_{k+1} = \overline{I}_k^T D_k \overline{I}_k$, we have

$$\left\| D_{k+1}^{-(1/2)} A_{k+1} D_{k+1}^{-(1/2)} \right\| = \sup_{\mathbf{v}} \frac{\mathbf{v}^T A_{k+1} \mathbf{v}}{\mathbf{v}^T D_{k+1} \mathbf{v}}$$
$$= \sup_{\mathbf{v}} \frac{\mathbf{v}^T \overline{I}_k^T S_k^T A_k S_k \overline{I}_k \mathbf{v}}{(\overline{I}_k \mathbf{v})^T D_k (\overline{I}_k \mathbf{v})}$$
$$\leq \sup_{\mathbf{v}} \frac{\mathbf{v}^T S_k^T A_k S_k \mathbf{v}}{\mathbf{v}^T D_k \mathbf{v}}.$$

Therefore, based on property (6.64) of φ_ν, we get

$$\mathbf{v}^T D_k^{-(1/2)} S_k^T A_k S_k D_k^{-(1/2)} \mathbf{v} \leq \sup_{t \in \left[0,\ \|D_k^{-(1/2)} A_k D_k^{-(1/2)}\|\right]} t(1 - t q_{\nu-1}(t))^2 \, \|\mathbf{v}\|^2$$
$$\leq \frac{\|D_k^{-(1/2)} A_k D_k^{-(1/2)}\|}{(2\nu+1)^2} \, \|\mathbf{v}\|^2.$$

That is, by recursion (with $D_0 = I$, $A_0 = A$), we end up with the estimate

$$\left\| D_{k+1}^{-(1/2)} A_{k+1} D_{k+1}^{-(1/2)} \right\| \leq \frac{\|A\|}{(2\nu+1)^{2(k+1)}}.$$

We conclude with the estimates

$$\|A_{k+1}\| \leq \left\| D_{k+1}^{1/2} \right\|^2 \left\| D_{k+1}^{-(1/2)} A_{k+1} D_{k+1}^{-(1/2)} \right\|$$
$$\leq \frac{\|A\|}{(2\nu+1)^{2(k+1)}} \left\| D_{k+1}^{1/2} \right\|^2 \simeq (2\nu+1)^{(d-2)(k+1)} \, \|A\|.$$

The latter inequality is based on the assumption that the composite aggregates are quasiuniform (see (6.70)). Thus the proof is complete. □

We use the main result regarding the relative spectral condition number of the ℓth-level V-cycle preconditioner B with respect to A given by Theorem 5.9, which we restate here.

Assume that smoothers M_j, interpolation matrices P_j, and respective coarse matrices related as $A_{j+1} = P_j^T A_j P_j$ are given. Each smoother M_j is such that $M_j^T + M_j - A_j$ is s.p.d.. Then, the following main identity holds

$$\mathbf{v}^T A \mathbf{v} \leq \mathbf{v}^T B \mathbf{v} = \inf_{(\mathbf{v}_k)} \left[\mathbf{v}_\ell^T A_\ell \mathbf{v}_\ell + \sum_{j<\ell} \left(M_j^T \mathbf{v}_j^f + A_j P_j \mathbf{v}_{j+1}\right)^T \right.$$
$$\left. \times \left(M_j^T + M_j - A_j\right)^{-1} \left(M_j^T \mathbf{v}_j^f + A_j P_j \mathbf{v}_{j+1}\right) \right]. \tag{6.72}$$

The inf here is taken over the components $(\mathbf{v}_k)$ of all possible decompositions of $\mathbf{v}$ defined as follows:

(i) Starting with $\mathbf{v}_0 = \mathbf{v}$.
(ii) For $k \geq 0$, $\mathbf{v}_k = \mathbf{v}_k^f + P_k \mathbf{v}_{k+1}$.

Introduce now the following averaging operators,

$$\overline{Q}_{k-1} = \left(I_{k-1}^T I_{k-1}\right)^{-1} I_{k-1}^T \; : \; \mathbb{R}^{n_0} \mapsto \mathbb{R}^{n_k}. \tag{6.73}$$

Note that $I_{k-1}\overline{Q}_{k-1}$ are ℓ_2-orthogonal projections.

We are interested in a particular recursive decomposition for any given fine-grid vector $\mathbf{v}$. Based on the characterization identity (6.72) utilizing an energy stable particular decomposition of the fine-grid vectors, we can get an upper bound of K_*, which is our goal. Introduce $\overline{Q}_{-1} = I$, and for $k \geq 0$, let $\mathbf{v}_k = \overline{Q}_{k-1}\mathbf{v} \in \mathbb{R}^{n_k}$. We have the recursive two-level decomposition

$$\mathbf{v}_k = \left(\overline{Q}_{k-1}\mathbf{v} - P_k \overline{Q}_k \mathbf{v}\right) + P_k \overline{Q}_k \mathbf{v} = \mathbf{v}_k^f + P_k \mathbf{v}_{k+1}.$$

In order to bound the relative condition number of the V-cycle preconditioner B with respect to A (due to estimate (6.72)), based on our choice of the smoother as in (6.67), it is sufficient to bound the expressions (i) and (ii) below:

(i) $\sum_{k<\ell} (\mathbf{v}_k^f)^T M_k \mathbf{v}_k^f = \sum_{k<\ell} (\overline{Q}_{k-1}\mathbf{v} - P_k \overline{Q}_k \mathbf{v})^T M_k (\overline{Q}_{k-1}\mathbf{v} - P_k \overline{Q}_k \mathbf{v})$,
(ii) $\sum_{k\leq\ell} \mathbf{v}_k^T A_k \mathbf{v}_k = \sum_{k<\ell} \mathbf{v}^T \overline{Q}_{k-1}^T A_k \overline{Q}_{k-1} \mathbf{v}$,

both in terms of $\mathbf{v}^T A \mathbf{v}$.

Estimating the first sum (i)

Recall that $P_k = S_k \overline{I}_k$, $S_k = I - D_k^{-1} A_k q_{\nu-1}(D_k^{-1} A_k)$, $I_k = \overline{I}_0 \overline{I}_1 \cdots \overline{I}_k$ and $D_k = (I_{k-1})^T I_{k-1}$. Note that (see (6.65)) $\|D_k^{1/2} S_k D_k^{-(1/2)}\| = \sup_{t\in[0,\, \|\overline{A}_k\|]} |\varphi_\nu(t)| = 1$. We start with the inequality

$$\begin{aligned}
\|(\overline{Q}_{k-1} - P_k \overline{Q}_k)\mathbf{v}\| &= \|(\overline{Q}_{k-1} - S_k \overline{I}_k \overline{Q}_k)\mathbf{v}\| \\
&= \|S_k(\overline{Q}_{k-1} - \overline{I}_k \overline{Q}_k)\mathbf{v} + (I - S_k)\overline{Q}_{k-1}\mathbf{v}\| \\
&\leq \|S_k(\overline{Q}_{k-1} - \overline{I}_k \overline{Q}_k)\mathbf{v}\| + \|(I - S_k)\overline{Q}_{k-1}\mathbf{v}\| \\
&\leq \left\|D_k^{-(1/2)}\right\| \left\|D_k^{1/2}(\overline{Q}_{k-1} - \overline{I}_k \overline{Q}_k)\mathbf{v}\right\| + \|(I - S_k)\overline{Q}_{k-1}\mathbf{v}\|.
\end{aligned}$$

Let $(0,\ b]$ be the interval that contains the eigenvalues of $\overline{A}_k = D_k^{-(1/2)} A_k D_k^{-(1/2)}$, which is used to construct the optimal polynomial $\varphi_\nu(t) = 1 - t q_{\nu-1}(t)$; that is, $b \geq \|\overline{A}_k\|$. Notice that

$$\begin{aligned}
I - S_k &= D_k^{-(1/2)} (\overline{A}_k q_{\nu-1}(\overline{A}_k)) D_k^{1/2} \\
&= D_k^{-(1/2)} \left(\overline{A}_k^{-(1/2)} (I - \varphi_\nu(\overline{A}_k))\right) \overline{A}_k^{1/2} D_k^{1/2}.
\end{aligned}$$

Based on estimate (6.66), we then get

$$\begin{aligned}
\|(I - S_k)\overline{Q}_{k-1}\mathbf{v}\| &\leq \|D_k^{-(1/2)}\| \max_{t\in(0,\,b]} \frac{1-\varphi_\nu(t)}{\sqrt{t}} \|\overline{Q}_{k-1}\mathbf{v}\|_{A_k} \\
&\leq C_\nu \frac{1}{\sqrt{b}} \left\|D_k^{-(1/2)}\right\| \|\overline{Q}_{k-1}\mathbf{v}\|_{A_k} \\
&\leq C_\nu \|D_k^{-(1/2)}\| \frac{1}{\|\overline{A}_k\|^{1/2}} \|\overline{Q}_{k-1}\mathbf{v}\|_{A_k}.
\end{aligned}$$

Thus we arrived at the estimate

$$\begin{aligned}
\|(\overline{Q}_{k-1} - P_k\overline{Q}_k)\mathbf{v}\| &\leq \left\|D_k^{-(1/2)}\right\| \|(I_{k-1}\overline{Q}_{k-1} - I_k\overline{Q}_k)\mathbf{v}\| \\
&\quad + \frac{C_\nu \left\|D_k^{-(1/2)}\right\|}{\|\overline{A}_k\|^{1/2}} \|\overline{Q}_{k-1}\mathbf{v}\|_{A_k}. \qquad (6.74)
\end{aligned}$$

The final bound on sum (i) is derived after an estimate of the terms in sum (ii) is obtained.

Estimating the second sum (ii)

Next, we bound $\|\overline{Q}_k\mathbf{v}\|_{A_{k+1}}$.

Because $\|A_k^{1/2} D_k^{-1} A_k^{1/2}\| = \|\overline{A}_k\|$, we have $\|A_k^{1/2} S_k A_k^{-(1/2)}\| = \|\varphi_\nu(A_k^{1/2} D_k^{-1} A_k^{1/2})\| \leq 1$ and similarly $\|D_k^{1/2} S_k D_k^{-(1/2)}\| = \|\varphi_\nu\big(D_k^{-(1/2)} A_k\ D_k^{-(1/2)}\big)\| = \|\varphi_\nu(\overline{A}_k)\| \leq 1$. The first estimate shows that

$$\mathbf{w}^T S_k^T A_k S_k \mathbf{w} \leq \mathbf{w}^T A_k \mathbf{w}.$$

Then, based on Lemma 6.30, we obtain

$$\begin{aligned}
\|\overline{Q}_k\mathbf{v}\|_{A_{k+1}} &= \|P_k\overline{Q}_k\mathbf{v}\|_{A_k} \\
&= \|S_k\overline{I}_k\overline{Q}_k\mathbf{v}\|_{A_k} \\
&\leq \|S_k(\overline{I}_k\overline{Q}_k - \overline{Q}_{k-1})\mathbf{v}\|_{A_k} + \|S_k\overline{Q}_{k-1}\mathbf{v}\|_{A_k} \\
&\leq \|S_k(\overline{I}_k\overline{Q}_k - \overline{Q}_{k-1})\mathbf{v}\|_{A_k} + \|\overline{Q}_{k-1}\mathbf{v}\|_{A_k} \\
&\leq \|A_k\|^{1/2}\|S_k(\overline{I}_k\overline{Q}_k - \overline{Q}_{k-1})\mathbf{v}\| + \|\overline{Q}_{k-1}\mathbf{v}\|_{A_k} \\
&\leq \|A_k\|^{1/2}\ \|D_k^{-(1/2)}\|\ \left\|D_k^{1/2}(\overline{I}_k\overline{Q}_k - \overline{Q}_{k-1})\mathbf{v}\right\| + \|\overline{Q}_{k-1}\mathbf{v}\|_{A_k} \\
&\leq \frac{\|A\|^{1/2}\ \|D_k^{1/2}\|}{(2\nu+1)^k} \left\|D_k^{-(1/2)}\right\| \|I_{k-1}(\overline{I}_k\overline{Q}_k - \overline{Q}_{k-1})\mathbf{v}\| + \|\overline{Q}_{k-1}\mathbf{v}\|_{A_k} \\
&\leq \kappa_D \frac{\|A\|^{1/2}}{(2\nu+1)^k} \|I_{k-1}(\overline{I}_k\overline{Q}_k - \overline{Q}_{k-1})\mathbf{v}\| + \|\overline{Q}_{k-1}\mathbf{v}\|_{A_k}. \qquad (6.75)
\end{aligned}$$

Here, κ_D is a uniform bound of the condition number of $D_k^{1/2}$ (see assumption (6.71)). We also have

$$\|\mathbf{v} - I_k\overline{Q}_k\mathbf{v}\|^2 = \|(I_{k-1}\overline{Q}_{k-1} - I_k\overline{Q}_k)\mathbf{v}\|^2 + \|\mathbf{v} - I_{k-1}\overline{Q}_{k-1}\mathbf{v}\|^2,$$

because $I_{k-1}^T I_{k-1} \overline{Q}_{k-1} = I_{k-1}^T$ and $I_k = I_{k-1} \overline{I}_k$, which implies that

$$(\mathbf{v} - I_{k-1}\overline{Q}_{k-1}\mathbf{v})^T (I_{k-1}\overline{Q}_{k-1} - I_k\overline{Q}_k)\mathbf{v} = (\mathbf{v} - I_{k-1}\overline{Q}_{k-1}\mathbf{v})^T I_{k-1}(\star) = 0.$$

Therefore,

$$\|(I_{k-1}\overline{Q}_{k-1} - I_k\overline{Q}_k)\mathbf{v}\| \leq \|\mathbf{v} - I_k\overline{Q}_k\mathbf{v}\|.$$

That is, if we bound $\|\mathbf{v} - I_k\overline{Q}_k\mathbf{v}\|$, the result will follow. Here, we use estimate (6.58), which was one of our main assumptions. It reads

$$\|\mathbf{v} - I_k\overline{Q}_k\mathbf{v}\|^2 \leq \sigma_a^2 \frac{(2\nu + 1)^{2(k+1)}}{\|A\|} \mathbf{v}^T A\mathbf{v}.$$

Then,

$$\|(I_{k-1}\overline{Q}_{k-1} - I_k\overline{Q}_k)\mathbf{v}\| \leq \sigma_a \frac{(2\nu + 1)^{k+1}}{\|A\|^{1/2}} \|\mathbf{v}\|_A. \tag{6.76}$$

Substituting the latter estimate in (6.75), leads to the following main recursive estimate,

$$\|\overline{Q}_k\mathbf{v}\|_{A_{k+1}} \leq \|\overline{Q}_{k-1}\mathbf{v}\|_{A_k} + \sigma_a \frac{(2\nu + 1)^k}{\|A\|^{1/2}} \kappa_D \frac{\|A\|^{1/2}}{(2\nu + 1)^k} \|\mathbf{v}\|_A.$$

That is, we proved the following main estimate with $\Delta = \sigma_a \kappa_D$,

$$\|\overline{Q}_k\mathbf{v}\|_{A_{k+1}} \leq \|\overline{Q}_{k-1}\mathbf{v}\|_{A_k} + \Delta \ \|\mathbf{v}\|_A \leq (1 + \Delta k) \ \|\mathbf{v}\|_A. \tag{6.77}$$

Thus the second sum is bounded as follows.

$$\sum_{l \leq \ell} \mathbf{v}_k^T A_k \mathbf{v}_k = \sum_{k \leq \ell} \|\overline{Q}_{k-1}\mathbf{v}\|_{A_k}^2 \leq C\ell^3 \ \mathbf{v}^T A\mathbf{v}. \tag{6.78}$$

Completing the bound of the first sum (i)

We showed (see estimate (6.74)) that

$$\|(\overline{Q}_{k-1} - P_k\overline{Q}_k)\mathbf{v}\| \leq \left\|D_k^{-(1/2)}\right\| \|(I_{k-1}\overline{Q}_{k-1} - I_k\overline{Q}_k)\mathbf{v}\| + \frac{C_\nu \|D_k^{-(1/2)}\|}{\|\overline{A}_k\|^{1/2}} \|\overline{Q}_{k-1}\mathbf{v}\|.$$

This estimate, together with (6.76) and (6.77), implies that

$$\|(\overline{Q}_{k-1} - P_k\overline{Q}_k)\mathbf{v}\| \leq \sigma_a \frac{(2\nu + 1)^k}{\|A\|^{1/2}} \|D_k^{-(1/2)}\| \ \|\mathbf{v}\|_A + \frac{C_\nu \|D_k^{-(1/2)}\|}{\|\overline{A}_k\|^{1/2}} (1 + \Delta k) \ \|\mathbf{v}\|_A.$$

We need to bound $\|M_k\|^{1/2}\|(\overline{Q}_{k-1} - P_k\overline{Q}_k)\mathbf{v}\|$. Recall that (by assumption) $M_k \simeq \|\overline{A}_k\|\ D_k$. This implies that

$$
\begin{aligned}
&\|M_k\|^{1/2}\|(\overline{Q}_{k-1} - P_k\overline{Q}_k)\mathbf{v}\| \\
&\leq \|\overline{A}_k\|^{1/2}\|D_k^{1/2}\|\,\|D_k^{-(1/2)}\|\left(\sigma_a\frac{(2\nu+1)^k}{\|A\|^{1/2}} + \frac{C_\nu}{\|\overline{A}_k\|^{1/2}}\ (1+\Delta k)\ \right)\|\mathbf{v}\|_A \\
&\leq \kappa_D\ \frac{\|A\|^{1/2}}{(2\nu+1)^k}\sigma_a\frac{(2\nu+1)^k}{\|A\|^{1/2}}\ \|\mathbf{v}\|_A + C_\nu\kappa_D(1+\Delta k)\ \|\mathbf{v}\|_A \\
&= \kappa_D\ [\sigma_a + C_\nu\ (1+\Delta k)]\ \|\mathbf{v}\|_A.
\end{aligned} \tag{6.79}
$$

Final estimates

In conclusion, we are ready to complete the proof of the following main result (given for $\nu = 1$ in [SA]).

Theorem 6.31. *Assume the following properties.*

- *The aggregates $\mathcal{A}_i^{(k)}$ at every coarse level k are quasiuniform in the sense of estimates* (6.70).
- *The approximation property* (6.58) *of the piecewise constant interpolants I_k (from coarse level $k+1$ all the way up to finest-level* 0*) holds.*
- *The choice of smoother is $M_k \simeq \|\overline{A}_k\|\ D_k$, where $D_k = I_{k-1}^T I_{k-1}$ and $\overline{A}_k = D_k^{-(1/2)} A_k D_k^{-(1/2)}$.*
- *The polynomials φ_ν are based on* (6.63) *with $b = \|\overline{A}_k\|$ at every level k. They are used in the construction of the smoothed interpolation matrices $P_k = \varphi_\nu(D_k^{-1}A_k)\overline{I}_k$, where $\overline{I}_k$ is the piecewise constant interpolant from coarse level $k+1$ to the next fine level k.*

Then, the resulting $V(1,1)$-cycle MG preconditioner B is nearly spectrally equivalent to A with $K_* \leq C\ell^3$, where K_* is the constant in (6.68).

Proof. It remains to use the estimates (6.79) and (6.78) for the particular decomposition $\mathbf{v}_k = (\overline{Q}_{k-1} - P_k\overline{Q}_k)\mathbf{v} + P_k\overline{Q}_k\mathbf{v}$ and $\mathbf{v}_{k+1} = \overline{Q}_k\mathbf{v}$. We have (see identity (6.72)),

$$
\begin{aligned}
\mathbf{v}^T B\mathbf{v} &\leq \left[\|P_\ell\overline{Q}_\ell\mathbf{v}\|_{A_\ell}^2 + 2\sum_k \|M_k\|\|(\overline{Q}_{k-1} - P_k\overline{Q}_k)\mathbf{v}\|^2 + 2\sum_k \|P_k\overline{Q}_k\mathbf{v}\|_{A_k}^2\right] \\
&\leq \left(2\sum_{k\leq\ell}\|\overline{Q}_{k-1}\mathbf{v}\|_{A_k}^2 + 2\kappa_D^2\ \sum_{k<\ell}(\sigma_a + C_\nu(1+k\Delta))^2\ \|\mathbf{v}\|_A^2\right) \\
&\leq C\left[\ell^3 + \sum_{k<\ell} k^2\right]\|\mathbf{v}\|_A^2 \\
&\leq C\ell^3\ \|\mathbf{v}\|_A^2.
\end{aligned}
$$

□

7

Domain Decomposition (DD) Methods

Domain decomposition (or DD) methods can generally be viewed as block versions of the Gauss–Seidel or Jacobi method that in addition may exploit overlap. If in the implementation, we use exact block inverses, the resulting algorithms are not generally of optimal complexity. For small blocks (corresponding to subdomains) to make the DD method of optimal order, we need a substantial in size coarse problem, and in order to have an overall optimal complexity of the method, we need an optimal solver for the coarse problem, which generally can be achieved by a multilevel method. For subdomain problems giving rise to large blocks, the coarse problem can be considered fixed. Then, in order to end up with an overall optimal complexity method, the subdomain problems have to be solved by an optimal method, which again can be a multilevel one. In summary, with DD-type methods to end up with an overall optimal complexity algorithm, we need in some of the components (such as subdomain or coarse-grid solutions) to employ some multilevel strategy.

This chapter also covers preconditioners based on domain embedding, auxiliary space methods, as well as preconditioners for problems with (multilevel) local refinement. In some cases, the subdomain problems allow for the use of fast (direct) elliptic solvers. We provide one such solver, as well.

7.1 Nonoverlapping blocks

In this section, we consider the following block structure of A

$$A = \begin{bmatrix} A_{0,0} & A_{0,b} \\ A_{b,0} & A_{b,b} \end{bmatrix},$$

where A_{00} is a block-diagonal matrix. Such a partitioning occurs in practice when A comes from a discretization of a PDE, where the dofs corresponding to index b form a separator (interface) Γ. For more details, see Section 1.8 (formula (1.19), in particular). That is, the domain is partitioned into a number of subdomains Ω_i by an interface Γ so that any entry a_{rs} of A with indices r, s corresponding to nodes $x_r \in \Omega_i$ and $x_s \in \Omega_j$ is zero for $i \neq j$.

P.S. Vassilevski, *Multilevel Block Factorization Preconditioners*,
doi: 10.1007/978-0-387-71564-3_7,

In order to achieve a stable block-form of A (cf. Section 3.4.1), we need an extension mapping

$$E = \begin{bmatrix} E_{0,b} \\ I \end{bmatrix}$$

that is almost "harmonic". The latter means that for a constant $\eta \geq 1$, we have

$$\mathbf{v}_b^T E^T A E \mathbf{v}_b \leq \eta \inf_{\mathbf{v}_0} \begin{bmatrix} \mathbf{v}_0 \\ \mathbf{v}_b \end{bmatrix}^T A \begin{bmatrix} \mathbf{v}_0 \\ \mathbf{v}_b \end{bmatrix} = \eta\, \mathbf{v}_b^T S \mathbf{v}_b, \text{ with } S = A_{bb} - A_{b,0} A_{0,0}^{-1} A_{0,b}.$$

If $\eta = 1$, we have that $E_{0,b} = -A_{0,0}^{-1} A_{0,b}$, which is sometimes called (A–) harmonic extension. In the latter case, $E^T A E = S$. The block-factorization preconditioner from Section 3.4.1 (modified to allow for nonsymmetric $M_{0,0}$) reads

$$B = \begin{bmatrix} I & 0 \\ A_{b,0} M_{0,0}^{-1} - E_{0,b}^T (I - A_{0,0} M_{0,0}^{-1}) & I \end{bmatrix} \begin{bmatrix} \overline{M}_0 & 0 \\ 0 & B_b \end{bmatrix} \times \begin{bmatrix} I & M_{0,0}^{-T} A_{0,b} - (I - M_{0,0}^{-T} A_{0,0}) E_{0,b} \\ 0 & I \end{bmatrix}.$$

Here

$$\overline{M}_0 = M_{0,0} \left(M_{0,0} + M_{0,0}^T - A_{0,0} \right)^{-1} M_{0,0}^T$$

is the symmetrized inexact subdomain solver, whereas B_b is a spectrally equivalent preconditioner to the interface Schur complement S, or equivalently, to $E^T A E$.

7.2 Boundary extension mappings based on solving special coarse problems

One way to construct computable extension mappings is by solving a coarse problem based on a bounded interpolation mapping which is identity at the boundary. Such a situation can arise if we coarsen the grid gradually away from the boundary Γ of a given domain Ω. An example is shown in Figure 7.1.

More specifically, let

$$P = \begin{bmatrix} \mathcal{P} \\ I \end{bmatrix} \begin{matrix} \} \ \mathcal{N}_f\text{–the set of fine dofs} \\ \} \ \mathcal{N}_c\text{–the set of coarse dofs} \end{matrix}$$

be the interpolation matrix, and assume that $\Gamma \subset \mathcal{N}_c$. Let

$$A = \begin{bmatrix} A_{0,0} & A_{0,b} \\ A_{b,0} & A_{b,b} \end{bmatrix} \begin{matrix} \} \ \Omega \setminus \Gamma \\ \} \ \Gamma \end{matrix}$$

be the fine-grid matrix. Similarly, let

$$A_c = P^T A P = \begin{bmatrix} A_{0,0}^c & A_{0,b}^c \\ A_{b,0}^c & A_{b,b}^c \end{bmatrix} \begin{matrix} \} \ \mathcal{N}_c \setminus \Gamma \\ \} \ \Gamma \end{matrix}$$

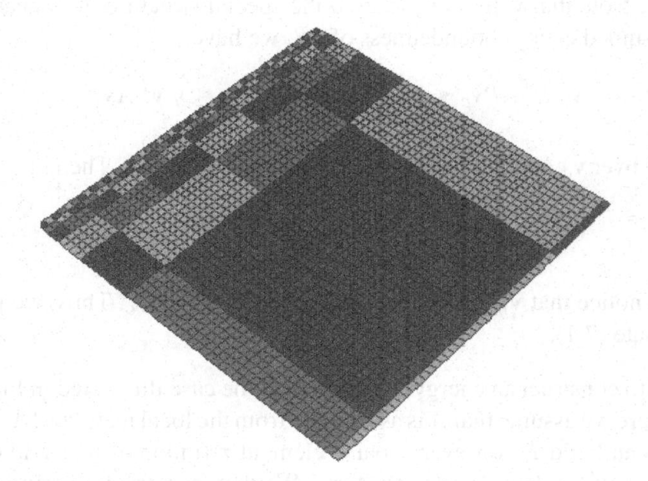

Fig. 7.1. Gradually coarsened mesh away from a boundary. The coarse degrees of freedom are the vertices of the agglomerated (coarse) elements. A typical coarse element has four or five coarse degrees of freedom.

be the coarse-grid matrix. The extension mapping of interest is defined by

$$E = P \begin{bmatrix} -\left(A^c_{0,0}\right)^{-1} A^c_{0,b} \\ I \end{bmatrix}.$$

The following result holds.

Theorem 7.1. *Assume that the interpolation matrix P is bounded in energy; that is, there is a restriction matrix $R: \mathbf{v} \mapsto \mathbf{v}_c = R\mathbf{v}$, such that*

$$\mathbf{v}_c^T P^T A P \mathbf{v}_c \le \eta \inf_{\mathbf{v}:\ R\mathbf{v}=\mathbf{v}_c} \mathbf{v}^T A \mathbf{v}.$$

Assuming that R and P are identity near Γ (see below), then the energy boundedness of PR implies energy boundedness of the extension mapping E; that is, we have

$$\mathbf{v}_b^T E^T A E \mathbf{v}_b \le \eta \inf_{\mathbf{v}:\ \mathbf{v}|_\Gamma=\mathbf{v}_b} \mathbf{v}^T A \mathbf{v}. \tag{7.1}$$

Proof. Because $\Gamma \subset \mathcal{N}_c$, it is natural to assume that R has similar structure to P, that is, that P and R act as the identity near Γ. More specifically, we consider R such that

$$R = \begin{bmatrix} * & 0 \\ 0 & I \end{bmatrix} \begin{matrix} \} \ \mathcal{N} \setminus \Gamma \\ \} \ \Gamma \end{matrix}.$$

The proof then proceeds as follows. Given $\mathbf{v}_b$, let $\mathbf{v}$ solve the minimization problem

$$\mathbf{v}^T A \mathbf{v} = \inf_{\mathbf{w}:\ \mathbf{w}|_\Gamma=\mathbf{v}_b} \mathbf{w}^T A \mathbf{w}.$$

Let $\mathbf{v}_c = R\mathbf{v}$. Note that $\mathbf{v}_c|_\Gamma = \mathbf{v}_b$ (due to the special structure of R near Γ). Then due to the assumed energy boundedness of PR, we have

$$\mathbf{v}_c^T P^T A P \mathbf{v}_c \le \eta \inf_{\mathbf{w}:\ R\mathbf{w}=\mathbf{v}_c} \mathbf{w}^T A \mathbf{w} \le \eta\ \mathbf{v}^T A \mathbf{v}.$$

Now take inf over $\mathbf{v}_c :\ \mathbf{v}_c|_\Gamma = \mathbf{v}_b$, on the left-hand side above. Then,

$$\inf_{\mathbf{v}_c|_\Gamma = \mathbf{v}_b} \mathbf{v}_c^T P^T A P \mathbf{v}_c \le \eta\ \mathbf{v}^T A \mathbf{v}.$$

It remains to notice that $\mathbf{v}_b^T E^T A E \mathbf{v}_b = \inf_{\mathbf{v}_c|_\Gamma = \mathbf{v}_b} \mathbf{v}_c^T P^T A P \mathbf{v}_c$. Thus, we proved the desired estimate (7.1). □

One way to construct an energy-bounded P in the case illustrated in Figure 7.1 is as follows. Here we assume that A is assembled from the local matrices $\{A_\tau\}$. Assume also that we can build P_τ for every coarse element τ (union of fine-grid elements), such that $\|A_\tau\| \|\mathbf{v}_\tau - P_\tau \mathbf{v}_\tau^c\|_\tau^2 \le \delta\ \mathbf{v}_\tau^T A_\tau \mathbf{v}_\tau$. We then construct a partition of unity diagonal matrices $\{Q_\tau\}$ where each Q_τ is nonzero only on τ and the set $\{Q_\tau\}$ satisfies $\sum_\tau Q_\tau = I$. Let $\widehat{P}_\tau$ be P_τ extended with zero rows outside τ. The global interpolation matrix is then defined as

$$P = \sum_\tau Q_\tau \left[0,\ \widehat{P}_\tau,\ 0\right].$$

Then, in precisely the same way as in Section 6.11, we show that the local estimates

$$\|A_\tau\| \|\mathbf{v}_\tau - P_\tau \mathbf{v}_\tau^c\|^2 \le \delta\ \mathbf{v}_\tau^T A_\tau \mathbf{v}_\tau$$

imply a global one

$$\|\mathbf{v} - P\mathbf{v}_c\|^2 \le \frac{\delta}{\eta \|A\|} \mathbf{v}^T A \mathbf{v}.$$

Here $\eta \le 1$ is such that

$$\|A_\tau\| \ge \eta\ \|A\|.$$

(Note that $\|A_\tau\| \le \|A\|$.) Then Theorem 6.20 shows that a two-grid preconditioner based on P and the Richardson smoother $M = (\|A\|/\omega)I$, $\omega \in (0, 2)$ (hence $\widetilde{M} = (\|A\|^2/\omega^2)(2(\|A\|/\omega)I - A)^{-1}$ is the symmetrized smoother) is spectrally equivalent to A. Finally, Corollary 3.24 shows that $\pi_{\widetilde{M}} = P\widetilde{M}_c^{-1} P^T \widetilde{M}$ is bounded in energy. The latter shows the result we needed. Indeed, because

$$P = \begin{bmatrix} * & 0 \\ 0 & I \end{bmatrix},$$

(i.e., P acts as the identity near Γ), we have that first

$$M_c = P^T M P = \begin{bmatrix} * & 0 \\ 0 & I \end{bmatrix}$$

acts as the identity near Γ, and hence,

$$M_c^{-1} P^T M = \begin{bmatrix} * & 0 \\ 0 & 0 \end{bmatrix}$$

also acts as the identity near Γ. Thus, if we define $R = M_c^{-1} P^T M$, then we have that R acts as the identity near Γ. Finally, because PR is bounded in energy (because $\pi_{\widetilde{M}}$ is bounded in energy, and $\widetilde{M} \simeq I$) the assumptions of Theorem 7.1 have been verified.

Note also that the elements τ are obtained by a standard "uniform" coarsening and a local P_τ bounded in A_τ-norm is feasible.

7.3 Weakly overlapping blocks

Here we assume that $A = P^T \widehat{A} P$; that is, A is obtained by a Galerkin (also called *RAP*) procedure from a bigger matrix $\widehat{A}$ based on a simple interpolation matrix P. In most of the applications, P has columns of the form

$$\begin{bmatrix} \vdots \\ 0 \\ 1 \\ 0 \\ \vdots \\ 0 \\ 1 \\ 0 \\ \vdots \end{bmatrix}.$$

That is, each column of P has at the most $m \geq 1$ nonzero entries equal to 1. Thus, we may say that every dof i of A is repeated $m = m(i)$ times in $\widehat{A}$. Assume that there is a set of indices "b" (further referred to as a separator boundary) and a set of indices "c" (further referred to as coarse dofs). The remaining dofs are denoted by "0" dofs. We can then reorder $\widehat{A}$ as follows

$$\widehat{A} = \begin{bmatrix} \widehat{A}_{0,0} & \widehat{A}_{0,b} & \widehat{A}_{0,c} \\ \widehat{A}_{b,0} & \widehat{A}_{b,b} & \widehat{A}_{b,c} \\ \widehat{A}_{c,0} & \widehat{A}_{c,b} & \widehat{A}_{c,c} \end{bmatrix} \begin{matrix} \} \text{ dofs outside the separator boundary} \\ \} \text{ dofs on the separator boundary} \\ \} \text{ coarse dofs} \end{matrix}.$$

Similarly, let

$$P = \begin{bmatrix} I & 0 & 0 \\ 0 & P_{0,b} & 0 \\ 0 & I & I \end{bmatrix}.$$

That is, only dofs on the separator boundary are allowed to have "multiple" copies in $\widehat{A}$.

An interesting case in practice is when the block

$$\mathcal{A} = \begin{bmatrix} \widehat{A}_{0,0} & \widehat{A}_{0,b} \\ \widehat{A}_{b,0} & \widehat{A}_{b,b} \end{bmatrix}$$

admits a block-diagonal structure (which is the case of "b" being a true separator), and the remaining diagonal block $\mathcal{B} = \widehat{A}_{c,c}$ has a small size (hence the name coarse dofs for the set "c").

We first use the exact inverse of $\widehat{A}$. It can be obtained by its exact factorization. Denoting $\mathcal{L} = [\widehat{A}_{c,0}, \widehat{A}_{c,b}]$, and similarly

$$\mathcal{R} = \begin{bmatrix} \widehat{A}_{0,c} \\ \widehat{A}_{b,c} \end{bmatrix},$$

we arrive at

$$\widehat{A} = \begin{bmatrix} I & 0 \\ \mathcal{L}\mathcal{A}^{-1} & I \end{bmatrix} \begin{bmatrix} \mathcal{A} & 0 \\ 0 & \mathcal{S} \end{bmatrix} \begin{bmatrix} I & \mathcal{A}^{-1}\mathcal{R} \\ 0 & I \end{bmatrix}.$$

Then the block preconditioner for A, or rather its inverse, takes the familiar form:

$$B^{-1} = (P^T P)^{-1} P^T \widehat{A}^{-1} P (P^T P)^{-1}. \tag{7.2}$$

Lemma 7.2. *The construction of B in (7.2) ensures that $A - B$ is positive semidefinite.*

Proof. Indeed, from $A = P^T \widehat{A} P$, we have $I = GG^T$ with $G = A^{-(1/2)} P^T \widehat{A}^{1/2}$, which implies that $I - G^T G$ is positive semidefinite. That is, $I - \widehat{A}^{1/2} P A^{-1} P^T \widehat{A}^{1/2}$ is positive semidefinite, or which is equivalent, $\widehat{A}^{-1} - P A^{-1} P^T$ is positive semidefinite. The latter finally implies that $(P^T P)^{-1} P^T \widehat{A}^{-1} P (P^T P)^{-1} - A^{-1} = B^{-1} - A^{-1}$ is positive semidefinite as well. □

Note that $\widehat{A}^{-1}$ involves the exact inverse of $\mathcal{A}$, which has a simple structure (i.e., block-diagonal). Also, it uses $\mathcal{S}^{-1}$, that is, the inverse of the exact Schur complement of $\widehat{A}$. The latter has a small size by assumption. We also note that $P^T P$ in the present setting is diagonal and $P(P^T P)^{-1}$ can be interpreted as a weight (or averaging) matrix. In the case of inexact inverses $\mathcal{M}^{-1}$ and $\mathcal{M}^{-T}$ of $\mathcal{A}$, we use the symmetrized preconditioner

$$\overline{\mathcal{M}} = \mathcal{M}(\mathcal{M} + \mathcal{M}^T - \mathcal{A})^{-1} \mathcal{M}^T$$

and a bounded extension mapping

$$\pi = \begin{bmatrix} I & 0 & E_{0,c} \\ 0 & I & E_{b,c} \\ 0 & 0 & I \end{bmatrix} = \begin{bmatrix} I & \mathcal{E} \\ 0 & I \end{bmatrix}, \quad \mathcal{E} = \begin{bmatrix} E_{0,c} \\ E_{b,c} \end{bmatrix}.$$

Introduce also the natural coarse matrix

$$\widehat{A}_c = \begin{bmatrix} \mathcal{E} \\ I \end{bmatrix}^T \widehat{A} \begin{bmatrix} \mathcal{E} \\ I \end{bmatrix}.$$

The resulting preconditioner inverse, B^{-1}, then reads:

$$B^{-1} = (P^T P)^{-1} P^T \begin{bmatrix} I & -\mathcal{M}^{-T}\mathcal{R} + (I - \mathcal{M}^{-T}\mathcal{A})\mathcal{E} \\ 0 & I \end{bmatrix} \begin{bmatrix} \overline{\mathcal{M}}^{-1} & 0 \\ 0 & \widehat{A}_c^{-1} \end{bmatrix}$$

$$\times \begin{bmatrix} I & 0 \\ -\mathcal{L}\mathcal{M}^{-1} + \mathcal{E}^T(I - \mathcal{A}\mathcal{M}^{-1}) & I \end{bmatrix} P(P^T P)^{-1}. \tag{7.3}$$

Consider first the following simple example. Let

$$A = \begin{bmatrix} \mathcal{A} & \underline{a}^T & 0 & \underline{p}^T \\ \underline{a} & d & \underline{b}^T & \underline{r}^T \\ 0 & \underline{b} & \mathcal{B} & \underline{q}^T \\ \underline{p} & \underline{r} & \underline{q} & \mathcal{C} \end{bmatrix}$$

and

$$P = \begin{bmatrix} I & 0 & 0 & 0 \\ 0 & I & 0 & 0 \\ 0 & I & 0 & 0 \\ 0 & 0 & I & 0 \\ 0 & 0 & 0 & I \end{bmatrix}.$$

Finally, let $d = \alpha + \beta$ and $\underline{r} = \underline{\eta} + \underline{\theta}$. Consider then the following extended matrix.

$$\widehat{A} = \begin{bmatrix} \mathcal{A} & \underline{a}^T & 0 & 0 & \underline{p}^T \\ \underline{a} & \alpha & 0 & 0 & \underline{\eta}^T \\ 0 & 0 & \beta & \underline{b}^T & \underline{\theta}^T \\ 0 & 0 & \underline{b} & \mathcal{B} & \underline{q}^T \\ \underline{p} & \underline{\eta} & \underline{\theta} & \underline{q} & \mathcal{C} \end{bmatrix}.$$

We can easily check that $A = P^T \widehat{A} P$. In the case of finite element matrices, we can naturally split $d = \alpha + \beta$ and $\underline{r} = \underline{\eta} + \underline{\theta}$ such that $\widehat{A}$, and hence the resulting blocks of $\widehat{A}$,

$$\begin{bmatrix} \mathcal{A} & \underline{a}^T \\ \underline{a} & \alpha \end{bmatrix} \text{ and } \begin{bmatrix} \beta & \underline{b}^T \\ \underline{b} & \mathcal{B} \end{bmatrix},$$

are symmetric positive semidefinite if A is symmetric positive semidefinite. These two major blocks of $\widehat{A}$ may be viewed as weakly overlapping in A because their shared dofs are from the set "b", which in practice has much smaller size than the nonoverlapping dofs "0" (giving rise to $\mathcal{A}$ and $\mathcal{B}$ in the present example). The above procedure of constructing $\widehat{A}$ and the respective M is referred to as a class of "Neumann–Neumann" methods. The variational relation between A and $\widehat{A}$ to specify unique value of the dofs on the separator "b" can be replaced by constraints. The latter procedure is sometimes called "tearing and interconnecting" and is a popular method in the finite element literature referred to as the FETI (finite element tearing and interconnecting) preconditioner (originated in [FETI] and in the presence of "coarse" dofs, commonly referred to as primal dofs; in [FETIdp0, FETIdp]). We have focused here on the possibility for inexact subdomain solvers (referring to $\mathcal{M}^{-1}$ and $\mathcal{M}^{-T}$ in (7.3)) as well as on the choice of the natural coarse matrix $\widehat{A}_c = E^T \widehat{A} E$. Detailed descriptions and finite element analysis of a variety of Neumann–Neumann and FETI (and FETI-DP) methods are found in [TW05].

7.4 Classical domain-embedding (DE) preconditioners

In this section, we also assume that $A = P^T \widehat{A} P$, in the sense that A can be viewed as a principal submatrix of $\widehat{A}$. Namely, let

$$\widehat{A} = \begin{bmatrix} A & R \\ L & T \end{bmatrix}.$$

Then with

$$P = \begin{bmatrix} I \\ 0 \end{bmatrix},$$

we get $A = P^T \widehat{A} P$. Note that this simple P may not be uniformly bounded in the $\widehat{A}$-norm. Namely, the constant η in the bound

$$\mathbf{v}^T A \mathbf{v} = \mathbf{v}^T P^T \widehat{A} P \mathbf{v} \le \eta \inf_{\mathbf{v}_0} \begin{bmatrix} \mathbf{v}_0 \\ \mathbf{v} \end{bmatrix}^T \widehat{A} \begin{bmatrix} \mathbf{v}_0 \\ \mathbf{v} \end{bmatrix}$$

may be mesh-dependent for a finite element matrix $\widehat{A}$. This is the case because a principal matrix of $\widehat{A}$ is not typically spectrally equivalent to a corresponding Schur complement $(A - RT^{-1}L)$ of $\widehat{A}$ (in a standard f.e. basis). In conclusion, the matrix embedding preconditioner (as defined in (7.2)) $(P^T P) P^T \widehat{A}^{-1} P (P^T P)^{-1} = P^T \widehat{A}^{-1} P$ will not generally be spectrally equivalent to A^{-1}.

That is why in the present case, we need an extension mapping

$$E = \begin{bmatrix} \mathcal{E} \\ I \end{bmatrix}$$

such that based on

$$\pi = \begin{bmatrix} I & \mathcal{E} \\ 0 & I \end{bmatrix}$$

the transformed matrix $\pi^T \widehat{A} \pi$ admits a stable two-by-two block form. Constructions of bounded extension mappings are bound in Section 7.2 and in Appendix D. The resulting "domain embedding" (DE), or rather, matrix-embedding preconditioner, takes the form

$$\begin{aligned} B^{-1} &= \begin{bmatrix} I \\ 0 \end{bmatrix}^T \pi^{-1} \widehat{A}^{-1} \pi^{-T} \begin{bmatrix} I \\ 0 \end{bmatrix} \\ &= [I, \ -\mathcal{E}]\, \widehat{A}^{-1} \begin{bmatrix} I \\ -\mathcal{E}^T \end{bmatrix}. \end{aligned} \tag{7.4}$$

It is clear also that we can use in (7.4) any available spectrally equivalent preconditioner $\widehat{B}$ for $\widehat{A}$ instead. Thus, we come up with the following inexact DE preconditioner,

$$B^{-1} = [I, \ -\mathcal{E}]\widehat{B}^{-1}\begin{bmatrix} I \\ -\mathcal{E}^T \end{bmatrix}. \tag{7.5}$$

The analysis of the resulting preconditioner is simple because with a bounded extension mapping E, we ensure that

$$\pi^T \widehat{A} \pi = \begin{bmatrix} A & A\mathcal{E} + R \\ \mathcal{E}^T A + L & E^T \widehat{A} E \end{bmatrix}$$

is spectrally equivalent to its block-diagonal part

$$\widehat{D} = \begin{bmatrix} A & 0 \\ 0 & E^T \widehat{A} E \end{bmatrix}.$$

More specifically, let

$$\mathbf{x}^T E^T \widehat{A} E \mathbf{x} \le \eta \inf_{\mathbf{v}} = \begin{bmatrix} \mathbf{v} \\ \mathbf{x} \end{bmatrix}^T \widehat{A} \begin{bmatrix} \mathbf{v} \\ \mathbf{x} \end{bmatrix}. \tag{7.6}$$

The above $\widehat{A}$-norm boundedness of E implies the following strengthened Cauchy–Schwarz inequality (simply examine the sign of the discriminant of the quadratic form,

$$Q(t) = \begin{bmatrix} t\mathbf{v} \\ \mathbf{x} \end{bmatrix}^T \widehat{A} \begin{bmatrix} t\mathbf{v} \\ \mathbf{x} \end{bmatrix} - \frac{1}{\eta}\mathbf{x}^T E^T \widehat{A} E \mathbf{x} \ge 0 \quad \text{for any } t \in \mathbb{R}),$$

$$\mathbf{v}^T (A\mathcal{E} + R)\mathbf{x} = \begin{bmatrix} \mathbf{v} \\ 0 \end{bmatrix}^T \widehat{A} \begin{bmatrix} 0 \\ \mathbf{x} \end{bmatrix} \le \sqrt{1 - \frac{1}{\eta}}\, (\mathbf{v}^T A \mathbf{v})^{1/2} (\mathbf{x}^T E^T \widehat{A} E \mathbf{x})^{1/2}.$$

Therefore, we have

$$\left(1 - \sqrt{1 - \frac{1}{\eta}}\right) \widehat{\mathbf{v}}^T \widehat{D} \widehat{\mathbf{v}} \le \widehat{\mathbf{v}}^T \widehat{A} \widehat{\mathbf{v}} \le \left(1 + \sqrt{1 - \frac{1}{\eta}}\right) \widehat{\mathbf{v}}^T \widehat{D} \widehat{\mathbf{v}}.$$

This shows that A^{-1} is spectrally equivalent to

$$\begin{bmatrix} I \\ 0 \end{bmatrix}^T (\pi^T \widehat{A} \pi)^{-1} \begin{bmatrix} I \\ 0 \end{bmatrix} = [I, \ -\mathcal{E}]\widehat{A}^{-1}\begin{bmatrix} I \\ -\mathcal{E}^T \end{bmatrix} = B^{-1}.$$

More specifically, the following bounds hold.

$$\left(1 - \sqrt{1 - \frac{1}{\eta}}\right) \mathbf{v}^T A \mathbf{v} \le \mathbf{v}^T B \mathbf{v} \le \left(1 + \sqrt{1 + \frac{1}{\eta}}\right) \mathbf{v}^T A \mathbf{v}.$$

Thus the following result is easily seen (cf., [V96]).

Theorem 7.3. *Let $\widehat{B}$ be a spectrally equivalent preconditioner to $\widehat{A}$ such that for two positive constants α and β, we have*

$$\alpha\, \widehat{\mathbf{v}}^T \widehat{B}\widehat{\mathbf{v}} \leq \widehat{\mathbf{v}}^T \widehat{A}\widehat{\mathbf{v}} \leq \beta\, \widehat{\mathbf{v}}^T \widehat{B}\widehat{\mathbf{v}}.$$

Consider the inexact DE preconditioner B defined in (7.5) based on an extension mapping

$$E = \begin{bmatrix} \mathcal{E} \\ I \end{bmatrix},$$

which satisfies the norm bound (7.6). Then the following spectral equivalence relations between A and B hold,

$$\frac{1-\sqrt{1-\frac{1}{\eta}}}{\beta}\, \mathbf{v}^T A\mathbf{v} \leq \mathbf{v}^T B\mathbf{v} \leq \frac{1+\sqrt{1-\frac{1}{\eta}}}{\alpha}\, \mathbf{v}^T A\mathbf{v}.$$

7.5 DE preconditioners without extension mappings

In some cases, A can be derived from a matrix

$$\widehat{A} = \begin{bmatrix} A+C & R \\ L & \mathcal{B} \end{bmatrix} \quad \text{where} \quad \begin{bmatrix} C & R \\ L & \mathcal{B} \end{bmatrix}$$

is symmetric positive semidefinite. More specifically, we assume that

$$A = \begin{bmatrix} \mathcal{A} & \underline{a}^T \\ \underline{a} & \alpha \end{bmatrix} \tag{7.7}$$

and let

$$R = \begin{bmatrix} 0 \\ \underline{r}^T \end{bmatrix} \quad \text{and} \quad C = \begin{bmatrix} 0 & 0 \\ 0 & \beta \end{bmatrix}.$$

That is, the extended matrix admits the form

$$\widehat{A} = \begin{bmatrix} \begin{bmatrix} \mathcal{A} & \underline{a}^T \\ \underline{a} & \alpha+\beta \end{bmatrix} & \begin{bmatrix} 0 \\ \underline{r}^T \end{bmatrix} \\ \begin{bmatrix} 0 & \underline{r} \end{bmatrix} & \mathcal{B} \end{bmatrix} = \begin{bmatrix} \mathcal{A} & \underline{a}^T & 0 \\ \underline{a} & \alpha+\beta & \underline{r}^T \\ 0 & \underline{r} & \mathcal{B} \end{bmatrix}. \tag{7.8}$$

Assuming now that the Schur complements $\underline{\alpha} = \alpha - \underline{a}\mathcal{A}^{-1}\underline{a}^T$ and $\underline{\beta} = \beta - \underline{r}^T\mathcal{B}^{-1}\underline{r}$ are related as in (7.9), then the same construction as in (7.4) or (7.5) gives a spectrally equivalent preconditioner B to A. Introduce the vectors

$$\widehat{\mathbf{v}} = \begin{bmatrix} \mathbf{v}_0 \\ \mathbf{v}_b \\ \mathbf{x} \end{bmatrix} \quad \text{and} \quad \mathbf{v} = \begin{bmatrix} \mathbf{v}_0 \\ \mathbf{v}_b \end{bmatrix}.$$

The analysis proceeds as follows. We first have

$$\inf_{\mathbf{x}} \begin{bmatrix} \mathbf{v} \\ \mathbf{x} \end{bmatrix}^T \widehat{A} \begin{bmatrix} \mathbf{v} \\ \mathbf{x} \end{bmatrix} = \mathbf{v}^T A\mathbf{v} + \mathbf{v}_b^T \underline{\beta}\mathbf{v}_b \geq \mathbf{v}^T A\mathbf{v},$$

where $\underline{\beta} = \beta - \underline{r}^T \mathcal{B}^{-1} \underline{r}$ as a Schur complement of the symmetric positive semidefinite matrix

$$\begin{bmatrix} \beta & \underline{r}^T \\ \underline{r} & \mathcal{B} \end{bmatrix}$$

is also positive semidefinite. Similarly, because by assumption $\underline{\beta}$ and $\underline{\alpha}$ are spectrally equivalent, we have that for some constant $\kappa > 0$,

$$\mathbf{v}_b^T \underline{\beta} \mathbf{v}_b \leq \kappa \mathbf{v}_b^T \underline{\alpha} \mathbf{v}_b, \tag{7.9}$$

which implies $\mathbf{v}_b^T \underline{\beta} \mathbf{v}_b \leq \kappa \mathbf{v}^T A \mathbf{v}$. The last inequality holds because $\underline{\alpha}$ is a Schur complement of the symmetric positive definite matrix A. Therefore,

$$\inf_{\mathbf{x}} \begin{bmatrix} \mathbf{v} \\ \mathbf{x} \end{bmatrix}^T \widehat{A} \begin{bmatrix} \mathbf{v} \\ \mathbf{x} \end{bmatrix} = \mathbf{v}^T A \mathbf{v} + \mathbf{v}_b^T \underline{\beta} \mathbf{v}_b \leq (1+\kappa) \mathbf{v}^T A \mathbf{v}.$$

Thus, we proved that A and the Schur complement $A + C - R\mathcal{B}^{-1} L$ of $\widehat{A}$ are spectrally equivalent. It is equivalent to say that A^{-1} and the exact DE preconditioner

$$B^{-1} = \begin{bmatrix} I \\ 0 \end{bmatrix}^T \widehat{A}^{-1} \begin{bmatrix} I \\ 0 \end{bmatrix},$$

which is a principal submatrix of $\widehat{A}^{-1}$, are spectrally equivalent. Finally, it is clear that we can instead use the inexact DE preconditioner

$$B^{-1} = \begin{bmatrix} I \\ 0 \end{bmatrix}^T \widehat{B}^{-1} \begin{bmatrix} I \\ 0 \end{bmatrix} \tag{7.10}$$

based on any given spectrally equivalent preconditioner $\widehat{B}$ for $\widehat{A}$ and still have spectral equivalence between A and B. That is, we proved the following main result.

Theorem 7.4. *Consider the matrix A given in* (7.7)*, which is embedded in matrix $\widehat{A}$ given in* (7.8)*. Assume that the two Schur complements on the separator set (denoted with index "b") satisfy* (7.9)*. Then the DE preconditioner B defined as in* (7.10) *is spectrally equivalent to A and the following bounds hold.*

$$\frac{\delta}{1+\kappa} \mathbf{v}^T B \mathbf{v} \leq \mathbf{v}^T A \mathbf{v} \leq \sigma\ \mathbf{v}^T B \mathbf{v}.$$

Here, the constants σ and δ are from the spectral equivalence relations between $\widehat{A}$ and $\widehat{B}$,

$$\delta\ \widehat{\mathbf{v}}^T \widehat{B} \widehat{\mathbf{v}} \leq \widehat{\mathbf{v}}^T \widehat{A} \widehat{\mathbf{v}} \leq \sigma\ \widehat{\mathbf{v}}^T \widehat{B} \widehat{\mathbf{v}},$$

and κ is from (7.9)*.*

We comment at the end that estimates of the form (7.9) are readily available in the finite element literature, and they represent the fact that the interface Schur complements $\underline{\alpha}$ and $\overline{\beta}$ define equivalent norms on the interface boundary (or separator) Γ under consideration. For second-order elliptic PDEs, this is an equivalent norm in the fractional-order Sobolev space $H_{0,0}^{1/2}(\Gamma)$. Finally, in the example of second-order elliptic PDEs, the finite element stiffness matrix A corresponds to a discrete problem with Neumann boundary conditions imposed on Γ. Thus the DE in the latter case is called DE through a Neumann boundary (perhaps first considered in [A78]).

7.6 Fast solvers for tensor product matrices

Here, we present a fast direct solver for special-type matrices that frequently appear in domain decomposition methods (e.g. as subdomain solvers). More specifically, we are interested in matrices

$$A = T \otimes I_{|B|} + I_{|T|} \otimes B,$$

where $I_{|B|}$ and $I_{|T|}$ are identity matrices of size $m = |B|$ (the size of B) and $|T|$ (the size of T), respectively. The matrix $T = (T_{r,s})_{r,s=1}^{n}$ is (block-)tridiagonal and $B = (b_{i,j})_{i,j=1}^{m}$ is (scalar-) tridiagonal. Generalizations for (block-)banded matrices are straightforward. The product $Q \otimes P$ stands for the block matrix $(p_{ij}Q)$ where $P = (p_{ij})$. We assume that T and B are s.p.d., but one of them is allowed to be only semidefinite. In other words, we assume that $\lambda_{\min}(T)I + B$ is s.p.d.

Assume that the eigendecomposition of T is computed so that we can utilize it for the solution of $A\mathbf{x} = \mathbf{b}$ as follows.

Let $T\mathbf{q}_k = \lambda_k \mathbf{q}_k$, $k = 1, 2, \ldots, |T|$. The eigenvectors $\mathbf{q}_k$ form an orthonormal system (i.e., $\mathbf{q}_k^T \mathbf{q}_l = \delta_{k,l}$). The system $A\mathbf{x} = \mathbf{b}$ can be rewritten as follows (letting $b_{1,0} = 0$ and $b_{m,m+1} = 0$),

$$T\mathbf{x}_i + b_{i,i-1}\mathbf{x}_{i-1} + b_{i,i}\mathbf{x}_i + b_{i,i+1}\mathbf{x}_{i+1} = \mathbf{b}_i, \qquad i = 1, \ldots, m. \tag{7.11}$$

The vectors $\mathbf{x}_i \in \mathbb{R}^{|T|}$, $i = 1, \ldots, m$ are unknown and $\mathbf{b}_i \in \mathbb{R}^n$, $i = 1, \ldots, m$ are given.

Using the orthogonal basis $\{\mathbf{q}_k\}$, we can expand both $\mathbf{x}_i$ and $\mathbf{b}_i$ as follows,

$$\mathbf{x}_i = \sum_k \eta_{i,k}\mathbf{q}_k,$$

$$\mathbf{b}_i = \sum_k \beta_{i,k}\mathbf{q}_k.$$

The coefficients $\beta_{i,k}$ are computed from the inner products

$$\beta_{i,k} = \mathbf{q}_k^T \mathbf{b}_i. \tag{7.12}$$

Substituting the above expressions for $\mathbf{x}_i$ and $\mathbf{b}_i$ in the original system (7.11), after rearranging the terms, we get

$$\sum_k ((\lambda_k + b_{i,i})\eta_{i,k} + b_{i,i-1}\eta_{i-1,k} + b_{i,i+1}\eta_{i+1,k})\,\mathbf{q}_k = \sum_k \beta_{i,k}\,\mathbf{q}_k, \quad i = 1, \ldots, m.$$

By comparing the coefficients in front of $\mathbf{q}_k$, we end up with the following system for the unknown coefficients $\{\eta_{i,k}\}$:

$$(\lambda_k + b_{i,i})\eta_{i,k} + b_{i,i-1}\eta_{i-1,k} + b_{i,i+1}\eta_{i+1,k} = \beta_{i,k}, \qquad i = 1, \ldots, m,$$

for $k = 1, \ldots, |T|$. Introducing the vectors $\boldsymbol{\eta}_k = (\eta_{i,k})_{i=1}^m \in \mathbb{R}^m$ and $\boldsymbol{\beta}_k = (\beta_{i,k})_{i=1}^m \in \mathbb{R}^m$, we end up with $|T|$ (decoupled) tridiagonal systems:

$$(\lambda_k I + B)\boldsymbol{\eta}_k = \boldsymbol{\beta}_k, \qquad k = 1, \ldots, |T|. \tag{7.13}$$

We can solve these systems for $|T|$ C_{tridiag} m flops.

Then, the solution vectors $\mathbf{x}_i$ are recovered from the formula $\mathbf{x}_i = \sum_k \eta_{i,k}\, \mathbf{q}_k$. The total cost of this evaluation is $2m|T|^2$ flops. The same cost, $2m|T|^2$ flops, is needed to compute the coefficients $\beta_{i,k} = \mathbf{q}_k^T \mathbf{b}_i$.

In summary, the standard method of separation of variables requires $\mathcal{O}(m|T|^2)$ flops in changing the basis plus C_{tridiag} $|T|m$ flops to compute the unknowns after the change of basis. The latter cost is an order of magnitude less than the cost of the actual change of basis.

In what follows, we describe a fast algorithm for separation of variables (FASV) originating in [V84]. It takes advantage of the fact that T is also (block-)tridiagonal. To explain the main idea, partition T into three blocks as follows,

$$T = \begin{bmatrix} T_1 & T_{1,0} & 0 \\ T_{0,1} & T_{0,0} & T_{0,2} \\ 0 & T_{2,0} & T_2 \end{bmatrix}.$$

Let $n = 2^\ell - 1$. Note now that $|T| = nN$ where N stands for the size of the blocks $T_{i,i}$ of the block-tridiagonal matrix T. We break T in the middle by using its $2^{\ell-1}$th row to define the blocks $T_{0,1}$, $T_{0,0}$, $T_{0,2}$. That is, $T_{0,0} = T_{2^{\ell-1},2^{\ell-1}}$ and $T_{0,1} = [0, \ldots, 0,\ T_{2^{\ell-1},2^{\ell-1}-1}]$, and similarly $T_{0,2} = [T_{2^{\ell-1},2^{\ell-1}+1},\ 0, \ldots, 0]$. Then, T_1 and T_2 are the major principal submatrices of T (which are also block-tridiagonal) but now with half the block size of the original matrix T.

The FASV exploits the following principal steps of Gaussian elimination realized on the basis of the standard separation of variables with paying attention to the nonzero pattern of the computed r.h.s. and utilizing the fact that we need to evaluate the sums only for specific components of the intermediate solutions. These observations are the key ingredients of the so-called "partial solution technique" developed by Y. A. Kuznetsov and A. M. Matsokin in [KM78]; see also [Ba78]. The method was further studied in [Ku85], and more recently in [KR96] and [RTa, RTb].

Use the above block partitioning of $T = T_{2^\ell - 1}$ and respective blocking of

$$\mathbf{x}_i = \begin{bmatrix} \mathbf{x}_i^{(1)} \\ \mathbf{x}_i^{(0)} \\ \mathbf{x}_i^{(2)} \end{bmatrix} \quad \text{and} \quad \mathbf{b}_i = \begin{bmatrix} \mathbf{b}_i^{(1)} \\ \mathbf{b}_i^{(0)} \\ \mathbf{b}_i^{(2)} \end{bmatrix}.$$

Let $\mathbf{x}_i = (\mathbf{x}_{i,\,j})_{j=1}^{n}$ and $\mathbf{b}_i = (\mathbf{b}_{i,\,j})_{j=1}^{n}$, $(n = 2^{\ell} - 1)$; then $\mathbf{x}_i^{(0)} = \mathbf{x}_{i,\,2^{\ell-1}}$, $\mathbf{x}_i^{(1)} = (\mathbf{x}_{i,\,j})_{j=1}^{2^{\ell-1}-1}$, and $\mathbf{x}_i^{(2)} = (\mathbf{x}_{i,\,j})_{j=2^{\ell-1}+1}^{2^{\ell}-1}$ and similarly, $\mathbf{b}_i^{(0)} = \mathbf{b}_{i,\,2^{\ell-1}}$, $\mathbf{b}_i^{(1)} = (\mathbf{b}_{i,\,j})_{j=1}^{2^{\ell-1}-1}$, and $\mathbf{b}_i^{(2)} = (\mathbf{b}_{i,\,j})_{j=2^{\ell-1}+1}^{2^{\ell}-1}$.
The following algorithm is of interest.

Algorithm 7.6.1. *Solve the systems*

$$T_1\mathbf{y}_i^{(1)} + b_{i,i-1}\mathbf{y}_{i-1}^{(1)} + b_{i,i}\mathbf{y}_i^{(1)} + b_{i,i+1}\mathbf{y}_{i+1}^{(1)} = \mathbf{b}_i^{(1)}, \qquad i = 1, \ldots, m,$$

and

$$T_2\mathbf{y}_i^{(2)} + b_{i,i-1}\mathbf{y}_{i-1}^{(2)} + b_{i,i}\mathbf{y}_i^{(2)} + b_{i,i+1}\mathbf{y}_{i+1}^{(2)} = \mathbf{b}_i^{(2)}, \qquad i = 1, \ldots, m,$$

and form

$$\mathbf{y}_i = \begin{bmatrix} \mathbf{y}_i^{(1)} \\ 0 \\ \mathbf{y}_i^{(2)} \end{bmatrix}.$$

Compute the residuals $\mathbf{r}_i = \mathbf{b}_i - (T\mathbf{y}_i + b_{i,i-1}\mathbf{y}_{i-1} + b_{i,i}\mathbf{y}_i + b_{i,i+1}\mathbf{y}_{i+1})$. *Notice that the residuals* $\mathbf{r}_i$ *have only one nonzero block component; namely,*

$$\mathbf{r}_{i,2^{\ell-1}} = \mathbf{b}_{i,2^{\ell-1}} - T_{2^{\ell-1},2^{\ell-1}-1}\mathbf{y}_{i,2^{\ell-1}-1}^{(1)} - T_{2^{\ell-1},2^{\ell-1}+1}\mathbf{y}_{i,2^{\ell-1}+1}^{(1)}.$$

Solve the residual equation

$$T\mathbf{z}_i + b_{i,i-1}\mathbf{z}_{i-1} + b_{i,i}\mathbf{z}_i + b_{i,i+1}\mathbf{z}_{i+1} = \mathbf{r}_i, \qquad i = 1, \ldots, m. \tag{7.14}$$

The desired solution is then $\mathbf{x}_i = \mathbf{y}_i + \mathbf{z}_i$.

Noticing that (by construction) $\mathbf{y}_{i,2^{\ell-1}} = 0$ *implies that* $\mathbf{x}_{i,2^{\ell-1}} = \mathbf{z}_{i,2^{\ell-1}}$, *we solve system (7.14) only for* $\mathbf{z}_{i,2^{\ell-1}}$. *By the method of separation of variables this is possible, because we have explicit formulas to compute* $\mathbf{z}_{i,j} = \sum_k \eta_{i,\,k}\mathbf{q}_{k,j}$.

The latter sums for a set of indices j *that form the middle block* $2^{\ell-1}$ *of* $\mathbf{z}_i$ *can be evaluated for* $2nm|N|$ *flops. Recall that* N *stands for the size of the block* $\mathbf{z}_{i,2^{\ell-1}}$ *(equal to the size of* $T_{0,0} = T_{2^{\ell-1},2^{\ell-1}}$*). To compute the coefficients* $\beta_{i,k} = \mathbf{q}_k^T\mathbf{r}_i$, *we use the fact that* $\mathbf{r}_i$ *has only one component that is nonzero. This reduces the cost from* $2n^2m$ *to only* $2nmN$ *flops, (where again* N *stands for the size of the nonzero component of* $\mathbf{r}_i$*.)*

After $\mathbf{x}_i^{(0)} = \mathbf{x}_{i,2^{\ell-1}} = \mathbf{z}_{i,2^{\ell-1}}$ *has been computed, the original problem decomposes into two decoupled pieces; namely, we have*

$$T_1\mathbf{x}_i^{(1)} + b_{i,i-1}\mathbf{x}_{i-1}^{(1)} + b_{i,i}\mathbf{x}_i^{(1)} + b_{i,i+1}\mathbf{x}_{i+1}^{(1)} = \mathbf{b}_i^{(1)} - T_{1,0}\mathbf{x}_i^{(0)}, \qquad i = 1, \ldots, m,$$

and

$$T_2\mathbf{x}_i^{(2)} + b_{i,i-1}\mathbf{x}_{i-1}^{(2)} + b_{i,i}\mathbf{x}_i^{(2)} + b_{i,i+1}\mathbf{x}_{i+1}^{(2)} = \mathbf{b}_i^{(2)} - T_{2,0}\mathbf{x}_i^{(0)}, \qquad i = 1, \ldots, m,$$

which have two times smaller block-size.

Applying the above algorithm recursively with respect to the block size of the blocks of T, we end up with the so-called fast algorithm for separation of variables (originating in [V84]).

We now give a fairly detailed description and motivation for the key steps that we have to take into account in order to implement FASV efficiently.

Algorithm 7.6.2 (FASV: Fast algorithm for separation of variables). *For every $k = 1, \ldots, \ell$, we introduce the principal submatrices of T,*

$$T_s^{(k)} = \begin{bmatrix} T_{(s-1)2^k+1,(s-1)2^k+1} & T_{(s-1)2^k+1,(s-1)2^k+2} & & & \\ T_{(s-1)2^k+2,(s-1)2^k+1} & T_{(s-1)2^k+2,(s-1)2^k+2} & T_{(s-1)2^k+2,(s-1)2^k+3} & & \\ & \ddots & & & \\ & & \ddots & \ddots & \ddots \\ & & & T_{s2^k-1,s2^k-2} & T_{s2^k-1,s2^k-1} \end{bmatrix},$$

for $s = 1, \ldots, 2^{\ell-k}$. Assume that all the eigenvalues and all the eigenvectors of $T_s^{(k)}$ have been precomputed. The eigenvectors use the same block-ordering as the matrices $T_s^{(k)}$. In FASV, the eigenvectors are needed only partially; namely, only the first, the last (i.e., the (2^k-1)th) and middle block-entry (i.e., the 2^{k-1}th) of each eigenvector need to be stored. Note that if T itself is separable, additional storage savings can be utilized.

(i) Forward recurrence. Assume that at step k, we have a r.h.s. $\mathbf{b}^{(k)}$ such that $\mathbf{b}_i^{(k)}$ have nonzero components at positions $s2^{k-1}$. Form the vectors $\mathbf{b}_i^{(k,s)}$ of length 2^k-1 by partitioning the $\mathbf{b}_i^{(k)}$, namely, $\mathbf{b}_{i,r}^{(k,s)} = \mathbf{b}_{i,(s-1)2^k+r}^{(k)}$, $r = 1, \ldots, 2^k-1$. Then, it is clear that only $\mathbf{b}_{i,2^{k-1}}^{(k,s)}$; that is, the middle component of each $\mathbf{b}_i^{(k,s)}$ will be nonzero.

1. Solve for the first, middle, and last components of $\mathbf{y}_i^{(k,s)}$, the systems

$$T_s^{(k)}\mathbf{y}_i^{(k,s)} + b_{i,i-1}\mathbf{y}_{i-1}^{(k,s)} + b_{i,i}\mathbf{y}_i^{(k,s)} + b_{i,i+1}\mathbf{y}_{i+1}^{(k,s)} = \mathbf{b}_i^{(k,s)}, \qquad i = 1, \ldots, m,$$

exploiting the fact that $\mathbf{b}_i^{(k,s)}$ has only one nonzero component (the one in the middle, 2^{k-1}th).

2. Form

$$\mathbf{y}_i^{(k)} = \begin{bmatrix} \mathbf{y}_i^{(k,1)} \\ 0 \\ \mathbf{y}_i^{(k,2)} \\ 0 \\ \vdots \\ 0 \\ \mathbf{y}_i^{(k,2^{\ell-k})} \end{bmatrix},$$

and compute the next residual

$$\mathbf{b}_i^{(k+1)} = \mathbf{b}_i^{(k)} - \left(T\mathbf{y}_i^{(k)} + b_{i,i-1}\mathbf{y}_{i-1}^{(k)} + b_{i,i}\mathbf{y}_i^{(k)} + b_{i,i+1}\mathbf{y}_{i+1}^{(k)}\right), \quad i = 1, \ldots, m.$$

We see (by induction) that $\mathbf{b}_i^{(k+1)}$ *has nonzero entries at positions* $s2^k$, $s = 1, \ldots, 2^{\ell-k} - 1$. *Those entries equal*

$$\mathbf{b}_{i,\, s2^k}^{(k+1)} = \mathbf{b}_{i,s2^k}^{(k)} - \left(T_{s2^k,s2^k-1}\mathbf{y}_{i,s2^k-1}^{(k)} + T_{s2^k,s2^k+1}\mathbf{y}_{i,s2^k+1}^{(k)}\right),$$

for $i = 1, \ldots, m$. *Noticing that* $\mathbf{y}_{i,s2^k-1}^{(k)} = \mathbf{y}_{i,2^k-1}^{(k,2s-1)}$ *(i.e., the last entry of* $\mathbf{y}_i^{(k,2s-1)}$*) and* $\mathbf{y}_{i,s2^k+1}^{(k)} = \mathbf{y}_{i,1}^{(k,2s)}$ *(the first entry of* $\mathbf{y}_i^{(k,2s)}$*), we see that the r.h.s.* $\mathbf{b}_i^{(k+1)}$ *are actually computable without full knowledge of* $\mathbf{y}_i^{(k,s)}$*; that is, we need only their first and last (i.e., the* $(2^k - 1)$*th) components. For computational efficiency (of the backward substitution), we also compute their middle component (the* 2^{k-1}*th one) here.*

(ii) Backward substitution. By construction, we have

$$\mathbf{b}^{(k+1)} = \mathbf{b}^{(k)} - A\mathbf{y}^{(k)}.$$

Therefore, letting $\mathbf{b}^{(\ell+1)} = 0$ *and* $\mathbf{b}^{(1)} = \mathbf{b}$, *we get*

$$A\left(\sum_k \mathbf{y}^{(k)}\right) = \mathbf{b}.$$

That is, the exact solution $\mathbf{x} = (\mathbf{x}_i)_{i=1}^m$ *has been decomposed as*

$$\mathbf{x}_i = \sum_k \mathbf{y}_i^{(k)}.$$

We recall that by construction $\mathbf{y}_{i,\, s2^k}^{(l)} = 0$ *for* $l \leq k$. *This, in particular implies that*

$$\mathbf{x}_{i,s2^k} = \sum_{l>k} \mathbf{y}_{i,\, s2^k}^{(l)}.$$

In the backward substitution steps for $k = \ell, \ldots, 1$, *we recover the exact solution* $\mathbf{x}_{i,s2^k}$ *utilizing the above formula.*

1. *Compute the r.h.s. vectors* $\mathbf{r}_i^{(k,s)}$ *that have nonzero components only at their first and last position, namely,*

$$\begin{aligned}
\mathbf{r}_{i,1}^{(k,s)} &= -T_{(s-1)2^k+1,(s-1)2^k}\mathbf{x}_{i,\,(s-1)2^k} \\
\mathbf{r}_{i,2^k-1}^{(k,s)} &= -T_{s2^k-1,s2^k}\mathbf{x}_{i,\, s2^k}.
\end{aligned}$$

2. *Solve, only for the middle component of* $\mathbf{z}_i^{(k,s)}$, *(for* $s = 1, \ldots, 2^{\ell-k}$*) the systems*

$$T_s^{(k)} \mathbf{z}_i^{(k,s)} + b_{i,i-1}\mathbf{z}_{i-1}^{(k,s)} + b_{i,i}\mathbf{z}_i^{(k,s)} + b_{i,i+1}\mathbf{z}_{i+1}^{(k,s)} = \mathbf{r}_i^{(k,s)}, \quad i = 1, \ldots, m,$$

3. *Recover* $\mathbf{x}_{i,\, s2^{k-1}}$ *as*

$$\mathbf{x}_{i,\, s2^{k-1}} = \mathbf{z}_{i,2^{k-1}}^{(k,s)} + \mathbf{y}_{i,2^{k-1}}^{(k,s)}.$$

We recall that the middle components (i.e., the 2^{k-1}*th) of* $\mathbf{y}_i^{(k,s)}$ *have been computed in the forward recurrence.*

Proposition 7.5. *We recall that* $|B| = m$ *stands for the size of* B, N *stands for the size of each block* T_{ii} *of* T, *and* n *is the number of blocks of* T. *That is, the size of* T *is* $|T| = nN$, *and the size of the overall problem is* $|T||B| = mnN$.

Following the steps of Algorithm 7.6.2, we can estimate the storage requirements and number of flops of a straightforward implementation of FASV as follows (the leading terms only).

Forward step (1) and backward step (2), can be implemented (using the sparsity of the r.h.s. and the fact that only certain components of the solutions are needed) for

$$\sum_k 2^{\ell-k}[8(2^k-1)Nm + 6(2^k-1)Nm] \simeq 14(nNm)\, \ell \textit{ flops}.$$

The corresponding (scalar) tridiagonal systems can be solved for

$$\sum_k 2^{\ell-k}\, C_{\text{tridiag}}\, (2^k-1)m \simeq C_{\text{tridiag}}\, (nNm)\, \ell \textit{ flops}.$$

That is, the FASV algorithm can be implemented with a cost

$$\simeq (14 + C_{\text{tridiag}})\, |T||B| \log n,$$

where n *is the block-size of* T. *Note that* $|T||B|$ *is the original problem size. The latter shows that FASV is nearly optimal direct solver.*

The storage requirements at every step k *for the needed* $3N$ *components of the* $(2^k-1)N$ *eigenvectors of the matrices* $T^{(k,s)}$ *equal*

$$2^{\ell-k}(2^k-1)N \times (3N) \simeq 3N^2 n.$$

If these components are stored in advance (for all ks), we would need storage $\simeq 3N^2 n\ell$, *which can be prohibitively large if* N *is of the same order as* $n \simeq m$. *However, if* T *is itself separable (as a sum of two tensor products of smaller matrices similarly to* A*) then the storage requirement reduces by an order of magnitude and hence is negligible. If* $N = 1$, *that is,* T *is also scalar tridiagonal (as* B*) the storage is negligible.*

The storage for all $2^{k-1}N$ *eigenvalues of* $T^{(k,s)}$ *gives* $\sum_k 2^{\ell-k}(2^k-1)N \simeq nN\, \ell$, *which is negligible.*

Finally, we need to store T *and* B, *and we also need two additional vector arrays, one for the r.h.s. and one for the solution. Customarily, the solution can overwrite the r.h.s. if the latter is not needed at the end.*

7.7 Schwarz methods

The classical (multiplicative) Schwarz methods are most commonly described as product iteration methods that exploit solutions corresponding to principal submatrices of A of relatively small size and for a large number of blocks, for optimal convergence, a global coarse solution is used in addition. For various algorithmic details, we refer to [DD]. In the present section, we give an equivalent formulation based on the relation we established in Section 3.2.1 between product iteration methods and certain (approximate) block-factorizations of A.

Given an $n \times n$ s.p.d. sparse matrix A, let Ω_k, $k = 1, \ldots, J$ be an overlapping partition of the set of indices $\{1, \ldots, n\}$. Also, let I_k be the extension by zero of vectors defined on the set Ω_k to a vector in $\mathbb{R}^n$. Denote by $A_k = I_k^T A I_k$ the principal submatrix of A corresponding to the index set Ω_k. Very often, as is customary in the DD literature, we call Ω_k subdomains.

Finally, let M_k be preconditioners for A_k such that $M_k^T + M_k - A_k$ are s.p.d. Recall (see Section 3.1.3) that the latter is equivalent to $\|I - A_k^{1/2} M_k^{-1} A_k^{1/2}\| < 1$. If $M_k = A_k$, the corresponding Schwarz preconditioner is said to exploit exact subdomain solutions. Note that our definition and subsequent analysis allow not only for inexact but also nonsymmetric subdomain and coarse-grid solvers.

Denote the local vector spaces $\mathbf{V}_k^{(0)} = \{\mathbf{v}|_{\Omega_k} : \mathbf{v} \in \mathbb{R}^n\}$, and form the following auxiliary subspaces of $\mathbb{R}^n$,

$$\mathbf{V}_k = \sum_{j \geq k} I_j \mathbf{V}_j^{(0)}.$$

Note that the vectors in $\mathbf{V}_k$ are zero outside

$$\widetilde{\Omega}_k \equiv \Omega_J \cup \Omega_{J-1} \cup \cdots \cup \Omega_k.$$

To define the overlapping Schwarz preconditioner B for A, we also need a coarse space $\mathbf{V}_0 = \text{Range}(P)$ for a given interpolation matrix $P : \mathbb{R}^m \mapsto \mathbb{R}^n$ where $m \leq n$. Let $A_0 = P^T A P$ and M_0 be a preconditioner for A_0 such that $M_0^T + M_0 - A_0$ is s.p.d. A typical case is $M_0 = A_0$. Alternatively, M_0 can be the downward part of a V-cycle multigrid based on A_0.

The overlapping Schwarz preconditioner B exploits solutions with M_k, M_k^T, $k = 1, \ldots, J$, and the coarse matrices M_0 and M_0^T, all being of smaller size compared to n (the size of A). The following recursive definition defines an overlapping Schwarz preconditioner with a coarse solution and inexact subdomain solutions (referring to actions of M_k^{-1} and M_k^{-T}).

Definition 7.6 (Multiplicative Schwarz preconditioner). *For $k = J, J-1, \ldots, 1$, let $\widetilde{I}_k$ be the extension by zero of vectors defined on $\widetilde{\Omega}_k = \Omega_J \cup \Omega_{J-1} \cdots \cup \Omega_k$ to vectors in $\mathbb{R}^n$. Note that $\widetilde{I}_0 = I$, and recall that $I_0 = P$.*

Let $\widetilde{A}_k = \widetilde{I}_k^T A \widetilde{I}_k$ be the principal submatrix of A corresponding to the auxiliary subdomain $\widetilde{\Omega}_k$ for $k > 0$, or the coarse matrix A_0, if $k = 0$.

Set $\widetilde{B}_J = \overline{M}_J \equiv M_J(M_J^T + M_J - A_J)^{-1}M_J^T$. *Assume that* $\widetilde{B}_{k+1}$, *for* $k < J$ *has been defined for vectors on* $\widetilde{\Omega}_{k+1}$ *only. In order to define* $\widetilde{B}_k$ *(for vectors defined on* $\widetilde{\Omega}_k$*), we first form*

$$\widehat{B}_k = \begin{bmatrix} M_k & 0 \\ \widetilde{I}_{k+1}^T A I_k & I \end{bmatrix} \begin{bmatrix} \left(M_k^T + M_k - A_k\right)^{-1} & 0 \\ 0 & \widetilde{B}_{k+1} \end{bmatrix} \begin{bmatrix} M_k^T & I_k^T A \widetilde{I}_{k+1} \\ 0 & I \end{bmatrix},$$

and then let

$$\widetilde{B}_k^{-1} = \widetilde{I}_k^T [I_k,\ \widetilde{I}_{k+1}] \widehat{B}_k^{-1} [I_k,\ \widetilde{I}_{k+1}]^T \widetilde{I}_k.$$

We notice that $[I_k,\ \widetilde{I}_{k+1}]\widehat{B}_k^{-1}[I_k,\ \widetilde{I}_{k+1}]^T$ *is s.p.d. when restricted to vectors defined on* $\widetilde{\Omega}_k$. *The latter is seen from the facts that* $\Omega_k \subset \widetilde{\Omega}_k$ *and* $\widetilde{\Omega}_{k+1} \subset \widetilde{\Omega}_k$, *hence for any vector* $\widetilde{\mathbf{v}}_k$ *defined on* $\widetilde{\Omega}_k$, *we have* $I_k^T \widetilde{I}_k \widetilde{\mathbf{v}}_k = \widetilde{\mathbf{v}}_k|_{\Omega_k}$, *the restriction to* Ω_k, *and* $\widetilde{I}_{k+1}^T \widetilde{I}_k \widetilde{\mathbf{v}}_k = \widetilde{\mathbf{v}}_k|_{\widetilde{\Omega}_{k+1}}$, *the restriction to* $\widetilde{\Omega}_{k+1}$. *That is,* $[I_k,\ \widetilde{I}_{k+1}]^T \widetilde{I}_k \widetilde{\mathbf{v}}_k = 0$ *implies* $\widetilde{\mathbf{v}}_k = 0$, *hence* $[I_k,\ \widetilde{I}_{k+1}]^T \widetilde{I}_k$ *has a full column rank. Therefore,* $\widetilde{B}_k$ *is a well-defined s.p.d. matrix.*

The following result holds as an application of the result in Section 3.2.1.

Lemma 7.7. *Consider the iteration matrix* $E_k = I - \widetilde{I}_k \widetilde{B}_k^{-1} \widetilde{I}_k^T A$. *Then, the following relation holds,*

$$E_k = (I - I_k M_k^{-T} I_k^T A) E_{k+1} (I - I_k M_k^{-1} I_k^T A).$$

Proof. We notice that $[I_k,\ \widetilde{I}_{k+1}]^T \widetilde{I}_k \widetilde{I}_k^T = [I_k,\ \widetilde{I}_{k+1}]^T$ (because $\Omega_k,\ \widetilde{\Omega}_{k+1} \subset \widetilde{\Omega}_k$). Therefore,

$$E_k = I - [I_k,\ \widetilde{I}_{k+1}] \widehat{B}_k^{-1} [I_k,\ \widetilde{I}_{k+1}]^T A.$$

Then based on the equivalence of the product iteration method exploiting solutions in the subspaces Range(I_k), Range($\widetilde{I}_{k+1}$), and Range(I_k), based on M_k, $\widetilde{B}_{k+1}$, and M_k^T respectively, and the block-factorization matrix $\widehat{B}_k$ on the other hand (as shown in Section 3.2.1), we have the identity

$$\begin{aligned} I &- [I_k,\ \widetilde{I}_{k+1}] \widehat{B}_k^{-1} [I_k,\ \widetilde{I}_{k+1}]^T A \\ &= (I - I_k M_k^{-T} I_k^T A)(I - \widetilde{I}_{k+1} \widetilde{B}_{k+1}^{-1} \widetilde{I}_{k+1}^T A)(I - I_k M_k^{-1} I_k^T A) \\ &= (I - I_k M_k^{-T} I_k^T A) E_{k+1} (I - I_k M_k^{-1} I_k^T A), \end{aligned}$$

which is the desired result. □

To analyze the Schwarz preconditioner, we need stable vector decomposition in the following sense. Let $\mathbf{v} = \sum_{j=0}^J I_j \mathbf{v}_j$ be such that

$$\sum_{j=0}^J \mathbf{v}_j^T I_j^T A I_j \mathbf{v}_j = \sum_{j=0}^J \mathbf{v}_j^T A_j \mathbf{v}_j \le \sigma\ \mathbf{v}^T A \mathbf{v}.$$

Then, we also need the partial sums restricted to $\widetilde{\Omega}_k$; that is, $\widetilde{\mathbf{v}}_k = \widetilde{I}_k^T \sum_{j=k}^J I_j \mathbf{v}_j$.

Let $\mathcal{I}(k)$ be the set of indices j such that $I_k^T A I_j$ is nonzero. Due to the sparsity of A, we can assume that $\max_{k \geq 1} |\mathcal{I}(k)| = \kappa$ is a bounded integer constant. It defines the maximal number of overlapping subdomains Ω_j with Ω_k. Introduce, for completeness, the set $\mathcal{I}(0) = \{0, 1, \ldots, J\}$.

In what follows, we estimate the term $(I_k^T A \widetilde{I}_{k+1} \widetilde{\mathbf{v}}_{k+1})^T A_k^{-1} (I_k^T A \widetilde{I}_{k+1} \widetilde{\mathbf{v}}_{k+1})$. We first notice that $\widetilde{I}_{k+1} \widetilde{I}_{k+1}^T I_j = I_j$, because $\Omega_j \subset \widetilde{\Omega}_{k+1}$ for $j \geq k+1$. Hence,

$$
\begin{aligned}
I_k^T A \widetilde{I}_{k+1} \widetilde{\mathbf{v}}_{k+1} &= I_k^T A \widetilde{I}_{k+1} \widetilde{I}_{k+1}^T \sum_{j=k+1}^{J} I_j \mathbf{v}_j \\
&= I_k^T A \sum_{j=k+1}^{J} I_j \mathbf{v}_j \\
&= I_k^T A \sum_{j>k,\ j \in \mathcal{I}(k)} I_j \mathbf{v}_j.
\end{aligned}
$$

Next, use the fact that $A_k = I_k^T A I_k$ and also use $\|X\| = \|X^T\| = 1$ for $X = A_k^{-(1/2)} I_k^T A^{1/2}$ to see that

$$
\mathbf{w}^T I_k A_k^{-1} I_k^T \mathbf{w} \leq \mathbf{w}^T A^{-1} \mathbf{w}.
$$

For $\mathbf{w} = A \sum_{j>k,\ j \in \mathcal{I}(k)} I_j \mathbf{v}_j$, we end up with the inequality

$$
\left(I_k^T A \widetilde{I}_{k+1} \widetilde{\mathbf{v}}_{k+1}\right)^T A_k^{-1} \left(I_k^T A \widetilde{I}_{k+1} \widetilde{\mathbf{v}}_{k+1}\right) \leq \left(\sum_{\substack{j \in \mathcal{I}(k) \\ j>k}} I_j \mathbf{v}_j \right)^T A \left(\sum_{\substack{j \in \mathcal{I}(k) \\ j>k}} I_j \mathbf{v}_j \right). \tag{7.15}
$$

We need next the following technical assumption

$$
\mathbf{v}_k^T (M_k^T + M_k - A_k) \mathbf{v}_k \geq \delta\ \mathbf{v}_k^T A_k \mathbf{v}_k. \tag{7.16}
$$

If M_k is s.p.d. preconditioner for A_k such that $\mathbf{v}_k^T M_k \mathbf{v}_k \geq \mathbf{v}_k^T A_k \mathbf{v}_k$ the above assumption holds with $\delta = 1$. In general, if M_k is a nonsymmetric matrix, we may prove the above inequality if M_k is a convergent splitting matrix for A_k. For example, if $\|I - A_k^{1/2} M_k^{-1} A_k^{1/2}\| \leq \varrho < 1$, we can show (see Lemma 7.8 below) that (7.16) holds with $\delta = ((1-\varrho)/(1+\varrho)) > 0$.

Lemma 7.8. *Assume that A is s.p.d. and let M provide an A-convergent iteration for solving systems with A. More specifically, let*

$$
\|I - A^{1/2} M^{-1} A^{1/2}\| \leq \varrho < 1.
$$

Then,

$$
\mathbf{v}^T (M^T + M - A) \mathbf{v} \geq \frac{1-\varrho}{1+\varrho}\ \mathbf{v}^T A \mathbf{v}. \tag{7.17}
$$

Proof. The result is seen from the inequalities $\|X - I\| \leq \varrho$ for $X = A^{1/2}M^{-1}A^{1/2}$. The norm inequality implies

$$\varrho^2 \, \|\mathbf{v}\|^2 \geq \mathbf{v}^T(I - X^T)(I - X)\mathbf{v} = \mathbf{v}^T\mathbf{v} - 2\mathbf{v}^T X\mathbf{v} + \|X\mathbf{v}\|^2. \tag{7.18}$$

That is, $2\|X\mathbf{v}\|\,\|\mathbf{v}\| \geq (1-\varrho^2)\,\|\mathbf{v}\|^2 + \|X\mathbf{v}\|^2$, or equivalently $2t \geq t^2 + (1-\varrho^2)$ for $t = \|X\mathbf{v}\|/\|\mathbf{v}\|$. This shows $(t-1)^2 \leq \varrho^2$, which implies $1+\varrho \geq t \geq 1-\varrho$. Using then $\|\mathbf{v}\| \geq (1/(1+\varrho))\,\|X\mathbf{v}\|$ in inequality (7.18) implies

$$2\mathbf{v}^T X\mathbf{v} \geq (1-\varrho^2)\,\|\mathbf{v}\|^2 + \|X\mathbf{v}\|^2 \geq \left[\frac{1-\varrho}{1+\varrho} + 1\right]\,\|X\mathbf{v}\|^2.$$

Letting $\mathbf{v} := X^{-1}\mathbf{v}$, we get

$$\mathbf{v}^T(X^{-T} + X^{-1})\mathbf{v} = 2\mathbf{v}^T X^{-1}\mathbf{v} \geq \left[1 + \frac{1-\varrho}{1+\varrho}\right]\,\|\mathbf{v}\|^2,$$

which is (7.17) (by letting $\mathbf{v} := A^{\frac{1}{2}}\mathbf{v}$). □

In summary, the following result can be formulated.

Theorem 7.9. *Assume that any* $\mathbf{v}$ *admits a decomposition* $\mathbf{v} = \sum_{j=0}^{J} I_j\mathbf{v}_j$, *which is stable; that is,*

$$\sum_{j=0}^{J} \mathbf{v}_j^T A_j \mathbf{v}_j \leq \sigma\, \mathbf{v}^T A\mathbf{v}. \tag{7.19}$$

Also, let the subdomain solvers M_j, $j > 0$ *and the coarse-grid solver* M_0 *provide* $\|.\|_{A_j}$ *convergent splittings for* A_j, *respectively. That is, we have*

$$\|I - A_j^{1/2}M_j^{-1}A_j^{1/2}\| \leq \varrho < 1.$$

The latter implies the following coercivity estimate,

$$\mathbf{w}_j^T(M_j^T + M_j - A_j)\mathbf{w}_j \geq \frac{1-\varrho}{1+\varrho}\,\mathbf{w}_j^T A_j\mathbf{w}_j,$$

as well as the following bounds for the symmetrized (subdomain or coarse-grid) solvers $\overline{M}_j = M_j(M_j^T + M_j - A_j)^{-1}M_j^T$,

$$\mathbf{w}_j^T A_j\mathbf{w}_j \leq \mathbf{w}_j^T\overline{M}_j\mathbf{w}_j \leq \frac{1}{1-\varrho^2}\,\mathbf{w}_j^T A_j\mathbf{w}_j.$$

Finally, let the subdomains Ω_k *satisfy the condition of bounded overlap, that is, that the sets* $\mathcal{I}(k) = \{j > 0;\ I_k^T A I_j \neq 0\}$ *have a bounded number of entries, in the sense that* $\max_{k>0} |\mathcal{I}(k)| \leq \kappa$ *for a bounded integer* κ. *Then, the following spectral equivalence result holds, for* $B = \widetilde{B}_0$,

$$\mathbf{v}^T A\mathbf{v} \leq \mathbf{v}^T B\mathbf{v} \leq 2\left[\sigma\left(\frac{1}{1-\varrho^2} + \kappa^2\,\frac{1+\varrho}{1-\varrho}\right) + 2\frac{1+\varrho}{1-\varrho}\right]\,\mathbf{v}^T A\mathbf{v}.$$

Proof. From the definition of $\widetilde{B}_k$, we have $XX^T = I$ for

$$X = \widetilde{B}_k^{1/2} \widetilde{I}_k^T [I_k,\ \widetilde{I}_{k+1}] \widehat{B}_k^{-(1/2)}.$$

Hence from $\|X\| = \|X^T\| = 1$, we get the inequality

$$\begin{aligned}
&[I_k \mathbf{v}_k + \widetilde{I}_{k+1} \widetilde{\mathbf{v}}_{k+1}]^T \widetilde{I}_k \widetilde{B}_k \widetilde{I}_k^T [I_k \mathbf{v}_k + \widetilde{I}_{k+1} \widetilde{\mathbf{v}}_{k+1}] \\
&= \begin{bmatrix} \mathbf{v}_k \\ \widetilde{\mathbf{v}}_{k+1} \end{bmatrix}^T [I_k,\ \widetilde{I}_{k+1}]^T \widetilde{I}_k \widetilde{B}_k \widetilde{I}_k^T [I_k,\ \widetilde{I}_{k+1}] \begin{bmatrix} \mathbf{v}_k \\ \widetilde{\mathbf{v}}_{k+1} \end{bmatrix} \\
&\leq \begin{bmatrix} \mathbf{v}_k \\ \widetilde{\mathbf{v}}_{k+1} \end{bmatrix}^T \widehat{B}_k \begin{bmatrix} \mathbf{v}_k \\ \widetilde{\mathbf{v}}_{k+1} \end{bmatrix}.
\end{aligned}$$

For any given $\mathbf{v}$, consider its stable decomposition $\mathbf{v} = \sum_{j \geq 0} I_j \mathbf{v}_j$ and the corresponding restricted partial sums $\widetilde{\mathbf{v}}_k = \widetilde{I}_k^T \sum_{j \geq k} I_j \mathbf{v}_j$. We have, $\widetilde{\mathbf{v}}_k = \widetilde{I}_k^T (I_k \mathbf{v}_k + \widetilde{I}_{k+1} \widetilde{\mathbf{v}}_{k+1})$, hence,

$$\begin{aligned}
\widetilde{\mathbf{v}}_k^T \widetilde{B}_k \widetilde{\mathbf{v}}_k &= \begin{bmatrix} \mathbf{v}_k \\ \widetilde{\mathbf{v}}_{k+1} \end{bmatrix}^T [I_k,\ \widetilde{I}_{k+1}]^T \widetilde{I}_k \widetilde{B}_k \widetilde{I}_k^T [I_k,\ \widetilde{I}_{k+1}] \begin{bmatrix} \mathbf{v}_k \\ \widetilde{\mathbf{v}}_{k+1} \end{bmatrix} \\
&\leq \begin{bmatrix} \mathbf{v}_k \\ \widetilde{\mathbf{v}}_{k+1} \end{bmatrix}^T \widehat{B}_k \begin{bmatrix} \mathbf{v}_k \\ \widetilde{\mathbf{v}}_{k+1} \end{bmatrix} \\
&= \widetilde{\mathbf{v}}_{k+1}^T \widetilde{B}_{k+1} \widetilde{\mathbf{v}}_{k+1} + (M_k^T \mathbf{v}_k + I_k^T A \widetilde{I}_{k+1} \widetilde{\mathbf{v}}_{k+1})^T (M_k^T + M_k - A_k)^{-1} \\
&\quad \times (M_k^T \mathbf{v}_k + I_k^T A \widetilde{I}_{k+1} \widetilde{\mathbf{v}}_{k+1}).
\end{aligned}$$

Therefore, by recursion, we get the inequality

$$\begin{aligned}
\mathbf{v}^T B \mathbf{v} &= \widetilde{\mathbf{v}}_0^T \widetilde{B}_0 \widetilde{\mathbf{v}}_0 \\
&\leq \sum_{k=0}^{J-1} \big(M_k^T \mathbf{v}_k + I_k^T A \widetilde{I}_{k+1} \widetilde{\mathbf{v}}_{k+1}\big)\big(M_k^T + M_k - A_k\big)^{-1}\big(M_k^T \mathbf{v}_k + I_k^T A \widetilde{I}_{k+1} \widetilde{\mathbf{v}}_{k+1}\big) \\
&\quad + \widetilde{\mathbf{v}}_J^T \widetilde{B}_J \widetilde{\mathbf{v}}_J \\
&= \sum_{k=0}^{J-1} \big(M_k^T \mathbf{v}_k + I_k^T A \widetilde{I}_{k+1} \widetilde{\mathbf{v}}_{k+1}\big)\big(M_k^T + M_k - A_k\big)^{-1}\big(M_k^T \mathbf{v}_k + I_k^T A \widetilde{I}_{k+1} \widetilde{\mathbf{v}}_{k+1}\big) \\
&\quad + \mathbf{v}_J^T \overline{M}_J \mathbf{v}_J. \qquad (7.20)
\end{aligned}$$

Now, use Cauchy–Schwarz inequality, the estimate for $\overline{M}_k$, the coercivity estimate for $M_k^T + M_k - A_k$, and the estimate (7.15) for the restricted partial sums, the Cauchy–Schwarz inequality using the bound on the cardinality of the sets $\mathcal{I}(k)$, to end up with the following upper bound,

$$\begin{aligned}
\mathbf{v}^T B \mathbf{v} &\leq 2 \sum_{k=0}^{J} \mathbf{v}_k^T \overline{M}_k \mathbf{v}_k + 2 \frac{1+\varrho}{1-\varrho} \sum_{k=0}^{J-1} \widetilde{\mathbf{v}}_{k+1}^T \widetilde{I}_{k+1}^T A I_k A_k^{-1} I_k^T A \widetilde{I}_{k+1} \widetilde{\mathbf{v}}_{k+1} \\
&\leq 2 \frac{1}{1-\varrho^2} \sum_{k=0}^{J} \mathbf{v}_k^T A_k \mathbf{v}_k + \frac{1+\varrho}{1-\varrho} \sum_{k=0}^{J-1} 2 \Bigg(\sum_{j>k,\, j \in \mathcal{I}(k)} I_j \mathbf{v}_j \Bigg)^T A \Bigg(\sum_{j>k,\, j \in \mathcal{I}(k)} I_j \mathbf{v}_j \Bigg)
\end{aligned}$$

$$= 2\frac{1}{1-\varrho^2}\sum_{k=0}^{J} \mathbf{v}_k^T A_k \mathbf{v}_k + 2\frac{1+\varrho}{1-\varrho}(\mathbf{v} - I_0\mathbf{v}_0)^T A(\mathbf{v} - I_0\mathbf{v}_0)$$

$$+ \frac{1+\varrho}{1-\varrho}\sum_{k=1}^{J-1} 2\left(\sum_{j>k,\ j\in\mathcal{I}(k)} I_j\mathbf{v}_j\right)^T A\left(\sum_{j>k,\ j\in\mathcal{I}(k)} I_j\mathbf{v}_j\right).$$

That is,

$$\mathbf{v}^T B\mathbf{v} \le 2\frac{1}{1-\varrho^2}\sum_{k=0}^{J} \mathbf{v}_k^T A_k \mathbf{v}_k + 4\frac{1+\varrho}{1-\varrho}(\mathbf{v}^T A\mathbf{v} + \mathbf{v}_0^T A_0\mathbf{v}_0)$$

$$+ \frac{1+\varrho}{1-\varrho}\sum_{k=1}^{J-1}\left(\sum_{j_1,\ j_2>k,\ j_1,\ j_2\in\mathcal{I}(k)} (\mathbf{v}_{j_1}^T A_{j_1}\mathbf{v}_{j_1} + \mathbf{v}_{j_2}^T A_{j_2}\mathbf{v}_{j_2})\right)$$

$$\le 2\frac{1}{1-\varrho^2}\sum_{k=0}^{J} \mathbf{v}_k^T A_k \mathbf{v}_k + 4\frac{1+\varrho}{1-\varrho}\left(\mathbf{v}^T A\mathbf{v} + \mathbf{v}_0^T A_0\mathbf{v}_0\right)$$

$$+ \frac{1+\varrho}{1-\varrho}\sum_{k=1}^{J-1} 2\kappa \sum_{j>k,\ j\in\mathcal{I}(k)} \mathbf{v}_j^T A_j\mathbf{v}_j$$

$$\le 2\left[\frac{1}{1-\varrho^2} + \kappa^2\frac{1+\varrho}{1-\varrho}\right]\sum_{k=0}^{J} \mathbf{v}_k^T A_k \mathbf{v}_k + 4\frac{1+\varrho}{1-\varrho}\mathbf{v}^T A\mathbf{v}.$$

Thus, based on the assumed stability of the decomposition $\mathbf{v} = \sum_{k=0}^{J} I_k\mathbf{v}_k$ the desired upper bound follows.

The lower bound $\mathbf{v}^T A\mathbf{v} \le \mathbf{v}^T B\mathbf{v}$ follows from the fact that the block-factorization preconditioner leads to an iteration matrix $E_{DD} = I - B^{-1}A$, which admits the following product form (proven in Lemma 7.7),

$$\begin{aligned} E_{DD} = {} & (I - I_0 M_0^{-T} I_0^T A)\cdots(I - I_k M_k^{-T} I_k^T A)\cdots(I - I_{J-1} M_{J-1}^{-T} I_{J-1}^T A) \\ & \times (I - I_J \overline{M}_J^{-1} I_J^T A)(I - I_{J-1} M_{J-1}^{-1} I_{J-1}^T A)\cdots(I - I_k M_k^{-1} I_k^T A)\cdots \\ & \times (I - I_0 M_0^{-1} I_0^T A). \end{aligned}$$

Noticing that

$$\begin{aligned} & (I - I_J M_J^{-T} I_J^T A)(I - I_J M_J^{-1} I_J^T A) \\ & = \left(I - I_J\left(M_J^{-T} + M_J^{-1} - M_J^{-T} I_J^T A I_J M_J^{-1}\right) I_J^T A\right) \\ & = \left(I - I_J\left(M_J^{-T} + M_J^{-1} - M_J^{-T} A_J M_J^{-1}\right) I_J^T A\right) \\ & = I - I_J \overline{M}_J^{-1} I_J^T A, \end{aligned}$$

we obtain that $AE_{DD} = \mathcal{E}^T A\mathcal{E}$, with

$$\mathcal{E}_{DD} = (I - I_J M_J^{-1} I_J^T A)\cdots(I - I_k M_k^{-1} I_k^T A)\cdots(I - I_0 M_0^{-1} I_0^T A). \tag{7.21}$$

Thus, we see that $A - AB^{-1}A = \mathcal{E}_{DD}^T A \mathcal{E}_{DD}$ is symmetric positive semidefinite. Hence, the positive definiteness of B implies that $\mathbf{v}^T A\mathbf{v} \leq \mathbf{v}^T B\mathbf{v}$. □

As a side result, we also proved the following identity.

Corollary 7.10. *The nonsymmetric Schwarz iteration matrix* $\mathcal{E}_{DD}$ *defined in* (7.21), *exploiting solutions with* M_k *in the subspaces Range*(I_k), $k = 0, \ldots, J$, *has a convergence factor equal to the square root of* $\varrho(E_{DD})$, *the convergence factor of* E_{DD}.

Noticing that inequality (7.20) holds as equality for special vectors (proven in the same way as in Lemma 5.8), we can formulate the following corollary.

Corollary 7.11. *The following characterization of the Schwarz preconditioner holds.*

$$\mathbf{v}^T B\mathbf{v} = \inf_{\mathbf{v}=\sum_{k=0}^{J} I_k \mathbf{v}_k} \left[\sum_{k=0}^{J-1} \left(M_k^T \mathbf{v}_k + I_k^T A \widetilde{I}_{k+1} \widetilde{\mathbf{v}}_{k+1}\right)^T \left(M_k^T + M_k - A_k\right)^{-1} \right.$$
$$\left. \times \left(M_k^T \mathbf{v}_k + I_k^T A \widetilde{I}_{k+1} \widetilde{\mathbf{v}}_{k+1}\right) + \mathbf{v}_J^T \overline{M}_J \mathbf{v}_J \right].$$

Here, $\widetilde{\mathbf{v}}_{k+1} = \widetilde{I}_{k+1}^T \sum_{j\geq k+1} I_j \mathbf{v}_j$ *are the partial sums* $\sum_{j\geq k+1} I_j \mathbf{v}_j$ *restricted to the union of subdomains* $\widetilde{\Omega}_{k+1} = \Omega_J \cup \cdots \cup \Omega_{k+1}$, *that is, to their support.*

7.8 Additive Schwarz preconditioners

Similar results (as in Theorem 7.9 and Corollary 7.11) hold for the additive Schwarz method defined by simply deleting the off-diagonal blocks of $\widehat{B}_k$ in Definition 7.6. For some pioneering works on additive Schwarz methods, we refer to [Li87], [MN85], and [DW87].

In this section, we present one specific version of the additive Schwarz method proposed in [CDS03]. It utilizes a nonoverlapping partition $\{\Omega_i\}$ of the degrees of freedom $\mathcal{N}$ and a overlapping one $\{\widetilde{\Omega}_i\}$ where each $\widetilde{\Omega}_i$ is obtained by extending each Ω_i by a few neighboring grid lines (or matrix graph level sets). The method is referred to as restricted additive Schwarz with harmonic overlap (or RASHO). It can be summarized as follows. Let I_k be the characteristic diagonal matrices that extend a local vector defined on Ω_k to a global vector with zero entries outside Ω_k. Let $\widetilde{A}_k$ be the principal submatrix of A corresponding to the extended subdomain $\widetilde{\Omega}_k$. Based on a two-by-two block partitioning of $\widetilde{A}_k$,

$$\widetilde{A}_k = \begin{bmatrix} A_k & U_k \\ L_k & X_k \end{bmatrix} \begin{matrix} \} \ \Omega_k \\ \} \ \widetilde{\Omega}_k \setminus \Omega_k \end{matrix},$$

we introduce the s.p.d. Schur complement matrices $S_k = A_k - U_k X_k^{-1} L_k$. Then,

$$B_{\text{RASHO}}^{-1} = \sum_k I_k S_k^{-1} I_k^T.$$

In practice, we do not have to explicitly form S_k (nor S_k^{-1}). The actions of S_k^{-1} are computable through the inverse actions of $\widetilde{A}_k^{-1}$. More specifically, we can use the relation

$$S_k^{-1}\mathbf{v}_k = \left(\widetilde{A}_k^{-1}\begin{bmatrix}\mathbf{v}_k\\0\end{bmatrix}\right)\bigg|_{\Omega_k}.$$

That is, we extend the r.h.s. vector $\mathbf{v}_k$ (defined on Ω_k only) to a vector defined on the extended domain $\widetilde{\Omega}_k$ with zero entries outside Ω_k, then apply $\widetilde{A}_k^{-1}$ to the extended vector, and finally restrict the result to Ω_k.

The RASHO preconditioner has some advantages over the more traditional Schwarz methods (with overlap) because it requires less communication in a parallel implementation. It can be analyzed in the same way as the traditional Schwarz methods as long as one can derive stable decompositions with components that are "$\widetilde{A}_k$"-harmonic in $\widetilde{\Omega}_k \backslash \Omega_k$.

To analyze the spectral equivalence properties of B_{RASHO} w.r.t. A, we also include a coarse space $\mathbf{V}_0$ such that for an interpolation matrix P, $P\mathbf{V}_0 \subset \mathbf{V}$. The role of the coarse space is such that for any $\mathbf{v} \in \mathbf{V}$, a coarse approximation $\mathbf{v}_0$ exists so that the difference $\mathbf{v} - \mathcal{P}\mathbf{v}_0$ can be decomposed as a sum of local components $\mathbf{v}_k$ supported in Ω_k.

In order to prove spectral equivalence, we need to construct a decomposition

$$\mathbf{v} = P\mathbf{v}_0 + \sum_k I_k \overline{\mathbf{v}}_k. \tag{7.22}$$

which is stable, that is, such that $\sum_k \|\overline{\mathbf{v}}_k\|^2_{S_k} \le \sigma\ \|\mathbf{v}\|^2_A$ and $\mathbf{v}_0^T A_0 \mathbf{v}_0 = \mathbf{v}_0^T P^T A P \mathbf{v}_0 \le \sigma\ \mathbf{v}^T A \mathbf{v}$.

Let $\widetilde{I}_k$ be the characteristic diagonal matrix that extends a vector defined on $\widetilde{\Omega}_k$ with zero entries outside $\widetilde{\Omega}_k$. We derive a stable decomposition as in (7.22), assuming that there is one suitable for the traditional Schwarz method; that is,

$$\mathbf{v} = P\mathbf{v}_0 + \sum_k \widetilde{I}_k \widetilde{\mathbf{w}}_k, \tag{7.23}$$

where $\sum_k \|\widetilde{I}_k \widetilde{\mathbf{w}}_k\|^2_A \le \widetilde{\sigma}\ \|\mathbf{v}\|^2_A$ and $\|P\mathbf{v}_0\|^2_A \le \widetilde{\sigma}\ \|\mathbf{v}\|^2_A$. Note that $\widetilde{\mathbf{w}}_k$ is supported in the extended subdomain $\widetilde{\Omega}_k$.

Assuming that a stable decomposition (7.23) exists, we construct a stable decomposition with components that are supported in the original subdomains Ω_k. Let the set $\mathcal{N}(k)$ consist of indices j such that Ω_j intersects $\widetilde{\Omega}_k$. We assume that $|\mathcal{N}(k)| \le \kappa$ for a fixed integer κ.

We decompose each component in (7.23) $\widetilde{I}_k \widetilde{\mathbf{w}}_k = \sum_{j \in \mathcal{N}(k)} \mathbf{v}_{k;\,j}$ where each $\mathbf{v}_{k;\,j}$ is now supported in Ω_j. The construction utilizes the local additive Schwarz operators $T_k = \sum_{j \in \mathcal{N}(k)} I_j S_j^{-1} I_j^T$. We note that T_k as a mapping from the vector space $\widetilde{\mathbf{V}}_k$ consisting of vectors supported in $\cup_{j \in \mathcal{N}(k)} \Omega_j$ into itself is s.p.d. and hence, invertible. Therefore, the following vectors are well defined

$$\mathbf{v}_{k;\,j} = I_j S_j^{-1} I_j^T T_k^{-1} \widetilde{I}_k \widetilde{\mathbf{w}}_k, \quad \text{for } j \in \mathcal{N}(k).$$

It is clear that

$$\sum_{j\in\mathcal{N}(k)} \mathbf{v}_{k;\ j} = \widetilde{I}_k\widetilde{\mathbf{w}}_k.$$

If we prove that $\|\mathbf{v}_{k;\ j}\|_A \le C\ \|\widetilde{I}_k\widetilde{\mathbf{w}}_k\|_A$, the desired result will then easily follow. Another realistic assumption is that T_k^{-1} (as a local additive Schwarz preconditioner) is spectrally equivalent to A restricted to the vector space $\widetilde{\mathbf{V}}_k$ (the vectors supported in $\cup_{j\in\mathcal{N}(k)}\Omega_j$). Thus, we can assume the uniform in $k \ge 1$ estimates, for vectors supported in $\cup_{j\in\mathcal{N}(k)}\Omega_j$,

$$\|\widetilde{I}_k\widetilde{\mathbf{w}}_k\|^2_{T_k^{-1}} \le \eta\ \|\widetilde{I}_k\widetilde{\mathbf{w}}_k\|_A^2.$$

By construction, letting $\mathbf{v}_{k;\ j} = I_j\overline{\mathbf{v}}_{k;\ j}$, $\overline{\mathbf{v}}_{k;\ j} = S_j^{-1}I_j^T T_k^{-1}\widetilde{I}_k\widetilde{\mathbf{w}}_k$, we have

$$\overline{\mathbf{v}}_{k;\ j}^T S_j\overline{\mathbf{v}}_{k;\ j} = \overline{\mathbf{v}}_{k;\ j}^T I_j^T T_k^{-1}\widetilde{I}_k\widetilde{\mathbf{w}}_k \le \|\overline{\mathbf{v}}_{k;\ j}\|_{S_j}\|S_j^{-(1/2)}I_j^T T_k^{-1}\widetilde{I}_k\widetilde{\mathbf{w}}_k\|.$$

That is,

$$\overline{\mathbf{v}}_{k;\ j}^T S_j\overline{\mathbf{v}}_{k;\ j} \le \|S_j^{-(1/2)}I_j^T T_k^{-1}\widetilde{I}_k\widetilde{\mathbf{w}}_k\|^2.$$

After a summation, we end up with the estimate

$$\begin{aligned}
\sum_{j\in\mathcal{N}(k)} \|\overline{\mathbf{v}}_{k;\ j}\|^2_{S_j} &\le \sum_{j\in\mathcal{N}(k)} (\widetilde{I}_k\widetilde{\mathbf{w}}_k)^T T_k^{-1}I_jS_j^{-1}I_j^T T_k^{-1}\widetilde{I}_k\widetilde{\mathbf{w}}_k \\
&= (\widetilde{I}_k\widetilde{\mathbf{w}}_k)^T T_k^{-1}\widetilde{I}_k\widetilde{\mathbf{w}}_k \\
&\le \eta\ \|\widetilde{I}_k\widetilde{\mathbf{w}}_k\|_A^2.
\end{aligned}$$

The final decomposition is then based on the components

$$\mathbf{v}_j = \sum_{k:\ j\in\mathcal{N}(k)} \mathbf{v}_{k;\ j} = I_jS_j^{-1}I_j^T \sum_{k:\ j\in\mathcal{N}(k)} T_k^{-1}\widetilde{I}_k\widetilde{\mathbf{w}}_k$$

that are supported in Ω_j. We also have

$$\sum_j \mathbf{v}_j = \sum_j \sum_{k:\ j\in\mathcal{N}(k)} \mathbf{v}_{k;\ j} = \sum_k \sum_{j\in\mathcal{N}(k)} \mathbf{v}_{k;\ j} = \sum_k \widetilde{I}_k\widetilde{\mathbf{w}}_k = \mathbf{v} - P\mathbf{v}_0.$$

Thus, from the main identity for additive Schwarz preconditioners, utilizing the particular decomposition derived above, we have the first desired estimate (with $\overline{\mathbf{v}}_j = S_j^{-1}I_j^T\sum_{k:\ j\in\mathcal{N}(k)} T_k^{-1}\widetilde{I}_k\widetilde{w}_k$):

$$\begin{aligned}
\mathbf{v}^T B_{\text{RASHO}}\mathbf{v} &= \min_{\mathbf{v}=P\mathbf{v}_0+\sum_j I_j\overline{\mathbf{v}}_j} \left(\mathbf{v}_0^T A_0\mathbf{v}_0 + \sum_j \overline{\mathbf{v}}_j^T S_j\overline{\mathbf{v}}_j\right) \\
&\le \mathbf{v}_0^T A_0\mathbf{v}_0 + \kappa \sum_k \sum_{j\in\mathcal{N}(k)} \|\overline{\mathbf{v}}_{k;\ j}\|^2_{S_j}
\end{aligned}$$

$$
\begin{aligned}
&\leq \mathbf{v}_0^T A_0 \mathbf{v}_0 + \kappa\ \eta \sum_k \|\widetilde{I}_k \widetilde{\mathbf{w}}_k\|_A^2 \\
&\leq \mathbf{v}_0^T A_0 \mathbf{v}_0 + \eta\ \kappa\ \widetilde{\sigma}\ \|\mathbf{v}\|_A^2 \\
&\leq \widetilde{\sigma}\ (1 + \eta\ \kappa)\ \|\mathbf{v}\|_A^2. \qquad (7.24)
\end{aligned}
$$

The estimate in the other direction uses the traditional additive Schwarz estimate (based on a bounded overlap assumption); namely, for any decomposition $\mathbf{v} = P\mathbf{v}_0 + \sum_k I_k \overline{\mathbf{v}}_k$,

$$
\mathbf{v}^T A \mathbf{v} \leq 2\mathbf{v}_0^T A_0 \mathbf{v}_0 + 2\kappa \sum_k \overline{\mathbf{v}}_k^T I_k^T A I_k \overline{\mathbf{v}}_k
$$

Without any additional assumptions, the analysis can proceed as follows. Use the fact that for relatively small subdomains Ω_k, $A_k = I_k^T A I_k$ and $\widetilde{A}_k = \widetilde{I}_k^T A \widetilde{I}_k$ are spectrally equivalent to their diagonals. Therefore, S_k as a Schur complement of $\widetilde{A}_k$ will be spectrally equivalent to the diagonal of A_k. In conclusion, we may assume that A_k is spectrally equivalent to S_k. This is reflected in the constant σ below. It is clear that σ will be a reasonable constant if $\text{diam}(\Omega_k)$ is relatively small. With the last assumption, we have the estimate

$$
\begin{aligned}
\mathbf{v}^T A \mathbf{v} &\leq 2\mathbf{v}_0^T A_0 \mathbf{v}_0 + 2\kappa\ \sigma \sum_k \overline{\mathbf{v}}_k^T S_k \overline{\mathbf{v}}_k \\
&\leq 2\ \max\{1,\ \kappa\ \sigma\} \left(\mathbf{v}_0^T A_0 \mathbf{v}_0 + \sum_k \overline{\mathbf{v}}_k^T S_k \overline{\mathbf{v}}_k \right).
\end{aligned}
$$

By taking the minimum over all possible decompositions, we arrive at the second desired estimate

$$
\begin{aligned}
\mathbf{v}^T A \mathbf{v} &\leq 2\ \max\{1,\ \kappa\ \sigma\} \min_{\mathbf{v} = P\mathbf{v}_0 + \sum_k I_k \overline{\mathbf{v}}_k} \left(\mathbf{v}_0^T A_0 \mathbf{v}_0 + \sum_k \overline{\mathbf{v}}_k^T S_k \overline{\mathbf{v}}_k \right) \\
&= 2\ \max\{1,\ \kappa\ \sigma\}\ \mathbf{v}^T B_{\text{RASHO}} \mathbf{v}. \qquad (7.25)
\end{aligned}
$$

In the model case of finite element matrices A coming from second-order elliptic equations, the precise dependence of the constants in the spectral equivalence relations between B_{RASHO} and A in terms of the maximal diameter H of the subdomains, the fine-grid mesh-size h, and the size of the overlap δ are studied in [CDS03]. Moreover, in [CDS03] it was shown that by solving local problems a computable w can be constructed such that the difference $w - u^*$ (u^* is the exact solution) can be decomposed as a sum of local functions that are harmonic in the extended subdomains. That is, RASHO can be implemented by keeping all iterates in terms of sum of components that are harmonic in the extended subdomains.

On the stability estimate (7.19) in a model finite element case

Given are a domain $\Omega \subset \mathbb{R}^d$, a plane polygon ($d = 2$), or a polytope ($d = 3$) and let $\mathcal{T}_H$ be a coarse triangulation of Ω. Assume that we are given a nonoverlapping partition $\{\Omega_j'\}_{j=1}^J$ of Ω with each Ω_j' being coarse-grid domains, that is, completely covered by elements from $\mathcal{T}_H$. By extending each Ω_j' to Ω_j, again coarse-grid domains, we get an overlapping domain decomposition of Ω. Let $\delta \simeq \text{distance}(\partial\Omega_j, \partial\Omega_j')$ be the size of the overlap. Let $\mathcal{T}_h$ be a triangulation of Ω obtained by a refinement of $\mathcal{T}_H$. Also, let $V = V_h$ be a finite element space associated with $\mathcal{T}_h$ and $\widetilde{V} = V_H$ be a coarse finite element space corresponding to $\mathcal{T}_H$. We assume that $V_H \subset V_h$. Consider finally the spaces V_j of finite element functions $\varphi \in V$ that are supported in $\overline{\Omega}_j$ and for convenience let $V_0 = V_H$. Then, we can write that $V = \sum_{j=0}^J V_j$. We may prove (cf., e.g., Dryja and Widlund [DW87]), that given a $v \in V$, the following stable decomposition exists,

$$v = \sum_{j=0}^{J} v_j, \qquad v_j \in V_j,$$

in the sense that there is a positive constant C independent of J and h, such that,

$$\sum_{j=0}^{J} \|v_j\|_1^2 \le C\|v\|_1^2. \tag{7.26}$$

Here $\|.\|_1$ stands for the norm in the Sobolev space $H^1(\Omega)$. The constant C satisfies $C = \mathcal{O}((H/\delta)^2)$ where δ is the size of the overlap. That is, if the overlap is generous ($\delta \simeq H$), C remains bounded uniformly in $H \to 0$. The proof is based on the construction of partition of unity functions $\theta_j \ge 0$ supported in $\overline{\Omega}_j$ such that $\|\nabla\theta_j\|_\infty \le C/\delta$. Partition of unity means that $\sum_j \theta_j = 1$. Let V_h be spanned by the Lagrangian (nodal) basis $\{\varphi_i\}$ associated with the nodes x_i of the triangulation $\mathcal{T}_h$. Let I_h stand for the nodal interpolation operator defined for any continuous function θ as $I_h\theta = \sum_{x_i} \theta(x_i)\,\varphi_i$. Then, for appropriately chosen coarse function v_H, consider the expansion

$$v - v_H = \sum_i I_h(\theta_i(v - v_H)).$$

Note that $v_i \equiv I_h(\theta_i(v - v_H)) \in V_i$. For the model case of second-order elliptic f.e. problems, we can easily show (exploiting the fact that the derivative of any piecewise linear function restricted to an element can be estimated by differences of its nodal values), that

$$|v_i|_1^2 \le C(\|\nabla\theta_i\|_\infty^2\, \|v - v_H\|_{0,\,\Omega_i}^2 + |v - v_H|_{1,\,\Omega_i}^2).$$

This estimate, after summation (based on bounded overlap assumption, i.e., that a domain Ω_i intersects a bounded number of subdomains Ω_j) shows the stability of the decomposition of $v - v_H$,

$$\sum_i |v_i|_1^2 \le C\left(\frac{1}{\delta^2}\,\|v - v_H\|_0^2 + |v - v_H|_1^2\right).$$

The desired result then follows by choosing the coarse space component v_H such that

$$H^{-1}\|v - v_H\|_0 + |v - v_H|_1 \le C\;|v|_1.$$

7.9 The domain decomposition paradigm of Bank and Holst

Here we present the DD paradigm of Bank and Holst [BH03]. It was introduced as a tool for parallel adaptive mesh generation. Here we take the linear algebra solver point of view. For the lowest-order (piecewise linear) finite elements the resulting matrix graph can be identified with the corresponding finite element mesh. The DD paradigm exploits several meshes. There is a final fine mesh that corresponds to the matrix A in question. It is assembled from a number of subdomain matrices A_i, $i = 1, 2, \ldots, p$. That is, the mesh domain Ω is composed of a set of subdomains Ω_i. We view these as sets of vertices of the mesh restricted to some (closed) geometric subdomains $\overline{\Omega}_i$ covered exactly by a number of finite elements from the final fine mesh. For our goal, the latter knowledge is not needed. We only need to know that there is a separator set $\Gamma = \cup_{i=1}^{p} \partial\Omega_i$, and every two subdomains Ω_i and Ω_j can have common nodes only from Γ.

Another ingredient is a global coarse mesh denoted by Ω_c and an associated coarse matrix A_c. This global coarse mesh is used only implicitly in the construction that follows. A main property of Ω_c is that it coincides with Ω on a strip around Γ. Denote this strip by Γ_δ. The width of the strip is assumed of order $2m + 1$ times the fine-grid mesh-size h. In geometric terms, we have $\delta \simeq (2m + 1)h$. This assumption is equivalent to the following property of A_c and A. Let

$$\mathbf{v}_c = \begin{bmatrix} 0 \\ \mathbf{v}_b \end{bmatrix} \begin{matrix} \} \; \Omega_c \setminus \Gamma \\ \} \; \Gamma \end{matrix} \quad \text{and similarly} \quad \mathbf{v} = \begin{bmatrix} 0 \\ \mathbf{v}_b \end{bmatrix} \begin{matrix} \} \; \Omega \setminus \Gamma \\ \} \; \Gamma \end{matrix} .$$

Then, $(A_c)^k \mathbf{v}_c$ and $A^k \mathbf{v}$, $0 \le k \le m$ are zero outside the strip Γ_δ, and also they coincide on Γ_δ.

The main ingredients of the paradigm are the partially coarse global meshes $\Omega^{(i)}$ and respective matrices $A^{(i)}$. Let the global coarse matrix A_c be assembled from the matrices $A_i^{(c)}$ coming from $\Omega_i^{(c)}$ (the part of Ω_c contained in Ω_i). Similarly, let A be assembled from the local (subdomain) matrices A_i. Then, the composite mesh matrices $A^{(i)}$ are assembled from A_i and $A_j^{(c)}$ for all $j \ne i$. Due to our assumption about the coarse mesh, $A^{(i)}$ coincides with both A and A_c on the strip Γ_δ.

To describe the linear system setting, we introduce vectors $\mathbf{v}_i$ defined on Ω_i. The vectors $\mathbf{v}_i$ and $\mathbf{v}_j$ do not necessarily match on $\Omega_i \cap \Omega_j \subset \Gamma$. We enforce continuity by proper constraints by simply identifying the values of $\mathbf{v}_i$ coming from different subdomains with a single (master) unique value on the common node on Γ. The latter is represented by the equation

$$\sum_i B_i \mathbf{v}_i = 0.$$

Here, the matrices B_i have entries equal to 0, -1, or 1. More specifically, the above equation rewritten entrywise, reads $(\mathbf{v}_{i(s)})_s - (\mathbf{v}_j)_s = 0$ for every master node s, coming from a unique subdomain $\Omega_{i(s)}$, a number of simple equations of the form for all indices $j \ne i(s)$ such that Ω_j and $\Omega_{i(s)}$ meet at the node s on Γ.

The linear system of equations then reads

$$\begin{bmatrix} A_1 & 0 & \dots & 0 & B_1^T \\ 0 & A_2 & 0 & & B_2^T \\ \vdots & \ddots & \ddots & \ddots & \vdots \\ 0 & & 0 & A_p & B_p^T \\ B_1 & B_2 & \dots & B_p & 0 \end{bmatrix} \begin{bmatrix} \mathbf{u}_1 \\ \mathbf{u}_2 \\ \vdots \\ \mathbf{u}_p \\ \boldsymbol{\lambda} \end{bmatrix} = \begin{bmatrix} \mathbf{f}_1 \\ \mathbf{f}_2 \\ \vdots \\ \mathbf{f}_p \\ 0 \end{bmatrix}.$$

The two-domain case

We describe the solution iterative method for the two-domain case first. Let $\mathbf{V} = \mathbf{V}_1 \oplus \mathbf{V}_2$ be the vector space of our interest. There is also an auxiliary space $\boldsymbol{\Lambda}$ of Lagrange multipliers. The matrix of the problem under consideration admits the saddle-point form,

$$\mathcal{A} = \begin{bmatrix} A_1 & 0 & B_1^T \\ 0 & A_2 & B_2^T \\ B_1 & B_2 & 0 \end{bmatrix}.$$

There are two coarse versions of this matrix, $\mathcal{A}_1$ and $\mathcal{A}_2$, corresponding to the spaces $\mathbf{V}^{(1)} = \mathbf{V}_1 \oplus P_2\mathbf{V}_2^c$ and $\mathbf{V}^{(2)} = P_1\mathbf{V}_1^c \oplus \mathbf{V}_2$, where $P_1\mathbf{V}_1^c \subset \mathbf{V}_1$ and $P_2\mathbf{V}_2^c \subset \mathbf{V}_2$, for two given interpolation matrices P_1 and P_2. Note the special form of the interpolation matrices P_1 and P_2; namely, they have an identity block corresponding to the strip Γ_δ (restricted to the respective subdomain Ω_i). For example,

$$P_1 = \begin{bmatrix} * & 0 \\ 0 & I \end{bmatrix} \begin{matrix} \} & \Omega_1 \setminus \Gamma_\delta \\ \} & \Gamma_\delta \cap \Omega_1 \end{matrix}. \tag{7.27}$$

We have

$$\mathcal{A}_1 = \begin{bmatrix} A_1 & 0 & B_1^T \\ 0 & P_2^T A_2 P_2 & P_2^T B_2^T \\ B_1 & B_2 P_2 & 0 \end{bmatrix} \quad \text{and} \quad \mathcal{A}_2 = \begin{bmatrix} P_1^T A_1 P_1 & 0 & P_1^T B_1^T \\ 0 & A_2 & B_2^T \\ B_1 P_1 & B_2 & 0 \end{bmatrix}.$$

We are interested in a iterative procedure (described below) for solving

$$\mathcal{A} \begin{bmatrix} \mathbf{x}_1 \\ \mathbf{x}_2 \\ \underline{\lambda} \end{bmatrix} = \begin{bmatrix} \mathbf{f}_1 \\ \mathbf{f}_2 \\ 0 \end{bmatrix}.$$

The matrices B_1 and B_2 are chosen in practice such that $B_1\mathbf{u}_1 = B_2\mathbf{u}_2$, which ensures that $\mathbf{u}_1$ and $\mathbf{u}_2$ coincide on the separator Γ (specified in (7.33) below).

Algorithm 7.9.1 (Bank–Holst DD paradigm). *Let* $(\mathbf{x}_1,\ \mathbf{x}_2,\ \underline{\lambda})$ *be a current iterate, such that the respective residual admits the form*

$$\mathbf{R} \equiv \mathbf{F} - \mathcal{A}\mathbf{X} = \begin{bmatrix} \mathbf{f}_1 \\ \mathbf{f}_2 \\ 0 \end{bmatrix} - \mathcal{A} \begin{bmatrix} \mathbf{x}_1 \\ \mathbf{x}_2 \\ \underline{\lambda} \end{bmatrix} = \begin{bmatrix} \mathbf{r}_1 \\ \mathbf{r}_2 \\ 0 \end{bmatrix}.$$

Let

$$Q_1 = \begin{bmatrix} P_1 & 0 & 0 \\ 0 & I & 0 \\ 0 & 0 & I \end{bmatrix}, \qquad Q_2 = \begin{bmatrix} I & 0 & 0 \\ 0 & P_2 & 0 \\ 0 & 0 & I \end{bmatrix}.$$

The next iterates $\mathbf{y}_1$, $\mathbf{y}_2$ *are defined as follows.*

1. *Solve for* $\mathbf{Y}_k$, $k = 1, 2$,

$$\mathcal{A}_k \mathbf{Y}_k = Q_k^T \mathbf{R}.$$

2. *Set*

$$\mathbf{y}_1 = \mathbf{x}_1 + [I,\ 0,\ 0]\, \mathbf{Y}_1, \qquad \mathbf{y}_2 = \mathbf{x}_2 + [0,\ I,\ 0]\, \mathbf{Y}_2.$$

In matrix notation, we have

$$\mathbf{y}_1 = \mathbf{x}_1 + [I,\ 0,\ 0]\, \mathcal{A}_1^{-1} Q_1^T (\mathbf{F} - \mathcal{A}\mathbf{X}),$$
$$\mathbf{y}_2 = \mathbf{x}_2 + [0,\ I,\ 0]\, \mathcal{A}_2^{-1} Q_2^T (\mathbf{F} - \mathcal{A}\mathbf{X}).$$

That is, the iteration matrix reads

$$\begin{bmatrix} [I,\ 0,\ 0](I - \mathcal{A}_1^{-1} Q_1^T \mathcal{A}) \\ [0,\ I,\ 0]\,(I - \mathcal{A}_2^{-1} Q_2^T \mathcal{A}) \\ \star \end{bmatrix}.$$

The following factorization holds.

$$\mathcal{A}_1 = \begin{bmatrix} I & 0 & 0 \\ 0 & I & 0 \\ B_1 A_1^{-1} & B_2 P_2 (P_2^T A_2 P_2)^{-1} & I \end{bmatrix}$$
$$\times \begin{bmatrix} A_1 & 0 & 0 \\ 0 & P_2^T A_2 P_2 & 0 \\ 0 & 0 & -B_1 A_1^{-1} B_1 - B_2 P_2 (P_2^T A_2 P_2)^{-1} P_2^T B_2^T \end{bmatrix}$$
$$\times \begin{bmatrix} I & 0 & A_1^{-1} B_1^T \\ 0 & I & (P_2^T A_2 P_2)^{-1} P_2^T B_2^T \\ 0 & 0 & I \end{bmatrix}.$$

A similar expression holds for $\mathcal{A}_2$. We need to compute the first row of $\mathcal{A}_1^{-1} Q_1^T \mathcal{A}$. We have, letting

$$S_1 = B_1 A_1^{-1} B_1^T + B_2 P_2 (P_2^T A_2 P_2)^{-1} P_2^T B_2^T,$$

and

$$S_2 = B_1 P_1 (P_1^T A_1 P_1)^{-1} P_1^T B_1^T + B_2 A_2^{-1} B_2,$$

$$
\begin{aligned}
&\left[I,\ 0,\ -A_1^{-1}B_1^T\right] \\
&\quad\times\begin{bmatrix} A_1^{-1} & 0 & 0 \\ 0 & (P_2^T A_2 P_2)^{-1} & 0 \\ 0 & 0 & -S_1^{-1} \end{bmatrix}\begin{bmatrix} I & 0 & 0 \\ 0 & I & 0 \\ -B_1 A_1^{-1} & -B_2 P_2 (P_2^T A_2 P_2)^{-1} & I \end{bmatrix} \\
&= \left[I,\ 0,\ -A_1^{-1}B_1^T\right]\begin{bmatrix} A_1^{-1} & 0 & 0 \\ 0 & (P_2^T A_2 P_2)^{-1} & 0 \\ S_1^{-1} B_1 A_1^{-1} & S_1^{-1} B_2 P_2 (P_2^T A_2 P_2)^{-1} & -S_1^{-1} \end{bmatrix} \\
&= \left[A_1^{-1}(I - B_1^T S_1^{-1} B_1 A_1^{-1}),\ -A_1^{-1} B_1^T S_1^{-1} B_2 P_2 (P_2^T A_2 P_2)^{-1},\ A_1^{-1} B_1^T S_1^{-1}\right].
\end{aligned}
$$

Now note that

$$
Q_1^T \mathcal{A} = \begin{bmatrix} A_1 & 0 & B_1^T \\ 0 & P_2^T A_2 & P_2^T B_2^T \\ B_1 & B_2 & 0 \end{bmatrix}
$$

Then, $[I,\ 0,\ 0]\mathcal{A}_1^{-1} Q_1^T \mathcal{A} = [M_1,\ M_2,\ M_3]$ where,

$$
\begin{aligned}
M_1 &= A_1^{-1}(I - B_1^T S_1^{-1} B_1 A_1^{-1}) A_1 + A_1^{-1} B_1^T S_1^{-1} B_1 \\
&= I \\
M_2 &= A_1^{-1} B_1^T S_1^{-1} B_2 - A_1^{-1} B^T 1 S_1^{-1} B_2 P_2 (P_2^T A_2 P_2)^{-1} P_2^T A_2 \\
&= A_1^{-1} B_1^T S_1^{-1} B_2 (I - P_2 (P_2^T A_2 P_2)^{-1} P_2^T A_2) \\
M_3 &= A_1^{-1}(I - B_1^T S_1^{-1} B_1 A_1^{-1}) B_1^T - A_1^{-1} B_1^T S_1^{-1} B_2 P_2 (P_2^T A_2 P_2)^{-1} P_2^T B_2^T \\
&= A_1^{-1} B_1^T - A_1^{-1} B_1^T S_1^{-1} \left[B_1 A_1^{-1} B_1^T + B_2 P_2 \left(P_2^T A_2 P_2\right)^{-1} P_2^T B_2^T\right] \\
&= A_1^{-1} B_1^T - A_1^{-1} B_1^T S_1^{-1} S_1 \\
&= 0.
\end{aligned}
$$

A similar expression holds for $[0,\ I,\ 0]\mathcal{A}_2^{-1} Q_2^T \mathcal{A}$. Thus, we showed the following formula for the iteration matrix $\mathcal{E}_{DD}$,

$$
\mathcal{E}_{DD} = \begin{bmatrix} 0 & -A_1^{-1} B_1^T S_1^{-1} B_2 \pi_2 & 0 \\ -A_2^{-1} B_2^T S_2^{-1} B_1 \pi_1 & 0 & 0 \\ 0 & 0 & I \end{bmatrix}.
$$

Here, $\pi_k = I - P_k (P_k^T A_k P_k)^{-1} P_k^T A_k$ is the A_k-projection onto the coarse space $\mathbf{V}_k^c$, $k = 1, 2$.

A norm estimate

We are interested in the following principal submatrix of $\mathcal{E}_{DD}$,

$$
E_{DD} = \begin{bmatrix} 0 & -A_1^{-1} B_1^T S_1^{-1} B_2 \pi_2 \\ -A_2^{-1} B_2^T S_2^{-1} B_1 \pi_1 & 0 \end{bmatrix}.
$$

For the following analysis, introduce the block-diagonal matrices $A = \text{diag}(A_i)$ and $\pi = \text{diag}(\pi_i)$.

Lemma 7.12. *The following inequality holds,*

$$\mathbf{w}^T A E_{DD} \mathbf{v} \leq (\mathbf{w}^T A \mathbf{w})^{1/2} \left(\|\pi_1 \mathbf{v}_1\|^2_{B_1^T S_2^{-1} B_1} + \|\pi_2 \mathbf{v}_2\|^2_{B_2^T S_1^{-1} B_2} \right)^{1/2},$$

for any

$$\mathbf{v} = \begin{bmatrix} \mathbf{v}_1 \\ \mathbf{v}_2 \end{bmatrix}, \quad \mathbf{w} = \begin{bmatrix} \mathbf{w}_1 \\ \mathbf{w}_2 \end{bmatrix} \text{ and } A = \begin{bmatrix} A_1 & 0 \\ 0 & A_2 \end{bmatrix}.$$

Proof. We have, using the Cauchy–Schwarz inequality,

$$\begin{aligned}
(\mathbf{w}^T A E_{DD} \mathbf{v})^2 &= \left(\mathbf{w}_1^T B_1^T S_1^{-1} B_2 \pi_2 \mathbf{v}_2 + \mathbf{w}_2^T B_2^T S_2^{-1} B_1 \pi_1 \mathbf{v}_1 \right)^2 \\
&\leq \left(\mathbf{w}_1^T A_1 \mathbf{w}_1 + \mathbf{w}_2^T A_2 \mathbf{w}_2 \right) \\
&\quad \times \left((\pi_2 \mathbf{v}_2)^T B_2^T S_1^{-1} B_1 A_1^{-1} B_1^T S_1^{-1} B_2 (\pi_2 \mathbf{v}_2) \right. \\
&\quad \left. + (\pi_1 \mathbf{v}_1)^T B_1^T S_2^{-1} B_2 A_2^{-1} B_2^T S_2^{-1} B_1 (\pi_1 \mathbf{v}_1) \right)
\end{aligned} \tag{7.28}$$

□

Lemma 7.13. *Let E_k^A be the A_k-harmonic extensions of vectors defined on $\Gamma = \partial\Omega_1 \cap \partial\Omega_2$ into the interior of Ω_k. Let $\delta > 0$ be such that*

$$\delta\, \mathbf{z}^T B_2 A_2^{-1} B_2^T \mathbf{z} \leq \mathbf{z}^T B_1 P_1 (P_1^T A_1 P_1)^{-1} P_1^T B_1^T \mathbf{z}, \quad \textit{for all } \mathbf{z}. \tag{7.29}$$

Similarly, we assume that $\delta > 0$ is such that

$$\delta\, \mathbf{z}^T B_1 A_1^{-1} B_1^T \mathbf{z} \leq \mathbf{z}^T B_2 P_2 (P_2^T A_2 P_2)^{-1} P_2^T B_2^T \mathbf{z}, \quad \textit{for all } \mathbf{z}. \tag{7.30}$$

Then, the following estimate holds,

$$\mathbf{w}^T A E_{DD} \mathbf{v} \leq \frac{1}{1+\delta} \|\mathbf{w}\|_A \left[\|E_2^A (\pi_1 \mathbf{v}_1)_\Gamma\|^2_{A_2} + \|E_1^A (\pi_2 \mathbf{v}_2)_\Gamma\|^2_{A_1} \right]^{1/2}. \tag{7.31}$$

Proof. We first comment that the estimates (7.29) and (7.30) hold if the Schur complements of A_1 and A_2 on Γ are spectrally equivalent in the case of the special coarsening around Γ. The special coarsening can ensure that the Schur complements of A_1 and its special coarse version $P_1^T A_1 P_1$ on Γ are spectrally equivalent, and similarly the Schur complements of A_2 and its respective coarse version $P_2^T A_2 P_2$ on Γ are spectrally equivalent. Therefore, we have that the mixed pairs of Schur complements of A_1 and $P_2^T A_2 P_2$, as well as A_2 and $P_1^T A_1 P_1$, are spectrally equivalent. We also have then that their respective inverses are spectrally equivalent. Then because B_1^T and $-B_2^T$ are simply matrices with identity and zero blocks (as in (7.33) below), we can see that the latter fact (about the inverses of the Schur complements) implies that the principal submatrices of the respective matrix inverses are spectrally equivalent, which in fact (based on the identity (7.34) below) represent the assumed inequalities (7.29) and (7.30).

Recall now the formulas

$$S_1 = B_1 A_1^{-1} B_1^T + B_2 P_2 \left(P_2^T A_2 P_2\right)^{-1} P_2^T B_2^T,$$
$$S_2 = B_1 P_1 \left(P_1^T A_1 P_1\right)^{-1} P_1^T B_1^T + B_2 A_2^{-1} B_2^T.$$

Consider the following problem for $\mathbf{u}_2$,

$$S_2 \mathbf{u}_2 = B_1 \pi_1 \mathbf{v}_1. \tag{7.32}$$

Estimate (7.31) is seen, by first noticing that for,

$$B_1^T = \begin{bmatrix} 0 \\ I \end{bmatrix} \begin{matrix} \Omega_1 \setminus \Gamma \\ \Gamma \end{matrix} \quad \text{and} \quad B_2^T = \begin{bmatrix} 0 \\ -I \end{bmatrix} \begin{matrix} \Omega_2 \setminus \Gamma \\ \Gamma \end{matrix}, \tag{7.33}$$

using the fact that P_1 is identity near Γ, (cf. (7.27), due to the special choice of the coarse mesh near Γ), implies

$$P_1^T B_1^T = \begin{bmatrix} 0 \\ I \end{bmatrix} \begin{matrix} \Omega_1 \setminus \Gamma \\ \Gamma \end{matrix}. \tag{7.34}$$

We now estimate the solution $\mathbf{u}_2$ of (7.32). We have (recalling that B_2^T is extension by zero in Ω_2),

$$\begin{aligned}
(1+\delta)\, \mathbf{u}_2^T B_2 A_2^{-1} B_2^T \mathbf{u}_2 &\le \mathbf{u}_2^T S_2 \mathbf{u}_2 \\
&= \mathbf{u}_2^T B_1 \pi_1 \mathbf{v}_1 \\
&= \mathbf{u}_2^T (\pi_1 \mathbf{v}_1)_\Gamma \\
&= -(B_2^T \mathbf{u}_2)^T E_2^A (\pi_1 \mathbf{v}_1)_\Gamma \\
&\le \left(\mathbf{u}_2^T B_2 A_2^{-1} B_2^T \mathbf{u}_2\right)^{1/2} \| E_2^A (\pi_1 \mathbf{v}_1)_\Gamma \|_{A_2}.
\end{aligned}$$

Thus, we proved that

$$\left(\mathbf{u}_2^T B_2 A_2^{-1} B_2^T \mathbf{u}_2\right)^{1/2} \le \frac{1}{1+\delta} \| E_2^A (\pi_1 \mathbf{v}_1)_\Gamma \|_{A_2}.$$

Because $\mathbf{u}_2 = S_2^{-1} B_1 \pi_1 \mathbf{v}_1$, we get that

$$\left((\pi_1 \mathbf{v}_1)^T B_1^T S_2^{-1} B_2 A_2^{-1} B_2^T S_2^{-1} B_1 (\pi_1 \mathbf{v}_1)\right)^{1/2} \le \frac{1}{1+\delta} \| E_2^A\, (\pi_1 \mathbf{v}_1)_\Gamma \|_{A_2}. \tag{7.35}$$

In the same way, we prove

$$\left((\pi_2 \mathbf{v}_2)^T B_2^T S_1^{-1} B_1 A_1^{-1} B_1^T S_1^{-1} B_2 (\pi_2 \mathbf{v}_2)\right)^{1/2} \le \frac{1}{1+\delta} \| E_1^A\, (\pi_2 \mathbf{v}_2)_\Gamma \|_{A_1}. \tag{7.36}$$

Substituting estimates (7.35)–(7.36) in (7.28), we arrive at the desired estimate

$$\mathbf{w}^T A E_{DD} \mathbf{v} \le \frac{1}{1+\delta} \|\mathbf{w}\|_A \left[\| E_2^A (\pi_1 \mathbf{v}_1)_\Gamma \|_{A_2}^2 + \| E_1^A (\pi_2 \mathbf{v}_2)_\Gamma \|_{A_1}^2 \right]^{1/2}. \qquad \square$$

Based on Lemma 7.30, because

$$\mathbf{w} = \begin{bmatrix} \mathbf{w}_1 \\ \mathbf{w}_2 \end{bmatrix}$$

can be arbitrary, by letting $\mathbf{w} := \pi \mathbf{w}$, using the inequality $\|\pi \mathbf{w}\|_A \leq \|\mathbf{w}\|_A$ and the identity $\pi^T A = A\pi$, we get the estimate

$$\mathbf{w}^T A \pi E_{DD} \mathbf{v} \leq \frac{1}{1+\delta} \|\mathbf{w}\|_A \left[\|E_2^A(\pi_1 \mathbf{v}_1)_\Gamma\|_{A_2}^2 + \|E_1^A(\pi_2 \mathbf{v}_2)_\Gamma\|_{A_1}^2 \right]^{1/2}.$$

Again, because $\mathbf{w}$ is arbitrary, by choosing $\mathbf{w}_1 = E_1^A(\mathbf{g}_2)_\Gamma$ and $\mathbf{w}_2 = E_2^A(\mathbf{g}_1)_\Gamma$, $\mathbf{g}_k|_\Gamma = (\pi E_{DD}\mathbf{v})_k|_\Gamma$, we get the convergence rate estimate formulated in the next theorem, letting $\mathbf{v} := \mathbf{x}_{k-1}$ and $\mathbf{x}_k := E_{DD}\mathbf{v}$.

Theorem 7.14.

$$|||\pi \mathbf{x}_k||| \leq \frac{1}{1+\delta} |||\pi \mathbf{x}_{k-1}|||.$$

Here $|||\cdot|||$ is defined for vectors

$$\mathbf{v} = \begin{bmatrix} \mathbf{v}_1 \\ \mathbf{v}_2 \end{bmatrix}$$

restricted to Γ, as follows,

$$|||\mathbf{v}|||^2 = \|E_2^A(\mathbf{v}_1)_\Gamma\|_{A_2}^2 + \|E_1^A(\mathbf{v}_2)_\Gamma\|_{A_1}^2.$$

An algorithm in the general case

The two-domain case analysis presented in the previous section cannot be generalized to the multidomain case without using the fact that the coarse problem has actually a certain approximation property. Also, the saddle-point formulation can be avoided due to the simple form of the constraint matrices B_i. In this section, we use the equivalent unconstrained setting of the problem. We assume that the global coarse problem on the mesh Ω_c defined by the matrix A_c admits the following weak approximation property. Introduce the coarse-grid correction operator $\pi_H = I - PA_c^{-1}P^T A$, where $A_c = P^T AP$, then the following estimate (a standard L_2-error estimate for finite element approximations) to hold is assumed,

$$\|\pi_H \mathbf{u}\|_0 \leq C\ H\ \|\mathbf{u}\|_A.$$

Here, H is the characteristic coarse mesh-size associated with the coarse mesh $\Omega_c = \Omega_H$. The corresponding fine-grid mesh-size is h associated with the mesh $\Omega = \Omega_h$. The following relation holds between the vector norm $\|\cdot\|$ (defined by $\|\mathbf{v}\|^2 = \mathbf{v}^T\mathbf{v}$) and the weak norm $\|\cdot\|_0$, (assuming two-dimensional geometric domain Ω),

$$\|\mathbf{v}\| \leq Ch^{-1}\ \|\mathbf{v}\|_0.$$

That is, combining both estimates, the following weak approximation property is assumed to hold,

$$\|\pi_H \mathbf{u}\| \le C \frac{H}{h} \|\mathbf{u}\|_A. \tag{7.37}$$

The matrix A corresponds to the problem that we are interested in, which formulated without Lagrange multipliers reads as follows,

$$\sum_k I_k A_k I_k^T \mathbf{u} = \mathbf{b}.$$

Here, I_k stands for zero extension of a vector defined on $\overline{\Omega}_k$ onto the entire domain Ω. In what follows, to stress the fact that a subdomain contains its boundary nodes (namely, that it is covered completely by a set of finite elements and contains all their degrees of freedom or nodes), we use overbars. Above, A_k stands for the subdomain matrix assembled from the individual element matrices corresponding to fine-grid elements contained in the subdomain $\overline{\Omega}_k$. The following decompositions of A are of interest.

$$A = I_k A_k I_k^T + I_k^{\text{ext}} A_k^{\text{ext}} \left(I_k^{\text{ext}}\right)^T.$$

Here, I_k^{ext} stands for zero extension of vectors defined on $\overline{\Omega \setminus \Omega_k}$ into (the interior of) Ω_k. Similarly, A_k^{ext} stands for the matrix assembled from the element matrices corresponding to the fine-grid elements contained in the subdomain complementary to Ω_k. We also need the global coarse matrix $\overline{A} = A_c = P^T A P$. Let $\overline{I}_k$ be the extension by zero of vectors defined on $\overline{\Omega}_k$ into the remaining part of the kth global coarsened-away mesh $\Omega_k^{(c)}$ (which outside $\overline{\Omega}_k$ coincides with Ω_c). We can also have the coarsened-away stiffness matrices

$$\overline{A}_k = \overline{I}_k A_k \overline{I}_k^T + \overline{I}_k^{\text{ext}} \overline{A}_k^{\text{ext}} \left(\overline{I}_k^{\text{ext}}\right)^T. \tag{7.38}$$

We have $\overline{A}_k = \overline{P}_k^T A \overline{P}_k$ for some interpolation matrices $\overline{P}_k$ that act as identity near $\overline{\Omega}_k$, and as $\overline{P}$ outside $\overline{\Omega}_k$.

Finally, we are interested in the Schur complements of $\overline{A}_k$ on $\overline{\Omega}_k$, which have the form

$$\overline{S}_k = A_k + J_k S_k^{\text{ext}} J_k^T.$$

Here, J_k is the trivial extension by zero from $\partial\Omega_k$ into the interior of Ω_k. We can use also the notation $\overline{P}_0 = P$ and $\overline{A}_0 = A_c$.

The iterative method of interest can be formulated as follows.

Algorithm 7.9.2 (A Neumann–Neumann algorithm). *Let $\{W_k\}$ be a set of global diagonal matrices with W_k having nonzero entries only on $\overline{\Omega}_k$. We assume that they provide "partition of unity;" that is,*

$$\sum_k W_k = I.$$

Consider the fine-grid problem $A\mathbf{u} = \mathbf{b}$. *Without loss of generality, we assume that* $\mathbf{b}$ *is nonzero only on* Γ. *We perform:*

(0) Global coarse-grid solution; that is, solve

$$A_c\overline{\mathbf{u}} = P^T\mathbf{b}.$$

(i) For every $k \geq 1$ *solution, in parallel, the coarsened-away problems*

$$\overline{A}_k\overline{\mathbf{u}}_k = \overline{P}_k^T(\mathbf{b} - AP\overline{\mathbf{u}}) = \begin{bmatrix} \star \\ 0 \\ 0 \end{bmatrix} \begin{matrix} \} \ \Omega_k \\ \} \ \partial\Omega_k \\ \} \ \text{everywhere else} \end{matrix}$$

Take the "good" part of $\overline{\mathbf{u}}_k$, *namely,* $\overline{I}_k^T\overline{\mathbf{u}}_k$. *The latter is a vector defined on* $\overline{\Omega}_k$.

(ii) "Average" the results to define a global conforming next iterate; that is, form

$$\sum_k W_k I_k \overline{I}_k^T \overline{\mathbf{u}}_k.$$

(iii) The next iterate is $\mathbf{u}_{\text{next}} = P\overline{\mathbf{u}} + \sum_k W_k I_k \overline{I}_k^T \overline{\mathbf{u}}_k$.

We comment on the crucial observation that the r.h.s. in item (i) is zero outside Ω_k. This is due to the fact that $\overline{P}_k$ coincides with P outside the subdomain $\Omega_k \setminus \Gamma_\delta$ of Ω_k. Hence $\overline{P}_k^T(\mathbf{b} - AP\overline{\mathbf{u}}) = P^T\mathbf{b} - A_c\overline{\mathbf{u}} = 0$ outside Ω_k.

Noting that W_k is the identity in the interior of $\overline{\Omega}_k$, $\overline{P}_k$ is the identity on $\overline{\Omega}_k$, and that the r.h.s. in item (i) is zero outside Ω_k because

$$\overline{I}_k^T \overline{A}_k^{-1} \overline{I}_k = \overline{S}_k^{-1},$$

we easily get that

$$\overline{I}_k^T \overline{\mathbf{u}}_k = \overline{S}_k^{-1} I_k^T W_k^T (\mathbf{b} - AP\overline{\mathbf{u}}).$$

Therefore,

$$\mathbf{u}_{\text{next}} = P\overline{\mathbf{u}} + \sum_k W_k I_k \overline{S}_k^{-1} I_k^T W_k^T (\mathbf{b} - AP\overline{\mathbf{u}}).$$

Consider the following mapping

$$M^{-1} = \sum_k W_k I_k \overline{S}_k^{-1} I_k^T W_k^T,$$

with the purpose of using $\overline{\pi} M^{-1} \overline{\pi}^T$ as a preconditioner for $\overline{\pi}^T A \overline{\pi}$, where

$$\overline{\pi} = I - PA_c^{-1}P^T A,$$

is the global coarse-grid correction operator. Mappings such as M above were originally analyzed in [Man93]; see also [DW95] and [DT91].

The method by Bank and Holst ignores the averaging matrices W_k. The reason is that the values of the vectors $\overline{\mathbf{u}}_k$ near the strip Γ_δ are actually negligible due to the special features of the composite meshes $\Omega_i^{(c)}$; that is, they contain a strip Γ_δ near the interface Γ being part of the fine-grid Ω_h with width $\delta = \mathcal{O}(H) \simeq (2m+1)h$.

7.9.1 Local error estimates

In this section, we show that the solution of problems such as shown in item (i) of Algorithm 7.9.2 are essentially local. More specifically, assuming that the coarse-mesh of size H is refined near $\Gamma = \cup\, \partial\Omega_k$ so that it coincides with the fine-mesh in a strip of width $\mathcal{O}(H)$ around Γ.

Consider a s.p.d. matrix A and let $\pi = I - P(P^TAP)^{-1}P^TA$ be an A-projection, for a given interpolation matrix P; that is, $\pi^2 = \pi$. It is clear also that $A\pi = \pi^TA$. In our application, P corresponds to the interpolation from the special global coarse space used in the Bank–Holst DD paradigm into the space associated with the composite coarsened-away mesh $\Omega_k^{(c)}$. The matrix $A := \overline{A}_k$ is any of the global composite-grid matrices. We notice that the r.h.s. in item (i) has a special form $\pi^T\mathbf{b}_k := (I - \overline{A}_k PA_c^{-1}P^T)\mathbf{b}_k$. Here, $\mathbf{b}_k = \overline{P}_k^T\mathbf{b}$.

In what follows, we omit the subscript k.

Consider now the following problem with a special r.h.s.,

$$A\mathbf{u} = \pi^T\mathbf{b}.$$

It is clear that $\mathbf{u} = \pi\mathbf{u}$, because $\pi^T\mathbf{b} = (\pi^2)^T\mathbf{b} = \pi^TA\mathbf{u} = A(\pi\mathbf{u})$. Thus, if we apply the CG method to solve the above problem with zero initial iterate, after $m \geq 1$ steps we get an approximation $\mathbf{u}_m$, which will satisfy the following estimate,

$$\|\mathbf{u} - \mathbf{u}_m\|_A \leq \min_{\varphi_m:\ \varphi_m(0)=1}\ \max_{t\in[0,\ \|A\|]} |\sqrt{t}\varphi_m(t)|\ \|\mathbf{u}\|,$$

where φ_m is a polynomial of degree m, which is normalized at the origin. Because $\mathbf{u} = \pi\ \mathbf{u}$ and in our application (assuming a two-dimensional domain), with $\|.\|_0$ being the integral L_2-norm and u being the finite element function corresponding to the vector $\mathbf{u}$, we end up with the estimate

$$\|\mathbf{u}\| \leq Ch^{-1}\ \|\pi_H u\|_0 \leq C\frac{H}{h}\ \|\mathbf{u}\|_A.$$

Here, we use the L_2-error estimate for the coarse-grid elliptic projection π_H (assuming full regularity); that is, $\|\pi_H u\|_0 \leq CH\ \|\mathbf{u}\|_A$. The final convergence rate estimate, for a proper polynomial φ_m, then takes the form

$$\|\mathbf{u} - \mathbf{u}_m\|_A \leq \frac{\|A\|^{1/2}}{2m+1}\, C\frac{H}{h}\ \|\mathbf{u}\|_A. \tag{7.39}$$

The best polynomial φ_m is defined through the Chebyshev polynomials of degree $2m+1$ (cf., Section 6.13.2) as follows; for $t \in [-1, 1]$,

$$T_{2m+1}(t) = (-1)^m(2m+1)t\varphi_m(\|A\|t^2).$$

Now, concentrate on the solution of problems such as shown in item (i). Note that the r.h.s. has the form $\pi^T\mathbf{b}$, where $A := \overline{A}_k$ and a π that comes from a P that interpolates from the global coarse space into the partially fine space (on a mesh-coarsened way from Ω_k). That is, this P is identity outside a subdomain $\Omega_k^{(0)}$ of Ω_k.

Under our refinement assumption, we have $\text{dist}(\Omega_k^{(0)}, \ \Omega \setminus \Omega_k) \simeq \mathcal{O}(H) \geq 2m_0 h$ for some integer m_0, which we assume is of order H/h. Thus, we have a problem

$$\overline{A}_k \overline{\mathbf{u}}_k = \overline{\mathbf{b}} = \pi^T \mathbf{b}.$$

The important observation is that $\overline{\mathbf{b}}$ is supported in $\Omega_k^{(0)}$. Applying m_0 standard CG iterations (with zero initial guess) to the last system leads to an approximation $\overline{\mathbf{u}}_k^{(m_0)}$. It is clear that $\overline{\mathbf{u}}_k^{(m_0)}$ is supported in Ω_k (because $A^{m_0}\overline{\mathbf{b}}$ is supported in Ω_k). More specifically, we have

$$\text{dist}\big(\text{support}(\overline{\mathbf{u}}_k^{(m_0)}), \ \Omega \setminus \Omega_k\big) \simeq \mathcal{O}(H) \geq m_0 h.$$

Applying the CG convergence rate estimate (7.39), we can get (because $\|A\| = \mathcal{O}(1)$) for an a priori chosen tolerance $\epsilon < 1$, the estimate

$$\|\overline{\mathbf{u}}_k - \overline{\mathbf{u}}_k^{(m_0)}\|_{\overline{A}_k} \leq \epsilon \ \|\overline{\mathbf{u}}_k\|_{\overline{A}_k}, \tag{7.40}$$

if we have chosen

$$m_0 \simeq \frac{H}{h} \frac{1}{\epsilon}.$$

We also want $m_0 h \simeq CH$; thus for $C \simeq 1/\epsilon$, the error estimate (7.40) is feasible.

In conclusion, because the solutions $\overline{\mathbf{u}}_k$ in item (i) are essentially local (due to the error estimate (7.40)) the averaging involved in Step (ii) does not really need to take place.

The actual counterpart of Algorithm 7.9.1 is an approximation of Algorithm 7.9.2. The global coarse matrix A_c is not used. Nevertheless, the r.h.s. for the composite grid problems involving $\overline{A}_k$ is kept orthogonal to the global coarse space. Thus, we get subdomain updates $\overline{I}_k^T \overline{\mathbf{u}}_k$ that are essentially supported in Ω_k in the sense that they can be approximated with local $\overline{\mathbf{u}}_k^0$ such that (see (7.38)) an estimate $\|\overline{I}_k^T(\overline{\mathbf{u}}_k - \overline{\mathbf{u}}_k^{(0)})\|_{A_k} \leq \|\overline{\mathbf{u}}_k - \overline{\mathbf{u}}_k^{(0)}\|_{\overline{A}_k} \leq \epsilon \ \|\overline{\mathbf{u}}_k\|_{\overline{A}_k}$ for an a priori chosen $\epsilon < 1$ holds. That is, the terms $\overline{\mathbf{u}}_k|_{\Omega_k \cap \Gamma_\delta} = (\overline{\mathbf{u}}_k - \overline{\mathbf{u}}_k^{(0)})|_{\Omega_k \cap \Gamma_\delta}$ are small, of relative order ϵ.

A perturbation analysis in the general case

The principal submatrix E_{DD} of the iteration matrix $\mathcal{E}_{DD}$, in the case of $p > 1$ subdomains, takes the form

$$E_{DD} = \begin{bmatrix} 0 & -E_1 B_2 \pi_2 & \ldots & -E_1 B_p \pi_p \\ -E_2 B_1 \pi_1 & 0 & & -E_2 B_p \pi_p \\ \vdots & & \ddots & \\ -E_p B_1 \pi_1 & -E_p B_2 \pi_2 & & 0 \end{bmatrix}, \tag{7.41}$$

where

$$\overline{A}_k = P_k^T A_k P_k$$
$$\pi_k = I - P_k \overline{A}_k^{-1} P_k^T A_k$$
$$\overline{B}_k = B_k P_k$$
$$S_k = B_k A_k^{-1} B_k^T + \sum_{j \neq k} \overline{B}_j \overline{A}_j^{-1} \overline{B}_j^T$$
$$E_k = A_k^{-1} B_k^T S_k^{-1}.$$

Let $\mathbf{w} = (\mathbf{w}_k)$, $\mathbf{v} = (\mathbf{v}_k)$ and $A = \text{diag}(A_k)$. We have

$$(\mathbf{w}^T A E_{DD} \mathbf{v})^2$$
$$= \left(\sum_k \mathbf{w}_k^T B_k^T S_k^{-1} \sum_{j \neq k} B_j \pi_j \mathbf{v}_j \right)^2$$
$$\leq \sum_k \mathbf{w}_k^T A_k \mathbf{w}_k \sum_k \left(\sum_{j \neq k} B_j \pi_j \mathbf{v}_j \right)^T S_k^{-1} B_k A_k^{-1} B_k^T S_k^{-1} \left(\sum_{j \neq k} B_j \pi_j \mathbf{v}_j \right).$$

Consider now the problem for $\mathbf{u}_k$,

$$S_k \mathbf{u}_k = \sum_{j \neq k} B_j \pi_j \mathbf{v}_j.$$

The above estimate then reads

$$(\mathbf{w}^T A E_{DD} \mathbf{v})^2 \leq \sum_k \mathbf{w}_k^T A_k \mathbf{w}_k \sum_k \mathbf{u}_k^T B_k A_k^{-1} B_k^T \mathbf{u}_k. \tag{7.42}$$

To estimate $\mathbf{u}_k$, we proceed as follows,

$$\mathbf{u}_k^T S_k \mathbf{u}_k = \mathbf{u}_k^T \left(\sum_{j \neq k} B_j \pi_j \mathbf{v}_j \right)$$
$$= \sum_{j \neq k} (B_j^T \mathbf{u}_k)^T \begin{bmatrix} \star \\ (\pi_j \mathbf{v}_j)|_{\partial \Omega_j} \end{bmatrix}$$
$$= \sum_{j \neq k} (\overline{B}_j^T \mathbf{u}_k)^T \begin{bmatrix} \star \\ (\pi_j \mathbf{v}_j)|_{\partial \Omega_j} \end{bmatrix}$$
$$\leq \left(\sum_{j \neq k} \mathbf{u}_k^T \overline{B}_j \overline{A}_j^{-1} \overline{B}_j^T \mathbf{u}_k \right)^{1/2} \left(\sum_{j \neq k} \| \overline{E}_j^H (\pi_j \mathbf{v}_j)_{\partial \Omega_j} \|_{\overline{A}_j}^2 \right)^{1/2}$$

Here, $\overline{E}_j^H$ stands for a coarse-grid extension of vectors on $\partial\Omega_j$ into the interior of Ω_j. Thus,

$$\mathbf{u}_k^T S_k \mathbf{u}_k \leq \sum_{j \neq k} \|\overline{E}_j^H (\pi_j \mathbf{v}_j)_{\partial\Omega_j}\|_{A_j}^2.$$

Assume now the estimate,

$$\delta_k \, \mathbf{z}^T B_k A_k^{-1} B_k^T \mathbf{z} \leq \sum_{j \neq k} \mathbf{z}^T \overline{B}_j \overline{A}_j^{-1} \overline{B}_j^T \mathbf{z}, \tag{7.43}$$

for any Lagrange multiplier vector $\mathbf{z}$ on Γ. The latter estimate implies

$$(1 + \delta_k) \, \mathbf{u}_k^T B_k A_k^{-1} B_k^T \mathbf{u}_k \leq \sum_{j \neq k} \|\overline{E}_j^H (\pi_j \mathbf{v}_j)_{\partial\Omega_j}\|_{A_j}^2,$$

which used in (7.42) leads to

$$\begin{aligned}
(\mathbf{w}^T A E_{DD} \mathbf{v})^2 &\leq \frac{1}{1 + \min_k \delta_k} \sum_k \mathbf{w}_k^T A_k \mathbf{w}_k \sum_k \sum_{j \neq k} \|\overline{E}_j^H (\pi_j \mathbf{v}_j)_{\partial\Omega_j}\|_{A_j}^2 \\
&\leq \frac{p-1}{1 + \min_k \delta_k} \sum_k \mathbf{w}_k^T A_k \mathbf{w}_k \sum_j \|\overline{E}_j^H (\pi_j \mathbf{v}_j)_{\partial\Omega_j}\|_{A_j}^2.
\end{aligned}$$

By choosing $\mathbf{w}_j = E_j^H (\pi E_{DD} \mathbf{v})_{\partial\Omega_j}$, we get the final estimate

$$\sum_j \|E_j^H (\pi_j (E_{DD} \mathbf{v})_j)_{\partial\Omega_j}\|_{A_j}^2 \leq \frac{p-1}{1 + \min_k \delta_k} \sum_j \|E_j^H (\pi_j \mathbf{v}_j)_{\partial\Omega_j}\|_{A_j}^2.$$

It is clear then that if $p = 2$, the method is convergent as a stationary iterative process. Note that here the norm is different from the norm in Theorem 7.14. Here we use coarse-grid extension mappings, whereas in Theorem 7.14, the norm involves fine-grid extension mappings.

Taking into account the constraints

Another observation is that if $\sum_{j \neq k} B_j \pi_j \mathbf{v}_j = -B_k \pi_k \mathbf{v}_k$, that is, $\pi \mathbf{v} = (\pi_k \mathbf{v}_k)$ satisfies the constraints, the estimate for $\mathbf{u}_k$ simplifies as follows.

$$\begin{aligned}
\mathbf{u}_k^T S_k \mathbf{u}_k &= \mathbf{u}_k^T \left(\sum_{j \neq k} B_j \pi_j \mathbf{v}_j \right) \\
&= -\mathbf{u}_k^T B_k \pi_k \mathbf{v}_k \\
&= -(B_k^T \mathbf{u}_k)^T \begin{bmatrix} \star \\ (\pi_k \mathbf{v}_k)|_{\partial\Omega_k} \end{bmatrix} \\
&\leq \left(\mathbf{u}_k^T B_k A_k^{-1} B_k^T \mathbf{u}_k\right)^{1/2} \|E_k^h (\pi_k \mathbf{v}_k)_{\partial\Omega_k}\|_{A_k}.
\end{aligned}$$

However, based on the argument leading to estimate (7.40), we can only assume that $\pi \mathbf{v} = (\pi_j \mathbf{v}_j)$ satisfies the constraint only approximately, that is, that

$\sum_j B_j \pi_j \mathbf{v}_j \approx 0$. Then, assuming the estimate

$$\sum_k \|S_k^{-1} \sum_j B_j \pi_j \mathbf{v}_j\|^2_{B_k A_k^{-1} B_k^T} \le \epsilon \sum_k \|S_k^{-1} B_k \pi_k \mathbf{v}_k\|^2_{B_k A_k^{-1} B_k^T}, \tag{7.44}$$

used first in (7.42), letting $S_k \widehat{\mathbf{u}}_k = -B_k \pi_k \mathbf{v}_k$ leads to

$$\begin{aligned}
&|\mathbf{w}^T A E_{DD} \mathbf{v}| \\
&\le \|\mathbf{w}\|_A \left[\left(\sum_k \widehat{\mathbf{u}}_k^T B_k A_k^{-1} B_k^T \widehat{\mathbf{u}}_k \right)^{1/2} + \left(\sum_k \left\| S_k^{-1} \sum_j B_j \pi_j \mathbf{v}_j \right\|^2_{B_k A_k^{-1} B_k^T} \right)^{1/2} \right] \\
&\le \|\mathbf{w}\|_A (1 + \sqrt{\epsilon}) \left(\sum_k \widehat{\mathbf{u}}_k^T B_k A_k^{-1} B_k^T \widehat{\mathbf{u}}_k \right)^{1/2}.
\end{aligned}$$

In a similar fashion as above, we obtain that

$$\|\widehat{\mathbf{u}}_k\|^2_{B_k A_k^{-1} B_k^T} = \|S_k^{-1} B_k \pi_k \mathbf{v}_k\|^2_{B_k A_k^{-1} B_k^T} \le \frac{1}{1+\delta_k} \|E_k^h \, \pi_k \mathbf{v}_k\|^2_{A_k}.$$

This finally shows the estimate

$$\sum_k \|E_k^h \, (\pi_k (E_{DD} \mathbf{v})_k)_{\partial \Omega_k} \|^2_{A_k} \le \frac{(1+\sqrt{\epsilon})^2}{1 + \min_k \delta_k} \sum_k \|E_k^h \, (\pi_k \mathbf{v}_k)_{\partial \Omega_k} \|^2_{A_k}.$$

Thus, the DD method of Bank and Holst will be convergent if ϵ is sufficiently small. The latter can be ensured by proper choice of the coarse mesh-size H. Note that H reflects the width $\delta = \mathcal{O}(H)$ of the strip Γ_δ covered by the fine mesh.

For a more precise finite element analysis we refer to [BV06].

7.10 The FAC method and related preconditioning

In the present section, we describe the fast adaptive composite-grid method (or FAC) proposed in [SMc84], [McT86], and its preconditioning versions [BEPS] and [ELV] suitable for solving discretization problems on meshes with patched local refinement (as shown in Figure 7.2). The method combines features of both a two-grid method and a domain decomposition method. Using matrix notation, the method can be summarized as follows.

Let A be the given matrix. There is a coarse version of A denoted by A_c. The fact that A corresponds to a discretization to a same problem as A_c but on a partially refined mesh can be expressed with the relation $A_c = P^T A P$ for an interpolation matrix P that has a major block being the identity; that is,

$$P = \begin{bmatrix} P_{FF} & P_{F,\Gamma} & 0 \\ 0 & I & 0 \\ 0 & 0 & I \end{bmatrix} \begin{matrix} \} \ \Omega_F \\ \} \ \Gamma \\ \} \ \Omega_C \setminus \Gamma. \end{matrix}$$

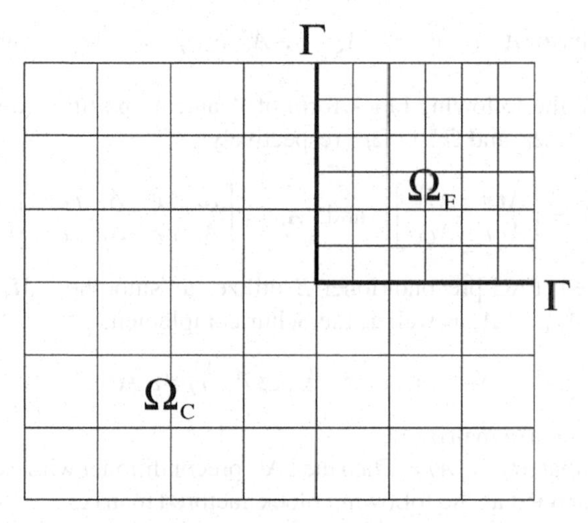

Fig. 7.2. Locally refined mesh; Ω_F consists of the top right corner fine-grid nodes (nodes on the interface Γ excluded). Nodes on Γ that are not coarse are not actual degrees of freedom.

The block-matrix

$$P_F = \begin{bmatrix} P_{FF} & P_{F,\Gamma} \\ 0 & I \end{bmatrix}$$

represents a "standard" interpolation matrix in a subdomain $\Omega_F \cup \Gamma$ of the original domain Ω. Here, Γ is a part of the boundary of the subdomain Ω_C where $\Omega = \Omega_F \cup \Omega_C$ is a direct decomposition. Here, we assume that the coarse–fine interface boundary Γ consists only of coarse dofs. We also need the set Ω_c^F consisting of the coarse dofs in Ω_F. Because Γ is assumed to be a separator, we have the following common DD block structure of A and A_c, corresponding to the ordering $\Omega_F,\ \Gamma,\ \Omega_{C_I} \equiv \Omega_C \backslash \Gamma$, and $\Omega_c^F,\ \Gamma,\ \Omega_{C_I}$, respectively.

$$A = \begin{bmatrix} A_{FF} & A_{F,\Gamma} & 0 \\ A_{\Gamma,F} & A_{\Gamma,\Gamma} & A_{\Gamma,C_I} \\ 0 & A_{C_I,\Gamma} & A_{C_I,C_I} \end{bmatrix} \quad \text{and} \quad A_c = \begin{bmatrix} A_{c,FF} & A_{c,F,\Gamma} & 0 \\ A_{c,\Gamma,F} & A_{c,\Gamma,\Gamma} & A_{c,\Gamma,C_I} \\ 0 & A_{c,C_I,\Gamma} & A_{c,C_I,C_I} \end{bmatrix}.$$

Due to the special form of P, we easily see that

$$\begin{aligned} A_c = P^T A P &= \begin{bmatrix} * & 0 & 0 \\ * & I & 0 \\ 0 & 0 & I \end{bmatrix} \begin{bmatrix} A_{FF} & A_{F,\Gamma} & 0 \\ A_{\Gamma,F} & A_{\Gamma,\Gamma} & A_{\Gamma,C_I} \\ 0 & A_{C_I,\Gamma} & A_{C_I,C_I} \end{bmatrix} \begin{bmatrix} * & * & 0 \\ 0 & I & 0 \\ 0 & 0 & I \end{bmatrix} \\ &= \begin{bmatrix} * & * & 0 \\ * & * & A_{\Gamma,C_I} \\ 0 & A_{C_I,\Gamma} & A_{C_I,C_I} \end{bmatrix} \begin{bmatrix} * & * & 0 \\ 0 & I & 0 \\ 0 & 0 & I \end{bmatrix} \\ &= \begin{bmatrix} * & * & 0 \\ * & * & A_{\Gamma,C_I} \\ 0 & A_{C_I,\Gamma} & A_{C_I,C_I} \end{bmatrix}. \end{aligned}$$

Therefore, we have $A_{c,\,C_I,\,\Gamma} = A_{C_I,\,\Gamma}$, $A_{c,\,\Gamma,\,C_I} = A_{\Gamma,\,C_I}$, and $A_{C_I,\,C_I} = A_{c,\,C_I,\,C_I}$.

We also need the following block form of A and A_c partitioned with respect to the ordering $\Omega_F \cup \Omega_C$ and $\Omega_c^F \cup \Omega_C$, respectively,

$$A = \begin{bmatrix} A_{FF} & A_{FC} \\ A_{CF} & A_{CC} \end{bmatrix} \quad \text{and} \quad A_c = \begin{bmatrix} A_{c,\,FF} & A_{c,\,FC} \\ A_{c,\,CF} & A_{c,\,CC} \end{bmatrix}.$$

The (symmetrized) FAC preconditioner B utilizes a "smoother" M_F coming from the major block A_{FF} of A, as well as the Schur complement

$$S_c = A_{c,\,CC} - A_{c,\,CF} A_{c,\,FF}^{-1} A_{c,\,FC}$$

coming from the coarse matrix A_c.

Assume first that $M_F = A_{FF}$. Then the FAC preconditioner with exact subdomain solutions, B, is defined as the following block-factored matrix,

$$B = \begin{bmatrix} A_{FF} & 0 \\ A_{CF} & I \end{bmatrix} \begin{bmatrix} I & A_{FF}^{-1} A_{FC} \\ 0 & S_c \end{bmatrix}. \tag{7.45}$$

Note that to implement the actions of B^{-1}, we need the inverse actions of S_c, which are readily available based on

$$A_c^{-1} = \begin{bmatrix} * & * \\ * & S_c^{-1} \end{bmatrix}.$$

That is, in order to solve a system with S_c we can instead solve a system with A_c, based on the formula $S_c^{-1} = [0,\ I] A_c^{-1} [0,\ I]^T$, where $[0,\ I]$ stands for restriction to $\Omega_C = \Omega \setminus \Omega_F$. Also, another important feature of B is that the interpolation matrix P is not explicitly used in the definition of B once the coarse matrix A_c (giving rise to S_c) is being given. The preconditioner B is efficient if there are efficient algorithms that compute the inverse actions of A_{FF} and A_c (the latter giving rise to inverse actions of S_c without actually having to form S_c).

For example, in the case of tensor product meshes (and separable variables PDEs), the matrices A_{FF} and A_c may allow for fast direct solvers as described in Section 7.6.

In the case when A_{FF} is not as easy for solving systems, we need to approximate it, and then it is natural to use smoothers M_F and the interpolation matrix (or rather its local block P_F) to define an efficient two-grid iteration process that takes advantage of special properties, such as data storage of A_{FF}, M_F, and P_F on uniform grids Ω_F and A_c as well. To illustrate the FAC method without exact inverses of A_{FF}, introduce the matrix

$$I_{\Omega_F} = \begin{bmatrix} I \\ 0 \end{bmatrix} \begin{matrix} \}\ \Omega_F \\ \}\ \Omega_C \end{matrix}.$$

Then, one step of the FAC method consists of performing the following composite iteration.

Algorithm 7.10.1 (FAC). *Let* $\mathbf{u}$ *be a current iterate for solving* $A\mathbf{x} = \mathbf{b}$. *Then:*

(i) *Solve for a subdomain correction* $\mathbf{u}_F$,

$$M_F \mathbf{u}_F = (I_{\Omega_F})^T (\mathbf{b} - A\mathbf{u})$$

and update $\mathbf{u} := \mathbf{u} + I_{\Omega_F} \mathbf{u}_F$.

(ii) *Solve for a coarse-grid correction* $\mathbf{u}_c$,

$$A_c \mathbf{u}_c = P^T (\mathbf{b} - A\mathbf{u}),$$

and update $\mathbf{u} := \mathbf{u} + P\mathbf{u}_c$.

(iii) *Optionally, to symmetrize the iteration, solve for one more subdomain correction* $\mathbf{u}_F$ *from*

$$M_F^T \mathbf{u}_F = (I_{\Omega_F})^T (\mathbf{b} - A\mathbf{u})$$

and update $\mathbf{u} := \mathbf{u} + I_{\Omega_F} \mathbf{u}_F$.

It is clear that the iteration matrix of the above process reads:

$$E_{FAC} = (I - I_{\Omega_F} M_F^{-T} (I_{\Omega_F})^T A)(I - P A_c^{-1} P^T A)(I - I_{\Omega_F} M_F^{-1} (I_{\Omega_F})^T A).$$

Equivalently, if we introduce the FAC preconditioner B_{FAC} from $I - B_{FAC}^{-1} A = E_{FAC}$, we obtain, letting $\overline{M}_F = M_F (M_F + M_F^T - A_{FF})^{-1} M_F^T$,

$$\begin{aligned}
B_{FAC}^{-1} &= [I_{\Omega_F},\ P] \begin{bmatrix} I & -M_F^{-T} (I_{\Omega_F})^T A P \\ 0 & I \end{bmatrix} \begin{bmatrix} \overline{M}_F^{-1} & 0 \\ 0 & A_c^{-1} \end{bmatrix} \\
&\quad \times \begin{bmatrix} I & 0 \\ -P^T A I_{\Omega_F} M_F^{-1} & I \end{bmatrix} \begin{bmatrix} (I_{\Omega_F})^T \\ P^T \end{bmatrix} \\
&= [I_{\Omega_F},\ (I - I_{\Omega_F} M_F^{-T} (I_{\Omega_F})^T A) P] \begin{bmatrix} \overline{M}_F^{-1} & 0 \\ 0 & A_c^{-1} \end{bmatrix} \\
&\quad \times \begin{bmatrix} (I_{\Omega_F})^T \\ P^T (I - A I_{\Omega_F} M_F^{-1} (I_{\Omega_F})^T) \end{bmatrix} \\
&= I_{\Omega_F} \overline{M}_F^{-1} (I_{\Omega_F})^T \\
&\quad + (I - I_{\Omega_F} M_F^{-T} (I_{\Omega_F})^T A) P A_c^{-1} P^T (I - A I_{\Omega_F} M_F^{-1} (I_{\Omega_F})^T).
\end{aligned} \tag{7.46}$$

The following identity holds (cf., Theorem 3.15 with $\mathcal{J} = I_{\Omega_F}$ and $\mathcal{D} = P^T A P$, $\mathcal{M} = M_F$, $\mathcal{A} = A_{FF}$),

$$\begin{aligned}
\mathbf{v}^T B_{FAC} \mathbf{v} = \min_{\mathbf{v} = I_{\Omega_F} \mathbf{v}_F + P \mathbf{v}_c} & \left[\mathbf{v}_c^T A_c \mathbf{v}_c + \left(M_F^T \mathbf{v}_F + (I_{\Omega_F})^T A P \mathbf{v}_c \right)^T \right. \\
& \left. \times \left(M_F + M_F^T - A_{FF} \right)^{-1} \left(M_F^T \mathbf{v}_F + (I_{\Omega_F})^T A P \mathbf{v}_c \right) \right].
\end{aligned} \tag{7.47}$$

Because the FAC method does not exploit smoothing on the entire domain, we cannot use directly our general two-grid result (cf. Theorem 3.25). However, based on (7.47), similar assumptions lead to spectral equivalence estimates between A and B_{FAC}. The main one is, if we can find a stable decomposition $\mathbf{v} = J_0\mathbf{v}_s + P\mathbf{v}_c$ such that $\mathbf{v}_c|_{\Omega_C} = \mathbf{v}|_{\Omega_C}$

$$\left(\text{hence } J_0\mathbf{v}_s = \begin{bmatrix} * \\ 0 \end{bmatrix} \begin{matrix} \} & \Omega_F \\ \} & \Omega_C \end{matrix}\right),$$

and such that the local symmetrized smoother $\overline{M}_F$ is spectrally equivalent to A_{FF} restricted to the subspace Range$((I_{\Omega_F})^T J_0)$, that is,

$$(\mathbf{v}_s)^T J_0^T I_{\Omega_F} \overline{M}_F (I_{\Omega_F})^T J_0 \mathbf{v}_s \leq \kappa \; (\mathbf{v}_s)^T J_0^T I_{\Omega_F} A_{FF} (I_{\Omega_F})^T J_0 \mathbf{v}_s,$$

with a bounded coarse component, that is, that

$$(P\mathbf{v}_c)^T A (P\mathbf{v}_c) \leq \eta \; \mathbf{v}^T A\mathbf{v}.$$

We also assume that M_F is properly scaled so that

$$\mathbf{w}_F^T (M_F + M_F^T - A_{FF})\mathbf{w}_F \geq \delta \; \mathbf{w}_F^T A_{FF} \mathbf{w}_F.$$

Then, based on the identity (7.47) and the above assumptions, we immediately get the following spectral equivalence between B_{FAC} and A,

$$\mathbf{v}^T A\mathbf{v} \leq \mathbf{v}^T B_{FAC}\mathbf{v} \leq [(1 + 2\delta^{-1})\eta + 2\kappa(1 + \sqrt{\eta})^2] \; \mathbf{v}^T A\mathbf{v}.$$

Thus, we proved the following main result.

Theorem 7.15. *Assume that every vector $\mathbf{v}$ admits a special decomposition $\mathbf{v} = J_0\mathbf{v}_s + P\mathbf{v}_c$ such that*

$$J_0\mathbf{v}_s = \begin{bmatrix} * \\ 0 \end{bmatrix} = I_{\Omega_F}(*)$$

is being localized in Ω_F, with the additional following properties.

(i) The symmetrized local smoother $\overline{M}_F$ is spectrally equivalent to A_{FF} restricted to the local subspace Range$(I_{\Omega_F}^T J_0)$; that is,

$$\mathbf{w}_F^T A_{FF}\mathbf{w}_F \leq \mathbf{w}_F^T \overline{M}_F \mathbf{w}_F \leq \kappa \; \mathbf{w}_F^T A_{FF} \mathbf{w}_F, \quad \textit{for all } \mathbf{w}_F = I_{\Omega_F}^T J_0 \mathbf{w}_s.$$

(ii) The coarse-grid extension $P\mathbf{v}_c$ of $\mathbf{v}_C$ in Ω_c^F (i.e., $\mathbf{v}_c|_{\Omega_C} = \mathbf{v}|_{\Omega_C} = \mathbf{v}_C$), is suboptimal; that is,

$$\mathbf{v}_c^T A_C \mathbf{v}_c = \mathbf{v}_c^T P^T A P\mathbf{v}_c \leq \eta \inf_{\mathbf{w}_F} \begin{bmatrix} \mathbf{w}_F \\ \mathbf{v}_C \end{bmatrix}^T A \begin{bmatrix} \mathbf{w}_F \\ \mathbf{v}_C \end{bmatrix}.$$

Assume also that the smoother M_F is scaled such that

$$\mathbf{v}_F^T (M_F + M_F^T - A_{FF})\mathbf{v}_F \geq \delta \; \mathbf{v}_F^T A_{FF} \mathbf{v}_F.$$

Then the FAC preconditioner B_{FAC} is spectrally equivalent to the composite grid matrix A and

$$\mathbf{v}^T A\mathbf{v} \leq \mathbf{v}^T B_{FAC}\mathbf{v} \leq \left[\left(1+\frac{2}{\delta}\right)\eta + 2\kappa(1+\sqrt{\eta})^2\right] \mathbf{v}^T A\mathbf{v}.$$

We now use the special features of B_{FAC} to specify the above assumptions. First, we assume that for any vector

$$\mathbf{v}_C = \begin{bmatrix} \mathbf{v}_\Gamma \\ \mathbf{v}_C^0 \end{bmatrix} \begin{matrix} \} & \Gamma \\ \} & \Omega_C \setminus \Gamma \end{matrix}$$

defined on Ω_C, we can find a "bounded coarse extension"

$$P\mathbf{v}_c = P\begin{bmatrix} \mathbf{v}_c^F \\ \mathbf{v}_C \end{bmatrix};$$

that is, $(P\mathbf{v}_c)^T A(P\mathbf{v}_c) \leq \sigma\ \mathbf{v}_C^T S_C \mathbf{v}_C$. Here, $S_C = A_{CC} - A_{CF}A_{FF}^{-1}A_{FC}$ is the Schur complement of A.

In the finite element case, the following decomposition of the global quadratic form holds,

$$\mathbf{v}^T A\mathbf{v} = \mathbf{v}_C^T A_C^{(N)} \mathbf{v}_C + \begin{bmatrix} \mathbf{v}_F \\ \mathbf{v}_\Gamma \end{bmatrix}^T A_F^{(N)} \begin{bmatrix} \mathbf{v}_F \\ \mathbf{v}_\Gamma \end{bmatrix}.$$

The matrices $A_C^{(N)}$ and $A_F^{(N)}$ are symmetric (semi)definite and defined for vectors on Ω_C and $\Omega_F \cup \Gamma$, respectively. Because in this case A, A_C, and S_C have a common major block, the estimate $(P\mathbf{v}_c)^T A(P\mathbf{v}_c) \leq \sigma\ \mathbf{v}_C^T S_C \mathbf{v}_C$ can be rewritten as

$$(P_F\mathbf{v}_c^F)^T A_F^{(N)} P_F \mathbf{v}_c^F \leq \sigma\ \mathbf{v}_\Gamma^T S_\Gamma \mathbf{v}_\Gamma.$$

Here, $\mathbf{v}_c^F|_\Gamma = \mathbf{v}_\Gamma \equiv \mathbf{v}_C|_\Gamma$. Also, $S_\Gamma = A_{\Gamma,\Gamma}^{(N)} - A_{\Gamma,F}^{(N)} A_{FF}^{-1} A_{F,\Gamma}^{(N)}$ is the Schur complement of the matrix

$$A_F^{(N)} = \begin{bmatrix} A_{FF} & A_{F,\Gamma}^{(N)} \\ A_{\Gamma,F}^{(N)} & A_{\Gamma,\Gamma}^{(N)} \end{bmatrix} \begin{matrix} \} & \Omega_F \\ \} & \Gamma \end{matrix},$$

which represents the contribution to A coming from the subdomain Ω_F. Note that $\mathbf{v}_C$ has the same interface component $\mathbf{v}_\Gamma$ (i.e., $\mathbf{v}_C|_\Gamma = \mathbf{v}_\Gamma$). We also have $\mathbf{v}_C^T S_C \mathbf{v}_C = \mathbf{v}_C^T A_C^{(N)} \mathbf{v}_C + \mathbf{v}_\Gamma^T S_\Gamma \mathbf{v}_\Gamma$.

In summary, we have the following main convergence result.

Theorem 7.16. *Assume:*

(i) *There is an energy bounded coarse extension mapping E from Γ into the interior of Ω_F, that is, $E\mathbf{v}_C = P\mathbf{v}_c^0$ for some vector $\mathbf{v}_c^0$, which coincides with $\mathbf{v}_C$ on Ω_C. The boundedness means that*

$$\mathbf{v}_C^T E^T A E \mathbf{v}_C \leq \|E\|^2 \inf_{\mathbf{w}_F} \begin{bmatrix} \mathbf{w}_F \\ \mathbf{v}_C \end{bmatrix}^T A \begin{bmatrix} \mathbf{w}_F \\ \mathbf{v}_C \end{bmatrix}.$$

(ii) The local two-grid method based on M_F, P_{FF} applied to A_{FF} is convergent in the sense that the respective two-grid preconditioner B_F^{TG} satisfies for any $\mathbf{v}_F$ the estimate:

$$(\mathbf{v}_F)^T B_F^{TG} \mathbf{v}_F \leq K_F^{TG} \, (\mathbf{v}_F)^T A_{FF} \mathbf{v}_F.$$

Finally, assume that M_F is properly scaled so that

$$\mathbf{w}_F^T (M_F + M_F^T - A_{FF}) \mathbf{w}_F \geq \delta \, \mathbf{w}_F^T A_{FF} \mathbf{w}_F. \tag{7.48}$$

Then the FAC preconditioner B_{FAC} is spectrally equivalent to the composite grid matrix A.

Proof. We recall the following main identity for B_F^{TG}.

$$\begin{aligned}(\mathbf{v}_F)^T B_F^{TG} \mathbf{v}_F = & \min_{\mathbf{v}_F = \mathbf{v}_s^F + P_{FF} \mathbf{v}_c^F} \left[\left(P_{FF} \mathbf{v}_c^F\right)^T A_{FF} P_{FF} \mathbf{v}_c^F \right. \\ & + \left(M_F^T \mathbf{v}_s^F + A_{FF} P_{FF} \mathbf{v}_c^F\right)^T \left(M_F + M_F^T - A_{FF}\right)^{-1} \\ & \left. \times \left(M_F^T \mathbf{v}_s^F + A_{FF} P_{FF} \mathbf{v}_c^F\right)\right]. \end{aligned} \tag{7.49}$$

Using the fact that we have a bounded extension mapping E that is in the range of the interpolation matrix P (i.e., $E\mathbf{v}_C = P\mathbf{v}_c^0$), we start with the decomposition $\mathbf{v} = E\mathbf{v}_C + I_{\Omega_F} \mathbf{v}_F^0 = P\mathbf{v}_c^0 + I_{\Omega_F} \mathbf{v}_F^0$. By assumption, we have the norm bound

$$\mathbf{v}_C^T E^T A E \mathbf{v}_C \leq \|E\|^2 \inf_{\mathbf{w}_F} \begin{bmatrix} \mathbf{w}_F \\ \mathbf{v}_C \end{bmatrix}^T A \begin{bmatrix} \mathbf{w}_F \\ \mathbf{v}_C \end{bmatrix}.$$

Now, use any local decomposition for $\mathbf{v}_F^0 = \mathbf{v}_s^F + P_{FF} \mathbf{v}_c^F$, to arrive at the global decomposition $\mathbf{v} = P\mathbf{v}_c + I_{\Omega_F} \mathbf{v}_s^F$ with

$$\mathbf{v}_c = \begin{bmatrix} \mathbf{v}_c^F \\ 0 \end{bmatrix} \begin{matrix} \} \; \Omega_c^F \\ \} \; \Omega_C \end{matrix} + \mathbf{v}_c^0.$$

Recall that Ω_c^F stands for the coarse dofs in Ω_F. Using the latter decomposition in the identity (7.47), the norm bound for $E\mathbf{v}_C$, the Cauchy–Schwarz inequality, estimate (7.48), the fact that $A^{-1} - I_{\Omega_F} A_{FF}^{-1} (I_{\Omega_F})^T$ is positive semidefinite, and the estimate $\|\mathbf{v}_F^0\|_{A_{FF}} = \|I_{\Omega_F} \mathbf{v}_F^0\|_A = \|\mathbf{v} - E\mathbf{v}_C\|_A \leq (1 + \|E\|) \, \|\mathbf{v}\|_A$, we arrive at the estimates

$$\begin{aligned} & \mathbf{v}^T B_{FAC} \mathbf{v} \\ & \quad \leq 2(1+\delta^{-1}) \, (E\mathbf{v}_C)^T A E \mathbf{v}_C + 2 \inf_{\mathbf{v}_F^0 = \mathbf{v}_s^F + P_{FF} \mathbf{v}_c^F} \left[(\mathbf{v}_c^F)^T P_{FF}^T A_{FF} P_{FF} \mathbf{v}_c^F \right. \\ & \quad \left. + (M_F^T \mathbf{v}_s^F + A_{FF} P_{FF} \mathbf{v}_c^F)^T (M_F + M_F^T - A_{FF})^{-1} (M_F^T \mathbf{v}_s^F + A_{FF} P_{FF} \mathbf{v}_c^F) \right] \\ & \quad = 2(1+\delta^{-1}) \, (E\mathbf{v}_C)^T A E \mathbf{v}_C + 2 (\mathbf{v}_F^0)^T B_F^{TG} \mathbf{v}_F^0 \\ & \quad \leq 2(1+\delta^{-1}) \|E\|^2 \, \mathbf{v}^T A \mathbf{v} + 2 K_F^{TG} \, (\mathbf{v}_F^0)^T A_{FF} \mathbf{v}_F^0 \\ & \quad \leq 2\left[(1+\delta^{-1}) \|E\|^2 + K_F^{TG} (1 + \|E\|)^2\right] \mathbf{v}^T A \mathbf{v}. \end{aligned}$$

Above, we used identity (7.49) for the local two-grid preconditioner B_F^{TG}. □

Multilevel FAC

We conclude the present section with the comment that the FAC method allows for multilevel extensions. We introduce the kth-level composite grid $\Omega^{(k)} = \Omega_C^{(\ell)} \cup \Omega_C^{(\ell-1)} \cup \cdots \cup \Omega_C^{(k+1)} \cup \Omega_F^{(k)}$ for $k = \ell - 1, \ldots, 0$ and $\Omega^{(\ell)} = \Omega_C^{(\ell)}$ (hence $\Omega_F^{(\ell)} = \emptyset$). Letting $\Omega_C^{(k)} = \Omega_C^{(\ell)} \cup \Omega_C^{(\ell-1)} \cup \cdots \cup \Omega_C^{(k+1)}$, we have the decomposition $\Omega^{(k)} = \Omega_C^{(k)} \cup \Omega_F^{(k)}$. The portion $\Omega_F^{(k)}$ of $\Omega^{(k)}$ is assumed to have some regular structure. In practice, $\Omega_F^{(k)}$ corresponds to a uniformly refined part of the computational domain.

The $k+1$st-level coarse composite matrix is $A^{(k+1)}$, and the interpolation matrix P_k that relates $A^{(k)}$ and $A^{(k+1)} = P_k^T A^{(k)} P_k$ has the special form

$$P_k = \begin{bmatrix} P_{FF}^{(k)} & P_{F,\Gamma_k}^{(k)} & 0 \\ 0 & I & 0 \\ 0 & 0 & I \end{bmatrix} \begin{matrix} \} \; \Omega_F^{(k)} \\ \} \; \Gamma_k \\ \} \; \Omega_{C_I}^{(k)} \equiv \Omega_C^{(k)} \setminus \Gamma_k. \end{matrix}$$

Here $\Gamma_k \subset \Omega^{(k)}$ is a separator so that $A^{(k)}$ admits the following DD block-form,

$$A^{(k)} = \begin{bmatrix} A_{FF}^{(k)} & A_{F,\Gamma_k}^{(k)} & 0 \\ A_{\Gamma_k,F}^{(k)} & A_{\Gamma_k,\Gamma_k}^{(k)} & A_{\Gamma_k,C_I}^{(k)} \\ 0 & A_{C_I,\Gamma_k}^{(k)} & A_{C_I,C_I}^{(k)} \end{bmatrix} \begin{matrix} \} \; \Omega_F^{(k)} \\ \} \; \Gamma_k \\ \} \; \Omega_{C_I}^{(k)}. \end{matrix}$$

The major block $A_{FF}^{(k)}$ of $A^{(k)}$ corresponds to the subdomain $\Omega_F^{(k)}$, which has some regular structure. We then assume that there is an easily invertible matrix $M_F^{(k)}$ to be used as a smoother for $A_{FF}^{(k)}$. As before, let $\overline{M}_F^{(k)} = M_F^{(k)}(M_F^{(k)} + M_F^{(k)^T} - A_{FF}^{(k)})^{-1} M_F^{(k)^T}$ stand for the symmetrized smoother, which is assumed s.p.d.

Definition 7.17 (Multilevel FAC preconditioner). *Starting with $B_{FAC}^{(\ell)} = A^{(\ell)}$, the coarsest matrix, we then define by recursion for $k = \ell - 1, \ldots, 0$, introducing*

$$I_F^k = \begin{bmatrix} I \\ 0 \end{bmatrix} \begin{matrix} \} \; \Omega_F^{(k)} \\ \} \; \Omega_C^{(k)} \end{matrix},$$

$$\begin{aligned} B_{FAC}^{(k)^{-1}} &= I_F^k \overline{M}_F^{(k)^{-1}} (I_F^k)^T + \left(I - I_F^k M_F^{(k)^{-T}} (I_F^k)^T A^{(k)}\right) P_k B_{FAC}^{(k+1)^{-1}} P_k^T \\ &\quad \times \left(I - A^{(k)} I_F^k M_F^{(k)^{-1}} (I_F^k)^T\right). \end{aligned}$$

Exploiting locality, assuming that the actions of $M_F^{(k)^{-1}}$ and $M_F^{(k)^{-T}}$ can be implemented in $\mathcal{O}(|\Omega_F^{(k)}|)$ flops, and based on the sparsity of P_k we can assume that the actions of P_k and P_k^T can also be implemented in $\mathcal{O}(|\Omega_F^{(k)}|)$ flops; it is easily seen then that one action of $B_{FAC}^{(0)^{-1}}$ can be implemented for $\mathcal{O}(|\Omega^{(\ell)}| + \sum_{k=1}^{\ell-1} |\Omega_F^{(k)}|)$ flops.

By construction, we have the product form of the FAC iteration matrix $E^{(k)}_{FAC} \equiv I - B^{(k)^{-1}}_{FAC} A^{(k)}$,

$$\begin{aligned}
E^{(k)}_{FAC} &= I - B^{(k)^{-1}}_{FAC} A^{(k)} \\
&= \left(I - I^k_F M^{(k)^{-T}}_F (I^k_F)^T A^{(k)}\right)\left(I - P_k B^{(k+1)^{-1}}_{FAC} P^T_k A^{(k)}\right) \\
&\quad \times \left(I - I^k_F M^{(k)^{-1}}_F (I^k_F)^T A^{(k)}\right).
\end{aligned}$$

We can show (by induction) based on the representation

$$B^{(k)^{-1}}_{FAC} = \left[I^k_F,\ P_k\right] \widehat{B}^{(k)^{-1}}_{FAC} \left[I^k_F,\ P_k\right]^T,$$

where

$$\begin{aligned}
\widehat{B}^{(k)}_{FAC} = &\begin{bmatrix} M^{(k)}_F & 0 \\ P^T_k A^{(k)} I^k_F & I \end{bmatrix} \\
&\times \begin{bmatrix} \left(M^{(k)}_F + M^{(k)^T}_F - A^{(k)}_{FF}\right)^{-1} & 0 \\ 0 & B^{(k+1)}_{FAC} \end{bmatrix} \\
&\times \begin{bmatrix} M^{(k)^T}_F & (I^k_F)^T A^{(k)} P_k \\ 0 & I \end{bmatrix},
\end{aligned}$$

that $\mathbf{v}^T A^{(k)} \mathbf{v} \le \mathbf{v}^T B^{(k)}_{FAC} \mathbf{v}$, hence $\|E^{(k)}_{FAC}\|_{A^{(k)}} < 1$. To prove the first fact, we first define $\widehat{B}^{(k)}_{TG}$ in the same way as $\widehat{B}^{(k)}_{FAC}$ with $B^{(k+1)}_{FAC}$ replaced by the exact coarse matrix $A^{(k+1)}$. Then define $B^{(k)}_{TG}$ as

$$B^{(k)^{-1}}_{TG} = \left[I^k_F,\ P_k\right] \widehat{B}^{(k)^{-1}}_{TG} \left[I^k_F,\ P_k\right]^T,$$

that is, in the same way as in (7.46). Note that because $[I^k_F,\ P_k][I^k_F,\ P_k]^T$ is s.p.d., and because $\widehat{B}^{(k)}_{FAC}$ is s.p.d., then $B^{(k)}_{FAC}$ is s.p.d. Similarly $B^{(k)}_{TG}$ is s.p.d. From the product representation of $I - B^{(k)^{-1}}_{TG} A^{(k)}$, we see that $A^{(k)}(I - B^{(k)^{-1}}_{TG} A^{(k)}) = (I - I^k_F M^{(k)^{-1}}_F (I^k_F)^T A^{(k)})^T (A^{(k)} - A^{(k)} P_k A^{(k+1)^{-1}} P^T_k A^{(k)})(I - I^k_F M^{(k)^{-1}}_F (I^k_F)^T A^{(k)}) = E^T_F (A^{(k)} - A^{(k)} P_k A^{(k+1)^{-1}} P^T_k A^{(k)}) E_F$. The latter matrix will be symmetric positive semidefinite if the middle term

$$A^{(k)} - A^{(k)} P_k A^{(k+1)^{-1}} P^T_k A^{(k)}$$

is symmetric positive semidefinite. From the relation $A^{(k+1)} = P^T_k A^{(k)} P_k$ and the fact that $\|X\| = \|X^T\| = 1$ used for $X = A^{(k)^{1/2}} P_k A^{(k+1)^{-(1/2)}}$, the middle term is seen to be symmetric positive semi definite. Thus,

$$\mathbf{v}^T B^{(k)}_{TG} \mathbf{v} \ge \mathbf{v}^T A^{(k)} \mathbf{v}.$$

Assuming now by induction that $B_{FAC}^{(k+1)} - A^{(k+1)}$ is positive semidefinite, we first observe that $\widehat{B}_{FAC}^{(k)} - \widehat{B}_{TG}^{(k)}$ is positive semidefinite, as well. Hence, the following inequalities are immediate.

$$\mathbf{v}^T B_{FAC}^{(k)^{-1}} \mathbf{v} \leq \mathbf{v}^T [I_F^{(k)},\ P_k] \widehat{B}_{TG}^{(k)^{-1}} [I_F^{(k)},\ P_k]^T \mathbf{v} = \mathbf{v}^T B_{TG}^{(k)^{-1}} \mathbf{v} \leq \mathbf{v}^T A^{(k)^{-1}} \mathbf{v}.$$

The latter confirms the induction assumption that $B_{FAC}^{(k)} - A^{(k)}$ is symmetric positive semidefinite.

Estimates from above for $B_{FAC}^{(k)}$ in terms of $A^{(k)}$ would require proving the existence of stable multilevel decompositions of vectors with terms supported in the refined regions $\Omega_F^{(k)}$. Such decompositions are typically derived based on stable decomposition of finite element functions that then give rise to corresponding stable decomposition of their coefficient vectors. For multilevel analysis of some versions of FAC, see [BP04] (and the references therein), and for some respective numerical results, see [BP04e].

7.11 Auxiliary space preconditioning methods

Given a vector space V and an s.p.d. matrix A operating on vectors from V, also let the (auxiliary) space $\widehat{V}$ together with another s.p.d. matrix $\widehat{A}$ be well understood in the sense that we know how to construct preconditioners $\widehat{B}$ for the matrix $\widehat{A}$. Then, we may want to take advantage of this fact to construct preconditioners B for A. To achieve this goal, we need some additional assumptions found in what follows.

Construction of the auxiliary space preconditioner

Assume that there is a computable mapping $\pi : \widehat{V} \mapsto V$, which relates the two spaces. Note that $V + \pi \widehat{V}$ gives an overlapping decomposition of the original space V. Assume that π is a bounded mapping; that is,

$$(\pi \widehat{\mathbf{v}})^T A \pi \widehat{\mathbf{v}} \leq \sigma\, \widehat{\mathbf{v}}^T \widehat{A} \widehat{\mathbf{v}}, \quad \text{for all } \widehat{\mathbf{v}} \in \widehat{V}. \tag{7.50}$$

We also assume that the space $\widehat{V}$ can well approximate elements from V in a norm $\|.\|_0$ weaker than $\|.\|_A$. That is, for any $\mathbf{v} \in V$, there is a $\widehat{\mathbf{v}} \in \widehat{V}$ such that:

- $\pi \widehat{\mathbf{v}}$ provides an approximation to $\mathbf{v}$; that is,

$$\|\mathbf{v} - \pi \widehat{\mathbf{v}}\|_0^2 \leq \delta\, \varrho(A)^{-1}\, \mathbf{v}^T A \mathbf{v}. \tag{7.51}$$

  Here,

$$\varrho(A) \geq \|A\| \sup_{\mathbf{v} \in V} \frac{\mathbf{v}^T \mathbf{v}}{\|\mathbf{v}\|_0^2} \geq \sup_{\mathbf{v} \in V} \frac{\mathbf{v}^T A \mathbf{v}}{\|\mathbf{v}\|_0^2} \quad \text{and} \quad \|A\| = \sup_{\mathbf{v} \in V} \frac{\mathbf{v}^T A \mathbf{v}}{\mathbf{v}^T \mathbf{v}}.$$

- The component $\widehat{\mathbf{v}}$ is stable; that is,

$$\widehat{\mathbf{v}}^T \widehat{A} \widehat{\mathbf{v}} \leq \eta\, \mathbf{v}^T A \mathbf{v}. \tag{7.52}$$

Note that the decomposition $\mathbf{v} = (\mathbf{v} - \pi\widehat{\mathbf{v}}) + \pi\widehat{\mathbf{v}}$ satisfies the condition that the first component $\mathbf{w} = \mathbf{v} - \pi\widehat{\mathbf{v}}$ is small (referring to $1 \ll \varrho(A)$), and the second component $\pi\widehat{\mathbf{v}}$ is bounded, both in terms of $\|\mathbf{v}\|_A$. The auxiliary space preconditioner exploits a smoother $\mathcal{M}$ intended to handle the oscillatory small component $\mathbf{v} - \pi\widehat{\mathbf{v}}$ and a preconditioner $\widehat{B}$ that is assumed readily available for the operator $\widehat{A}$.

Consider then the following transformation matrix $[I,\ \pi]$. Our scheme of two-by-two block-factorization preconditioners for A utilizing the decomposition based on $[I,\ \pi]$ leads to the following (inverse of) preconditioner $B^{-1} = [I,\ \pi]\overline{B}^{-1}[I,\ \pi]^T$ defined by

$$\overline{B} = \begin{bmatrix} I & 0 \\ \pi^T A\mathcal{M}^{-1} & I \end{bmatrix} \begin{bmatrix} \mathcal{M}(\mathcal{M}+\mathcal{M}^T - A)^{-1}\mathcal{M}^T & 0 \\ 0 & \widehat{B} \end{bmatrix} \begin{bmatrix} I & \mathcal{M}^{-T}A\pi \\ 0 & I \end{bmatrix}, \tag{7.53}$$

where $\widehat{B}$ is a given preconditioner to $\widehat{A}$ and $\mathcal{M}$ is a simple preconditioner ("smoother") for A. More explicitly, we get

$$B^{-1} = \mathcal{M}^{-T}(\mathcal{M}+\mathcal{M}^T - A)\mathcal{M}^{-1} + (I - \mathcal{M}^{-T}A)\pi\widehat{B}^{-1}\pi^T(I - A\mathcal{M}^{-1}). \tag{7.54}$$

Introduce the symmetrized smoother $\overline{\mathcal{M}} = \mathcal{M}(\mathcal{M}+\mathcal{M}^T - A)^{-1}\mathcal{M}^T$. To analyze the preconditioner B, we need the additional assumptions on the smoother $\mathcal{M}$ (or $\overline{\mathcal{M}}$),

$$0 \le \mathbf{v}^T\overline{\mathcal{M}}\mathbf{v} \le \kappa\ \varrho(A)\ \|\mathbf{v}\|_0^2, \tag{7.55}$$

and

$$\theta\ \mathbf{v}^T(\mathcal{M}+\mathcal{M}^T - A)\mathbf{v} \ge \mathbf{v}^T A\mathbf{v}. \tag{7.56}$$

Note that the r.h.s. inequality in (7.55) is trivially satisfied with $\mathcal{M} = (\|A\|/\omega)\ I$, $\omega \in (0, 2)$, and $\kappa = 1/(\omega(2-\omega))$. We have

$$\overline{\mathcal{M}} = \mathcal{M}(\mathcal{M}+\mathcal{M}^T - A)^{-1}\mathcal{M}^T = \frac{\|A\|^2}{\omega^2}\left(\frac{2\|A\|}{\omega}\ I - A\right)^{-1}.$$

Therefore,

$$\frac{\|A\|^2}{\omega^2}\ \mathbf{w}^T\left(\frac{2\|A\|}{\omega}\ I - A\right)^{-1}\mathbf{w} \le \frac{\|A\|}{\omega(2-\omega)}\ \mathbf{w}^T\mathbf{w} \le \frac{1}{\omega(2-\omega)}\ \varrho(A)\ \|\mathbf{w}\|_0^2.$$

The left-hand side of (7.55) means that $\mathcal{M}$ is a convergent smoother for A in the A-inner product. That is, $\mathbf{v}^T(I - \mathcal{M}^{-1}A)^T A(I - \mathcal{M}^{-1}A)\mathbf{v} \le \varrho\ \mathbf{v}^T A\mathbf{v}$ for some $\varrho < 1$, which is true for $\mathcal{M} = (\|A\|/\omega)\ I$ with $\omega \in (0, 2)$. Simply note that the eigenvalues of $I - \mathcal{M}^{-1}A$ are in $(-1, 1)$. Finally, assumption (7.56) can be ensured for any $\mathcal{M}$ by proper scaling. For the particular example $\mathcal{M} = (\|A\|/\omega)\ I$, it holds with $\theta = \omega/(2-\omega)$.

For the preconditioner $\widehat{B}$ to $\widehat{A}$, we need the equivalence relations,

$$\widehat{\mathbf{v}}^T\pi^T A\pi\widehat{\mathbf{v}} \le \widehat{\mathbf{v}}^T\widehat{B}\widehat{\mathbf{v}} \le \widehat{\sigma}\ \widehat{\mathbf{v}}^T\widehat{A}\widehat{\mathbf{v}}, \quad \text{for all } \widehat{\mathbf{v}}. \tag{7.57}$$

The above inequality includes the assumed boundedness (7.50) of the operator π. The estimate (7.57) can be obtained if $\widehat{B}$ is a spectrally equivalent preconditioner to $\widehat{A}$, that is, if

$$\sigma\, \widehat{\mathbf{v}}^T \widehat{A}\widehat{\mathbf{v}} \leq \widehat{\mathbf{v}}^T \widehat{B}\widehat{\mathbf{v}} \leq \widehat{\sigma}\, \widehat{\mathbf{v}}^T \widehat{A}\widehat{\mathbf{v}}.$$

The latter relations, together with the boundedness of π, that is, estimate (7.50), imply the equivalence relations (7.57).

We first have (from the assumption $\widehat{\mathbf{v}}^T\pi^T A\pi\widehat{\mathbf{v}} \leq \widehat{\mathbf{v}}^T\widehat{B}\widehat{\mathbf{v}}$) based on Theorem 3.16 that

$$0 \leq \mathbf{v}^T(B - A)\mathbf{v}.$$

Next, consider the special decomposition $\mathbf{v} = \mathbf{w} + \pi\widehat{\mathbf{v}}$, or equivalently,

$$\mathbf{v} = \mathbf{w} + \pi\widehat{\mathbf{v}} = [I,\ \pi]\begin{bmatrix} \mathbf{w} \\ \widehat{\mathbf{v}} \end{bmatrix}.$$

Therefore, for the upper bound, we get

$$\begin{aligned}
\mathbf{v}^T B\mathbf{v} &\leq \begin{bmatrix} \mathbf{w} \\ \widehat{\mathbf{v}} \end{bmatrix}^T \overline{B} \begin{bmatrix} \mathbf{w} \\ \widehat{\mathbf{v}} \end{bmatrix} \\
&= (\mathcal{M}^T\mathbf{w} + A\pi\widehat{\mathbf{v}})^T(\mathcal{M} + \mathcal{M}^T - A)^{-1}(\mathcal{M}^T\mathbf{w} + A\pi\widehat{\mathbf{v}}) + \widehat{\mathbf{v}}^T\widehat{B}\widehat{\mathbf{v}} \\
&\leq 2\,\mathbf{w}^T\overline{\mathcal{M}}\mathbf{w} + 2\,\widehat{\mathbf{v}}^T\pi^T A(\mathcal{M} + \mathcal{M}^T - A)^{-1}A\pi\widehat{\mathbf{v}} + \widehat{\mathbf{v}}^T\widehat{B}\widehat{\mathbf{v}}.
\end{aligned}$$

Then, based on (7.55) and (7.56), using the estimate $\mathbf{w}^T A\mathbf{w} \leq \varrho(A)\ \|\mathbf{w}\|_0^2$ and the boundedness of π, we arrive at the final estimates:

$$\begin{aligned}
\mathbf{v}^T B\mathbf{v} &\leq 2\,\kappa\varrho(A)\ \|\mathbf{w}\|_0^2 + 2\theta\ \widehat{\mathbf{v}}^T\pi^T A\pi\widehat{\mathbf{v}} + \widehat{\mathbf{v}}^T\widehat{B}\widehat{\mathbf{v}} \\
&\leq 2\kappa\varrho(A)\ \|\mathbf{w}\|_0^2 + (2\theta + 1)\ \widehat{\mathbf{v}}^T\widehat{B}\widehat{\mathbf{v}} \\
&\leq 2\kappa\varrho(A)\ \|\mathbf{w}\|_0^2 + (2\theta + 1)\widehat{\sigma}\ \widehat{\mathbf{v}}^T\widehat{A}\widehat{\mathbf{v}} \\
&\leq [2\kappa\delta + (1 + 2\theta)\widehat{\sigma}\eta]\ \mathbf{v}^T A\mathbf{v}.
\end{aligned}$$

We used in the last line above that $\|\mathbf{w}\|_0^2 \leq \delta(\varrho(A))^{-1}\ \mathbf{v}^T A\mathbf{v}$ and $\widehat{\mathbf{v}}^T\widehat{B}\widehat{\mathbf{v}} \leq \eta\ \mathbf{v}^T A\mathbf{v}$, that is, the assumptions that $\mathbf{w} = \mathbf{v} - \pi\widehat{\mathbf{v}}$ is small in the sense of estimate (7.51) and that the second component $\widehat{\mathbf{v}}$ in the decomposition is stable in the sense of (7.52). The last estimate completes the proof that the auxiliary space preconditioner B is spectrally equivalent to A.

A somewhat simpler auxiliary space preconditioner is the additive one (proposed in Xu [Xu96b]). For a symmetric smoother $\mathcal{M}$, (e.g., $\mathcal{M} = (\|A\|/\omega)\ I$) it takes the form

$$B_{\text{add}}^{-1} = \mathcal{M}^{-1} + \pi\widehat{B}^{-1}\pi^T. \tag{7.58}$$

It can be analyzed analogously (under the same assumption). We have

$$I = B_{\text{add}}^{1/2}[I,\ \pi]\begin{bmatrix} \mathcal{M}^{-1} & 0 \\ 0 & \widehat{B}^{-1} \end{bmatrix}[I,\ \pi]^T B_{\text{add}}^{1/2}.$$

Therefore,

$$\left(\begin{bmatrix} \mathcal{M}^{-1} & 0 \\ 0 & \widehat{B}^{-1} \end{bmatrix}^{-(1/2)} [I,\ \pi]^T B_{\text{add}}^{1/2}\right)\left(\begin{bmatrix} \mathcal{M}^{-1} & 0 \\ 0 & \widehat{B}^{-1} \end{bmatrix}^{-(1/2)} [I,\ \pi]^T B_{\text{add}}^{1/2}\right)^T - I$$

is symmetric negative semidefinite; that is,

$$[I,\ \pi]^T B_{\text{add}}[I,\ \pi] - \begin{bmatrix} \mathcal{M} & 0 \\ 0 & \widehat{B} \end{bmatrix}$$

is symmetric negative semidefinite. This shows, that for any decomposition

$$\mathbf{v} = \mathbf{w} + \pi\widehat{\mathbf{v}} = [I,\ \pi]\begin{bmatrix} \mathbf{w} \\ \widehat{\mathbf{v}} \end{bmatrix},$$

we have

$$\begin{aligned}
\mathbf{v}^T B_{\text{add}}\mathbf{v} &= \begin{bmatrix} \mathbf{w} \\ \widehat{\mathbf{v}} \end{bmatrix}^T [I,\ \pi]^T B_{\text{add}}[I,\ \pi]\begin{bmatrix} \mathbf{w} \\ \widehat{\mathbf{v}} \end{bmatrix} \\
&\leq \begin{bmatrix} \mathbf{w} \\ \widehat{\mathbf{v}} \end{bmatrix}^T \begin{bmatrix} \mathcal{M} & 0 \\ 0 & \widehat{B} \end{bmatrix}\begin{bmatrix} \mathbf{w} \\ \widehat{\mathbf{v}} \end{bmatrix} \\
&= \mathbf{w}^T \mathcal{M}\mathbf{w} + \widehat{\mathbf{v}}^T \widehat{B}\widehat{\mathbf{v}} \\
&= 2\ \mathbf{w}^T \mathcal{M}(2\mathcal{M})^{-1}\mathcal{M}\mathbf{w} + \widehat{\mathbf{v}}^T \widehat{B}\widehat{\mathbf{v}} \\
&\leq 2\ \mathbf{w}^T \mathcal{M}(2\mathcal{M} - A)^{-1}\mathcal{M}\mathbf{w} + \widehat{\sigma}\ \widehat{\mathbf{v}}^T \widehat{A}\widehat{\mathbf{v}} \\
&= 2\ \mathbf{w}^T \mathcal{M}(\mathcal{M} + \mathcal{M}^T - A)^{-1}\mathcal{M}^T\mathbf{w} + \widehat{\sigma}\ \widehat{\mathbf{v}}^T \widehat{A}\widehat{\mathbf{v}} \\
&= 2\ \mathbf{w}^T\overline{M}\mathbf{w} + \widehat{\sigma}\ \widehat{\mathbf{v}}^T \widehat{A}\widehat{\mathbf{v}} \\
&\leq 2\varrho(A)\kappa\ \|\mathbf{w}\|_0^2 + \widehat{\sigma}\ \widehat{\mathbf{v}}^T \widehat{A}\widehat{\mathbf{v}}.
\end{aligned}$$

Now, use the estimates for the components of the decomposition $\mathbf{v} = \mathbf{v} + \pi\widehat{\mathbf{v}}$, $\|\mathbf{w}\|_0^2 = \|\mathbf{v} - \pi\widehat{\mathbf{v}}\|_0^2 \leq \delta(\varrho(A))^{-1}\ \mathbf{v}^T A\mathbf{v}$ and $\widehat{\mathbf{v}}^T \widehat{A}\widehat{\mathbf{v}} \leq \eta\ \mathbf{v}^T A\mathbf{v}$, to arrive at the lower bound

$$\mathbf{v}^T A\mathbf{v} \geq \frac{1}{\widehat{\sigma}\eta + 2\kappa\delta}\ \mathbf{v}^T B_{\text{add}}\mathbf{v}.$$

The estimate from above is similarly derived. We have

$$\begin{aligned}
\mathbf{v}^T B_{\text{add}}^{-1}\mathbf{v} &= \mathbf{v}^T \mathcal{M}^{-1}\mathbf{v} + \mathbf{v}^T \pi\widehat{A}^{-1}\pi^T\mathbf{v} \\
&\leq \frac{2\theta}{1+\theta}\ \mathbf{v}^T A^{-1}\mathbf{v} + \widehat{\sigma}\ \mathbf{v}^T A^{-1}\mathbf{v} \\
&= \left(\widehat{\sigma} + \frac{2\theta}{1+\theta}\right)\ \mathbf{v}^T A^{-1}\mathbf{v}.
\end{aligned}$$

Here we used that $(\widehat{A}^{-(1/2)}\pi^T A^{1/2})(\widehat{A}^{-(1/2)}\pi^T A^{1/2})^T - \widehat{\sigma}I$ is negative semidefinite (by assumption), hence $(\widehat{A}^{-(1/2)}\pi^T A^{1/2})^T(\widehat{A}^{-(1/2)}\pi^T A^{1/2}) - \widehat{\sigma}I$ is negative semidefinite, which implies that $\pi \widehat{A}^{-1}\pi^T - \widehat{\sigma} A^{-1}$ is negative semidefinite. Also (again by assumption) $\mathcal{M} + \mathcal{M}^T - (1 + (1/\theta))\ A$ is positive semidefinite, therefore $\mathbf{v}^T \mathcal{M} \mathbf{v} \geq ((1+\theta)/2\theta)\ \mathbf{v}^T A \mathbf{v}$.

To summarize, the following main result holds.

Theorem 7.18. *Consider two vector spaces V and $\widehat{V}$ and let $\pi : \widehat{V} \mapsto V$ relate the auxiliary space $\widehat{V}$ with V. Let the given s.p.d. matrix A of main interest, acting on vectors from V, be related to a s.p.d. matrix $\widehat{A}$ defined on $\widehat{V}$. Finally, let $\widehat{A}$ admit a spectrally equivalent preconditioner $\widehat{B}$ that is properly scaled, such as in* (7.57). *Under the assumptions that any $\mathbf{v} \in V$ admits a decomposition $\mathbf{v} = \mathbf{w} + \pi \widehat{\mathbf{v}}$ where the first component $\mathbf{w}$ is small in the sense of* (7.51)*, and the second component $\widehat{\mathbf{v}}$ is stable in the sense of estimate* (7.52)*, provided that A also allows for a smoother $\mathcal{M}$ (e.g., $\mathcal{M} = (\|A\|/\omega)\ I$, for any $\omega \in (0, 2)$), which satisfies* (7.55) *and* (7.56)*, then both the additive auxiliary space preconditioner B_{add} defined in* (7.58) *and the multiplicative one B defined in* (7.53) *are spectrally equivalent to A.*

An H(curl) auxiliary space preconditioner

Examples of auxiliary space preconditioners for finite element problems rely on important finite element decompositions. The power of the auxiliary space preconditioning method can be illustrated with the following $\mathbf{H}(\text{curl})$ example. A main result in [HX06] shows that vector functions in $H(\text{curl})$-conforming Nédélec spaces $\mathbf{V}_h$ (cf., Section B.6 in the appendix for definition of the f.e. spaces) allow for decompositions based on three components,

$$\mathbf{v}_h = \mathbf{w}_h + \nabla z_h + \mathbf{\Pi}_h \mathbf{z}_h, \tag{7.59}$$

where we have two auxiliary spaces, S_h and $\mathbf{S}_h$. S_h is a scalar finite element space suitable for Poisson-like problems and $\mathbf{S}_h = (S_h)^3$ is a vector one. The mapping $\mathbf{\Pi}_h$ relates the dofs in $\mathbf{S}_h$ with the dofs in the original space $\mathbf{V}_h$.

Note that all spaces use the original triangulation (set of elements) $\mathcal{T}_h$. The component $\mathbf{w}_h$ is small, in the sense that $\|\mathbf{w}_h\|_0 \leq Ch\ \|\mathbf{v}_h\|_{\mathbf{H}(\text{curl})}$, and the remaining components are stable, namely, $\|\nabla z_h\|_0 \leq C\ \|\mathbf{v}_h\|_{\mathbf{H}(\text{curl})}$ and $\|\mathbf{z}_h\|_1 \leq C\ \|\mathbf{v}_h\|_{\mathbf{H}(\text{curl})}$. The additive auxiliary space preconditioner for the original $\mathbf{H}(\text{curl})$ form,

$$(\mathbf{A}_h \mathbf{v}_h,\ \mathbf{v}_h) \equiv (\text{curl}\ \mathbf{v}_h,\ \text{curl}\ \mathbf{v}_h) + (\mathbf{v}_h,\ \mathbf{v}_h), \qquad \mathbf{v}_h \in \mathbf{V}_h,$$

utilizes the subspaces $\mathbf{V}_h$, ∇S_h, and $\mathbf{\Pi}_h \mathbf{S}_h$ of $\mathbf{V}_h$, and respective operators $\mathbf{\Lambda}_h$, B_h, and $\mathbf{B}_h$, coming from the symmetric Gauss–Seidel smoother for the original $\mathbf{H}(\text{curl})$ operator (matrix) $\mathbf{A}_h$, a MG preconditioner for the Laplace-like form $(\mathbf{A}_h \nabla z_h,\ \nabla z_h) = (\nabla z_h,\ \nabla z_h)$ with $z_h \in S_h$, and similarly, a MG preconditioner for the (vector) Laplace-like form $(\mathbf{A}_h \mathbf{\Pi}_h \mathbf{z}_h,\ \mathbf{\Pi}_h \mathbf{z}_h) \simeq (\nabla \mathbf{z}_h,\ \nabla \mathbf{z}_h)$, $\mathbf{z}_h \in \mathbf{S}_h$. Introducing the matrix representation G_h of the embedding $\nabla S_h \subset \mathbf{V}_h$, that is, the matrix representation of the mapping of the dofs in S_h into dofs of $\mathbf{V}_h$ (based on $\nabla S_h \subset \mathbf{V}_h$), the additive auxiliary space preconditioner B_{add} takes the form

$$B_{\text{add}}^{-1} = \mathbf{\Lambda}_h^{-1} + G_h\ B_h^{-1} G_h^T + \mathbf{\Pi}_h \mathbf{B}_h^{-1} \mathbf{\Pi}_h^T.$$

The fact that B_{add} is an optimal preconditioner for $A = \mathbf{A}_h$ follows from the properties of the decomposition (7.59) and the fact that with $\|.\|_0$ being the L_2-norm, we have then $\varrho(A) \simeq h^{-2}$.

The performance of both additive and multiplicative auxiliary space preconditioners for $H(\text{curl})$-type problems is documented in detail in [KV06i] and [KV06ii].

General approaches for constructing auxiliary space preconditioners were studied in [Nep91a], [Nep91b], [Xu96b], and [BPZ]. Applications to $H(\text{div})$-bilinear forms were considered in [RVWb] and based on auxiliary meshes for $H(\text{curl})$ in [KPVa].

Concluding remarks for this chapter

For more details on DD methods, we refer to the books [DD], [Wo00], and [TW05].

8

Preconditioning Nonsymmetric and Indefinite Matrices

This chapter describes an approach of preconditioning nonsymmetric and indefinite matrices that can be treated as perturbations of symmetric positive definite ones. Namely, we assume that $A = A_0 + R$ where A_0 is a s.p.d. matrix and R can be treated as a perturbation of A. More specifically, the main assumption is that in a space complementary to a coarse space of a fixed size, R has a small norm. This is made more precise in what follows.

An additive version of the approach described in what follows was originally considered by Yserentant [Y86]. The Schur complement preconditioner that we present next is found in [V92a]. An equivalent finite element type preconditioner was considered in [Xu92b].

The assumptions made in the present chapter are verified for finite element (nonsymmetric and possibly indefinite) matrices corresponding to general second-order elliptic bilinear forms in Appendix B; see in particular, Theorems B.3 and B.4 for the perturbation approach described in Section 8.2.

8.1 An abstract setting

Partition the set of indices (dofs) into a fixed small set of "c" dofs and a large complementary set of "f" dofs. Let

$$P = \begin{bmatrix} \mathcal{P} \\ I \end{bmatrix} \begin{matrix} \} \text{ “}f\text{” dofs} \\ \} \text{ “}c\text{” dofs} \end{matrix}$$

be a given interpolation matrix. Let $A_c = P^T A P$ be the coarse matrix, which we assume to be invertible.

Consider the transformation matrix

$$\pi = \begin{bmatrix} I & \mathcal{P} \\ 0 & I \end{bmatrix},$$

P.S. Vassilevski, *Multilevel Block Factorization Preconditioners*,
doi: 10.1007/978-0-387-71564-3_8,

and the corresponding transformed matrix

$$\widehat{A} = \pi^T A \pi = \begin{bmatrix} A_{ff} & A_{fc} + A_{ff}\mathcal{P} \\ A_{cf} + \mathcal{P}^T A_{ff} & A_c \end{bmatrix}.$$

Because A_c is assumed of small size, we are actually interested in the reduced Schur complement

$$W_f = A_{ff} - (A_{fc} + A_{ff}\mathcal{P})A_c^{-1}(A_{cf} + \mathcal{P}^T A_{ff}).$$

Our goal is to construct an efficient preconditioner to W_f based on preconditioners for the same Schur complement W_f^0 coming from the transformed form $\widehat{A}_0 = \pi^T A_0 \pi$ of the given s.p.d. matrix A_0. The first main assumption is the following Gärding inequality, valid for two constants $\gamma_0 \in [0, 1)$ and $\delta_0 > 0$,

$$\mathbf{v}^T A \mathbf{v} \geq (1 - \gamma_0)\mathbf{v}^T A_0 \mathbf{v} - c_0 \|\mathbf{v}\|_0^2. \tag{8.1}$$

Here, $\|\mathbf{v}\|_0$ is a given norm weaker than $\sqrt{\mathbf{v}^T A_0 \mathbf{v}}$. In a typical finite element application, $\|\mathbf{v}\|_0$ comes from the integral L_2-norm of the finite element function v that $\mathbf{v}$ represents. Then, in d dimensions ($d = 2$, or $d = 3$), $\|\mathbf{v}\|_0^2 \simeq h^d \mathbf{v}^T \mathbf{v}$, where $h \mapsto 0$ is the mesh-size.

Our second main assumption is that for any given $\mathbf{v}$, the solutions of $A\mathbf{x} = \mathbf{v}$ and $A_c \mathbf{x}_c = P^T \mathbf{v}$ are close in the weaker norm. That is, for some small $\delta > 0$, we have

$$\|\mathbf{x} - P\mathbf{x}_c\|_0^2 \leq \delta\, \mathbf{x}^T A_0 \mathbf{x},$$

which is the same as

$$\|(A^{-1}\mathbf{v} - PA_c^{-1}P^T)A\mathbf{x}\|_0^2 \leq \delta\, \mathbf{x}^T A_0 \mathbf{x}. \tag{8.2}$$

Equivalently, we may say that the coarse-grid projection $\pi_A = PA_c^{-1}P^T A$ satisfies

$$\|(I - \pi_A)\mathbf{v}\|_0^2 \leq \delta\, \mathbf{v}^T A_0 \mathbf{v}. \tag{8.3}$$

In what follows we assume δ is sufficiently small such that $\delta c_0 < 1 - \gamma_0$.

For more details, in the case of second-order elliptic equations, see Theorem B.7 in the appendix.

Using inequality (8.3) in (8.1) for $\mathbf{v} = (I - \pi_A)\mathbf{v}$ leads to

$$\mathbf{v}^T A \mathbf{v} \geq (1 - \gamma_0 - \delta c_0)\mathbf{v}^T A_0 \mathbf{v}. \tag{8.4}$$

That is (for $\delta > 0$ sufficiently small), A becomes coercive in terms of A_0 in a subspace complementary to the coarse space; namely, for $\mathbf{v} \in$ Range $(I - \pi_A)$.

Next, we find a representation of the solution to the equation $\mathbf{v} = (I - \pi_A)\mathbf{v}$. We have $0 = \pi_A \mathbf{v} = PA_c^{-1}P^T A\mathbf{v}$, which is equivalent to $P^T A \mathbf{v} = 0$. Recall the transformation matrix

$$\pi = \begin{bmatrix} I & \mathcal{P} \\ 0 & I \end{bmatrix}.$$

Because π is invertible, we can seek $\mathbf{v} = \pi\overline{\mathbf{v}}$, where

$$\overline{\mathbf{v}} = \begin{bmatrix} \overline{\mathbf{v}}_f \\ \mathbf{v}_c \end{bmatrix}.$$

Then,

$$\mathbf{v} = \begin{bmatrix} \overline{\mathbf{v}}_f + \mathcal{P}\mathbf{v}_c \\ \mathbf{v}_c \end{bmatrix}.$$

From the equation

$$0 = P^T A\mathbf{v} = P^T A\left(P\mathbf{v}_c + \begin{bmatrix} \overline{\mathbf{v}}_f \\ 0 \end{bmatrix}\right),$$

we get

$$A_c\mathbf{v}_c + P^T A \begin{bmatrix} \overline{\mathbf{v}}_f \\ 0 \end{bmatrix} = 0.$$

That is,

$$A_c\mathbf{v}_c + (A_{cf} + \mathcal{P}^T A_{ff})\overline{\mathbf{v}}_f = 0.$$

This shows that

$$\widehat{A}\overline{\mathbf{v}} = \begin{bmatrix} W_f\overline{\mathbf{v}}_f \\ 0 \end{bmatrix}.$$

Therefore, $\mathbf{v}^T A\mathbf{v} = \overline{\mathbf{v}}^T\pi^T A\pi\overline{\mathbf{v}} = \overline{\mathbf{v}}^T\widehat{A}\overline{\mathbf{v}} = \overline{\mathbf{v}}_f^T W_f\overline{\mathbf{v}}_f$. Thus, we proved the following coercivity estimate for the Schur complement W_f of A,

$$\overline{\mathbf{v}}_f^T W_f\overline{\mathbf{v}}_f \geq (1 - \gamma_0 - \delta c_0)\overline{\mathbf{v}}^T\pi^T A_0\pi\overline{\mathbf{v}}.$$

Now, because $\pi^T A_0\pi$ is s.p.d. for its Schur complement W_f^0, we have the estimate

$$\overline{\mathbf{v}}^T\pi^T A_0\pi\overline{\mathbf{v}} \geq \min_{\mathbf{v}_c} \begin{bmatrix} \overline{\mathbf{v}}_f \\ \mathbf{v}_c \end{bmatrix}^T \widehat{A}_0 \begin{bmatrix} \overline{\mathbf{v}}_f \\ \mathbf{v}_c \end{bmatrix} = \overline{\mathbf{v}}_f^T W_f^0\overline{\mathbf{v}}_f. \tag{8.5}$$

Thus, we finally arrive at the main coercivity estimate for the Schur complements W_f and W_f^0:

$$\overline{\mathbf{v}}_f^T W_f\overline{\mathbf{v}}_f \geq (1 - \gamma_0 - \delta c_0)\overline{\mathbf{v}}_f^T W_f^0\overline{\mathbf{v}}_f. \tag{8.6}$$

In order to complete the proof that W_f^0 is a good s.p.d. preconditioner for the coercive nonsymmetric matrix W_f, we need some bounds from above. A natural assumption is as follows,

$$\mathbf{v}^T A\mathbf{w} \leq (1 + \Delta)\left(\mathbf{v}^T A_0\mathbf{v}\right)^{1/2}\left(\mathbf{w}^T A_0\mathbf{w}\right)^{1/2}. \tag{8.7}$$

This can be equivalently stated that A_0 is the principal (leading) term in A.

The above inequality used for $\mathbf{w} = \pi\overline{\mathbf{w}}$ and $\mathbf{v} = \pi\overline{\mathbf{v}}$, shows the boundedness of $\widehat{A}$ in terms of $\widehat{A}_0$. Next, for $\overline{\mathbf{w}}$:

$$\widehat{A}\overline{\mathbf{w}} = \begin{bmatrix} W_f \overline{\mathbf{w}}_f \\ 0 \end{bmatrix}$$

used in the boundedness estimate for $\widehat{A}$ in terms of $\widehat{A}_0$, we get

$$\begin{aligned}
\overline{\mathbf{v}}_f^T W_f \overline{\mathbf{w}}_f &\leq (1+\Delta)(\overline{\mathbf{v}}^T \widehat{A}_0 \overline{\mathbf{v}})^{1/2} (\overline{\mathbf{w}}^T \widehat{A}_0 \overline{\mathbf{w}})^{1/2} \\
&\leq \frac{1+\Delta}{\sqrt{1-\gamma_0-\delta c_0}} (\overline{\mathbf{v}}^T \widehat{A}_0 \overline{\mathbf{v}})^{1/2} (\overline{\mathbf{w}}^T \widehat{A} \overline{\mathbf{w}})^{1/2} \\
&= \frac{1+\Delta}{\sqrt{1-\gamma_0-\delta c_0}} (\overline{\mathbf{v}}^T \widehat{A}_0 \overline{\mathbf{v}})^{1/2} \left(\overline{\mathbf{w}}_f^T W_f \overline{\mathbf{w}}_f\right)^{1/2}.
\end{aligned}$$

We first used inequality (8.4) because $\mathbf{w} = \pi\overline{\mathbf{w}}$ is in the proper subspace and then the fact that $\overline{\mathbf{w}}^T \widehat{A}\overline{\mathbf{w}} = \overline{\mathbf{w}}_f^T W_f \overline{\mathbf{w}}_f$. The left-hand side is independent of $\mathbf{v}_c$ by taking the minimum over $\mathbf{v}_c$ based on (8.5), thus we arrive at the estimate

$$\overline{\mathbf{v}}_f^T W_f \overline{\mathbf{w}}_f \leq \frac{1+\Delta}{\sqrt{1-\gamma_0-\delta c_0}} \left(\overline{\mathbf{v}}_f^T W_f^0 \overline{\mathbf{v}}_f\right)^{1/2} \left(\overline{\mathbf{w}}_f^T W_f \overline{\mathbf{w}}_f\right)^{1/2}. \tag{8.8}$$

Finally, letting $\overline{\mathbf{v}}_f = \overline{\mathbf{w}}_f$ in (8.8) implies

$$\left(\overline{\mathbf{w}}_f^T W_f \overline{\mathbf{w}}_f\right)^{1/2} \leq \frac{1+\Delta}{\sqrt{1-\gamma_0-\delta c_0}} \left(\overline{\mathbf{w}}_f^T W_f^0 \overline{\mathbf{w}}_f\right)^{1/2}.$$

Then using the last inequality back in (8.8) leads to the final estimate, which bounds W_f in terms of W_f^0:

$$\overline{\mathbf{v}}_f^T W_f \overline{\mathbf{w}}_f \leq \frac{(1+\Delta)^2}{1-\gamma_0-\delta c_0} \left(\overline{\mathbf{v}}_f^T W_f^0 \overline{\mathbf{v}}_f\right)^{1/2} \left(\overline{\mathbf{w}}_f^T W_f^0 \overline{\mathbf{w}}_f\right)^{1/2}. \tag{8.9}$$

To summarize:

Theorem 8.1. *Let $A = A_0 + R$ be a nonsymmetric and possibly indefinite matrix and A_0 be its principal s.p.d. part in the sense of inequalities* (8.1) *and* (8.7). *The norm $\|.\|_0$ is assumed weaker than $\sqrt{(.)^T A_0(.)}$ such that an error estimate* (8.3) *holds for sufficiently small $\delta > 0$ and proper coarse subspace Range(P),*

$$P = \begin{bmatrix} \mathcal{P} \\ I \end{bmatrix}.$$

Then, A and its principal s.p.d. part A_0 are spectrally equivalent in a subspace complementary to the coarse space. More precisely, based on

$$\pi = \begin{bmatrix} I & \mathcal{P} \\ 0 & I \end{bmatrix},$$

after transforming A and A_0 to $\pi^T A \pi$ and $\pi^T A_0 \pi$, respectively, the transformed matrices have Schur complements W_f and W_f^0 that satisfy the coercivity and boundedness estimates (8.6) *and* (8.9). *Therefore, using W_f^0 as a preconditioner for solving systems with W_f in a preconditioned GCG-type method (see Appendix A) will have a convergence rate no worse than*

$$\left(1 - \frac{(1-\gamma_0 - \delta c_0)^4}{(1+\Delta)^4}\right)^{1/2}.$$

8.2 A perturbation point of view

At this point, we emphasize that we actually need an inequality of the type (8.7) for special $\mathbf{w}$; namely, for $\pi_A \mathbf{w} = 0$. If we assume that $A - A_0 = R$ can be bounded above as follows,

$$\mathbf{v}^T (A - A_0)\mathbf{w} \leq \sigma \left(\mathbf{v}^T A_0 \mathbf{v}\right)^{1/2} \|\mathbf{w}\|_0,$$

which implies a similar lower bound estimate

$$\mathbf{v}^T (A - A_0)\mathbf{w} \geq -\sigma \left(\mathbf{v}^T A_0 \mathbf{v}\right)^{1/2} \|\mathbf{w}\|_0,$$

for $\mathbf{w}$ such that $\mathbf{w} = (I - \pi_A)\mathbf{w}$. These estimates imply estimates that demonstrate that A can be viewed as a perturbation of A_0 in the subspace $\pi_A \mathbf{w} = 0$. More specifically, we have

$$\mathbf{v}^T A\mathbf{w} \leq (1 + \sqrt{\delta}\sigma)(\mathbf{v}^T A_0 \mathbf{v})^{1/2} (\mathbf{w}^T A_0 \mathbf{w})^{1/2}, \quad \text{for any } \mathbf{v} \text{ and any } \mathbf{w} : \pi_A \mathbf{w} = 0. \tag{8.10}$$

Similarly,

$$\mathbf{w}^T A\mathbf{w} \geq (1 - \sqrt{\delta}\sigma)\, \mathbf{w}^T A_0 \mathbf{w}, \quad \text{for any } \mathbf{w} :\ \pi_A \mathbf{w} = 0. \tag{8.11}$$

Then, the coercivity estimate for W_f in terms of W_f^0 reads

$$\overline{\mathbf{v}}_f^T W_f \overline{\mathbf{v}}_f \geq (1 - \sqrt{\delta}\sigma) \overline{\mathbf{v}}_f^T W_f^0 \overline{\mathbf{v}}_f. \tag{8.12}$$

Following the analysis from the preceding section, we end up with the following sharper boundedness estimate for W_f in terms of W_f^0,

$$\overline{\mathbf{v}}_f^T W_f \overline{\mathbf{w}}_f \leq \frac{(1 + \sqrt{\delta}\sigma)^2}{1 - \sqrt{\delta}\sigma} \left(\overline{\mathbf{v}}_f^T W_f^0 \overline{\mathbf{v}}_f\right)^{1/2} \left(\overline{\mathbf{w}}_f^T W_f^0 \overline{\mathbf{w}}_f\right)^{1/2}. \tag{8.13}$$

We can actually prove more than coercivity and boundedness estimates. From the inequality valid for $\mathbf{w} :\ \pi_A \mathbf{w} = 0$,

$$\mathbf{v}^T (A_0 - A)\mathbf{w} \leq \sigma\sqrt{\delta} (\mathbf{v}^T A_0 \mathbf{v})^{1/2} (\mathbf{w}^T A_0 \mathbf{w})^{1/2}, \tag{8.14}$$

used for special $\mathbf{v} = \pi \overline{\mathbf{v}}$; namely such that

$$\pi^T A_0 \mathbf{v} = \begin{bmatrix} W_f^0 \overline{\mathbf{v}}_f \\ 0 \end{bmatrix},$$

and because $\pi_A \mathbf{w} = 0$ with $\mathbf{w} = \pi \overline{\mathbf{w}}$, we get $\mathbf{v}^T A \mathbf{w} = \overline{\mathbf{v}}_f^T W_f \overline{\mathbf{w}}_f$, and therefore,

$$\overline{\mathbf{v}}_f^T (W_f^0 - W_f) \overline{\mathbf{w}}_f \leq \sigma \sqrt{\delta} \left(\overline{\mathbf{v}}_f^T W_f^0 \overline{\mathbf{v}}_f \right)^{1/2} \left(\overline{\mathbf{w}}^T \pi^T A_0 \pi \overline{\mathbf{w}} \right)^{1/2}. \tag{8.15}$$

Finally, from the same estimate (8.14), used for $\mathbf{v} = \pi \overline{\mathbf{v}} = \mathbf{w} = \pi \overline{\mathbf{w}}$, we also have

$$\overline{\mathbf{w}}^T \pi^T A_0 \pi \overline{\mathbf{w}} \leq \overline{\mathbf{w}}_f^T W_f \overline{\mathbf{w}}_f + \sigma \sqrt{\delta}\, \overline{\mathbf{w}}^T \pi^T A_0 \pi \overline{\mathbf{w}}.$$

That is, together with the boundedness estimate (8.13) (used for $\overline{\mathbf{v}}_f = \overline{\mathbf{w}}_f$), we arrive at

$$(1 - \sigma \sqrt{\delta})\, \overline{\mathbf{w}}^T \pi^T A_0 \pi \overline{\mathbf{w}} \leq \overline{\mathbf{w}}_f^T W_f \overline{\mathbf{w}}_f \leq \frac{(1 + \sigma \sqrt{\delta})^2}{1 - \sigma \sqrt{\delta}} \overline{\mathbf{w}}_f^T W_f^0 \overline{\mathbf{w}}_f.$$

Substituting this inequality in (8.15) the following main perturbation estimate is obtained.

$$\overline{\mathbf{v}}_f^T (W_f^0 - W_f) \overline{\mathbf{w}}_f \leq \sigma \sqrt{\delta}\, \frac{1 + \sigma \sqrt{\delta}}{1 - \sigma \sqrt{\delta}} \left(\overline{\mathbf{v}}_f^T W_f^0 \overline{\mathbf{v}}_f \right)^{1/2} \left(\overline{\mathbf{w}}_f^T W_f^0 \overline{\mathbf{w}}_f \right)^{1/2}. \tag{8.16}$$

The latter estimate is equivalent to the fact that the norm of

$$I - \left(W_f^0\right)^{-(1/2)} W_f \left(W_f^0\right)^{-(1/2)}$$

is less than $\sigma \sqrt{\delta}((1 + \sigma \sqrt{\delta})/(1 - \sigma \sqrt{\delta})) < 1$. The latter can be equivalently stated in the following compact form.

Theorem 8.2. *For a constant $\delta_f \leq \sigma \sqrt{\delta}((1 + \sigma \sqrt{\delta})/(1 - \sigma \sqrt{\delta})) = \mathcal{O}(\sqrt{\delta})$, the following main deviation estimate holds,*

$$\left\| \left(W_f^{-1} - (W_f^0)^{-1}\right) \overline{\mathbf{v}}_f \right\|_{W_f^0} \leq \delta_f \| \overline{\mathbf{v}}_f \|_{(W_f^0)^{-1}}.$$

Remark 8.3. It is clear that W_f^0 can be replaced by an accurate preconditioner S_f^0, such as a corresponding Schur complement coming from a few ($\nu \geq 1$) V-cycles applied to the s.p.d. matrix A_0 (see the next section for more details). Then we have an estimate

$$(1 - \varrho^\nu) \overline{\mathbf{v}}_f^T S_f^0 \overline{\mathbf{v}}_f \leq \overline{\mathbf{v}}_f^T W_f^0 \overline{\mathbf{v}}_f \leq \overline{\mathbf{v}}_f^T S_f^0 \overline{\mathbf{v}}_f,$$

where $\varrho \in (0, 1)$ is, for example, the convergence factor of the V-cycle MG. Then it is clear that a result similar to that of Theorem 8.2 will hold with W_f^0 replaced by S_f^0 and a different constant $\delta_f := \varrho^\nu + \delta_f/(1 - \varrho^\nu)$, which can be made arbitrary close to $\sigma \sqrt{\delta}((1 + \sigma \sqrt{\delta})/(1 - \sigma \sqrt{\delta}))$ by increasing ν. Thus, we can use the variable-step preconditioned CG (conjugate gradient) method (see Chapter 10) to solve systems with W_f using S_f^0 as a preconditioner. The convergence rate will be close to that of the V-cycle MG perturbed by the value of $\sqrt{\delta}$.

8.3 Implementation

A main example for nonsymmetric and possibly indefinite matrices A comes from discretized second-order elliptic PDEs, where A_0 would refer to the principal symmetric and definite part of the PDE, and the remaining part would come from the terms with lower-order (first and zeroth) derivatives. It is then straightforward to prove an estimate of the form (8.1) for constants $\gamma_0 \in [0, 1)$ and $\delta \geq 0$ where the norm $\|.\|_0$ comes from the integral L_2-norm of functions. It is clear then if we use a coarse space, we should be able to prove an estimate of the form (8.2) with $\delta \mapsto 0$ when the coarse mesh gets smaller and smaller (for more details see Theorem B.7 in the appendix). Note that we need δ (or the coarse mesh-size) to be sufficiently small to compensate for the coefficient $c_0/(1-\gamma_0)$, which is mesh-independent. The quantity $c_0/(1-\gamma_0)$ depends only on the coefficients of the underlined PDE. Hence, δ (or the coarse mesh-size) can be considered fine mesh independent and therefore fixed. In the case when the lower-order derivative terms of the differential operator are dominating, such as the convection–diffusion operator $-\epsilon\Delta u + \underline{b} \cdot \nabla u$ for small $\epsilon > 0$, or the Helmholtz operator $-\Delta u - k^2 u$ for large k, the coarse mesh that reflects the coefficient δ will become practically unacceptably fine. The terms with lower-order derivatives in these operators cannot be treated as perturbations of the principal elliptic part and therefore require other approaches that, however (so far), have limited partial success (cf., [BW97], [HP97], [BL97], [BL00]).

It is also clear that we do not have to work with A_0 and its Schur complement. We can instead use any spectrally equivalent preconditioner M_0 for A_0 and work with its corresponding Schur complement.

To implement the actions of W_f, we have to solve a coarse problem with A_c, which is of small size. We solve, for a given $\overline{\mathbf{v}}_f$, the coarse-grid equation for $\mathbf{v}_c$,

$$A_c \mathbf{v}_c = -P^T A \begin{bmatrix} \overline{\mathbf{v}}_f \\ 0 \end{bmatrix}.$$

Then,

$$W_f \overline{\mathbf{v}}_f = [I, 0]\widehat{A}\overline{\mathbf{v}} = [I, 0]\pi^T A \pi \overline{\mathbf{v}} = [I, 0] A \begin{bmatrix} \overline{\mathbf{v}}_f + \mathcal{P}\mathbf{v}_c \\ \mathbf{v}_c \end{bmatrix}.$$

The inverse actions of the Schur complement S_f^0 of $\pi^T M_0 \pi$ are computed as follows.

$$\left(S_f^0\right)^{-1} \overline{\mathbf{v}}_f = [I,\ 0]\pi^{-1} M_0^{-1} \pi^{-T} \begin{bmatrix} I \\ 0 \end{bmatrix} \overline{\mathbf{v}}_f = [I,\ -\mathcal{P}] M_0^{-1} \begin{bmatrix} I \\ -\mathcal{P}^T \end{bmatrix} \overline{\mathbf{v}}_f.$$

Here, we used the fact that the inverse of a Schur complement of a matrix is a principal submatrix of the inverse of the given matrix.

In summary, the implementation of an iterative method based on the Schur complements of $\pi^T A \pi$ and $\pi^T M_0 \pi$ requires only actions of A_c^{-1}, A, M_0^{-1}, $\mathcal{P}$, and $\mathcal{P}^T$; that is, the iterations can be implemented in terms of the original matrix, the coarse matrix, the original s.p.d. preconditioner M_0 (its inverse action), and the interpolation matrix P and its transpose.

9

Preconditioning Saddle-Point Matrices

In the present chapter, we consider problems with symmetric (nonsingular) matrices that admit the following two-by-two block form,

$$A = \begin{bmatrix} \mathcal{A} & \mathcal{B}^T \\ \mathcal{B} & -\mathcal{C} \end{bmatrix}.$$

Here, $\mathcal{A}$ and $\mathcal{C}$ are symmetric and positive semidefinite, which makes A indefinite. A common case in practice is $\mathcal{C} = 0$. Matrices A of the above form are often called saddle-point matrices. In the following sections, we study the construction of preconditioners that exploit the above block structure of A. We consider both cases of definite and indefinite preconditioners.

There is extensive literature devoted to the present topic. We mention only the more recent surveys [AN03] and [BGL].

9.1 Basic properties of saddle-point matrices

We assume that $\mathcal{B}^T$ has full column rank; that is, $\mathcal{B}\mathcal{B}^T$ is invertible. Then the following result is easily seen.

Theorem 9.1. *A necessary and sufficient condition for A to be invertible is that $\mathcal{A} + \mathcal{B}^T\mathcal{B}$ be invertible.*

Proof. Indeed, if there is a nonzero vector $\mathbf{w}$ such that simultaneously $\mathcal{A}\mathbf{w} = 0$ and $\mathcal{B}\mathbf{w} = 0$ then

$$A \begin{bmatrix} \mathbf{w} \\ 0 \end{bmatrix} = \begin{bmatrix} 0 \\ 0 \end{bmatrix},$$

that is, A has a nontrivial null space. Thus, $\mathcal{A} + \mathcal{B}^T\mathcal{B}$ being only semidefinite implies that A is singular.

The converse is also true. Assume that $\mathcal{A} + \mathcal{B}^T\mathcal{B}$ is invertible. We show then that A is invertible. Because $\mathcal{C}$ is positive semidefinite, for a sufficiently small constant

P.S. Vassilevski, *Multilevel Block Factorization Preconditioners*,
doi: 10.1007/978-0-387-71564-3_9,

$\delta > 0$ we can guarantee that $I - \delta C$ is s.p.d. Consider the system

$$\begin{bmatrix} \mathcal{A} & \mathcal{B}^T \\ \mathcal{B} & -\mathcal{C} \end{bmatrix} \begin{bmatrix} \mathbf{w} \\ \mathbf{x} \end{bmatrix} = 0.$$

We can transform it as follows. Multiply the second block row with $\delta \mathcal{B}^T$ and add it to the first block row, and then multiply the second block row with $I - \delta C$; we get

$$\begin{bmatrix} \mathcal{A} + \delta \mathcal{B}^T \mathcal{B} & \mathcal{B}^T (I - \delta \mathcal{C}) \\ (I - \delta \mathcal{C}) \mathcal{B} & -(I - \delta \mathcal{C}) \mathcal{C} \end{bmatrix} \begin{bmatrix} \mathbf{w} \\ \mathbf{x} \end{bmatrix} = 0.$$

Note that $\mathcal{A}$ and $\mathcal{B}$ having only the zero vector as a common null vector implies that $\mathcal{A} + \delta \mathcal{B}^T \mathcal{B}$ is invertible, and hence, we get the following reduced problem for $\mathbf{x}$,

$$[(I - \delta \mathcal{C})\mathcal{C} + (I - \delta \mathcal{C})\mathcal{B}(\mathcal{A} + \delta \mathcal{B}^T \mathcal{B})^{-1} \mathcal{B}^T (I - \delta \mathcal{C})]\mathbf{x} = 0.$$

Observe now that the matrix $(I - \delta \mathcal{C})\mathcal{C}$ is symmetric positive semidefinite, hence $(I - \delta \mathcal{C})\mathcal{B}(\mathcal{A} + \delta \mathcal{B}^T \mathcal{B})^{-1} \mathcal{B}^T (I - \delta \mathcal{C})\mathbf{x} = 0$. That is, $\mathcal{B}^T (I - \delta \mathcal{C})\mathbf{x} = 0$, and due to the full-column rank of $\mathcal{B}^T$, we get $(I - \delta \mathcal{C})\mathbf{x} = 0$. Finally, because $I - \delta \mathcal{C}$ is invertible, we get $\mathbf{x} = 0$. Then $\mathbf{w} = -(\mathcal{A} + \delta \mathcal{B}^T \mathcal{B})^{-1} \mathcal{B}^T (I - \delta \mathcal{C})\mathbf{x} = 0$. Thus, we proved that A has only a trivial null space that completes the proof. □

In some applications $\mathcal{B}^T$ is rank deficient. This is the case, for example, when $\mathcal{B}$ corresponds to a discrete divergence operator. Then $\mathbf{1}^T \mathcal{B} = 0$ for any constant vector $\mathbf{1}$ (assuming that essential boundary conditions were imposed). Strictly speaking, A may be singular then (the case for $\mathcal{C} = 0$). To avoid nonuniqueness of the solution we add additional constraints, namely, for all vectors $\mathbf{q}_k$, $k = 1, \ldots, m$, providing a basis of the null space of $\mathcal{B}^T$, we impose

$$\mathbf{q}_k^T \mathbf{x} = 0.$$

We may assume that $\mathcal{Q} = [\mathbf{q}_1, \ldots, \mathbf{q}_m]$ is orthogonal; that is, $\mathcal{Q}^T \mathcal{Q} = I$.

The original problem

$$A \begin{bmatrix} \mathbf{w} \\ \mathbf{x} \end{bmatrix} = \begin{bmatrix} \mathbf{f} \\ \mathbf{g} \end{bmatrix}$$

with a possibly nonunique solution is transformed to one with a unique solution

$$\begin{bmatrix} \mathcal{A} & 0 & \mathcal{B}^T \\ 0 & 0 & \mathcal{Q}^T \\ \mathcal{B} & \mathcal{Q} & -\mathcal{C} \end{bmatrix} \begin{bmatrix} \mathbf{w} \\ \boldsymbol{\lambda} \\ \mathbf{x} \end{bmatrix} = \begin{bmatrix} \mathbf{f} \\ 0 \\ \mathbf{g} \end{bmatrix}. \tag{9.1}$$

Here, $\boldsymbol{\lambda} \in \mathbb{R}^m$ is the vector of the so-called Lagrange multipliers. The new saddle-point matrix has a full-column rank off-diagonal block

$$\begin{bmatrix} \mathcal{B}^T \\ \mathcal{Q}^T \end{bmatrix}$$

ensured by construction. Indeed, $(\mathcal{B}\mathcal{B}^T+\mathcal{Q}\mathcal{Q}^T)\mathbf{x} = 0$ implies $\mathcal{B}^T\mathbf{x} = 0$ and $\mathcal{Q}^T\mathbf{x} = 0$. The first equation $\mathcal{B}^T\mathbf{x} = 0$ implies that $\mathbf{x} = \mathcal{Q}\sigma \in \text{Null}(\mathcal{B}^T)$ for some $\sigma \in \mathbb{R}^m$. Then, the second equation $0 = \mathcal{Q}^T\mathbf{x} = \mathcal{Q}^T(\mathcal{Q}\sigma)$ shows that $\sigma = 0$. That is, $\mathbf{x} = 0$. This proves that $\mathcal{B}\mathcal{B}^T + \mathcal{Q}\mathcal{Q}^T$ is invertible, which is equivalent to

$$\begin{bmatrix} \mathcal{B}^T \\ \mathcal{Q}^T \end{bmatrix}$$

having full-column rank. Finally, notice that the principal block of the expanded saddle-point matrix is

$$\begin{bmatrix} \mathcal{A} & 0 \\ 0 & 0 \end{bmatrix},$$

which is symmetric positive semidefinite. The result of Theorem 9.1 then tells us that the new problem has a unique solution if (and only if)

$$\begin{bmatrix} \mathcal{A} & 0 \\ 0 & 0 \end{bmatrix} + \begin{bmatrix} \mathcal{B}^T \\ \mathcal{Q}^T \end{bmatrix} \begin{bmatrix} \mathcal{B}^T \\ \mathcal{Q}^T \end{bmatrix}^T = \begin{bmatrix} \mathcal{A}+\mathcal{B}^T\mathcal{B} & \mathcal{B}^T\mathcal{Q} \\ \mathcal{Q}^T\mathcal{B} & \mathcal{Q}^T\mathcal{Q} \end{bmatrix} = \begin{bmatrix} \mathcal{A}+\mathcal{B}^T\mathcal{B} & 0 \\ 0 & I \end{bmatrix}$$

is invertible, that is if (and only if) $\mathcal{A} + \mathcal{B}^T\mathcal{B}$ is invertible. In practice, we do not form the expanded (three-by-three) system (9.1) explicitly. During the computation, the vectors $\mathbf{x}$ are considered (and kept explicitly) orthogonal to $\mathbf{q}_1, \dots, \mathbf{q}_m$.

Consider the special case $\mathcal{C} = 0$. If the second r.h.s. component $\mathbf{g}$ is orthogonal to null$(\mathcal{B}^T)$, we can prove that $\boldsymbol{\lambda} = 0$. Indeed, the last equation of the three-by-three system implies that $\mathcal{Q}^T(\mathcal{B}\mathbf{w}+\mathcal{Q}\boldsymbol{\lambda}) = \mathcal{Q}^T\mathbf{g} = 0$. Because $\mathcal{Q}^T\mathcal{B} = 0$, we have $\boldsymbol{\lambda} = 0$.

Therefore, as we proved above, the following result holds in the case of rank-deficient $\mathcal{B}^T$ and $\mathcal{C} = 0$.

Theorem 9.2. *Consider the saddle-point problem*

$$\begin{bmatrix} \mathcal{A} & \mathcal{B}^T \\ \mathcal{B} & 0 \end{bmatrix} \begin{bmatrix} \mathbf{w} \\ \mathbf{x} \end{bmatrix} = \begin{bmatrix} \mathbf{f} \\ \mathbf{g} \end{bmatrix}.$$

Here, $\mathcal{A}$ is symmetric positive semidefinite. Let $\{\mathbf{q}_k\}_{k=1}^m$ form a basis of null$(\mathcal{B}^T)$. The above saddle-point problem for any given $\mathbf{f}$ and $\mathbf{g}$: $\mathbf{q}_k^T\mathbf{g} = 0$, $k = 1, \dots, m$, will have a unique solution $(\mathbf{w},\ \mathbf{x})$ with second component $\mathbf{x}$ satisfying the constraints $\mathbf{q}_k^T\mathbf{x} = 0$, $k = 1, \dots, m$, if and only if $\mathcal{A} + \mathcal{B}^T\mathcal{B}$ is invertible.

The case of general positive semidefinite $\mathcal{C}$ is considered in the final theorem.

Theorem 9.3. *Consider the saddle-point problem*

$$\begin{bmatrix} \mathcal{A} & \mathcal{B}^T \\ \mathcal{B} & -(I - \mathcal{Q}\mathcal{Q}^T)\mathcal{C} \end{bmatrix} \begin{bmatrix} \mathbf{w} \\ \mathbf{x} \end{bmatrix} = \begin{bmatrix} \mathbf{f} \\ \mathbf{g} \end{bmatrix}. \tag{9.2}$$

Here, $\mathcal{A}$ and $\mathcal{C}$ are symmetric positive semidefinite and $\mathcal{Q} = [\mathbf{q}_1, \dots, \mathbf{q}_m]$ is such that $\{\mathbf{q}_k\}_{k=1}^m$ form an orthogonal basis of null$(\mathcal{B}^T)$. The above saddle-point problem for any given $\mathbf{f}$, and $\mathbf{g}$ orthogonal to null$(\mathcal{B}^T)$, has a unique solution $(\mathbf{w},\ \mathbf{x})$ with second component $\mathbf{x}$ satisfying the constraints $\mathbf{q}_k^T\mathbf{x} = 0$, $k = 1, \dots, m$, if and only if $\mathcal{A} + \mathcal{B}^T\mathcal{B}$ is invertible.

Proof. We showed that the necessary and sufficient condition for the expanded saddle-point problem (9.1) to have a unique solution is $\mathcal{A}+\mathcal{B}^T\mathcal{B}$ to be s.p.d. We show next that (9.1) in the case $\mathcal{Q}^T\mathbf{g} = 0$ having a unique solution $(\mathbf{w},\ \mathbf{x})$ with $\mathcal{Q}^T\mathbf{x} = 0$ is equivalent to (9.2) having a unique solution $(\mathbf{w},\ \mathbf{x})$ such that $\mathcal{Q}^T\mathbf{x} = 0$ (if $\mathcal{Q}^T\mathbf{g} = 0$).

It is clear that $\mathcal{Q}^T\mathbf{g} = 0$ is a necessary condition for solvability of (9.2). This is seen by multiplying its second equation by $\mathcal{Q}^T$ using the fact that $\mathcal{Q}^T B = 0$ and $\mathcal{Q}^T(I - \mathcal{Q}\mathcal{Q}^T) = 0$.

Consider the expanded system (9.1) with $\mathbf{g}:\ \ \mathcal{Q}^T\mathbf{g} = 0$. Multiplying its third equation $\mathcal{B}\mathbf{w} + \mathcal{Q}\boldsymbol{\lambda} - \mathcal{C}\mathbf{x} = \mathbf{g}$ with the projection $I - \mathcal{Q}\mathcal{Q}^T$, based on the facts that $\mathcal{Q}^T\mathcal{B} = 0$, $\mathcal{Q}^T\mathcal{Q} = I$ and $\mathcal{Q}^T\mathbf{g} = 0$, leads to the desired second equation $\mathcal{B}\mathbf{w} - (I - \mathcal{Q}\mathcal{Q}^T)\mathcal{C}\mathbf{x} = \mathbf{g}$ of (9.2). This and the first equation of (9.1) show that $(\mathbf{w},\ \mathbf{x})$ with $\mathcal{Q}^T\mathbf{x} = 0$ give a solution to the problem (9.2).

On the other hand, any solution $(\mathbf{w},\ \mathbf{x})$ of (9.2) such that $\mathcal{Q}^T\mathbf{x} = 0$ together with $\boldsymbol{\lambda} = \mathcal{Q}^T\mathcal{C}\mathbf{x}$, provides a solution to (9.1) that we know is unique if $\mathcal{A} + \mathcal{B}^T\mathcal{B}$ is s.p.d. Thus, $(\mathbf{w}, \mathbf{x})$ with $\mathcal{Q}^T\mathbf{x} = 0$ must be the unique solution to (9.2). □

Remark 9.4. Based on the above result, we can in principle transform a problem

$$A\begin{bmatrix}\mathbf{w}\\ \mathbf{x}\end{bmatrix} = \begin{bmatrix}\mathbf{f}\\ \mathbf{g}\end{bmatrix}$$

to the following equivalent one,

$$\begin{bmatrix}\mathcal{A}+\delta\mathcal{B}^T\mathcal{B} & \mathcal{B}^T(I-\delta\mathcal{C})\\ (I-\delta\mathcal{C})\mathcal{B} & -(I-\delta\ \mathcal{C})\mathcal{C}\end{bmatrix}\begin{bmatrix}\mathbf{w}\\ \mathbf{x}\end{bmatrix} = \begin{bmatrix}\mathbf{f}+\delta\mathcal{B}^T\mathbf{g}\\ (I-\delta\ \mathcal{C})\mathbf{g}\end{bmatrix}.$$

The transformed matrix is again of saddle-point type. It has the property that $\mathcal{B}^T(I - \delta\ \mathcal{C})$ has a full-column rank, and finally, its first block $\mathcal{A}+\delta\ \mathcal{B}^T\mathcal{B}$ is invertible. Thus, at least in theory, we can assume without loss of generality that the original matrix A has invertible first block $\mathcal{A}$.

9.2 S.p.d. preconditioners

Consider the following saddle-point problem

$$A\mathbf{X} = \mathbf{F},$$

where

$$\mathbf{X} = \begin{bmatrix}\mathbf{v}\\ \mathbf{x}\end{bmatrix} \quad \text{and} \quad \mathbf{F} = \begin{bmatrix}\mathbf{w}\\ \mathbf{b}\end{bmatrix}.$$

It is typical to derive a priori estimates for the solution $\mathbf{X}$ in a norm $\|.\|_D$ whereas the r.h.s. $\mathbf{F}$ is taken in the dual space, that is, in the norm $\|\mathbf{F}\|_{D^{-1}} = \max_{\mathbf{Y}}(\mathbf{Y}^T\mathbf{F}/\|\mathbf{Y}\|_D)$. The a priori estimate reads:

$$\|\mathbf{X}\|_D \le \kappa\ \|\mathbf{F}\|_{D^{-1}}. \tag{9.3}$$

From this a priori estimate used for $\mathbf{X} = A^{-1}\mathbf{F}$, we get the first spectral relation between A and D

$$\frac{1}{\kappa^2}\mathbf{X}^T D\mathbf{X} \leq \mathbf{X}^T A^T D^{-1} A\mathbf{X}. \tag{9.4}$$

The latter estimate represents coercivity of $A^T D^{-1} A$ in terms of D.

Note that the norm $\|.\|_D^2$ is typically a sum of squares of two norms; that is, D is a block-diagonal s.p.d. matrix with blocks $\mathcal{M}$ and $\mathcal{D}$. Then, for

$$\mathbf{V} = \begin{bmatrix} \mathbf{v} \\ \mathbf{x} \end{bmatrix},$$

we have $\|\mathbf{V}\|_D^2 = \mathbf{v}^T \mathcal{M}\mathbf{v} + \mathbf{x}^T \mathcal{D}\mathbf{x}$.

Assume now, that we can construct a block-diagonal matrix M that is spectrally equivalent to D. The coercivity estimate (9.4) implies a similar estimate with D replaced by M; namely, we have

$$\mathbf{F}^T M\mathbf{F} \leq \widehat{\kappa}^2\ \mathbf{F}^T A^T M^{-1} A\mathbf{F}. \tag{9.5}$$

Because $A^T = A$, the above estimate, by letting $\mathbf{F} = M^{-(1/2)}\mathbf{V}$, is equivalent to

$$\mathbf{V}^T\mathbf{V} \leq \widehat{\kappa}^2\ \mathbf{V}^T (M^{-(1/2)} A M^{-(1/2)})^2 \mathbf{V}.$$

That is, the absolute value of the eigenvalues of $M^{-(1/2)} A M^{-(1/2)}$ is bounded from below by $1/\widehat{\kappa}$. Or equivalently, the eigenvalues of $M^{-(1/2)} A M^{-(1/2)}$ stay away from the origin. The latter property is a main reason to use block-diagonal s.p.d. preconditioners M that come from the norm $\|.\|_D$ in which the saddle-point problem is well posed.

To complete the spectral equivalence relations between A and M, we need an estimate from above for the absolute value of the eigenvalues of $M^{-(1/2)} A M^{-(1/2)}$. This comes from the following (assumed) boundedness estimate of A in terms of D; namely,

$$\mathbf{V}^T A\mathbf{W} \leq \sigma\ (\mathbf{V}^T D\mathbf{V})^{1/2} (\mathbf{W}^T D\mathbf{W})^{1/2}. \tag{9.6}$$

Because M is spectrally equivalent to D, a similar estimate holds with D replaced by M,

$$\mathbf{V}^T A\mathbf{W} \leq \widehat{\sigma}\ (\mathbf{V}^T M\mathbf{V})^{1/2} (\mathbf{W}^T M\mathbf{W})^{1/2}. \tag{9.7}$$

Thus, the eigenvalues of $M^{-(1/2)} A M^{-(1/2)}$ (because $A^T = A$) are bounded in absolute value by $\widehat{\sigma}$.

To summarize, we have the following result.

Theorem 9.5. *The construction of block-diagonal s.p.d. preconditioners for the symmetric saddle-point (indefinite) block-matrix A can be reduced to the construction of preconditioners M for the block-diagonal matrix D. The matrix D first defines a norm* *in which the saddle-point problem is well posed; namely, an a priori estimate* (9.3) *holds for any r.h.s.* **F**, *and second, the saddle-point operator A is bounded in terms of D in the sense that a boundedness estimate such as* (9.6) *holds. The thus-constructed s.p.d. preconditioners M when used in a MINRES iterative method (cf., Appendix A) will exhibit a rate of convergence dependent only on the constants involved in the spectral equivalence relations* (9.5) *and* (9.7).

9.2.1 Preconditioning based on "inf–sup" condition

We assume here that $\mathcal{C} = 0$. The case of nonzero (symmetric positive semidefinite) $\mathcal{C}$ is treated as a perturbation of the case $\mathcal{C} = 0$ at the end of this section.

Introduce the block-diagonal matrix

$$D = \begin{bmatrix} \mathcal{M} & 0 \\ 0 & \mathcal{D} \end{bmatrix}.$$

We also need the Schur complements $\mathcal{S} = \mathcal{B}\mathcal{A}^{-1}\mathcal{B}^T$ and $\mathcal{S}_{\mathcal{A}_\mathcal{D}} = \mathcal{B}\mathcal{A}_\mathcal{D}^{-1}\mathcal{B}^T$ with $\mathcal{A}_\mathcal{D} = \mathcal{A} + \mathcal{B}^T\mathcal{D}^{-1}\mathcal{B}$.

We have the following important result.

Lemma 9.6. *The a priori estimate* (9.3) *implies the following well-known LBB (Ladyzhenskaya–Babuška–Brezzi) or simply "inf–sup" condition*

$$\frac{1}{\kappa} \le \inf_{\mathbf{y}} \sup_{\mathbf{w}} \frac{\mathbf{y}^T\mathcal{B}\mathbf{w}}{\|\mathbf{w}\|_\mathcal{M}\|\mathbf{y}\|_\mathcal{D}}. \tag{9.8}$$

Proof. For any $\mathbf{y}$, solve the saddle-point problem

$$\begin{aligned} \mathcal{A}\mathbf{v} \quad + \mathcal{B}^T\mathbf{x} \quad &= 0, \\ \mathcal{B}\mathbf{v} \quad\quad\quad &= -\mathcal{D}\mathbf{y}. \end{aligned}$$

The solution components are $\mathbf{x} = \mathcal{S}^{-1}\mathcal{D}\mathbf{y}$ and $\mathbf{v} = -\mathcal{A}^{-1}\mathcal{B}^T\mathcal{S}^{-1}\mathcal{D}\mathbf{y}$. The assumed a priori estimate (9.3) rewritten as in (9.4) implies that

$$\mathbf{v}^T\mathcal{M}\mathbf{v} + \mathbf{x}^T\mathcal{D}\mathbf{x} \le \kappa^2\, \mathbf{y}^T\mathcal{D}\mathbf{y}.$$

In particular, we have $\mathbf{v}^T\mathcal{M}\mathbf{v} \le \kappa^2\, \mathbf{y}^T\mathcal{D}\mathbf{y}$. Consider finally,

$$\begin{aligned} \sup_{\mathbf{w}} \frac{(\mathbf{y}^T\mathcal{B}\mathbf{w})^2}{\mathbf{w}^T\mathcal{M}\mathbf{w}} &\ge \frac{(\mathbf{y}^T\mathcal{B}\mathbf{v})^2}{\mathbf{v}^T\mathcal{M}\mathbf{v}} \\ &= \frac{(\mathbf{y}^T\mathcal{B}\mathcal{A}^{-1}\mathcal{B}^T\mathcal{S}^{-1}\mathcal{D}\mathbf{y})^2}{\mathbf{v}^T\mathcal{M}\mathbf{v}} \\ &= \frac{(\mathbf{y}^T\,\mathcal{D}\mathbf{y})^2}{\mathbf{v}^T\mathcal{M}\mathbf{v}} \\ &\ge \frac{(\mathbf{y}^T\,\mathcal{D}\mathbf{y})^2}{\kappa^2\mathbf{y}^T\mathcal{D}\mathbf{y}} \\ &= \frac{1}{\kappa^2}\,\mathbf{y}^T\mathcal{D}\mathbf{y}. \end{aligned}$$

The latter estimate is the desired "inf–sup" (or LBB) condition. □

The converse is also true in the following sense.

Theorem 9.7. *The "inf–sup" condition* (9.8) *and boundedness of A in terms of D as in* (9.6) *imply the a priori estimate for* $\mathbf{AX} = \mathbf{F}$ *in the pair of norms generated by* $\mathcal{A}_{\mathcal{D}} = \mathcal{A} + \mathcal{B}^T\mathcal{D}^{-1}\mathcal{B}$ *and* $\mathcal{D}$.

Proof. The boundedness of A in terms of D (9.6) used for

$$\mathbf{V} = \mathbf{W} = \begin{bmatrix} \mathbf{w} \\ 0 \end{bmatrix}$$

implies that

$$\mathbf{w}^T\mathcal{A}\mathbf{w} \le \sigma\ \mathbf{w}^T\mathcal{M}\mathbf{w}. \tag{9.9}$$

The boundedness of A in terms of D (9.6) now used for

$$\mathbf{V} = \mathbf{W} = \begin{bmatrix} \mathbf{w} \\ \mathbf{y} \end{bmatrix}$$

implies

$$\mathbf{w}^T\mathcal{A}\mathbf{w} + 2\mathbf{y}^T\mathcal{B}\mathbf{w} \le \sigma\ (\mathbf{w}^T\mathcal{M}\mathbf{w} + \mathbf{y}^T\mathcal{D}\mathbf{y}).$$

Substituting $\mathbf{y} := t\mathbf{y}$ for any real number t, we get that the quadratic form $Q(t) \equiv \sigma\ t^2\ \mathbf{y}^T\mathcal{D}\mathbf{y} - 2t\ \mathbf{y}^T\mathcal{B}\mathbf{w} + \mathbf{w}^T(\sigma\ \mathcal{M} - \mathcal{A})\mathbf{w}$ is nonnegative. This shows, based on (9.9), that the discriminant $D \equiv (\mathbf{y}^T\mathcal{B}\mathbf{w})^2 - \sigma\ \mathbf{y}^T\mathcal{D}\mathbf{y}\ \mathbf{w}^T(\sigma\ \mathcal{M} - \mathcal{A})\mathbf{w}$ is nonpositive; that is, we have

$$(\mathbf{y}^T\mathcal{B}\mathbf{w})^2 \le \sigma\ \mathbf{y}^T\mathcal{D}\mathbf{y}\ \mathbf{w}^T(\sigma\ \mathcal{M} - \mathcal{A})\mathbf{w}. \tag{9.10}$$

Letting $\mathbf{y} = \mathcal{D}^{-1}\mathcal{B}\mathbf{w}$ in the last estimate gives

$$\mathbf{w}^T\mathcal{B}^T\mathcal{D}^{-1}\mathcal{B}\mathbf{w} \le \sqrt{\sigma}\ (\mathbf{w}^T(\sigma\ \mathcal{M} - \mathcal{A})\mathbf{w})^{1/2}(\mathbf{w}^T\mathcal{B}^T\mathcal{D}^{-1}\mathcal{B}\mathbf{w})^{1/2},$$

or

$$\mathbf{w}^T\mathcal{B}^T\mathcal{D}^{-1}\mathcal{B}\mathbf{w} \le \sigma\ \mathbf{w}^T(\sigma\ \mathcal{M} - \mathcal{A})\mathbf{w}.$$

Equivalently,

$$\mathbf{w}^T(\sigma\ \mathcal{A} + \mathcal{B}^T\mathcal{D}^{-1}\mathcal{B})\mathbf{w} \le \sigma^2\ \mathbf{w}^T\mathcal{M}\mathbf{w}. \tag{9.11}$$

The "inf–sup" condition (9.8) implies then that

$$\begin{aligned} \frac{1}{\kappa}\|\mathbf{y}\|_{\mathcal{D}} &\le \sup_{\mathbf{w}} \frac{\mathbf{w}^T\mathcal{B}^T\mathbf{y}}{\|\mathbf{w}\|_{\mathcal{M}}} \\ &\le \sigma\ \sup_{\mathbf{w}} \frac{\mathbf{w}^T\mathcal{B}^T\mathbf{y}}{(\mathbf{w}^T(\sigma\ \mathcal{A} + \mathcal{B}^T\mathcal{D}^{-1}\mathcal{B})\mathbf{w})^{1/2}} \\ &\le \frac{\sigma}{\min\{1,\ \sqrt{\sigma}\}} \frac{\mathbf{w}^T\mathcal{B}^T\mathbf{y}}{(\mathbf{w}^T(\mathcal{A} + \mathcal{B}^T\mathcal{D}^{-1}\mathcal{B})\mathbf{w})^{1/2}}. \end{aligned} \tag{9.12}$$

That is, we have an "inf–sup" condition also when $\|.\|_{\mathcal{M}}$ is replaced by the norm generated by $\mathcal{A}_{\mathcal{D}} = \mathcal{A} + \mathcal{B}^T\mathcal{D}^{-1}\mathcal{B}$ with a different "inf–sup" constant

$$\widehat{\kappa} = \kappa \; \frac{\sigma}{\min\{1, \; \sqrt{\sigma}\}}.$$

Consider now the saddle-point problem $A\mathbf{X} = \mathbf{F}$ with a general r.h.s.

$$\mathbf{F} = \begin{bmatrix} \mathbf{f} \\ \mathbf{g} \end{bmatrix}$$

and let

$$\mathbf{X} = \begin{bmatrix} \mathbf{v} \\ \mathbf{x} \end{bmatrix}$$

be its solution. We transform the system $A\mathbf{X} = \mathbf{F}$ by multiplying its second block equation with $\mathcal{B}^T\mathcal{D}^{-1}$ and adding the result to the first one to arrive at

$$\begin{array}{rcl} (\mathcal{A}+ \; \mathcal{B}^T\mathcal{D}^{-1}\mathcal{B})\mathbf{v} + \mathcal{B}^T\mathbf{x} & = & \mathbf{f} + \mathcal{B}^T\mathcal{D}^{-1}\mathbf{g}, \\ \mathcal{B}\mathbf{v} & = & \mathbf{g}. \end{array} \tag{9.13}$$

We can estimate the first block of the r.h.s. above as follows.

$$\begin{aligned} \|\mathbf{f} + \mathcal{B}^T\mathcal{D}^{-1}\mathbf{g}\|_{\mathcal{A}_{\mathcal{D}}^{-1}} &\leq \|\mathbf{f}\|_{\mathcal{A}_{\mathcal{D}}^{-1}} + \|\mathcal{A}_{\mathcal{D}}^{-(1/2)}\mathcal{B}^T\mathcal{D}^{-(1/2)}\| \|\mathbf{g}\|_{\mathcal{D}^{-1}} \\ &= \|\mathbf{f}\|_{\mathcal{A}_{\mathcal{D}}^{-1}} + \|\mathcal{D}^{-(1/2)}\mathcal{B}\mathcal{A}_{\mathcal{D}}^{-(1/2)}\| \|\mathbf{g}\|_{\mathcal{D}^{-1}} \\ &\leq \|\mathbf{f}\|_{\mathcal{A}_{\mathcal{D}}^{-1}} + \|\mathbf{g}\|_{\mathcal{D}^{-1}}. \end{aligned} \tag{9.14}$$

Also with $\mathcal{S}_{\mathcal{A}_{\mathcal{D}}} = \mathcal{B}\mathcal{A}_{\mathcal{D}}^{-1}\mathcal{B}^T$, the following reduced equation for $\mathbf{x}$ is obtained from (9.13),

$$-\mathcal{S}_{\mathcal{A}_{\mathcal{D}}}\mathbf{x} = -\mathcal{B}\mathcal{A}_{\mathcal{D}}^{-1}(\mathbf{f} + \mathcal{B}^T\mathcal{D}^{-1}\mathbf{g}) + \mathbf{g}.$$

From the "inf–sup" condition (9.12), with $\widehat{\kappa} = \kappa \; (\sigma/(\min\{\sqrt{\sigma}, \; 1\}))$, we easily obtain

$$\frac{1}{\widehat{\kappa}^2}\mathbf{x}^T\mathcal{D}\mathbf{x} \leq \mathbf{x}^T\mathcal{B}\mathcal{A}_{\mathcal{D}}^{-1}\mathcal{B}^T\mathbf{x} = \mathbf{x}^T\mathcal{S}_{\mathcal{A}_{\mathcal{D}}}\mathbf{x}. \tag{9.15}$$

Therefore,

$$\begin{aligned} \frac{1}{\widehat{\kappa}^2}\mathbf{x}^T\mathcal{D}\mathbf{x} &\leq \mathbf{x}^T\mathcal{S}_{\mathcal{A}_{\mathcal{D}}}\mathbf{x} \\ &= -\mathbf{x}^T\mathbf{g} + \mathbf{x}^T\mathcal{B}\mathcal{A}_{\mathcal{D}}^{-1}(\mathbf{f} + \mathcal{B}^T\mathcal{D}^{-1}\mathbf{g}) \\ &\leq \|\mathbf{g}\|_{\mathcal{D}^{-1}}\|\mathbf{x}\|_{\mathcal{D}} + \|\mathbf{x}\|_{\mathcal{D}}\|\mathcal{D}^{-(1/2)}\mathcal{B}\mathcal{A}_{\mathcal{D}}^{-(1/2)}\| \|(\mathbf{f} + \mathcal{B}^T\mathcal{D}^{-1}\mathbf{g})\|_{\mathcal{A}_{\mathcal{D}}^{-1}} \\ &\leq \|\mathbf{x}\|_{\mathcal{D}}\big(2\|\mathbf{g}\|_{\mathcal{D}^{-1}} + \|\mathbf{f}\|_{\mathcal{A}_{\mathcal{D}}^{-1}}\big). \end{aligned}$$

Thus we arrive at the first a priori estimate,

$$\frac{1}{\widehat{\kappa}^2}\,\|\mathbf{x}\|_{\mathcal{D}} \leq 2\|\mathbf{g}\|_{\mathcal{D}^{-1}} + \|\mathbf{f}\|_{\mathcal{A}_{\mathcal{D}}^{-1}}. \tag{9.16}$$

Finally, because

$$\begin{aligned}\mathbf{w}^T\mathcal{B}^T\mathbf{x} &\leq \|\mathcal{A}_{\mathcal{D}}^{1/2}\mathbf{w}\|\,\|\mathcal{A}_{\mathcal{D}}^{-(1/2)}\mathcal{B}^T\mathcal{D}^{-(1/2)}\|\,\|\mathbf{x}\|_{\mathcal{D}}\\ &= \|\mathcal{A}_{\mathcal{D}}^{1/2}\mathbf{w}\|\,\|\mathcal{D}^{-(1/2)}\mathcal{B}\mathcal{A}_{\mathcal{D}}^{-(1/2)}\|\,\|\mathbf{x}\|_{\mathcal{D}}\\ &\leq \|\mathbf{w}\|_{\mathcal{A}_{\mathcal{D}}}\|\mathbf{x}\|_{\mathcal{D}},\end{aligned}$$

from the first equation of (9.13), we get

$$\begin{aligned}\mathbf{v}^T\mathcal{A}_{\mathcal{D}}\mathbf{v} &= -\mathbf{v}^T\mathcal{B}^T\mathbf{x} + \mathbf{v}^T(\mathbf{f} + \mathcal{B}^T\mathcal{D}^{-1}\mathbf{g})\\ &\leq \|\mathbf{v}\|_{\mathcal{A}_{\mathcal{D}}}\|\mathbf{x}\|_{\mathcal{D}} + \|\mathbf{v}\|_{\mathcal{A}_{\mathcal{D}}}\|\mathbf{f} + \mathcal{B}^T\mathcal{D}^{-1}\mathbf{g}\|_{\mathcal{A}_{\mathcal{D}}^{-1}}.\end{aligned}$$

That is,

$$\|\mathbf{v}\|_{\mathcal{A}_{\mathcal{D}}} \leq \|\mathbf{x}\|_{\mathcal{D}} + \|\mathbf{f} + \mathcal{B}^T\mathcal{D}^{-1}\mathbf{g}\|_{\mathcal{A}_{\mathcal{D}}^{-1}}.$$

Based on (9.14) and the proven estimate (9.16) for $\|\mathbf{x}\|_{\mathcal{D}}$, we then arrive at the other desired a priori estimate,

$$\|\mathbf{v}\|_{\mathcal{A}_{\mathcal{D}}} \leq \|\mathbf{x}\|_{\mathcal{D}} + \|\mathbf{f} + \mathcal{B}^T\mathcal{D}^{-1}\mathbf{g}\|_{\mathcal{A}_{\mathcal{D}}^{-1}} \leq (\widehat{\kappa}^2 + 1)\|\mathbf{f}\|_{\mathcal{A}_{\mathcal{D}}^{-1}} + (2\widehat{\kappa}^2 + 1)\|\mathbf{g}\|_{\mathcal{D}^{-1}}.\ \square$$

The following result (see (9.15)) also holds.

Corollary 9.8. *Assume the "inf–sup" condition* (9.8) *with* $\mathcal{M} = \mathcal{A}_{\mathcal{D}} = \mathcal{A} + \mathcal{B}^T\mathcal{D}^{-1}\mathcal{B}$. *Then, the Schur complement* $\mathcal{S}_{\mathcal{A}_{\mathcal{D}}} = \mathcal{B}(\mathcal{A}_{\mathcal{D}})^{-1}\mathcal{B}^T$ *of the transformed saddle-point problem* (9.13) *is spectrally equivalent to* $\mathcal{D}$*; namely, we have*

$$\frac{1}{\kappa^2}\,\mathbf{x}^T\mathcal{D}\mathbf{x} \leq \mathbf{x}^T\mathcal{S}_{\mathcal{A}_{\mathcal{D}}}\mathbf{x} \leq \mathbf{x}^T\mathcal{D}\mathbf{x}.$$

In particular, if $\mathcal{D}$ *is well conditioned then* $\mathcal{S}_{\mathcal{A}_{\mathcal{D}}}$ *is also well conditioned.*

We also need the following auxiliary result.

Lemma 9.9. *Assume that an a priori estimate* (9.3) *in the pair of norms* $\|.\|_{\mathcal{M}}$, $\|.\|_{\mathcal{D}}$ *holds (i.e.,* $\|.\|_D^2 = \|.\|_{\mathcal{M}}^2 + \|.\|_{\mathcal{D}}^2$*), together with a boundedness estimate* (9.6)*. Then,* $\mathcal{M}$ *is spectrally equivalent to* $\mathcal{A}_{\mathcal{D}}$*.*

Proof. The boundedness estimate (9.6) implies estimate (9.11), which shows $\min\{1,\sigma\}\ \mathbf{w}^T\mathcal{A}_{\mathcal{D}}\mathbf{w} \leq \sigma^2\ \mathbf{w}^T\mathcal{M}\mathbf{w}$. The latter represents one side of the desired spectral equivalence between $\mathcal{A}_{\mathcal{D}}$ and $\mathcal{M}$. To prove an estimate in the other direction, consider the saddle-point problem

$$\begin{aligned}\mathcal{A}\mathbf{v} + \mathcal{B}^T\mathbf{x} &= \mathcal{A}\mathbf{f}\\ \mathcal{B}\mathbf{v} \qquad\quad &= \mathcal{B}\mathbf{f}.\end{aligned}$$

It is clear that $\mathbf{x} = 0$ and $\mathbf{v} = \mathbf{f}$. Based on the assumed a priori estimate (9.3), we have for some κ,

$$\kappa^{-2}\, \mathbf{v}^T \mathcal{M}\mathbf{v} \leq (\mathcal{A}\mathbf{f})^T \mathcal{M}^{-1}(\mathcal{A}\mathbf{f}) + (\mathcal{B}\mathbf{f})^T \mathcal{D}^{-1}\mathcal{B}\mathbf{f}.$$

This estimate, together with the boundedness estimate (9.9) $\mathbf{v}^T \mathcal{A}\mathbf{v} \leq \sigma\ \mathbf{v}^T \mathcal{M}\mathbf{v}$, imply for $\mathbf{v} = \mathbf{f}$ that

$$\kappa^{-2}\, \mathbf{f}^T \mathcal{M}\mathbf{f} \leq \sigma\ \mathbf{f}^T \mathcal{A}\mathbf{f} + (\mathcal{B}\mathbf{f})^T \mathcal{D}^{-1}\mathcal{B}\mathbf{f} \leq \max\{1,\ \sigma\}\ \mathbf{f}^T \mathcal{A}_{\mathcal{D}}\mathbf{f}.$$

This is the desired second spectral relation between $\mathcal{M}$ and $\mathcal{A}_{\mathcal{D}}$. □

We conclude with the following main result that characterizes a s.p.d. block-diagonal preconditioner D for the saddle-point matrix A.

Theorem 9.10. *Consider the saddle-point matrix A (with $\mathcal{C} = 0$) and a block-diagonal matrix D with blocks $\mathcal{M}$ and $\mathcal{D}$. The saddle-point problem $A\mathbf{X} = \mathbf{F}$ is well posed and bounded in the D-norm so that estimates* (9.3) *and* (9.6) *hold (or equivalently the spectral equivalence relations* (9.4) *and* (9.6) *hold) if and only if:*

1. *The pair $(\mathcal{M},\ \mathcal{D})$ ensures an "inf–sup" condition of the form* (9.8) *for the block $\mathcal{B}$.*
2. *$\mathcal{M}$ is spectrally equivalent to $\mathcal{A}_{\mathcal{D}} = \mathcal{A} + \mathcal{B}^T\mathcal{D}^{-1}\mathcal{B}$.*

Proof. The fact that an a priori estimate and boundedness estimate imply that $\mathcal{M}$ is spectrally equivalent to $\mathcal{A}_{\mathcal{D}}$ is given by Lemma 9.9. Also, that an a priori estimate (9.3) implies the "inf–sup" condition (9.8) is given by Lemma 9.6.

Assume now that $(\mathcal{M},\ \mathcal{D})$ ensures an "inf–sup" condition of the form (9.8) and that $\mathcal{A}_{\mathcal{D}}$ is spectrally equivalent to $\mathcal{M}$. If we show that A is bounded in terms of the pair of norms $\|.\|_{\mathcal{D}}$ and $\|.\|_{\mathcal{M}}$, then from Theorem 9.7 we get that for some $\overline{\kappa}$ an a priori estimate $\overline{\kappa}^{-1}\ \|\mathbf{X}\|_D \leq \|\mathbf{F}\|_{D^{-1}}$ holds for $A\mathbf{X} = \mathbf{F}$ and $\|.\|_D^2 = \|.\|_{\mathcal{A}_{\mathcal{D}}}^2 + \|.\|_{\mathcal{D}}^2$. Then, because $\mathcal{A}_{\mathcal{D}}$ is spectrally equivalent to $\mathcal{M}$, a similar a priori estimate holds in the pair of norms $\|.\|_{\mathcal{M}}$ and $\|.\|_{\mathcal{D}}$, which is in fact the desired result. Thus, the proof is complete if we show that A is bounded in terms of the block-diagonal matrix D with blocks $\mathcal{M}$ and $\mathcal{D}$. The latter is seen as follows. We have that $\mathcal{B}^T\mathcal{D}^{-1}\mathcal{B} = \mathcal{A}_{\mathcal{D}} - \mathcal{A}$ is positive semidefinite, and hence,

$$(\mathbf{v}^T \mathcal{A}\mathbf{w})^2 \leq \mathbf{v}^T \mathcal{A}\mathbf{v}\mathbf{w}^T \mathcal{A}\mathbf{w} \leq \mathbf{v}^T \mathcal{A}_{\mathcal{D}}\mathbf{v}\mathbf{w}^T \mathcal{A}_{\mathcal{D}}\mathbf{w}.$$

Also,

$$\mathbf{x}^T \mathcal{B}\mathbf{w} \leq \|\mathbf{x}\|_{\mathcal{D}}\|\mathcal{D}^{-(1/2)}\mathcal{B}\mathbf{w}\| = \|\mathbf{x}\|_{\mathcal{D}}(\mathbf{w}^T(\mathcal{A}_{\mathcal{D}} - \mathcal{A})\mathbf{w})^{1/2} \leq \|\mathbf{x}\|_{\mathcal{D}}\|\mathbf{w}\|_{\mathcal{A}_{\mathcal{D}}}.$$

For any two vectors

$$\mathbf{V} = \begin{bmatrix} \mathbf{v} \\ \mathbf{x} \end{bmatrix} \quad \text{and} \quad \mathbf{W} = \begin{bmatrix} \mathbf{w} \\ \mathbf{y} \end{bmatrix},$$

combining the last two estimates gives

$$\begin{aligned}
\mathbf{V}^T A\mathbf{W} &= \mathbf{v}^T(\mathcal{A}\mathbf{w} + \mathcal{B}^T\mathbf{y}) + \mathbf{x}^T\mathcal{B}\mathbf{w} \\
&\leq (\mathbf{v}^T\mathcal{A}_{\mathcal{D}}\mathbf{v})^{1/2}(\mathbf{w}^T\mathcal{A}_{\mathcal{D}}\mathbf{w})^{1/2} + \|\mathbf{x}\|_{\mathcal{D}}\|\mathbf{w}\|_{\mathcal{A}_{\mathcal{D}}} + \mathbf{y}^T\mathcal{B}\mathbf{v} \\
&\leq (\|\mathbf{v}\|^2_{\mathcal{A}_{\mathcal{D}}} + \|\mathbf{x}\|^2_{\mathcal{D}})^{1/2}(\|\mathbf{w}\|^2_{\mathcal{A}_{\mathcal{D}}} + \|\mathbf{y}\|^2_{\mathcal{D}})^{1/2} + \|\mathbf{v}\|_{\mathcal{A}_{\mathcal{D}}}\|\mathbf{y}\|_{\mathcal{D}} \\
&\leq 2(\|\mathbf{v}\|^2_{\mathcal{A}_{\mathcal{D}}} + \|\mathbf{x}\|^2_{\mathcal{D}})^{1/2}(\|\mathbf{w}\|^2_{\mathcal{A}_{\mathcal{D}}} + \|\mathbf{y}\|^2_{\mathcal{D}})^{1/2} \\
&= 2\|\mathbf{V}\|_D\|\mathbf{W}\|_D.
\end{aligned}$$

□

Consider at the end the more general case

$$A = \begin{bmatrix} \mathcal{A} & \mathcal{B}^T \\ \mathcal{B} & -\mathcal{C} \end{bmatrix}.$$

Denote by A_0 the matrix with zero $\mathcal{C}$ block. Because then

$$A = A_0 + \begin{bmatrix} 0 \\ I \end{bmatrix} \mathcal{T}^{-1}[0,\ I]$$

with $\mathcal{T} = (-\mathcal{C})^{-1}$, based on the Sherman–Morrison formula (see Proposition 3.5), we have

$$A^{-1} = A_0^{-1} - A_0^{-1}\begin{bmatrix} 0 \\ I \end{bmatrix}\left(\mathcal{T} + [0,\ I]A_0^{-1}\begin{bmatrix} 0 \\ I \end{bmatrix}\right)^{-1}[0,\ I]A_0^{-1}.$$

Because $[0,\ I]A_0^{-1}[{}^0_I] = -\mathcal{S}^{-1} = -(\mathcal{B}\mathcal{A}^{-1}\mathcal{B}^T)^{-1}$ (the inverse of the Schur complement of A_0), the above formula for A^{-1} takes the following form,

$$A^{-1} = A_0^{-1} - A_0^{-1}\begin{bmatrix} 0 \\ I \end{bmatrix}(-\mathcal{C}^{-1} - \mathcal{S}^{-1})^{-1}[0,\ I]A_0^{-1},$$

which can be rewritten as

$$A^{-1} = A_0^{-1} + A_0^{-1}\begin{bmatrix} 0 \\ I \end{bmatrix}\mathcal{C}^{1/2}(I + \mathcal{C}^{1/2}\mathcal{S}^{-1}\mathcal{C}^{1/2})^{-1}\mathcal{C}^{1/2}[0,\ I]A_0^{-1}. \tag{9.17}$$

The latter expression makes sense also for singular $\mathcal{C}$.

The following result holds for A.

Corollary 9.11. *Let $(\mathcal{M},\ \mathcal{D})$ satisfy the same properties for A_0 as in Theorem 9.10. Assume also that $\mathcal{C}$ is bounded in terms of $\mathcal{D}$. Then, the block-diagonal matrix D with blocks $\mathcal{M}$ and $\mathcal{D}$ provides a uniform preconditioner for A in the sense that the spectral equivalence relations* (9.4) *and* (9.6) *hold.*

Proof. The fact that $\mathcal{C}$ is bounded in terms of $\mathcal{D}$ and the boundedness assumption on A_0 in terms of D implies that A is bounded in terms of D, which proves (9.6).

By assumption, we have

$$\|D^{1/2}A_0^{-1}D^{1/2}\| \leq \kappa.$$

Also by assumption there is a $\sigma > 0$ such that

$$\|\mathcal{D}^{-(1/2)}\mathcal{C}\mathcal{D}^{-(1/2)}\| \leq \sigma.$$

Also, because $\mathcal{C}^{1/2}\mathcal{S}^{-1}\mathcal{C}^{1/2}$ is symmetric positive semidefinite, we have

$$\|\mathcal{D}^{-(1/2)}\mathcal{C}^{1/2}(I+\mathcal{C}^{1/2}\mathcal{S}^{-1}\mathcal{C}^{1/2})^{-1}\mathcal{C}^{1/2}\mathcal{D}^{-(1/2)}\| \leq \|\mathcal{D}^{-(1/2)}\mathcal{C}\mathcal{D}^{-(1/2)}\| \leq \sigma.$$

To prove (9.4), we use the representation of A^{-1} in (9.17). We have

$$\begin{aligned}\|D^{1/2}A^{-1}D^{1/2}\| &\leq \|D^{1/2}A_0^{-1}D^{1/2}\| + \|D^{1/2}A_0^{-1}D^{1/2}\|\|\mathcal{D}^{-(1/2)} \\ &\quad \times \mathcal{C}^{1/2}(I+\mathcal{C}^{1/2}\mathcal{S}^{-1}\mathcal{C}^{1/2})^{-1}\mathcal{C}^{1/2}\mathcal{D}^{-(1/2)}\|\|D^{1/2}A_0^{-1}D^{1/2}\| \\ &\leq \kappa + \kappa^2\,\|\mathcal{D}^{-(1/2)}\mathcal{C}\mathcal{D}^{-(1/2)}\| \\ &\leq \kappa + \kappa^2\,\sigma.\end{aligned}$$

Thus, for the solution $\mathbf{X}$ of the problem $A\mathbf{X} = \mathbf{F}$, the following a priori estimate holds.

$$\|\mathbf{X}\|_D = \|D^{1/2}A^{-1}\mathbf{F}\| \leq \|D^{1/2}A^{-1}D^{1/2}\|\ \|\mathbf{F}\|_{D^{-1}}.$$

Because we proved that boundedness of $\|D^{1/2}A^{-1}D^{1/2}\|$, the proof is complete. □

The following result provides a natural norm for well posedness of the saddle-point problem of our main interest in the case of s.p.d. major block $\mathcal{A}$.

Proposition 9.12. *Let $\mathcal{A}$ be s.p.d. and $\mathcal{C}$ be positive semidefinite. Consider the saddle-point problem*

$$\begin{bmatrix}\mathcal{A} & \mathcal{B}^T \\ \mathcal{B} & -\mathcal{C}\end{bmatrix}\begin{bmatrix}\mathbf{w} \\ \mathbf{x}\end{bmatrix} = \begin{bmatrix}\mathbf{f} \\ \mathbf{g}\end{bmatrix}.$$

This problem is well posed in the pair of norms $\mathcal{M} = \mathcal{A}$ and $\mathcal{D} = \mathcal{S}_A \equiv \mathcal{C}+\mathcal{B}\mathcal{A}^{-1}\mathcal{B}^T$ and A is bounded in terms of

$$D_A = \begin{bmatrix}\mathcal{M} & 0 \\ 0 & \mathcal{S}_A\end{bmatrix}.$$

Proof. After eliminating $\mathbf{w} = \mathcal{A}^{-1}(\mathbf{f} - \mathcal{B}^T\mathbf{x})$ the reduced problem for $\mathbf{x}$ reads

$$-\mathcal{S}_A\mathbf{x} = \mathbf{g} - \mathcal{B}\mathcal{A}^{-1}\mathbf{f}.$$

Then,

$$\begin{aligned}\|\mathbf{x}\|_{\mathcal{S}_A} &\leq \|\mathbf{g}\|_{\mathcal{S}_A^{-1}} + \|\mathcal{S}_A^{-(1/2)}\mathcal{B}\mathcal{A}^{-(1/2)}\|\,\|\mathbf{f}\|_{\mathcal{A}^{-1}} \\ &= \|\mathbf{g}\|_{\mathcal{S}_A^{-1}} + \|\mathcal{A}^{-(1/2)}\mathcal{B}^T\mathcal{S}_A^{-(1/2)}\|\,\|\mathbf{f}\|_{\mathcal{A}^{-1}} \\ &\leq \|\mathbf{g}\|_{\mathcal{S}_A^{-1}} + \|\mathbf{f}\|_{\mathcal{A}^{-1}}.\end{aligned}$$

The second a priori estimate is obtained from $\mathcal{A}\mathbf{w} = \mathbf{f} - \mathcal{B}^T\mathbf{x}$, which gives

$$\|\mathbf{w}\|_{\mathcal{A}} \leq \|\mathbf{f}\|_{\mathcal{A}^{-1}} + \|\mathcal{A}^{-(1/2)}\mathcal{B}^T\mathbf{x}\| \leq \|\mathbf{f}\|_{\mathcal{A}^{-1}} + \|\mathbf{x}\|_{\mathcal{S}_A}.$$

Thus, the two estimates combined show the desired final a priori estimate

$$\left\| \begin{bmatrix} \mathbf{w} \\ \mathbf{x} \end{bmatrix} \right\|_{D_A} \leq \sqrt{8} \left\| \begin{bmatrix} \mathbf{f} \\ \mathbf{g} \end{bmatrix} \right\|_{D_A^{-1}}.$$

The boundedness of A in terms of D_A is also easily seen. For any

$$\mathbf{V} = \begin{bmatrix} \mathbf{v} \\ \mathbf{y} \end{bmatrix} \quad \text{and} \quad \mathbf{W} = \begin{bmatrix} \mathbf{w} \\ \mathbf{x} \end{bmatrix}$$

using the inequality $\mathbf{z}^T\mathcal{B}\mathbf{u} = \mathbf{u}^T\mathcal{B}^T\mathbf{z} \leq \|\mathcal{A}^{-(1/2)}\mathcal{B}^T\mathbf{z}\| \|\mathbf{u}\|_{\mathcal{A}} \leq \|\mathbf{z}\|_{\mathcal{S}_A}\|\mathbf{u}\|_{\mathcal{A}}$ and the Cauchy–Schwarz inequality, we have

$$\begin{aligned}
\mathbf{W}^T A\mathbf{V} &= \mathbf{w}^T\mathcal{A}\mathbf{v} + \mathbf{w}^T\mathcal{B}^T\mathbf{y} + \mathbf{x}^T\mathcal{B}\mathbf{v} - \mathbf{x}^T\mathcal{C}\mathbf{y} \\
&\leq \|\mathbf{w}\|_{\mathcal{A}}\|\mathbf{v}\|_{\mathcal{A}} + \|\mathbf{x}\|_{\mathcal{S}_A}\|\mathbf{y}\|_{\mathcal{S}_A} + \|\mathbf{w}\|_{\mathcal{A}}\|\mathbf{y}\|_{\mathcal{S}_A} + \|\mathbf{x}\|_{\mathcal{S}_A}\|\mathbf{v}\|_{\mathcal{A}} \\
&\leq 2\,(\|\mathbf{w}\|^2_{\mathcal{A}} + \|\mathbf{x}\|^2_{\mathcal{S}_A})^{1/2}(\|\mathbf{v}\|^2_{\mathcal{A}} + \|\mathbf{y}\|^2_{\mathcal{S}_A})^{1/2} \\
&= 2\,\|\mathbf{V}\|_{D_A}\|\mathbf{W}\|_{D_A}.
\end{aligned}$$

□

Concluding remarks

Block-diagonal preconditioners for solving saddle-point problems were considered in [RW92], [SW93], and [VL96]; see also the book [ESW06].

9.3 Transforming A to a positive definite matrix

In this section, we outline an approach originally proposed in Bramble and Pasciak (1988) ([BP88]), which utilizes a preconditioner $\mathcal{M}$ for the major block $\mathcal{A}$ such that $\mathcal{A}-\mathcal{M}$ is symmetric positive definite and transforms the original saddle-point matrix A to a positive definite one $\widehat{A}$. The transformation reads

$$\widehat{A} \equiv \begin{bmatrix} \mathcal{A}\mathcal{M}^{-1} - I & 0 \\ \mathcal{B}\mathcal{M}^{-1} & -I \end{bmatrix} \begin{bmatrix} \mathcal{A} & \mathcal{B}^T \\ \mathcal{B} & -\mathcal{C} \end{bmatrix} = \begin{bmatrix} \mathcal{A}\mathcal{M}^{-1}\mathcal{A} - \mathcal{A} & (\mathcal{A}\mathcal{M}^{-1} - I)\mathcal{B}^T \\ \mathcal{B}(\mathcal{M}^{-1}\mathcal{A} - I) & \mathcal{C} + \mathcal{B}\mathcal{M}^{-1}\mathcal{B}^T \end{bmatrix}. \tag{9.18}$$

We first notice that the Schur complement $\widehat{\mathcal{S}}$ of $\widehat{A}$ equals the negative Schur complement $\mathcal{S} = \mathcal{C} + \mathcal{B}\mathcal{A}^{-1}\mathcal{B}^T$ of A. Indeed, we readily see that $\widehat{A}$ admits the following block-factorization,

$$\widehat{A} = \begin{bmatrix} I & 0 \\ * & I \end{bmatrix} \begin{bmatrix} * & 0 \\ 0 & \mathcal{S} \end{bmatrix} \begin{bmatrix} I & * \\ 0 & I \end{bmatrix}.$$

The same result is seen by direct computation,

$$\begin{aligned}\widehat{\mathcal{S}} &= \mathcal{C} + \mathcal{B}\mathcal{M}^{-1}\mathcal{B}^T - \mathcal{B}(\mathcal{M}^{-1}\mathcal{A} - I)\mathcal{A}^{-1}(\mathcal{A}\mathcal{M}^{-1} - I)^{-1}(\mathcal{A}\mathcal{M}^{-1} - I)\mathcal{B}^T \\ &= \mathcal{C} + \mathcal{B}\mathcal{A}^{-1}\mathcal{B}^T.\end{aligned}$$

Because $\mathcal{A}\mathcal{M}^{-1}\mathcal{A} - \mathcal{A}$ is s.p.d. (due to the choice of $\mathcal{M}$) and $\widehat{\mathcal{S}} = \mathcal{S}$ is also s.p.d., we conclude that the transformed matrix $\widehat{\mathrm{A}}$ is s.p.d., as well.

There are several possible choices that naturally ensure $\mathbf{v}^T\mathcal{M}\mathbf{v} \le \mathbf{v}^T\mathcal{A}\mathbf{v}$. One construction is found in Section 7.3 (see Lemma 7.2). With this choice, in some cases, we may even be able to construct $\mathcal{M}$ so that $\mathcal{M}^{-1}$ is explicitly available and sparse, for example, by letting the subdomains used in the FETI method be of size comparable to a single fine-grid element and without coarse-grid. The latter is the case if $\mathcal{A}$ is assembled from respective element matrices $\{\mathcal{A}_\tau\}$ that are invertible. Another possible choice is based on the element-by-element construction of $\mathcal{M}$ described in Section 4.7. To actually end up with an $\mathcal{M}$ such that $\mathcal{M}^{-1}$ is efficiently computable, we may have to apply the procedure recursively and thus end up with a multilevel block-factored $\mathcal{M}$. In the latter case, $\mathcal{M}^{-1}$ is not feasible in an explicit (sparse) form.

In either case, the actual $\mathcal{M}$ has to be scaled by some $\theta \in (0, 1)$; that is, $\mathcal{M} := \theta\mathcal{M}$. The latter is needed to ensure that the resulting $\mathcal{A} - \mathcal{M}$ is positive definite (not only semidefinite).

If $\mathcal{M}$ provides a convergent splitting for $\mathcal{A}$, then the block form of the transformed matrix $\widehat{\mathrm{A}}$ is stable. More specifically, the following main result holds.

Theorem 9.13. *Assume that $\mathcal{M}$ provides a convergent splitting for $\mathcal{A}$, such that for a constant $\gamma \in [0, 1)$, we have*

$$0 \le \mathbf{w}^T(\mathcal{A}^{1/2}\mathcal{M}^{-1}\mathcal{A}^{1/2} - I)\mathbf{w} \le \gamma^2\, \mathbf{w}^T\mathbf{w}. \tag{9.19}$$

Then the following strengthened Cauchy–Schwarz inequality holds,

$$\mathbf{y}^T\mathcal{B}(\mathcal{M}^{-1}\mathcal{A} - I)\mathbf{w} \le \gamma\, (\mathbf{w}^T(\mathcal{A}\mathcal{M}^{-1}\mathcal{A} - \mathrm{A})\mathbf{w})^{1/2}(\mathbf{y}^T(\mathcal{C} + \mathcal{B}\mathcal{M}^{-1}\mathcal{B}^T)\mathbf{y})^{1/2}.$$

Note that $\mathcal{B}(\mathcal{M}^{-1}\mathcal{A} - I)$ is the strictly block lower-triangular part of $\widehat{\mathrm{A}}$ and $\mathcal{A}\mathcal{M}^{-1}\mathcal{A} - \mathrm{A}$ and $\mathcal{C} + \mathcal{B}\mathcal{M}^{-1}\mathcal{B}^T$ are the principal blocks on the main diagonal of $\widehat{\mathrm{A}}$. In other words, the two-by-two block form of $\widehat{\mathrm{A}}$ in (9.18) is stable.

Proof. The proof uses the standard Cauchy–Schwarz inequality. Letting $\mathcal{E} = \mathcal{A}^{1/2}\mathcal{M}^{-1}\mathcal{A}^{1/2} - I$, we have

$$\begin{aligned}\mathbf{y}^T\mathcal{B}(\mathcal{M}^{-1}\mathcal{A} - I)\mathbf{w} &= (\mathcal{A}^{-(1/2)}\mathcal{B}^T\mathbf{y})^T(\mathcal{E}\mathcal{A}^{1/2}\mathbf{w}) \\ &\le (\mathbf{y}^T\mathcal{B}\mathcal{A}^{-1}\mathcal{B}^T\mathbf{y})^{1/2}(\mathbf{w}^T\mathcal{A}^{1/2}\mathcal{E}^2\mathcal{A}^{1/2}\mathbf{w})^{1/2} \\ &\le (\mathbf{y}^T\mathcal{B}\mathcal{M}^{-1}\mathcal{B}^T\mathbf{y})^{1/2}\gamma\, (\mathbf{w}^T\mathcal{A}^{1/2}\mathcal{E}\mathcal{A}^{1/2}\mathbf{w})^{1/2} \\ &\le \gamma\, (\mathbf{y}^T(\mathcal{C} + \mathcal{B}\mathcal{M}^{-1}\mathcal{B}^T)\mathbf{y})^{1/2}(\mathbf{w}^T(\mathcal{A}\mathcal{M}^{-1}\mathcal{A} - \mathcal{A})\mathbf{w})^{1/2}. \quad \square\end{aligned}$$

Thus, what is left is that we need in general a good preconditioner for $\widehat{\mathcal{S}} \equiv \mathcal{C} + \mathcal{B}\mathcal{M}^{-1}\mathcal{B}^T$ in order to complete the construction of preconditioners for $\widehat{\mathrm{A}}$.

If $\mathcal{M}^{-1}$ is explicitly available and sparse, this is feasible by the methods described in the preceding sections for s.p.d. matrices. To ensure both, $\mathcal{M}$ being efficient for $\mathcal{A}$ (with a good γ in estimate (9.19)) and $\mathcal{M}^{-1}$ being sparse, this basically means that $\mathcal{A}$ is well conditioned. This is the case, for example, for matrices A coming from mixed finite element discretizations of second-order elliptic PDEs. Otherwise, $\mathcal{M}^{-1}$ is not sparse and the second block on the diagonal of $\widehat{A}$, $\widehat{\mathcal{S}}$, is also not feasible. But as it often happens, $\widehat{\mathcal{S}}$ turns out to be well conditioned then. In that case, good approximations to $\widehat{\mathcal{S}}^{-1}$ can be derived by iterations, which generally leads to "inner–outer" iterative methods because the actions of $\mathcal{M}^{-1}$ can typically come from another iterative procedure.

Final comments

In certain applications coming from mixed finite element discretizations, it is feasible to consider a least squares form of the saddle-point problem. A more feasible approach is to change the discretization procedure and then use least squares with proper weights in the norms (sometimes using certain "negative" Sobolev space norms). For some original papers utilizing least squares approaches, see [BPL], [CLMM], and [CPV].

9.4 (Inexact) Uzawa and distributive relaxation methods

Consider the saddle-point problem of the form

$$A = \begin{bmatrix} \mathcal{A} & \mathcal{B}^T \\ \mathcal{B} & -\mathcal{C} \end{bmatrix}. \tag{9.20}$$

Here, $\mathcal{A}$ is s.p.d. and $\mathcal{C}$ is symmetric positive semidefinite.

9.4.1 Distributive relaxation

Given a transformation matrix G such that AG is easier to handle (i.e., block triangular with s.p.d. blocks on the diagonal, or simply being s.p.d.), we can use it to define a smoothing procedure. Such an idea originated in [BD79] (see also [Wi89] and [Wi90]), and was referred to as "distributive" relaxation or transforming smoothers.

In what follows, we assume that AG is s.p.d. Consider an initial iterate $\mathbf{x}_0$ for $A\mathbf{x} = \mathbf{b}$. Letting $\mathbf{x} = G\mathbf{y}$, we get $AG\mathbf{y} = \mathbf{b}$. Finally, let D be an s.p.d. matrix such that the transformed iteration matrix $I - D^{-1}(AG)$ corresponds to a convergent method. In terms of the original variables, we have the iteration

$$\mathbf{x} = \mathbf{x}_0 + GD^{-1}(\mathbf{b} - A\mathbf{x}_0).$$

The respective iteration matrix reads

$$E = I - GD^{-1}A.$$

We notice that in the inner product defined by the s.p.d. matrix $G^{-T}A$, E is symmetric positive semidefinite. We have, using the fact that $AG = G^T A$,

$$\mathbf{v}^T G^{-T} A E \mathbf{w} = \mathbf{v}^T G^{-T} A (I - G D^{-1} A)\mathbf{w} = \mathbf{v}^T (G^{-T} A - A D^{-1} A)\mathbf{w},$$

which is a symmetric form. We also have

$$\mathbf{v}^T G^{-T} A E \mathbf{v} = \mathbf{v}^T G^{-T} [AG - (AG) D^{-1} (AG)] G^{-1} \mathbf{v} \geq 0.$$

The latter shows the positive semidefiniteness of E in the $G^{-T}A$-inner product.

9.4.2 The Bramble–Pasciak transformation

Now, consider the Bramble–Pasciak transformation matrix (see Section 9.3),

$$G = \begin{bmatrix} \mathcal{M}^{-1}\mathcal{A} - I & \mathcal{M}^{-1}\mathcal{B}^T \\ 0 & -I \end{bmatrix}.$$

Here $\mathcal{M}$ is a given s.p.d. matrix such that for a constant $\gamma \in (0, 1)$,

$$\mathbf{u}^T \mathcal{M} \mathbf{u} \leq \gamma \ \mathbf{u}^T \mathcal{A} \mathbf{u}, \quad \text{for all } \mathbf{u}. \tag{9.21}$$

Note that at this point, we do not assume that $\mathcal{M}$ is necessarily spectrally equivalent to $\mathcal{A}$. Nevertheless, a $\gamma \in (0, 1)$ uniformly bounded away from unity can be found even for $\mathcal{M}$ that is not spectrally equivalent to $\mathcal{A}$. A simple example is $\mathcal{M} = \gamma \ \lambda_{\min}(\mathcal{A}) \ I$.

Compute the transformed matrix

$$AG = \begin{bmatrix} \mathcal{A}\mathcal{M}^{-1}\mathcal{A} - \mathcal{A} & (\mathcal{A}\mathcal{M}^{-1} - I)\mathcal{B}^T \\ \mathcal{B}(\mathcal{M}^{-1}\mathcal{A} - I) & \mathcal{B}\mathcal{M}^{-1}\mathcal{B}^T + \mathcal{C} \end{bmatrix}. \tag{9.22}$$

As we showed earlier (see Section 9.3), AG is s.p.d., because the main block $\mathcal{A}\mathcal{M}^{-1}\mathcal{A} - \mathcal{A}$ is s.p.d., and the Schur complement $\mathcal{S}$ of AG equals the (negative) Schur complement $\mathcal{S}_A$ of the original saddle-point matrix A, hence is s.p.d.

Next, compute $G^{-T}A$ explicitly. We have

$$\begin{aligned}
G^{-T} &= \begin{bmatrix} \mathcal{A}\mathcal{M}^{-1} - I & 0 \\ \mathcal{B}\mathcal{M}^{-1} & -I \end{bmatrix}^{-1} \\
&= \left(\begin{bmatrix} I & 0 \\ \mathcal{B}\mathcal{M}^{-1}(\mathcal{A}\mathcal{M}^{-1} - I)^{-1} & I \end{bmatrix} \begin{bmatrix} \mathcal{A}\mathcal{M}^{-1} - I & 0 \\ 0 & -I \end{bmatrix} \right)^{-1} \\
&= \begin{bmatrix} (\mathcal{A}\mathcal{M}^{-1} - I)^{-1} & 0 \\ 0 & -I \end{bmatrix} \begin{bmatrix} I & 0 \\ -\mathcal{B}\mathcal{M}^{-1}(\mathcal{A}\mathcal{M}^{-1} - I)^{-1} & I \end{bmatrix} \\
&= \begin{bmatrix} (\mathcal{A}\mathcal{M}^{-1} - I)^{-1} & 0 \\ \mathcal{B}\mathcal{M}^{-1}(\mathcal{A}\mathcal{M}^{-1} - I)^{-1} & -I \end{bmatrix}.
\end{aligned}$$

Then,

$$
\begin{aligned}
G^{-T}A &= \begin{bmatrix} (\mathcal{A}\mathcal{M}^{-1}-I)^{-1} & 0 \\ \mathcal{B}\mathcal{M}^{-1}(\mathcal{A}\mathcal{M}^{-1}-I)^{-1} & -I \end{bmatrix} \begin{bmatrix} \mathcal{A} & \mathcal{B}^T \\ \mathcal{B} & -\mathcal{C} \end{bmatrix} \\
&= \begin{bmatrix} (\mathcal{A}\mathcal{M}^{-1}-I)^{-1}\mathcal{A} & (\mathcal{A}\mathcal{M}^{-1}-I)^{-1}\mathcal{B}^T \\ \mathcal{B}(\mathcal{M}^{-1}(\mathcal{A}\mathcal{M}^{-1}-I)^{-1}\mathcal{A}-I) & \mathcal{C}+\mathcal{B}\mathcal{M}^{-1}(\mathcal{A}\mathcal{M}^{-1}-I)^{-1}\mathcal{B}^T \end{bmatrix} \\
&= \begin{bmatrix} (\mathcal{M}^{-1}-\mathcal{A}^{-1})^{-1} & (\mathcal{A}\mathcal{M}^{-1}-I)^{-1}\mathcal{B}^T \\ \mathcal{B}(\mathcal{M}^{-1}\mathcal{A}-I)^{-1} & \mathcal{C}+\mathcal{B}(\mathcal{A}-\mathcal{M})^{-1}\mathcal{B}^T \end{bmatrix}.
\end{aligned}
$$

Use the inequality $\mathbf{w}^T(\mathcal{X}-\mathcal{X}^2)\mathbf{w} \leq \gamma\ \mathbf{w}^T(I-\mathcal{X})\mathbf{w}$, for $\mathcal{X} = \mathcal{A}^{-(1/2)}\mathcal{M}\mathcal{A}^{-(1/2)}$ based on (9.21) to show that

$$\mathbf{z}^T\mathcal{M}(\mathcal{A}-\mathcal{M})^{-1}\mathcal{M}\mathbf{z} \leq \gamma\ \mathbf{z}^T(\mathcal{M}^{-1}-\mathcal{A}^{-1})^{-1}\mathbf{z}.$$

Therefore, based on the Cauchy–Schwarz inequality, we get

$$
\begin{aligned}
(\mathbf{v}^T\mathcal{B}(\mathcal{M}^{-1}\mathcal{A}-I)^{-1}\mathbf{z})^2 &= (\mathbf{v}^T\mathcal{B}(\mathcal{A}-\mathcal{M})^{-1}\mathcal{M}\mathbf{z})^2 \\
&\leq \mathbf{v}^T\mathcal{B}(\mathcal{A}-\mathcal{M})^{-1}\mathcal{B}^T\mathbf{v}\ \mathbf{z}^T\mathcal{M}(\mathcal{A}-\mathcal{M})^{-1}\mathcal{M}\mathbf{z} \\
&\leq \gamma\ \mathbf{v}^T\mathcal{B}(\mathcal{A}-\mathcal{M})^{-1}\mathcal{B}^T\mathbf{v}\ \mathbf{z}^T(\mathcal{M}^{-1}-\mathcal{A}^{-1})^{-1}\mathbf{z} \\
&\leq \gamma\ \mathbf{v}^T(\mathcal{C}+\mathcal{B}(\mathcal{A}-\mathcal{M})^{-1}\mathcal{B}^T)\mathbf{v}\ \mathbf{z}^T(\mathcal{M}^{-1}-\mathcal{A}^{-1})^{-1}\mathbf{z}.
\end{aligned}
$$

The latter represents a strengthened Cauchy–Schwarz inequality for $G^{-T}A$ in the sense that the off-diagonal block of $G^{-T}A$ is dominated by the principal block-diagonal part of $G^{-T}A$. Therefore, $G^{-T}A$ is spectrally equivalent to its block-diagonal part

$$\begin{bmatrix} (\mathcal{M}^{-1}-\mathcal{A}^{-1})^{-1} & 0 \\ 0 & \mathcal{C}+\mathcal{B}(\mathcal{A}-\mathcal{M})^{-1}\mathcal{B}^T \end{bmatrix}.$$

The latter block-diagonal matrix can be further simplified based on the inequalities $\mathbf{w}^T\mathcal{A}^{-1}\mathbf{w} \leq \mathbf{w}^T(\mathcal{A}-\mathcal{M})^{-1}\mathbf{w} \leq (1/(1-\gamma))\ \mathbf{w}^T\mathcal{A}^{-1}\mathbf{w}$ (using again (9.21)). That is, $G^{-T}A$ is spectrally equivalent to the block-diagonal matrix

$$\begin{bmatrix} (\mathcal{M}^{-1}-\mathcal{A}^{-1})^{-1} & 0 \\ 0 & \mathcal{C}+\mathcal{B}\mathcal{A}^{-1}\mathcal{B}^T \end{bmatrix}.$$

Finally, using again (9.21), we have

$$\mathbf{w}^T\mathcal{M}\mathbf{w} \leq \mathbf{w}^T(\mathcal{M}^{-1}-\mathcal{A}^{-1})^{-1}\mathbf{w} \leq \frac{1}{1-\gamma}\ \mathbf{w}^T\mathcal{M}\mathbf{w},$$

which proves at the end the following result.

Lemma 9.14. *Assume that* (9.21) *holds. The matrix* $\mathcal{G}^{-T}\mathcal{A}$ *is spectrally equivalent to the block-diagonal one,*

$$\begin{bmatrix} \mathcal{M} & 0 \\ 0 & \mathcal{C}+\mathcal{B}\mathcal{A}^{-1}\mathcal{B}^T \end{bmatrix}. \tag{9.23}$$

The constants in the spectral equivalence relations depend only on $\gamma \in (0, 1)$ *and they deteriorate if* γ *gets close to unity.*

As a corollary, we have that if $\mathcal{M}$ is spectrally equivalent to $\mathcal{A}$ then $G^{-T}A$ is spectrally equivalent to

$$D_A \equiv \begin{bmatrix} \mathcal{A} & 0 \\ 0 & \mathcal{C} + \mathcal{B}\mathcal{A}^{-1}\mathcal{B}^T \end{bmatrix}.$$

The latter matrix is typically the one used to define a norm in which the saddle-point problem $A\mathbf{x} = \mathbf{b}$ is well posed (cf., Proposition 9.12), that is, to have an a priori estimate of the form

$$\|\mathbf{x}\|_{D_A} \leq \sigma \ \|\mathbf{b}\|_{D_A^{-1}}.$$

To find a simple $\mathcal{M}$ (i.e., $\mathcal{M}^{-1}$ explicitly given sparse matrix) that is spectrally equivalent to $\mathcal{A}$, this is feasible if $\mathcal{A}$ itself is well conditioned. This is the case, for example, for matrices A coming from a mixed f.e. method applied to second-order scalar elliptic PDEs. Then $\mathcal{A}$ is a mass matrix and $\mathcal{M}$ can be chosen to be its properly scaled diagonal part. In general, $\mathcal{M}$ can be obtained by an optimal $V(1, 1)$-cycle multigrid applied to $\mathcal{A}$ and scaled so that (9.21) to hold.

9.4.3 A note on two-grid analysis

Our goal is to study the convergence of the two-grid method utilizing the so-called "distributive relaxation" based on the transformation matrix G (defined below) and a standard coarse grid correction.

We assume that $\mathcal{M}$ is spectrally equivalent to $\mathcal{A}$ such that for two uniform constants $\gamma < 1 < \gamma^{-1} \leq \kappa$, the following estimates hold,

$$\mathbf{w}^T\mathcal{M}\mathbf{w} \leq \gamma \ \mathbf{w}^T\mathcal{A}\mathbf{w}, \qquad \mathbf{w}^T\mathcal{A}\mathbf{w} \leq \kappa \ \mathbf{w}^T\mathcal{M}\mathbf{w}, \quad \text{for all } \mathbf{w}. \tag{9.24}$$

We study the convergence in the natural "energy" norm $\|.\|_{G^{-T}A}$, which we showed is spectrally equivalent to the norm defined from the block-diagonal matrix (9.23) or

$$D_A = \begin{bmatrix} \mathcal{A} & 0 \\ 0 & \mathcal{S}_A \end{bmatrix}$$

(because we assumed that $\mathcal{M}$ is spectrally equivalent to $\mathcal{A}$). Let

$$\mathbf{x} = \begin{bmatrix} \mathbf{u} \\ \mathbf{p} \end{bmatrix}$$

and define $\|\mathbf{x}\|^2 = \|\mathbf{u}\|^2 + \|\mathbf{p}\|^2$. The interpolation matrix P is assumed to be block-diagonal; that is,

$$P = \begin{bmatrix} \mathcal{P} & 0 \\ 0 & \mathcal{Q} \end{bmatrix}.$$

In practice, we have the property that the (negative) Schur complement $\mathcal{S}_A = \mathcal{C} + \mathcal{B}\mathcal{A}^{-1}\mathcal{B}^T$ of A is spectrally equivalent to a sparse matrix (either a mass matrix for

Stokes problems, cf., e.g., [LQ86], or a matrix defining a discrete counterpart of H^1-Sobolev norm, cf., e.g., [RVWa]).

Define the coarse grid matrix

$$A_c = P^T A P = \begin{bmatrix} \mathcal{A}_c & \mathcal{B}_c^T \\ \mathcal{B}_c & -\mathcal{C}_c \end{bmatrix}.$$

It is clear that A_c is also a saddle-point. Define next the "distributive relaxation" matrix $M_{\text{distr}} = DG^{-1}$ where D is a "standard" smoother (e.g., Richardson as in (9.25) below) for $AG = G^T A$. Then, $M_{TR} \equiv G^{-T} M_{\text{distr}} = G^{-T} D G^{-1}$ can be used as a smoother for the transformed s.p.d. matrix $A_{TR} = G^{-T} A$.

We are interested in the two-grid method defined by the product iteration matrix

$$E^m (I - \pi_A),$$

where $E = I - M_{\text{distr}}^{-1} A = I - GD^{-1}A$ is the smoothing iteration matrix, $m \geq 1$ is the number of smoothing iterations, and $\pi_A = I - PA_c^{-1}P^T A$ is the standard coarse-grid projection onto the space Range(P).

To be specific, we define the following smoother D coming from AG, letting $\kappa = \lambda_{\max}(\mathcal{M}^{-1}\mathcal{A}) > 1$,

$$D = 2 \begin{bmatrix} (\kappa - 1)\kappa \, \mathcal{M} & 0 \\ 0 & \|\mathcal{S}_A\| \kappa \, I \end{bmatrix}. \tag{9.25}$$

It is easily seen from (9.22) that D is a convergent smoother for AG; that is, we have $\mathbf{V}^T AG\mathbf{V} \leq \mathbf{V}^T D\mathbf{V}$ for any $\mathbf{V}$. Hence $G^{-T}DG^{-1}$ is a convergent smoother for $G^{-T}A$. Next, we compute $M_{TR} = G^{-T}DG^{-1}$. We have

$$\begin{aligned} M_{TR} &= \begin{bmatrix} (\mathcal{A}\mathcal{M}^{-1} - I)^{-1} & 0 \\ \mathcal{B}(\mathcal{A} - \mathcal{M})^{-1} & -I \end{bmatrix} D \begin{bmatrix} (\mathcal{M}^{-1}\mathcal{A} - I)^{-1} & (\mathcal{A} - \mathcal{M})^{-1}\mathcal{B}^T \\ 0 & -I \end{bmatrix} \\ &= \begin{bmatrix} 2\kappa(\kappa - 1)\, \mathcal{M}(\mathcal{A} - \mathcal{M})^{-1}\mathcal{M}(\mathcal{A} - \mathcal{M})^{-1}\mathcal{M} & \mathcal{Y}^T \\ \mathcal{Y} & \mathcal{X} \end{bmatrix} \end{aligned}$$

where

$$\begin{aligned} \mathcal{X} &= 2\kappa \|\mathcal{S}_A\|\, I + 2\kappa(\kappa - 1)\, \mathcal{B}(\mathcal{A} - \mathcal{M})^{-1}\mathcal{M}(\mathcal{A} - \mathcal{M})^{-1}\mathcal{B}^T, \\ \mathcal{Y} &= 2\kappa(\kappa - 1)\, \mathcal{B}(\mathcal{A} - \mathcal{M})^{-1}\mathcal{M}(\mathcal{A} - \mathcal{M})^{-1}\mathcal{M}. \end{aligned}$$

Because $\mathcal{M}$ is spectrally equivalent to $\mathcal{A}$, it is clear then that $M_{TR} = G^{-T}DG^{-1}$ is bounded above (in terms of inner product) by the block-diagonal matrix

$$\begin{bmatrix} \mathcal{M} & 0 \\ 0 & \|\mathcal{S}_A\|\, I \end{bmatrix}$$

times a constant δ depending only on κ. That is, we proved that

$$\|M_{TR}\| = \|G^{-T}DG^{-1}\| \leq \delta \, \|D_A\|. \tag{9.26}$$

Note now that

$$E = I - GD^{-1}A = I - GD^{-1}G^T(G^{-T}A) = I - M_{TR}^{-1}A_{TR}.$$

To analyze the convergence of the two-grid method in the natural energy norm coming from $A_{TR} = G^{-T}A$, we utilize the classical approach due to Hackbusch (cf., [H82], [H85], [H94]) based on establishing a so-called "smoothing property" and an "approximation property".

The following "smoothing property" is standard. We have

$$\begin{aligned}
\|A_{TR}^{1/2}(I - M_{TR}^{-1}A_{TR})^m \mathbf{e}\| &\leq \|A_{TR}^{1/2}(I - M_{TR}^{-1}A_{TR})^m A_{TR}^{-(1/2)}(A_{TR}^{1/2}M_{TR}^{-1}A_{TR}^{1/2})^{1/2}\| \\
&\quad \times \|(A_{TR}^{1/2}M_{TR}^{-1}A_{TR}^{1/2})^{-(1/2)}A_{TR}^{1/2}\mathbf{e}\| \\
&\leq \max_{t\in[0,1]} (1-t)^m t^{1/2}\ \|(A_{TR}^{1/2}M_{TR}^{-1}A_{TR}^{1/2})^{-(1/2)}A_{TR}^{1/2}\mathbf{e}\| \\
&= \max_{t\in[0,1]} (1-t)^m t^{1/2}\ \|M_{TR}^{1/2}\mathbf{e}\| \\
&\leq \frac{1}{\sqrt{2m}}\ \|M_{TR}^{1/2}\mathbf{e}\| \\
&\leq \frac{\|M_{TR}\|^{1/2}}{\sqrt{2m}}\ \|\mathbf{e}\|.
\end{aligned}$$

Consider finally the estimate

$$\begin{aligned}
\|E_{TG}\mathbf{e}\|_{A_{TR}} &= \|A_{TR}^{1/2}E^m(I - \pi_A)\mathbf{e}\| \\
&\leq \frac{\|M_{TR}\|^{1/2}}{\sqrt{2m}}\ \|(I - \pi_A)\mathbf{e}\| \\
&\leq \frac{\sqrt{\delta}}{2m}\ \|D_A\|^{1/2}\|(I - \pi_A)\mathbf{e}\|.
\end{aligned}$$

Here we used estimate (9.26).

Assume now the following "approximation property"

$$\|D_A\|^{1/2}\|(I - \pi_A)\mathbf{e}\| \leq \eta_a\ \|\mathbf{e}\|_{D_A}. \tag{9.27}$$

Recall that D_A defines the norm in which the saddle-point problem is well posed (Proposition 9.12), hence the above estimate is a natural one. In the application of mixed finite element discretizations, such an approximation property typically requires L_2-error estimates that can be obtained by duality argument.

The convergence of the two-grid method then follows (recalling that $A_{TR} = G^{-T}A$ is spectrally equivalent to D_A), because for sufficiently large $m \geq 1$, we have that the bound provided by the estimate

$$\|E_{TR}\mathbf{e}\|_{A_{TR}} \leq \frac{\sqrt{\delta}}{\sqrt{2m}}\ \eta_s\ \lambda_{\max}^{1/2}(A_{TR}^{-1}D_A)\ \|\mathbf{e}\|_{A_{TR}}$$

will be strictly less than unity.

We summarize as follows.

Theorem 9.15. *Let the coarse-grid projection* $\pi_A = PA_c^{-1}P^TA$, $A_c = P^TAP$, *possess an "approximation property" as in (9.27). Assume also that* $\mathcal{M}$ *used in the Bramble–Pasciak transformation matrix* G *is spectrally equivalent to* $\mathcal{A}$ *(as in (9.24)). Consider the distributive relaxation* $M_{\text{distr}} = DG^{-1}$, *where* D *is defined in (9.25). Then the two-grid method giving rise to the iteration matrix* $E_{TG} = (I - M_{\text{distr}}^{-1}A)^m(I - \pi_A)$ *has a convergence factor measured in the* $A_{TR} = G^{-1}A$*-norm that behaves as* $m^{-(1/2)}$ *where* $m \geq 1$ *is the number of smoothing steps. This estimate in particular implies uniform two-grid convergence for sufficiently large* m.

9.4.4 Inexact Uzawa methods

The distributive relaxation-based, two-grid method presented earlier is suitable in practice if the major block $\mathcal{A}$ of A is well conditioned. In the present section, we present inexact Uzawa algorithms suitable for more general $\mathcal{A}$. In particular, we assume that $\mathcal{S}_A$ is well conditioned. We comment that the Schur complement $\mathcal{S}_A$ can be guaranteed to be well conditioned (or rather spectrally equivalent to a matrix $\mathcal{D}$) by proper transformation of the saddle-point matrix A, namely, by adding the second block row of A multiplied by $\mathcal{B}^T\mathcal{D}^{-1}$ to the first block row of A (see Corollary 9.8).

The Uzawa algorithm (originating in [AHU]) is referred to the iteration process for solving

$$A\begin{bmatrix}\mathbf{u}\\ \mathbf{p}\end{bmatrix} = \begin{bmatrix}\mathbf{f}\\ \mathbf{g}\end{bmatrix},$$

described in what follows. Namely, for a given $\mathcal{M}^{-1}$, an approximate inverse to $\mathcal{A}$, and a suitable parameter τ from a current approximation $\mathbf{u}_k$, $\mathbf{p}_k$, we compute the next one as follows:

$$\begin{aligned}\mathbf{u}_{k+1} &= \mathbf{u}_k + \mathcal{M}^{-1}(\mathbf{f} - \mathcal{A}\mathbf{u}_k - \mathcal{B}^T\mathbf{p}_k),\\ \mathbf{p}_{k+1} &= \mathbf{p}_k + \tau\,(\mathcal{B}\mathbf{u}_{k+1} - \mathbf{g}).\end{aligned}$$

We consider a symmetrized version of the above Uzawa algorithm, namely:

$$\begin{aligned}\mathbf{u}_{k+1/2} &= \mathbf{u}_k + \mathcal{M}^{-1}(\mathbf{f} - \mathcal{A}\mathbf{u}_k - \mathcal{B}^T\mathbf{p}_k),\\ \mathbf{p}_{k+1} &= \mathbf{p}_k + \tau\,(\mathcal{B}\mathbf{u}_{k+1/2} - \mathbf{g}),\\ \mathbf{u}_{k+1} &= \mathbf{u}_{k+1/2} + \mathcal{M}^{-1}(\mathbf{f} - \mathcal{A}\mathbf{u}_{k+1/2} - \mathcal{B}^T\mathbf{p}_{k+1}).\end{aligned} \tag{9.28}$$

In general, we may consider $\mathcal{M}^{-1}$ to be a nonlinear mapping, for example, one obtained by auxiliary (inner) CG-type iterations that provide approximations to the inverse action of $\mathcal{A}$. We use the notation $\mathcal{M}^{-1}[.]$ to indicate the latter fact.

Assume now the following estimate

$$\|\mathbf{u} - \mathcal{M}^{-1}[\mathcal{A}\mathbf{u}]\|_{\mathcal{A}} \leq \delta\ \|\mathbf{u}\|_{\mathcal{A}},$$

or equivalently, (letting $\mathbf{u} = \mathcal{A}^{-1}\mathbf{v}$),

$$\|\mathcal{A}^{-1}\mathbf{v} - \mathcal{M}^{-1}[\mathbf{v}]\|_{\mathcal{A}} \leq \delta\ \|\mathbf{v}\|_{\mathcal{A}^{-1}}. \tag{9.29}$$

In the analysis to follow, $\delta \in [0, 1)$ is assumed sufficiently small. The construction of the respective inexact Uzawa algorithm and the analysis to follow is based on [AV91].

Based on the exact factorization of A^{-1},

$$A^{-1} = \begin{bmatrix} I & -\mathcal{A}^{-1}\mathcal{B}^T \\ 0 & I \end{bmatrix} \begin{bmatrix} \mathcal{A}^{-1} & 0 \\ \mathcal{S}_A^{-1}\mathcal{B}\mathcal{A}^{-1} & -\mathcal{S}_A^{-1} \end{bmatrix},$$

we can define the mapping $\mathbf{r} \mapsto B[\mathbf{r}]$ as an approximate inverse to A by replacing the actions of $\mathcal{A}^{-1}$ with $\mathcal{M}^{-1}[.]$ and $\mathcal{S}_A^{-1}$ by an appropriate constant τ. More specifically, the actions of $B[\cdot]$ are computed based on the following approximate inverse,

$$\begin{bmatrix} 2I - \mathcal{M}^{-1}\mathcal{A} & -\mathcal{M}^{-1}\mathcal{B}^T \\ 0 & I \end{bmatrix} \begin{bmatrix} \mathcal{M}^{-1} & 0 \\ \tau\, \mathcal{B}\mathcal{M}^{-1} & -\tau \end{bmatrix},$$

as implemented in the following algorithm.

Algorithm 9.4.1 (A nonlinear approximate inverse). *Given* $\mathbf{r} = \begin{bmatrix} \mathbf{f} \\ \mathbf{g} \end{bmatrix}$ *compute:*

1. $\mathbf{u}_0 = \mathcal{M}^{-1}[\mathbf{f}]$;
2. $\mathbf{q}_0 = \mathcal{B}\mathbf{u}_0 - \mathbf{g}$;
3. $\mathbf{p} = \tau\, \mathbf{q}_0$;
4. $\mathbf{v} = \mathbf{f} - \mathcal{A}\mathbf{u}_0 - \mathcal{B}^T\mathbf{p}$;
5. $\mathbf{u}_1 = \mathcal{M}^{-1}[\mathbf{v}]$;
6. $\mathbf{u} = \mathbf{u}_0 + \mathbf{u}_1$.

Then,

$$B[\mathbf{r}] = \begin{bmatrix} \mathbf{u} \\ \mathbf{p} \end{bmatrix}.$$

We notice that one step of the symmetrized (inexact) Uzawa algorithm (9.28) with $\mathbf{u}_k = 0$, $\mathbf{p}_k = 0$ reduces to the above Algorithm 9.4.1, which defines a nonlinear approximate inverse (or preconditioner) of A.

We have the representation

$$\begin{aligned} AB[\mathbf{r}] &= \begin{bmatrix} \mathcal{A} & \mathcal{B}^T \\ \mathcal{B} & -\mathcal{C} \end{bmatrix} \begin{bmatrix} \mathbf{u} \\ \mathbf{p} \end{bmatrix} = \begin{bmatrix} \mathcal{A}\mathbf{u} + \mathcal{B}^T\mathbf{p} \\ \mathcal{B}\mathbf{u} - \mathcal{C}\mathbf{p} \end{bmatrix} \\ &= \begin{bmatrix} \mathcal{A}(\mathbf{u}_0 + \mathbf{u}_1) + \mathcal{B}^T\mathbf{p} \\ \mathcal{B}(\mathbf{u}_0 + \mathbf{u}_1) - \mathcal{C}\mathbf{p} \end{bmatrix} \\ &= \begin{bmatrix} \mathbf{f} \\ \mathbf{g} \end{bmatrix} + \begin{bmatrix} \mathcal{A}(\mathbf{u}_0 + \mathcal{M}^{-1}[\mathbf{v}]) - \mathbf{f} + \mathcal{B}^T\mathbf{p} \\ \mathcal{B}(\mathbf{u}_0 + \mathcal{M}^{-1}[\mathbf{v}]) - \mathcal{C}\mathbf{p} - \mathbf{g} \end{bmatrix} \\ &= \begin{bmatrix} \mathbf{f} \\ \mathbf{g} \end{bmatrix} + \begin{bmatrix} (\mathcal{A}\mathbf{u}_0 + \mathcal{M}^{-1}[\mathbf{v}]) - \mathbf{v} - \mathcal{A}\mathbf{u}_0 - \mathcal{B}^T\mathbf{p} + \mathcal{B}^T\mathbf{p} \\ \mathcal{B}(\mathbf{u}_0 + \mathcal{M}^{-1}[\mathbf{v}]) - \mathbf{g} - \mathcal{C}\mathbf{p} \end{bmatrix} \\ &= \begin{bmatrix} \mathbf{f} \\ \mathbf{g} \end{bmatrix} + \begin{bmatrix} (\mathcal{A}\mathcal{M}^{-1}[\mathbf{v}]) - \mathbf{v} \\ \mathcal{B}(\mathbf{u}_0 + \mathcal{M}^{-1}[\mathbf{v}]) - \mathbf{g} - \mathcal{C}\mathbf{p} \end{bmatrix}. \end{aligned}$$

We can first show that for a proper $\tau > 0$, there is a $\delta_0 \in [0, 1)$ such that the following estimate holds,

$$\|\mathcal{S}_A^{-1}\mathbf{q} - \tau\mathbf{q}\|_{\mathcal{S}_A} \leq \delta_0 \, \|\mathbf{q}\|_{\mathcal{S}_A^{-1}}. \tag{9.30}$$

We can choose $\tau = 2/(\lambda_{\min}[\mathcal{S}_A] + \lambda_{\max}[\mathcal{S}_A])$ and have

$$\delta_0^2 = \frac{\lambda_{\max}[\mathcal{S}_A] - \lambda_{\min}[\mathcal{S}_A]}{\lambda_{\max}[\mathcal{S}_A] + \lambda_{\min}[\mathcal{S}_A]} < 1.$$

These are "good" constants if $\mathcal{S}_A$ is well conditioned. Later, we consider a choice of τ as a nonlinear mapping, which makes the resulting inexact Uzawa algorithm fairly parameter free.

Our next goal is to estimate the deviation $AB[\mathbf{r}] - \mathbf{r}$ in an appropriate norm. We choose the norm defined from

$$D_A^{-1} = \begin{bmatrix} \mathcal{A}^{-1} & 0 \\ 0 & \mathcal{S}_A^{-1} \end{bmatrix}.$$

We first estimate $\|\mathbf{v}\|_{\mathcal{A}^{-1}}$. We have, recalling that $\mathbf{v} = \mathbf{f} - \mathcal{A}\mathbf{u}_0 - \mathcal{B}^T\mathbf{p} = (\mathbf{f} - \mathcal{A}\mathcal{M}^{-1}[\mathbf{f}]) - \mathcal{B}^T\mathbf{p}$,

$$\begin{aligned} \|\mathbf{v}\|_{\mathcal{A}^{-1}} &\leq \|\mathbf{f} - \mathcal{A}\mathcal{M}^{-1}[\mathbf{f}]\|_{\mathcal{A}^{-1}} + \|\mathcal{B}^T\mathbf{p}\|_{\mathcal{A}^{-1}} \\ &\leq \delta \, \|\mathbf{f}\|_{\mathcal{A}^{-1}} + \|\mathcal{A}^{-(1/2)}\mathcal{B}^T\mathbf{p}\| \\ &\leq \delta \, \|\mathbf{f}\|_{\mathcal{A}^{-1}} + \|\mathbf{p}\|_{\mathcal{S}_A}. \end{aligned}$$

For the term $\|\mathbf{p}\|_{\mathcal{S}_A}$, we have

$$\begin{aligned} \|\mathbf{p}\|_{\mathcal{S}_A} &= \|\tau \, (\mathcal{B}\mathcal{M}^{-1}[\mathbf{f}] - \mathbf{g})\|_{\mathcal{S}_A} \\ &= \left\| (\tau - \mathcal{S}_A^{-1})(\mathcal{B}\mathcal{M}^{-1}[\mathbf{f}] - \mathbf{g}) + \mathcal{S}_A^{-1}(\mathcal{B}\mathcal{M}^{-1}[\mathbf{f}] - \mathbf{g}) \right\|_{\mathcal{S}_A} \\ &\leq (1 + \delta_0) \, \|\mathcal{B}\mathcal{M}^{-1}[\mathbf{f}] - \mathbf{g}\|_{\mathcal{S}_A^{-1}} \\ &\leq (1 + \delta_0) \left\| \mathcal{S}_A^{-(1/2)}\mathcal{B}\mathcal{A}^{-(1/2)} \right\| \|\mathcal{M}^{-1}[\mathbf{f}]\|_{\mathcal{A}} + (1 + \delta_0)\|\mathbf{g}\|_{\mathcal{S}_A^{-1}} \\ &\leq (1 + \delta_0)(1 + \delta) \, \|\mathbf{f}\|_{\mathcal{A}^{-1}} + (1 + \delta_0)\|\mathbf{g}\|_{\mathcal{S}_A^{-1}}. \end{aligned}$$

We used above the fact that $\|\mathcal{X}\| = \|\mathcal{X}^T\| \leq 1$ for $\mathcal{X} = \mathcal{S}_A^{-(1/2)}\mathcal{B}\mathcal{A}^{-(1/2)}$.

Next, we estimate $\mathcal{B}(\mathbf{u}_0 + \mathcal{M}^{-1}[\mathbf{v}]) - \mathbf{g} - \mathcal{C}\mathbf{p}$, which we rearrange as $\mathcal{B}(\mathbf{u}_0 + \mathcal{M}^{-1}[\mathbf{v}]) - \mathbf{g} - (\mathcal{C}\mathbf{p} + \mathcal{B}\mathcal{A}^{-1}\mathcal{B}^T\mathbf{p} + \mathbf{q}_0) - \mathbf{g} + \mathcal{B}\mathcal{A}^{-1}\mathcal{B}^T\mathbf{p} - \mathbf{q}_0$. Because $\mathbf{p} = \tau\mathbf{q}_0$ and $\mathbf{q}_0 = \mathcal{B}\mathbf{u}_0 - \mathbf{g}$, $\mathbf{v} = \mathbf{f} - \mathcal{A}\mathbf{u}_0 - \mathcal{B}^T\mathbf{p}$, we arrive at

$$\begin{aligned} \mathcal{B}(\mathbf{u}_0 + \mathcal{M}^{-1}[\mathbf{v}]) - \mathbf{g} - \mathcal{C}\mathbf{p} &= (\mathcal{S}_A\tau\mathbf{q}_0 - \mathbf{q}_0) + \mathcal{B}(\mathcal{M}^{-1}[\mathbf{v}] + \mathcal{A}^{-1}\mathcal{B}^T\mathbf{p}) \\ &= (\mathcal{S}_A\tau\mathbf{q}_0 - \mathbf{q}_0) + \mathcal{B}(\mathcal{M}^{-1}[\mathbf{v}] - \mathcal{A}^{-1}\mathbf{v}) \\ &\quad + \mathcal{B}\mathcal{A}^{-1}(\mathbf{f} - \mathcal{A}\mathbf{u}_0) \\ &= (\mathcal{S}_A\tau\mathbf{q}_0 - \mathbf{q}_0) + \mathcal{B}(\mathcal{M}^{-1}[\mathbf{v}] - \mathcal{A}^{-1}\mathbf{v}) \\ &\quad + \mathcal{B}\mathcal{A}^{-1}(\mathbf{f} - \mathcal{A}\mathcal{M}^{-1}[\mathbf{f}]). \end{aligned} \tag{9.31}$$

Estimating the term

$$\|(\mathcal{S}_A \tau \mathbf{q}_0 - \mathbf{q}_0)\|_{\mathcal{S}_A^{-1}} \leq \delta_0 \|\mathbf{q}_0\|_{\mathcal{S}_A^{-1}} \leq \delta_0 \ \|\mathcal{B}\mathcal{M}^{-1}[\mathbf{f}]\|_{\mathcal{S}_A^{-1}} + \delta_0 \|\mathbf{g}\|_{\mathcal{S}_A^{-1}}$$

and using again the norm bound $\|\mathcal{S}_A^{-(1/2)} \mathcal{B} \mathcal{A}^{-(1/2)}\| \leq 1$, leads to

$$\|(\mathcal{S}_A \tau \mathbf{q}_0 - \mathbf{q}_0)\|_{\mathcal{S}_A^{-1}} \leq \delta_0 (1+\delta) \ \|\mathbf{f}\|_{\mathcal{A}^{-1}} + \delta_0 \|\mathbf{g}\|_{\mathcal{S}_A^{-1}}.$$

From the identity (9.31), we can see that

$$\begin{aligned}
\|\mathcal{B}(\mathbf{u}_0 + \mathcal{M}^{-1}[\mathbf{v}]) - \mathbf{g} - \mathcal{C}\mathbf{p}\|_{\mathcal{S}_A^{-1}} &\leq \delta_0 [(1+\delta)\|\mathbf{f}\|_{\mathcal{A}^{-1}} + \|\mathbf{g}\|_{\mathcal{S}_A^{-1}}] \\
&\quad + \delta \ (\|\mathbf{v}\|_{\mathcal{A}^{-1}} + \|\mathbf{f}\|_{\mathcal{A}^{-1}}) \\
&\leq [(1+\delta)[\delta + \delta_0(1+\delta)] + \delta^2] \|\mathbf{f}\|_{\mathcal{A}^{-1}} \\
&\quad + [\delta(1+\delta_0) + \delta_0] \|\mathbf{g}\|_{\mathcal{S}_A^{-1}} \\
&\leq [\delta_0(1+\delta) + C\delta][\|\mathbf{f}\|_{\mathcal{A}^{-1}} + \|\mathbf{g}\|_{\mathcal{S}_A^{-1}}].
\end{aligned}$$

Thus, we proved the following main result.

Theorem 9.16. *The nonlinear inexact Uzawa algorithm 9.4.1 is convergent if* $\delta > 0$ *in* (9.29) *is sufficiently small, for any* $\delta_0 \in [0, 1)$ *from estimate* (9.30)*; that is, there is a constant* $q \in [0, 1)$ *such that,*

$$\|AB[\mathbf{r}] - \mathbf{r}\|_{D_A^{-1}} \leq q \ \|\mathbf{r}\|_{D_A^{-1}}.$$

Note that the result in Theorem 9.16 holds with no condition on δ_0 to be small enough. However, the choice of the parameter τ (in (9.30)) remains a bit unclear. To resolve this issue, we may want to replace τ with another (generally) nonlinear mapping $\mathcal{T}[\cdot]$, which is close to $\mathcal{S}_A^{-1}$ in the sense that the following estimate holds for a tolerance $\delta_0 \in [0, 1)$,

$$\|\mathcal{S}_A^{-1}\mathbf{q} - \mathcal{T}[\mathbf{q}]\|_{\mathcal{S}_A} \leq \delta_0 \ \|\mathbf{q}\|_{\mathcal{S}_A^{-1}}.$$

We notice that the proof above does not change if we consider τ to be a nonlinear mapping (with the proper understanding of Step 3 in Algorithm 9.4.1).

The mapping $\mathcal{T}[\mathbf{q}]$ is computed by performing a few steps of the following steepest descent-type algorithm (as proposed and analyzed in [AV92]). Define the nonlinear mapping $\widetilde{\mathcal{S}}[\cdot] = \mathcal{C} + \mathcal{B}\mathcal{M}^{-1}[\mathcal{B}(\cdot)]$. We have

$$\begin{aligned}
\mathbf{q}^T \widetilde{\mathcal{S}}[\mathbf{q}] &= \mathbf{q}^T (\widetilde{\mathcal{S}}[\mathbf{q}] - \mathcal{S}_A \mathbf{q})) + \mathbf{q}^T \mathcal{S}_A \mathbf{q} \\
&= \mathbf{q}^T \mathcal{B}\mathcal{A}^{-(1/2)} (\mathcal{A}^{1/2} (\mathcal{M}^{-1}[\mathcal{B}^T \mathbf{q}] - \mathcal{A}^{-1}\mathcal{B}^T \mathbf{q})) + \mathbf{q}^T \mathcal{S}_A \mathbf{q} \\
&\geq (1-\delta) \ \|\mathbf{q}\|_{\mathcal{S}_A}^2.
\end{aligned}$$

Similarly, $\mathbf{q}^T \widetilde{\mathcal{S}}[\mathbf{q}] \leq (1+\delta) \ \mathbf{q}^T \mathcal{S}_A \mathbf{q}$.

Algorithm 9.4.2 (Steepest descent definition of T).

(0) Initiate: choose $\mathbf{x}_0$ *and compute* $\mathbf{r}_0 = \mathbf{q} - \widetilde{\mathcal{S}}[\mathbf{x}_0]$.

(i) Iterate: for $k \geq 1$, compute:

1. $\widetilde{\mathbf{r}}_{k-1} = \widetilde{\mathcal{S}}[\mathbf{r}_{k-1}]$;
2. $\alpha_{k-1} = \dfrac{\mathbf{r}_{k-1}^T \mathbf{r}_{k-1}}{\mathbf{r}_{k-1}^T \widetilde{\mathbf{r}}_{k-1}}$;
3. $\mathbf{x}_k = \mathbf{x}_k + \alpha_{k-1}\, \mathbf{r}_{k-1}$;
4. $\mathbf{r}_k = \mathbf{q} - \widetilde{\mathcal{S}}[\mathbf{x}_k]$.

After sufficiently many steps $k \geq 1$, we let $\mathcal{T}[\mathbf{q}] := \mathbf{x}_k$.

Theorem 9.17. *Under the coercivity and boundedness properties of $\widetilde{S}[\cdot]$ for δ sufficiently small, the above steepest descent algorithm can terminate for a $\delta_0 \in [0, 1)$ in the sense that if we define $\mathcal{T}[\mathbf{q}] \equiv \mathbf{x}_k$ for a sufficiently large k, an estimate of the form $\|\mathbf{q} - \widetilde{\mathcal{S}}[\mathcal{T}[\mathbf{q}]]\|_{\mathcal{S}_A^{-1}} \leq \delta_0\, \|\mathbf{q}\|_{\mathcal{S}_A^{-1}}$ holds for a $\delta_0 \in [0, 1)$. More specifically, the following estimate holds,*

$$\|\mathbf{q} - \widetilde{\mathcal{S}}[\mathcal{T}[\mathbf{q}]]\|_{\mathcal{S}_A^{-1}} \equiv \|\mathbf{r}_k\|_{\mathcal{S}_A^{-1}} \leq \left(q^k + \frac{1}{1-q}\frac{2\delta}{1-\delta}\right) \|\mathbf{q}\|_{\mathcal{S}_A^{-1}},$$

where

$$q = \sqrt{1 - \frac{1-2\delta}{1-\delta^2}\left(\frac{2\sqrt{\kappa}}{1+\kappa}\right)^2 + \frac{3\delta}{1-\delta}}.$$

The constant κ stands for the condition number of $\mathcal{S}_A$, whereas $\delta \in [0, 1)$, generally, sufficiently small, is such that

$$(1-\delta)\, \mathbf{q}^T \mathcal{S}_A \mathbf{q} \leq \mathbf{q}^T \widetilde{\mathcal{S}}[\mathbf{q}] \leq (1+\delta)\, \mathbf{q}^T \mathcal{S}_A \mathbf{q}.$$

Proof. We have

$$(1-\delta)\, \|\mathbf{x}_k\|^2_{\mathcal{S}_A} \leq \mathbf{x}_k^T \widetilde{\mathcal{S}}[\mathbf{x}_k] \leq \|\mathbf{x}_k\|_{\mathcal{S}_A^{-1}}\, \|\mathbf{q} - \mathbf{r}_k\|_{\mathcal{S}_A^{-1}}.$$

That is,

$$(1-\delta)\, \|\mathbf{x}_k\|_{\mathcal{S}_A} \leq \left(\|\mathbf{r}_0\|_{\mathcal{S}_A^{-1}} + \|\mathbf{r}_k\|_{\mathcal{S}_A^{-1}}\right). \tag{9.32}$$

Next,

$$\begin{aligned}
\|\mathbf{r}_k\|_{\mathcal{S}_A^{-1}} &= \|\mathbf{q} - \widetilde{\mathcal{S}}[\mathbf{x}_k]\|_{\mathcal{S}_A^{-1}} \\
&= \|\mathbf{q} - \widetilde{\mathcal{S}}[\mathbf{x}_{k-1}] + \widetilde{\mathcal{S}}[\mathbf{x}_{k-1}] - \widetilde{\mathcal{S}}[\mathbf{x}_k]\|_{\mathcal{S}_A^{-1}} \\
&\leq \|\mathbf{r}_{k-1} - \alpha_{k-1}\, \mathcal{S}_A \mathbf{r}_{k-1}\|_{\mathcal{S}_A^{-1}} + \|\mathcal{S}_A \mathbf{x}_{k-1} - \widetilde{\mathcal{S}}[\mathbf{x}_{k-1}]\|_{\mathcal{S}_A^{-1}} \\
&\quad + \|\mathcal{S}_A \mathbf{x}_k - \widetilde{\mathcal{S}}[\mathbf{x}_k]\|_{\mathcal{S}_A^{-1}} \\
&\leq \|\mathbf{r}_{k-1} - \alpha_{k-1}\, \mathcal{S}_A \mathbf{r}_{k-1}\|_{\mathcal{S}_A^{-1}} + \delta\, \|\mathbf{x}_{k-1}\|_{\mathcal{S}_A} + \delta\, \|\mathbf{x}_k\|_{\mathcal{S}_A} \\
&\leq \|\mathbf{r}_{k-1} - \alpha_{k-1}\, \mathcal{S}_A \mathbf{r}_{k-1}\|_{\mathcal{S}_A^{-1}} + 2\delta\, \|\mathbf{x}_{k-1}\|_{\mathcal{S}_A} + \delta\, \alpha_{k-1}\, \|\mathbf{r}_{k-1}\|_{\mathcal{S}_A}.
\end{aligned}$$

Use now the inequality

$$\alpha_{k-1}\ \|\mathbf{r}_{k-1}\|_{\mathcal{S}_A} \leq \frac{\|\mathbf{r}_{k-1}\|_{\mathcal{S}_A^{-1}}\|\mathbf{r}_{k-1}\|^2_{\mathcal{S}_A}}{\mathbf{r}^T_{k-1}\widetilde{\mathcal{S}}[\mathbf{r}_{k-1}]} \leq \frac{1}{1-\delta}\ \|\mathbf{r}_{k-1}\|_{\mathcal{S}_A^{-1}}$$

and the bound (9.32) to arrive at

$$\|\mathbf{r}_k\|_{\mathcal{S}_A^{-1}} \leq \|\mathbf{r}_{k-1} - \alpha_{k-1}\ \mathcal{S}_A\mathbf{r}_{k-1}\|_{\mathcal{S}_A^{-1}} + \frac{\delta}{1-\delta}\ (2\|\mathbf{r}_0\|_{\mathcal{S}_A^{-1}} + 3\|\mathbf{r}_{k-1}\|_{\mathcal{S}_A^{-1}}).$$

It is clear that the term $\|\mathbf{r}_{k-1} - \alpha_{k-1}\ \mathcal{S}_A\mathbf{r}_{k-1}\|^2_{\mathcal{S}_A^{-1}} = \|\mathbf{r}_{k-1}\|^2_{\mathcal{S}_A^{-1}} - 2\alpha_{k-1}\|\mathbf{r}_{k-1}\|^2 + \alpha^2_{k-1}\mathbf{r}^T_{k-1}\mathcal{S}_A\mathbf{r}_{k-1} \leq \|\mathbf{r}_{k-1}\|^2_{\mathcal{S}_A^{-1}} - 2\alpha_{k-1}\|\mathbf{r}_{k-1}\|^2 + \alpha^2_{k-1}(1/(1-\delta))\ \mathbf{r}^T_{k-1}\widetilde{\mathcal{S}}[\mathbf{r}_{k-1}]$ can be estimated by $\overline{q}^2\ \|\mathbf{r}_{k-1}\|^2_{\mathcal{S}_A^{-1}}$ with a $\overline{q} \in [0, 1)$, similarly to the steepest descent algorithm. More specifically, we have

$$\begin{aligned}
&\|\mathbf{r}_{k-1} - \alpha_{k-1}\ \mathcal{S}_A\mathbf{r}_{k-1}\|^2_{\mathcal{S}_A^{-1}} \\
&\quad \leq \|\mathbf{r}_{k-1}\|^2_{\mathcal{S}_A^{-1}} - 2\alpha_{k-1}\|\mathbf{r}_{k-1}\|^2 + \alpha^2_{k-1}\frac{1}{1-\delta}\ \mathbf{r}^T_{k-1}\widetilde{\mathcal{S}}[\mathbf{r}_{k-1}] \\
&\quad \leq \|\mathbf{r}_{k-1}\|^2_{\mathcal{S}_A^{-1}} - \left[2 - \frac{1}{1-\delta}\right]\alpha_{k-1}\|\mathbf{r}_{k-1}\|^2 \\
&\quad \leq \|\mathbf{r}_{k-1}\|^2_{\mathcal{S}_A^{-1}}\frac{1-2\delta}{1-\delta}\ \frac{(\mathbf{r}^T_{k-1}\mathbf{r}_{k-1})^2}{(1-\delta)\mathbf{r}^T_{k-1}\mathcal{S}_A\mathbf{r}_{k-1}}.
\end{aligned}$$

Now use the Kantorovich's inequality (Proposition G.1)

$$\mathbf{r}^T_{k-1}\mathcal{S}_A^{-1}\mathbf{r}_{k-1}\mathbf{r}^T_{k-1}\mathcal{S}_A\mathbf{r}_{k-1} \leq \left(\frac{\kappa+1}{2\sqrt{\kappa}}\right)^2\ (\mathbf{r}^T_{k-1}\mathbf{r}_{k-1})^2,$$

where $\kappa = (\lambda_{\max}[\mathcal{S}_A])/(\lambda_{\min}[\mathcal{S}_A])$ is the condition number of $\mathcal{S}_A$, to arrive at the desired estimate

$$\|\mathbf{r}_{k-1} - \alpha_{k-1}\ \mathcal{S}_A\mathbf{r}_{k-1}\|_{\mathcal{S}_A^{-1}} \leq \overline{q}\ \|\mathbf{r}_{k-1}\|_{\mathcal{S}_A^{-1}},$$

with

$$\overline{q} = \sqrt{1 - \frac{1-2\delta}{1-\delta^2}\left(\frac{2\sqrt{\kappa}}{\kappa+1}\right)^2} \simeq \frac{\kappa-1}{\kappa+1} < 1, \quad \text{when } \delta \mapsto 0.$$

Thus, we have

$$\|\mathbf{r}_k\|_{\mathcal{S}_A^{-1}} \leq \left[\overline{q} + \frac{3\delta}{1-\delta}\right]\ \|\mathbf{r}_{k-1}\|_{\mathcal{S}_A^{-1}} + \frac{2\delta}{1-\delta}\ \|\mathbf{q}\|_{\mathcal{S}_A^{-1}}.$$

The desired result then follows by recursion, assuming that $q = \overline{q} + (3\delta/(1-\delta)) < 1$ (valid for a sufficiently small δ). □

In order to match estimate (9.30) (with $\mathcal{T}[\cdot]$ in place of τ), we can use the following corollary easily obtained from Theorem 9.17. We first show that $\widetilde{\mathcal{S}}[\cdot] = \mathcal{C} + \mathcal{B}\mathcal{M}^{-1}[\mathcal{B}^T(\cdot)]$ and $\mathcal{S}_A = \mathcal{C} + \mathcal{B}\mathcal{A}^{-1}\mathcal{B}^T$ are close in the following sense,

$$\begin{aligned} \|\mathcal{S}_A^{-1}(\widetilde{\mathcal{S}}[\mathbf{v}] - \mathcal{S}_A\mathbf{v})\|_{\mathcal{S}_A} &\leq \|\mathcal{S}_A^{-(1/2)}\mathcal{B}\mathcal{A}^{-(1/2)}\| \|\mathcal{A}^{1/2)}(\mathcal{M}^{-1}[\mathcal{B}^T\mathbf{v}] - \mathcal{A}^{-1}(\mathcal{B}\mathbf{v}))\| \\ &\leq \delta \ \|\mathcal{A}^{-(1/2)}\mathcal{B}^T\mathbf{v}\| \\ &\leq \delta \ \|\mathbf{v}\|_{\mathcal{S}_A}. \end{aligned} \tag{9.33}$$

We used the fact that $\mathcal{M}^{-1}[\cdot]$ is close to $\mathcal{A}^{-1}$ (estimate (9.29)) and that $\|\mathcal{X}\| = \|\mathcal{X}^T\| \leq 1$ for $\mathcal{X} = \mathcal{S}_A^{-(1/2)}\mathcal{B}\mathcal{A}^{-(1/2)}$.

To prove the desired analogue of (9.30), we proceed as follows. Letting

$$\overline{\delta}_0 = q^k + \frac{1}{1-q}\frac{2\delta}{1-\delta}, \qquad \delta_0 = \overline{\delta}_0 + (1+\overline{\delta}_0)\,\frac{\delta}{1-\delta},$$

and $\mathcal{T}[\mathbf{q}] = \mathbf{x}_k$, using estimates (9.32) and (9.33),

$$\begin{aligned} \|\mathcal{S}_A^{-1}\mathbf{q} - \mathcal{T}[\mathbf{q}]\|_{\mathcal{S}_A} &\leq \overline{\delta}_0 \ \|\mathbf{q}\|_{\mathcal{S}_A^{-1}} + \|\mathbf{x}_k - \mathcal{S}_A^{-1}\widetilde{\mathcal{S}}[\mathbf{x}_k]\|_{\mathcal{S}_A} \\ &\leq \overline{\delta}_0 \ \|\mathbf{q}\|_{\mathcal{S}_A^{-1}} + \delta \ \|\mathbf{x}_k\|_{\mathcal{S}_A} \\ &\leq \overline{\delta}_0 \ \|\mathbf{q}\|_{\mathcal{S}_A^{-1}} + \frac{\delta}{1-\delta}\left(\|\mathbf{q}\|_{\mathcal{S}_A^{-1}} + \|\mathbf{r}_k\|_{\mathcal{S}_A^{-1}}\right) \\ &\leq \left(\overline{\delta}_0 + (1+\overline{\delta}_0)\,\frac{\delta}{1-\delta}\right) \ \|\mathbf{q}\|_{\mathcal{S}_A^{-1}} \\ &= \delta_0 \ \|\mathbf{q}\|_{\mathcal{S}_A^{-1}}. \end{aligned}$$

Implementations of inner–outer methods of the above type for solving saddle-point problems are found in [ChV99]. Other related results that provide inexact Uzawa-type methods are found in [LQ86], [BWY], [EG94], and [BPV97].

9.5 A constrained minimization approach

A monotone subspace minimization scheme

In the present section, we consider the saddle-point problem,

$$\begin{bmatrix} \mathcal{A} & \mathcal{B}^T \\ \mathcal{B} & 0 \end{bmatrix} \begin{bmatrix} \mathbf{u} \\ \mathbf{x} \end{bmatrix} = \begin{bmatrix} \mathbf{f} \\ 0 \end{bmatrix},$$

recast as the following equivalent constrained minimization problem for $\mathbf{u}$.

Find the solution $\mathbf{u}$ of

$$\mathcal{J}(\mathbf{u}) \equiv \frac{1}{2}\,\mathbf{u}^T\mathcal{A}\mathbf{u} - \mathbf{f}^T\mathbf{u} \mapsto \min$$

subject to the equality constraint

$$\mathcal{B}\mathbf{u} = 0.$$

Consider an overlapping partitioning $\{\Omega_i\}$ of the set of indices of the vector $\mathbf{u}$ corresponding to the first block $\mathcal{A}$ of A. Let $\mathcal{I}_i$ be the characteristic diagonal matrix corresponding to Ω_i. The latter means that

$$\mathcal{I}_i \mathbf{v}_i = \begin{bmatrix} 0 \\ \mathbf{v}_i \\ 0 \end{bmatrix} \} \;\; \Omega_i.$$

That is, $\mathcal{I}_i$ extends a local vector $\mathbf{v}_i$ defined on Ω_i by zero outside Ω_i. Let $\mathcal{Q}_i^T$ be a restriction matrix onto the support of $\mathcal{B}\mathcal{I}_i$. We assume that $\mathcal{Q}_i$ is such that $\mathcal{Q}_i^T \mathcal{B}\mathcal{I}_i \mathbf{u}_i = 0$ implies $\mathcal{B}\mathcal{I}_i \mathbf{u}_i = 0$. Define then $\mathcal{B}_i = \mathcal{Q}_i^T \mathcal{B}\mathcal{I}_i$ and let $\mathcal{A}_i = \mathcal{I}_i^T \mathcal{A}\mathcal{I}_i$. The assumption on $\mathcal{Q}_i$ implies that if $\mathcal{B}_i \mathbf{u}_i = 0$ then $\mathcal{B}\mathcal{I}_i \mathbf{u}_i = 0$, that is $\mathcal{I}_i \text{null}(\mathcal{B}_i) \subset \text{null}(\mathcal{B})$.

We assume that the local saddle-point matrices

$$A_i = \begin{bmatrix} \mathcal{A}_i & \mathcal{B}_i^T \\ \mathcal{B}_i & 0 \end{bmatrix}$$

are invertible.

Consider the following local constrained minimization problem. Let $\mathbf{u}^0$ be a current approximation to the original (global) problem. Solve for a local correction $\mathcal{I}_i \mathbf{u}_i$ such that

$$\mathcal{J}(\mathbf{u}^0 + I_i \mathbf{u}_i) \mapsto \min \text{ subject to } \mathcal{B} I_i \mathbf{u}_i = 0. \tag{9.34}$$

Because $\mathcal{J}(\mathbf{u}^0 + I_i \mathbf{u}_i) = \mathcal{J}(\mathbf{u}^0) + \frac{1}{2}\, \mathbf{u}_i^T \mathcal{I}_i^T \mathcal{A}\mathcal{I}_i \mathbf{u}_i - (\mathcal{I}_i^T(\mathbf{f} - \mathcal{A}\mathbf{u}^0))^T \mathbf{u}_i$, it is clear that (9.34) is equivalent to the following local minimization problem,

$$\mathcal{J}_i(\mathbf{u}_i) \equiv \frac{1}{2}\, \mathbf{u}_i^T \mathcal{A}_i \mathbf{u}_i - (\mathcal{I}_i^T(\mathbf{f} - \mathcal{A}\mathbf{u}^0)^T \mathbf{u}_i \mapsto \min \text{ subject to } \mathcal{B}_i \mathbf{u}_i = 0. \tag{9.35}$$

Equivalently, to determine $\mathbf{u}_i$, we can instead solve the local saddle-point problem

$$\begin{bmatrix} \mathcal{A}_i & \mathcal{B}_i^T \\ \mathcal{B}_i & 0 \end{bmatrix} \begin{bmatrix} \mathbf{u}_i \\ \mathbf{x}_i \end{bmatrix} = \begin{bmatrix} \mathcal{I}_i^T(\mathbf{f} - \mathcal{A}\mathbf{u}^0) \\ 0 \end{bmatrix}. \tag{9.36}$$

Moreover, assume that the sets $\mathcal{I}_i(\text{null}(\mathcal{B}_i))$ provide an (overlapping) partition of $\text{null}(\mathcal{B})$. In other words, we assume that $\text{null}(\mathcal{B})$ allows for a basis, locally supported with respect to the partition $\{\Omega_i\}$, which we formulate in the following assumption,

(n) any $\mathbf{v} \in \text{null}(\mathcal{B})$ admits a decomposition

$$\mathbf{v} = \sum_i \mathcal{I}_i \mathbf{v}_i, \;\; \mathbf{v}_i \in \text{null}(\mathcal{B}_i).$$

For a given interpolation matrix $\mathcal{P}$, define

$$\mathcal{A}_c = \mathcal{P}^T \mathcal{A}\mathcal{P}.$$

We also need a restriction matrix $\mathcal{Q}_c^T$ such that $\mathcal{Q}_c^T \mathcal{B}\mathcal{P}_c \mathbf{u}_c = 0$ implies $\mathcal{B}\mathcal{P}_c \mathbf{u}_c = 0$. Define then $\mathcal{B}_c = \mathcal{Q}_c^T \mathcal{B}\mathcal{P}_c$. The assumptions on $\mathcal{B}_i$ and $\mathcal{B}_c$ constructed on the basis of the restriction matrices $\mathcal{Q}_i$ and $\mathcal{Q}_c$ that $\mathcal{B}_i\mathbf{u}_i$ implies $\mathcal{B}\mathcal{I}_i\mathbf{u}_i = 0$ and $\mathcal{B}_c\mathbf{u}_c = 0$ implies $\mathcal{B}\mathbf{u}_c = 0$ are naturally met for mixed finite element discretization matrices based on Raviart–Thomas spaces, and $\mathcal{Q}_i^T$ and $\mathcal{Q}_c^T$ are restrictions onto the associated discontinuous (piecewise constants) spaces. For more details, see Sections B.4 and B.5 in the appendix.

Because $\mathcal{A}$ is sparse, the products $\mathcal{I}_j^T\mathcal{A}\mathcal{I}_i$ can be nonzero for a finite number of indices. We assume that the number of indices j for which $\mathcal{I}_j^T\mathcal{A}\mathcal{I}_i$ is nonzero for any i is bounded by an integer $\kappa \geq 1$.

Similarly to the local problems (9.34)–(9.35), in order to find a coarse correction, we can solve the following coarse subspace constrained minimization problem,

$$\mathcal{J}(\mathbf{u}^0 + \mathcal{P}\mathbf{u}_c) \mapsto \min \text{ subject to } \mathcal{B}\mathcal{P}\mathbf{u}_c = 0. \tag{9.37}$$

It is clear that it can be rewritten as

$$\mathcal{J}_c(\mathbf{u}_c) \equiv \frac{1}{2}\,\mathbf{u}_c^T\mathcal{A}_c\mathbf{u}_c - (\mathcal{P}^T(\mathbf{f} - \mathcal{A}\mathbf{u}^0))^T\mathbf{u}_c \mapsto \min \text{ subject to } \mathcal{B}_c\mathbf{u}_c = 0,$$

which leads to the following coarse saddle-point problem

$$\begin{bmatrix} \mathcal{A}_c & \mathcal{B}_c^T \\ \mathcal{B}_c & 0 \end{bmatrix}\begin{bmatrix} \mathbf{u}_c \\ \mathbf{x}_c \end{bmatrix} = \begin{bmatrix} \mathcal{P}^T(\mathbf{f} - \mathcal{A}\mathbf{u}^0) \\ 0 \end{bmatrix}. \tag{9.38}$$

The following subspace minimization-type algorithm is of interest.

Algorithm 9.5.1 (A subspace minimization algorithm).

- *For a given iterate* $\mathbf{u}^0$, *set* $\mathbf{u} = \mathbf{u}^0$ *and perform the following subspace correction steps running over all sets* Ω_i,

$$\mathcal{J}(\mathbf{u} + \mathcal{I}_i\mathbf{u}_i) \mapsto \min,$$

subject to $\mathcal{B}\mathcal{I}_i\mathbf{u}_i = 0$, *or equivalently* $\mathcal{B}_i\mathbf{u}_i = 0$. *Then, update* $\mathbf{u} := \mathbf{u} + \mathcal{I}_i\mathbf{u}_i$.
- *Compute a "coarse subspace correction." For a given initial coarse approximation* $\mathbf{u}_c^{(0)}$ *(e.g.,* $\mathbf{u}_c^{(0)} = 0$*) such that* $\mathcal{B}_c\mathbf{u}_c^{(0)} = 0$,

  1. *First form*

$$\mathbf{f}_c = \mathcal{P}^T(\mathbf{f} - \mathcal{A}\,\mathbf{u}) + \mathcal{A}_c\mathbf{u}_c^{(0)}.$$

  2. *Then, solve the coarse constrained minimization problem:*

$$\mathcal{J}_c(\mathbf{u}_c) \equiv \frac{1}{2}\,\mathbf{u}_c^T\mathcal{A}_c\mathbf{u}_c - \mathbf{f}_c^T\mathbf{u}_c \mapsto \min$$

  subject to $\mathcal{B}_c\mathbf{u}_c = 0$.
- *The new iterate is*

$$\mathbf{u}^{new} = \mathbf{u} + \mathcal{P}(\mathbf{u}_c - \mathbf{u}_c^{(0)}).$$

The above subspace correction technique based on solving local saddle-point problems was originally used in [M92i] and [M92ii] as an overlapping Schwarz method.

The above two-grid scheme can be generalized in a straightforward manner to a multilevel one. This is done in a later chapter devoted to inequality constrained quadratic minimization problems.

Lemma 9.18. *The following monotonicity property holds.*

$$\mathcal{J}(\mathbf{u}^{\text{new}}) = \mathcal{J}(\mathbf{u}) + \mathcal{J}_c(\mathbf{u}_c) - \mathcal{J}_c(\mathbf{u}_c^{(0)}) \leq \mathcal{J}(\mathbf{u}).$$

Proof. By a straightforward computation, we have

$$\begin{aligned}
\mathcal{J}(\mathbf{u}^{\text{new}}) &= \frac{1}{2}(\mathbf{u} + \mathcal{P}(\mathbf{u}_c - \mathbf{u}_c^{(0)}))^T \mathcal{A}(\mathbf{u} + \mathcal{P}(\mathbf{u}_c - \mathbf{u}_c^{(0)})) \\
&\quad - \mathbf{f}^T (\mathbf{u} + \mathcal{P}(\mathbf{u}_c - \mathbf{u}_c^{(0)})) \\
&= \frac{1}{2}\, \mathbf{u}^T \mathcal{A}\mathbf{u} - \mathbf{f}^T \mathbf{u} - \mathbf{f}^T \mathcal{P}(\mathbf{u}_c - \mathbf{u}_c^{(0)}) \\
&\quad + (\mathbf{u}_c - \mathbf{u}_c^{(0)})^T \mathcal{P}^T \mathcal{A}\mathbf{u} + \frac{1}{2}\, (\mathbf{u}_c - \mathbf{u}_c^{(0)})^T \mathcal{A}_c (\mathbf{u}_c - \mathbf{u}_c^{(0)}) \\
&= \mathcal{J}(\mathbf{u}) - (\mathbf{u}_c - \mathbf{u}_c^{(0)})^T (\mathcal{P}^T (\mathbf{f} - \mathcal{A}\mathbf{u}) + \mathcal{A}_c \mathbf{u}_c^{(0)}) \\
&\quad + (\mathbf{u}_c - \mathbf{u}_c^{(0)})^T \mathcal{A}_c \mathbf{u}_c^{(0)} + \frac{1}{2}\, \mathbf{u}_c^T \mathcal{A}_c \mathbf{u}_c - \mathbf{u}_c^{(0)^T} \mathcal{A}_c \mathbf{u}_c + \frac{1}{2} \mathbf{u}_c^{(0)^T} \mathcal{A}_c \mathbf{u}_c^{(0)} \\
&= \mathcal{J}(\mathbf{u}) - \mathbf{f}_c^T (\mathbf{u}_c - \mathbf{u}_c^{(0)}) + \frac{1}{2}\, \mathbf{u}_c^T \mathcal{A}_c \mathbf{u}_c - \frac{1}{2}\, \mathbf{u}_c^{(0^T} \mathcal{A}_c \mathbf{u}_c^{(0)} \\
&= \mathcal{J}(\mathbf{u}) + \mathcal{J}_c(\mathbf{u}_c) - \mathcal{J}_c(\mathbf{u}_c^{(0)}).
\end{aligned}$$

Then, because $\mathcal{J}_c(\mathbf{u}_c) \leq \mathcal{J}_c(\mathbf{u}_c^{(0)})$, the desired monotonicity property follows. □

Convergence rate analysis

We are interested in the quadratic functional

$$\mathcal{J}(\mathbf{u}) = \frac{1}{2}\, \mathbf{u}^T \mathcal{A}\mathbf{u} - \mathbf{f}^T \mathbf{u}$$

for $\mathbf{u} \in \mathcal{K}$. In our case, $\mathcal{K} = \text{null}(\mathcal{B})$ is a linear space. The following characterization result holds.

Lemma 9.19. *Let* $\mathbf{u}$ *be the solution of the constrained minimization problem* $\mathcal{J}(\mathbf{u}) \mapsto \min$ *subject to* $\mathbf{u} \in \mathcal{K}$*. Then, for any* $\mathbf{g} \in \mathcal{K}$

$$\mathbf{g}^T (\mathcal{A}\mathbf{u} - \mathbf{f}) = 0.$$

Proof. For any $\mathbf{g} \in \mathcal{K}$ and any real t, $t\mathbf{g}$ is also in $\mathcal{K}$. Then, from $\mathcal{J}(\mathbf{u} + t\mathbf{g}) \geq \mathcal{J}(\mathbf{u})$, we obtain

$$0 \leq \mathcal{J}(\mathbf{u} + t\mathbf{g}) - \mathcal{J}(\mathbf{u}) = t^2 \frac{1}{2} \mathbf{g}^T \mathcal{A}\mathbf{g} + t\mathbf{g}^T (\mathcal{A}\mathbf{u} - \mathbf{f}).$$

Varying $t \mapsto 0$ with positive and negative values shows that in order to maintain the nonnegativity of the expression, we must have $\mathbf{g}^T (\mathcal{A}\mathbf{u} - \mathbf{f}) = 0$. □

In what follows, we describe a simplified version of a main result in [BTW]. We make the following assumption.

(d) The space $\mathcal{K}$ is decomposed into subspaces $\mathcal{K}_i$, $i = 1, \ldots, m, m+1$ such that

$$\mathbf{u} = \mathbf{u}_{m+1} + \sum_{i=1}^{m} \widehat{\mathbf{u}}_i.$$

In our application, $\mathbf{u}_{m+1} = \mathcal{P}\mathbf{u}_c$, and $\widehat{\mathbf{u}}_i = \mathcal{I}_i\mathbf{u}_i$, for $i \leq m$ where $\mathbf{u}_i$ belongs to the local spaces null$(\mathcal{B}_i)$. We assume that every $\mathbf{z} \in \mathcal{K}$ allows for a decomposition $\mathbf{z} = \mathbf{z}_{m+1} + \sum_{i \leq m} \mathbf{z}_i$ such that $\mathbf{z}_i \in \mathcal{K}_i$ and

$$\|\mathbf{z}_{m+1}\|_{\mathcal{A}}^2 + \sum_{i \leq m} \|\mathbf{z}_i\|_{\mathcal{A}}^2 \leq C_1^2 \, \|\mathbf{z}\|_{\mathcal{A}}^2.$$

In the particular application of our main interest, the above estimate takes the form, using the fact that $\mathcal{A}_i = \mathcal{I}_i^T \mathcal{A} \mathcal{I}_i$ and $\mathcal{A}_c = \mathcal{P}^T \mathcal{A} \mathcal{P}$, $\mathbf{z}_{m+1} = \mathcal{P}\mathbf{y}_c$ and $\mathbf{z}_i = \mathcal{I}_i \mathbf{y}_i$,

$$\|\mathbf{y}_c\|_{\mathcal{A}_c}^2 + \sum_{i \leq m} \|\mathbf{y}_i\|_{\mathcal{A}_i}^2 \leq C_1^2 \, \|\mathbf{z}\|_{\mathcal{A}}^2.$$

Using the fact that the spaces $\mathcal{K}_i$ for $i \leq m$ are local, the following estimate is straightforward; for any $\mathbf{w} = \mathbf{w}_{m+1} + \sum_{i \leq m} \mathbf{w}_i$,

$$\begin{aligned}
\|\mathbf{w}\|_{\mathcal{A}}^2 &\leq 2\mathbf{w}_{m+1}^T \mathcal{A}\mathbf{w}_{m+1} + 2 \sum_{i,j \leq m} \mathbf{w}_j^T \mathcal{A} \mathbf{w}_i \\
&= 2\mathbf{w}_{m+1}^T \mathcal{A}\mathbf{w}_{m+1} + 2 \sum_{i \leq m} \sum_{\{j \leq m:\ \Omega_j \cap \Omega_i \neq \emptyset\}} \mathbf{w}_j^T \mathcal{A} \mathbf{w}_i \\
&\leq 2\mathbf{w}_{m+1}^T \mathcal{A}\mathbf{w}_{m+1} + 2\kappa \sum_{i \leq m} \|\mathbf{w}_i\|_{\mathcal{A}}^2 \\
&\leq 2\kappa \sum_{i=1}^{m+1} \|\mathbf{w}_i\|_{\mathcal{A}}^2.
\end{aligned}$$

Recall that $\kappa \geq 1$ stands for the maximum number of subdomains Ω_j that intersect any given subdomain Ω_i in the sense that $\mathcal{I}_j^T \mathcal{A} \mathcal{I}_i \neq 0$, which due to assumed locality is a bounded number. In a similar fashion, we prove that for any two decompositions $\mathbf{w} = \mathbf{w}_{m+1} + \sum_{i \leq m} \mathbf{w}_i$ and $\mathbf{g} = \mathbf{g}_{m+1} + \sum_{i \leq m} \mathbf{g}_i$, we have the estimate

$$\begin{aligned}
&\sum_{i=1}^{m+1} \mathbf{g}_i^T \mathcal{A} \sum_{j=i+1}^{m+1} \mathbf{w}_j \\
&\quad = \mathbf{g}_{m+1}^T \mathcal{A}\mathbf{w}_{m+1} + \mathbf{g}_{m+1}^T \mathcal{A} \sum_{j=1}^{m} \mathbf{w}_j + \sum_{i=1}^{m} \mathbf{g}_i^T \mathcal{A}\mathbf{w}_{m+1} + \sum_{i=1}^{m} \mathbf{g}_i^T \mathcal{A} \sum_{j=i+1}^{m} \mathbf{w}_j \\
&\quad \leq \|\mathbf{g}_{m+1}\|_{\mathcal{A}} \|\mathbf{w}_{m+1}\|_{\mathcal{A}} + \|\mathbf{g}_{m+1}\|_{\mathcal{A}} \left\| \sum_{j=1}^{m} \mathbf{w}_j \right\|_{\mathcal{A}} + \|\mathbf{w}_{m+1}\|_{\mathcal{A}} \left\| \sum_{i=1}^{m} \mathbf{g}_i \right\|_{\mathcal{A}} \\
&\qquad + \sum_{i=1}^{m} \|\mathbf{g}_i\|_{\mathcal{A}} \sum_{\{j > i,\ j \leq m:\ \Omega_j \cap \Omega_i \neq \emptyset\}} \|\mathbf{w}_j\|_{\mathcal{A}}.
\end{aligned}$$

That is,

$$\sum_{i=1}^{m+1} \mathbf{g}_i^T \mathcal{A} \sum_{j=i+1}^{m+1} \mathbf{w}_j \leq \left[2\mathbf{w}_{m+1}^T \mathcal{A} \mathbf{w}_{m+1} + 2\kappa \sum_{i\leq m} \|\mathbf{w}_i\|_{\mathcal{A}}^2\right]^{1/2}$$
$$\times \left[2\mathbf{g}_{m+1}^T \mathcal{A} \mathbf{g}_{m+1} + 2\kappa \sum_{i\leq m} \|\mathbf{g}_i\|_{\mathcal{A}}^2\right]^{1/2}$$
$$\leq 2\kappa \left(\sum_{i=1}^{m+1} \|\mathbf{w}_i\|_{\mathcal{A}}^2\right)^{1/2} \left(\sum_{i=1}^{m+1} \|\mathbf{g}_i\|_{\mathcal{A}}^2\right)^{1/2} \tag{9.39}$$

For a given current iterate $\mathbf{u}^0$ consider Algorithm 9.5.1 and let

$$\mathbf{u}^{\frac{i}{m+1}} = \mathbf{u}^{\frac{i-1}{m+1}} + \mathcal{I}_i \mathbf{u}_i \quad \text{for } i \leq m,$$

and

$$\mathbf{u}^{\frac{i}{m+1}} = \mathbf{u}^{\text{new}} = \mathbf{u}^{\frac{m}{m+1}} + \mathcal{P}\mathbf{u}_c \quad \text{for } i = m+1.$$

Thus letting $\mathcal{I}_{m+1} = \mathcal{P}$ and $\mathbf{u}_{m+1} = \mathbf{u}_c$, $\mathbf{u}_i$ solves the problem,

$$\mathcal{J}(\mathbf{u}^{((i-1)/(m+1))} + \mathcal{I}_i \mathbf{u}_i) = \min_{\mathbf{g}_i \in \mathcal{K}_i} \mathcal{J}(\mathbf{u}^{((i-1/)(m+1))} + \mathbf{g}_i).$$

The latter problem is equivalent to the local one

$$\mathcal{J}_i(\mathbf{u}_i) = \frac{1}{2}\, \mathbf{u}_i^T \mathcal{A}_i \mathbf{u}_i - \left(\mathcal{I}_i^T \left(\mathbf{f} - \mathcal{A}\mathbf{u}^{\frac{i-1}{m+1}}\right)\right)^T \mathbf{u}_i \mapsto \min \text{ subject to } \mathcal{I}_i \mathbf{u}_i \in \mathcal{K}_i.$$

Then, due to Lemma 9.19, we have

$$\mathbf{g}_i^T (\mathcal{A}(\mathbf{u}^{((i-1)/(m+1))} + \mathcal{I}_i \mathbf{u}_i) - \mathbf{f}) = 0, \quad \text{for all } \mathbf{g}_i \in \mathcal{K}_i. \tag{9.40}$$

We have the identity

$$\mathcal{J}(\mathbf{w}) - \mathcal{J}(\mathbf{u}) = (\mathcal{A}\mathbf{u} - \mathbf{f})^T (\mathbf{w} - \mathbf{u}) + \frac{1}{2}\, \|\mathbf{w} - \mathbf{u}\|_{\mathcal{A}}^2. \tag{9.41}$$

Letting $\mathbf{w} = \mathbf{u}^{((i-1)/(m+1))}$, $\mathbf{u} = \mathbf{u}^{i/m}$ and from (9.40) used for $\mathbf{g}_i = -\mathcal{I}_i \mathbf{u}_i$, we get

$$\mathcal{J}(\mathbf{u}^{((i-1)/(m+1))}) - \mathcal{J}(\mathbf{u}^{i/m}) = \frac{1}{2}\, \|\mathcal{I}_i \mathbf{u}_i\|_{\mathcal{A}}^2.$$

Therefore,

$$\mathcal{J}(\mathbf{u}) - \mathcal{J}(\mathbf{u}^{\text{new}}) = \sum_{i=1}^{m+1} (\mathcal{J}(\mathbf{u}^{((i-1)/(m+1))}) - \mathcal{J}(\mathbf{u}^{i/(m+1)})) \geq \frac{1}{2} \sum_{i=1}^{m+1} \|\mathcal{I}_i \mathbf{u}_i\|_{\mathcal{A}}^2. \tag{9.42}$$

In particular, we obtain the monotonicity (which we already proved in Lemma 9.18),

$$\mathcal{J}(\mathbf{u}^0) \geq \mathcal{J}(\mathbf{u}^{\text{new}}).$$

Let $\mathbf{u}$ be the exact solution of the constrained minimization problem $\mathcal{J}(\mathbf{u}) \mapsto \min$ subject to $\mathbf{u} \in \mathcal{K}$. Use the assumed stable decomposition (d) for

$$\mathbf{u} - \mathbf{u}^0 = \mathbf{z}_{m+1} + \sum_{i \leq m} \mathbf{z}_i.$$

Then,

$$\begin{aligned}
(\mathcal{A}\mathbf{u}^{\text{new}} - \mathbf{f})^T \, (\mathbf{u}^{\text{new}} - \mathbf{u}) &= (\mathcal{A}\mathbf{u}^{\text{new}} - \mathbf{f})^T \left(\sum_{i=1}^{m+1} \mathcal{I}_i \mathbf{u}_i + \mathbf{u}^0 - \mathbf{u} \right) \\
&= \sum_{i=1}^{m+1} (\mathcal{A}\mathbf{u}^{\text{new}} - \mathbf{f})^T \, (\mathcal{I}_i \mathbf{u}_i - \mathbf{z}_i).
\end{aligned}$$

Use now (9.40) for $\mathbf{g}_i = \mathbf{z}_i - \mathcal{I}_i \mathbf{u}_i$, which then reads $(\mathbf{z}_i - \mathcal{I}_i \mathbf{u}_i)^T \, (\mathcal{A}\mathbf{u}^{i/(m+1)} - \mathbf{f}) = 0$. Because $\mathbf{u}^{\text{new}} - \mathbf{u}^{i/(m+1)} = \sum_{j>i} \mathcal{I}_j \mathbf{u}_j$, we get, based on (9.39) used for $\mathbf{w}_j = \mathcal{I}_j \mathbf{u}_j$ and $\mathbf{g} = \sum_{i=1}^{m+1} \mathbf{g}_i$ with $\mathbf{g}_i = \mathcal{I}_i \mathbf{u}_i - \mathbf{z}_i$,

$$\begin{aligned}
(\mathcal{A}\mathbf{u}^{\text{new}} - \mathbf{f})^T \, (\mathbf{u}^{\text{new}} - \mathbf{u}) &= \sum_{i=1}^{m+1} (\mathcal{A}\mathbf{u}^{\text{new}} - \mathbf{f} - (\mathcal{A}\mathbf{u}^{i/(m+1)} - \mathbf{f}))^T \, (\mathcal{I}_i \mathbf{u}_i - \mathbf{z}_i) \\
&= \sum_{i=1}^{m+1} \sum_{j>i} (\mathcal{A}\mathcal{I}_j \mathbf{u}_j)^T \, (\mathcal{I}_i \mathbf{u}_i - \mathbf{z}_i) \\
&\leq 2\kappa \left(\sum_j \|\mathcal{I}_j \mathbf{u}_j\|_{\mathcal{A}}^2 \right)^{1/2} \left(\sum_i \|\mathcal{I}_i \mathbf{u}_i - \mathbf{z}_i\|_{\mathcal{A}}^2 \right)^{1/2} \\
&\leq 2\kappa \left(\sum_j \|\mathcal{I}_j \mathbf{u}_j\|_{\mathcal{A}}^2 \right)^{1/2} \\
&\quad \times \left(\left(\sum_i \|\mathcal{I}_i \mathbf{u}_i\|_{\mathcal{A}}^2 \right)^{1/2} + C_1 \, \|\mathbf{u}^0 - \mathbf{u}\|_{\mathcal{A}} \right).
\end{aligned} \tag{9.43}$$

In the last line, we used the triangle inequality and the assumed stability estimate $\sum_i \|\mathbf{z}_i\|_{\mathcal{A}}^2 \leq C_1^2 \, \|\mathbf{u}^0 - \mathbf{u}\|_{\mathcal{A}}^2$. Now use the estimate (9.42) in (9.43) to arrive at

$$\begin{aligned}
(\mathcal{A}\mathbf{u}^{\text{new}} - \mathbf{f})^T \, (\mathbf{u}^{\text{new}} - \mathbf{u}) \leq 4\kappa \, (\mathcal{J}(\mathbf{u}^0) - \mathcal{J}(\mathbf{u}^{\text{new}})) \\
+ 2\kappa\sqrt{2} \, C_1 \, \sqrt{\mathcal{J}(\mathbf{u}^0) - \mathcal{J}(\mathbf{u}^{\text{new}})} \, \|\mathbf{u}^0 - \mathbf{u}\|_{\mathcal{A}}.
\end{aligned} \tag{9.44}$$

At this point, use the fact that $\mathbf{u}$ is the exact solution, hence due to Lemma 9.19 $(\mathcal{A}\mathbf{u} - \mathbf{f})^T(\mathbf{u}^0 - \mathbf{u}) = 0$. Then from identity (9.41), we obtain,

$$\mathcal{J}(\mathbf{u}^0) - \mathcal{J}(\mathbf{u}) = (\mathcal{A}\mathbf{u} - \mathbf{f})^T(\mathbf{u}^0 - \mathbf{u}) + \frac{1}{2}\,\|\mathbf{u}^0 - \mathbf{u}\|_{\mathcal{A}}^2 \geq \frac{1}{2}\,\|\mathbf{u}^0 - \mathbf{u}\|_{\mathcal{A}}^2.$$

The latter estimate combined with (9.44) leads to

$$\begin{aligned}\mathcal{J}(\mathbf{u}^{\text{new}}) - \mathcal{J}(\mathbf{u}) &\leq (\mathcal{A}\mathbf{u}^{\text{new}} - \mathbf{f})^T\;(\mathbf{u}^{\text{new}} - \mathbf{u})\\ &\leq 4\kappa\;(\mathcal{J}(\mathbf{u}^0) - \mathcal{J}(\mathbf{u}^{\text{new}}))\\ &\quad + 4\sqrt{2}\kappa\;C_1\;\sqrt{\mathcal{J}(\mathbf{u}^0) - \mathcal{J}(\mathbf{u}^{\text{new}})}\;\sqrt{2}\;\sqrt{\mathcal{J}(\mathbf{u}^0) - \mathcal{J}(\mathbf{u})}.\end{aligned}$$

Introduce $d_0 = \mathcal{J}(\mathbf{u}^0) - \mathcal{J}(\mathbf{u}) \geq 0$ and $d_{\text{new}} = \mathcal{J}(\mathbf{u}^{\text{new}}) - \mathcal{J}(\mathbf{u}) \geq 0$ and let $\mu \in (0, 1)$. In terms of d_0 and d_{new}, the latter inequality reads

$$\begin{aligned}d_{\text{new}} &\leq 4\kappa\;(d_0 - d_{\text{new}}) + 8\kappa\;C_1\;\sqrt{d_0 - d_{\text{new}}}\sqrt{d_0}\\ &\leq \left(4\kappa\; + \frac{16\kappa^2 C_1^2}{\mu}\right)(d_0 - d_{\text{new}}) + \mu d_0\\ &\leq C_*\mu^{-1}(d_0 - d_{\text{new}}) + \mu d_0.\end{aligned}$$

Here, $C_* = 4\kappa\; + 16\kappa^2 C_1^2$. Then the latter inequality reads

$$d_{\text{new}} \leq \left(1 - \frac{\mu(1-\mu)}{\mu + C_*}\right)\;d_0.$$

Consider the scalar function $g(\mu) \equiv \mu(1-\mu)/(\mu+C_*)$. Based on $(\mu+C_*)^2 g'(\mu) = (1-2\mu)(\mu+C_*) - \mu(1-\mu) = 0$, that is, $\mu^2 + 2C_*\mu - C_* = 0$, we see that with the choice $\mu = \mu_* \equiv -C_* + \sqrt{C_*^2 + C_*} = C_*/(C_* + \sqrt{C_*^2 + C_*}) \in (0, 1)$, we get the following expression for the maximum of $g(\mu)$:

$$\begin{aligned}\max_{\mu\in[0,1]} \frac{\mu(1-\mu)}{\mu + C_*} &= 1 - 2\mu_* = 1 + 2C_* - 2\sqrt{C_*^2 + C_*}\\ &= \frac{1}{1 + 2C_* + 2\sqrt{C_*^2 + C_*}} \in (0, 1).\end{aligned}$$

That is, we proved the following convergence rate estimate

$$d_{\text{new}} \leq\; 2\mu_*\;d_0 = \frac{2C_*}{C_* + \sqrt{C_*^2 + C_*}}\;d_0 < d_0.$$

Equivalently,

$$d_{\text{new}} \leq \left(1 - \frac{1}{1 + 2C_* + 2\sqrt{C_*^2 + C_*}}\right)\;d_0 = \left(1 - \frac{1}{(\sqrt{1 + C_*} + \sqrt{C_*})^2}\right)\;d_0.$$

The following main result then holds.

Theorem 9.20. *Under the assumption (d) providing for any* $\mathbf{z} \in \mathcal{K}$, *a stable decomposition* $\mathbf{z} = \mathcal{P}\mathbf{z}_c + \sum_i \mathcal{I}_i \mathbf{z}_i$, *in the sense that for a constant* C_1, *the estimate*

$$\|\mathcal{P}\mathbf{z}_c\|_{\mathcal{A}}^2 + \sum_{i=1}^{m} \|\mathcal{I}_i \mathbf{z}_i\|_{\mathcal{A}}^2 \leq C_1^2 \; \|\mathbf{z}\|_{\mathcal{A}}^2$$

holds. The matrices $\mathcal{I}_i$ *and* $\mathcal{P}$ *are such that there are restriction matrices* $\mathcal{Q}_i^T$ *and* $\mathcal{Q}_c^T$ *with the property* $\mathcal{Q}_i^T \mathcal{B}\mathcal{I}_i \mathbf{u}_i = 0$ *and* $\mathcal{Q}_c^T \mathcal{B}\mathcal{P}\mathbf{u}_c = 0$ *imply* $\mathcal{B}\mathcal{I}_i \mathbf{u}_i = 0$ *and* $\mathcal{B}\mathcal{P}\mathbf{u}_c = 0$. *Define* $\mathcal{B}_i = \mathcal{Q}_i^T B_i \mathcal{I}_i$ *and* $\mathcal{B}_c = \mathcal{Q}_c^T \mathcal{B}\mathcal{P}$, $\mathcal{A}_i = \mathcal{I}_i^T \mathcal{A}\mathcal{I}_i$ *and* $\mathcal{A}_c = \mathcal{P}^T \mathcal{A}\mathcal{P}$. *We also assume that the local saddle-point problems based on*

$$A_i = \begin{bmatrix} \mathcal{A}_i & \mathcal{B}_i^T \\ \mathcal{B}_i & 0 \end{bmatrix}$$

and the coarse one,

$$\mathcal{A}_c = \begin{bmatrix} \mathcal{A}_c & \mathcal{B}_c \\ \mathcal{B}_c & 0 \end{bmatrix},$$

are solvable. Finally, we assume that the interaction matrices $\mathcal{I}_j^T \mathcal{A}\mathcal{I}_i$ *for any* i *are nonzero for a bounded number of indices* j *denoted by* $\kappa \geq 1$. *Let* $\mathbf{u}^k$ *be obtained by applying* $k \geq 1$ *steps of Algorithm 9.5.1. Then the following geometric convergence in the* $\mathcal{A}$*-norm holds,*

$$\frac{1}{2} \; \|\mathbf{u}^k - \mathbf{u}\|_{\mathcal{A}}^2 \leq \mathcal{J}(\mathbf{u}^k) - \mathcal{J}(\mathbf{u}) \leq (\varrho_{TR})^k \; (\mathcal{J}(\mathbf{u}^0) - \mathcal{J}(\mathbf{u})),$$

with

$$\varrho_{TR} \equiv 1 - \frac{1}{1 + 2C_* + 2\sqrt{C_*^2 + C_*}}, \quad \text{and} \quad C_* = 4\kappa + 16\kappa^2 \; C_1^2.$$

We comment at the end that the main ingredient in the proof is establishing a proper stable decomposition. The latter can be verified in practice for a number of finite element discretization problems (e.g., for mixed methods for second-order elliptic PDEs) based on proper stable decompositions for finite element functions. For more details, we refer to Section F.3 in the appendix.

Final remarks

Other approaches to solving saddle-point problems exploiting distributive relaxation are found in [Wi89] and [Wi90]. For the use of indefinite smoothers or preconditioners, see [Van86], [ELLV], [BS97], and [SZ03].

10

Variable-Step Iterative Methods

In this chapter, we consider iterative methods that exploit preconditioners or approximate inverses that are of fixed quality but may change from step to step (hence the name variable-step). These are generally represented by nonlinear mappings. Examples of such mappings are given by solutions obtained by a few steps of variational (or CG) methods applied to some auxiliary problems.

In particular we consider block-preconditioners for matrices

$$A = \begin{bmatrix} \mathcal{A} & \mathcal{R} \\ \mathcal{L} & \mathcal{B} \end{bmatrix}$$

that exploit linear preconditioners (or smoothers) for $\mathcal{A}$ and variable-step preconditioners for the Schur complement $\mathcal{S} = \mathcal{B} - \mathcal{L}\mathcal{A}^{-1}\mathcal{R}$ of A. The procedure can also be applied recursively to define variable-step multilevel preconditioners as well as to AMLI-cycle MG with variable-step recursive calls to coarse levels.

We use the terminology introduced in the original paper [AV91]. Y. Saad has introduced the name "flexible" preconditioning (in [Sa93]; see also [Sa03]), which is more popular today. Other papers that deal with the topic of variable-step/flexible preconditioning are [VV94], [GY99], [Not0b], [SSa], and [SSb].

The variable-step multilevel preconditioners were originally proposed in [AV94] (additive versions), and later in [JK02] the multiplicative case was analyzed. Here, we also introduce and analyze the variable-step AMLI-cycle MG method.

10.1 Variable-step (nonlinear) preconditioners

Let $D[.]$ be generally a nonlinear mapping that is close to $\mathcal{D}^{-1}$. We assume that $\mathcal{D}$ is s.p.d. and impose that the following measure of the deviation of $D[.]$ from $\mathcal{D}^{-1}$ be small, in the sense that

$$\|D[\mathbf{x}] - \mathcal{D}^{-1}\mathbf{x}\|_{\mathcal{D}} \leq \delta \; \|\mathcal{D}^{-1}\mathbf{x}\|_{\mathcal{D}}, \quad \text{for all } \mathbf{x}. \tag{10.1}$$

P.S. Vassilevski, *Multilevel Block Factorization Preconditioners*,
doi: 10.1007/978-0-387-71564-3_10,

Consider the following preconditioner,

$$B[.] = \begin{bmatrix} I & -\mathcal{M}^{-T}\mathcal{R} \\ 0 & I \end{bmatrix} \begin{bmatrix} \overline{\mathcal{M}}^{-1} & 0 \\ 0 & D[.] \end{bmatrix} \begin{bmatrix} I & 0 \\ -\mathcal{L}\mathcal{M}^{-1} & I \end{bmatrix}. \tag{10.2}$$

To be specific, we define the actions of $B[\mathbf{v}]$ as follows. Let

$$\mathbf{v} = \begin{bmatrix} \mathbf{w} \\ \mathbf{x} \end{bmatrix}$$

Then

$$\begin{aligned} B[\mathbf{v}] &= \begin{bmatrix} I & -\mathcal{M}^{-T}\mathcal{R} \\ 0 & I \end{bmatrix} \begin{bmatrix} \overline{\mathcal{M}}^{-1}\mathbf{w} \\ D[\mathbf{x} - \mathcal{L}\mathcal{M}^{-1}\mathbf{w}] \end{bmatrix} \\ &= \begin{bmatrix} \overline{\mathcal{M}}^{-1}\mathbf{w} - \mathcal{M}^{-T}\mathcal{R}D[\mathbf{x} - \mathcal{L}\mathcal{M}^{-1}\mathbf{w}] \\ D[\mathbf{x} - \mathcal{L}\mathcal{M}^{-1}\mathbf{w}] \end{bmatrix}. \end{aligned}$$

We prove a bound for the deviation of $B[.]$ from the corresponding linear preconditioner

$$\mathrm{B}^{-1} = \begin{bmatrix} I & -\mathcal{M}^{-T}\mathcal{R} \\ 0 & I \end{bmatrix} \begin{bmatrix} \overline{\mathcal{M}}^{-1} & 0 \\ 0 & \mathcal{D}^{-1} \end{bmatrix} \begin{bmatrix} I & 0 \\ -\mathcal{L}\mathcal{M}^{-1} & I \end{bmatrix}.$$

We have

$$B[\mathbf{v}] - \mathrm{B}^{-1}\mathbf{v} = \begin{bmatrix} -\mathcal{M}^{-T}\mathcal{R} \\ I \end{bmatrix} (D[\mathbf{x} - \mathcal{L}\mathcal{M}^{-1}\mathbf{w}] - \mathcal{D}^{-1}(\mathbf{x} - \mathcal{L}\mathcal{M}^{-1}\mathbf{w})).$$

Therefore,

$$B(B[\mathbf{v}] - \mathrm{B}^{-1}\mathbf{v}) = \begin{bmatrix} 0 \\ \mathcal{D} \end{bmatrix} (D[\mathbf{x} - \mathcal{L}\mathcal{M}^{-1}\mathbf{w}] - \mathcal{D}^{-1}(\mathbf{x} - \mathcal{L}\mathcal{M}^{-1}\mathbf{w})),$$

which implies

$$\begin{aligned} \|B[\mathbf{v}] - \mathrm{B}^{-1}\mathbf{v}\|_{\mathrm{B}} &= \|D[\mathbf{x} - \mathcal{L}\mathcal{M}^{-1}\mathbf{w}] - \mathcal{D}^{-1}(\mathbf{x} - \mathcal{L}\mathcal{M}^{-1}\mathbf{w})\|_{\mathcal{D}} \\ &\leq \delta\ \|\mathcal{D}^{-1}(\mathbf{x} - \mathcal{L}\mathcal{M}^{-1}\mathbf{w})\|_{\mathcal{D}}. \end{aligned}$$

Finally, noticing that

$$\|\mathrm{B}^{-1}\mathbf{v}\|_{\mathrm{B}}^2 = \mathbf{v}^T\mathrm{B}^{-1}\mathbf{v} = \mathbf{w}^T\overline{\mathcal{M}}^{-1}\mathbf{w} + (\mathbf{x} - \mathcal{L}\mathcal{M}^{-1}\mathbf{w})^T\mathcal{D}^{-1}(\mathbf{x} - \mathcal{L}\mathcal{M}^{-1}\mathbf{w}),$$

we end up with the following result.

Theorem 10.1. *Consider the nonlinear preconditioner $B[.]$ defined in* (10.2) *and the corresponding linear one B^{-1}, which differ by their Schur complements $D[.]$ and $\mathcal{D}^{-1}$, respectively. Then, under the assumption* (10.1) *of small deviation of $D[.]$ from $\mathcal{D}^{-1}$, the same estimate of the deviation between $B[.]$ and B^{-1} holds:*

$$\|B[\mathbf{v}] - \mathrm{B}^{-1}\mathbf{v}\|_{\mathrm{B}} \leq \delta\ \|\mathrm{B}^{-1}\mathbf{v}\|_{\mathrm{B}}. \tag{10.3}$$

In a following section, we describe a procedure to improve on the variable-step (nonlinear) preconditioner defined in (10.2) by applying the CG-like procedure described in the next section.

10.2 Variable-step preconditioned CG method

In the present section, we describe a somewhat standard preconditioned CG (conjugate gradient) method where the preconditioner $B[.]$ is a nonlinear mapping that is assumed to approximate the inverse of a linear one B, which we assume is s.p.d. A more general case (that includes nonsymmetric and possibly indefinite matrices) was considered in [AV91].

We now formulate an algorithm that can be used to provide iterated approximate inverses to A on the basis of a given initial (nonlinear) mapping $B[\cdot]$ that approximates a given s.p.d. matrix B. Our main application is $\text{B} = A$.

For any $\nu \geq 1$, choose a fixed sequence of integers $\{m_k\}_{k=0}^{\nu}$, $0 \leq m_k \leq m_{k-1} + 1 \leq k-1$. A typical choice is $m_k = 0$. We define the ν-times iterated nonlinear preconditioner $B_\nu[\mathbf{v}] = \mathbf{u}_{\nu+1}$ where $\mathbf{u}_{\nu+1}$ is the $\nu+1$st iterate obtained by the following variable-step preconditioned CG (conjugate gradient) procedure.

Algorithm 10.2.1 (Variable-step preconditioned CG). *For a given* $\mathbf{v}$, *define* $B_\nu[\mathbf{v}] = \mathbf{u}_{\nu+1}$, *where* $\mathbf{u}_{\nu+1}$ *is computed as follows.*

1. *Let* $\mathbf{v}_0 = \mathbf{v}$ *and* $\mathbf{u}_0 = 0$. *Compute* $\mathbf{r}_0 = B[\mathbf{v}_0]$ *and let* $\mathbf{d}_0 = \mathbf{r}_0$. *Then let*

$$\mathbf{u}_1 = \frac{\mathbf{d}_0^T \mathbf{v}_0}{\mathbf{d}_0^T A \mathbf{d}_0} \mathbf{d}_0 \quad \textit{and} \quad \mathbf{v}_1 = \mathbf{v}_0 - \frac{\mathbf{d}_0^T \mathbf{v}_0}{\mathbf{d}_0^T A \mathbf{d}_0} A \mathbf{d}_0.$$

2. *For* $k = 1, \ldots, \nu$, *compute* $\mathbf{r}_k = B[\mathbf{v}_k]$ *and then based on* $\{\mathbf{d}_j\}_{j=k-1-m_k}^{k}$ *form*

$$\mathbf{d}_k = \mathbf{r}_k - \sum_{j=k-1-m_k}^{k-1} \frac{\mathbf{r}_k^T A \mathbf{d}_j}{\mathbf{d}_j^T A \mathbf{d}_j} \mathbf{d}_j.$$

Then the next iterate is

$$\mathbf{u}_{k+1} = \mathbf{u}_k + \frac{\mathbf{d}_k^T \mathbf{v}_k}{\mathbf{d}_k^T A \mathbf{d}_k} \mathbf{d}_k,$$

and the corresponding residual equals

$$\mathbf{v}_{k+1} = \mathbf{v} - A\mathbf{u}_k = \mathbf{v}_k - \frac{\mathbf{d}_k^T \mathbf{v}_k}{\mathbf{d}_k^T A \mathbf{d}_k} A \mathbf{d}_k.$$

3. *Finally, we let* $B_\nu[\mathbf{v}] = \mathbf{u}_{\nu+1}$.

Here, $B[\cdot]$ approximates the inverse of $\mathrm{B} = A$, with accuracy $\delta \in [0, 1)$, that is,

$$\|A^{-1}\mathbf{v} - B[\mathbf{v}]\|_A \leq \delta\ \|\mathbf{v}\|_{A^{-1}}. \tag{10.4}$$

We have the following convergence result for Algorithm 10.2.1.

Theorem 10.2. *Consider Algorithm 10.2.1 for a given* $\mathbf{v}$. *Assume that the s.p.d. matrix* A *has an approximate inverse* $B[\cdot]$ *satisfying* (10.4). *In Algorithm 10.2.1, define the kth step search direction* $\mathbf{d}_k$ *to be A-orthogonal to the* $m_k + 1$ *most recent search directions. The integers* $\{m_k\}$ *satisfy* $0 \leq m_k \leq m_{k-1} + 1 \leq k - 1$. *That is, we have*

$$\mathbf{d}_k = \mathbf{r}_k - \sum_{j=k-1-m_k}^{k-1} \beta_{k,j}\ \mathbf{d}_j,$$

with $\beta_{k,j} = (\mathbf{r}_k^T A\mathbf{d}_j)/(\mathbf{d}_j^T A\mathbf{d}_j)$. *Recall, that* $\mathbf{v}_0 = \mathbf{v}$, $\mathbf{v}_k = \mathbf{v}_{k-1} - \alpha_{k-1}\ A\mathbf{d}_{k-1}$ *for* $\alpha_{k-1} = (\mathbf{d}_{k-1}^T\mathbf{v}_{k-1})/(\mathbf{d}_{k-1}^T A\mathbf{d}_{k-1})$, *and* $\mathbf{r}_k = B[\mathbf{v}_k]$. *Also with* $\mathbf{u}_0 = 0$, *Algorithm 10.2.1 computes* $\mathbf{u}_k = \mathbf{u}_{k-1} + \alpha_{k-1}\ \mathbf{d}_{k-1}$, *and the kth step iterated approximate inverse* $B_k[.]$ *is defined as* $B_k[\mathbf{v}] = \mathbf{u}_{k+1}$. *Note that* $B_0[\mathbf{v}] = \alpha_0 B[\mathbf{v}]$; *that is,* $B_0[\mathbf{v}]$ *differs from* $B[\mathbf{v}]$ *by a scalar factor.*

The following convergence rate estimate holds,

$$\|\mathbf{v}_k\|_{A^{-1}} \leq \delta\ \|\mathbf{v}_{k-1}\|_{A^{-1}}.$$

Equivalently, because $\mathbf{v}_{k+1} = \mathbf{v} - A\mathbf{u}_{k+1} = \mathbf{v} - AB_k[\mathbf{v}]$, *the following deviation estimate between* $A^{-1}\mathbf{v}$ *and* $B_k[\mathbf{v}]$ *holds,*

$$\|A^{-1}\mathbf{v} - B_k[\mathbf{v}]\|_A = \|\mathbf{v} - AB_k[\mathbf{v}]\|_{A^{-1}} \leq \delta^{k+1}\ \|\mathbf{v}\|_{A^{-1}}.$$

Proof. Assuming by induction that $\mathbf{v}_{k-1}$ is orthogonal to $\mathbf{d}_j$ for $k-2-m_{k-1} \leq j < k-1$, we have then that

$$\mathbf{v}_k^T\mathbf{d}_j = \mathbf{v}_{k-1}^T\mathbf{d}_j - \alpha_{k-1}\mathbf{d}_{k-1}^T A\mathbf{d}_j = 0, \quad \text{for all } j < k-1 \quad \text{and} \quad j \geq k-2-m_{k-1}.$$

For $j = k-1$, we also have

$$\mathbf{v}_k^T\mathbf{d}_{k-1} = \mathbf{v}_{k-1}^T\mathbf{d}_{k-1} - \alpha_{k-1}\ \mathbf{d}_{k-1}^T A\mathbf{d}_{k-1} = 0$$

due to the choice of α_{k-1}. That is, $\mathbf{v}_k$ is orthogonal to $\mathbf{d}_j$ for all $j\ :\ k > j \geq k-1-m_k \geq k-2-m_{k-1}$, which confirms the induction assumption.
Note then, that

$$\begin{aligned}
\alpha_{k-1} &= \frac{1}{\mathbf{d}_{k-1}^T A\mathbf{d}_{k-1}}\ \mathbf{v}_{k-1}^T\left(\mathbf{r}_{k-1} - \sum_{j=k-2-m_{k-1}}^{k-2} \beta_{k-1,j}\ \mathbf{d}_j\right) \\
&= \frac{1}{\mathbf{d}_{k-1}^T A\mathbf{d}_{k-1}}\ \mathbf{v}_{k-1}^T\mathbf{r}_{k-1} \\
&= \frac{\mathbf{v}_{k-1}^T B[\mathbf{v}_{k-1}]}{\mathbf{d}_{k-1}^T A\mathbf{d}_{k-1}}.
\end{aligned}$$

Another observation is that

$$\|\mathbf{r}_k\|_A^2 = \|\mathbf{d}_k\|_A^2 + \sum_{j=k-1-m_k}^{k-1} \beta_{k,j}^2 \, \|\mathbf{d}_j\|_A^2 \geq \|\mathbf{d}_k\|_A^2.$$

Now we are ready to proceed with the convergence rate estimate in the case when $B[.]$ provides an approximate inverse to A as in (10.4). We have

$$\|\mathbf{v} - AB_k[\mathbf{v}]\|_{A^{-1}} = \|\mathbf{v} - A\mathbf{u}_{k+1}\|_{A^{-1}} = \|\mathbf{v}_{k+1}\|_{A^{-1}}.$$

Also,

$$\|\mathbf{v}_{k+1}\|_{A^{-1}}^2 = \|\mathbf{v}_k - \alpha_k \, A\mathbf{d}_k\|_{A^{-1}}^2 = \|\mathbf{v}_k\|_{A^{-1}}^2 - \left(\frac{\mathbf{v}_k^T \mathbf{d}_k}{\|\mathbf{d}_k\|_A}\right)^2.$$

Because $\mathbf{v}_k^T \mathbf{d}_k = \mathbf{v}_k^T B[\mathbf{v}_k]$ and $\|\mathbf{d}_k\|_A \leq \|\mathbf{r}_k\|_A = \|B[\mathbf{v}_k]\|_A$, the following estimate is seen.

$$\|\mathbf{v}_{k+1}\|_{A^{-1}}^2 \leq \|\mathbf{v}_k\|_{A^{-1}}^2 - \left(\frac{\mathbf{v}_k^T B[\mathbf{v}_k]}{\|B[\mathbf{v}_k]\|_A}\right)^2 = \min_{\alpha \in \mathbb{R}} \|\mathbf{v}_k - \alpha \, AB[\mathbf{v}_k]\|_{A^{-1}}^2. \tag{10.5}$$

Estimate (10.4) upon expanding reads,

$$\|\mathbf{v}\|_{A^{-1}}^2 - 2\mathbf{v}^T B[\mathbf{v}] + \|B[\mathbf{v}]\|_A^2 \leq \delta^2 \, \|\mathbf{v}\|_{A^{-1}}^2,$$

or equivalently

$$2\mathbf{v}^T B[\mathbf{v}] \geq (1 - \delta^2) \, \|\mathbf{v}\|_{A^{-1}}^2 + \|B[\mathbf{v}]\|_A^2.$$

Based on the Cauchy–Schwarz inequality $a^2 + b^2 \geq 2ab$, we also get

$$\mathbf{v}^T B[\mathbf{v}] \geq \sqrt{1 - \delta^2} \, \|\mathbf{v}\|_{A^{-1}} \|B[\mathbf{v}]\|_A.$$

Using the last estimate for $\mathbf{v} := \mathbf{v}_k$, in (10.5) gives the desired convergence rate estimate:

$$\|\mathbf{v}_{k+1}\|_{A^{-1}}^2 \leq \|\mathbf{v}_k\|_{A^{-1}}^2 - (1 - \delta^2) \, \|\mathbf{v}_k\|_{A^{-1}}^2 = \delta^2 \, \|\mathbf{v}_k\|_{A^{-1}}^2.$$ □

Based on the convergence property of the above variable-step preconditioned CG method,

$$\|\mathbf{v}_{k+1}\|_{A^{-1}} \leq \min_{\alpha} \|\mathbf{v}_k - \alpha AB[\mathbf{v}_k]\|_{A^{-1}} \leq \Delta \|\mathbf{v}_k\|_{A^{-1}} \leq \cdots \leq \Delta^{k+1} \, \|\mathbf{v}_0\|_{A^{-1}},$$

we obtain an improved estimate for the iterated approximate inverse $B_\nu[\cdot]$ for A, that is,

$$\|A^{-1}\mathbf{v} - B_\nu[\mathbf{v}]\|_A \leq \Delta^{\nu+1} \, \|A^{-1}\mathbf{v}\|_A.$$

That is, by enlarging ν, we can always achieve that $\Delta_\nu \equiv \Delta^{\nu+1} \leq \delta$. Because $\Delta \leq 1 - (2/(1 + C_\delta \kappa))$, we get

$$(\nu + 1) \log \left(1 - \frac{2}{1 + C_\delta \kappa}\right) \leq \log \delta.$$

That is,

$$\nu + 1 \geq \frac{\log \delta^{-1}}{\log \frac{C_\delta \kappa + 1}{C_\delta \kappa - 1}}. \tag{10.6}$$

The following modification of Theorem 10.2 holds due to R. Blaheta [Bl02].

Corollary 10.3. *Under the main assumptions of Theorem 10.2, assume that the variable-step (nonlinear preconditioner) $B[.]$ is close to a fixed s.p.d. matrix B^{-1} such that*

$$\|B[\mathbf{v}] - B^{-1}\mathbf{v}\|_{B^{-1}} \leq \delta \; \|\mathbf{v}\|_{B^{-1}}.$$

Let $\kappa \geq 1$ be an upper bound of the condition number of $B^{-1}A$. Then the following convergence rate estimate holds,

$$\|\mathbf{v}_k\|_{A^{-1}} \leq \overline{\delta} \; \|\mathbf{v}_{k-1}\|_{A^{-1}}, \quad \textit{with } \overline{\delta} = \sqrt{1 - \frac{1 - \delta^2}{\kappa}}.$$

Proof. The following coercivity of $B[\cdot]$ is established in the same way as in the proof of Theorem 10.2. We have

$$\mathbf{v}^T B[\mathbf{v}] \geq \sqrt{1 - \delta^2} \; \|\mathbf{v}\|_{B^{-1}} \|B[\mathbf{v}]\|_B.$$

The latter implies

$$\mathbf{v}^T B[\mathbf{v}] \geq \sqrt{\frac{1 - \delta^2}{\kappa}} \; \|\mathbf{v}\|_{A^{-1}} \|B[\mathbf{v}]\|_A.$$

The remainder of the proof is the same as of Theorem 10.2, replacing $1 - \delta^2$ with $(1 - \delta^2)/\kappa$. □

At the end, we present one more corollary to Theorem 10.2 which is due to Notay (cf., [Not0b]).

Corollary 10.4. *Under the main assumptions of Theorem 10.2 assume in addition that the variable-step (nonlinear preconditioner) $B[.]$ is close to a fixed s.p.d. matrix B^{-1} such that*

$$\|B[\mathbf{v}] - B^{-1}\mathbf{v}\|_{B^{-1}} \leq \delta \; \|\mathbf{v}\|_{B^{-1}}.$$

Let $\kappa \geq 1$ be an upper bound of the condition number of $B^{-1}A$. Then the following convergence rate estimate holds,

$$\|\mathbf{v}_k\|_{A^{-1}}^2 \leq \left(1 - \frac{4\kappa(1 - \delta)^2}{((\kappa - 1)\delta^2 + (1 - \delta)^2 + \kappa)^2}\right) \; \|\mathbf{v}_{k-1}\|_{A^{-1}}^2.$$

Note that in the case of $\delta = 0$, the above estimate reduces to the familiar steepest descent convergence result estimate

$$\|\mathbf{v}_k\|^2_{A^{-1}} \leq \left(\frac{\kappa - 1}{\kappa + 1}\right)^2 \|\mathbf{v}_{k-1}\|^2_{A^{-1}}.$$

Proof. In the proof of Theorem 10.2, we derived the estimate (10.5)

$$\|\mathbf{v}_{k+1}\|^2_{A^{-1}} \leq \min_{\alpha \in \mathbb{R}} \|\mathbf{v}_k - \alpha\, AB[\mathbf{v}_k]\|^2_{A^{-1}}.$$

We have (letting $\mathbf{v} = \mathbf{v}_k$)

$$\begin{aligned}
\|\mathbf{v} - \alpha AB[\mathbf{v}]\|^2_{A^{-1}} &= \|\mathbf{v} - \alpha A\mathrm{B}^{-1}\mathbf{v} + \alpha A(\mathrm{B}^{-1}\mathbf{v} - B[\mathbf{v}]\|^2_{A^{-1}} \\
&= \|\mathbf{v} - \alpha A\mathrm{B}^{-1}\mathbf{v}\|^2_{A^{-1}} + 2\alpha\, (\mathbf{v} - \alpha A\mathrm{B}^{-1}\mathbf{v})^T(\mathrm{B}^{-1}\mathbf{v} - B[\mathbf{v}]) \\
&\quad + \alpha^2\, \|\mathrm{B}^{-1}\mathbf{v} - B[\mathbf{v}]\|^2_A \\
&\leq \|\mathbf{v} - \alpha A\mathrm{B}^{-1}\mathbf{v}\|^2_{A^{-1}} \\
&\quad + 2\alpha\, \|\mathbf{v} - \alpha A\mathrm{B}^{-1}\mathbf{v}\|_{\mathrm{B}^{-1}} \|\mathrm{B}^{-1}\mathbf{v} - B[\mathbf{v}]\|_{\mathrm{B}} \\
&\quad + \alpha^2\, \|\mathrm{B}^{-1}\mathbf{v} - B[\mathbf{v}]\|^2_A.
\end{aligned}$$

Let $\mathbf{r} = \mathrm{B}^{-1}\mathbf{v} - B[\mathbf{v}]$. Then,

$$\|\mathbf{r}\|^2_A \leq \lambda_{\max}(\mathrm{B}^{-1}A)\, \|\mathbf{r}\|^2_{\mathrm{B}} \leq \delta\, \lambda_{\max}(\mathrm{B}^{-1}A)\, \|\mathbf{v}\|^2_{\mathrm{B}^{-1}}.$$

We also have,

$$\|\mathbf{v} - \alpha A\mathrm{B}^{-1}\mathbf{v}\|^2_{\mathrm{B}^{-1}} \leq \|I - \alpha\, \mathrm{B}^{-(1/2)}A\mathrm{B}^{-(1/2)}\|\, \|\mathbf{v}\|^2_{\mathrm{B}^{-1}}.$$

Thus,

$$\begin{aligned}
\|\mathbf{v} - \alpha AB[\mathbf{v}]\|^2_{A^{-1}} \leq{}& \|\mathbf{v} - \alpha A\mathrm{B}^{-1}\mathbf{v}\|^2_{A^{-1}} + 2\alpha\, \delta\, \|I - \alpha\, \mathrm{B}^{-(1/2)}A\mathrm{B}^{-(1/2)}\|\, \|\mathbf{v}\|^2_{\mathrm{B}^{-1}} \\
&+ \alpha^2\delta^2\, \lambda_{\max}(\mathrm{B}^{-1}A)\|\mathbf{v}\|^2_{\mathrm{B}^{-1}}.
\end{aligned}$$

Because for $\alpha \geq 0$

$$\beta = \|I - \alpha\, \mathrm{B}^{-(1/2)}A\mathrm{B}^{-(1/2)}\| = \max\{|1 - \alpha\lambda_{\min}(\mathrm{B}^{-1}A)|,\ |1 - \alpha\lambda_{\max}(\mathrm{B}^{-1}A)|\}, \tag{10.7}$$

the last estimate reads,

$$\begin{aligned}
\|\mathbf{v} - \alpha AB[\mathbf{v}]\|^2_{A^{-1}} &\leq \|\mathbf{v} - \alpha A\mathrm{B}^{-1}\mathbf{v}\|^2_{A^{-1}} + (\alpha^2\lambda_{\max}(\mathrm{B}^{-1}A)\, \delta^2 + 2\alpha\beta\delta)\|\mathbf{v}\|^2_{\mathrm{B}^{-1}} \\
&= \mathbf{v}^T(A^{-1} - (2\alpha - \alpha^2\lambda_{\max}(\mathrm{B}^{-1}A)\delta^2 - 2\alpha\beta\delta)\mathrm{B}^{-1} \\
&\quad + \alpha^2\, \mathrm{B}^{-1}A\mathrm{B}^{-1})\mathbf{v} \\
&\leq \max_{t \in [\lambda_{\min}(\mathrm{B}^{-1}A),\ \lambda_{\max}(\mathrm{B}^{-1}A)]} Q(t)\, \mathbf{v}^T A^{-1}\mathbf{v},
\end{aligned}$$

where

$$Q(t) = 1 - (2\alpha - \alpha^2 \lambda_{\max}(\mathrm{B}^{-1}A)\delta^2 - 2\alpha\beta\delta)t + \alpha^2\, t^2. \tag{10.8}$$

Now choose

$$\alpha = \frac{2}{(1-\delta)\lambda_{\min}(\mathrm{B}^{-1}A) + \frac{\lambda_{\max}(\mathrm{B}^{-1}A)+\delta^2(\lambda_{\max}(\mathrm{B}^{-1}A)-\lambda_{\min}(\mathrm{B}^{-1}A))}{1-\delta}}.$$

This choice ensures that

$$\alpha \in \left(0,\ \frac{2}{\lambda_{\min}(\mathrm{B}^{-1}A) + \lambda_{\max}(\mathrm{B}^{-1}A)}\right),$$

which implies (see (10.7)) that

$$\beta = 1 - \alpha\ \lambda_{\min}(\mathrm{B}^{-1}A).$$

We also have

$$\begin{aligned}
&2\alpha - \alpha^2 \lambda_{\max}(\mathrm{B}^{-1}A)\delta^2 - 2\alpha\beta\delta \\
&\quad = 2\alpha - \alpha^2 \lambda_{\max}(\mathrm{B}^{-1}A)\delta^2 - 2\alpha(1 - \alpha\ \lambda_{\min}(\mathrm{B}^{-1}A))\delta \\
&\quad = \alpha\ (2(1-\delta) - \alpha\lambda_{\max}(\mathrm{B}^{-1}A)\delta^2 + 2\alpha\lambda_{\min}(\mathrm{B}^{-1}A)\delta) \\
&\quad = \alpha\ (\lambda_{\min}(\mathrm{B}^{-1}A) + \lambda_{\max}(\mathrm{B}^{-1}A)).
\end{aligned}$$

To see the last equality, it is equivalent to show that

$$\begin{aligned}
2(1-\delta) &= \alpha\ (\lambda_{\min}(\mathrm{B}^{-1}A) + \lambda_{\max}(\mathrm{B}^{-1}A) - \delta(2\lambda_{\min}(\mathrm{B}^{-1}A) - \delta\ \lambda_{\max}(\mathrm{B}^{-1}A))) \\
&= \alpha\ (\lambda_{\max}(\mathrm{B}^{-1}A) + \delta^2\ (\lambda_{\max}(\mathrm{B}^{-1}A) - \lambda_{\min}(\mathrm{B}^{-1}A)) \\
&\quad + \lambda_{\min}(\mathrm{B}^{-1}A)\ (1-\delta)^2).
\end{aligned}$$

The latter equation coincides exactly with the definition of α,

$$\alpha = \frac{2}{\lambda_{\min}(\mathrm{B}^{-1}A)(1-\delta) + \frac{\lambda_{\max}(\mathrm{B}^{-1}A)+\delta^2\ (\lambda_{\max}(\mathrm{B}^{-1}A)-\lambda_{\min}(\mathrm{B}^{-1}A))}{1-\delta}}.$$

With the choice of α we have made, the quadratic form $Q(t)$ from (10.8) simplifies to

$$Q(t) = 1 - \alpha\ (\lambda_{\min}(\mathrm{B}^{-1}A) + \lambda_{\max}(\mathrm{B}^{-1}A))t + t^2\alpha^2.$$

It is clear then that for $t \in [\lambda_{\min}(\mathrm{B}^{-1}A),\ \lambda_{\max}(\mathrm{B}^{-1}A)]$, we have

$$\begin{aligned}
Q(t) &= 1 - \alpha^2\lambda_{\min}(\mathrm{B}^{-1}A)\lambda_{\max}(\mathrm{B}^{-1}A) + \alpha^2(t - \lambda_{\min}(\mathrm{B}^{-1}A))(t - \lambda_{\min}(\mathrm{B}^{-1}A)) \\
&\le 1 - \alpha^2\lambda_{\min}(\mathrm{B}^{-1}A)\lambda_{\max}(\mathrm{B}^{-1}A).
\end{aligned}$$

Therefore, the final estimate reads,

$$\begin{aligned}\|\mathbf{v} - \alpha AB[\mathbf{v}]\|_{A^{-1}}^2 &\leq \left[1 - \alpha^2 \lambda_{\min}(\mathrm{B}^{-1}A)\lambda_{\max}(\mathrm{B}^{-1}A)\right]\mathbf{v}^T A^{-1}\mathbf{v} \\ &= \left(1 - \frac{4\kappa(1-\delta)^2}{((\kappa-1)\delta^2 + (1-\delta)^2 + \kappa)^2}\right) \|\mathbf{v}\|_{A^{-1}}^2. \end{aligned}$$ □

10.3 Variable-step multilevel preconditioners

Consider a sequence of matrices $\{A_k\}$ related in the following hierarchical fashion,

$$A_k = \begin{bmatrix} \mathcal{A}_k & \mathcal{R}_k \\ \mathcal{L}_k & A_{k+1} \end{bmatrix}.$$

If the above two-by-two block form of A_k is stable, which in particular means that A_k is spectrally equivalent to its block-diagonal part, and hence, A_{k+1} is spectrally equivalent to the exact Schur complement $\mathcal{S}_k = A_{k+1} - \mathcal{L}_k \mathcal{A}_k^{-1} \mathcal{R}_k$, the following two-level preconditioner is viable.

$$B_{TL} = \begin{bmatrix} \mathcal{M}_k & 0 \\ \mathcal{L}_k & I \end{bmatrix} \begin{bmatrix} \left(\mathcal{M}_k + \mathcal{M}_k^T - \mathcal{A}_k\right)^{-1} & 0 \\ 0 & A_{k+1} \end{bmatrix} \begin{bmatrix} \mathcal{M}_k^T & \mathcal{R}_k \\ 0 & I \end{bmatrix},$$

where $\mathcal{M}_k$ comes from a convergent splitting for $\mathcal{A}_k$ such that

$$\|I - \mathcal{A}_k^{1/2} \mathcal{M}_k^{-1} \mathcal{A}_k^{1/2}\| < 1.$$

We recall also the symmetrized preconditioners

$$\overline{\mathcal{M}}_k = \mathcal{M}_k \left(\mathcal{M}_k + \mathcal{M}_k^T - \mathcal{A}_k\right)^{-1} \mathcal{M}_k^T.$$

The following recursive multilevel procedure can be utilized to define a multilevel factorization variable-step preconditioner of guaranteed quality.

Let $B_{k+1}[.]$ be a given (defined by induction) variable-step preconditioner for A_{k+1} and consider the better quality preconditioner obtained by $\nu = \nu_k \geq 0$ steps of the variable-step preconditioned CG algorithm 10.2.1, $B_{k+1}^{(\nu)}$. Then the kth-level variable-step preconditioner approximating A_k^{-1}, is defined as follows.

$$B_k[.] = \begin{bmatrix} I & -\mathcal{M}_k^{-T} \mathcal{R}_k \\ 0 & I \end{bmatrix} \begin{bmatrix} \overline{\mathcal{M}}_k^{-1} & 0 \\ 0 & B_{k+1}^{(\nu)}[.] \end{bmatrix} \begin{bmatrix} I & 0 \\ -\mathcal{L}_k \mathcal{M}_k^{-1} & I \end{bmatrix}.$$

Based on the result of the preceding section, we can guarantee a fixed quality of the preconditioner at every level k by properly choosing $\nu = \nu_k$ at every level. In particular, if we knew that there is a fixed (linear) multilevel preconditioner with a guaranteed quality then, the nonlinear one will also have a guaranteed quality, as well, because the estimate for $\nu = \nu_k$ given in (10.6) will be level-independent.

The nonlinear preconditioner has a potential advantage of being a parameter (to estimate) free one.

10.4 Variable-step AMLI-cycle MG

Here we construct recursively nonlinear approximate inverses to A_k used in an AMLI-cycle MG.

We first define a variable-step AMLI-cycle MG.

Definition 10.5 (Variable–step AMLI–cycle MG). *Let A_k, P_k, and M_k for $k = 0, \ldots, \ell$ where $A_{k+1} = P_k^T A_k P_k$, be the parameters of a MG hierarchy. Introduce also the symmetrized smoothers $\overline{M}_k = M_k(M_k^T + M_k - A_k)^{-1} M_k^T$. The AMLI-cycle also exploits a sequence of integers $\nu_k \geq 0$, $k = 0,\ 1, \ldots, \ell$.*

At the coarsest level, set $B_\ell = A_\ell^{-1}$. Then, assume that $B_{k+1}[\cdot]$ for some $k < \ell$ has been defined as an approximate inverse to A_{k+1}. On its basis construct an iterated one, $B_{k+1}^{(\nu_k)}[\cdot]$ implemented as in Algorithm 10.2.1, letting $B[\cdot] = B_{k+1}[\cdot]$ as input and $B_{k+1}^{(\nu_k)}[\cdot] = B_{\nu_k}[\cdot]$ as output. If $\nu_k = 0$, we simply let $B_{k+1}^{(0)}[\cdot] = B_{k+1}[\cdot]$; that is, we do not use Algorithm 10.2.1.

Then, define first

$$\overline{B}_k[\cdot] = \begin{bmatrix} I & -M_k^{-T} A_k P_k \\ 0 & I \end{bmatrix} \begin{bmatrix} \overline{M}_k^{-1} & 0 \\ 0 & B_{k+1}^{(\nu_k)}[\cdot] \end{bmatrix} \begin{bmatrix} I & 0 \\ -P_k^T A_k M_k^{-1} & I \end{bmatrix},$$

and then for the approximate inverse of A_k let

$$\begin{aligned} B_k[\cdot] &= [I,\ P_k]\, \overline{B}_k[\cdot] \begin{bmatrix} I \\ P_k^T \end{bmatrix} \\ &= \begin{bmatrix} I, & (I - M_k^{-T} A_k) P_k \end{bmatrix} \begin{bmatrix} \overline{M}_k^{-1} & 0 \\ 0 & B_{k+1}^{(\nu_k)}[\cdot] \end{bmatrix} \begin{bmatrix} I \\ P_k^T (I - A_k M_k^{-1}) \end{bmatrix} \\ &= \overline{M}_k^{-1} + (I - M_k^{-T} A_k) P_k B_{k+1}^{(\nu_k)} \left[P_k^T (I - A_k M_k^{-1})(\cdot) \right]. \end{aligned}$$

The following monotonicity property holds (similarly to Theorem 10.1).

Lemma 10.6. *Consider the (linear) MG preconditioner B_k defined as follows,*

$$B_k^{-1} = \overline{M}_k^{-1} + (I - M_k^{-T} A_k) P_k B_{k+1}^{-1} P_k^T (I - A_k M_k^{-1}).$$

The deviation $B_k^{-1}\mathbf{v} - B_k[\mathbf{v}]$ does not increase from level $k+1$ to level k; that is, we have

$$\| B_k^{-1}\mathbf{v} - B_k[\mathbf{v}] \|_{B_k} \leq \| B_{k+1}^{-1} \overline{\mathbf{v}} - B_{k+1}[\overline{\mathbf{v}}] \|_{B_{k+1}}. \tag{10.9}$$

Here, $\overline{\mathbf{v}} = P_k^T (I - A_k M_k^{-1})\mathbf{v}$, and it can be estimated as

$$\| \overline{\mathbf{v}} \|_{B_{k+1}^{-1}} \leq \| \mathbf{v} \|_{B_k^{-1}}. \tag{10.10}$$

Proof. We have

$$\begin{aligned}\|\overline{\mathbf{v}}\|^2_{B^{-1}_{k+1}} &= \left\|B^{-(1/2)}_{k+1} P^T_k \left(I - A_k M^{-1}_k\right)\mathbf{v}\right\|^2 \\ &\le \mathbf{v}^T \overline{M}_k \mathbf{v} + \mathbf{v}^T \left(I - M^{-T}_k A_k\right) P_k B^{-1}_{k+1} P^T_k \left(I - A_k M^{-1}_k\right)\mathbf{v} \\ &= \|\mathbf{v}\|^2_{B^{-1}_k},\end{aligned}$$

which proves (10.10). It also shows that $\|B^{-(1/2)}_{k+1} P^T_k (I - A_k M^{-1}_k) B^{1/2}_k\| \le 1$. For the deviation in question, we have

$$\begin{aligned}\|B^{-1}_k \mathbf{v} - B_k[\mathbf{v}]\|^2_{B_k} &= \left\|\left(I - M^{-T}_k A_k\right) P_k \left(B^{-1}_{k+1}\overline{\mathbf{v}} - B_{k+1}[\overline{\mathbf{v}}]\right)\right\|_{B_k} \\ &\le \left\|B^{1/2}_k \left(I - M^{-T}_k A_k\right) P_k B^{-(1/2)}_{k+1}\right\| \left\|B^{-1}_{k+1}\overline{\mathbf{v}} - B_{k+1}[\overline{\mathbf{v}}]\right\|_{B_{k+1}} \\ &= \left\|B^{-(1/2)}_{k+1} P^T_k \left(I - A_k M^{-1}_k\right) B^{1/2}_k\right\| \left\|B^{-1}_{k+1}\overline{\mathbf{v}} - B_{k+1}[\overline{\mathbf{v}}]\right\|_{B_{k+1}} \\ &\le \left\|B^{-1}_{k+1}\overline{\mathbf{v}} - B_{k+1}[\overline{\mathbf{v}}]\right\|_{B_{k+1}},\end{aligned}$$

which proves the desired monotonicity property (10.9). □

We are now ready to prove our main result (see [NV07]).

Theorem 10.7. *Given an integer parameter* $k_0 \ge 1$ *and another integer* $\nu \ge 1$. *Let* $\nu_k = \nu$ *for* $k = sk_0$, $s = 1, \ldots, [\ell/k_0]$, *and* $\nu_k = 0$ *otherwise. Consider for a given MG hierarchy of matrices* $\{A_k\}$, *interpolation matrices* $\{P_k\}$ *such that* $A_{k+1} = P^T_k A_k P_k$, *and smoothers* $\{M_k\}$. *They define fixed-length symmetric* $V(1,1)$*-cycle MG matrices* $B^{(k+k_0)\mapsto k}_{MG}$, *from any coarse-level* $k + k_0$ *to level* k *with exact solution at level* $k + k_0$. *Assume that the convergence factor of such fixed length V-cycles are uniformly in* $k \ge 0$ *bounded by a* $\delta_{k_0} \in [0, 1)$. *Let* ν *and* k_0 *be related such that the inequality*

$$(1 - (1 - \delta^2)(1 - \delta_{k_0}))^{\nu/2} \le \delta \tag{10.11}$$

has a solution $\delta \in (0, 1)$. *A sufficient condition for this is*

$$\nu > \frac{1}{1 - \delta_{k_0}}. \tag{10.12}$$

Then the variable-step AMLI-cycle MG as defined in Definition 10.5 for the sequence $\{\nu_k\}$ *above, provides an approximate inverse for* A_k *with guaranteed quality* δ; *that is, we have the uniform deviation estimate*

$$\|A^{-1}_k \mathbf{v} - B^{(\nu)}_k[\mathbf{v}]\|_{A_k} \le \delta\ \|\mathbf{v}\|_{A^{-1}_k}.$$

Proof. We first show that (10.12) implies the existence of a $\delta \in [0, 1)$, which solves (10.11). Indeed, letting $\overline{\delta} = \delta^{2/\nu}$ inequality (10.11) reads

$$1 - (1 - \overline{\delta}^{\nu})(1 - \delta_{k_0}) \le \overline{\delta},$$

or equivalently,

$$\varphi(\overline{\delta}) \equiv 1 - (1 - \delta_{k_0})\,(1 + \overline{\delta} + \cdots + \overline{\delta}^{\nu-1}) \le 0.$$

Because $\varphi(1) = 1 - (1 - \delta_{k_0})\nu < 0$ (due to (10.12)) and $\varphi(0) = 1 - (1 - \delta_{k_0}) = \delta_{k_0} > 0$, there is a $\overline{\delta} \in [0, 1)$ such that $\varphi(\overline{\delta}) = 0$. Hence any $\delta \in [\overline{\delta}^{2/\nu}, 1)$ will satisfy (10.11).

Applying Lemma 10.6 recursively, we end up with the deviation estimate

$$\|B_k^{-1}\mathbf{v} - B_k[\mathbf{v}]\|_{B_k} \le \|B_{k+k_0}^{-1}\overline{\mathbf{v}} - B_{k+k_0}^{(\nu)}[\overline{\mathbf{v}}]\|_{B_{k+k_0}}$$

for a vector $\overline{\mathbf{v}}$ such that

$$\|\overline{\mathbf{v}}\|_{B_{k+k_0}^{-1}} \le \|\mathbf{v}\|_{B_k^{-1}}. \tag{10.13}$$

Because at level $k + k_0$ we use an exact solution (in the definition of B_k^{-1}) (i.e., $B_{k+k_0} = A_{k+k_0}$), the above estimate becomes

$$\|B_k^{-1}\mathbf{v} - B_k[\mathbf{v}]\|_{B_k} \le \|A_{k+k_0}^{-1}\overline{\mathbf{v}} - B_{k+k_0}^{(\nu)}[\overline{\mathbf{v}}\|_{A_{k+k_0}}. \tag{10.14}$$

Assume now by induction that there is a $\delta \in [0, 1)$ such that

$$\|A_{k+k_0}^{-1}\overline{\mathbf{v}} - B_{k+k_0}^{(\nu)}[\overline{\mathbf{v}}]\|_{A_{k+k_0}} \le \delta\ \|\overline{\mathbf{v}}\|_{A_{k+k_0}^{-1}}.$$

The last estimate, together with (10.14) and (10.13), implies that

$$\|B_k^{-1}\mathbf{v} - B_k[\mathbf{v}]\|_{B_k^{-1}} \le \delta\ \|\mathbf{v}\|_{B_k^{-1}}.$$

By assumption, the k_0th-length V-cycle has a certain quality, such as

$$\left\|(B_k^{-1} - A_k^{-1})\mathbf{v}\right\|_{A_k} \le \delta_{k_0}\ \|\mathbf{v}\|_{A_k^{-1}}.$$

That is, $\kappa_{k_0} = 1/(1 - \delta_{k_0})$ is a (uniform) upper bound on the condition number of B_k with respect to A_k. Corollary 10.3 implies then the following convergence estimate for the iterated nonlinear mapping $B_k^{(\nu)}[\cdot]$,

$$\|A_k^{-1}\mathbf{v} - B_k^{(\nu)}[\mathbf{v}]\|_{A_k} \le \left(1 - \frac{1-\delta^2}{\kappa_{k_0}}\right)^{\nu/2} \|\mathbf{v}\|_{A_k^{-1}}.$$

To confirm the induction assumption, we need the inequality

$$\left(1 - \frac{1-\delta^2}{\kappa_{k_0}}\right)^{\nu/2} \le \delta,$$

which as we already shown has a solution for $\nu > \kappa_{k_0} = 1/(1 - \delta_{k_0})$. Thus the proof is complete. □

Theorem 10.7 assumes that $\nu > 1/(1-\delta_{k_0})$, but increasing k_0 may deteriorate the k_0th-length $V(1,1)$-cycle in general; that is, δ_{k_0} can get closer to unity and hence ν needs to be chosen sufficiently large. However, if ν gets too large then the complexity of the AMLI-cycle may become unacceptable. To address both the quality of the variable-step AMLI-cycle MG and its complexity, consider now the example of matrices A_k coming from second-order finite element elliptic equations posed on a domain $\Omega \subset \mathbb{R}^d$, $d = 2$ or $d = 3$. In that case, δ_{k_0} has the following asymptotic behavior (cf., Section 5.6.2),

$$K_{MG}^{(k+k_0)\mapsto k} \le \kappa_{k_0} = \frac{1}{1-\delta_{k_0}} \simeq \begin{cases} k_0^2, & d = 2, \\ 2^{k_0}, & d = 3. \end{cases} \tag{10.15}$$

Then, the following result holds.

Corollary 10.8. *Consider the variable-step AMLI-cycle as defined in Theorem 10.7 for matrices A_k coming from second-order elliptic finite element equations on uniformly refined meshes and M_k being the Gauss–Seidel smoother, or any other smoother giving rise to $\overline{M}_k$ that is spectrally equivalent to the diagonal of A_k. The second-order elliptic PDE is assumed to have coefficients that vary smoothly within each element from the coarsest triangulation $\mathcal{T}_H = \mathcal{T}_\ell$. Assume that $h_{k+1} = 2h_k$ where $h_0 = h$ is the finest mesh-size and $h_\ell = H$ is the coarsest mesh-size. Finally, assume that the number of dofs at level k are $n_k = 2^d\ n_{k+1}$, where $d = 2$ or $d = 3$ is the dimension of the domain (where the PDE is posed). Then, we can select $\nu < 2^{dk_0}$ for k_0 sufficiently large so that the inequality* (10.11) *has a solution $\delta \in (0,1)$. This choice of ν guarantees uniform quality of the variable-step preconditioner $B_k[\cdot]$ and at the same time ensures its optimal complexity.*

Proof. We have the following asymptotic inequality for ν coming from (10.12) as $\delta_{k_0} \mapsto 1$ based on (10.15),

$$\nu > \frac{1}{(1-\delta_{k_0})} \simeq \begin{cases} k_0^2, & d = 2, \\ 2^{k_0}, & d = 3. \end{cases}$$

From complexity restrictions, we have (estimated in the same way as in Section 5.6.4) $\nu < 2^{dk_0}$. It is clear then that in both cases, $d = 2$ and $d = 3$, we can choose ν, for k_0 sufficiently large, such that the variable-step AMLI-cycle MG are of fixed quality δ (from (10.11)) and at the same time have optimal complexity. □

We remark at the end that because δ_{k_0} is bounded independently of possible jumps in the coefficients of the PDE (which are assumed to vary smoothly within the elements of the initial coarse triangulation $\mathcal{T}_H$) the resulting $\delta \in (0,1)$ will also be coefficient independent.

11

Preconditioning Nonlinear Problems

11.1 Problem formulation

We are interested in the following nonlinear operators,

$$A[.] = A_0 + B[.],$$

where $B[.]$ is treated as a perturbation to the linear one A_0 similarly to the case of nonsymmetric and possibly indefinite matrices we considered in Chapter 8. This means that in a norm $\|.\|_0$ coming from an inner product $(\cdot,\ \cdot)_0$, we have

$$(A[\mathbf{v}],\ \mathbf{v})_0 \geq (1 - \gamma_0)(A_0\mathbf{v},\ \mathbf{v})_0 - c_0\|\mathbf{v}\|_0^2. \tag{11.1}$$

The linear operator A_0 is assumed coercive in the norm $\|.\|_0$; that is, $\|\mathbf{v}\|_0^2 \leq \Delta(A_0\mathbf{v},\ \mathbf{v})_0$ for a constant $\Delta > 0$. We also assume that A_0 is $(\cdot,\ \cdot)_0$-symmetric which together with its coercivity imples that $\|\mathbf{v}\|_{A_0} \equiv ((A_0\mathbf{v},\ \mathbf{v})_0)^{1/2}$ is a norm stronger than $\|.\|_0$. In the analysis to follow we make use of a third norm $\|.\|$ which is assumed stronger than $\|.\|_{A_0}$. To avoid technical details we assume that $B[.]$ is positive; that is

$$(B[\mathbf{v}],\ \mathbf{v})_0 \geq 0. \tag{11.2}$$

Then the estimate (11.1) is trivially satisfied with $\gamma_0 = c_0 = 0$.

We are interested in the solution of the nonlinear problem $A[\mathbf{u}] = \mathbf{f}$ which we assume is uniquely solvable. We make some assumptions about differentiability of A in a neighborhood of the exact solution $\mathbf{u}^\star$ of $A[\mathbf{u}] = \mathbf{f}$, as well as on the approximation of a coarse-grid solution $P\mathbf{u}_c$ defined variationally by the identity $(A[P\mathbf{u}_c],\ P\mathbf{y})_0 = (\mathbf{f},\ P\mathbf{y})_0$ for any coarse vector $\mathbf{y}$.

More specifically, the nonlinear mapping $B[.]$ is assumed differentiable in the sense that for some $\sigma > 0$, for any $\mathbf{g}$ uniformly with respect $\mathbf{v}_0$ in a ball near the exact solution $A[\mathbf{u}^*] = \mathbf{f}$, we have

$$B[\mathbf{v}_0 + \mathbf{g}] = B[\mathbf{v}_0] + B'(\mathbf{v}_0)\mathbf{g} + \mathcal{O}(\|\mathbf{g}\|^{1+\sigma}).$$

P.S. Vassilevski, *Multilevel Block Factorization Preconditioners*,
doi: 10.1007/978-0-387-71564-3_11,

Here, $B'(\mathbf{v}_0)$ is the derivative of $B[.]$ at $\mathbf{v}_0$. More specifically, for the derivative A' (or B'), we assume that for any $\mathbf{v}_0$ in a ball near the exact solution $\mathbf{u}^*$ of $A[\mathbf{u}] = \mathbf{f}$, we have the estimate

$$(A[\mathbf{v}_0 + \mathbf{g}] - A[\mathbf{v}_0] - A'(\mathbf{v}_0)\mathbf{g},\ \mathbf{v})_0 \leq \gamma\ \|\mathbf{g}\|\|\mathbf{v}\|_0.$$

Here, $\|.\|$ is the norm (introduced above) that is stronger than $\|.\|_{A_0}$. The latter inequality implies

$$\|A[\mathbf{v}_0 + \mathbf{g}] - A[\mathbf{v}_0] - A'(\mathbf{v}_0)\mathbf{g}\|_0 \leq \gamma\ \|\mathbf{g}\|. \tag{11.3}$$

Because $A[.] = A_0 + B[.]$, we also have

$$(B[\mathbf{v}_0 + \mathbf{g}] - B[\mathbf{v}_0] - B'(\mathbf{v}_0)\mathbf{g},\ \mathbf{v})_0 \leq \gamma\ \|\mathbf{g}\|\|\mathbf{v}\|_0. \tag{11.4}$$

In some cases (as in semilinear second-order elliptic PDEs), we can actually prove a stronger estimate,

$$(B[\mathbf{v}_0 + \mathbf{g}] - B[\mathbf{v}_0] - B'(\mathbf{v}_0)\mathbf{g},\ \mathbf{v})_0 \leq L\|\mathbf{g}\|\|\mathbf{g}\|_0\|\mathbf{v}\|_0. \tag{11.5}$$

Then for $\|\mathbf{g}\|_0$ sufficiently small, we can achieve $L\|\mathbf{g}\| \leq \gamma < 1$. Finally, we assume that the derivative B' is continuous; that is,

$$((B'(\mathbf{u}_0) - B'(\mathbf{v}_0))\mathbf{g},\ \mathbf{v})_0 \leq L\|\mathbf{u}_0 - \mathbf{v}_0\|\|\mathbf{g}\|_0\|\mathbf{v}\|_0,$$

for any $\mathbf{v}_0$ and $\mathbf{u}_0$ close to the exact solution $\mathbf{u}^\star$ of $A[\mathbf{u}] = \mathbf{f}$. This in particular implies

$$\|(A'(\mathbf{u}_0) - A'(\mathbf{v}_0))\mathbf{g}\|_0 \leq L\|\mathbf{u}_0 - \mathbf{v}_0\|\|\mathbf{g}\|_0. \tag{11.6}$$

The following error estimate is our next assumption. Consider the coarse problem, for any given $\mathbf{v}_0$ in a small neighborhood of the exact solution $A[\mathbf{u}^*] = \mathbf{f}$,

$$(A[P\mathbf{x}],\ P\mathbf{y})_0 = (A[\mathbf{v}_0],\ P\mathbf{y})_0, \quad \text{for all } \mathbf{y}. \tag{11.7}$$

Then, we assume that for a small $\delta > 0$, which gets smaller with increasing the size of the coarse problem,

$$\|\mathbf{v}_0 - P\mathbf{x}\|_0 \leq \delta\ \|A[\mathbf{v}_0]\|_0.$$

We also assume an error estimate in the stronger norm $\|.\|$, namely, for a small $\alpha < 1$,

$$\|\mathbf{v}_0 - P\mathbf{x}\| \leq \delta^\alpha\ \|A[\mathbf{v}_0]\|_0. \tag{11.8}$$

We prove next an a priori estimate for the solution $P\mathbf{x}$ of the coarse problem. Because the norm $\|.\|_0$ is weaker than $\|.\|_{A_0}$, we have

$$\begin{aligned}
\Delta^{-1}\|P\mathbf{x}\|_0^2 &\leq (A_0 P\mathbf{x},\ P\mathbf{x})_0 \\
&\leq (A_0 P\mathbf{x} + B[P\mathbf{x}],\ P\mathbf{x})_0 \\
&= (A[P\mathbf{x}],\ P\mathbf{x})_0 \\
&= (A[\mathbf{v}_0],\ P\mathbf{x})_0 \\
&\leq \|A[\mathbf{v}_0]\|_0\|P\mathbf{x}\|_0.
\end{aligned}$$

Therefore,

$$\|P\mathbf{x}\|_0 \leq \Delta \ \|A[\mathbf{v}_0]\|_0.$$

All the assumptions made in the present section can be verified for a certain class of semilinear second-order elliptic PDEs and proper choice of the norms $\|.\|_0$ and $\|.\|$, cf. [BVW03] or Section B.2.

11.2 Choosing an accurate initial approximation

Under the assumption made in the previous section, consider the linearized problem

$$(A_0 + B'(P\mathbf{x}))\mathbf{u} = \mathbf{r} \equiv A[\mathbf{v}_0] - B[P\mathbf{x}] + B'(P\mathbf{x})P\mathbf{x}. \tag{11.9}$$

It approximates the nonlinear problem $A[\mathbf{v}] = A[\mathbf{v}_0]$. Due to (11.3) and the stronger error estimate (11.8), we have

$$\|\mathbf{r}\|_0 = \|A[\mathbf{v}_0] - B[P\mathbf{x}] + B'(P\mathbf{x})P\mathbf{x}\|_0 \leq \gamma \|\mathbf{v}_0 - P\mathbf{x}\| \leq \gamma \ \delta^\alpha \ \|A[\mathbf{v}_0]\|_0. \tag{11.10}$$

The difference $\mathbf{v}_0 - \mathbf{u}$ solves the linear system

$$\begin{aligned} A'(P\mathbf{x})(\mathbf{v}_0 - \mathbf{u}) &= A'(P\mathbf{x})\mathbf{v}_0 - A[\mathbf{v}_0] + B[P\mathbf{x}] - B'(P\mathbf{x})P\mathbf{x} \\ &= A'(P\mathbf{x})(\mathbf{v}_0 - P\mathbf{x}) - A[\mathbf{v}_0] + A'(P\mathbf{x})P\mathbf{x} \\ &\quad + A[P\mathbf{x}] - A'(P\mathbf{x})P\mathbf{x} \\ &= -A[\mathbf{v}_0] + A[P\mathbf{x}] + A'(P\mathbf{x})(\mathbf{v}_0 - P\mathbf{x}). \end{aligned}$$

Therefore, based on (11.3), (11.6), and the error estimates for $P\mathbf{x} - \mathbf{v}_0$,

$$\begin{aligned} \|A'(P\mathbf{x})(\mathbf{v}_0 - \mathbf{u})\|_0 &\leq \|(A'(P\mathbf{x}) - A'(\mathbf{v}_0))(\mathbf{v}_0 - P\mathbf{x})\|_0 \\ &\quad + \| - A[\mathbf{v}_0] + A[P\mathbf{x}] + A'(\mathbf{v}_0)(\mathbf{v}_0 - P\mathbf{x})\|_0 \\ &\leq \gamma \|\mathbf{v}_0 - P\mathbf{x}\| + L \ \|\mathbf{v}_0 - P\mathbf{x}\| \|\mathbf{v}_0 - P\mathbf{x}\|_0 \\ &\leq \delta^\alpha (\gamma + L\delta) \ \|A[\mathbf{v}_0]\|_0. \end{aligned}$$

The linear system (11.9) can be solved by $\nu \geq 0$ iterations thus ending up with a sufficiently accurate approximation $\mathbf{u}_0$ to $\mathbf{u}$ such that (using (11.10))

$$\|A'(P\mathbf{x})(\mathbf{u}_0 - \mathbf{u})\|_0 \leq C \frac{1}{1+\nu} \|\mathbf{r}\|_0 \leq C \ \frac{1}{1+\nu} \ \|A[\mathbf{v}_0]\|_0.$$

Note that here, we need an iterative method that reduces the residual in the weaker norm $\|.\|_0$ with an optimal rate. In the application of semilinear second-order elliptic equations giving rise to s.p.d. matrices $A'(P\mathbf{x}) = A_0 + B'(P\mathbf{x})$, we may use the cascadic MG to get optimal convergence for the residuals by increasing the number of smoothing steps (for details, see Section 5.8.1).

Recall that $A'(P\mathbf{x})$ is coercive, hence, we have the estimate $\|\mathbf{g}\|_0^2 \leq \Delta\ (A'(P\mathbf{x})\mathbf{g}, \mathbf{g})_0 \leq \Delta\ \|A'(P\mathbf{x})\mathbf{g}\|_0 \|\mathbf{g}\|_0$. That is,

$$\|\mathbf{g}\|_0 \leq \Delta\ \|A'(P\mathbf{x})\mathbf{g}\|_0.$$

Using this coercivity estimate, and because $P\mathbf{x}$ and $\mathbf{v}_0$ are close, then $A'(\mathbf{v}_0)$ and $A'(P\mathbf{x})$ are also close (see (11.6)). Therefore, for $\gamma = L\Delta\ \delta^\alpha \|A[\mathbf{v}_0]\|_0$, we have

$$\begin{aligned}
\|A'(\mathbf{v}_0)(\mathbf{v}_0 - \mathbf{u}_0)\|_0 &\leq \|(A'(\mathbf{v}_0) - A'(P\mathbf{x}))(\mathbf{v}_0 - \mathbf{u}_0)\|_0 + \|A'(P\mathbf{x}))(\mathbf{v}_0 - \mathbf{u}_0)\|_0 \\
&\leq \|A'(P\mathbf{x})(\mathbf{v}_0 - \mathbf{u}_0)\|_0 + L\ \|\mathbf{v}_0 - P\mathbf{x}\|\,\|\mathbf{v}_0 - \mathbf{u}_0\|_0 \\
&\leq \|A'(P\mathbf{x})(\mathbf{v}_0 - \mathbf{u}_0)\|_0 + L\delta^\alpha \|A[\mathbf{v}_0]\|_0 \Delta\ \|A'(P\mathbf{x})(\mathbf{v}_0 - \mathbf{u}_0\|_0 \\
&\leq (1 + L\Delta\ \delta^\alpha \|A[\mathbf{v}_0]\|_0)\ \|A'(P\mathbf{x})(\mathbf{v}_0 - \mathbf{u}_0\|_0 \\
&\leq (1+\gamma)\|A'(P\mathbf{x})(\mathbf{v}_0 - \mathbf{u})\|_0 + (1+\gamma)\|A'(P\mathbf{x})(\mathbf{u} - \mathbf{u}_0)\|_0 \\
&\leq (1+\gamma)C\frac{1}{1+\nu}\|\mathbf{r}\|_0 + (1+\gamma)\gamma\delta^\alpha \|A[\mathbf{v}_0]\|_0 \\
&\leq (1+\gamma)\max\left\{\frac{C}{1+\nu},\ \gamma\delta^\alpha\right\}\ \|A[\mathbf{v}_0]\|_0 \\
&\leq \eta < 1,
\end{aligned}$$

for any $\eta < 1$ chosen a priori.

In summary, consider the nonlinear problem,

$$A[\mathbf{v}_0] = \mathbf{f}.$$

Here, $\mathbf{f}$ is given and $\mathbf{v}_0$ unknown. We can find a sufficiently accurate approximation $\mathbf{u}_0$ to $\mathbf{v}_0$ in the following steps. First by solving the coarse nonlinear problem (11.7), we obtain $P\mathbf{x}$. Then, we can form the linearized fine-grid problem (11.9) and solve it approximately by $\nu \geq 0$ iterations thus ending up with $\mathbf{u}_0$. Then, for any a priori chosen $\eta < 1$, if the coarse problem is sufficiently accurate (hence, we have sufficiently small δ in the error estimates), and if ν is sufficiently large, the approximation $\mathbf{u}_0$ will be sufficiently close to the unknown solution $\mathbf{v}_0$ in the sense that the following estimate holds:

$$\|A'(\mathbf{v}_0)(\mathbf{v}_0 - \mathbf{u}_0)\|_0 \leq (1+\gamma)\max\left\{\frac{C}{1+\nu},\ \gamma\delta^\alpha\right\}\ \|\mathbf{f}\|_0 \leq \eta < 1. \tag{11.11}$$

The thus-constructed approximation $\mathbf{u}_0$ can be used as a sufficiently accurate initial guess in an inexact Newton method that we present in the following section.

General two-level discretization schemes for certain finite element problems were presented in [Xu96a]. They provide accurate approximations from a coarse space, and at the fine-level, we need to solve a linearized problem only.

11.3 The inexact Newton algorithm

Consider the nonlinear equation,

$$A[\mathbf{u}] = \mathbf{f}.$$

The nonlinear mapping $A[.]$ is considered as a mapping from a given (infinite-dimensional) space $\mathcal{X}$ equipped with a strong norm $\|.\|$ to another (infinite-dimensional) space $\mathcal{Y}$ equipped with a weaker norm $\|.\|_0$. We assume that $\mathcal{X} \subset \mathcal{Y}$. We recall that a discrete counterpart of $A[\cdot]$ is sometimes denoted by F.

Let $\mathbf{u}_0$ be an accurate initial approximation to $\mathbf{u}^\star$, the exact solution of the above problem.

Algorithm 11.3.1 (Modified inexact Newton method).

- *For* $n = 0, 1, \ldots,$ *until convergence, compute the inexact Newton correction* $\mathbf{s}_n$ *such that*

$$A'(\mathbf{u}_0)\mathbf{s}_n = \mathbf{f} - A[\mathbf{u}_n] + \mathbf{r}_n \quad \textit{where } \|\mathbf{r}_n\|_0 \leq \eta \|A[\mathbf{u}_n] - \mathbf{f}\|_0. \tag{11.12}$$

- *Then, set*

$$\mathbf{u}_{n+1} = \mathbf{u}_n + \mathbf{s}_n. \tag{11.13}$$

We make now the following main assumptions.

(A1) The mapping $A[.]$ acting from the space $(\mathcal{X}, \|.\|) \mapsto (\mathcal{X}, \|.\|_0)$ is invertible in a small neighborhood of a given $\mathbf{f} \in (\mathcal{X}, \|.\|_0)$. Let $\mathbf{u}^\star$ be the exact solution of $A[\mathbf{u}] = \mathbf{f}$.

(A2) The mapping $A[.]$ is differentiable in a neighborhood of $\mathbf{u}^\star$ and $(A'(\mathbf{u}))^{-1}$ exists and is uniformly bounded for any $\mathbf{u}$ in a neighborhood of $\mathbf{u}^\star$; that is,

$$\|(A'(\mathbf{u}))^{-1}\mathbf{v}\| \leq \mu \ \|\mathbf{v}\|_0.$$

Note that if $\mathbf{v} \in (\mathcal{X}, \|.\|)$, then $\mathbf{v} \in (\mathcal{X}, \|.\|_0)$. Therefore, we also have,

$$\|\mathbf{v}\| \leq \mu \ \|A'(\mathbf{u})\mathbf{v}\|_0. \tag{11.14}$$

(A3) For any $\epsilon > 0$ there is a $\delta > 0$ such that the derivative A' satisfies the estimates,

$$\|A[\mathbf{u}] - A[\mathbf{u}^\star] - A'(\mathbf{u}^\star)(\mathbf{u} - \mathbf{u}^\star)\|_0 \leq \epsilon \ \|\mathbf{u} - \mathbf{u}^\star\|,$$

and

$$\begin{aligned} \|(\mathbf{v} - A'(\mathbf{u}^\star)(A'(\mathbf{u}))^{-1})\mathbf{v}\|_0 &\leq \epsilon \ \|\mathbf{v}\|_0, \quad \text{all } \mathbf{v} \in \mathcal{Y}, \\ \|(\mathbf{v} - (A'(\mathbf{u}))^{-1}A'(\mathbf{u}^\star))\mathbf{v}\| &\leq \epsilon \ \|\mathbf{v}\|, \quad \text{all } \mathbf{v} \in \mathcal{X}, \end{aligned} \tag{11.15}$$

whenever $\|\mathbf{u} - \mathbf{u}^\star\| < \delta$. Note that (11.15) and (11.14) imply (with $\epsilon := \epsilon\mu$)

$$\|((A'(\mathbf{u}))^{-1} - (A'(\mathbf{u}^\star))^{-1})\mathbf{v}\| \ \leq \epsilon \ \|\mathbf{v}\|_0, \quad \text{all } \mathbf{v} \in \mathcal{X}.$$

It is clear that without loss of generality, we may assume that $\mathbf{f} = 0$, otherwise we can consider the shifted nonlinear operator $A[\mathbf{u}] := A[\mathbf{u}] - \mathbf{f}$. The derivative of the shifted nonlinear operator does not change, nor do the assumptions (A2)–(A3).

The following main result holds (in the spirit of [DES] as modified in [BVW03]).

Theorem 11.1. *Let the assumptions (A1)–(A3) hold and let* η, t *satisfying* $0 \le \eta < t < 1$ *be given. Then, there is an* $\delta > 0$ *such that if* $\|\mathbf{u}_0 - \mathbf{u}^*\|_0 < \delta$, *then the sequence of iterates* $\{\mathbf{u}_k\}$ *generated by* (11.12)–(11.13), *converges to* $\mathbf{u}^*$. *Moreover, the convergence is linear in the sense that*

$$\|\mathbf{u}_{k+1} - \mathbf{u}^*\|_* \le t\|\mathbf{u}^k - \mathbf{u}^*\|_*, \tag{11.16}$$

where $\|\mathbf{v}\|_* = \|A'(\mathbf{u}^*)\mathbf{v}\|_0$, *provided the initial iterate satisfies the (stronger) estimate*

$$\mu\|A'(\mathbf{u}^*)(\mathbf{u}_0 - \mathbf{u}^*\|_0 < \delta. \tag{11.17}$$

Here μ *is such that*

$$\|\mathbf{v}\| \le \mu\|A'(\mathbf{u})\mathbf{v}\|_0,$$

for any $\mathbf{u}$ *in a neighborhood of* $\mathbf{u}^\star$ *(see* (11.14)*).*

Proof. Because $0 < \eta < t$, there is a $\epsilon > 0$ such that

$$\epsilon + \mu\epsilon(\epsilon + 1) + (\epsilon + 1)\eta(1 + \mu\epsilon) < t. \tag{11.18}$$

Based on the properties of A' and $(A')^{-1}$, (A2), and (A3), now choose a $\delta > 0$ sufficiently small such that for any $\mathbf{v}: \ \|\mathbf{v} - \mathbf{u}^\star\| < \delta$,

$$\|I - (A'(\mathbf{v}))^{-1}A'(\mathbf{u}^\star)\| \le \epsilon, \tag{11.19}$$

$$\|I - A'(\mathbf{u}^\star)(A'(\mathbf{v}))^{-1}\| \le \epsilon, \tag{11.20}$$

and

$$\|A[\mathbf{v}] - A[\mathbf{u}^\star] - A'(\mathbf{u}^\star)(\mathbf{v} - \mathbf{u}^\star)\|_0 \le \epsilon \ \|\mathbf{v} - \mathbf{u}^\star\|. \tag{11.21}$$

The norms in (11.19) and (11.20) are the corresponding operator norms induced by (11.15).

Note now that if we choose the initial iterate $\mathbf{u}_0$ such that (11.17) holds, then we also have $\|\mathbf{u}_0 - \mathbf{u}^\star\| \le \mu \ \|A'(\mathbf{u}^\star)(\mathbf{u}_0 - \mathbf{u}^\star)\|_0 < \delta$. The proof proceeds then by induction. Because $A'(\mathbf{u}_0)$ is invertible (by assumption (A2)), the system $A'(\mathbf{u}_0)\mathbf{s} = -A[\mathbf{u}_0]$ has a solution, and hence, we can find a $\mathbf{s}_0$ such that $A'(\mathbf{u}_0)\mathbf{s}_0 = -A[\mathbf{u}_0] + \mathbf{r}_0$ with $\|\mathbf{r}_0\|_0 \le \eta \ \|A[\mathbf{u}_0]\|_0$. We then define, with $G = A'(\mathbf{u}_0)$,

$$\mathbf{u}_1 = \mathbf{u}_0 - G^{-1}(\mathbf{r}_0 - A[\mathbf{u}_0]).$$

Next,

$$\mathbf{u}_1 - \mathbf{u}^\star = \mathbf{u}_0 - \mathbf{u}^\star - G^{-1}A[\mathbf{u}_0] + (G^{-1} - (A'(\mathbf{u}^\star))^{-1})\mathbf{r}_0 + (A'(\mathbf{u}^\star))^{-1}\mathbf{r}_0.$$

Replace now $A[\mathbf{u}_0]$ by $A'(\mathbf{u}^\star)(\mathbf{u}_0 - \mathbf{u}^\star) + [A[\mathbf{u}_0] - A[\mathbf{u}^\star] - A'(\mathbf{u}^\star)(\mathbf{u}_0 - \mathbf{u}^\star)]$. Then,

$$\begin{aligned}\mathbf{u}_1 - \mathbf{u}_0 &= \mathbf{u}_0 - \mathbf{u}^\star - G^{-1}A'(\mathbf{u}^\star)(\mathbf{u}_0 - \mathbf{u}^\star)\\ &\quad - G^{-1}[A[\mathbf{u}_0] - A[\mathbf{u}^\star] - A'(\mathbf{u}^\star)(\mathbf{u}_0 - \mathbf{u}^\star)]\\ &\quad + (G^{-1} - (A'(\mathbf{u}^\star))^{-1})\mathbf{r}_0 + (A'(\mathbf{u}^\star))^{-1}\mathbf{r}_0. \end{aligned} \tag{11.22}$$

Thus, we end up with the identity

$$\begin{aligned}A'(\mathbf{u}^\star)(\mathbf{u}_1 - \mathbf{u}_0) &= (I - A'(\mathbf{u}^\star)G^{-1})A'(\mathbf{u}^\star)(\mathbf{u}_0 - \mathbf{u}^*)\\ &\quad - A'(\mathbf{u}^\star)G^{-1}[A[\mathbf{u}_0] - A[\mathbf{u}^\star] - A'(\mathbf{u}^\star)(\mathbf{u}_0 - \mathbf{u}^\star)]\\ &\quad + (A'(\mathbf{u}^\star)G^{-1} - I)\mathbf{r}_0 + \mathbf{r}_0.\end{aligned}$$

Therefore,

$$\begin{aligned}\|\mathbf{u}_1 - \mathbf{u}_0\|_* &\le \|I - A'(\mathbf{u}^\star)G^{-1}\|\,\|\mathbf{u}_0 - \mathbf{u}^\star\|_* + \|A'(\mathbf{u}^\star)G^{-1}\|\ \epsilon\mu\ \|\mathbf{u}_0 - \mathbf{u}^\star\|_*\\ &\quad + \|A'(\mathbf{u}^\star)G^{-1}\mathbf{r}_0\|_0\\ &\le \epsilon\ \|\mathbf{u}_0 - \mathbf{u}^\star\|_* + \mu\epsilon(\epsilon+1)\ \|\mathbf{u}_0 - \mathbf{u}^\star\|_* + (\epsilon+1)\|\mathbf{r}_0\|_0. \end{aligned} \tag{11.23}$$

Because,

$$\begin{aligned}\|\mathbf{r}_0\|_0 &\le \eta\ \|A[\mathbf{u}_0]\|_0\\ &= \eta\ \|A'(\mathbf{u}^\star)(\mathbf{u}_0 - \mathbf{u}^\star) + [A[\mathbf{u}_0] - A[\mathbf{u}^\star] - A'(\mathbf{u}^\star)(\mathbf{u}_0 - \mathbf{u}^\star)]\|_0\\ &\le \eta\ (\|A'(\mathbf{u}^\star)(\mathbf{u}_0 - \mathbf{u}^\star)\|_0 + \|A[\mathbf{u}_0] - A[\mathbf{u}^\star] - A'(\mathbf{u}^\star)(\mathbf{u}_0 - \mathbf{u}^\star)\|_0)\\ &\le \eta(\|\mathbf{u}_0 - \mathbf{u}^\star\|_* + \epsilon\|\mathbf{u}_0 - \mathbf{u}^\star\|)\\ &\le \eta(1 + \epsilon\mu)\ \|\mathbf{u}_0 - \mathbf{u}^\star\|_*,\end{aligned}$$

we have from (11.23) that

$$\begin{aligned}\|\mathbf{u}_1 - \mathbf{u}_0\|_* &\le [\epsilon + \mu\epsilon(\epsilon+1) + (\epsilon+1)\eta(1+\mu\epsilon)]\|\mathbf{u}_0 - \mathbf{u}^*\|_*\\ &\le t\ \|\mathbf{u}_0 - \mathbf{u}^*\|_*.\end{aligned}$$

Then it is clear that

$$\|\mathbf{u}_k - \mathbf{u}^*\| \le \mu\ \|\mathbf{u}_k - \mathbf{u}^*\|_* \le \mu t^k\ \|\mathbf{u}_0 - \mathbf{u}^*\|_* \le \mu\ \|\mathbf{u}_0 - \mathbf{u}^*\|_* \le \delta.$$

That is, all iterates remain in the δ neighborhood of $\mathbf{u}^\star$, and the induction argument can be repeated. Thus the proof is complete. □

We remark at the end that the computation of a sufficiently accurate initial iterate $\mathbf{u}_0$ was considered in the preceding section under additional assumptions that $A[.]$ was semilinear and that the linearized problems can be solved by an optimal-order iterative method that reduces the residuals in the $\|.\|_0$- norm. Such an iterative method can be a cascadic MG or a W-cycle MG with sufficiently many smoothing steps provided the underlined linearized PDE is regular enough. Then all the bounds in Theorem (11.1) are mesh-independent and the resulting iteration method defined by Algorithm 11.3.1 has an optimal complexity. For more details, we refer to Section B.2 (or see [BVW03]). Other choices of norms $\|.\|_0$ and $\|.\|$ were considered in [KPV].

12

Quadratic Constrained Minimization Problems

This chapter deals with solving quadratic minimization problems defined from a s.p.d. matrix A subject to box inequality constraints that model Signorini's problems in contact mechanics. In particular, we investigate the use of preconditioners B for A incorporated in the commonly used projection methods. The latter methods are also quadratic minimization problems involving the preconditioner B to define the quadratic functional. To make the projection methods computationally feasible (for more general than diagonal B) an equivalent dual formulation is introduced that involves the inverse actions of B (and not the actions of B). For the special case when the constrained set involves a small subset of the unknowns a reduced problem formulation is introduced and analyzed. Our presentation of these topics is based on the results by J. Schoeberl in [Sch98] and [Sch01]. We conclude the chapter with a multilevel FAS (full approximation scheme) based on monotone smoothers (such as projected Gauss–Seidel) providing a monotonicity proof from [IoV04].

12.1 Problem formulation

For a convex set K and a quadratic functional based on a symmetric positive definite matrix A,

$$J(\mathbf{v}) = \frac{1}{2}\mathbf{v}^T A\mathbf{v} - \mathbf{b}^T\mathbf{v},$$

solve the following optimization problem.

$$\text{Find } \mathbf{u} \in K : \; J(\mathbf{u}) = \min_{\mathbf{v}\in K} J(\mathbf{v}). \tag{12.1}$$

Lemma 12.1. *The solution* $\mathbf{u}$ *of the above problem is characterized by the variational inequality,*

$$\mathbf{u}^T A(\mathbf{v} - \mathbf{u}) \geq \mathbf{b}^T(\mathbf{v} - \mathbf{u}), \quad \textit{for all } \mathbf{v} \in K. \tag{12.2}$$

Proof. The proof follows from the minimization property of $\mathbf{u}$ and convexity of K by considering for any $\mathbf{v}$ the element $t\mathbf{v} + (1 - t)\mathbf{u} \in K$ for $t \in (0,\ 1)$. We have

P.S. Vassilevski, *Multilevel Block Factorization Preconditioners*,
doi: 10.1007/978-0-387-71564-3_12,

$J(\mathbf{u}) \leq J(t\mathbf{v} + (1-t)\mathbf{u})$, which leads first to

$$\frac{1}{2}t^2\mathbf{v}^T A\mathbf{v} + t(1-t)\mathbf{v}^T A\mathbf{u} + \frac{1}{2}(1-t)^2\mathbf{u}^T A\mathbf{u} - \mathbf{b}^T(t\mathbf{v} + (1-t)\mathbf{u})$$
$$\geq \frac{1}{2}\mathbf{u}^T A\mathbf{u} - \mathbf{b}^T\mathbf{u}.$$

Equivalently, we have

$$\frac{1}{2}t^2\mathbf{v}^T A\mathbf{v} + t(1-t)\mathbf{v}^T A\mathbf{u} - t\mathbf{b}^T\mathbf{v} \geq \frac{1}{2}\, t(2-t)\mathbf{u}^T A\mathbf{u} - t\mathbf{b}^T\mathbf{u},$$

which leads to

$$\frac{1}{2}t\mathbf{v}^T A\mathbf{v} + (1-t)\mathbf{v}^T A\mathbf{u} - \mathbf{b}^T\mathbf{v} \geq \frac{1}{2}\,(2-t)\mathbf{u}^T A\mathbf{u} - \mathbf{b}^T\mathbf{u},$$

and by letting $t \mapsto 0$ we end up with,

$$\mathbf{v}^T A\mathbf{u} - \mathbf{b}^T\mathbf{v} \geq \mathbf{u}^T A\mathbf{u} - \mathbf{b}^T\mathbf{u},$$

which is the desired result. □

12.1.1 Projection methods

We are interested in the following projection method. Given a symmetric positive definite matrix B, a preconditioner to A, define the projection

$$P_B\mathbf{v} : \mathbf{V} \mapsto K,$$

as the solution of the minimal distance problem:

$$\|P_B\mathbf{v} - \mathbf{v}\|_B = \min_{\mathbf{w}\in K} \|\mathbf{w} - \mathbf{v}\|_B. \tag{12.3}$$

Let $\alpha_{\min}$ and $\alpha_{\max}$ be the spectral bounds

$$\alpha_{\min}\mathbf{v}^T B\mathbf{v} \leq \mathbf{v}^T A\mathbf{v} \leq \alpha_{\max}\, \mathbf{v}^T B\mathbf{v}.$$

For a proper parameter $\tau > 0$ consider the following iteration process.

Algorithm 12.1.1 (Projection Iteration). *Given $\mathbf{v}_0 \in K$, for $k = 0, 1, \ldots,$ until convergence, compute:*

1. $\widehat{\mathbf{v}}_k = \mathbf{v}_k + \tau B^{-1}(\mathbf{b} - A\mathbf{v}_k)$,
2. $\mathbf{v}_{k+1} = P_B\widehat{\mathbf{v}}_k$.

We study next the convergence of this algorithm. Assume that $(1/\tau)B - A$ is positive semidefinite (or nonnegative). That is, let $1 \geq \tau\alpha_{\max}$.

The following result has been proved by Schöberl in [Sch98].

Theorem 12.2. *The following convergence rate holds, for $k \geq 0$,*

$$J(\mathbf{v}_k) - J(\mathbf{u}) \leq \varrho^k\,(J(\mathbf{v}_0) - J(\mathbf{u})), \quad \|\mathbf{v}_k - \mathbf{u}\|_A^2 \leq 2\varrho^k(J(\mathbf{v}_0) - J(\mathbf{u})),$$

where

$$\varrho \le 1 - \frac{\tau\alpha_{\min}}{2} \simeq 1 - \frac{1}{2\kappa}, \quad \kappa = \frac{\alpha_{\max}}{\alpha_{\min}} \simeq cond(B^{-1}A).$$

Proof. We have, from the first step of Algorithm 12.1.1,

$$\mathbf{b} = A\mathbf{v}_k + \frac{1}{\tau}B(\widehat{\mathbf{v}}_k - \mathbf{v}_k),$$

and

$$\begin{aligned}
J(\mathbf{v}_{k+1}) &= \frac{1}{2}\mathbf{v}_{k+1}^T A\mathbf{v}_{k+1} - \mathbf{b}^T\mathbf{v}_{k+1} \\
&= \frac{1}{2}\big[(\mathbf{v}_{k+1} - \mathbf{v}_k)^T A(\mathbf{v}_{k+1} - \mathbf{v}_k) + 2(\mathbf{v}_{k+1} - \mathbf{v}_k)^T A\mathbf{v}_k + \mathbf{v}_k^T A\mathbf{v}_k\big] \\
&\quad - \mathbf{b}^T\mathbf{v}_k - \left(A\mathbf{v}_k + \frac{1}{\tau}B(\widehat{\mathbf{v}}_k - \mathbf{v}_k)\right)^T(\mathbf{v}_{k+1} - \mathbf{v}_k) \\
&= J(\mathbf{v}_k) + \frac{1}{2}(\mathbf{v}_{k+1} - \mathbf{v}_k)^T A(\mathbf{v}_{k+1} - \mathbf{v}_k) - \frac{1}{\tau}(B(\widehat{\mathbf{v}}_k - \mathbf{v}_k))^T(\mathbf{v}_{k+1} - \mathbf{v}_k).
\end{aligned}$$

Using the fact that $\frac{1}{\tau}B - A$ is nonnegative, we arrive at

$$\begin{aligned}
J(\mathbf{v}_{k+1}) &\le J(\mathbf{v}_k) + \frac{1}{2\tau}(\mathbf{v}_{k+1} - \mathbf{v}_k)^T B(\mathbf{v}_{k+1} - \mathbf{v}_k) - \frac{1}{\tau}(B(\widehat{\mathbf{v}}_k - \mathbf{v}_k))^T(\mathbf{v}_{k+1} - \mathbf{v}_k) \\
&= J(\mathbf{v}_k) + \frac{1}{2\tau}(\mathbf{v}_{k+1} - \widehat{\mathbf{v}}_k)^T B(\mathbf{v}_{k+1} - \mathbf{v}_k) - \frac{1}{2\tau}(\widehat{\mathbf{v}}_k - \mathbf{v}_k)^T B(\mathbf{v}_{k+1} - \mathbf{v}_k) \\
&= J(\mathbf{v}_k) + \frac{1}{2\tau}(\mathbf{v}_{k+1} - \widehat{\mathbf{v}}_k)^T B((\mathbf{v}_{k+1} - \widehat{\mathbf{v}}_k) + (\widehat{\mathbf{v}}_k - \mathbf{v}_k)) \\
&\quad - \frac{1}{2\tau}(\widehat{\mathbf{v}}_k - \mathbf{v}_k)^T B((\widehat{\mathbf{v}}_k - \mathbf{v}_k) + (\mathbf{v}_{k+1} - \widehat{\mathbf{v}}_k)) \\
&= J(\mathbf{v}_k) + \frac{1}{2\tau}\big[(\mathbf{v}_{k+1} - \widehat{\mathbf{v}}_k)^T B(\mathbf{v}_{k+1} - \widehat{\mathbf{v}}_k) - (\widehat{\mathbf{v}}_k - \mathbf{v}_k)^T B(\widehat{\mathbf{v}}_k - \mathbf{v}_k)\big].
\end{aligned}$$

A principal step in the proof is the following estimate for $\mathbf{v}_{k+1} = P_B\widehat{\mathbf{v}}_k$,

$$\begin{aligned}
\|P_B\widehat{\mathbf{v}}_k - \widehat{\mathbf{v}}_k\|_B^2 &\le \|P_{[\mathbf{u},\,\mathbf{v}_k]}\widehat{\mathbf{v}}_k - \widehat{\mathbf{v}}_k\|_B^2 \\
&\le -\tau\alpha_{\min}\,(\mathbf{u} - \mathbf{v}_k)^T B(\widehat{\mathbf{v}}_k - \mathbf{v}_k) + \|\mathbf{v}_k - \widehat{\mathbf{v}}_k\|_B^2. \qquad (12.4)
\end{aligned}$$

Here, $P_{[\mathbf{u},\,\mathbf{v}_k]}$ stands for the projection on the segment with endpoints $\mathbf{u}$ and $\mathbf{v}_k$ (a convex subset of K). That is, we have

$$\begin{aligned}
\|\mathbf{v}_{k+1} - \widehat{\mathbf{v}}_k\|_B^2 &\le \min_{t\in[0,1]} \|t\mathbf{u} + (1-t)\mathbf{v}_k - \widehat{\mathbf{v}}_k\|_B^2 \\
&\le t^2\|\mathbf{u} - \mathbf{v}_k\|_B^2 - 2t(\widehat{\mathbf{v}}_k - \mathbf{v}_k)^T B(\mathbf{u} - \mathbf{v}_k) + \|\widehat{\mathbf{v}}_k - \mathbf{v}_k\|_B^2. \qquad (12.5)
\end{aligned}$$

Due to the variational representation of the solution $\mathbf{u}$, we have $(\mathbf{b}-A\mathbf{u})^T(\mathbf{u}-\mathbf{v}_k) \geq 0$, hence,

$$\begin{aligned}(\widehat{\mathbf{v}}_k - \mathbf{v}_k)^T B(\mathbf{u} - \mathbf{v}_k) &= \tau(\mathbf{b} - A\mathbf{v}_k)^T(\mathbf{u} - \mathbf{v}_k) \\ &= \tau(\mathbf{b} - A\mathbf{u})^T(\mathbf{u} - \mathbf{v}_k) + \tau\|\mathbf{u} - \mathbf{v}_k\|_A^2 \\ &\geq \tau\alpha_{\min}\ \|\mathbf{u} - \mathbf{v}_k\|_B^2. \end{aligned} \tag{12.6}$$

This shows,

$$\|\mathbf{v}_{k+1} - \mathbf{v}_k\|_B^2 \leq \min_{t\in[0,1]} \left[t^2\|\mathbf{u} - \mathbf{v}_k\|_B^2 - 2t\tau\alpha_{\min}\ \|\mathbf{u} - \mathbf{v}_k\|_B^2 + \|\widehat{\mathbf{v}}_k - \mathbf{v}_k\|_B^2\right].$$

The last expression indicates that choosing $t = \tau\alpha_{\min} \leq 1$ is appropriate. Estimate (12.5) with this choice of t implies the desired one, (12.4),

$$\begin{aligned}\|\mathbf{v}_{k+1} - \mathbf{v}_k\|_B^2 &\leq (\tau\alpha_{\min})^2\|\mathbf{u} - \mathbf{v}_k\|_B^2 \\ &\quad - 2\tau\alpha_{\min}(\widehat{\mathbf{v}}_k - \mathbf{v}_k)^T B(\mathbf{u} - \mathbf{v}_k) + \|\widehat{\mathbf{v}}_k - \mathbf{v}_k\|_B^2 \\ &\leq \|\widehat{\mathbf{v}}_k - \mathbf{v}_k\|_B^2 - \tau\alpha_{\min}(\widehat{\mathbf{v}}_k - \mathbf{v}_k)^T B(\mathbf{u} - \mathbf{v}_k),\end{aligned}$$

where we have used the estimate (12.6). Thus we proved the following estimate,

$$J(\mathbf{v}_{k+1}) \leq J(\mathbf{v}_k) + \frac{1}{2\tau}(-\tau\alpha_{\min}\ (\mathbf{u} - \mathbf{v}_k)^T B(\widehat{\mathbf{v}}_k - \mathbf{v}_k)).$$

Finally,

$$\begin{aligned}J(\mathbf{v}_{k+1}) &\leq J(\mathbf{v}_k) + \frac{1}{2\tau}\ (-\tau\alpha_{\min})\ (\mathbf{u} - \mathbf{v}_k)^T B(\widehat{\mathbf{v}}_k - \mathbf{v}_k) \\ &= J(\mathbf{v}_k) - \frac{\tau\alpha_{\min}}{2}\left((\mathbf{b} - A\mathbf{u})^T(\mathbf{u} - \mathbf{v}_k) + \|\mathbf{u} - \mathbf{v}_k\|_A^2\right) \\ &= J(\mathbf{v}_k) - \frac{\tau\alpha_{\min}}{2}\left(\mathbf{b}^T\mathbf{u} - \mathbf{b}^T\mathbf{v}_k + \mathbf{v}_k^T A\mathbf{v}_k - \mathbf{u}^T A\mathbf{v}_k\right) \\ &\leq J(\mathbf{v}_k) - \frac{\tau\alpha_{\min}}{2}\left(\mathbf{b}^T\mathbf{u} - \mathbf{b}^T\mathbf{v}_k - \frac{1}{2}\mathbf{u}^T A\mathbf{u} + \frac{1}{2}\mathbf{v}_k^T A\mathbf{v}_k\right) \\ &= \left(1 - \frac{\tau\alpha_{\min}}{2}\right) J(\mathbf{v}_k) + \frac{\tau\alpha_{\min}}{2} J(\mathbf{u}) \\ &= \varrho J(\mathbf{v}_k) + (1 - \varrho)J(\mathbf{u}).\end{aligned}$$

Therefore,

$$J(\mathbf{v}_{k+1}) - J(\mathbf{u}) \leq \varrho(J(\mathbf{v}_k) - J(\mathbf{u})).$$

The iterates are estimated as follows,

$$\begin{aligned}\|\mathbf{v}_k - \mathbf{u}\|_A^2 &= 2(J(\mathbf{v}_k) - J(\mathbf{u}) - (A\mathbf{u} - \mathbf{b})^T(\mathbf{v}_k - \mathbf{u})) \\ &\leq 2(J(\mathbf{v}_k) - J(\mathbf{u})) \\ &\leq 2\varrho^k(J(\mathbf{v}_0) - J(\mathbf{u})).\end{aligned}$$

□

12.1.2 A modified projection method

We present here a modification of Algorithm 12.1.1 proposed in [DJ05] that results in a better convergence rate estimate. Let B be a s.p.d. preconditioner for A such that

$$\alpha \, \mathbf{v}^T B \mathbf{v} \leq \mathbf{v}^T A \mathbf{v} \leq \mathbf{v}^T B \mathbf{v}. \tag{12.7}$$

Consider then the following algorithm.

Algorithm 12.1.2 (Modified Projection Iteration). *Consider the original problem*

$$J(\mathbf{v}) = \frac{1}{2} \, \mathbf{v}^T A \mathbf{v} - \mathbf{f}^T \mathbf{v} \mapsto \min \quad \textit{subject to } \mathbf{v} \in K. \tag{12.8}$$

Given $\mathbf{v}_0 \in K$, *for* $k = 0, 1, \ldots,$ *until convergence, compute* $\mathbf{v}_{k+1}$ *by solving the constrained minimization problem:*

$$J_k(\mathbf{x}) = J(\mathbf{x}) + \frac{1}{2} \, \|\mathbf{x} - \mathbf{v}_k\|_{B-A}^2 \mapsto \min \quad \textit{subject to } \mathbf{x} \in K.$$

We have

$$\begin{aligned}
J_k(\mathbf{x}) &= J(x) + \frac{1}{2} \, \|\mathbf{x} - \mathbf{v}_k\|_{B-A}^2 \\
&= \frac{1}{2} \, \mathbf{x}^T A \mathbf{x} - \mathbf{f}^T \mathbf{x} + \frac{1}{2} \, (\mathbf{x} - \mathbf{v}_k)^T (B - A)(\mathbf{x} - \mathbf{v}_k) \\
&= \frac{1}{2} \, (\mathbf{x} - \mathbf{v}_k)^T B (\mathbf{x} - \mathbf{v}_k) + \frac{1}{2} \, \mathbf{x}^T A \mathbf{x} - \mathbf{f}^T \mathbf{x} \\
&\quad - \frac{1}{2} \, \mathbf{x}^T A \mathbf{x} - \frac{1}{2} \, \mathbf{v}_k^T A \mathbf{v}_k + \mathbf{x}^T A \mathbf{v}_k \\
&= \frac{1}{2} \, (\mathbf{x} - \mathbf{v}_k)^T B (\mathbf{x} - \mathbf{v}_k) - \mathbf{x}^T (\mathbf{f} - A \mathbf{v}_k) - \frac{1}{2} \, \mathbf{v}_k^T A \mathbf{v}_k.
\end{aligned}$$

That is,

$$J_k(\mathbf{x}) = \frac{1}{2} \, (\mathbf{x} - \mathbf{v}_k)^T B (\mathbf{x} - \mathbf{v}_k) + \mathbf{x}^T (A \mathbf{v}_k - \mathbf{f}) - \frac{1}{2} \, \mathbf{v}_k^T A \mathbf{v}_k \tag{12.9}$$

is a quadratic functional with quadratic term that involves only B. Hence J_k is computationally similar to the quadratic functional involved in the minimal distance problem $\|\mathbf{v} - P_K \mathbf{v}\|_B = \min_{\mathbf{x} \in K} \|\mathbf{v} - \mathbf{x}\|_B$ that defines the projection P_K.

Theorem 12.3. *The following convergence result holds for the iterates computed by Algorithm 12.1.2,*

$$J(\mathbf{v}_k) - J(\mathbf{v}^*) \leq (1 - \alpha)^k \, (J(\mathbf{v}_0) - J(\mathbf{v}^*)),$$

where $\mathbf{v}^*$ *is the exact solution of* (12.8) *and* $\alpha \in (0, 1]$ *is from the spectral equivalence relation* (12.7).

Proof. Using the convexity of K and that $\alpha \in [0, 1]$ gives

$$\begin{aligned} J(\mathbf{v}_{k+1}) - J(\mathbf{v}^*) &\le J_k(\mathbf{v}_{k+1}) - J(\mathbf{v}^*) \\ &\le \min_{t\in[0,1]} J_k(\mathbf{v}_k + t(\mathbf{v}^* - \mathbf{v}_k)) - J(\mathbf{v}^*) \\ &\le J_k(\mathbf{v}_k + \alpha(\mathbf{v}^* - \mathbf{v}_k)) - J(\mathbf{v}^*). \end{aligned}$$

Use the left-hand side of (12.7) and the identity (12.9) for $\mathbf{x} = \mathbf{v}_k + \alpha(\mathbf{v}^* - \mathbf{v}_k)$ to arrive at the estimates

$$\begin{aligned} &J_k(\mathbf{v}_k + \alpha(\mathbf{v}^* - \mathbf{v}_k)) - J(\mathbf{v}^*) \\ &\quad = \frac{\alpha^2}{2} (\mathbf{v}^* - \mathbf{v}_k)^T B(\mathbf{v}^* - \mathbf{v}_k) + \left(\mathbf{v}_k + \alpha(\mathbf{v}^* - \mathbf{v}_k)\right)^T (A\mathbf{v}_k - \mathbf{f}) \\ &\quad\quad - \frac{1}{2} \mathbf{v}_k^T A\mathbf{v}_k - J(\mathbf{v}^*) \\ &\quad \le \frac{\alpha}{2} (\mathbf{v}^* - \mathbf{v}_k)^T A(\mathbf{v}^* - \mathbf{v}_k) + \alpha\, (\mathbf{v}^* - \mathbf{v}_k)^T (A\mathbf{v}_k - \mathbf{f}) + \mathbf{v}_k^T (A\mathbf{v}_k - \mathbf{f}) \\ &\quad\quad - \frac{1}{2} \mathbf{v}_k^T A\mathbf{v}_k - J(\mathbf{v}^*) \\ &\quad = \frac{\alpha}{2} (\mathbf{v}^* - \mathbf{v}_k)^T A(\mathbf{v}^* - \mathbf{v}_k) + \alpha\, (\mathbf{v}^* - \mathbf{v}_k)^T (A\mathbf{v}_k - \mathbf{f}) + J(\mathbf{v}_k) - J(\mathbf{v}^*) \\ &\quad = \frac{\alpha}{2} \left((\mathbf{v}^*)^T A\mathbf{v}^* + \mathbf{v}_k^T A\mathbf{v}_k - 2\mathbf{v}_k^T A\mathbf{v}^*\right) + \alpha\, (\mathbf{v}^*)^T A\mathbf{v}_k - \alpha\, (\mathbf{v}^*)^T \mathbf{f} \\ &\quad\quad + \alpha\mathbf{v}_k^T \mathbf{f} - \alpha\, \mathbf{v}_k^T A\mathbf{v}_k + J(\mathbf{v}_k) - J(\mathbf{v}^*) \\ &\quad = \frac{\alpha}{2} (\mathbf{v}^*)^T A\mathbf{v}^* - \alpha\, (\mathbf{v}^*)^T \mathbf{f} - \frac{\alpha}{2} \mathbf{v}_k^T A\mathbf{v}_k + \alpha\, \mathbf{v}_k^T \mathbf{f} + J(\mathbf{v}_k) - J(\mathbf{v}^*) \\ &\quad = (1 - \alpha)\, (J(\mathbf{v}_k) - J(\mathbf{v}^*)). \end{aligned}$$

That is, we have

$$J(\mathbf{v}_{k+1}) - J(\mathbf{v}^*) \le (1 - \alpha)\, (J(\mathbf{v}_k) - J(\mathbf{v}^*)),$$

which is the desired result. □

12.2 Computable projections

The problem of computing the actions of projection $P_B\mathbf{v}$ for a given B, is again a constrained minimization problem; namely,

$$\frac{1}{2}\mathbf{w}^T B\mathbf{w} - (B\mathbf{v})^T \mathbf{w} \mapsto \min, \tag{12.10}$$

over $\mathbf{w} \in K$. Note that if B is a diagonal matrix $\operatorname{diag}(d_i)$, $d_i > 0$, and $K = \{\mathbf{w} : \mathbf{w}_i \le \mathbf{g}_i, \text{for } i \in \Gamma\}$, for a given index set Γ, then the above minimization problem decouples and reduces to a number of one-dimensional quadratic constrained minimization problems; namely,

$$\sum_i \left(\frac{1}{2} d_i \mathbf{w}_i^2 - d_i \mathbf{v}_i \mathbf{w}_i\right) \mapsto \min \tag{12.11}$$

subject to $\{\mathbf{w}_i \leq \mathbf{g}_i\}$ where $i \in \Gamma$.

The latter one-dimensional problems are trivially solved; we have $\mathbf{w}_i = \mathbf{v}_i$ for indices outside Γ, and for $i \in \Gamma$, either $\mathbf{w}_i = \mathbf{v}_i$ if $\mathbf{v}_i \leq \mathbf{g}_i$ or $\mathbf{w}_i = \mathbf{g}_i$.

12.3 Dual problem approach

The solution of the projection problem (12.10) for a general (nondiagonal) B may be as difficult as the original problem. We may want to apply a projection method (with a simpler matrix) to solve the constrained minimization problem (12.10), for example, based on a (block-)diagonal preconditioner to B.

12.3.1 Dual problem formulation

There is one more difficulty with problem (12.10). Typically, for a general B defined by an algorithm (such as multigrid) we do not have the actions of B on vectors available, rather we have the inverse actions of B; that is, $B^{-1}\mathbf{v}$ is easily computable. The dual method was used in [Sch98] to reformulate (12.10) to involve B^{-1}.

Lemma 12.4. *Given is the original problem,*

$$\frac{1}{2}\mathbf{w}^T B\mathbf{w} - (B\mathbf{v})^T \mathbf{w} \mapsto \min, \tag{12.12}$$

subject to $\mathbf{w} \in K = \{\mathbf{w}_i \leq \mathbf{g}_i,\ i \in \Gamma\}$. *Then the following formulation is equivalent to the original one, in the sense that the solution of the* B*-projection problem is given by* $P_B\mathbf{v} = \mathbf{v} - B^{-1} I_\Gamma \mathbf{q}$, *where also,*

$$I_\Gamma \mathbf{v}_\Gamma = \begin{bmatrix} 0 \\ \mathbf{v}_\Gamma \end{bmatrix} \begin{matrix} \} & \Omega \setminus \Gamma \\ \} & \Gamma \end{matrix}$$

is the trivial extension of vectors $\mathbf{v}_\Gamma$ *defined on* Γ *by zero in the rest of* Ω, *and* $\mathbf{q}$ *solves the following (dual) problem,*

$$\frac{1}{2}\mathbf{q}^T (I_\Gamma)^T B^{-1} I_\Gamma \mathbf{q} - \mathbf{q}^T (I_\Gamma)^T (\mathbf{v} - I_\Gamma \mathbf{g}) \mapsto \min, \tag{12.13}$$

subject to $\mathbf{q} \geq 0$. *Here,* $\mathbf{q}$ *and* $\mathbf{g}$ *are vectors defined only on* Γ.

Proof. The derivation of the dual quadratic minimization problem (12.13) is given in what follows in full detail.

Consider the Lagrangian

$$\mathcal{L}(\mathbf{w}, \underline{\lambda}) = J(\mathbf{w}) - \underline{\lambda}^T \left(\mathbf{g} - I_\Gamma^T \mathbf{w}\right),$$

of the following constrained minimization problem,

$$J(\mathbf{w}) = \frac{1}{2}\mathbf{w}^T B\mathbf{w} - \mathbf{b}^T\mathbf{w} \mapsto \min,$$

subject to $\mathbf{w} \in K = \{\mathbf{w}_i \leq \mathbf{g}_i,\ i \in \Gamma\}$. In our application $\mathbf{b} = B\mathbf{v}$. Note that $\underline{\lambda}$ is defined only on Γ.

The (well-known) Karush–Kuhn–Tucker (KKT) conditions (cf., e.g., [SW97]), in the present setting, take the form:

1.
$$0 = \frac{\partial \mathcal{L}}{\partial \mathbf{w}} = \frac{\partial J(\mathbf{w})}{\partial \mathbf{w}} + I_\Gamma \underline{\lambda} = B\mathbf{w} - \mathbf{b} + I_\Gamma \underline{\lambda}.$$
2. $\mathbf{g} - I_\Gamma^T \mathbf{w} \geq 0$ (componentwise).
3. $\underline{\lambda} \geq 0$ (componentwise).
4. $\underline{\lambda}_i(\mathbf{g}_i - \mathbf{w}_i) = 0,\ i \in \Gamma$.

We now rewrite the above conditions to involve only actions of B^{-1}.

Let $\mathbf{w} = I_\Gamma \mathbf{w}_\Gamma + \mathbf{w}_0$ with $\mathbf{w}_0 = 0$ on Γ; that is, $I_\Gamma^T \mathbf{w}_0 = 0$. Let $\mathbf{y} = -\mathbf{g} + \mathbf{w}_\Gamma$. We have $\mathbf{y} \leq 0$. Then the first condition takes the form

$$B(I_\Gamma(\mathbf{y} + \mathbf{g}) + \mathbf{w}_0) = \mathbf{b} - I_\Gamma \underline{\lambda}.$$

That is, after multiplying with $I_\Gamma^T B^{-1}$ and using the fact that $I_\Gamma^T I_\Gamma = I$ and $I_\Gamma^T \mathbf{w}_0 = 0$ we end up with

$$\mathbf{y} + \mathbf{g} = I_\Gamma^T B^{-1}\mathbf{b} - I_\Gamma^T B^{-1} I_\Gamma \underline{\lambda},$$

or equivalently,

$$I_\Gamma^T B^{-1} I_\Gamma \underline{\lambda} + \mathbf{y} = I_\Gamma^T B^{-1}\mathbf{b} - \mathbf{g}. \tag{12.14}$$

The second condition simply reads,

$$\mathbf{y} \leq 0 \quad \text{(componentwise)}. \tag{12.15}$$

We also have,

$$\underline{\lambda} \geq 0 \quad \text{(componentwise)}, \tag{12.16}$$

and

$$\underline{\lambda}_i \mathbf{y}_i = 0,\ i \in \Gamma. \tag{12.17}$$

Consider now the dual quadratic minimization problem

$$J^\star(\mathbf{q}) = \frac{1}{2}\mathbf{q}^T I_\Gamma^T B^{-1} I_\Gamma \mathbf{q} - \mathbf{q}^T I_\Gamma^T (B^{-1}\mathbf{b} - I_\Gamma \mathbf{g}) \mapsto \min,$$

subject to the constraints $\mathbf{q}_i \geq 0$. Its Lagrangian reads,

$$\mathcal{L}^\star(\mathbf{q},\ \underline{\mu}) = J^\star(\mathbf{q}) - \underline{\mu}^T \mathbf{q}.$$

The corresponding Karush–Kuhn–Tucker conditions read:

1. $$0 = \frac{\partial \mathcal{L}^\star}{\partial \mathbf{q}} = \frac{\partial J^\star(\mathbf{q})}{\partial \mathbf{q}} - \underline{\mu} = I_\Gamma^T B^{-1} I_\Gamma \mathbf{q} - I_\Gamma^T (B^{-1}\mathbf{b} - I_\Gamma \mathbf{g}) - \underline{\mu}.$$
   This is exactly Equation (12.14) with $\underline{\lambda} = \mathbf{q}$ and $\mathbf{y} = -\underline{\mu}$.
2. $\mathbf{q} \geq 0$ (componentwise). This is exactly condition (12.16), $\underline{\lambda} \geq 0$, if $\underline{\lambda} = \mathbf{q}$.
3. $\underline{\mu} \geq 0$ (componentwise), which is inequality (12.15) with $\mathbf{y} = -\underline{\mu}$.
4. $\underline{\mu}_i \mathbf{q}_i = 0$, $i \in \Gamma$. This is exactly condition (12.17) with $\underline{\lambda} = \mathbf{q}$ and $\mathbf{y} = -\underline{\mu}$.

Thus problem (12.13) provides a solution $\mathbf{q}$ which gives $\mathbf{w} = B^{-1}(\mathbf{b} - I_\Gamma \mathbf{q}) = \mathbf{v} - B^{-1} I_\Gamma \mathbf{q}$ as the solution to the original constrained minimization problem (12.12). □

We again stress the fact that solving the dual problem and recovering $P_B \mathbf{v} = B^{-1}(\mathbf{b} - I_\Gamma \mathbf{q})$ do not involve actions of B; only actions of B^{-1} are required.

12.3.2 Reduced problem formulation

If we introduce the Schur complement S_B of B on Γ, that is, if

$$B^{-1} = \begin{bmatrix} \star & \star \\ \star & (S_B)^{-1} \end{bmatrix} \begin{matrix} \} & \Omega \setminus \Gamma \\ \} & \Gamma \end{matrix},$$

then problem (12.13) can be reformulated in the following reduced form,

$$\frac{1}{2} \mathbf{q}^T (S_B)^{-1} \mathbf{q} - \mathbf{q}^T ((I_\Gamma)^T \mathbf{v} - \mathbf{g}) \mapsto \min, \qquad (12.18)$$

subject to $\mathbf{q} \geq 0$. This is a reduced problem and S_B has in general a better condition number $\kappa(S_B)$ than the condition number $\kappa(B)$ of B. Typically, for matrices A coming from second-order elliptic finite element equations, assuming that B is spectrally equivalent to A, the behavior is, from $\mathcal{O}(h^{-2})$ conditioning for B, it is reduced to $\mathcal{O}(h^{-1})$ for its Schur complement S_B. Here, $h \mapsto 0$ is the mesh size. Then, we may use the projection method with diagonal matrix for defining the projection to solve the reduced problem (12.18). Based on Theorem 12.2, we get that problem (12.18) can be solved in $\mathcal{O}(\kappa(S_B))$ iterations. The cost of each iteration is proportional to the cost of one action of S_B^{-1}. To make the method efficient we must choose B such that the actions of S_B^{-1} are inexpensive to compute, for example, proportional to $|\Gamma| = \mathcal{O}(h^{-d+1})$ where $d = 2$ or 3 is the dimension of Ω. One possibility is to consider B^{-1} defined as follows. Let $\mathbf{V}_\Gamma$ be a subspace of $\mathbf{V}$ and $\mathbf{V}_0$ be a subspace of $\mathbf{V}$ with vectors vanishing on Γ. Let I_0 and $\mathcal{I}_\Gamma$ be extensions of vectors from $\mathbf{V}_0$ and $\mathbf{V}_\Gamma$ into vectors of full dimension. More specifically let I_0 be the trivial extension with zero on Γ; that is, let

$$I_0 = \begin{bmatrix} I \\ 0 \end{bmatrix} \begin{matrix} \} & \Omega \setminus \Gamma \\ \} & \Gamma \end{matrix}.$$

In general $\mathcal{I}_\Gamma$ can be viewed as an interpolation (or rather extension) mapping. Define the respective subspace matrices $A_0 = (I_0)^T A I_0$ and $A_\Gamma = (\mathcal{I}_\Gamma)^T A \mathcal{I}_\Gamma$. Note that $\mathbf{V} = \mathbf{V}_0 + \mathbf{V}_\Gamma$ may be an overlapping decomposition. To be specific in what follows we introduce a "finite element" terminology. For so-called contact problems in mechanics (cf., [HHNL]) Γ is considered to be part of the boundary of the domain Ω where the corresponding PDE is posed. We consider then $\mathbf{V}_\Gamma$ to be a proper coarse subspace of $\mathbf{V}$ corresponding to a coarse mesh $\mathcal{N}_H$ gradually coarsened away from Γ (and being not coarsened in a neighborhood of Γ). That is, in particular, this implies that $\Gamma \subset \mathcal{N}_H$. We assume that the "interpolation" mapping $\mathcal{I}_\Gamma$ is bounded in energy in the sense that for a constant $\eta \geq 1$ we have

$$(\mathcal{I}_\Gamma \mathbf{v}_\Gamma)^T A (\mathcal{I}_\Gamma \mathbf{v}_\Gamma) \leq \eta \min_{\mathbf{v}_0 :\, \mathbf{v}_0|_{\mathcal{N}_H}=0} (\mathbf{v}_0 + \mathcal{I}_\Gamma \mathbf{v}_\Gamma)^T A (\mathbf{v}_0 + \mathcal{I}_\Gamma \mathbf{v}_\Gamma), \quad \text{for all } \mathbf{v}_\Gamma. \tag{12.19}$$

Note that above we also have $\mathbf{v}_0 = 0$ on Γ (because $\Gamma \subset \mathcal{N}_H$). In other words, by defining

$$(R_\Gamma)^T = \begin{bmatrix} 0 \\ I \end{bmatrix} \begin{matrix} \} \; \Omega \setminus \mathcal{N}_H \\ \} \; \mathcal{N}_H \supset \Gamma \end{matrix},$$

the above norm boundedness (12.19) can be rewritten as

$$\mathbf{v}^T (\mathcal{I}_\Gamma R_\Gamma)^T A (\mathcal{I}_\Gamma R_\Gamma) \mathbf{v} \leq \eta \; \mathbf{v}^T A \mathbf{v}.$$

We now define B. Let B_0 and B_Γ be given preconditioners for A_0 and A_Γ (e.g., corresponding MG methods for the spaces $\mathbf{V}_0$ and $\mathbf{V}_\Gamma$). Then, consider

$$B^{-1} = [I_0, \; \mathcal{I}_\Gamma] \begin{bmatrix} B_0^{-1} & 0 \\ 0 & B_\Gamma^{-1} \end{bmatrix} [I_0, \; \mathcal{I}_\Gamma]^T. \tag{12.20}$$

The inverse action of B requires actions of B_0^{-1} and B_Γ^{-1}. The Schur complement S_B of B on Γ is defined from

$$(S_B)^{-1} = \begin{bmatrix} 0 \\ I_\Gamma \end{bmatrix}^T B^{-1} \begin{bmatrix} 0 \\ I_\Gamma \end{bmatrix} = \begin{bmatrix} 0 \\ I_\Gamma \end{bmatrix}^T (\mathcal{I}_\Gamma) B_\Gamma^{-1} (\mathcal{I}_\Gamma)^T \begin{bmatrix} 0 \\ I_\Gamma \end{bmatrix}.$$

Here,

$$\begin{bmatrix} 0 \\ I_\Gamma \end{bmatrix}$$

gets a vector defined on Γ and extends it by zero in the rest of Ω (the fine grid mesh $\mathcal{N}_h$).

Remark 12.5. Based on our general two-level results in "additive" form (cf., Section 3.2.8) it is straightforward to prove that B and A are spectrally equivalent with bounds depending on the spectral equivalence bounds between A_0 and B_0, the coarse matrix A_Γ and its preconditioner B_Γ (which may equal A_Γ), and the norm bound η from (12.19). The same holds, if we consider B^{-1} that has multiplicative form, as defined next.

Let $B_0 = D_0 + L_0$ and $B_0^T = D_0 + U_0$ be given preconditioners to A_0 such that $B_0 + B_0^T - A_0$ is positive definite. This in particular means that $D_0^T = D_0$ and $U_0^T = L_0$. The multiplicative version of B then reads,

$$B^{-1} = [I_0, \mathcal{I}_\Gamma] \begin{bmatrix} I & -(D_0 + U_0)^{-1} I_0^T A \mathcal{I}_\Gamma \\ 0 & I \end{bmatrix}$$
$$\times \begin{bmatrix} (D_0 + U_0)^{-1}(2D_0 + L_0 + U_0 - A_0)(D_0 + L_0)^{-1} & 0 \\ 0 & B_\Gamma^{-1} \end{bmatrix}$$
$$\times \begin{bmatrix} I & 0 \\ -\mathcal{I}_\Gamma^T A I_0 (D_0 + L_0)^{-1} & I \end{bmatrix} [I_0, \mathcal{I}_\Gamma]^T. \quad (12.21)$$

Note that in the applications we may have $D_0 = 0$ and $L_0^T = U_0$ be two nonsymmetric preconditioners to A_0. One example could be an L_0 defined from a downward (nonsymmetric) V-cycle multigrid applied to A_0. Then U_0 will correspond to an upward V-cycle multigrid with a smoother applied in reverse order at every level. Alternatively, we may have $L_0 = U_0 = 0$ and hence $D_0 = D_0^T$ be a given positive definite preconditioner to A_0.

The inverse actions of the multiplicative B are computed by the following algorithm, which can be seen as one step of a product subspace iteration method.

Algorithm 12.3.1 (Multiplicative preconditioner). *Introduce the subspace residual iteration matrices* $E_0 = I - AI_0(D_0 + L_0)^{-1} I_0^T$, $\widehat{E}_0 = I - AI_0(D_0 + L_0)^{-T} I_0^T$, *and* $E_\Gamma = I - A\mathcal{I}_\Gamma B_\Gamma^{-1} \mathcal{I}_\Gamma^T$.

Given $\mathbf{v}$, *we compute* $\mathbf{x} = B^{-1}\mathbf{v}$ *in the following steps.*

- *Forward elimination:*
  1. *Compute* $\mathbf{y}_0 = (D_0 + L_0)^{-1} I_0^T \mathbf{v}$.
  2. *Compute residual* $\mathbf{v} - AI_0\mathbf{y}_0$, *restrict it to* $\mathbf{V}_\Gamma$, *and solve with* B_Γ; *that is,*

$$\begin{aligned} \mathbf{x}_\Gamma &= B_\Gamma^{-1} \mathcal{I}_\Gamma^T (\mathbf{v} - AI_0\mathbf{y}_0) \\ &= B_\Gamma^{-1} \mathcal{I}_\Gamma^T (\mathbf{v} - AI_0(D_0 + L_0)^{-1} I_0^T \mathbf{v}) \\ &= B_\Gamma^{-1} \mathcal{I}_\Gamma^T (I - AI_0(D_0 + L_0)^{-1} I_0^T)\mathbf{v} \\ &= B_\Gamma^{-1} \mathcal{I}_\Gamma^T E_0 \mathbf{v}. \end{aligned}$$

So far, we have computed the solution $\mathbf{y}_0$, $\mathbf{x}_\Gamma$ *of*

$$\begin{bmatrix} (D_0 + L_0)\mathbf{y}_0 \\ B_\Gamma \mathbf{x}_\Gamma \end{bmatrix} = \begin{bmatrix} I & 0 \\ -\mathcal{I}_\Gamma^T A I_0 (D_0 + L_0)^{-1} & I \end{bmatrix} [I_0, \mathcal{I}_\Gamma]^T \mathbf{v}.$$

- *Backward recurrence:*
  1. *Solve for* $\mathbf{x}_0$ *the equation*

$$(D_0 + L_0)^T \mathbf{x}_0 + I_0^T A \mathcal{I}_\Gamma \mathbf{x}_\Gamma = (2D_0 + L_0 + L_0^T - A_0)\mathbf{y}_0.$$

That is, compute

$$\begin{aligned}\mathbf{x}_0 &= -(D_0 + L_0)^{-T} I_0^T A \mathcal{I}_\Gamma \mathbf{x}_\Gamma + \mathbf{y}_0 + (D_0 + L_0)^{-T} (D_0 + L_0 - A_0)\mathbf{y}_0 \\ &= \mathbf{y}_0 + (D_0 + L_0)^{-T} I_0^T (\mathbf{v} - A I_0 \mathbf{y}_0) - (D_0 + L_0)^{-T} I_0^T A \mathcal{I}_\Gamma \mathbf{x}_\Gamma \\ &= \mathbf{y}_0 + (D_0 + L_0)^{-T} I_0^T (\mathbf{v} - A(I_0 \mathbf{y}_0 + \mathcal{I}_\Gamma \mathbf{x}_\Gamma)).\end{aligned}$$

In this way, we have computed the expression

$$\begin{bmatrix}\mathbf{x}_0 \\ \mathbf{x}_\Gamma\end{bmatrix} = \begin{bmatrix}(D_0 + L_0)^{-T}((2D_0 + L_0 + L_0^T - A_0)\mathbf{y}_0 - I_0^T A \mathcal{I}_\Gamma \mathbf{x}_\Gamma) \\ \mathbf{x}_\Gamma\end{bmatrix}.$$

The latter represents the product

$$\begin{aligned}\pi \equiv &\begin{bmatrix} I & -(D_0 + U_0)^{-1} I_0^T A \mathcal{I}_\Gamma \\ 0 & I \end{bmatrix} \\ &\times \begin{bmatrix}(D_0 + U_0)^{-1}(2D_0 + L_0 + U_0 - A_0)(D_0 + L_0)^{-1} & 0 \\ 0 & B_\Gamma^{-1}\end{bmatrix} \\ &\times \begin{bmatrix} I & 0 \\ -\mathcal{I}_\Gamma^T A I_0 (D_0 + L_0)^{-1} & I\end{bmatrix} [I_0,\ \mathcal{I}_\Gamma]^T \mathbf{v}.\end{aligned}$$

Indeed, we have

$$\begin{aligned}\pi = &\begin{bmatrix} I & -(D_0 + U_0)^{-1} I_0^T A \mathcal{I}_\Gamma \\ 0 & I \end{bmatrix} \\ &\times \begin{bmatrix}(D_0 + U_0)^{-1}(2D_0 + L_0 + U_0 - A_0)(D_0 + L_0)^{-1} I_0^T \mathbf{v} \\ B_\Gamma^{-1} \mathcal{I}_\Gamma^T (\mathbf{v} - A I_0 (D_0 + L_0)^{-1} I_0^T \mathbf{v})\end{bmatrix} \\ = &\begin{bmatrix} I & -(D_0 + U_0)^{-1} I_0^T A \mathcal{I}_\Gamma \\ 0 & I \end{bmatrix} \\ &\times \begin{bmatrix}(D_0 + U_0)^{-1}(2D_0 + L_0 + U_0 - A_0)\mathbf{y}_0 \\ \mathbf{x}_\Gamma\end{bmatrix} \\ = &\begin{bmatrix} I & -(D_0 + U_0)^{-1} I_0^T A \mathcal{I}_\Gamma \\ 0 & I \end{bmatrix} \\ &\times \begin{bmatrix}(D_0 + U_0)^{-1}((D_0 + U_0)\mathbf{x}_0 + I_0^T A \mathcal{I}_\Gamma \mathbf{x}_\Gamma) \\ \mathbf{x}_\Gamma\end{bmatrix} \\ = &\begin{bmatrix} I & -(D_0 + U_0)^{-1} I_0^T A \mathcal{I}_\Gamma \\ 0 & I \end{bmatrix} \begin{bmatrix}\mathbf{x}_0 + (D_0 + U_0)^{-1} I_0^T A \mathcal{I}_\Gamma \mathbf{x}_\Gamma \\ \mathbf{x}_\Gamma\end{bmatrix} \\ = &\begin{bmatrix}\mathbf{x}_0 \\ \mathbf{x}_\Gamma\end{bmatrix}.\end{aligned}$$

2. *Compute the solution* $\mathbf{x} = B^{-1}\mathbf{v} = [I_0,\ \mathcal{I}_\Gamma]\pi = I_0 \mathbf{x}_0 + \mathcal{I}_\Gamma \mathbf{x}_\Gamma$. *We have*

$$\begin{aligned}\mathbf{x} &= I_0 \mathbf{x}_0 + \mathcal{I}_\Gamma \mathbf{x}_\Gamma \\ &= I_0 \mathbf{y}_0 + \mathcal{I}_\Gamma \mathbf{x}_\Gamma + I_0 (D_0 + L_0)^{-T} I_0^T (\mathbf{v} - A(I_0 \mathbf{y}_0 + \mathcal{I}_\Gamma \mathbf{x}_\Gamma)) \\ &= I_0 (D_0 + L_0)^{-T} I_0^T \mathbf{v} + E_0^T (I_0 \mathbf{y}_0 + \mathcal{I}_\Gamma \mathbf{x}_\Gamma)).\end{aligned}$$

The final residual equals

$$\begin{aligned}
\mathbf{v} - A\mathbf{x} &= (I - AB^{-1})\mathbf{v} \\
&= \widehat{E}_0(\mathbf{v} - A(I_0\mathbf{y}_0 + \mathcal{I}_\Gamma \mathbf{x}_\Gamma)) \\
&= \widehat{E}_0\big(\mathbf{v} - AI_0(L_0 + D_0)^{-1}I_0^T\mathbf{v} - A\mathcal{I}_\Gamma B_\Gamma^{-1}\mathcal{I}_\Gamma^T E_0\mathbf{v}\big) \\
&= \widehat{E}_0\big(I - A\mathcal{I}_\Gamma B_\Gamma^{-1}\mathcal{I}_\Gamma^T\big)E_0\mathbf{v} \\
&= \widehat{E}_0 E_\Gamma E_0\mathbf{v}.
\end{aligned}$$

That is, we have the product representation of the residual iteration matrix,

$$I - AB^{-1} = \widehat{E}_0 E_\Gamma E_0.$$

The projection algorithm: A summary

Here we present the overall projection algorithm to solve the original constrained minimization problem (12.1) based on the spectrally equivalent preconditioner B defined in (12.20).

Algorithm 12.3.2 (Composite projection method).

- *Given current iterate* $\mathbf{v}_k \in K$, $k = 0, 1, \ldots$, *compute the following.*
- $\widehat{\mathbf{v}}_k = \mathbf{v}_k + \tau B^{-1}(\mathbf{b} - A\mathbf{v}_k)$ *for a properly chosen iteration parameter* τ, *that is,* $(1/\tau)B - A$ *positive semidefinite.*
- *Compute the next iterate* $\mathbf{v}_{k+1} = P_B\widehat{\mathbf{v}}_k$, *where the actions of* P_B *are computed by solving the projection minimization problem in its dual form* (12.13) *or rather in its reduced form* (12.18)*. This is done by iterations again using a projection method with a simple diagonal matrix* D_Γ *as a preconditioner for* S_B*. Alternatively, we may use any other conventional method for solving constrained minimization problems, because this is a problem of relatively small size.*

12.4 A monotone two-grid scheme

To define a two-grid scheme we need a coarse space and a smoothing procedure. The coarse space we consider satisfies an important (special) property. The smoothing procedure is monotone (described in what follows).

Consider a coarse space $\mathbf{V}_c \subset \mathbf{V}$ and let P be an interpolation matrix that has the form

$$P = \begin{bmatrix} W \\ I \end{bmatrix}.$$

The degrees of freedom x_i corresponding to the constraint set $i \in \Gamma$ are denoted by C. To carry around both i and a degree of freedom x_i helps to associate the vectors $\mathbf{v} = (v_i)$ with an actual finite element grid $\mathcal{N}_h = \{x_i\}$, corresponding to a finite element discretization of a respective PDE. One main example of the quadratic constrained minimization problems we consider in the present chapter comes from contact problems in mechanics.

The identity block of P defines a natural embedding of the coarse degrees of freedom x_{i_c} into the fine degrees of freedom x_i. We also assume that constrained degrees of freedom $x_i \in C$ (or equivalently $i \in \Gamma$) are all present on the coarse grid. This implies that

$$(P\mathbf{v}_c)_\Gamma = \mathbf{v}_c|_\Gamma. \tag{12.22}$$

12.4.1 Projected Gauss–Seidel

We next describe the projected Gauss–Seidel method. Mathematically it can be described as a sequence of one-dimensional minimization problems. Consider the functional $J(\mathbf{v})$. Given a current iterate $\mathbf{v} = (v_i)$ which satisfies the constraints, we vary only a single component v_i (the remaining ones v_j, $j \neq i$ are kept fixed). This leads to a scalar quadratic function $\varphi(v_i) = J(v_i\mathbf{e}_i + \mathbf{v}^0)$ where $\mathbf{v}^0 = \sum_{j \neq i} v_j\mathbf{e}_j$, and $\{\mathbf{e}_i\}$ are the unit coordinate vectors. If $i \in \Gamma$ then we have to satisfy the constraint $v_i \leq g_i$. Thus a problem of finding the minimum of a quadratic function subject to a simple inequality constraint is obtained. More specifically, with $x = v_i$, $a = \mathbf{e}_i^T A\mathbf{e}_i > 0$, $b = \mathbf{e}_i^T(\mathbf{b} - A\mathbf{v}^0)$, a constant $c = J(\mathbf{v}^0) = \frac{1}{2}(\mathbf{v}^0)^T A\mathbf{v}^0 - \mathbf{b}^T\mathbf{v}^0$ and $d = g_i$, we have to solve

$$\varphi(x) = \frac{1}{2}ax^2 - bx + c \mapsto \min$$
$$\text{subject to } x \leq d.$$

The solution is $x = b/a$ if $b/a \leq d$, or $x = d$ otherwise. The new iterate then is $\mathbf{v} := \mathbf{v}^0 + v_i\mathbf{e}_i$ with $v_i = x$. After a loop over all indices i we complete the projected Gauss–Seidel cycle. This procedure used iteratively is referred to as the projected Gauss–Seidel method. We can also develop block versions of this method or even use overlapping blocks, by solving small-size constrained minimization problems for every block. An important property of the projected Gauss–Seidel is that every intermediate iterate decreases the functional, and hence after a complete cycle, we have that $J(\mathbf{v}) \leq J(\mathbf{v}^{\text{initial}})$; that is, it is a monotone method.

12.4.2 Coarse-grid solution

Let $\mathbf{v}^{m-1}$, $m \geq 1$, be a current iterate for solving our fine-grid problem

$$J(\mathbf{v}) = \frac{1}{2}\,\mathbf{v}^T A\mathbf{v} - \mathbf{b}^T\mathbf{v} \mapsto \min, \quad \mathbf{v} \in \mathbb{R}^n, \tag{12.23}$$

subject to the inequality constraints

$$v_i \leq g_i \quad \text{for all } i \in \Gamma.$$

Here Γ is a given subset of the index set $\{0, 1, 2, \ldots, n-1\}$ and $\mathbf{g} = (g_i)$ is a given vector defined for indices in Γ. Finally, A is a given symmetric positive definite matrix.

After performing a few steps of the projected Gauss–Seidel (or any other monotone smoothing scheme) we end up with an intermediate iterate $\mathbf{v}^{m-(1/2)} = (v_i^{m-(1/2)})$ which satisfies the constraints $v_i^{m-(1/2)} \leq g_i$, $i \in \Gamma$, and we also have,

$$J(\mathbf{v}^{m-(1/2)}) \leq J(\mathbf{v}^{m-1}).$$

The next iterate $\mathbf{v}^m$ is sought as $\mathbf{v}^m = \mathbf{v}^{m-(1/2)} + P_C\mathbf{y}_C$ where $\mathbf{y}_C$ is a coarse-grid vector such that the resulting coarse-grid energy functional is minimized. More specifically, because Γ is a subset of the set of coarse-grid degrees of freedom, due to (12.22), we can solve the following coarse-grid minimization problem.

Find $\mathbf{y}_C$ such that

$$\frac{1}{2}(\mathbf{v}^{m-(1/2)} + P\mathbf{y}_C)^T A(\mathbf{v}^{m-(1/2)} + P\mathbf{y}_C) - \mathbf{b}^T(\mathbf{v}^{m-(1/2)} + P\mathbf{y}_C) \mapsto \min \quad (12.24)$$

subject to the constraints $(\mathbf{y}_C)_i \leq g_i - \mathbf{v}_i^{m-(1/2)}$ for $i \in \Gamma$.

Let $A_C = P^T AP$ and $\mathbf{b}_C = P^T(\mathbf{b} - A\mathbf{v}^{m-(1/2)})$. Then (12.24) is equivalent to the following coarse quadratic constrained minimization problem,

$$J_C(\mathbf{y}_C) = \frac{1}{2}\mathbf{y}_C^T A_C \mathbf{y}_C - \mathbf{b}_C^T \mathbf{y}_C \mapsto \min$$

subject to the inequality constraints

$$(\mathbf{y}_C)_i \leq g_i - (\mathbf{v}^{m-(1/2)})_i \quad \text{for all } i \in \Gamma.$$

The two-grid iteration method can be summarized as follows.

Algorithm 12.4.1 (Two-grid minimization method). *Given an iterate $\mathbf{v}^{m-1}$, compute the next iterate $\mathbf{v}^m$ performing the following steps.*

- *Step 1: Compute $\mathbf{v}^{m-(1/2)} = \mathbf{v}^{m-1} + \widetilde{\mathbf{y}^1} + \widetilde{\mathbf{y}^2} + \cdots + \widetilde{\mathbf{y}^n}$. Here $\widetilde{\mathbf{y}^i}$, $1 \leq i \leq n$, are corrections spanned by the unit coordinate vectors $\mathbf{e}_i$, produced by the projected Gauss–Seidel algorithm with initial approximation $\mathbf{v}^{m-1}$.*
- *Step 2: Compute $\mathbf{v}^m = \mathbf{v}^{m-(1/2)} + P\mathbf{y}_C$, where $\mathbf{y}_C \in V_C$ and V_C is the coarse-grid vector space. The coarse-grid correction $\mathbf{y}_C$ solves the coarse quadratic minimization problem,*

$$J_C(\mathbf{y}_C) = \frac{1}{2}\mathbf{y}_C^T A_C \mathbf{y}_C - \mathbf{b}_C^T \mathbf{y}_C \mapsto \min$$

subject to the inequality constraints

$$(\mathbf{y}_C)_i \leq g_i - (\mathbf{v}^{m-(1/2)})_i \quad \textit{for all } i \in \Gamma.$$

The following main result holds.

Theorem 12.6. *The Algorithm 12.4.1 provides a monotone scheme; that is, for any two consecutive iterates, $\mathbf{v}^{m-1}$ and $\mathbf{v}^m$, produced by the algorithm, we have*

$$J(\mathbf{v}^m) \leq J(\mathbf{v}^{m-1}).$$

Proof. Given the iterate $\mathbf{v}^{m-1}$ and applying the projected Gauss–Seidel algorithm in Step 1, we get a new intermediate iterate $\mathbf{v}^{m-(1/2)}$ for which we have:

1. The new intermediate iterate satisfies the inequality constraints due to the projection operation in the projected Gauss–Seidel algorithm:

$$(\mathbf{v}^{m-(1/2)})_i \leq g_i \quad \text{for all } i \in \Gamma. \tag{12.25}$$

2. The value of the functional at the new intermediate iterate is less than the one at the previous one because the projected Gauss–Seidel algorithm provides a monotone scheme:

$$J(\mathbf{v}^{m-(1/2)}) \leq J(\mathbf{v}^{m-1}).$$

At the next step we look for a correction $\mathbf{y}_C \in V_C$ such that

$$(\mathbf{v}^{m-(1/2)} + P\mathbf{y}_C)_i \leq g_i \quad \text{for all } i \in \Gamma,$$

and

$$J(\mathbf{v}^{m-(1/2)} + P\mathbf{y}_C) \leq J(\mathbf{v}^{m-(1/2)}).$$

Simplifying the the expression gives,

$$\begin{aligned} J(\mathbf{v}^{m-(1/2)} + P\mathbf{y}_C) &= \frac{1}{2}(\mathbf{v}^{m-(1/2)})^T A\mathbf{v}^{m-(1/2)} + \frac{1}{2}\mathbf{y}_C^T P^T A P\mathbf{y}_C \\ &\quad + \mathbf{y}_C^T P^T A\mathbf{v}^{m-(1/2)} - \mathbf{b}^T(\mathbf{v}^{m-(1/2)} + P\mathbf{y}_C) \\ &= J(\mathbf{v}^{m-(1/2)}) + J_C(\mathbf{y}_C). \end{aligned} \tag{12.26}$$

It is clear that it is equivalent to solve the coarse-grid constraint minimization problem,

$$J_C(\mathbf{y}_C) \equiv \frac{1}{2}\mathbf{y}_C^T A_C \mathbf{y}_C - \mathbf{b}_C^T \mathbf{y}_C \mapsto \min$$

subject to

$$(\mathbf{y}_C)_i \leq g_i - (\mathbf{v}^{m-(1/2)})_i \quad \text{for all } i \in \Gamma.$$

Note that here we use the fact that the constraints are exactly present on the coarse level by our assumption on P, namely, that, $(P\mathbf{y}_C)|_\Gamma = \mathbf{y}_C|_\Gamma$. It is clear then that if we choose the correction $\mathbf{y}_C = \mathbf{y}_C^{opt}$ where $\mathbf{y}_C^{opt}$ is the solution of the above constraint minimization problem, we have that

$$J\left(\mathbf{v}^{m-(1/2)} + P\mathbf{y}_C^{opt}\right) \leq J(\mathbf{v}^{m-(1/2)}). \tag{12.27}$$

The latter is true because we may choose $\mathbf{y}_C = 0$ and satisfy the constraints due to the inequality (12.25). We then have

$$J_C\left(\mathbf{y}_C^{opt}\right) \leq J_C(0) = 0.$$

Therefore from (12.26) we have $J_C(\mathbf{y}_C^{opt}) = J(\mathbf{v}^{m-(1/2)} + P\mathbf{y}_C^{opt}) - J(\mathbf{v}^{m-(1/2)})$ which implies (12.27). Thus from (12.4.2) and (12.27) because $\mathbf{v}^m = \mathbf{v}^{m-(1/2)} + P\mathbf{y}_C^{opt}$ the proof is complete; that is, we have

$$J(\mathbf{v}^m) \leq J(\mathbf{v}^{m-(1/2)}) \leq J(\mathbf{v}^{m-1}).$$ □

12.5 A monotone FAS constrained minimization algorithm

Because we can treat problem (12.23) as a nonlinear one, we could attempt to solve it by applying the classical full approximation scheme (or simply FAS) by Achi Brandt (cf., [AB77], [BC83]).

Let $\{A_k\}_{k=0}^{\ell}$, $\{P_k\}$ be a MG hierarchy of matrices that satisfy $A_{k+1} = P_k^T A_k P_k$ with $A_0 = A$ being the given fine-grid s.p.d. matrix that defines the original quadratic functional $J(\mathbf{v}) = \frac{1}{2}\, \mathbf{v}^T A \mathbf{v} - \mathbf{b}^T \mathbf{v}$. Let $\Gamma_k = \Gamma$ be the constraint sets at all levels; that is, the main assumption is that the constraint set does not change from level to level and hence that each P_k is an identity on Γ as assumed in (12.22). Let $\mathcal{N}_k$ be the kth-level set of degrees of freedom. We have $\mathcal{N}_{k+1} \subset \mathcal{N}_k$ and for all k, $\Gamma \subset \mathcal{N}_k$. Finally, let $\mathbf{V}_k$ be the kth-level coarse vector space.

A corresponding FAS algorithm in the present context takes the following form.

Algorithm 12.5.1 (FAS constrained minimization algorithm). *Consider the problem* (12.23) *with* $\mathbf{b}_0 = \mathbf{b}$ *and* $\mathbf{g}^0 = \mathbf{g}$ *given. Let* ℓ *be the coarsest level.*

(0) For $k \geq 0$ *let* $\mathbf{v}_k^0 \in \mathbf{V}_k$ *be a current iterate at level* k *satisfying the constraints* $(\mathbf{v}_k^0)_i \leq (\mathbf{g}^k)_i$, $i \in \Gamma$. *Let*

$$J_k(\mathbf{y}) = \frac{1}{2}\mathbf{y}^T A_k \mathbf{y} - \mathbf{b}_k^T \mathbf{y},$$

be the kth-level quadratic functional.

(1) If $k < \ell$ *apply* $\nu_1 \geq 1$ *projected Gauss–Seidel smoothing iterations with initial iterate* $\mathbf{v}_k^0$. *Denote the resulting iterate obtained after a full cycle of projected Gauss–Seidel by* $\mathbf{v}_k$. *Go to Step (3).*

(2) Else (i.e., if $k = \ell$*), then solve the corresponding constrained minimization problem exactly. Denote the resulting solution by* $\mathbf{v}_k$. *Set* $k := k - 1$ *and go to Step (4).*

(3) Based on a coarse-grid constrained minimization problem correct the kth-level iterate $\mathbf{v}_k$. *Define* $\mathbf{g}_{k+1}$ *and* $\mathbf{b}_{k+1}$ *for the next level coarse-grid problem as follows.*

- *Set* $\mathbf{g}^{k+1} = \mathbf{g}^k = \cdots = \mathbf{g}$ *and choose as initial approximation at level* $k+1$, $\mathbf{v}_{k+1}^0 = \mathbf{v}_k|_{\mathcal{N}_{k+1}}$.
- *Set* $\mathbf{b}_{k+1} = P_k^T(\mathbf{b}_k - A_k\mathbf{v}_k) + A_{k+1}\mathbf{v}_{k+1}^0$. *Set* $k := k+1$ *and go to Step (1).*

(4) Update level k iterate $\mathbf{v}_k$,

$$\mathbf{v}_k^{new} = \mathbf{v}_k + P_k\left(\mathbf{v}_{k+1} - \mathbf{v}_{k+1}^0\right).$$

(5) Apply $\nu_2 \geq 1$ projected Gauss–Seidel iterations with initial iterate $\mathbf{v}_k$. The resulting iterate is also denoted by $\mathbf{v}_k$. Set $k := k-1$. If $k \geq 0$, go to Step (4), otherwise one V-cycle is completed.

We can show that the above algorithm is well defined, that is, that all the intermediate iterates in Algorithm 12.5.1 satisfy the appropriate constraints which is done in what follows.

The resulting iterate $\mathbf{v}_k$ after the application of the projected Gauss–Seidel algorithm in Step 1 of Algorithm 12.5.1 satisfies the constraints $(\mathbf{v}_k)_i \leq (\mathbf{g}^k)_i,\ i \in \Gamma$. This is true due to the projection procedure in the projected Gauss–Seidel algorithm.

In Step 2 we again have that the resulting iterate satisfies the same constraints because we use exact coarse-grid solution.

In Step 3 the constraint set does not change from level k to level $k+1$ (by assumption). Thus, $\mathbf{v}^0_{k+1}$ satisfies the constraints because it is a restriction of $\mathbf{v}_k$ and the latter one satisfies the constraints.

In Step 4 we have to show that $(\mathbf{v}^{\text{new}}_k)_i \leq (\mathbf{g}^k)_i,\ i \in \Gamma$. Indeed, due to the main property of the interpolation matrix $P \equiv P_k$, (12.22),

$$\begin{aligned}
(\mathbf{v}^{\text{new}}_k)|_\Gamma &= \mathbf{v}_k\big|_\Gamma + \big(P_k\big(\mathbf{v}_{k+1} - \mathbf{v}^0_{k+1}\big)\big)\big|_\Gamma \\
&= \mathbf{v}_k|_\Gamma + \mathbf{v}_{k+1}|_\Gamma - \mathbf{v}^0_{k+1}\big|_\Gamma \\
&= \mathbf{v}_k|_\Gamma + \mathbf{v}_{k+1}|_\Gamma - \mathbf{v}_k|_\Gamma \\
&= \mathbf{v}_{k+1}|_\Gamma \\
&\leq \mathbf{g}^{k+1} = \mathbf{g}^k.
\end{aligned}$$

In Step 5 the resulting iterate $\mathbf{v}_k$ satisfies the appropriate constraints again due to the properties of the projected Gauss–Seidel.

The following main fact easily follows from the construction of the FAS iterates.

Theorem 12.7. *Algorithm 12.5.1 provides a monotone scheme.*

Proof. It is sufficient to prove that

$$J_k\big(\mathbf{v}^{\text{new}}_k\big) \leq J_k(\mathbf{v}_k).$$

Denote for brevity $P = P_k$. Based on the definition of $A_{k+1} = P^T A_k P$, J_k and J_{k+1}, and $\mathbf{b}_{k+1} = P^T(\mathbf{b}_k - A_k\mathbf{v}_k) + A_{k+1}\mathbf{v}^0_{k+1}$, we can derive the identity,

$$\begin{aligned}
J_k\big(\mathbf{v}^{\text{new}}_k\big) &= J_k\big(\mathbf{v}_k + P\big(\mathbf{v}_{k+1} - \mathbf{v}^0_{k+1}\big)\big) \\
&= \frac{1}{2}\big(\mathbf{v}_k + P\big(\mathbf{v}_{k+1} - \mathbf{v}^0_{k+1}\big)\big)^T A_k\big(\mathbf{v}_k + P\big(\mathbf{v}_{k+1} - \mathbf{v}^0_{k+1}\big)\big) \\
&\quad - (\mathbf{b}_k)^T\big(\mathbf{v}_k + P\big(\mathbf{v}_{k+1} - \mathbf{v}^0_{k+1}\big)\big) \\
&= J_k(\mathbf{v}_k) + \frac{1}{2}\big(\mathbf{v}_{k+1} - \mathbf{v}^0_{k+1}\big)^T A_{k+1}\big(\mathbf{v}^0_{k+1} - \mathbf{v}_{k+1}\big) \\
&\quad - (\mathbf{b}_k - A_k\mathbf{v}_k)^T P\big(\mathbf{v}_{k+1} - \mathbf{v}^0_{k+1}\big)
\end{aligned}$$

$$\begin{aligned}
&= J_k(\mathbf{v}_k) + \frac{1}{2}(\mathbf{v}_{k+1})^T A_{k+1}\mathbf{v}_{k+1} \\
&\quad - \left(P^T\left(\mathbf{b}_k - A_k\mathbf{v}_k\right) + A_{k+1}\mathbf{v}^0_{k+1}\right)^T \mathbf{v}_{k+1} \\
&\quad + \frac{1}{2}\left(\mathbf{v}^0_{k+1}\right)^T A_{k+1}\mathbf{v}^0_{k+1} + (\mathbf{b}_k - A_k\mathbf{v}_k)^T P\mathbf{v}^0_{k+1} \\
&= J_k(\mathbf{v}_k) + J_{k+1}(\mathbf{v}_{k+1}) + \frac{1}{2}\left(\mathbf{v}^0_{k+1}\right)^T A_{k+1}\mathbf{v}^0_{k+1} + (\mathbf{b}_k - A_k\mathbf{v}_k)^T P\mathbf{v}^0_{k+1}.
\end{aligned}$$

We also have,

$$\begin{aligned}
-J_{k+1}\left(\mathbf{v}^0_{k+1}\right) &= -\frac{1}{2}\left(\mathbf{v}^0_{k+1}\right)^T A_{k+1}\mathbf{v}^0_{k+1} + (\mathbf{b}_{k+1})^T \mathbf{v}^0_{k+1} \\
&= -\frac{1}{2}\left(\mathbf{v}^0_{k+1}\right)^T A_{k+1}\mathbf{v}^0_{k+1} + \left(P^T\left(\mathbf{b}_k - A_k\mathbf{v}_k\right) + A_{k+1}\mathbf{v}^0_{k+1}\right)^T \mathbf{v}^0_{k+1} \\
&= \frac{1}{2}\left(\mathbf{v}^0_{k+1}\right)^T A_{k+1}\mathbf{v}^0_{k+1} + (\mathbf{b}_k - A_k\mathbf{v}_k)^T P\mathbf{v}^0_{k+1}.
\end{aligned}$$

The latter two identities imply the following main one,

$$J_k\left(\mathbf{v}^{\text{new}}_k\right) = J_k(\mathbf{v}_k) + J_{k+1}(\mathbf{v}_{k+1}) - J_{k+1}\left(\mathbf{v}^0_{k+1}\right).$$

Now, having in mind that in Algorithm 12.5.1 the vector $\mathbf{v}_{k+1}$ reduces the functional J_{k+1} (assumed by induction, true at the coarsest level, and because we use a monotone smoother), that is,

$$J_{k+1}(\mathbf{v}_{k+1}) \le J_{k+1}\left(\mathbf{v}^0_{k+1}\right),$$

we arrive at the final desired inequality,

$$J_k\left(\mathbf{v}^{\text{new}}_k\right) \le J_k(\mathbf{v}_k). \qquad \square$$

Part III

Appendices

A

Generalized Conjugate Gradient Methods

In this chapter we summarize a general approach for solving a nonsymmetric and possibly indefinite system of equations. We first present a general variational approach and then consider some special cases that lead to the most popular methods of generalized conjugate gradient type.

A.1 A general variational setting for solving nonsymmetric problems

A common approach to solve the nonsymmetric problem in question is to minimize a certain norm of the current residual $\mathbf{r} = \mathbf{b} - A\mathbf{x}_k$ where $\mathbf{x}_k$ is a current kth-step iterate, for $k \geq 0$. We assume an arbitrary initial iterate $\mathbf{x}_0$. Given are two inner products $(\cdot, \ \cdot)$ and $\langle \cdot, \ \cdot \rangle$. Based on a current set of search vectors $\{\mathbf{d}_k\}_{j\geq 0}^k$, typically a $\langle \cdot, \ \cdot \rangle$-orthogonal system, we compute the next iterate $\mathbf{x}_{k+1} = \mathbf{x}_k + \sum_{j=0}^k \alpha_j^{(k)} \mathbf{d}_j$ such that

$$\|\mathbf{r}_{k+1}\| = \|\mathbf{r}_k - A\mathbf{x}_{k+1}\| \mapsto \min$$

over the set of coefficients $\{\alpha_j^{(k)}\}_{j=0}^k$. The solution

$$\boldsymbol{\alpha}_k = \begin{bmatrix} \alpha_0^{(k)} \\ \alpha_1^{(k)} \\ \vdots \\ \alpha_k^{(k)} \end{bmatrix}$$

solves the Gram system

$$\Lambda_k \boldsymbol{\alpha}_k = \mathbf{g}_k = \begin{bmatrix} (\mathbf{r}_k, \ A\mathbf{d}_0) \\ (\mathbf{r}_k, \ A\mathbf{d}_1) \\ \vdots \\ (\mathbf{r}_k, \ A\mathbf{d}_k) \end{bmatrix}. \tag{A.1}$$

The Gram matrix Λ_k has entries

$$(\Lambda_k)_{j,l} = (A\mathbf{d}_j,\ A\mathbf{d}_l),\ j, l = 0, \ldots,\ k.$$

The next search vector $\mathbf{d}_{k+1}$ is computed from $\mathbf{r}_{k+1}$ and the current set of search vectors $\{\mathbf{d}\}_{j=0}^{k}$. Due to the minimization procedure, we notice that if $\mathbf{r}_{k+1} \neq 0$ then the the set of $\{\mathbf{d}_j\}_{j=0}^{k} \cup \{\mathbf{r}_{k+1}\}$ is linearly independent. Hence, we can define

$$\mathbf{d}_{k+1} = \mathbf{r}_{k+1} - \sum_{j<k+1} \beta_j^{(k)}\ \mathbf{d}_j \tag{A.2}$$

such that $\mathbf{d}_{k+1}$ are $\langle \cdot\ ,\ \cdot \rangle$-orthogonal to all previous $\mathbf{d}_j$. Note that we have the option to choose the inner product $\langle \cdot\ ,\ \cdot \rangle$. The choice A.2 is very natural (and most popular in practice) because it implies that $\mathbf{d}_{k+1} = P_{k+1}(A)\mathbf{r}_0$ for a proper polynomial P_{k+1} of degree $k+1$.

In the special case, most commonly used in practice,

$$\langle \cdot\ ,\ \cdot \rangle = (A\cdot\ ,\ A\cdot), \tag{A.3}$$

a mathematically equivalent choice of the search vectors, seen from the equality

$$\mathbf{r}_{k+1} = \mathbf{r}_k - A\sum_{j=0}^{k} \alpha_j \mathbf{d}_j = -\alpha_k^{(k)} A\mathbf{d}_k + P_k(A)\mathbf{r}_0,$$

and the fact that $\alpha_k^{(k)} \neq 0$, is the Arnoldi construction of $\mathbf{d}_{k+1}$. Namely,

$$\mathbf{d}_{k+1} = A\mathbf{d}_k - \sum_{j=0}^{k} \beta_j^{(k)} \mathbf{d}_j, \tag{A.4}$$

such that $\langle \mathbf{d}_{k+1}, \mathbf{d}_j \rangle = 0$ for $j = 0, \ldots, k$. Then,

$$\beta_j^{(k)} = \frac{\langle A\mathbf{d}_k,\ \mathbf{d}_j \rangle}{\langle \mathbf{d}_j,\ \mathbf{d}_j \rangle}.$$

In the special least squares choice of inner products (A.3), some simplifications occur (cf. [EES83], [Ax87]). The major simplification is that the Gram matrix Λ_k becomes diagonal. We also have proved by induction that $(\mathbf{r}_k,\ A\mathbf{d}_j) = 0$ for $j < k$, hence $(\mathbf{r}_k,\ A\mathbf{r}_k) = (\mathbf{r}_k,\ A\mathbf{d}_k) + \sum_{j<k} \beta_j^{(k-1)}\ (\mathbf{r}_k,\ A\mathbf{d}_j) = (\mathbf{r}_k,\ A\mathbf{r}_k)$. Thus $\alpha_j^{(k)} = 0$ for $j < k$, and

$$\alpha_k \equiv \alpha_k^{(k)} = \frac{(\mathbf{r}_k,\ A\mathbf{d}_k)}{(A\mathbf{d}_k,\ A\mathbf{d}_k)} = \frac{(\mathbf{r}_k,\ A\mathbf{r}_k)}{(A\mathbf{d}_k,\ A\mathbf{d}_k)}.$$

We easily see, for $i < k$,

$$(\mathbf{r}_k,\ A\mathbf{r}_i) = (\mathbf{r}_k, A\mathbf{d}_k) + \sum_{j<k} \beta_j^{(k-1)} (\mathbf{r}_k,\ A\mathbf{d}_j) = 0,$$

We formulate the resulting algorithm, sometimes referred to as the generalized conjugate gradient, least squares (or GCG-LS) method. In [EES83] it was named the GCR (generalized conjugate residual) method, and in one of the most popular papers [SS86], the GMRES method. In the latter paper, the Arnoldi process (A.4) was used to compute the search directions and the inner products $\langle \cdot\, ,\, \cdot\rangle$, $(\cdot\, ,\, \cdot)$ were not related as in (A.3). Thus the GMRES method requires solving systems (A.1) with the Gram matrix Λ_k which differs from the one on the previous step $k-1$ by its last row and column. The latter allows for efficient solution of the system, for example, by Householder orthogonal transformations as demonstrated in [Wal88]. The GCR method with Arnoldi construction of search vectors was considered in [YJ80].

The presentation of the method based on two general inner products and the name GCG-LS is due to Axelsson ([Ax87]).

We summarize the algorithm in the case (A.3).

Algorithm A.1.1 (GCG-LS algorithm). *Given the system* $A\mathbf{x} = \mathbf{b}$ *and a general inner product* $(\cdot\, ,\, \cdot)$.

- *Initiate: let* $\mathbf{x}_0$ *be arbitrary; compute* $\mathbf{r}_0 = \mathbf{b} - A\mathbf{x}_0$, *and let* $\mathbf{d}_0 = \mathbf{r}_0$.
- *For* $k = 0, \ldots,$ *until convergence, compute:*
  1. $\alpha = \frac{(\mathbf{r}_k,\, A\mathbf{d}_k)}{(A\mathbf{d}_k,\, A\mathbf{d}_k)} = \frac{(\mathbf{r}_k,\, A\mathbf{r}_k)}{(A\mathbf{d}_k,\, A\mathbf{d}_k)}$;
  2. $\mathbf{x}_{k+1} = \mathbf{x}_k - \alpha\mathbf{d}_k$;
  3. $\mathbf{r}_{k+1} = \mathbf{r}_k - \alpha A\mathbf{d}_k$.
- *Compute the next search vector as*

$$\mathbf{d}_{k+1} = \mathbf{r}_{k+1} - \sum_{j=0}^{k} \frac{(A\mathbf{r}_{k+1},\, A\mathbf{d}_j)}{(A\mathbf{d}_j,\, A\mathbf{d}_j)}\, \mathbf{d}_j.$$

The convergence of the method is seen from the minimization property of the method. We have

$$\begin{aligned}\|\mathbf{r}_{k+1}\| &= \min_{\alpha} \|\mathbf{r}_k - \alpha A\mathbf{d}_k\| \\ &= \min_{\alpha_0,\, \ldots,\, \alpha_k} \left\| \mathbf{r}_0 - \sum_{j=0}^{k} \alpha_j A\mathbf{d}_j \right\| \\ &= \min_{P_k} \|(I - AP_k(A))\mathbf{r}_0\|,\end{aligned}$$

for any polynomial P_k of degree k. In other words,

$$\|\mathbf{r}_{k+1}\| = \min_{P_{k+1}:\, P_{k+1}(0)=1} \|P_{k+1}(A)\mathbf{r}_0\|.$$

In the case when A is symmetric in the $\langle\cdot, \cdot\rangle$ inner product substantial simplification occurs because the Arnoldi process truncates; that is, it reduces to the Lanczos algorithm

$$\mathbf{d}_{k+1} = A\mathbf{d}_k - \beta_k^{(k)}\mathbf{d}_k - \beta_{k-1}^{(k)}\mathbf{d}_{k-1}.$$

Practical choices for $\langle \mathbf{v}, \mathbf{w} \rangle$ are $\mathbf{w}^T \mathbf{v}$ and $\mathbf{v}^T A^2 \mathbf{w}$. The choice $\langle \mathbf{v}, \mathbf{w} \rangle = \mathbf{w}^T A^2 \mathbf{v}$ equals to $(A\cdot, A\cdot)$ for $(\cdot, \cdot) = (\cdot)^T(\cdot)$. In that case the Gram matrix Λ_k is diagonal and the respective system (A.1) has solution $\alpha_j^{(k)} = 0$ for $j < k$. The only nonzero coefficient $\alpha_k \equiv \alpha_k^{(k)}$ is computed as in Algorithm A.1.1. The resulting method was most probably first considered in [Ch78]. If we use the choice $\langle \mathbf{v}, \mathbf{w} \rangle = (\mathbf{v}, \mathbf{w}) = \mathbf{w}^T \mathbf{v}$, we then end up with a tridiagonal Gram matrix Λ_k. The solution of systems with Λ_k can efficiently be implemented based on orthogonal transformations as originally proposed in [PS75]. The latter resulting method has the popular name MINRES. A complete presentation of these and other CG-type methods is found in Saad [Sa03].

A.2 A quick CG guide

A.2.1 The CG algorithm

The popular CG (conjugate gradient) method is the fundamental tool for solving linear systems of equations $A\mathbf{x} = \mathbf{b}$ with a s.p.d. matrix A. A computational form of the algorithm is as follows.

Algorithm A.2.1 (CG algorithm). *Given vectors* $\mathbf{x}_0$ *(an initial approximation), residual vector* $\mathbf{r} = \mathbf{b} - A\mathbf{x}_0$, *and an initial search direction* $\mathbf{p} = \mathbf{r}$, *and an auxiliary vector* $\mathbf{g} = A\mathbf{p}$, *and also, given a tolerance* $\epsilon > 0$ *and a maximal number of iterations allowed,* $\max_{iter}$, *we perform the following steps.*

(0) *Set iter* $= 0$ *and compute* $\mathbf{r} = \mathbf{b} - A\mathbf{x}_0$, $\mathbf{p} = \mathbf{r}$, *and* $\mathbf{g} = A\mathbf{p}$.
Form the inner products $\delta_{old} = \mathbf{r}^T\mathbf{r}$, *and* $\gamma = \mathbf{p}^T\mathbf{g} = \mathbf{p}^T A\mathbf{p}$.
If $\delta_{old} \leq \epsilon^2 \mathbf{b}^T\mathbf{b}$ *go to (ix).*

(i) *Compute a step length,*

$$\alpha = \frac{\delta_{old}}{\gamma} = \frac{\mathbf{r}^T\mathbf{r}}{\mathbf{p}^T A\mathbf{p}}.$$

(ii) *Compute the next iterate,*

$$\mathbf{x} := \mathbf{x} + \alpha\mathbf{p}.$$

(iii) *Compute the next residual* $\mathbf{r} = \mathbf{b} - A\mathbf{x}$ *as follows,*

$$\mathbf{r} := \mathbf{r} - \alpha\mathbf{g} = \mathbf{r} - \alpha A\mathbf{p}.$$

(iv) *Compute the norm square of the new residual* $\mathbf{r}$ *by forming the inner product* $\delta = \mathbf{r}^T\mathbf{r}$.

(v) *Check for convergence, that is, if* $\delta < \epsilon^2\, \delta_{old}$, *or if the number of iterations iter has reached the prescribed maximal value* $\max_{iter}$. *If one of these conditions is satisfied, go to Step (ix). Otherwise, set iter* $:=$ *iter* $+ 1$ *and go to Step (vi).*

(vi) *Compute* $\beta = \delta/\delta_{old}$ *and set* $\delta_{old} = \delta$.

(vii) Form a new search direction

$$\mathbf{p} := \mathbf{r} + \beta\mathbf{p}.$$

(viii) Compute $\mathbf{g} = A\mathbf{p}$ *and* $\gamma = \mathbf{p}^T\mathbf{g} = \mathbf{p}^T A\mathbf{p}$, *and then go to Step (i).*
(ix) End.

A.2.2 Preconditioning

We can see that the problem $A\mathbf{x} = \mathbf{b}$ can be rewritten for any given s.p.d. matrix B (called a preconditioner) as $B^{-(1/2)}AB^{-(1/2)}(B^{1/2}\mathbf{x}) = B^{-(1/2)}\mathbf{b}$. If we formally apply the CG algorithm A.2.1 to the thus-transformed system and make appropriate change of variables to get an algorithm in terms of the original unknowns, we end up with the popular preconditioned CG (or PCG) algorithm below.

Algorithm A.2.2 (PCG algorithm). *Given vectors* $\mathbf{x}_0$ *(an initial approximation), residual vector* $\mathbf{r} = \mathbf{b} - A\mathbf{x}_0$, *preconditioned residual* $\widetilde{\mathbf{r}} = B^{-1}\mathbf{r}$ *and an initial search direction* $\mathbf{p} = \widetilde{\mathbf{r}}$, *and an auxiliary vector* $\mathbf{g} = A\mathbf{p}$, *and also, given a tolerance* $\epsilon > 0$ *and a maximal number of iterations allowed,* $\max_{iter}$, *we perform the following steps.*

(0) Set iter $= 0$ *and compute* $\mathbf{r} = \mathbf{b} - A\mathbf{x}_0$, $\widetilde{\mathbf{r}} = B^{-1}\mathbf{r}$, $\mathbf{p} = \widetilde{\mathbf{r}}$, *and* $\mathbf{g} = A\mathbf{p}$.
Form the inner products $\delta_{old} = \mathbf{r}^T\widetilde{\mathbf{r}}$ *and* $\gamma = \mathbf{p}^T\mathbf{g} = \mathbf{p}^T A\mathbf{p}$.
If $\delta_{old} \leq \epsilon^2\mathbf{b}^T B^{-1}\mathbf{b}$ *go to (ix).*

(i) Compute a step length

$$\alpha = \frac{\delta_{old}}{\gamma} = \frac{\mathbf{r}^T\widetilde{\mathbf{r}}}{\mathbf{p}^T A\mathbf{p}}.$$

(ii) Compute the next iterate

$$\mathbf{x} := \mathbf{x} + \alpha\mathbf{p}.$$

(iii) Compute the next residual $\mathbf{r} = \mathbf{b} - A\mathbf{x}$ *as follows,*

$$\mathbf{r} := \mathbf{r} - \alpha\mathbf{g} = \mathbf{r} - \alpha A\mathbf{p}.$$

(p) Preconditioning step: compute

$$\widetilde{\mathbf{r}} = B^{-1}\mathbf{r}.$$

(iv) Compute the preconditioned norm square of the new residual $\mathbf{r}$ *by forming the inner product* $\delta = \mathbf{r}^T\widetilde{\mathbf{r}} = \mathbf{r}^T B^{-1}\mathbf{r}$.

(v) Check for convergence, that is, if $\delta < \epsilon^2\,\delta_{old}$, *or if the number of iterations iter has reached the prescribed maximal value* $\max_{iter}$. *If one of these conditions is satisfied, go to Step (ix). Otherwise set iter := iter + 1 and go to Step (vi).*

(vi) Compute $\beta = \delta/\delta_{old}$ *and set* $\delta_{old} = \delta$.

(vii) Form a new search direction

$$\mathbf{p} := \widetilde{\mathbf{r}} + \beta \mathbf{p}.$$

(viii) Compute $\mathbf{g} = A\mathbf{p}$ *and* $\gamma = \mathbf{p}^T \mathbf{g} = \mathbf{p}^T A\mathbf{p}$*, and then go to Step (i).*
(ix) End.

If we compare both algorithms, A.2.1 and A.2.2, they differ by Step (p) in the latter, where we have to apply the preconditioner in order to compute the preconditioned residual. They also differ in the initial step (o), where in Algorithm A.2.2 we need one more inverse action of the preconditioner B. In terms of storage, the PCG algorithm A.2.2 requires one more vector to store the preconditioned residual.

A.2.3 Best polynomial approximation property of CG

The CG exploits a number of properties which make it the efficient and popular method it is, among which is its minimization property of the energy norm of the error in a space spanned by powers of the matrix A times the initial error.

Proposition A.1. *The CG method has the following main property. Let* $\mathbf{x}_k$ *be the kth iterate and* $\mathbf{e}_k = A^{-1}\mathbf{b} - \mathbf{x}_k$ *be the corresponding error. We have, defining* $\|\mathbf{v}\|_A = (\mathbf{v}^T A\mathbf{v})^{1/2}$,

$$\|\mathbf{e}_k\|_A \leq \min_{p_k} \|p_k(A)\mathbf{e}_0\|_A,$$

where the minimum is taken over all polynomials $p_k = p_k(t)$ *of degree k normalized at the origin; that is,* $p_k(0) = 1$.

For a proof, see any appropriate text on numerical methods, or [H94], for example.

The latter error estimate has the property that if we replace A with $B^{-(1/2)} A B^{-(1/2)}$, and $\mathbf{e}_k := B^{1/2}\mathbf{e}_k$, we obtain,

$$\|\mathbf{e}_k\|_A \leq \min_{p_k} \|p_k(B^{-(1/2)} A B^{-(1/2)})\mathbf{e}_0\|_A.$$

That is, the norm of the error does not change but the argument of the polynomial p_k does; it is the preconditioned matrix $B^{-(1/2)} A B^{-(1/2)}$, which may have a much more favorable spectrum than A for proper choice of the preconditioner B.

A.2.4 A decay rate estimate for A^{-1}

Here we present a result originally proved in [DMS84] (already mentioned in Section 6.12).

Let A be s.p.d. with spectrum contained in $[\alpha,\ \beta] \subset \mathbb{R}^+$.

The result in question is based on two observations.

- *First observation.* Let $B = (b_{ij})$ be any matrix. Then for any vector norm $\|.\|$ such that $\|\mathbf{e}_i\| = 1$ where $\mathbf{e}_i$ is the ith unit coordinate vector, then

$$|b_{ij}| \leq \|B\|.$$

  This follows from $b_{ij} = \mathbf{e}_i^T B\mathbf{e}_j \leq \|B\| \|\mathbf{e}_i\| \|\mathbf{e}_j\| = \|B\|$.

- *Second observation.* Let $B = A^{-1}$ and let $\|.\|$ be defined for any $\mathbf{v} = (v_i) \in \mathbb{R}^n$ as $\|\mathbf{v}\|^2 = \mathbf{v}^T\mathbf{v} = \sum_i v_i^2$; that is, $\|.\|$ is the standard Euclidean vector norm. Obviously, $\|\mathbf{e}_i\| = 1$.
  Then, for any polynomial p_k of degree $k \geq 0$, consider the matrix $B - p_k(A)$. For any entry b_{ij} of $B = A^{-1}$ with indices (i, j) outside the nonzero sparsity pattern of $p_k(A)$, we have $(p_k(A))_{ij} = 0$ and hence

$$|b_{ij}| \leq \|B - p_k(A)\| = \|A^{-1} - p_k(A)\| \leq \sup_{\lambda \in [\alpha, \beta]} |\lambda^{-1} - p_k(\lambda)|.$$

Note that we have the flexibility to choose the coefficients of the polynomial p_k (because this will not change the sparsity pattern of $p_k(A)$). Therefore,

$$|b_{ij}| \leq \inf_{p_k} \sup_{\lambda \in [\alpha, \beta]} |\lambda^{-1} - p_k(\lambda)|.$$

Thus, the following simple upper bound holds

$$|b_{ij}| \leq \frac{1}{\alpha} \inf_{p_k} \sup_{\lambda \in [\alpha, \beta]} |1 - \lambda p_k(\lambda)| = \frac{1}{\alpha} \frac{2q^{k+1}}{1 + q^{2(k+1)}}, \qquad q = \frac{\sqrt{\kappa} - 1}{\sqrt{\kappa} + 1}, \qquad \kappa = \frac{\beta}{\alpha}.$$

In the last estimate we use

$$p_k : 1 - t p_k(t) = \frac{T_{k+1}\left(\frac{\alpha+\beta-2t}{\beta-\alpha}\right)}{T_{k+1}\left(\frac{\alpha+\beta}{\beta-\alpha}\right)},$$

where T_{k+1} is the well-known Chebyshev polynomial of degree $k + 1$.

In conclusion, if A is well conditioned (i.e., $\kappa = \beta/\alpha$ is a nice number) and (i, j) is away from the sparsity pattern of A, say at distance k, for large k the entry $b_{ij} = (A^{-1})_{ij} \simeq q^{k+1}$; that is, its value decays geometrically, with $k \mapsto \infty$.

B

Properties of Finite Element Matrices. Further Details

This chapter summarizes some additional properties of finite element matrices arising in the finite element discretization of second-order elliptic PDEs. The topics covered include:

- Piecewise linear Lagrangian basis functions; element stiffness matrices and global matrix conditioning; mass matrices and equivalent L_2-norms. Gärding inequality, duality argument, L_2-error estimates, and weak approximation property. This material, in particular, supplements Chapter 8.
- A semilinear second-order elliptic PDE (supplements Chapter 11).
- Mixed finite elements for second-order elliptic PDEs; the space $H(\text{div})$, and the related "inf–sup" condition. The (computable) Fortin projection. This material, in particular, supplements Chapter 9.
- Nonconforming finite elements and Stokes problem (supplements part of Chapter 9).
- Maxwell equations and $H(\text{curl})$-problems (can be viewed as an addition to Section 7.11).

B.1 Piecewise linear finite elements

Consider the following second-order elliptic operator

$$\mathcal{L}u \equiv -\sum_{r=1}^{d} \frac{\partial}{\partial x_r}\left(\sum_{s=1}^{d} a_{r,s}(\mathbf{x})\frac{\partial u}{\partial x_s}\right) + \sum_{i=1}^{d} b_i(\mathbf{x})\frac{\partial u}{\partial x_i} + c(\mathbf{x})u.$$

Here, the coefficient matrix $\mathcal{A}(\mathbf{x}) = \{a_{i,j}(\mathbf{x})\}_{i,j=1}^{d}$ is assumed symmetric positive definite uniformly in $\mathbf{x} \in \Omega$ where Ω is a plane polygon ($d = 2$) or a 3D polytope ($d = 3$). The vector field $\mathbf{b} = (b_i(\mathbf{x}))_{i=1}^{d}$ and the low-order term coefficient $c = c(\mathbf{x})$ are also given bounded measurable functions in Ω. We associate with $\mathcal{L}$ the following boundary value problem posed variationally. For a given function $f \in L_2(\Omega)$, find a weak solution u, that is, $u \in L_2(\Omega)$, $\partial u/\partial x_i \in L_2(\Omega)$, such that for any sufficiently smooth function φ, we have

$$a(u, \varphi) \equiv \sum_{i,j=1}^{d} \left(a_{i,j} \frac{\partial u}{\partial x_i}, \frac{\partial \varphi}{\partial x_j} \right) + \left(\sum_{i=1}^{d} b_i \frac{\partial u}{\partial x_i} + cu, \varphi \right) = (f, \varphi). \tag{B.1}$$

We use the common notation $(.,.)$ for the $L_2(\Omega)$-inner product. The standard $L_2(\Omega)$ norm is denoted by $\|.\|_0 = \sqrt{(.\ .)}$. Here, u and φ are assumed vanishing on $\partial\Omega$. The corresponding space of functions is the well-known Sobolev space $H_0^1(\Omega)$. The norm $\|v\|_1$ for any $v \in H^1(\Omega)$ is defined as $\|v\|_1^2 = \|v\|_0^2 + \sum_{i=1}^{d} \|\partial u/\partial x_i\|_0^2$. For functions that vanish on $\partial\Omega$ the following seminorm gives an equivalent norm

$$|v|_1^2 = \sum_{i=1}^{d} \left\| \frac{\partial u}{\partial x_i} \right\|_0^2.$$

Inhomogeneous Dirichlet boundary conditions are similarly treated. We first find a function u_0 (explicitly) that satisfies the boundary conditions and then the difference $u - u_0$ will satisfy the homogeneous ones. This is easily achievable on a discrete level (by approximating the boundary data to belong to the discrete space).

The finite element method of interest refers to the following Ritz–Galerkin procedure. We construct a discrete (i.e., finite-dimensional) space $V = V_h$ where the parameter $h \mapsto 0$ and then V approximates the continuous (infinite-dimensional) space $H_0^1(\Omega)$. The functions in V are simple piecewise polynomials, that is why they admit certain approximation properties. More specifically, let $\{\tau\}$ be a set of simple-shaped polygons (or polytopes in 3D), called elements, which provide a nonoverlapping partition of Ω. There is a requirement that every two elements either share exactly a single common vertex, or a single common face (or a single common edge in 3D) or their intersection is empty. With this property the set of elements $\mathcal{T} = \{\tau\}$ is called the triangulation of Ω. Then h is typically referred to the maximal diameter of τ when τ runs over the elements in $\mathcal{T}$. The functions $v \in V$ restricted to any element τ are polynomials of a given fixed degree p. For the class of problems we consider, the functions in V should belong to $H^1(\Omega)$. This is guaranteed if we can construct V such that for any $v \in V$ the formula for integration by parts is valid. Namely, for any two neighboring elements τ_1 and τ_2 sharing a common face, and any smooth function φ vanishing outside $\tau_1 \cup \tau_2$, we have

$$\left(\frac{\partial v}{\partial x_i}, \varphi \right) = -\left(v, \frac{\partial \varphi}{\partial x_i} \right) + \int_{\partial\tau_1 \cap \partial\tau_2} [v]\, \varphi\, \mathbf{n}_i \, d\varrho.$$

Here $[v]$ stands for the jump of v across the common face $\partial\tau_1 \cap \partial\tau_2$ and $\mathbf{n}_i$ is the ith component of a unit vector $\mathbf{n}$ normal to that face. That is, in order to have the formula of integration by parts valid we must ensure that $[v] = 0$. This imposes continuity of the finite element functions $v \in V$.

Lagrangian piecewise linear basis

From now on, we consider the model 2D case and triangular elements τ. The extension to 3D is straightforward. A simple way to construct V being the space of piecewise linear continuous functions is to construct a basis for V. Let $\{\mathbf{x}_i\}$ be the set of vertices of all triangles τ. Introduce also the (geometrical) coordinates (x_i, y_i) of the vertex $\mathbf{x}_i$. For any element τ consider its three vertices $\mathbf{x}_{i_1}$, $\mathbf{x}_{i_2}$, and $\mathbf{x}_{i_3}$ with respective coordinates (x_{i_s}, y_{i_s}), $s = 1, 2, 3$. For each i_s, we define a basis function $\varphi_{i_s} = \varphi_{i_s}(x, y)$ locally, for now only on the element τ, as the solution of the following linear equation

$$\begin{vmatrix} x & y & \varphi_{i_s} & 1 \\ x_{i_1} & y_{i_1} & \delta_{i_1,\, i_s} & 1 \\ x_{i_2} & y_{i_2} & \delta_{i_2,\, i_s} & 1 \\ x_{i_3} & y_{i_3} & \delta_{i_3,\, i_s} & 1 \end{vmatrix} = 0. \tag{B.2}$$

Here, $\delta_{q,r} = 0$ if $q \neq r$ and 1 otherwise. It is clear that $\varphi_{i_s}(x_{i_r}, y_{i_r}) = \delta_{r,s}$. Simply let $x = x_{i_r}$ and $y = y_{i_r}$ in (B.2). Then $\varphi_{i_s} = \delta_{i_r,\, i_s}$ solves (B.2) because the determinant has two identical rows then. It is also clear that φ_{i_s} is a linear function of the form,

$$\varphi_{i_s} = a_s(x - x_{i_s}) + b_s(y - y_{i_s}) + 1, \qquad a_s = \frac{\partial \varphi_{i_s}}{\partial x} \text{ and } b_s = \frac{\partial \varphi_{i_s}}{\partial y}. \tag{B.3}$$

It is also trivial to see that φ_{i_s} being defined on all elements that share vertex $\mathbf{x}_{i_s}$ is actually continuous across the element boundaries. This is the case because on every common edge of two such elements φ_{i_s} it is uniquely defined as a linear function (effectively of one variable along that edge) that takes value one at vertex $\mathbf{x}_{i_s}$ and vanishes at the other endpoint of that edge. Because φ_{i_s} vanishes on the boundary of $\cup\{\tau :\ \mathbf{x}_{i_s}$ is a vertex of $\tau\}$, φ_{i_s} can be extended by zero in the rest of Ω and still be continuous. The set of functions φ_{i_s} is easily seen to be linearly independent and span the space of continuous piecewise linear functions V. If we want to satisfy homogeneous Dirichlet boundary conditions, we simply remove the functions φ_{i_s} for nodes $\mathbf{x}_{i_s} \in \partial\Omega$ from the basis.

We can easily compute the partial derivatives a_s and b_s of the basis function in (B.2). We have

$$0 = \begin{vmatrix} 1 & 0 & \dfrac{\partial \varphi_{i_s}}{\partial x} & 0 \\ x_{i_1} & y_{i_1} & \delta_{i_1,\, i_s} & 1 \\ x_{i_2} & y_{i_2} & \delta_{i_2,\, i_s} & 1 \\ x_{i_3} & y_{i_3} & \delta_{i_3,\, i_s} & 1 \end{vmatrix} = \frac{\partial \varphi_{i_s}}{\partial x} \begin{vmatrix} x_{i_1} & y_{i_1} & 1 \\ x_{i_2} & y_{i_2} & 1 \\ x_{i_3} & y_{i_3} & 1 \end{vmatrix} + \begin{vmatrix} y_{i_1} & \delta_{i_1,\, i_s} & 1 \\ y_{i_2} & \delta_{i_2,\, i_s} & 1 \\ y_{i_3} & \delta_{i_3,\, i_s} & 1 \end{vmatrix}, \tag{B.4}$$

and

$$0 = \begin{vmatrix} 0 & 1 & \dfrac{\partial \varphi_{i_s}}{\partial y} & 0 \\ x_{i_1} & y_{i_1} & \delta_{i_1,\, i_s} & 1 \\ x_{i_2} & y_{i_2} & \delta_{i_2,\, i_s} & 1 \\ x_{i_3} & y_{i_3} & \delta_{i_3,\, i_s} & 1 \end{vmatrix} = \frac{\partial \varphi_{i_s}}{\partial y} \begin{vmatrix} x_{i_1} & y_{i_1} & 1 \\ x_{i_2} & y_{i_2} & 1 \\ x_{i_3} & y_{i_3} & 1 \end{vmatrix} - \begin{vmatrix} x_{i_1} & \delta_{i_1,\, i_s} & 1 \\ x_{i_2} & \delta_{i_2,\, i_s} & 1 \\ x_{i_3} & \delta_{i_3,\, i_s} & 1 \end{vmatrix}. \tag{B.5}$$

That is, the basis function φ_{i_s} takes the form (see (B.3)),

$$\varphi_{i_s}(x,y) = 1 - \frac{\begin{vmatrix} y_{i_1} & \delta_{i_1,\,i_s} & 1 \\ y_{i_2} & \delta_{i_2,\,i_s} & 1 \\ y_{i_3} & \delta_{i_3,\,i_s} & 1 \end{vmatrix}}{\begin{vmatrix} x_{i_1} & y_{i_1} & 1 \\ x_{i_2} & y_{i_2} & 1 \\ x_{i_3} & y_{i_3} & 1 \end{vmatrix}}(x - x_{i_s}) + \frac{\begin{vmatrix} x_{i_1} & \delta_{i_1,\,i_s} & 1 \\ x_{i_2} & \delta_{i_2,\,i_s} & 1 \\ x_{i_3} & \delta_{i_3,\,i_s} & 1 \end{vmatrix}}{\begin{vmatrix} x_{i_1} & y_{i_1} & 1 \\ x_{i_2} & y_{i_2} & 1 \\ x_{i_3} & y_{i_3} & 1 \end{vmatrix}}(y - y_{i_s}). \quad \text{(B.6)}$$

We finally note that the determinant

$$\begin{vmatrix} x_{i_1} & y_{i_1} & 1 \\ x_{i_2} & y_{i_2} & 1 \\ x_{i_3} & y_{i_3} & 1 \end{vmatrix}$$

is nonzero if the nodes $\{\mathbf{x}_{i_r}\}_{r=1}^{3}$ form a nondegenerated triangle.

Again, we note that the above construction is general and can be applied in 3D as well. For basis functions on tetrahedral elements, we will end up with 4-by-4 determinants in place of (B.2).

In what follows, we derive an explicit expression (see (B.7)) for the three-by-three element matrix A_τ computed from the Laplace bilinear form restricted to a general triangle τ; that is, $A_\tau = (\int_\tau \nabla\varphi_{i_r} \cdot \nabla\varphi_{i_s}\, d\mathbf{x})_{r,s=1}^{3}$. Consider the following triangle $\overline{\tau}$ with vertices $(0,0)$, $(0,1)$ and (X,Y). For a general triangle τ, we can use the transformation

$$\begin{bmatrix} x \\ y \end{bmatrix} = \begin{bmatrix} x_{i_1} \\ y_{i_1} \end{bmatrix} + (x_{i_2} - x_{i_1})\, Q \begin{bmatrix} \overline{x} \\ \overline{y} \end{bmatrix}.$$

Here Q is a rotation, that is, an orthogonal matrix $Q^T Q = I$. Then the basis functions on τ are given by $\varphi_{i_r}(\mathbf{x}) = \overline{\varphi}_r((1/(x_{i_2} - x_{i_1}))\, Q^T(\mathbf{x} - \mathbf{x}_{i_1}))$ where $\overline{\varphi}_r(\overline{\mathbf{x}})$, $r = 1, 2, 3$ are the Lagrangian basis functions associated with the vertices of the "reference" triangle $\overline{\tau}$. The respective element integrals are related as follows,

$$\int_\tau \nabla\varphi_{i_r} \cdot \nabla\varphi_{i_s}\, dxdy = \int_{\overline{\tau}} \overline{\nabla}\overline{\varphi}_r \cdot \overline{\nabla}\overline{\varphi}_s\, d\overline{x}d\overline{y}.$$

We use here the fact that Q is orthogonal. In other words, the element matrices for the 2D Poisson equation do not change if the triangle is translated, rotated, and replaced by a geometrically similar one. Thus, without loss of generality, we can compute the element matrices for the triangle with vertices $(x_{i_1},\ y_{i_1}) = (0,0)$, $(x_{i_2},\ y_{i_2}) = (0,1)$ and $(x_{i_3},\ y_{i_3}) = (X,Y)$. Let the angles of τ associated with the vertices $(0,0)$ and $(0,1)$ be α and β. Then, $X = \cot\alpha/(\cot\alpha + \cot\beta)$ and $Y = 1/(\cot\alpha + \cot\beta)$. Based on the expressions for the derivatives of φ_{i_s}, $s = 1, 2, 3$, given by (B.4) and (B.5), we readily get

$$\nabla\varphi_{i_1} = \begin{bmatrix} -1 \\ \frac{-1+X}{Y} \end{bmatrix}, \quad \nabla\varphi_{i_2} = \begin{bmatrix} 1 \\ \frac{-X}{Y} \end{bmatrix}, \quad \text{and} \quad \nabla\varphi_{i_3} = \begin{bmatrix} 0 \\ \frac{1}{Y} \end{bmatrix}.$$

Then, because the gradients are constant vectors, we have $\int_\tau \nabla\varphi_{i_r} \cdot \nabla\varphi_{i_s}\, dxdy = |\tau|\ \nabla\varphi_{i_r} \cdot \nabla\varphi_{i_s}$ with $|\tau| = (\sin\alpha \sin\beta)/(2\sin(\alpha+\beta))$ being the area of τ.

The resulting 3-by-3 element matrix equals

$$A_\tau = |\tau| \begin{bmatrix} 1+\left(\frac{-1+X}{Y}\right)^2 & -1+\frac{(1-X)X}{Y^2} & -\frac{1-X}{Y^2} \\ -1+\frac{X(1-X)}{Y^2} & 1+\frac{X^2}{Y^2} & -\frac{X}{Y^2} \\ -\frac{1-X}{Y^2} & -\frac{X}{Y^2} & \frac{1}{Y^2} \end{bmatrix}.$$

We first notice that

$$\frac{|\tau|}{Y} = \frac{\sin\alpha \sin\beta}{2\sin(\alpha+\beta)}(\cot\alpha + \cot\beta) = \frac{1}{2}.$$

Also, we have

$$\frac{X}{Y} = \cot\alpha \quad \text{and} \quad \frac{X^2+Y^2}{Y} = X\frac{X}{Y} + Y = \frac{\cot^2\alpha + 1}{\cot\alpha + \cot\beta} = \cot\alpha + \cot\gamma,$$

where $\gamma = \pi - \alpha - \beta$ is the third angle of τ (at the vertex (X, Y)). The latter follows from the well-known identity

$$\cot\alpha\cot\beta + \cot\alpha\cot\gamma + \cot\beta\cot\gamma = 1.$$

Thus, we end up with the following expression for A_τ,

$$A_\tau = \frac{1}{2}\begin{bmatrix} \cot\beta + \cot\gamma & -\cot\gamma & -\cot\beta \\ -\cot\gamma & \cot\gamma + \cot\alpha & -\cot\alpha \\ -\cot\beta & -\cot\alpha & \cot\alpha + \cot\beta \end{bmatrix}. \tag{B.7}$$

Element matrices and assembling

Once having a basis $\{\varphi_i\}$, we can derive the discrete system for the finite element solution of (B.1). Namely, we seek $u_h = \sum_i u_h(\mathbf{x}_i)\varphi_i \in V$ such that

$$a(u_h, \varphi) = (f, \varphi) \quad \text{for all } \varphi \in V.$$

It is sufficient above to have φ run over the basis $\{\varphi_i\}$ of V. Upon expanding, we get

$$\int_\Omega \left[\sum_{\mathbf{x}_k\in\mathcal{N}} u_h(\mathbf{x}_k) \sum_{r,s=1}^{2} a_{r,s}(\mathbf{x}) \frac{\partial\varphi_k}{\partial x_r}\frac{\partial\varphi_i}{\partial x_s} + \sum_r \mathbf{b}_r(\mathbf{x})\frac{\partial\varphi_k}{\partial x_r}\varphi_i + c(\mathbf{x})\varphi_k\varphi_i\right] d\mathbf{x}$$
$$= (f, \varphi_i).$$

Splitting the integral on Ω as a sum over $\tau \in \mathcal{T}$, we get

$$a(u_h, \varphi_i) = \sum_{\tau:\, \mathbf{x}_i\in\tau} \sum_{\mathbf{x}_k\in\tau} u_h(\mathbf{x}_k) \left[\int_\tau \sum_{r,s=1}^{2} a_{r,s}(\mathbf{x})\frac{\partial\varphi_k}{\partial x_r}\frac{\partial\varphi_i}{\partial x_s}\, d\mathbf{x} \right.$$
$$\left. + \int_\tau \sum_r \mathbf{b}_r(\mathbf{x})\frac{\partial\varphi_k}{\partial x_r}\varphi_i\, d\mathbf{x} + \int_\tau c(\mathbf{x})\varphi_k\varphi_i\, d\mathbf{x}\right]$$
$$= (f, \varphi_i).$$

For general coefficients $a_{r,s}(\mathbf{x})$, $\mathbf{b}_r(\mathbf{x})$, and $c(\mathbf{x})$ a suitable approximation is to assume that they are constants (generally, different) on every element τ. That is, we assume $a_{r,s}(\mathbf{x}) \simeq a_{r,s}(\mathbf{x}_\tau)$, $\mathbf{b}_r(\mathbf{x}) \simeq \mathbf{b}_r(\mathbf{x}_\tau)$, and $c(\mathbf{x}) \simeq c(\mathbf{x}_\tau)$ for a node $\mathbf{x}_\tau \in \tau$. Other (more accurate) approximations are also possible. Then the discrete bilinear form takes the (approximate) form

$$
\begin{aligned}
a(u_h, \varphi_i) \simeq & \sum_{\tau:\, \mathbf{x}_i \in \tau} \sum_{\mathbf{x}_k \in \tau} u_h(\mathbf{x}_k) \left[\sum_{r,s=1}^{2} a_{r,s}(\mathbf{x}_\tau) \int_\tau \frac{\partial \varphi_k}{\partial x_r} \frac{\partial \varphi_i}{\partial x_s} \, d\mathbf{x} \right. \\
& \left. + \sum_r \mathbf{b}_r(\mathbf{x}_\tau) \int_\tau \frac{\partial \varphi_k}{\partial x_r} \varphi_i \, d\mathbf{x} + c(\mathbf{x}_\tau) \int_\tau \varphi_k \varphi_i \, d\mathbf{x} \right] \\
= & \, (f, \varphi_i) \\
\simeq & \sum_{\tau:\, \mathbf{x}_i \in \tau} f(\mathbf{x}_\tau) \int_\tau \varphi_i \, d\mathbf{x}.
\end{aligned}
$$

Introduce now the three-by-three element matrices

$$
\begin{aligned}
A_\tau &= \left(\sum_{r,s=1}^{2} a_{r,s}(\mathbf{x}_\tau) \int_\tau \frac{\partial \varphi_k}{\partial x_r} \frac{\partial \varphi_i}{\partial x_s} \, d\mathbf{x} \right)_{\mathbf{x}_k,\, \mathbf{x}_i \in \tau}, \\
B_\tau &= \left(\sum_r \mathbf{b}_r(\mathbf{x}_\tau) \int_\tau \frac{\partial \varphi_k}{\partial x_r} \varphi_i \, d\mathbf{x} \right)_{\mathbf{x}_k,\, \mathbf{x}_i \in \tau}, \\
C_\tau &= c(\mathbf{x}_\tau) \left(\int_\tau \varphi_k \varphi_i \, d\mathbf{x} \right)_{\mathbf{x}_k,\, \mathbf{x}_i \in \tau}.
\end{aligned}
$$

Note that B_τ is nonsymmetric and we should take care when computing its entries (otherwise we may end up computing B_τ^T instead). Because the derivatives of the basis functions are constant on τ, some simplification occurs. We have

$$
\begin{aligned}
A_\tau &= \left(\sum_{r,s=1}^{2} a_{r,s}(\mathbf{x}_\tau) |\tau| \frac{\partial \varphi_k}{\partial x_r} \frac{\partial \varphi_i}{\partial x_s} \right)_{\mathbf{x}_k,\, \mathbf{x}_i \in \tau}, \\
B_\tau &= \left(\sum_r \mathbf{b}_r(\mathbf{x}_\tau) \frac{\partial \varphi_k}{\partial x_r} \int_\tau \varphi_i \, d\mathbf{x} \right)_{\mathbf{x}_k,\, \mathbf{x}_i \in \tau}, \\
C_\tau &= c(\mathbf{x}_\tau) \left(\int_\tau \varphi_k \varphi_i \, d\mathbf{x} \right)_{\mathbf{x}_k,\, \mathbf{x}_i \in \tau}.
\end{aligned}
$$

Here (and in what follows), $|\tau|$ stands for the area of τ. Because $|\tau| = \mathcal{O}(h^2)$ and $\partial\varphi_k/\partial x_r = \mathcal{O}(h^{-1})$, it is clear then that the entries of A_τ are order $\mathcal{O}(1)$, the entries of B_τ are $\mathcal{O}(h)$, and the entries of C_τ are $\mathcal{O}(h^2)$.

The global matrix is the sum of three matrices A, B, and C, which are assembled from the respective element matrices in the following sense. Let $\mathbf{v}_\tau \in \mathbb{R}^3$ be the restriction of $\mathbf{v}$ to τ; that is, $\mathbf{v}_\tau = (v(\mathbf{x}_s))_{\mathbf{x}_s\in\tau}$. Then, "assembly" refers to the following representation of the global quadratic form as a sum of local quadratic forms,

$$\mathbf{v}^T A\mathbf{w} = \sum_\tau \mathbf{v}_\tau^T A_\tau \mathbf{w}_\tau,$$

$$\mathbf{v}^T B\mathbf{w} = \sum_\tau \mathbf{v}_\tau^T B_\tau \mathbf{w}_\tau,$$

$$\mathbf{v}^T C\mathbf{w} = \sum_\tau \mathbf{v}_\tau^T C_\tau \mathbf{w}_\tau.$$

The global (or assembled) matrix is called a stiffness matrix, and the element matrices are sometimes called element stiffness matrices.

Introducing the vector of unknowns $\mathbf{u} = (u_h(\mathbf{x}_k))_{\mathbf{x}_k\in\mathcal{N}}$, and the right-hand side vector

$$\mathbf{f} = (f_i)_{\mathbf{x}_i\in\mathcal{N}}, \qquad f_i = \sum_{\tau:\ \mathbf{x}_i\in\tau} f(\mathbf{x}_\tau)\int_\tau \varphi_i\ d\mathbf{x},$$

we end up with the following linear systems of equations,

$$(A+B+C)\mathbf{u} = \mathbf{f}.$$

The element matrices that form the lower-order term matrix C,

$$M_\tau = \left(\int_\tau \varphi_k\varphi_i\ d\mathbf{x}\right)_{\mathbf{x}_k,\mathbf{x}_i\in\tau},$$

are referred to as element mass matrices, and they play an important role in analyzing the conditioning of the global stiffness matrix. We have the following explicit form of M_τ,

$$M_\tau = \frac{|\tau|}{12}\begin{bmatrix} 2 & 1 & 1 \\ 1 & 2 & 1 \\ 1 & 1 & 2 \end{bmatrix}.$$

We notice that M_τ is equivalent to $D_\tau = (|\tau|/3)I$ in the sense that for any vector $\mathbf{v}_\tau$, we have

$$\frac{1}{4}\,\mathbf{v}_\tau^T D_\tau \mathbf{v}_\tau \le \mathbf{v}_\tau^T M_\tau \mathbf{v}_\tau \le \mathbf{v}_\tau^T D_\tau \mathbf{v}_\tau.$$

Let D be the diagonal matrix assembled from the element (diagonal) matrices D_τ and M be the assembled mass matrix. Then, we also have the global equivalence result

$$\frac{1}{4}\,\mathbf{v}^T D\mathbf{v} \le \mathbf{v}^T M\mathbf{v} \le \mathbf{v}^T D\mathbf{v}.$$

Note that $\mathbf{v}^T M\mathbf{v} = (v,\ v) = \|v\|_0^2$, where $v = \sum v_i\varphi_i$ is the finite element function that corresponds to the coefficient vector $\mathbf{v}$. Finally assume that the triangulation $\{\tau\}$ is quasiuniform in the sense that for two positive constants ν and μ,

$$\nu h^2 \le |\tau| \le \mu\, h^2.$$

Then, $\mathbf{v}^T D\mathbf{v} = \sum_\tau |\tau|\mathbf{v}_\tau^T\mathbf{v}_\tau \le \mu h^2 m_0 \mathbf{v}^T\mathbf{v}$, where $m_0 \ge 1$ is the maximal number of elements that share a common vertex. Similarly $\mathbf{v}^T D\mathbf{v} \ge \nu\ h^2\mathbf{v}^T\mathbf{v}$. That is, the following estimate holds,

$$4\mu m_0 \ge \frac{h^2\ \mathbf{v}^T\mathbf{v}}{\mathbf{v}^T M\mathbf{v}} \ge \nu.$$

We often use the inner product $(\mathbf{v},\ \mathbf{w})_0 = \mathbf{w}^T M\mathbf{v} = (v,\ w)$. The above estimates show that the vector inner product $\mathbf{w}^T\mathbf{v}$ and the one generated by the mass matrix, $(\mathbf{w},\ \mathbf{v})_0$, are equivalent up to a scaling factor proportional to h^{-2}. In 3D a similar result holds; the scaling factor then is h^{-3}.

Matrix conditioning

Assume for the time being that $B = C = 0$. The conditioning of the stiffness matrix A is studied in what follows. Because we have assumed that the coefficient matrix $\mathcal{A}(\mathbf{x}) = \{a_{r,s}(\mathbf{x})\}_{r,s=1}^2$ is s.p.d., uniformly w.r.t. $\mathbf{x}$, there are two constants μ_τ and ν_τ such that for all $\mathbf{x} \in \tau$,

$$\nu_\tau\ \underline{\xi}^T\underline{\xi} \le \sum_{r,s=1}^2 a_{r,s}(\mathbf{x})\xi_r\xi_s \le \mu_\tau\underline{\xi}^T\underline{\xi},\quad \text{for all } \underline{\xi} \in \mathbb{R}^2.$$

This estimate immediately implies that for the matrix

$$A_\tau^{(0)} = \left(\int_\tau \sum_{r=1}^2 \frac{\partial\varphi_k}{\partial x_r}\frac{\partial\varphi_i}{\partial x_r}\,d\mathbf{x}\right)_{\mathbf{x}_k,\ \mathbf{x}_i\in\tau},$$

the following spectral equivalence relations hold,

$$\nu_\tau\ \mathbf{v}_\tau^T A_\tau^{(0)}\mathbf{v}_\tau \le \mathbf{v}_\tau^T A_\tau\mathbf{v}_\tau \le \mu_\tau\ \mathbf{v}_\tau^T A_\tau^{(0)}\mathbf{v}_\tau,$$

and after summation over $\tau \in \mathcal{T}$, we arrive at the final estimate,

$$\min_\tau \nu_\tau\ \mathbf{v}^T A^{(0)}\mathbf{v} \le \mathbf{v}^T A\mathbf{v} \le \max_\tau \mu_\tau\ \mathbf{v}^T A^{(0)}\mathbf{v}.$$

Here $A^{(0)}$ is the matrix obtained by assembling the element matrices $A_\tau^{(0)}$, that is, corresponding to the Laplace operator $\mathcal{L} = -\Delta$.

We would like to compute an estimate of $\varrho(A) \simeq \|A\| \ \sup_{\mathbf{v}} \ \mathbf{v}^T\mathbf{v}/\|\mathbf{v}\|_0^2$. Because $\|A\| = \mathcal{O}(1)$ (because the entries of A are $\mathcal{O}(1)$ and A is sparse) it is sufficient to find a good estimate of $\sup_{\mathbf{v}} \mathbf{v}^T\mathbf{v}/\|\mathbf{v}\|_0^2$. The latter, as we already proved, is of order $\mathcal{O}(h^{-2})$. That is, for finite element matrices for second order elliptic PDEs discretized on quasiuniform triangulations, we have

$$\varrho(A) = \sup_{\mathbf{v}} \frac{\mathbf{v}^T A\mathbf{v}}{\|\mathbf{v}\|_0^2} \simeq \|A\| \ \sup_{\mathbf{v}} \ \frac{\mathbf{v}^T\mathbf{v}}{\|\mathbf{v}\|_0^2} \simeq \mathcal{O}(h^{-2}).$$

We can also show that $\min_{\mathbf{v}} \mathbf{v}^T A\mathbf{v}/\|\mathbf{v}\|_0^2 = \mathcal{O}(1)$. This is seen from the Friedrich's inequality valid for any function $v \in H_0^1(\Omega)$

$$\|v\|_0^2 \leq C_F \int\limits_\Omega |\nabla v|^2 \ d\mathbf{x}.$$

In conclusion, the following main result has been proved.

Proposition B.1. *For finite element matrices A coming from second-order elliptic PDEs discretized on quasiuniform triangulations, we have Cond(A)* $= \mathcal{O}(h^{-2})$. *This estimate used for the Laplace operator is sometimes referred to the following "inverse inequality" valid for all* $v \in V$,

$$|v|_1^2 \equiv \int\limits_\Omega |\nabla v|^2 \ d\mathbf{x} = \mathbf{v}^T A^{(0)}\mathbf{v} \leq Ch^{-2}\mathbf{v}^T M\mathbf{v} = Ch^{-2}(\mathbf{v}, \ \mathbf{v})_0 = Ch^{-2}\|v\|_0^2.$$

The familiar Gärding inequality is immediately seen for a general (nonsymmetric and possibly indefinite) operator $\mathcal{L}$ and respective stiffness matrix

$$\mathbf{v}^T(A+B+C)\mathbf{v} \geq \mathbf{v}^T A\mathbf{v} - c_0 \ \|\mathbf{v}\|_0^2 - \sum_\tau \|\mathbf{b}_\tau^T \mathcal{A}^{-1}\mathbf{b}_\tau\|_\infty^{1/2} \left(\mathbf{v}_\tau^T A_\tau \mathbf{v}_\tau\right)^{1/2} \left(\mathbf{v}_\tau^T M_\tau \mathbf{v}_\tau\right)^{1/2}.$$

Here, we used the representation

$$\begin{aligned}
\mathbf{w}_\tau^T B_\tau \mathbf{v}_\tau &= \int\limits_\tau (\mathbf{b}^T(\mathbf{x})\nabla v) w \ d\mathbf{x} \\
&= \int\limits_\tau (\mathcal{A}^{-(1/2)}\mathbf{b})^T(\mathcal{A}^{1/2}\nabla v) w \ d\mathbf{x} \\
&\leq \int\limits_\tau (\mathbf{b}^T\mathcal{A}^{-1}\mathbf{b})^{1/2}((\nabla v)^T\mathcal{A}(\mathbf{x})\nabla v)^{1/2}|w| \ d\mathbf{x} \\
&\leq \max_{\mathbf{x}\in\tau} \ (\mathbf{b}^T(\mathbf{x})\mathcal{A}^{-1}(\mathbf{x})\mathbf{b}(\mathbf{x}))^{1/2} \int\limits_\tau ((\nabla v)^T\mathcal{A}(\mathbf{x})\nabla v)^{1/2}|w| \ d\mathbf{x}.
\end{aligned}$$

Furthermore, we have

$$\begin{aligned}\mathbf{w}_\tau^T B_\tau \mathbf{v}_\tau &\le \frac{1}{\sqrt{\nu_\tau}} \max_{\mathbf{x}\in\tau} \|\mathbf{b}(\mathbf{x})\| \left(\mathbf{v}_\tau^T A_\tau \mathbf{v}_\tau\right)^{1/2}\left(\mathbf{w}_\tau^T M_\tau \mathbf{w}_\tau\right)^{1/2}\\ &\le \delta_0 \left(\mathbf{v}_\tau^T A_\tau \mathbf{v}_\tau\right)^{1/2}\left(\mathbf{w}_\tau^T M_\tau \mathbf{w}_\tau\right)^{1/2}.\end{aligned}$$

We have assumed that

$$\frac{1}{\sqrt{\nu_\tau}} \max_{\mathbf{x}\in\tau} \|\mathbf{b}(\mathbf{x})\| \le \delta_0.$$

Then the Gärding inequality takes the form

$$\begin{aligned}\mathbf{v}^T(A+B+C)\mathbf{v} &\ge \mathbf{v}^T A\mathbf{v} - c_0 \|\mathbf{v}\|_0^2 - \delta_0 \sum_\tau \left(\mathbf{v}_\tau^T A_\tau \mathbf{v}_\tau\right)^{1/2}\left(\mathbf{v}_\tau^T M_\tau \mathbf{v}_\tau\right)^{1/2}\\ &\ge \mathbf{v}^T A\mathbf{v} - c_0 \|\mathbf{v}\|_0^2 - \delta_0(\mathbf{v}^T A\mathbf{v})^{1/2}\|\mathbf{v}\|_0\\ &\ge \mathbf{v}^T A\mathbf{v} - \left(c_0 \frac{\sqrt{C_F}}{\min_\tau \sqrt{\nu_\tau}} + \delta_0\right)\left(\mathbf{v}^T A\mathbf{v}\right)^{1/2}\|\mathbf{v}\|_0. \qquad \text{(B.8)}\end{aligned}$$

It is clear that we can set $\delta_0 = 0$ if $\mathbf{b} = 0$. Also, we can choose $c_0 = 0$ if $c(\mathbf{x}) \ge 0$. A similar estimate holds from above

$$\begin{aligned}\mathbf{w}^T(A+B+C)\mathbf{v} &\le \mathbf{w}^T A\mathbf{v} + c_0 \|\mathbf{v}\|_0\|\mathbf{w}\|_0 + \delta_0(\mathbf{v}^T A\mathbf{v})^{1/2}\|\mathbf{w}\|_0\\ &\le \mathbf{w}^T A\mathbf{v} + \left(\delta_0 + \frac{\sqrt{C_F}}{\min_\tau \sqrt{\nu_\tau}} c_0\right)(\mathbf{v}^T A\mathbf{v})^{1/2}\|\mathbf{w}\|_0.\end{aligned}$$

Combining both estimates show, with $\sigma = \delta_0 + c_0(\sqrt{C_F}/(\min_\tau \sqrt{\nu_\tau}))$, that

$$\mathbf{w}^T(B+C)\mathbf{v} \le \sigma\, (\mathbf{v}^T A\mathbf{v})^{1/2}\|\mathbf{w}\|_0. \qquad \text{(B.9)}$$

L_2-error estimates

The following error estimate has been proved in Schatz and Wang [SW96].

Theorem B.2. *For every $\epsilon > 0$ there is a mesh-size $h_0 = h_0(\epsilon)$ such that for every finer mesh-size $h < h_0$ and corresponding f.e. space $V_h \subset H_0^1(\Omega)$, if $u_h \in V_h$ and $u \in H_0^1(\Omega)$ are such that*

$$a(u - u_h,\ \varphi) = 0, \quad \textit{for all } \varphi \in V_h,$$

then the following error bound holds,

$$\|u - u_h\|_0 \le \epsilon^{1/2}\ \|u - u_h\|_1.$$

Based on this result, given a fixed small ϵ, if we select a (fixed) coarse mesh-size H such that $H < h_0(\epsilon)$ and assuming that $V_H \subset V_h$, (i.e., $\mathcal{T}_h$ is a refinement of $\mathcal{T}_H$), if

$a(u - u_H, \varphi) = 0$, for all $\varphi \in V_H$, we would then have $a(u_h - u_H, \varphi) = 0$ for all $\varphi \in V_H$, and hence

$$\|u_H - u_h\|_0 \leq \epsilon^{1/2} \|u_H - u_h\|_1. \tag{B.10}$$

For now, we do not discuss the uniqueness of the solution u_h of the finite element problem of our main interest; namely, for a given $f \in L_2(\Omega)$ find $u_h \in V_h$ such that

$$a(u_h, \varphi) = (f, \varphi), \quad \text{for all } \varphi \in V_h. \tag{B.11}$$

We prove later, that (B.11) is actually uniquely solvable for a sufficiently small mesh h (see further Lemma B.5).

The above general L_2-error estimate implies the following main perturbation result.

Theorem B.3. *Given $\epsilon > 0$. Let $h_0(\epsilon)$ be the fixed mesh-size for which Theorem B.2 holds. Consider a coarse f.e. space V_H with a fixed H such that $H < h_0(\epsilon)$. Finally, let V_h be any f.e. space such that $V_H \subset V_h$. Then, the vectors $\mathbf{w}$ in the linear space*

$$\underline{\theta}^T (A + B + C)\mathbf{w} = 0 \ \ \textit{for all} \ \ \underline{\theta} \textit{ being the fine-grid coefficient vector of a } \theta \in V_H,$$

and arbitrary $\mathbf{v}$, satisfy the following perturbation estimate

$$\mathbf{v}^T (B + C)\mathbf{w} \leq \sigma \sqrt{C\epsilon} \, (\mathbf{v}^T A\mathbf{v})^{1/2} (\mathbf{w}^T A\mathbf{w})^{1/2}.$$

Proof. To prove the estimate, we notice that $\underline{\theta}^T (A + B + C)\mathbf{w} = 0$ for all $\theta \in V_H$, or using finite element notation $a(w, \theta) = 0$, for all $\theta \in V_H$, implies $\|\mathbf{w}\|_0 = \|w\|_0 \leq \epsilon^{1/2} \|w\|_1 = \epsilon^{1/2} (\mathbf{w}^T A^{(0)}\mathbf{w})^{1/2} \leq (C\epsilon)^{1/2} (\mathbf{w}^T A\mathbf{w})^{1/2}$, $C = 1/\min_\tau \nu_\tau$. The rest follows from the Gärding inequality (B.9). □

Because the adjoint operator of $\mathcal{L}$ (for functions in $H_0^1(\Omega)$) equals

$$\mathcal{L}^\star u = -\sum_{r=1}^d \frac{\partial}{\partial x_r}\left(\sum_{s=1}^d a_{r,s}(\mathbf{x})\frac{\partial u}{\partial x_s}\right) - \sum_{i=1}^d b_i(\mathbf{x})\frac{\partial u}{\partial x_i} + (c(\mathbf{x}) - \operatorname{div} \mathbf{b})u,$$

a dual result similar to that in Theorem B.3 holds, namely, we have the following.

Theorem B.4. *Given is $\epsilon > 0$. Let $h_0(\epsilon)$ be the fixed mesh-size for which Theorem B.2 holds. Consider a coarse f.e. space V_H with a fixed H such that $H < h_0(\epsilon)$. Finally, let V_h be any f.e. space such that $V_H \subset V_h$. Then, the vectors $\mathbf{w}$ in the linear space*

$$\mathbf{w}^T (A + B + C)\underline{\theta} = 0 \ \ \textit{for all} \ \ \underline{\theta} \textit{ being the fine-grid coefficient vector of a } \theta \in V_H,$$

and arbitrary $\mathbf{v}$, satisfy the perturbation estimate

$$\mathbf{v}^T (B^T + C)\mathbf{w} = \mathbf{w}^T (B + C)\mathbf{v} \leq \sigma \sqrt{C\epsilon} \, (\mathbf{v}^T A\mathbf{v})^{1/2} (\mathbf{w}^T A\mathbf{w})^{1/2}.$$

Here, we assume that the vector function $\mathbf{b} = \mathbf{b}(\mathbf{x})$ is sufficiently smooth (i.e., its divergence exists and is bounded).

Theorems B.3 and B.4 verify the main assumption needed in Chapter 8.

Duality argument

Here, we consider the bilinear form $a(.,.)$ corresponding to the operator $\mathcal{L}$, as well as the adjoint operator $\mathcal{L}^\star$ for functions in $H_0^1(\Omega)$. We assume that they are uniquely invertible for functions $f \in L_2(\Omega)$. This is the case because the lower-order terms in L and L^* containing the coefficients $\mathbf{b}(\mathbf{x})$ and $c(\mathbf{x})$ define compact operators for functions in $H_0^1(\Omega)$. Moreover, if we assume uniqueness, that is,

$$a(v,\ \varphi) = 0, \quad \text{for all } \varphi \in H_0^1(\Omega) \quad \text{implies } v = 0,$$

then the following "inf–sup" estimate holds (cf. [BPL]):

$$\|v\|_1 \leq C \sup_{w \in H_0^1(\Omega)} \frac{a(v,\ w)}{\|w\|_1}.$$

That is, uniqueness implies the a priori estimate,

$$\|u\|_1 \leq C\sqrt{C_F}\ \|f\|_0,$$

for the solution $u \in H_0^1(\Omega)$ of the variationally posed problem $a(u,\ \varphi) = (f,\ \varphi)$ for all $\varphi \in H_0^1(\Omega)$. Introduce now the Sobolev space $H^2(\Omega)$. This is the space of functions in $L_2(\Omega)$ that have all their derivatives up to order two also belonging to $L_2(\Omega)$. The norm in $H^2(\Omega)$, further denoted by $\|.\|_2$, is by definition the square root of the sum of the squares of the L_2-norms of the function and its derivatives (up to order two). Assume that the adjoint problem, for any $f \in L_2(\Omega)$, has a solution w,

$$\mathcal{L}^\star w = f, \tag{B.12}$$

which satisfies the (full) regularity estimate,

$$\|w\|_2 \leq C\|f\|_0.$$

The regularity is determined by the principal elliptic part of the operators $\mathcal{L}$ (and $\mathcal{L}^\star$) if the coefficients $\mathbf{b}_i(\mathbf{x})$ and $c_0(\mathbf{x})$ are sufficiently smooth. For smooth coefficients $a_{r,s}(\mathbf{x})$, and domains Ω being convex polygons such regularity estimates are known in the literature.

The regularity estimate, combined with the "duality argument" (due to Aubin and Nitsche) gives the following well-known L_2-estimate for the finite element solution $u_h \in V_h$ given by

$$a(u - u_h,\ \varphi) = 0, \quad \text{for all } \varphi \in V_h.$$

Consider problem (B.12) with $f = u - u_h$. We then have, for any $\varphi_h \in V_h$, $a(u - u_h,\ \varphi_h) = 0$. Therefore,

$$\begin{aligned}(f,\ f) &= a(u - u_h,\ w)\\ &= a(u - u_h,\ w - \varphi_h)\\ &\leq C\|u - u_h\|_1\|w - \varphi_h\|_1.\end{aligned}$$

Then because $\varphi_h \in V_h$ is arbitrary we can get $\|w - \varphi_h\|_1 \leq Ch\|w\|_2$ and using the regularity estimate, we finally arrive at

$$\|f\|_0^2 \leq C\|u - u_h\|_1 Ch\|w\|_2 \leq Ch\|u - u_h\|_1\|f\|_0.$$

That is, we have ($f = u - u_h$) the desired error bound

$$\|u - u_h\|_0 \leq Ch\|u - u_h\|_1. \tag{B.13}$$

Based on the same argument as above we can compare two finite element solutions $u_h \in V_h$ and $u_H \in V_H$, where $V_H \subset V_h$. We have

$$\|u_H - u_h\|_0 \leq CH\|u_H - u_h\|_1. \tag{B.14}$$

We have not yet shown that the discrete problem has a unique solution u_h for sufficiently small h. It is clear that the discrete problem, at worst for a zero r.h.s., $f = 0$, may have a nonzero solution (because the number of equations equals the number of unknowns). If $f = 0$ then the continuous solution is $u = 0$ and $a(u_h, \varphi) = 0$ for all $\varphi \in V_h$. Based on the Gärding inequality, used for $u - u_h$, and the general error estimate $\|u - u_h\|_0 \leq C\sqrt{\epsilon}\ \|u - u_h\|_1$ (or in the presence of full regularity, we can let $\epsilon = h^2$), we arrive at the coercivity estimate, for any $h < h_0 = h_0(\epsilon)$,

$$a(u - u_h,\ u - u_h) \geq \left(\min_{\tau} \nu_\tau\ - \sigma\ \sqrt{\epsilon}\right) \|u - u_h\|_1^2. \tag{B.15}$$

This coercivity estimate immediately implies, because $u = 0$ and $a(u_h,\ u_h) = 0$, that $0 \geq (\min_\tau \nu_\tau - \sigma\ C\sqrt{\epsilon})\ \|u_h\|_1^2$, and if ϵ is sufficiently small, we must have $u_h = 0$. That is, we have the following result.

Lemma B.5. *The discrete problem* (B.11) *is uniquely solvable if h is sufficiently small.*

Based on the coercivity estimate (B.15) and the boundedness of the bilinear form, we also get, for any $\varphi \in V_h$,

$$C\|u-u_h\|_1^2 \leq a(u-u_h,\ u-u_h) = a(u-u_h,\ u-\varphi) \leq C\|u-u_h\|_1 \inf_{\varphi\in V_h} \|u-\varphi\|_1.$$

That is, we proved the following result.

Lemma B.6. *If the mesh-size is sufficiently small, under the full regularity assumption, the following main error estimate holds*

$$h^{-1}\ \|u - u_h\|_0 + \|u - u_h\|_1 \leq C \inf_{\varphi\in V_h} \|u - \varphi\|_1 \leq Ch\ \|u\|_2 \leq Ch\ \|f\|_0. \tag{B.16}$$

Theorem B.7. *Consider the matrix $L_h = A + B + C$ and a coarse version of it L_H obtained by finite element discretization of the problem $a(u,\ \varphi) = (f,\ \varphi)$, $\varphi \in H_0^1(\Omega)$, corresponding to two finite element spaces V_h and V_H where $V_H \subset V_h$. Let P be the interpolation matrix that relates the coefficient vector $\mathbf{v}_c$ of a coarse function $v_H \in V_H$ expanded in terms of the coarse (Lagrangian) basis, and its* *coefficient vector* $\mathbf{v}$ *corresponding to the expansion of* $v_H \in V_h$ *in terms of the fine-grid Lagrangian basis of* V_h*; that is, let* $\mathbf{v} = P\mathbf{v}_c$*. Consider then the coarse-grid correction matrix* $I - \pi = I - P(L_H)^{-1}P^T L_h$*. It appears naturally, if we compare the solutions* u_h *and* u_H *of the problems* $L_h\mathbf{u}_h = \mathbf{f}_h$ *and* $L_H\mathbf{u}_H = P^T\mathbf{f}_h$*. We have* $\mathbf{u}_h - P\mathbf{u}_H = \mathbf{u}_h - PL_H^{-1}P^T\mathbf{f}_h = \mathbf{u}_h - PL_H^{-1}P^T L_h\mathbf{u}_h = (I - \pi)\mathbf{u}_h$*. Then the following estimate holds*

$$\|(I - \pi)\mathbf{u}_h\|_0 = \|\mathbf{u}_h - P\mathbf{u}_H\|_0 \leq C\epsilon^{1/2} \left(\mathbf{u}_h^T A_h \mathbf{u}_h\right)^{1/2}.$$

Under the assumption of full regularity of problem (B.12)*, we can let* $\epsilon = H^2$ *and hence, have the estimate*

$$\|(I - \pi)\mathbf{u}_h\|_0 = \|\mathbf{u}_h - P\mathbf{u}_H\|_0 \leq CH \left(\mathbf{u}_h^T A_h \mathbf{u}_h\right)^{1/2}.$$

Proof. Let u_h and u_H be the finite element functions corresponding to the coefficient vectors $\mathbf{u}_h$ and $\mathbf{u}_H$. Also, let f be the finite element function corresponding to the vector $M_h^{-1}L_h\mathbf{u}_h$ (M_h is the mass matrix). That is, let $f = \sum_{\mathbf{x}_i \in \mathcal{N}_h} f_i \varphi_i^h$ where f_i is the ith entry of the vector $M_h^{-1}L_h\mathbf{u}_h$ and $\{\varphi_i^h\}_{\mathbf{x}_i \in \mathcal{N}_h}$ is the nodal basis of V_h. Consider the variationally posed second-order elliptic problem,

$$a(u, \varphi) = (f, \varphi).$$

Its finite element discretization based on V_h takes the form $L_h\mathbf{u}_h = M_h(M_h^{-1}L_h\mathbf{u}_h) =: \mathbf{f}_h$, and similarly for V_H, $L_H\mathbf{u}_H = P^T M_h(M_h^{-1}L_h\mathbf{u}_h) = P^T\mathbf{f}_h$. That is, u_H and u_h are finite element solutions of the above continuous problem. Using then the error estimate (B.10), we get

$$\|(I - \pi)\mathbf{u}_h\|_0 = \|u_h - u_H\|_0 \leq C\sqrt{\epsilon}\|u_h - u_H\|_1. \tag{B.17}$$

If we show that $\|u_h - u_H\|_1 \leq C\|u_h\|_1$, then $\|(I - \pi)\mathbf{u}_h\|_0 \leq C\sqrt{\epsilon}\|u_h\|_1$ which rewritten in terms of matrices and vectors will give the desired result.

From the Gärding inequality (B.9), we have

$$\mathbf{v}^T L_h\mathbf{v} \geq \mathbf{v}^T A\mathbf{v} - \sigma\ (\mathbf{v}^T A\mathbf{v})^{1/2}\|\mathbf{v}\|_0.$$

Letting $\mathbf{v} = \mathbf{u}_h - P\mathbf{u}_H$ and using the estimate $\|\mathbf{v}\|_0 \leq C\epsilon^{1/2}\ (\mathbf{v}^T A\mathbf{v})^{1/2}$, we arrive at the following coercivity estimate,

$$\mathbf{v}^T L_h\mathbf{v} \geq (1 - \sigma\ C\sqrt{\epsilon})\ \mathbf{v}^T A\mathbf{v}. \tag{B.18}$$

On the other hand, using the fact the u_H is a finite element solution, together with the $H_0^1 \times H_0^1$– boundedness of the bilinear form $a(.,.)$, because $\mathbf{v}^T A\mathbf{v} \simeq \|u_h - u_H\|_1^2$, we arrive at the following upper bound,

$$\begin{aligned}
\mathbf{v}^T L_h\mathbf{v} &= a(u_h - u_H,\ u_h - u_H)\\
&= a(u_h - u_H,\ u_h)\\
&\leq C\ \|u_h - u_H\|_1\|u_h\|_1\\
&\leq C\ (\mathbf{v}^T A\mathbf{v})^{1/2}\|u_h\|_1.
\end{aligned}$$

Combining the last estimate with the preceding one (B.18), for sufficiently small ϵ (or equivalently, sufficiently small H), we arrive at the desired estimate (needed in (B.17))

$$\|u_h - u_H\|_1 \leq C(\mathbf{v}^T A\mathbf{v})^{1/2} \leq C\|u_h\|_1.$$

□

Theorem B.8. *Under the full regularity assumption, we have the following stronger approximation property in matrix–vector form (with* $\|.\|$: $\|\mathbf{w}\| = \sqrt{\mathbf{w}^T\mathbf{w}}$*),*

$$\|(I - PL_H^{-1}P^T L_h)\mathbf{v}\| \leq C\frac{H^2}{h^2}\frac{\|L_h\mathbf{v}\|}{\|L_h\|}. \tag{B.19}$$

Proof. To prove the required estimate construct $f \in V_h \subset L_2(\Omega)$ as in the proof of Theorem B.7. That is, f has coefficient vector $\mathbf{f}$ equal to $M_h^{-1}L_h\mathbf{u}_h$ (where M_h is the mass matrix). Use now the optimal L_2-error estimate $\|u_h - u_H\|_0 \leq CH^2\|f\|_0$. Then because $\|f\|_0^2 \simeq h^d\ \|\mathbf{f}\|^2$ ($d = 2$ or $d = 3$), we have $\|f\|_0^2 = (M_h^{-1}L_h\mathbf{u}_h)^T M_h(M_h^{-1}L_h\mathbf{u}_h) \leq Ch^{-d}\|L_h\mathbf{u}_h\|^2$. Therefore,

$$h^{d/2}\ \|(I - \pi)\mathbf{u}_h\| \leq C\|(I - \pi)\mathbf{u}_h\|_0 \leq C\frac{H^2}{h^{d/2}}\ \|L_h\mathbf{u}_h\|.$$

That is,

$$\|(I - \pi)\mathbf{u}_h\| \leq C\ \frac{H^2}{h^2}\ h^{2-d}\ \|L_h\mathbf{u}_h\|,$$

which completes the proof because $\mathbf{v} := \mathbf{u}_h$ can be arbitrary and it is easily seen (looking at element matrix level, e.g.) that $\|L_h\| \simeq h^{d-2}$. □

Corollary B.9. *In the full regularity case, for* $H = 2h$*, or more generally for* $H \leq Ch$*, we have the uniform estimate*

$$\|L_h\|\ \|(L_h^{-1} - PL_H^{-1}P^T)\mathbf{v}\| \leq C\|\mathbf{v}\|.$$

Note that $\|L_h\|\|L_h^{-1}\mathbf{v}\|$ *behaves as the condition number of* L_h *(for some vectors* $\mathbf{v}$*). That is,* $\|L_h\|L_h^{-1}$ *is not uniformly bounded in* $h \mapsto 0$*, whereas the difference* $\|L_h\|\ (L_h^{-1} - PL_H^{-1}P^T)$ *is uniformly bounded.*

B.2 A semilinear second-order elliptic PDE

Consider the following semilinear elliptic problem

$$\mathcal{L}u \equiv -\Delta u + b(\mathbf{x},\ u) = f(\mathbf{x}),\ \mathbf{x} \in \Omega,$$

subject to Dirichlet boundary conditions $u = 0$ on $\partial\Omega$. The coefficient $b = b(\mathbf{x},\ \cdot)$ is a given scalar function, and $f \in L_2(\Omega)$ is a given r.h.s. Under appropriate assumptions on b and Ω (cf., e.g., [Sh95]) the above problem has a (unique) solution $u^\star \in H^2(\Omega) \cap H_0^1(\Omega)$.

For the purpose of verifying the assumptions on the inexact Newton method we made in Section 11.3, we further assume that b is smooth; namely, its partial derivative $b_u(\mathbf{x}, \cdot)$ exists and is nonnegative. Moreover, we also assume that $b_u(., u)$ is Lipschitz; that is,

$$|b_u(\mathbf{x}, u) - b_u(\mathbf{x}, v)| \leq L\ |u - v| \text{ uniformly in } \mathbf{x} \in \Omega. \tag{B.20}$$

Introduce a finite element space $V = V_h \subset H_0^1(\Omega)$ of piecewise linear basis functions on a given triangulation $\mathcal{T}_h$. Consider the discrete nonlinear problem; find $u_h \in V_h$ such that $L_h(u_h) = f_h$ posed variationally,

$$(L_h(u_h), \varphi) \equiv (\nabla u_h, \nabla \varphi) + (b(\cdot, u_h), \varphi) = (f, \varphi), \quad \text{for all } \varphi \in V_h. \tag{B.21}$$

We additionally assume that Ω is a convex polygon (or polytope in 3D). Thus the Poisson problem $-\Delta u = f$ for $f \in L_2(\Omega)$ has a $H^2(\Omega) \cap H_0^1(\Omega)$ solution that satisfies $\|u\|_2 \leq C\|f\|_0$. Note that the same holds for the linearized equation, for any fixed $u_0 \in H^2(\Omega) \cap H_0^1(\Omega)$ (note that then u_0 is a continuous function, cf., e.g., [BS96])

$$-\Delta u + b_u(\mathbf{x}, u_0)u = f(\mathbf{x}). \tag{B.22}$$

That is, we have $\|u\|_2 \leq C\|b_u(., u_0)\|_{\max}\ \|u\|_0 + C\|f\|_0 \leq C(\|b_u(., u_0)\|_{\max} + 1)\ \|f\|_0$. Define now two Banach spaces $\mathcal{X}_h$ and $\mathcal{Y}_h$. The first one, $\mathcal{X}_h$, is V_h equipped with the norm

$$\|v\|_{\mathcal{X}_h} = \|v\| \equiv \max\{\|v\|_1, \|v\|_{\max}\},$$

and the second one, $\mathcal{Y}_h$, is simply V_h equipped with the L_2-norm $\|v\|_0$.

We are interested first in the properties of the linear mapping $L_h'(u_0)$ and its inverse.

Lemma B.10. *The finite element linear operator defined variationally*

$$(L_h'(u_0)\ v, \varphi) = (\nabla v, \nabla \varphi) + (b_u(., u_0)\ v, \varphi) \quad \textit{for all } \varphi \in V_h,$$

has the following properties.

(i) For any $u_0, v_0 \in \mathcal{X}_h$ such that $\|u_0 - v_0\| \leq \delta < 1$, and any $v \in \mathcal{X}_h$,

$$\left\|\left(I - \left(L_h'(u_0)\right)^{-1} L_h'(v_0)\right)v\right\| \leq C\delta\left[1 + \|b_u(., u_0)\|_{\max}\right]\|v\|.$$

(ii) For any $u_0, v_0 \in \mathcal{X}_h$ such that $\|u_0 - v_0\| \leq \delta < 1$ and any $g \in V_h$, we have

$$\left\|\left(I - L_h'(v_0)\left(L_h'(u_0)\right)^{-1}\right) g\right\|_0 \leq C\delta\ \|g\|_0.$$

(iii) For any $u_0, v_0 \in \mathcal{X}_h$ such that $\|u_0 - v_0\| \leq \delta < 1$ and any $\varphi \in V_h$, we have

$$(L_h(v_0) - L_h(u_0) - L_h'(u_0)(u_0 - v_0), \varphi) \leq C\delta\ \|v_0 - u_0\|\ \|\varphi\|_0.$$

Proof. To prove (ii), let $v = (L_h'(u_0))^{-1} g$, or equivalently, solve for $v \in V_h$ the linear finite element problem

$$(\nabla v,\ \nabla\varphi) + (b_u(.,\ u_0)v,\ \varphi) = (g,\ \varphi), \text{ for all } \varphi \in V_h.$$

Because $b_u(.,\ u_0) \geq 0$, we have the a priori estimate $\|v\|_0 \leq C_F \|g\|_0$. We then have

$$\begin{aligned}\|g - L_h'(v_0)(L_h'(u_0))^{-1} g\|_0 &= \|L_h'(u_0)v - L_h'(v_0)v\|_0 \\ &= \|(b_u(.,\ u_0) - b_u(.,\ v_0))v\|_0 \\ &\leq L\|u_0 - v_0\|_{\max}\ \|v\|_0 \\ &\leq LC_F\delta\ \|g\|_0.\end{aligned}$$

The latter is estimate (ii).

To prove (iii), we have, for any $\varphi \in V_h$

$$\begin{aligned}&\left(L_h(v_0) - L_h(u_0) - L_h'(u_0)(u_0 - v_0),\ \varphi\right) \\ &\quad = (b(.,\ v_0) - b(.,\ u_0) - b_u(.,\ u_0)(u_0 - v_0),\ \varphi) \\ &\quad \leq L\ \|u_0 - v_0\|_{\max} \|u_0 - v_0\|_0 \|\varphi\|_0 \\ &\quad \leq L\delta\ C_F\ \|u_0 - v_0\|\ \|\varphi\|_0\end{aligned}$$

which verifies (iii).

To prove (i), solve the following discrete problem for $\psi \in V_h$,

$$(L_h'(u_0)\psi,\ \varphi) = (L_h'(v_0)v,\ \varphi), \quad \text{for all } \varphi \in V_h.$$

We have that $((L_h'(u_0))^{-1} L_h'(v_0) - I)v = \psi - v \in V_h$ solves

$$\left(L_h'(u_0)(\psi - v),\ \varphi\right) = \left((L_h'(v_0) - L_h'(u_0))v,\ \varphi\right), \quad \text{for all } \varphi \in V_h.$$

Consider the solution w of the continuous problem $(-\Delta + b_u(.,\ u_0))w = g \equiv (L_h'(v_0) - L_h'(u_0))v \in L_2(\Omega)$. We have the error estimate

$$\begin{aligned}\|w - (\psi - v)\|_1 &\leq Ch\ \|w\|_2 \\ &\leq Ch(\|g\|_0 + C\|b_u(.,\ u_0)\|_{\max}) \\ &\leq Ch\ (\|b_u(.,\ u_0)\|_{\max} + \|b_u(.,\ v_0) - b_u(.,\ u_0)\|_{\max} \|v\|_0) \\ &\leq Ch(\|b_u(.,\ u_0)\|_{\max} + \|v_0 - u_0\|_{\max}\ \|v\|_0).\end{aligned}$$

Introduce the nodal interpolation operator

$$I_h w = \sum_{\mathbf{x}_i \in \mathcal{N}_h} w(\mathbf{x}_i)\varphi_i \in V_h$$

where $\{\varphi_i\}_{\mathbf{x}_i \mathcal{N}_h}$ is the nodal (Lagrangian) basis of V_h. The following familiar estimate holds $\|I_h w - w\|_1 \leq Ch\|w\|_2$ as well as $\|I_h w - w\|_{\max} \leq Ch\|w\|_2$. Therefore, $\|I_h w - (\psi - v)\|_1 \leq \|I_h w - w\|_1 + \|w - (\psi - v)\|_1 \leq Ch\ \|w\|_2 \leq Ch(\|b_u(.,\ u_0)\|_{\max} + \|v_0 - u_0\|_{\max} \|v\|_0)$. The desired result follows from the fact that $\|w\|_\infty \leq C\|w\|_2$

(see, e.g., [BS96]) and the inverse inequality $\|\varphi\|_{\max} \le Ch^{-\alpha}\ \|\varphi\|_1$ (see Section G.3 in the appendix) for $\varphi \in V_h$ and an $\alpha < 1$. More specifically,

$$\begin{aligned} \|\psi - v\|_{\max} &\le \|I_h w\|_{\max} + \|I_h w - (\psi - v)\|_{\max} \\ &\le \|I_h w\|_{\max} + Ch^{-\alpha}\ \|I_h w - (\psi - v)\|_1 \le C\ \|w\|_2. \end{aligned}$$

That is, because $\|w\|_0 \le C_F^{-1}\|g\|_0 \le C\|v_0 - u_0\|_{\max}\|v\|_0$, we finally arrive at

$$\begin{aligned} \|\psi - v\| &\le C\|w\|_2 \\ &\le C(\|b_u(.,\ u_0)\|_{\max}\|w\|_0 + \|g\|_0) \\ &\le C(\|b_u(.,\ u_0)\|_{\max}\|w\|_0 + \|v_0 - u_0\|_{\max}\|v\|_0) \\ &\le C\delta\ (1 + \|b_u(.,\ u_0)\|_{\max})\ \|v\|_0. \end{aligned}$$

□

Now, we provide precise estimates for the algorithm to compute accurate initial approximations for solving the (fine-grid) nonlinear problem described earlier in Section 11.2. For this purpose, consider two finite element spaces $V_H \subset V_h$. The mesh H will be sufficiently small but fixed, that is, independent of $h \mapsto 0$. Then the fine-grid nonlinear problem can be treated as a "perturbation" of the linearized one in the following sense. Solve the coarse-grid nonlinear problem

$$(L_H(u_H^\star),\ \varphi) = (f,\ \varphi) \quad \text{for all } \varphi \in V_H.$$

Consider then $g = b(.,\ u_H^\star) - b_u(.,\ u_H^\star)u_H^\star$ and the fine-grid linear problem

$$\left(L_h^{'}(u_H^\star)\overline{u}_h^\star,\ \varphi\right) = (g,\ \varphi) \quad \text{for all } \varphi \in V_h. \tag{B.23}$$

Rewrite the fine-grid nonlinear problem $(L_h^{'}(u_h^\star),\ \varphi) = (f,\ \varphi)$ for all $\varphi \in V_h$ as

$$\left(L_h^{'}(u_H^\star)u_h^\star,\ \varphi\right) = (f + b_u(.,\ u_H^\star)u_h^\star - b(.,\ u_h^\star),\ \varphi) \quad \text{for all } \varphi \in V_h.$$

Therefore, the difference $u_h^\star - \overline{u}_h^\star$ solves

$$\begin{aligned} \left(L_h^{'}(u_H^\star)(u_h^\star - \overline{u}_h^\star),\ \varphi\right) = (b(.,\ u_h^\star) - b(.,\ u_H^\star) \\ - b_u(.,\ u_H^\star)(u_h^\star - u_H^\star),\ \varphi) \quad \text{for all } \varphi \in V_h. \end{aligned}$$

We have the estimate

$$\left(L_h^{'}(u_H^\star)(u_h^\star - \overline{u}_h^\star),\ \varphi\right) \le L\ \|u_h^\star - u_H^\star\|_{\max}\|u_h^\star - u_H^\star\|_0\|\varphi\|_0 \quad \text{for all } \varphi \in V_h.$$

Hence

$$\left\|L_h^{'}(u_H^\star)(u_h^\star - \overline{u}_h^\star)\right\|_0 \le L\ \|u_h^\star - u_H^\star\|_{\max}\|u_h^\star - u_H^\star\|_0.$$

For $h < H \le h_0$ and an $\alpha \in (0, 1)$, we have the error estimate $\|u_h^\star - u_H^\star\|_{\max}\|u_h^\star - u_H^\star\|_0 \le CH^{1+\alpha}\|u^\star\|_2^2$. Finally, solve the linear problem (B.23) approximately (e.g., by a few iterations of an optimal MG method), with a guaranteed residual reduction. That is, for a small tolerance $\varrho \in (0, 1)$, we compute an accurate u_h^0 that satisfies

$$\|L_h'(u_H^\star)(\overline{u}_h^\star - u_h^0)\|_0 = \|g - L_h'(u_H^\star)u_h^0\|_0 \le \varrho\ \|g\|_0.$$

The final estimate that we need reads using (ii) for $u_0 = u_H^\star$ and $v_0 = u_h^\star$, and $g = L_h'(u_H^\star)(\overline{u}_h^\star - u_h^0)$,

$$\begin{aligned}
\|L_h'(u_h^\star)(\overline{u}_h^\star - u_h^0)\|_0 &\le (1 + C\delta)\ \|L_h'(u_H^\star)(\overline{u}_h^\star - u_h^0)\|_0 \\
&\le 2\big[\|L_h'(u_H^\star)(\overline{u}_h^\star - u_h^\star)\|_0 + \|L_h'(u_H^\star)(\overline{u}_h^\star - u_h^0)\|_0\big] \\
&\le 2\big[CH^{1+\alpha}\|u^\star\|_2^2 + \varrho\|g\|_0\big] \\
&\le \frac{\epsilon}{\mu}.
\end{aligned}$$

The quantity ϵ/μ for the purpose of the present analysis is a prescribed small tolerance (for details cf. Section 11.2).

The last estimate can be guaranteed if H is sufficiently small, and (after H has been chosen and fixed) by choosing the tolerance ϱ sufficiently small, because then g is fixed and $\|u^\star\|_2^2$ is just a constant.

B.3 Stable two-level HB decomposition of finite element spaces

We analyze the stability property of the two-level HB decomposition of finite element spaces V_h corresponding to a triangulation $\mathcal{T}_h$ obtained by a refinement of a coarse one $\mathcal{T}_H$ and its corresponding coarse f.e. space $V_H \subset V_h$.

B.3.1 A two-level hierarchical basis and related strengthened Cauchy–Schwarz inequality

In what follows we consider a triangle T that is refined into four geometrically similar triangles τ_s, $s = 1, 2, 3, 4$. We can associate with the vertices of T the three standard basis functions $\varphi_i^{(H)}$, $i = 4, 5, 6$, whereas with the midpoints of its edges, the piecewise linear basis functions $\varphi_i^{(h)}$, $i = 1, 2, 3$, associated with the fine triangles τ_s. It is clear that the two sets of functions are linearly independent. For example if we have (on T)

$$c_4\varphi_4^{(H)} + c_5\varphi_5^{(H)} + c_6\varphi_6^{(H)} = c_1\varphi_1^{(h)} + c_2\varphi_2^{(h)} + c_3\varphi_3^{(h)},$$

for some coefficients c_s, $s = 1, \ldots, 6$, because $\varphi_i^{(h)}$, $i = 1, 2, 3$ vanish at the vertices of the coarse triangle T, we have that $c_4 = c_5 = c_6 = 0$. That is, we have then

$$c_1\varphi_1^{(h)} + c_2\varphi_2^{(h)} + c_3\varphi_3^{(h)} = 0,$$

which implies that $c_1 = c_2 = c_3 = 0$ because $\varphi_i^{(h)}$ are linearly independent.

Consider the two finite-dimensional spaces $V_{H,\,T} = \text{span}\ \{\varphi_4^{(H)},\ \varphi_5^{(H)},\ \varphi_6^{(H)}\}$ and $V_{h,\,T}^f = \text{span}\ \{\varphi_1^{(h)},\ \varphi_2^{(h)},\ \varphi_3^{(h)}\}$. We proved that they are linearly independent. Therefore, the following estimate holds

$$\max_{\substack{v_H \in V_{H,T}, v_H \neq \text{Const} \\ v_h^{(f)} \in V_{h,T}^f}} \frac{(\nabla v_H,\ \nabla v_h^{(f)})_T}{|v_h^{(f)}|_{1,\,T}\,|v_H|_{1,\,T}} \leq \gamma_T < 1. \tag{B.24}$$

It is clear that $\gamma_T \in [0, 1)$ depends only on the angles of T and not on its size. The following explicit expression has been derived in [MM82] for a triangle T with angles α_1, α_2, and α_3,

$$1 - \gamma_T^2 = \frac{5}{8} - \frac{1}{8}\sqrt{4d - 3},$$

where $d = \sum_i \cos^2 \alpha_i$. Because $d < 3$ ($d = 3$ for degenerated triangle $\alpha_1 = \alpha_2 = 0$ and $\alpha_3 = \pi$), we have the following uniform upper bound,

$$\gamma_T^2 < \frac{3}{4}.$$

Our goal is to apply the above estimate to the global matrix A corresponding to a triangulation $\mathcal{T}_h$ and its coarse version A_c corresponding to a triangulation $\mathcal{T}_H$ with $H = 2h$. That is, we assume that the fine-grid triangles τ are obtained by refining the coarse triangles T into four geometrically similar ones. The vertices of the coarse triangles form the coarse-grid and are denoted "c"-nodes, whereas the fine-grid nodes that are not coarse are denoted "f"-nodes. We also have the Galerkin relation $A_c = P^T A P$ for a proper (linear) interpolation matrix P (cf. Section 1.2).

To do this, consider first A_T, the 6×6 matrix corresponding to the triangle T computed with respect to the fine-grid piecewise linear basis functions $\varphi_s^{(h)}$ associated with the six fine-grid node $\mathbf{x}_s$, $s = 1, 2, \ldots, 6$. Alternatively, we can use the two-level hierarchical basis (or HB) $\{\varphi_4^{(H)},\ \varphi_5^{(H)},\ \varphi_6^{(H)}\} \cup \{\varphi_1^{(h)},\ \varphi_2^{(h)},\ \varphi_3^{(h)}\}$ introduced earlier. This gives rise to another matrix (called the two-level HB matrix) $\overline{A}_T$. We show below (see (B.26)) that it can be obtained from A_T by a proper transformation. The following observation is in order. Let $\mathbf{v}$ be the coefficient vector of a f.e. function v expanded in terms of the standard nodal basis $\varphi_i^{(h)}$. Let P be the interpolation matrix that implements the embedding $V_H \subset V_h$ in terms of coefficient vectors; that is, if $v \in V_H$ has a coefficient vector w.r.t. the coarse basis $\varphi_{i_c}^{(H)}$, then $P\mathbf{v}_c$ will be its coefficient vector w.r.t. to the fine-grid nodal basis $\varphi_i^{(h)}$. It is clear that P admits the following form

$$P = \begin{bmatrix} W \\ I \end{bmatrix} \begin{matrix} \}\ \text{"f"-nodes} \\ \}\ \text{"c"-nodes} \end{matrix}.$$

This is seen from the fact that the interpolation at the coarse nodes $\mathbf{x}_i$ is an identity because the coarse nodes are the subset of all nodes on the fine-grid. Then, if $v_c = \sum_i v_i^c \varphi_i^{(h)}$ for every coarse node $\mathbf{x}_i$, we have $v_i^c = v_c(\mathbf{x}_i)$. In other words, the ith entry of the vector $P\mathbf{v}_c$ equals a corresponding entry of $\mathbf{v}_c$ for any coarse node $\mathbf{x}_i$. That is, P is the identity at the coarse nodes. The same argument applies to P restricted to an individual coarse element T. We have

$$P_T = \begin{bmatrix} W_T \\ I \end{bmatrix}.$$

We can rewrite the strengthened Cauchy–Schwarz inequality (see (B.24)),

$$\left(\nabla\varphi_h^{(f)},\ \nabla\varphi_H\right)_T \leq \gamma_T\ |\nabla\varphi_h^{(f)}|_{1,\,T}\ |\nabla\varphi_H|_{1,\,T},$$

in terms of the local stiffness matrix A_T and coefficient vectors

$$\mathbf{v}^f = \begin{bmatrix} \mathbf{v}_f \\ 0 \end{bmatrix}$$

and $P\mathbf{v}_c$ of $\varphi_h^{(f)}$ and $\varphi_H = v_c \in V_H$, respectively. The first vector has zero entries at the "c" nodes because $\varphi_h^{(f)}$ vanishes at the coarse nodes. We have for their restrictions to T,

$$\left(\mathbf{v}_T^f\right)^T A_T P\mathbf{v}_{c,\,T} \leq \gamma_T \left(\left(\mathbf{v}_T^f\right)^T A_T \mathbf{v}_T^f\right)^{1/2} \left(\mathbf{v}_{c,\,T}^T P_T^T A_T P_T \mathbf{v}_{c,\,T}\right)^{1/2}.$$

We note that $A_{c,\,T} = P_T^T A_T P_T$ is the coarse element matrix. Introducing

$$J = \begin{bmatrix} I \\ 0 \end{bmatrix},$$

with zero block corresponding to the "c" nodes, and letting J_T be the restriction of J to T, we finally arrive at the local strengthened Cauchy–Schwarz inequality of our main interest

$$\mathbf{v}_{f,\,T}^T J_T^T A_T P_T \mathbf{v}_{c,\,T} \leq \gamma_T \left(\mathbf{v}_{f,\,T}^T J_T^T A_T J_T \mathbf{v}_{f,\,T}\right)^{1/2} \left(\mathbf{v}_{c,\,T}^T P_T^T A_T P_T \mathbf{v}_{c,\,T}\right)^{1/2}.$$

Introduce the transformed matrix

$$\overline{A}_T = [J_T,\ P_T]^T A_T [J_T,\ P_T] = \begin{bmatrix} J_T^T A_T A_T & J_T^T A_T P_T \\ P_T^T A_T J_T & P_T^T A_T P_T \end{bmatrix} = \begin{bmatrix} A_{ff,\,T} & \overline{A}_{fc,\,T} \\ \overline{A}_{cf,\,T} & A_{c,\,T} \end{bmatrix}.$$

We note (due to the special form of J_T) that $A_{ff,\,T}$ is actually a principal submatrix of A_T (corresponding to the "f" nodes in T) and also that $P_T^T A_T P_T$ is the coarse element matrix $A_{c,\,T}$. This is one reason to have $\overline{A}_T$ referred to as the two-level hierarchical basis (or HB) matrix. Consider its Schur complement $\overline{S}_T = A_{c,\,T} - \overline{A}_{cf,\,T} (A_{ff,\,T})^{-1} \overline{A}_{fc,\,T}$. We know (cf., Lemma 3.3) that the strengthened Cauchy–Schwarz inequality can equivalently be stated as

$$\begin{aligned}
\mathbf{v}_{c,\,T}^T A_{c,\,T}\mathbf{v}_{c,\,T} &\leq \frac{1}{1-\gamma_T^2}\,\mathbf{v}_{c,\,T}^T \overline{S}_T \mathbf{v}_{c,\,T} \\
&= \frac{1}{1-\gamma_T^2} \min_{\mathbf{v}_{f,\,T}} \begin{bmatrix} \mathbf{v}_{f,\,T} \\ \mathbf{v}_{c,\,T} \end{bmatrix}^T \overline{A}_T \begin{bmatrix} \mathbf{v}_{f,\,T} \\ \mathbf{v}_{c,\,T} \end{bmatrix} \\
&= \frac{1}{1-\gamma_T^2} \min_{\mathbf{v}_{f,\,T}} \begin{bmatrix} \mathbf{v}_{f,\,T} \\ \mathbf{v}_{c,\,T} \end{bmatrix}^T [J_T,\; P_T]^T A_T \,[J_T,\; P_T] \begin{bmatrix} \mathbf{v}_{f,\,T} \\ \mathbf{v}_{c,\,T} \end{bmatrix} \\
&= \frac{1}{1-\gamma_T^2} \min_{\mathbf{v}_T = J_T\mathbf{v}_{f,\,T} + P_T\mathbf{v}_{c,\,T}} \mathbf{v}_T^T A_T \mathbf{v}_T. \qquad \text{(B.25)}
\end{aligned}$$

Finally, note that the f.e. function $v \in V_h$ (restricted to T) that has standard nodal coefficient vector $\mathbf{v}_T$ if expanded in the two-level HB $\{\varphi_4^{(H)},\; \varphi_5^{(H)},\; \varphi_6^{(H)}\} \cup \{\varphi_1^{(h)},\; \varphi_2^{(h)},\; \varphi_3^{(h)}\}$ will have coefficient vector

$$\begin{bmatrix} \mathbf{v}_{f,\,T} \\ \mathbf{v}_{c,\,T} \end{bmatrix}$$

coming from the representation $\mathbf{v}_T = J_T\mathbf{v}_{f,\,T} + P_T\mathbf{v}_{c,\,T}$. That is, we have then

$$\mathbf{v}_T^T A_T \mathbf{v}_T = \begin{bmatrix} \mathbf{v}_{f,\,T} \\ \mathbf{v}_{c,\,T} \end{bmatrix}^T \overline{A}_T \begin{bmatrix} \mathbf{v}_{f,\,T} \\ \mathbf{v}_{c,\,T} \end{bmatrix},$$

or equivalently

$$\overline{A}_T = [J_T,\; P_T]^T A_T [J_T,\; P_T]. \qquad \text{(B.26)}$$

Observe that A_T and $\overline{A}_T$ are different. However, their first pivot blocks $A_{ff,T}$ and $\overline{A}_{ff,\,T}$ coincide. We show below that their Schur complements are also the same. Use now the hierarchical decomposition of any fine-grid vector $\mathbf{v}$,

$$\mathbf{v} = \begin{bmatrix} \overline{\mathbf{v}}_f \\ 0 \end{bmatrix} + P\mathbf{v}_c,$$

where $\mathbf{v}_c$ is the restriction of $\mathbf{v}$ to the "c" nodes. Using the matrix form of

$$P = \begin{bmatrix} W \\ I \end{bmatrix},$$

we get

$$\mathbf{v} = \begin{bmatrix} I & W \\ 0 & I \end{bmatrix} \begin{bmatrix} \overline{\mathbf{v}}_f \\ \mathbf{v}_c \end{bmatrix}.$$

That is, if

$$\mathbf{v} = \begin{bmatrix} \mathbf{v}_f \\ \mathbf{v}_c \end{bmatrix},$$

we will have $\mathbf{v}_f = \overline{\mathbf{v}}_f + W\mathbf{v}_c$.

Introduce now the element Schur complement S_T of A_T. We have

$$\mathbf{v}_{c,\,T}^T S_T \mathbf{v}_{c,\,T} = \min_{\mathbf{v}_{f,\,T}:\ \mathbf{v}_T = \begin{bmatrix} \mathbf{v}_{f,\,T} \\ \mathbf{v}_{c,\,T} \end{bmatrix}} \mathbf{v}_T^T A_T \mathbf{v}_T. \tag{B.27}$$

It is clear that by using the hierarchical decomposition instead (restricted to T), we also have

$$\begin{aligned}
\mathbf{v}_{c,\,T}^T S_T \mathbf{v}_{c,\,T} &= \min_{\mathbf{v}_T = \begin{bmatrix} \overline{\mathbf{v}}_{f,\,T} \\ 0 \end{bmatrix} + P_T \mathbf{v}_{c,\,T}} \mathbf{v}_T^T A_T \mathbf{v}_T \\
&= \min_{\mathbf{v}_T = \begin{bmatrix} \mathbf{v}_{f,\,T} \\ \mathbf{v}_{c,\,T} \end{bmatrix}} \mathbf{v}_T^T \begin{bmatrix} I & W_T \\ 0 & I \end{bmatrix}^T A_T \begin{bmatrix} I & W_T \\ 0 & I \end{bmatrix} \mathbf{v}_T \\
&= \min_{\mathbf{v}_T = \begin{bmatrix} \mathbf{v}_{f,\,T} \\ \mathbf{v}_{c,\,T} \end{bmatrix}} \mathbf{v}_T^T \overline{A}_T \mathbf{v}_T.
\end{aligned}$$

Here, we used the relation

$$\overline{A}_T = \begin{bmatrix} I & W_T \\ 0 & I \end{bmatrix}^T A_T \begin{bmatrix} I & W_T \\ 0 & I \end{bmatrix}.$$

Thus, we showed that S_T, the Schur complement of A_T equals the Schur complement $\overline{S}_T$ of the two-level hierarchical matrix $\overline{A}_T$. Therefore, we can rewrite (B.25) in the following final form relating the coarse element matrix $A_{c,\,T}$ and the Schur complement S_T of the local (fine-grid) matrix A_T. We have

$$\mathbf{v}_{c,\,T}^T A_{c,\,T} \mathbf{v}_{c,\,T} \le \frac{1}{1-\gamma_T^2}\, \mathbf{v}_{c,\,T}^T S_T \mathbf{v}_{c,\,T}. \tag{B.28}$$

Based on the estimate for $\lambda_T \equiv 1/(1-\gamma_T^2)$, the following local-to-global analysis is immediate, utilizing the minimization property (B.27) of the local Schur complements S_T. We have

$$\begin{aligned}
\mathbf{v}_c^T A_c \mathbf{v}_c &= \sum_T \mathbf{v}_{c,\,T}^T A_{c,\,T} \mathbf{v}_{c,\,T} \\
&\le \sum_T \lambda_T\, \mathbf{v}_{c,\,T}^T S_T \mathbf{v}_{c,\,T} \\
&\le \sum_T \lambda_T \begin{bmatrix} \mathbf{v}_{f,\,T} \\ \mathbf{v}_{c,\,T} \end{bmatrix}^T A_T \begin{bmatrix} \mathbf{v}_{f,\,T} \\ \mathbf{v}_{c,\,T} \end{bmatrix} \\
&= \sum_T \lambda_T\, \mathbf{v}_T^T A_T \mathbf{v}_T \\
&\le \max_T \lambda_T \sum_T \mathbf{v}_T^T A_T \mathbf{v}_T \\
&= \max_T \lambda_T\, \mathbf{v}^T A \mathbf{v}.
\end{aligned}$$

Because $\lambda_T < 4$, we have the uniform (global) estimate

$$\mathbf{v}_c^T P^T A P \mathbf{v}_c \leq 4 \min_{\mathbf{v}=J\mathbf{v}_f+P\mathbf{v}_c} \mathbf{v}^T A \mathbf{v},$$

where

$$J = \begin{bmatrix} I \\ 0 \end{bmatrix}.$$

The latter is equivalent (cf., Lemma 3.3) to the (global) strengthened Cauchy–Schwarz inequality,

$$\mathbf{v}_f^T J^T A P \mathbf{v}_c \leq \gamma \left(\mathbf{v}_f^T J^T A J \mathbf{v}_f\right)^{1/2} \left(\mathbf{v}_c^T P^T A P \mathbf{v}_c\right)^{1/2}. \tag{B.29}$$

Here $\gamma = \max_T \gamma_T$. For the particular case of piecewise linear triangular elements, we have

$$\gamma = \max_T \gamma_T < \frac{\sqrt{3}}{2}. \tag{B.30}$$

B.3.2 On the MG convergence uniform w.r.t. the mesh and jumps in the PDE coefficients

We start with the observation that $\gamma = \max_{T \in \mathcal{T}_H} \gamma_T^2$ stays away from unity is general. Note that the local γ_T depends only on the shape of T and not on its size. That is, if we keep the elements at every refinement level geometrically similar to a finite number of coarse elements, then the global $\gamma = \max_T \gamma_T$ will be uniformly bounded away from unity. This verifies one of two of the sufficient conditions (see (ii) in Section 3.3) that imply two-grid convergence. With the simple choice of

$$J = \begin{bmatrix} I \\ 0 \end{bmatrix},$$

if we can also show that $J^T A J$ and the symmetrized smoother $\widetilde{M}$ restricted to Range(J) (i.e., $J^T \widetilde{M} J$) are spectrally equivalent (see condition (i) in Section 3.3) then we have a mesh-independent two-grid convergence estimate result which follows from Theorem 3.25. This was a result originally shown by R. Bank and T. Dupont in 1980 ([BD80]) for $H = 2h$ and finite element problems corresponding to second-order self-adjoint elliptic problems.

For the simple J, we have that $A_{ff} = J^T A J$ corresponds to a principal submatrix of A where we have deleted all rows and columns of A corresponding to the vertices of the coarse triangles. We first show that A_{ff} is spectrally equivalent to its diagonal part, independent of the mesh-size h. To prove this result, we notice that A_{ff} is assembled from the local matrices $\overline{A}_{ff,\,T}$. It is clear (see, e.g., (B.7)) that the local s.p.d. matrices $\overline{A}_{ff,\,T}$ are spectrally equivalent to its diagonal part with constants of spectral equivalence that depend on the angles of T only. Hence, if we keep the triangles T geometrically similar to a fixed initial set of triangles, then the resulting matrices A_{ff} will be spectrally equivalent to its diagonal part with mesh-independent constants. Therefore if M is a smoother such that the symmetrized one $\widetilde{M}$ is spectrally equivalent to the diagonal D of A (such as the Gauss–Seidel, cf., Proposition 6.12), we have then that A_{ff} and $J^T \widetilde{M} J$ are both spectrally equivalent to $D_{ff} = J^T D J$ which is the desired result.

In conclusion, we showed that the two-grid method for A based on M and P has a uniformly bounded convergence factor. Moreover, because all estimates we derived are based on properties of the local matrices A_T the result does not change if we replace them by any constant $\omega_T > 0$ times A_T, that is, if we consider problems that give rise to local matrices $\omega_T A_T$ (and A_T referring to the ones corresponding to the Laplace bilinear form). Thus, we have also proved uniform convergence of the two-grid method for finite element problems coming from second-order elliptic bilinear forms $a(u, \varphi) = \int_\Omega \omega(\mathbf{x}) \nabla u \cdot \nabla \varphi \, d\mathbf{x}$ for given polygonal domain Ω (the local analysis holds in 3D as well) where the coefficient $\omega(\mathbf{x})$ is piecewise constant over the coarse elements from $\mathcal{T}_H$, no matter how large the jumps of ω across the coarse element boundaries might be. This is one of the main attractive features of the (two- and multilevel) HB methods. We showed here that the same holds for the standard two-grid method. We also proved that this extends to the multilevel case, namely, to the AMLI-cycle MG methods (see Section 5.6.3). Results regarding V-cycle MG convergence for such problems are not as easy to obtain. We refer to [JW94] and [WX94], or to [XZh07] for a more recent treatment of the topic.

B.4 Mixed methods for second-order elliptic PDEs

In this section, we consider the mixed finite element method applied to second-order elliptic problems

$$\mathcal{L}u \equiv -\sum_{r=1}^{d} \frac{\partial}{\partial x_r} \left(\sum_{s=1}^{d} a_{r,s}(\mathbf{x}) \frac{\partial u}{\partial x_s} \right) = f.$$

Here $f \in L_2(\Omega)$ and $u \in H_0^1(\Omega)$. In the mixed finite element method, we first introduce a new unknown (vector) function $\underline{\sigma} = -\mathcal{A}(\mathbf{x})\nabla u$, where $\mathcal{A}(\mathbf{x}) = (a_{r,s}(\mathbf{x}))_{r,s=1}^{d}$ and rewrite the problem $\mathcal{L}u = f$ as the following system of first-order PDEs posed variationally, for appropriate test functions $\underline{\eta}$ and φ,

$$\begin{aligned} (\mathcal{A}^{-1}\underline{\sigma}, \ \underline{\eta}) + (\nabla u, \ \underline{\eta}) &= 0, \\ -(\operatorname{div} \underline{\sigma}, \ \varphi) \qquad\qquad &= -(f, \varphi). \end{aligned}$$

If we use integration by parts, assuming that $\underline{\eta}$ is sufficiently smooth (in order to perform integration by parts) to arrive at $(\nabla u, \ \underline{\eta}) = -(u, \ \operatorname{div} \underline{\eta})$ using the fact that $u \in H_0^1(\Omega)$, the above system admits the symmetric form,

$$\begin{aligned} (\mathcal{A}^{-1}\underline{\sigma}, \ \underline{\eta}) - (u, \ \operatorname{div} \underline{\eta}) &= 0, \\ -(\operatorname{div} \underline{\sigma}, \ \varphi) \qquad\qquad &= -(f, \varphi). \end{aligned} \tag{B.31}$$

Because we assume $f \in L_2(\Omega)$ and $\nabla u \in (L_2(\Omega))^d$ a natural space for $\underline{\sigma}$ is $H(\mathrm{div})$. The latter is defined as the vector functions $\underline{\eta} \in (L_2(\Omega))^d$ that also have divergence in $L_2(\Omega)$. Therefore the test functions above $\underline{\eta}$ and φ satisfy $\underline{\eta} \in H(\mathrm{div})$ and $\varphi \in L_2(\Omega)$. The system (B.31) does not contain any derivatives of u, therefore a minimal space for u is $L_2(\Omega)$.

Note that inhomogeneous Dirichlet boundary conditions for $\mathcal{L}u = f$, $u = g$ on $\partial\Omega$, are easily handled by the mixed method. The corresponding mixed problem takes the form, introducing $\mathbf{n}$ the unit vector normal to $\partial\Omega$ pointing outward Ω,

$$\begin{aligned} (\mathcal{A}^{-1}\underline{\sigma},\ \underline{\eta}) - (u,\ \mathrm{div}\ \underline{\eta}) &= -\int\limits_{\partial\Omega} g\ \underline{\eta} \cdot \mathbf{n}\, d\varrho, \\ -(\mathrm{div}\ \underline{\sigma},\ \varphi) &= -(f, \varphi). \end{aligned} \tag{B.32}$$

Mixed finite elements for second-order elliptic problems

The discretization of (B.32) requires two finite element spaces, a vector one, $\mathbf{R} = \mathbf{R}_h$, and a scalar one, $W = W_h$ which are subspaces of $H(\mathrm{div})$ and $L_2(\Omega)$, respectively. The discrete problem takes then the following form.

Find $\underline{\sigma}_h \in \mathbf{R}_h$ and $u_h \in W_h$ such that

$$\begin{aligned} (\mathcal{A}^{-1}\underline{\sigma}_h,\ \underline{\eta}_h) - (u_h,\ \mathrm{div}\ \underline{\eta}) &= -\int\limits_{\partial\Omega} g\ \underline{\eta}_h \cdot \mathbf{n}\, d\varrho \quad \text{for all } \underline{\eta}_h \in \mathbf{R}_h, \\ -(\mathrm{div}\ \underline{\sigma}_h,\ \chi_h) &= -(f,\ \chi_h) \quad\quad\quad\ \text{for all } \chi_h \in W_h. \end{aligned} \tag{B.33}$$

For triangular elements τ a popular pair of spaces is the lowest-order Raviart–Thomas spaces $\mathbf{R}_h$ and W_h. They are associated with a common triangulation $\mathcal{T}_h = \{\tau\}$. Any function $\underline{\eta} \in \mathbf{R}_h$ restricted to a triangle τ has the form

$$\begin{bmatrix} a + cx \\ b + cy \end{bmatrix};$$

that is, it has three degrees of freedom (dofs); namely, the coefficients a, b, and c. We can define locally three basis functions $\underline{\eta}_{e_k}$, associated with the three edges e_{i_k}, $k = 1, 2, 3$ of any given triangle τ. Thus, globally, the number of the basis functions will equal the number of the edges of all triangles in $\mathcal{T}$. The basis functions are defined from the following edge-based integral conditions, for all three edges e_l of τ,

$$\int\limits_{e_l} \underline{\eta}_{e_k} \cdot \mathbf{n}_{e_l}\, d\varrho = \delta_{e_k,\ e_l}, \qquad l = 1, 2, 3.$$

These conditions ensure that $\underline{\eta}_e$ associated with a particular edge e is supported in the (possibly) two neighboring elements τ_e^+ and τ_e^- that share e. (One of these triangles may be empty if e is a boundary edge). Also, we automatically guarantee that $\underline{\eta}_e \cdot \mathbf{n}_e$ is continuous across e because it happens (see below Remark B.11) that $\underline{\eta}_e \cdot \mathbf{n}_e$ is constant on e and due to the integral condition $\int_e \underline{\eta}_e \cdot \mathbf{n}_e = 1$. Therefore the constant in question is uniquely specified (from the possibly two neighboring elements that share e). Thus, the basis functions belong to $H(\text{div})$, because their normal components are continuous. This is a characterization of $H(\text{div})$ (for piecewise smooth vector functions). To prove it, consider any piecewise polynomial vector function $\underline{\eta}$ with continuous normal components $\underline{\eta} \cdot \mathbf{n}_e$, for any $e \in \mathcal{E}$ (where $\mathcal{E}$ is the set of all edges). By integration by parts, for any sufficiently smooth function φ vanishing on $\partial\Omega$, we get

$$(\nabla\varphi,\ \underline{\eta}) = \sum_\tau \int_\tau (\nabla\varphi) \cdot \underline{\eta}\ d\mathbf{x}$$
$$= -\sum_\tau \int_\tau \varphi \text{ div } \underline{\eta}\ d\mathbf{x} + \sum_{e\in\mathcal{E}} \int_e [\underline{\eta} \cdot \mathbf{n}_e]\ \varphi\ d\varrho = -(\varphi,\ \text{div } \underline{\eta}),$$

because the jump terms $[\underline{\eta} \cdot \mathbf{n}_e]$ are zero. This shows that div $\underline{\eta}$ is well defined as a function of $L_2(\Omega)$ (due to the density of $H_0^1(\Omega)$ in $L_2(\Omega)$).

The basis functions χ_τ of W, for every τ are trivially constructed because those are piecewise constant functions; that is, for any given τ, $\chi_\tau = 1$ on τ and $\chi_\tau = 0$ outside τ.

We can derive the following more explicit equations for computing the basis functions of $\mathbf{R}$. For any triangle τ with vertices $\mathbf{x}_{i_k}, k = 1, 2, 3$, consider the midpoints $\mathbf{x}_{i_r,i_s} = \frac{1}{2}(\mathbf{x}_{i_r} + \mathbf{x}_{i_s})$. $r = 1,\ s = 2, r = 2,\ s = 3$, and $r = 1,\ s = 3$. Then, define for each midpoint $\mathbf{x}_{i_r,i_s}$, or equivalently, any edge $e = (\mathbf{x}_{i_r},\ \mathbf{x}_{i_s})$, a basis function

$$\underline{\eta}_e = \underline{\eta}_{(\mathbf{x}_{i_r},\ \mathbf{x}_{i_s})} = \begin{bmatrix} a_e + c_e \frac{1}{h}(x - x_{i_r,i_s}) \\ b_e + c_e \frac{1}{h}(y - y_{i_r,i_s}) \end{bmatrix}.$$

The conditions for determining the three coefficients $a_e, b_e,\ c_e$ read, for all three edges $e' = (i_k, i_l)$,

$$\delta_{e,\ e'} \frac{1}{|e'|} = \mathbf{n}_{e'} \cdot \begin{bmatrix} a_e \\ b_e \end{bmatrix} + c_e \frac{1}{h} \frac{1}{|e'|} \int_{e'} \mathbf{n}_{e'} \cdot \begin{bmatrix} x - x_{i_r,i_s} \\ y - y_{i_r,i_s} \end{bmatrix} d\varrho.$$

The midpoint $\mathbf{x}_{(i_k,i_l)}$ has geometric coordinates $x_{i_k,i_l} = \frac{1}{2}(x_{i_k} + x_{i_l})$ and $y_{i_k,i_l} = \frac{1}{2}(y_{i_k} + y_{i_l})$. Now, because

$$\mathbf{n}_{e'} \cdot \begin{bmatrix} x - x_{i_k,i_l} \\ y - y_{i_k,i_l} \end{bmatrix} = 0$$

for (x, y) on edge e' (because $\mathbf{n}_{e'}$ is normal to that edge), we get the somewhat simplified system of three equations (for the three edges e' of τ) and three unknowns, $(a_e,\ b_e,\ c_e)$,

$$\delta_{e,\ e'} \frac{1}{|e'|} = \mathbf{n}_{e'} \cdot \begin{bmatrix} a_e \\ b_e \end{bmatrix} + c_e \mathbf{n}_{e'} \cdot \begin{bmatrix} (x_{i_k,i_l} - x_{i_r,i_s})/h \\ (y_{i_k,i_l} - y_{i_r,i_s})/h \end{bmatrix}.$$

Here, $\delta_{e,\ e'} = 0$ if $e \neq e'$ and $\delta_{e,\ e} = 1$.

Remark B.11. A general observation is that a_e, b_e, and c_e are order $\mathcal{O}(h^{-1})$ because $1/|e^{'}| = \mathcal{O}(h^{-1})$. Also, div $\underline{\eta}_e = (2/h)c_e = \mathcal{O}(h^{-2})$. Also, we have

$$\mathbf{n}_e \cdot \underline{\eta}_e = \mathbf{n}_e \cdot \begin{bmatrix} a_e \\ b_e \end{bmatrix} + \frac{c_e}{h}\mathbf{n}_e \cdot \begin{bmatrix} x - x_{i_r,i_s} \\ y - y_{i_r,i_s} \end{bmatrix}$$

and the latter term vanishes for (x, y) on e. That is,

$$\mathbf{n}_e \cdot \underline{\eta}_e = \mathbf{n}_e \cdot \begin{bmatrix} a_e \\ b_e \end{bmatrix} = \text{Const on } e.$$

Because the degrees of freedom for the space $\mathbf{R}_h$ are associated with the set $\mathcal{E}$ it is convenient to have global numbering of the edges, that is, $e_1,\ e_2,\ e_3, \dots, e_{n_e}$. Compute now for any element τ, having edges $e_{i_1}, e_{i_2},\ e_{i_3}$, the element matrices,

$$A_\tau = \left(\int\limits_\tau \underline{\eta}^T_{e_{i_k}} \mathcal{A}(\mathbf{x})\, \underline{\eta}_{e_{i_l}}\ d\mathbf{x} \right)_{k,l=1}^{3},$$

$$B_\tau = -|\tau| \left[\text{div } \eta_{e_{i_1}}, \text{div } \eta_{e_{i_2}},\ \text{div } \eta_{e_{i_3}} \right].$$

Here $|\tau|$ is the area of τ and we used the fact that the div η_{i_l} for $l = 1, 2, 3$, are constants on τ. Based on Remark B.11, we see that the entries of both A_τ and B_τ are order $\mathcal{O}(1)$.

Then after the usual assembly, we end up with the global matrices A and B,

$$\mathbf{w}^T A \mathbf{v} = \sum_{\tau \in \mathcal{T}} \mathbf{w}_\tau^T A_\tau \mathbf{v}_\tau, \quad \text{and} \quad \mathbf{y}^T B \mathbf{v} = \sum_{\tau \in \mathcal{T}} \mathbf{y}_\tau^T B_\tau \mathbf{v}_\tau.$$

Note that $\mathbf{v}_\tau,\ \mathbf{w}_\tau \in \mathbb{R}^3$, whereas $\mathbf{y}_\tau$ is just a scalar. Finally, the finite element problem (B.33) takes the following matrix–vector form

$$\begin{bmatrix} A & B^T \\ B & 0 \end{bmatrix} \begin{bmatrix} \mathbf{v} \\ \mathbf{x} \end{bmatrix} = \begin{bmatrix} \mathbf{g} \\ \mathbf{f} \end{bmatrix}. \tag{B.34}$$

Here $\mathbf{v} = (\mathbf{v}_e)_{e \in \mathcal{E}}$ and $\mathbf{x} = (\xi_\tau)_{\tau \in \mathcal{T}}$. The entries $\mathbf{v}_e$ and $\mathbf{x}_\tau$ are scalars and they are the coefficients in the expansion of $\underline{\sigma}_h = \sum_{e \in \mathcal{E}} \mathbf{v}_e\ \underline{\eta}_e$ and $u_h = \sum_{\tau \in \mathcal{T}} \xi_\tau\ \chi_\tau$. The r.h.s. $\mathbf{g} = (\mathbf{g}_e)_{e \in \mathcal{E}}$ has entries $\mathbf{g}_e = -\int_e g\ \underline{\eta} \cdot \mathbf{n}_e\ d\varrho$ that might be possibly nonzero only for boundary edges e. Finally, for the second block of the r.h.s., $\mathbf{f}$, we have $\mathbf{f} = (\mathbf{f}_\tau)_{\tau \in \mathcal{T}}$ with $\mathbf{f}_\tau = -\int_\tau f\ d\mathbf{x}$.

We introduce next the so-called Fortin projection (originated in [F77]). It is a useful tool in proving then "inf–sup" condition as well as in deriving error estimates in the mixed f.e. method.

Definition B.12 (Fortin projection). *Consider, for any sufficiently smooth vector-function $\underline{\eta}$, the following edge-based defined projection $\pi = \pi_h :\ \underline{\eta} \mapsto \pi\underline{\eta} \in \mathbf{R}_h$ with dofs equal to*

$$\int\limits_e (\pi\underline{\eta}) \cdot \mathbf{n}_e\ d\varrho = \int\limits_e (\underline{\eta}) \cdot \mathbf{n}_e\ d\varrho.$$

The projection $\pi = \pi_h$ defines a unique function

$$\pi \underline{\eta} = \sum_{e \in \mathcal{E}} \left(\int_e \underline{\eta} \cdot \mathbf{n}_e \, d\varrho \right) \underline{\eta}_e \in \mathbf{R}. \tag{B.35}$$

It satisfies the important "commutativity" property

$$(\text{div } \underline{\eta}, \ \chi) = (\text{div } \pi \underline{\eta}, \ \chi), \quad \text{for any } \chi \in W_h.$$

The latter is true because both expressions equal (using the divergence theorem),

$$\sum_e \int_e \underline{\eta} \cdot \mathbf{n}_e \, [\chi]_e \, d\varrho.$$

We used here the fact that the jump term $[\chi]_e$ is constant on e for any $\chi \in W$ (because χ is a piecewise constant function). Introducing the L_2-projection $Q_h : L_2(\Omega) \mapsto W_h$, one can rewrite the above commutativity property as follows,

$$Q_h \text{ div } = \text{div } \pi_h. \tag{B.36}$$

We note that both π_h and Q_h are easily computable based on local operations. The actions of π_h are seen from the expression (B.35) whereas Q_h is computable from the following explicit expression,

$$Q_h \chi = \sum_\tau \frac{1}{|\tau|} \int_\tau \chi \, d\mathbf{x} \, \chi_\tau.$$

We can easily check that $(Q_h \chi, \ \theta) = (\chi, \ \theta)$ for any $\theta \in W_h$.

In what follows, we verify an "inf–sup" estimate.

We assume, for simplicity, that the domain Ω is such that the Poisson equation

$$-\Delta u = \chi \text{ in } \Omega \quad \text{and} \quad u = 0 \text{ on } \partial\Omega,$$

for any given $\chi \in L_2(\Omega)$ admits full regularity (some minimal regularity is generally sufficient); that is, we have

$$\|u\|_2 \leq C \|\chi\|_0.$$

This is the case for Ω being a convex polygon. Due to the identity $\|\nabla u\|^2 = (\chi, \ u) \leq C_F \|\chi\|_0 \|\nabla u\|_0$ (using Schwarz and Friedrich's inequalities), we also have $\|u\|_1 \leq C_F \|\chi\|_0$. Consider then $\mathbf{w} = \pi_h(\nabla u) \in \mathbf{R}_h$. We first notice (because div $\pi_h = Q_h$div) that

$$\|\text{div } \mathbf{w}\|_0 = \|Q_h \nabla u\|_0 \leq \|\nabla u\|_0 \leq C_F \|\chi\|_0.$$

Also, by the definition of π_h, we have $\int_e (\nabla u) \cdot \mathbf{n}_e \, d\varrho = \int_e \mathbf{w} \cdot \mathbf{n}_e \, d\varrho$. Therefore (based on Remark B.11),

$$\|\mathbf{w}\|_0^2 \simeq \sum_{e \in \mathcal{E}} \left(\int_e (\nabla u) \cdot \mathbf{n}_e \, d\varrho \right)^2 \|\underline{\eta}_e\|^2_{\tau_e^- \cup \tau_e^+} \leq C \sum_{e \in \mathcal{E}} \left(\int_e (\nabla u) \cdot \mathbf{n}_e \, d\varrho \right)^2.$$

Use now the following inequality for any function $\psi \in H^2(\tau_e^- \cup \tau_e^+)$,

$$\begin{aligned} \left(\int_e (\nabla \psi) \cdot \mathbf{n}_e \, d\varrho \right)^2 &\leq Ch \int_e (\nabla \psi \cdot \mathbf{n}_e)^2 \, d\varrho \\ &\leq C \left(\|\nabla \psi\|^2_{0,\, \tau_e^- \cup \tau_e^+} + h^2 \|\psi\|^2_{2,\, \tau_e^- \cup \tau_e^+} \right). \end{aligned}$$

This estimate is proved by first verifying the result on the unit-size domain (based on a trace theorem result) and then changing the variables to get the domain of size h which gives rise to the above powers of h. Therefore,

$$\|\mathbf{w}\|_0^2 \leq C \|\nabla u\|_0^2 + Ch^2 \|u\|_2^2 \leq (C + Ch^2) \, \|\chi\|_0^2.$$

That is, we proved

$$\|\mathbf{w}\|^2_{H(\text{div})} \leq \beta \, \|\chi\|_0^2.$$

The latter estimate shows the first "inf–sup" estimate

$$\|\chi\|_0 \leq \beta \sup_{\mathbf{w} \in \mathbf{R}} \frac{(\chi, \text{ div } \mathbf{w})}{\|\mathbf{w}\|_{H(\text{div})}}. \tag{B.37}$$

However, another choice of norms is also possible. It gives a different "inf–sup" condition. Consider the identity $(\chi, \text{ div } \mathbf{w}) = -\sum_{e \in \mathcal{E}} \int_e [\chi]_e \, \mathbf{w} \cdot \mathbf{n}_e \, d\varrho$. Choose now

$$\mathbf{w} = -\sum_{e \in \mathcal{E}} [\chi]_e \, \underline{\eta}_e.$$

Then, $(\chi, \text{ div } \mathbf{w}) = \sum_{e \in \mathcal{E}} [\chi]_e^2$. Again based on Remark B.11, we have

$$\|\mathbf{w}\|_0^2 \simeq \sum_{e \in \mathcal{E}} [\chi]_e^2 \, \|\underline{\eta}_e\|^2_{\tau_e^- \cup \tau_e^+} \leq C \sum_{e \in \mathcal{E}} [\chi]_e^2,$$

and

$$\|\text{div } \mathbf{w}\|_0^2 \simeq \sum_{e \in \mathcal{E}} [\chi]_e^2 \, \|\text{div } \underline{\eta}_e\|^2_{\tau_e^- \cup \tau_e^+} \leq Ch^{-2} \sum_{e \in \mathcal{E}} [\chi]_e^2.$$

This shows, that for a mesh-independent constant β,

$$\|\mathbf{w}\|_0^2 + h^2\|\text{div } \mathbf{w}\|_0^2 \le \beta \sum_{e\in\mathcal{E}} [\chi]_e^2 = \beta\ (\chi,\ \text{div } \mathbf{w}),$$

which implies the following alternative "inf–sup" estimate

$$\begin{aligned}\|\chi\|_{1,\,h} \equiv \left(\sum_{e\in\mathcal{E}} [\chi]_e^2\right)^{1/2} &\le \beta \sup_{\mathbf{w}\in\mathbf{R}} \frac{(\chi,\ \text{div } \mathbf{w})}{\left(\|\mathbf{w}\|_0^2 + h^2\|\text{div } \mathbf{w}\|_0^2\right)^{1/2}} \\ &\le \beta \sup_{\mathbf{w}\in\mathbf{R}} \frac{(\chi,\ \text{div } \mathbf{w})}{\|\mathbf{w}\|_0}.\end{aligned} \tag{B.38}$$

We next show that the norm $\|.\|_{1,\,h}$ is stronger than the L_2-norm.

Lemma B.13. *The following inequality holds, for any* $\chi \in W_h$,

$$\|\chi\|_0 \le C\|\chi\|_{1,h}.$$

Proof. The space W_h is associated with triangular elements τ and let $|\tau| \simeq \mathcal{O}(h^2)$. We refine the triangles (by connecting the midpoints of the edges of every triangle) and we do this twice. The resulting triangulation is denoted $\mathcal{T}_{h/4}$. Next, we introduce the space of piecewise linear functions $W_{h/4}$ associated with $\mathcal{T}_{h/4}$. Let $\mathcal{N}_{h/4}$ denote the set of all vertices of the triangles in $\mathcal{T}_{h/4}$. Define the mapping $P:\ W_h \mapsto W_{h/4}$ as follows. On the strictly interior triangle $\tau^0 \subset \tau$, $\tau \in \mathcal{T}_h$, and $\tau^0 \in \mathcal{T}_{h/4}$, we let $P\chi = \chi$. At all remaining nodes in $\mathcal{N}_{h/4}$ that are shared by two or more triangles from $\mathcal{T}_h$ we let $P\chi$ be a simple arithmetic average of the values of χ coming from the triangles that share that node. Finally, define the "cut-off" function $\theta \in W_{h/4}$ which is 1 on all strictly interior nodes in $\mathcal{N}_{h/4}$ and zero on $\partial\Omega$. It is clear that

$$\|\chi\|_0 \simeq \|\theta(P\chi)\|_0.$$

Based on the chain rule $\nabla(\theta P\chi) = \theta\nabla(P\chi) + (P\chi)\nabla\theta$, and noticing that $\nabla\theta$ is zero outside a strip Ω_h near $\partial\Omega$ of width $\mathcal{O}(h)$, and that within the strip $|\nabla\theta| \le Ch^{-1}$, we obtain the estimate,

$$\begin{aligned}|\theta P\chi|_1^2 &\le C|P\chi|_1^2 + \frac{C}{h^2}\int_{\Omega_h} \chi^2(\mathbf{x})\,d\mathbf{x} \\ &\le C\|\chi\|_{1,h}^2 + C\sum_{e=\partial\Omega\cap\tau} \chi_\tau^2 \\ &\le C\|\chi\|_{1,h}^2.\end{aligned} \tag{B.39}$$

We used that the $|.|_1$ (semi)norm of a finite element function $\varphi = P\chi$ is simply a square root of the sum of squares of differences $(\varphi(\mathbf{x}_i) - \varphi(\mathbf{x}_j))$, for any neighboring pair of nodes $\mathbf{x}_i,\ \mathbf{x}_j \in \mathcal{N}_{h/4}$ (i.e., belonging to a common fine-grid element in $\mathcal{T}_{h/4}$). And because all such differences can be expressed as a (fixed) sum of differences of the form $\chi_{\tau_1} - \chi_{\tau_2}$ for neighboring elements $\tau_1,\ \tau_2 \in \mathcal{T}_h$, the estimate $|P\chi|_1^2 \le C\|\chi\|_{1,h}^2$ is seen.

The desired norm bound follows from Friedrich's inequality and estimate (B.39)

$$\|\chi\|_0^2 \simeq \|\theta P\chi\|_0^2 \le C_F\ |\theta P\chi|_1^2 \le C\|\chi\|_{1,\ h}^2. \qquad \square$$

We complete the present section by summarizing the result for the "inf–sup" conditions that we can exploit for the saddle-point matrix in (B.34).

Theorem B.14. *The pair of lowest-order Raviart–Thomas spaces* $\mathbf{R}_h,\ W_h$, *ensure respective "inf–sup" conditions if equipped with the following norms.*

- $\mathbf{R}_h$ *equipped with the* $H(div)$*-norm and* W_h *with the* L_2*-norm; or*
- $\mathbf{R}_h$ *equipped with the* $(L_2(\Omega))^2$*-norm* $\|.\|_0$, *whereas* W_h *equipped with the* $\|.\|_{1,h}$*-norm (defined in* (B.38)*).*

Also, the saddle-point operator (matrix) in (B.34) *is bounded in both pairs of norms.*

Proof. To prove the boundedness of the saddle-point operator, in both pairs of norms, we first notice that

$$\mathbf{w}^T A\mathbf{v} \le C\|\mathbf{w}\|_0\ \|\mathbf{v}\|_0,$$

and second, let $\mathbf{v}$ and $\mathbf{x}$ be the coefficient vectors of $\underline{v} \in \mathbf{R}_h$ and $\chi \in W_h$, respectively. Then,

$$\mathbf{x}^T B\mathbf{v} = (\text{div}\ \underline{v},\ \chi) \le \|\text{div}\ \underline{v}\|_0\|\chi\|_0,$$

which proves boundedness of the saddle-point operator in the first pair of norms $\|.\|_{H(\text{div})},\ \|.\|_0$. On the other hand, because (using integration by parts)

$$\mathbf{x}^T B\mathbf{v} = (\text{div}\ \underline{v},\ \chi) = \sum_{e\in\mathcal{E}} \int_e \underline{v}\cdot\mathbf{n}_e\ [\chi]_e\ d\varrho \le C\|\underline{v}\|_0\ \|\chi\|_{1,h},$$

the boundedness of the saddle-point operator, in the second pair of norms, $\|.\|_0,\ \|.\|_{1,h}$, also follows. $\square$

Corollary B.15. *The pair of norms in* $\|.\|_{H(div)}$ *and* $\|.\|_{L_2(\Omega)}$ *used for the finite element spaces* $\mathbf{R}_h$ *and* W_h *give rise to the following block-diagonal matrix,*

$$\begin{bmatrix} A + B^T M^{-1} B & 0 \\ 0 & M \end{bmatrix}.$$

Here, $M = \text{diag}(|\tau|)_{\tau\in\mathcal{T}_h} \simeq h^2\ I$ *is the diagonal mass matrix corresponding to the space* W_h *of piecewise constant functions. Due to our choice of basis* $\{\underline{\eta}_e\}_{e\in\mathcal{E}_h}$ *of* $\mathbf{R}_h$, *we have* $A + B^T M^{-1} B \simeq I + h^{-2} B^T B$ *(cf. Remark B.11). The other pair of norms,*

$\|.\|_{(L_2(\Omega))^2}$ *and* $\|.\|_{1,h}$, *give rise again to a block-diagonal matrix*

$$\begin{bmatrix} A & 0 \\ 0 & D \end{bmatrix},$$

where $A \simeq I$ *and* D *corresponds now to a discrete (cell–centered) Laplace operator* $-\Delta_h$.

The first pair of norms, $\|.\|_{H(\text{div})}$, $\|.\|_{L_2(\Omega)}$, require constructing preconditioners for the Raviart–Thomas space $\mathbf{R}_h$ and the $H(\text{div})$ bilinear form. In Appendix F we prove a general MG convergence result for the weighted $H(\text{div})$-bilinear form $(\mathbf{u},\ \mathbf{v}) + \tau\ (\text{div}\ \mathbf{u},\ \text{div}\ \mathbf{v})$. For the purpose of that analysis, consider the mass (Gram) matrix $\mathbf{G}$ computed from the Raviart–Thomas space $\mathbf{R}_h$. Consider also the diagonal matrix $\mathbf{D} = ((\text{div}\ \underline{\eta}_e,\ \text{div}\ \underline{\eta}_e))_{e\in\mathcal{E}_h}$. Because on a given element τ that shares an edge (face in 3D) $e \in \mathcal{E}_h$, div $\underline{\eta}_e$ is constant, we have the identity

$$|\tau|\ |\text{div}\ \underline{\eta}_e|^2 = \int\limits_\tau |\text{div}\ \underline{\eta}_e|^2\ d\mathbf{x} = |\text{div}\ \underline{\eta}_e| \int\limits_e \underline{\eta}_e \cdot \mathbf{n}_e\ d\varrho = |\text{div}\ \underline{\eta}_e|.$$

Note that for $e' \subset \partial\tau$, $\int_{e'} \underline{\eta}_e \cdot \mathbf{n}_e\ d\varrho = 1$ or 0. Therefore, we have that $|\text{div}\ \underline{\eta}_e| = 1/|\tau|$. Note that div $\underline{\eta}_e \neq 0$ for any basis function $\underline{\eta}_e$. Indeed, if we assume that div $\underline{\eta}_e = 0$ on an element τ such that $e \subset \partial\tau$, then by the divergence theorem, we will have $0 = \int_{\partial\tau} \underline{\eta}_e \cdot \mathbf{n}\ d\varrho = \int_e \underline{\eta}_e \cdot \mathbf{n}\ d\varrho = 1$, or -1, which is a contradiction.

Therefore, the above diagonal matrix $\mathbf{D}$ is spectrally equivalent to the scaled mass matrix h^{-2} $\mathbf{G}$. Note that the part of the stiffness matrix that comes from the (div ., div .) form is only semidefinite (with a large null space), however, its diagonal $\mathbf{D}$ is s.p.d. Thus, we showed the following result.

Proposition B.16. *The diagonal of* $\mathbf{A}$ *coming from the parameter-dependent* $H(\text{div})$*-bilinear form* $(\mathbf{u},\ \mathbf{v}) + \tau\ (\text{div}\ \mathbf{u},\ \text{div}\ \mathbf{v})$ *and the lowest-order Raviart–Thomas space* $\mathbf{R}_h$ *is spectrally equivalent to the weighted mass matrix* $(1 + \tau h^{-2})\ \mathbf{G}$.

The choice of the second pair of norms $\|.\|_0$, $\|.\|_{1,h}$ to define block-diagonal preconditioners for the saddle-point matrix in (B.34) was considered in [RVWa]. Multigrid methods for spaces of discontinuous functions (giving rise to generalizations to the norm $\|.\|_{1,h}$) are found in [GK03].

An alternative to the block-diagonal preconditioning approach for the saddle-point operator in (B.34) is to explore preconditioning by multigrid methods in a div-free subspace. Assume that we have two nested pairs of finite element spaces, a coarse one $\mathbf{R}_H$, W_H, and a fine one $\mathbf{R}_h$, W_h coming from two nested triangulations $\mathcal{T}_H$ and $\mathcal{T}_h$. That is, the elements in $\mathcal{T}_h$ are obtained by refining the elements T of $\mathcal{T}_H$. We notice that $W_H \subset W_h$ and $\mathbf{R}_H \subset \mathbf{R}_h$. Therefore, from these embeddings, we have interpolation matrices $\mathcal{Q}$ and $\mathcal{P}$ defined naturally. The coarse saddle-point matrix reads

$$\begin{bmatrix} A_c & B_c^T \\ B_c & 0 \end{bmatrix},$$

and due to the nestedness of the f.e. spaces, we have $A_c = \mathcal{P}^T A \mathcal{P}$ and $B_c = \mathcal{Q}^T B \mathcal{P}$. Also, if a coarse function $\underline{v}_H \in \mathbf{R}_H$ satisfies

$$(\text{div } \underline{v}_H, \ \chi_H) = 0 \quad \text{for all } \chi_H \in W_H,$$

this implies that (pointwise) div $\underline{v}_H = 0$. These facts, translated into a matrix–vector form can be interpreted, first as $B_c \mathbf{v}_c = 0$, and second as $B\mathcal{P}\mathbf{v}_c = 0$. This gives us the opportunity to use the constrained minimization approach (see Section 9.5) for solving the saddle-point problem (B.34) if $\mathbf{f} = 0$. More details are given further in Section F.3.

B.5 Nonconforming elements and Stokes problem

We are concerned in the present section with the following mixed system

$$\begin{aligned} -\Delta \underline{\sigma} - \nabla p &= \mathbf{f}, \\ \text{div } \underline{\sigma} \qquad &= 0, \end{aligned}$$

for $\underline{\sigma} \in (H_0^1(\Omega))^d$ and $p \in L_2(\Omega)$. Here $d = 2$ or 3. The problem is seen to be well posed if rewritten variationally,

$$\begin{aligned} (\nabla \underline{\sigma}, \ \nabla \underline{\theta}) + (p, \ \text{div } \underline{\theta}) &= (\mathbf{f}, \ \underline{\theta})), \quad && \text{for all } \underline{\theta} \in (H_0^1(\Omega))^d \\ (\text{div } \underline{\sigma}, \ \chi) &= 0 && \text{for all } \chi \in L_2(\Omega). \end{aligned}$$

Note that there are no boundary conditions imposed on p. Also, because $(1, \ \text{div } \underline{\theta}) = \int_{\partial\Omega} \mathbf{n} \cdot \underline{\theta} \ d\varrho = 0$ for $\underline{\theta} \in (H_0^1(\Omega))^d$ which implies $(p + C, \ \text{div } \underline{\theta}) = (p, \ \text{div } \underline{\theta})$ for any constant C, it is clear that p is determined up to an additive constant.

We introduce now the popular P_1 nonconforming triangular element (also called the Crouzeix–Raviart element). The finite element space V_h (of scalar functions) consists of piecewise linear basis functions that are continuous at the midpoints $\mathbf{x}_{m_e}$ of every edge $e \in \mathcal{E}_h$ of the triangles $\tau \in \mathcal{T}_h$. That is, the functions from V_h are not generally in $H^1(\Omega)$. However, a certain integration by parts formula still holds. Consider the vector function space $\mathbf{V}_h = (V_h)^2$. Let W_h be again the space of piecewise constant functions (w.r.t. the triangles $\tau \in \mathcal{T}_h$). For any $\underline{\eta} \in \mathbf{V}_h$ and a $\chi \in W_h$, we have

$$\begin{aligned} \sum_{\tau \in \mathcal{T}_h} \int_\tau \text{div } \underline{\eta} \ \chi \ d\mathbf{x} &= - \sum_{\tau \in \mathcal{T}_h} \int_{\partial\tau} \underline{\eta} \cdot \mathbf{n} \ \chi \ d\varrho \\ &= - \sum_{\tau \in \mathcal{T}_h} \sum_{e \subset \partial\tau} \int_e \underline{\eta} \cdot \mathbf{n}_e \ \chi \ d\varrho \\ &= \sum_{\tau \in \mathcal{T}_h} \sum_{e \subset \partial\tau} |e| \ \underline{\eta}(\mathbf{x}_{m_e}) \cdot \mathbf{n}_e \ \chi \, . \end{aligned}$$

Here we used the fact that $\underline{\eta}$ is a linear function on any edge e, therefore the midpoint integration rule is exact for such functions. Thus, becaues $\underline{\eta}$ is continuous at the midpoints $\mathbf{x}_{m_e}$, the following integration by parts formula holds,

$$\sum_{\tau\in\mathcal{T}_h}\int_\tau \text{div}\ \underline{\eta}\ \chi\ d\ \mathbf{x} = \sum_{e\in\mathcal{E}_h}\int_e \underline{\eta}\cdot\mathbf{n}_e\ [\chi]_e\ d\varrho. \tag{B.40}$$

We can also define a Fortin projection $\pi = \pi_h$ for any sufficiently smooth scalar function and for each individual component of any (sufficiently smooth) vector function. The projection π_h defines a function $\pi_h\eta \in V_h$, where the degrees of freedom (dofs) of $\pi_h\eta$ (one dof per edge $e \in \mathcal{E}$) are specified from the equations

$$\int_e \pi_h\eta\ d\varrho = \int_e \eta\ d\varrho \quad \text{for all } e \in \mathcal{E}.$$

This means that $(\pi_h\eta)(\mathbf{x}_{m_e}) = 1/|e| \int_e \eta\ d\varrho$. Then, for any sufficiently smooth vector function $\underline{\eta}$ and any $\chi \in W_h$, we have

$$(\text{div}\ \underline{\eta},\ \chi) = \sum_{\tau\in\mathcal{T}_h}\int_\tau (\text{div}\ \underline{\eta})\ \chi\ d\ \mathbf{x} = \sum_{\tau\in\mathcal{T}_h}\int_\tau (\text{div}\ \pi_h\underline{\eta})\ \chi\ d\ \mathbf{x}, \tag{B.41}$$

because both last expressions equal $\sum_{e\in\mathcal{E}_h}\int_e \underline{\eta}\cdot\mathbf{n}_e\ [\chi]_e\ d\varrho$, (see (B.40)). Introduce the piecewise divergence operator div_h; that is, for any piecewise smooth $\underline{\eta}$, $\text{div}_h\underline{\eta} \in L_2(\Omega)$ is simply equal to div $\underline{\eta}$, well-defined, piecewise on every element $\tau \in \mathcal{T}_h$. Also, let $Q_h : L_2(\Omega) \mapsto W_h$ be the L_2-projection. We can then rewrite the commutativity property (B.41) in the operator form,

$$Q_h\text{div} = \text{div}_h\pi_h. \tag{B.42}$$

The above product of operators is applied to smooth functions.

We are interested in the finite element problem which reads as follows.

Find $\underline{\sigma}_h \in \mathbf{V}_h$ and $u_h \in W_h$, such that

$$\begin{array}{lll}\sum_{\tau\in\mathcal{T}_h}\int_\tau \nabla\underline{\sigma}_h\cdot\nabla\underline{\theta}\ d\mathbf{x} & +(u_h,\ \text{div}_h\ \underline{\theta}) = (\mathbf{f},\ \underline{\theta}) & \text{for all } \underline{\theta}\in\mathbf{V}_h,\\ (\text{div}_h\ \underline{\sigma}_h,\ \chi) & = 0 & \text{for all } \chi\in W_h.\end{array} \tag{B.43}$$

Again, we stress the fact that u_h is determined up to an additive constant if $\mathbf{V}_h$ consist of functions vanishing at the midpoints of the edges on the boundary of Ω. Introduce the stiffness matrices $\mathbf{A}$ and B which are computed by assembling the element matrices

$$\mathbf{A}_\tau = \begin{bmatrix} A_\tau & 0\\ 0 & A_\tau\end{bmatrix} \quad \text{and} \quad B_\tau = \left[B_\tau^x,\ B_\tau^y\right].$$

The element matrices are computed based on the scalar (edge-based) piecewise linear nonconforming basis functions $\{\varphi_{e_1},\ \varphi_{e_2}, \ldots, \varphi_{e_{n_e}}\}$ and the trivially constructed basis functions $\{\chi_\tau\}_{\tau\in\mathcal{T}_h}$ of W_h. More specifically, for any triangle τ with edges e_{i_1}, e_{i_2}, and e_{i_3}, we have

$$A_\tau = \left(\int_\tau \nabla \varphi_{e_{i_l}} \cdot \nabla \varphi_{e_{i_k}} \, d\mathbf{x} \right)_{l,k=1}^{3}, \qquad B_\tau^x = |\tau| \left(\frac{\partial \varphi_{e_{i_1}}}{\partial x}, \frac{\partial \varphi_{e_{i_2}}}{\partial x}, \frac{\partial \varphi_{e_{i_3}}}{\partial x} \right),$$

and similarly,

$$B_\tau^y = |\tau| \left(\frac{\partial \varphi_{e_{i_1}}}{\partial y}, \frac{\partial \varphi_{e_{i_2}}}{\partial y}, \frac{\partial \varphi_{e_{i_3}}}{\partial y} \right).$$

Note that A_τ corresponds to a nonconforming (scalar) Laplace element matrix. Introduce also the coefficient vectors

$$\mathbf{V} = \begin{bmatrix} \mathbf{V}^x \\ \mathbf{V}^y \end{bmatrix} \quad \text{of } \underline{\sigma}_h = \begin{bmatrix} \sigma_h^x \\ \sigma_h^y \end{bmatrix}$$

and $\mathbf{x}$ of u_h, and the r.h.s. vector

$$\mathbf{F} = \begin{bmatrix} \mathbf{F}^x \\ \mathbf{F}^y \end{bmatrix}$$

coming from the r.h.s. of the continuous problem

$$\mathbf{f} = \begin{bmatrix} f^x \\ f^y \end{bmatrix}.$$

More specifically, we have $\mathbf{F}^x = ((f^x, \varphi_e))_{e\in\mathcal{E}_h}$ and $\mathbf{F}^y = ((f^y, \varphi_e))_{e\in\mathcal{E}_h}$.

The discrete problem (B.43) takes the following saddle-point matrix-vector form

$$\begin{bmatrix} \mathbf{A} & B^T \\ B & 0 \end{bmatrix} \begin{bmatrix} \mathbf{V} \\ \mathbf{x} \end{bmatrix} = \begin{bmatrix} \mathbf{F} \\ 0 \end{bmatrix}. \tag{B.44}$$

We show next an "inf–sup" condition for the discrete saddle-point operator coming from the finite element discretization of the Stokes problem.

Denote by $\mathcal{E}^0 = \mathcal{E}_h^0$ the set of all interior edges. Note first that π_h is bounded in the L_2-norm

$$\begin{aligned} \|\pi_h \eta\|_0^2 &\simeq \sum_{e\in\mathcal{E}^0} |\tau_e^- \cup \tau_e^+| \left(\frac{1}{e} \int_e \eta \, d\varrho \right)^2 \\ &\le C \sum_\tau \int_{\tau_e^+ \cup \tau_e^-} (\eta^2 + h^2 |\nabla \eta|^2) \, d\mathbf{x} \le C \|\eta\|_1^2. \end{aligned}$$

Also, using the fact that the square of the L_2-norm of the gradient of a finite element function can be bounded by a sum of squared differences, we easily obtain the next estimate

$$\sum_{\tau} \int_{\tau} |\nabla \pi_h \eta|^2 \, d\mathbf{x} \simeq \sum_{\tau \in \mathcal{T}_h} \sum_{e_1,\, e_2 \subset \partial \tau} \left(\frac{1}{e_1} \int_{e_1} \eta \, d\varrho - \frac{1}{e_2} \int_{e_2} \eta \, d\varrho \right)^2 \leq C |\eta|_1^2.$$

The second inequality is seen easily for smooth functions η, and in general, by continuity. Thus, we have

$$|\pi_h \eta|_{1,h}^2 \equiv \sum_{\tau \in \mathcal{T}_h} \int_{\tau} |\nabla \pi_h \eta|^2 \, d\mathbf{x} \leq C \|\eta\|_1^2. \tag{B.45}$$

We use the same estimate for vector functions $\eta \in (H_0^1(\Omega))^2$. Assume now that the continuous "inf–sup" condition holds (cf., e.g., [B01])

$$|\chi|_0 \equiv \inf_{C = \text{const}} \|\chi - C\|_0 \leq \beta \sup_{\underline{v} \in (H_0^1(\Omega))^2} \frac{(\chi,\, \text{div}\, \underline{v})}{|\underline{v}|_1}.$$

Recall (B.42) which reads $(\chi,\ \text{div}\ \underline{v}) = (\chi,\ Q_h \text{div}\ \underline{v}) = (\chi,\ \text{div}_h\ \pi_h \underline{v})$ for $\chi \in W_h$. This commutativity property, the boundedness of π_h (shown in (B.45)) used in the continuous "inf–sup" condition implies the desired discrete one; that is, we have

$$|\chi|_0 \leq \beta \sup_{\underline{v} \in (H_0^1(\Omega))^2} \frac{|\pi_h \underline{v}|_{1,h}}{|\underline{v}|_1} \sup_{\underline{\eta} \in \mathbf{V}_h} \frac{(\chi,\, \text{div}_h \underline{\eta})}{|\underline{\eta}|_{1,h}^2} \leq C\beta \sup_{\underline{\eta} \in \mathbf{V}_h} \frac{(\chi,\, \text{div}_h \underline{\eta})}{|\underline{\eta}|_{1,h}^2}. \tag{B.46}$$

The boundedness of the discrete (finite element) Stokes operator in the pair of norms $\|.\|_{1,h},\ |.|_0$ is trivially seen. To summarize, we have the following main result.

Theorem B.17. *The discrete Stokes operator in* (B.44) *is well posed in the pair of norms* $|.|_{1,h},\ |.|_0$ *that give rise to the block-diagonal matrix*

$$\begin{bmatrix} \mathbf{A} & 0 \\ 0 & M \end{bmatrix}.$$

Here, $M = \text{diag}(|\tau|)_{\tau \in \mathcal{T}_h} \simeq h^2\, I$, *is the diagonal mass matrix coming from the space* W_h *of piecewise constants.*

The s.p.d. matrix $\mathbf{A}$ (a pair of discrete nonconforming finite element Laplacian) can be preconditioned very efficiently by MG methods that can be constructed based on the fact that the nonconforming space contains the conforming one on the same mesh. On coarse levels one can use conforming spaces.

Alternatively, the discrete problem (B.44) can be treated as a constrained minimization one as in Section 9.5. For the purpose of defining a coarse space needed for efficiency of the constrained minimization algorithm analyzed there, consider two nested triangulations: a coarse one, $\mathcal{T}_H = \{T\}$, and a refinement of it, $\mathcal{T}_h = \{\tau\}$.

Introduce also the coarse and fine pairs of finite element spaces $\mathbf{V}_H,\ W_H$ and $\mathbf{V}_h,\ W_h$. They lead to the fine-grid matrices $\mathbf{A}$ and B and to the coarse ones $\mathbf{A}_c$ and B_c, respectively. We notice that $W_H \subset W_h$ but $\mathbf{V}_H \not\subset \mathbf{V}_h$. That is why we need to define an interpolation matrix $\mathcal{P}$ that maps a coarse coefficient vector $\mathbf{v}_c$ into a coefficient vector $\mathcal{P}\mathbf{v}_c$ of the f.e. function from $\mathbf{V}_h$. The interpolation mapping for the second component of the solution $\mathcal{Q}$ is naturally defined from the embedding $W_H \subset W_h$ simply as piecewise constant interpolation.

Our goal next is to define a $\mathcal{P}$ such that $B_c\mathbf{v}_c = 0$ implies $B\mathcal{P}\mathbf{v}_c = 0$ (a condition needed in Section 9.5). For every coarse edge $e_H = \partial T^+ \cap \partial T^-,\ T^+,\ T^- \in \mathcal{T}_H$, define $\underline{v}_H^+$ and $\underline{v}_H^-$ as the traces of a given coarse function $\underline{v}_H \in \mathbf{V}_H$ coming from the two neighboring coarse elements T^+ and T^-. Also, for any coarse element T considered as a fine-grid subdomain (union of four similar fine-grid triangles) introduce the local spaces $\mathbf{V}_h(T)$ of functions from $\mathbf{V}_h$ restricted to T and similarly, let $W_h(T)$ be the restriction of the functions from W_h to T. Finally, let $\mathbf{V}_h^0(T)$ be the subspace of $\mathbf{V}_h(T)$ of functions that vanish at the midpoints of (fine-grid) edges $e \subset \partial T$.

Given $\underline{v}_H \in \mathbf{V}_H$, consider then for any $T \in \mathcal{T}_H$ the following local problems.

Find $\underline{v}_h \in \mathbf{V}_h(T)$ and $u_h \in W_h(T)$ such that

$$
\begin{aligned}
&\sum_{\tau \subset T} \int_\tau (\nabla \underline{v}_h) \cdot (\nabla \underline{\eta})\, d\mathbf{x} + \sum_{\tau \subset T} \int_\tau u_h \mathrm{div}\ \underline{\eta}\, d\mathbf{x} \\
&\quad = \sum_{\tau \subset T} \int_\tau (\nabla \underline{v}_H) \cdot (\nabla \underline{\eta})\, d\mathbf{x}, \quad \text{for all } \underline{\eta} \in \mathbf{V}_h^0(T), \\
&\sum_{\tau \subset T} \int_\tau \mathrm{div}_h\ \underline{v}_h \chi\, d\mathbf{x} = \sum_{\tau \subset T} \int_\tau (\mathrm{div}_H\ \underline{v}_H)\ \chi\, d\mathbf{x}, \quad \text{for all } \chi \in W_h(T),
\end{aligned}
\tag{B.47}
$$

subject to the Dirichlet boundary conditions on every coarse edge $e_H \subset \partial T$, and any $e \subset e_H, e \in \mathcal{E}_h$,

$$
\frac{1}{|e|} \int_e \underline{v}_h\, d\varrho = \text{Const} = \frac{1}{|e_H|} \int_{e_H} \underline{v}_H^+\, d\varrho = \frac{1}{|e_H|} \int_{e_H} \underline{v}_H^-\, d\varrho. \tag{B.48}
$$

That is, $\underline{v}_h$ is constant on every fine-grid edge e contained in a given coarse edge e_H. For edges e_H on the boundary $\partial\Omega$ this expression is seen to be zero. The inhomogeneous Dirichlet problems (B.47)–(B.48) are solvable because $\sum_{\tau \subset T} \int_\tau (\mathrm{div}_h\ \underline{v}_h - \mathrm{div}_H\ \underline{v}_H)\ d\mathbf{x} = 0$, which is equivalent to $\int_{\partial T} \underline{v}_h \cdot \mathbf{n}\ d\varrho = \int_{\partial T} \underline{v}_H \cdot \mathbf{n}\ d\varrho \equiv \int_{\partial T} \underline{v}_H^+ \cdot \mathbf{n}\ d\varrho = \int_{\partial T} \underline{v}_H^- \cdot \mathbf{n}\ d\varrho$. Because the boundary conditions on the coarse edges $e_H \in \mathcal{E}_H$ are consistent, the local problems (B.47)–(B.48) define a global function $\underline{v}_h \in \mathbf{V}_h$.

The matrix representation of the mapping $\underline{v}_H \in \mathbf{V}_H \mapsto \underline{v}_h \in \mathbf{V}_h$ defines our interpolation matrix $\mathcal{P}$. It is clear that (by construction) $\mathcal{P}$ satisfies the property "$B_c\mathbf{v}_c = 0$ implies $B\mathcal{P}\mathbf{v}_c = 0$". Also, because $W_H \subset W_h$, we have that $(\mathrm{div}_h\ \underline{v}_h,\ \chi) = (\mathrm{div}_H\ \underline{v}_H,\ \chi)$ for any $\chi \in W_H$ which in a matrix–vector form translates to $\mathcal{Q}^T B\mathcal{P}\mathbf{v}_c = B_c\mathbf{v}_c$. That is, $B_c = \mathcal{Q}^T B\mathcal{P}$.

The overall method (Algorithm 9.5.1 described in Section 9.5) explores solving small fine-grid constrained minimization problems in addition to a global coarse constrained minimization one. We note that here the coarse matrix used in the algorithm is $\mathcal{P}^T \mathbf{A} \mathcal{P}$ (which is different from $\mathbf{A}_c$ provided by the coarse discretization).

B.6 F.e. discretization of Maxwell's equations

An $H(\text{curl})$ formulation for the electric field

Without entering into much detail, the time-dependent Maxwell equations are described by five vector fields **E**, **H**, **D**, **B**, and **J** plus one scalar function ϱ which are related as follows

$$\begin{aligned}
\text{curl } \mathbf{E} &= -\frac{\partial \mathbf{B}}{\partial t}, \\
\text{curl } \mathbf{H} &= \frac{\partial \mathbf{D}}{\partial t} + \mathbf{J}, \\
\text{div } \mathbf{D} &= \varrho, \\
\text{div } \mathbf{B} &= 0, \\
\text{div } \mathbf{J} &= -\frac{\partial \varrho}{\partial t}.
\end{aligned}$$

Here **D** and **H** are the densities of the electric and magnetic flux and under some assumptions about linearity the following relations hold for some known positive coefficients ϵ and μ,

$$\mathbf{D} = \epsilon\, \mathbf{E} \quad \text{and} \quad \mathbf{H} = \mu^{-1}\mathbf{B}.$$

Theoretically, these equations are solved on $\mathbb{R}^3$. In practice, this is done on a bounded domain Ω imposing boundary conditions such as

$$\mathbf{E} \times \mathbf{n} = 0 \quad \text{and} \quad \mathbf{B} \cdot \mathbf{n} = 0 \text{ on } \partial\Omega.$$

Using the above relations, assuming that **J** is known, we end up with the following system

$$\begin{aligned}
\text{curl } \mathbf{E} &= -\frac{\partial \mathbf{B}}{\partial t}, \\
\text{curl } \mu^{-1}\mathbf{B} &= \frac{\partial \epsilon \mathbf{E}}{\partial t} + \mathbf{J}.
\end{aligned}$$

After a time discretization $t_{n+1} - t_n = \Delta t$, the following reduced problem for $\mathbf{E}_{n+1}$ is obtained

$$\epsilon \mathbf{E}_{n+1} + (\Delta t)^2 \text{ curl } \mu^{-1} \text{curl } \mathbf{E}_{n+1} = \epsilon \mathbf{E}_n + (\Delta t)\ (\text{curl } \mu^{-1}\mathbf{B}_n - \mathbf{J}_{n+1}).$$

Letting $\mathbf{u} = \mathbf{E}_{n+1}$, $\epsilon = \beta$ and $\alpha = (\Delta t)^2 \mu^{-1}$, we have the following second-order PDE

$$\text{curl } \alpha \text{ curl } \mathbf{u} + \beta \mathbf{u} = \mathbf{f}$$

subject to $\mathbf{u} \times \mathbf{n} = 0$ on $\partial\Omega$. We consider it in a "weak sense"; that is, for any test function $\mathbf{v}$ using integration by parts and the boundary conditions, we arrive at the identity

$$(\alpha \text{ curl } \mathbf{u}, \text{ curl } \mathbf{v}) + (\beta \mathbf{u}, \mathbf{v}) = (\mathbf{f}, \mathbf{v}). \tag{B.49}$$

The space of vector functions from $L_2(\Omega)$ that have a well-defined curl in the L_2-sense is denoted by $H(\text{curl}, \Omega)$ and if the functions satisfy the boundary condition $\mathbf{u} \times \mathbf{n} = 0$ on $\partial\Omega$ it is denoted by $H_0(\text{curl}, \Omega)$.

We now describe the f.e. discretization process in more detail. Introduce a triangulation $\mathcal{T}_h$ of Ω consisting of tetrahedral elements T. A popular choice of finite element space is the lowest-order Nédélec space

$$\mathbf{Q}_h = \{\boldsymbol{\varphi} \in H_0(\text{curl}, \Omega) : \boldsymbol{\varphi}|_T = \mathbf{a} + \mathbf{b} \times \mathbf{x}, \ T \in \mathcal{T}_h\}.$$

Here, $\mathbf{a}, \mathbf{b} \in \mathbb{R}^3$ depend on T. The latter definition more specifically means that the tangential components $\boldsymbol{\varphi} \cdot \mathbf{t}_e$ of any $\boldsymbol{\varphi} = \mathbf{a} + \mathbf{b} \times \mathbf{x}$ are continuous on the edges e of all Ts. Note that every T has 6 edges and there are 6 degrees of freedom (the coefficients $\mathbf{a}$ and $\mathbf{b}$ of $\boldsymbol{\varphi}$ restricted to T) to be specified. A natural choice then seems to be the quantities

$$\int_e \boldsymbol{\varphi} \cdot \mathbf{t}_e \, d\varrho$$

for all 6 edges e of T. Let $\mathcal{E}_h^0$ be the set of interior edges (w.r.t. Ω). To define a Lagrangian basis $\{\boldsymbol{\varphi}_e\}_{e \in \mathcal{E}_h^0}$ of $\mathbf{Q}_h$, we let $\boldsymbol{\varphi}_e = \boldsymbol{\varphi}_e(\mathbf{x}) = \mathbf{a}_e + 1/|e| \, \mathbf{b}_e \times (\mathbf{x}_{m_e} - \mathbf{x})$ where $\mathbf{x}_{m_e}$ is the midpoint of the edge e. Then the Lagrangian condition $\int_{e'} \boldsymbol{\varphi}_e \cdot \mathbf{t}_{e'} \, d\varrho = \delta_{e,e'}$ reads, for $e' = e$,

$$\frac{1}{|e|} = \mathbf{a}_e \cdot \mathbf{t}_e,$$

and for $e' \neq e$, based on the vector identity $(\mathbf{p} \times \mathbf{q}) \cdot \mathbf{r} = \mathbf{p} \cdot (\mathbf{q} \times \mathbf{r})$,

$$\begin{aligned} 0 &= \left(\mathbf{a}_e + \frac{1}{|e|} \mathbf{b}_e \times (\mathbf{x}_{m_e} - \mathbf{x}_{m_{e'}})\right) \cdot \mathbf{t}_{e'} \\ &= \mathbf{a}_e \cdot \mathbf{t}_{e'} + \frac{1}{|e|} ((\mathbf{x}_{m_e} - \mathbf{x}_{m_{e'}}) \times \mathbf{t}_{e'}) \cdot \mathbf{b}_e. \end{aligned}$$

Note that $\mathbf{a}_e, \mathbf{b}_e = \mathcal{O}(h^{-1})$ and hence curl $\boldsymbol{\varphi}_e = 1/|e| \ (-2\mathbf{b}_e)$ is order $\mathcal{O}(h^{-2})$, whereas $\|\boldsymbol{\varphi}_e\|_0^2 \simeq \mathcal{O}(|T|h^{-2}) = \mathcal{O}(h)$. This shows that the entries of the element stiffness matrix $\mathbf{A}_T = (a(\boldsymbol{\varphi}_{e'}, \boldsymbol{\varphi}_e))_{e, e' \subset \partial T}$ are of order $\mathcal{O}(\alpha \ h^{-1} + \beta h)$. Finally, notice that for $\mathbf{x} \in e'$, $\boldsymbol{\varphi}_e(\mathbf{x}) \cdot \mathbf{t}_{e'} = (\mathbf{a}_e + 1/|e| \, \mathbf{b}_e \times (\mathbf{x}_{m_e} - \mathbf{x}_{m_{e'}})) \cdot \mathbf{t}_{e'} = \text{const}$, because then $\mathbf{x} - \mathbf{x}_{m_{e'}}$ is parallel to $\mathbf{t}_{e'}$ and hence $\mathbf{b}_e \times (\mathbf{x}_{m_{e'}} - \mathbf{x}_e)$ is orthogonal to $\mathbf{t}_{e'}$. This shows the global continuity of the tangential components of the basis functions $\boldsymbol{\varphi}_e$.

A direct construction of the Lagrangian basis $\{\boldsymbol{\varphi}_e\}$ of $\mathbf{Q}_h$ is as follows (e.g., [SLC]). Introduce the scalar piecewise linear basis functions $\varphi_{\mathbf{x}_k}$ associated with every vertex $\mathbf{x}_k \in \mathcal{N}_h$. A compact definition is based on the nodes $\mathbf{x}_k^*$ which are the orthogonal projections of $\mathbf{x}_k$ onto the opposite (to $\mathbf{x}_k$) face F_k of T. More specifically, we have

$$\varphi_{\mathbf{x}_k}(\mathbf{x}) = \frac{1}{|\mathbf{x}_k - \mathbf{x}_k^\star|^2}(\mathbf{x}_k - \mathbf{x}_k^\star) \cdot (\mathbf{x} - \mathbf{x}_k^\star).$$

We easily compute the gradient of the basis functions, namely, $\nabla\varphi_k = 1/(|\mathbf{x}_k - \mathbf{x}_k^\star|^2)\,(\mathbf{x}_k - \mathbf{x}_k^\star)$. Then, because $\mathbf{t}_e = \epsilon_e\, 1/(|\mathbf{x}_k - \mathbf{x}_l|)(\mathbf{x}_k - \mathbf{x}_l)$ ($\epsilon_e = 1$ or -1) and $(\mathbf{x}_k - \mathbf{x}_k^\star) \cdot (\mathbf{x}_l - \mathbf{x}_k^\star) = 0$ (noting that $\mathbf{x}_l,\ \mathbf{x}_k^\star \in F_k$), we have

$$\nabla\varphi_{\mathbf{x}_k} \cdot \mathbf{t}_e = \frac{1}{|\mathbf{x}_k - \mathbf{x}_k^\star|^2}\,(\mathbf{x}_k - \mathbf{x}_k^\star) \cdot \epsilon_e\,\frac{1}{|\mathbf{x}_k - \mathbf{x}_l|}(\mathbf{x}_k - \mathbf{x}_l) = \frac{\epsilon_e}{|\mathbf{x}_k - \mathbf{x}_l|}. \tag{B.50}$$

Similarly,

$$\nabla\varphi_{\mathbf{x}_l} \cdot \mathbf{t}_e = -\frac{\epsilon_e}{|\mathbf{x}_k - \mathbf{x}_l|}. \tag{B.51}$$

The Nédélec (vector) basis function $\boldsymbol{\varphi}_e$, for a given edge $e = (\mathbf{x}_k,\ \mathbf{x}_l)$ is then defined as

$$\boldsymbol{\varphi}_e = \varphi_{\mathbf{x}_k}\ \nabla\varphi_{\mathbf{x}_l} - \varphi_{\mathbf{x}_l}\ \nabla\varphi_{\mathbf{x}_k}.$$

We notice that $\boldsymbol{\varphi}_e \cdot \mathbf{t}_{e'} = 0$ for any edge e' different from e. This is true because either $\nabla\varphi_{\mathbf{x}_l} \cdot \mathbf{t}_{e'} = 0$ or $\varphi_{\mathbf{x}_k}(\mathbf{x}) = 0$ for $\mathbf{x}$ on e'. Also, for $\mathbf{x}$ on e, using (B.50) and (B.51), we finally obtain

$$\begin{aligned}\boldsymbol{\varphi}_e \cdot \mathbf{t}_e &= \varphi_{\mathbf{x}_k}\nabla\varphi_l \cdot \mathbf{t}_e - \varphi_{\mathbf{x}_l}\nabla\varphi_k \cdot \mathbf{t}_e\\ &= -\frac{\epsilon_e}{|\mathbf{x}_k - \mathbf{x}_l|}(\varphi_{\mathbf{x}_k} + \varphi_{\mathbf{x}_l})\\ &= -\frac{\epsilon_e}{|\mathbf{x}_k - \mathbf{x}_l|}.\end{aligned}$$

We used the fact that the scalar basis functions $\{\varphi_{\mathbf{x}_i}\}_{\mathbf{x}_i \in \mathcal{N}_h}$ sum up to unity and the only ones that are nonzero on e are $\varphi_{\mathbf{x}_k}$ and $\varphi_{\mathbf{x}_l}$; that is why $\varphi_{\mathbf{x}_k} + \varphi_{\mathbf{x}_l} = 1$ on e.

We are interested in solving the following variational problem,

$$\text{Find } \mathbf{u} \in \mathbf{Q}_h\ :\quad (\alpha\,\text{curl}\,\mathbf{u}, \text{curl}\,\mathbf{v}) + (\beta\,\mathbf{u}, \mathbf{v}) = (\mathbf{f}, \mathbf{v}) \qquad \text{for all } \mathbf{v} \in \mathbf{Q}_h\,.$$

Once having a basis $\{\boldsymbol{\varphi}_e\}_{e \in \mathcal{E}_h^0}$ of $\mathbf{Q}_h$, we can compute the stiffness matrix of (B.49). It consists of two parts, a weighted mass matrix $\mathbf{G} = ((\beta\ \boldsymbol{\varphi}_{e'},\ \boldsymbol{\varphi}_e))_{e,\, e' \in \mathcal{E}_h^0}$ and a curl-term $\mathbf{C} = ((\alpha\ \text{curl}\ \boldsymbol{\varphi}_{e'},\ \text{curl}\ \boldsymbol{\varphi}_e))_{e,\, e' \in \mathcal{E}_h^0}$. Compute also the r.h.s. vector $\underline{\mathbf{f}} = ((\mathbf{f},\ \boldsymbol{\varphi}_e))_{e \in \mathcal{E}_h^0}$ and expand the unknown solution $\mathbf{u} = \sum_{e \in \mathcal{E}_h^0} u_e\ \boldsymbol{\varphi}_e$ in terms of the basis of $\mathbf{Q}_h$. The coefficient vector $\underline{\mathbf{u}} = (u_e)_{e \in \mathcal{E}_h^0}$ is the unknown vector. After the discretization, we end up with the system of linear equations, letting $\mathbf{A} = \mathbf{C} + \mathbf{G}$,

$$\mathbf{A}\underline{\mathbf{u}} = \underline{\mathbf{f}}.$$

Each unknown (degree of freedom) is associated with an edge e from $\mathcal{E}_h^0$. The problem can get very large when $h \mapsto 0$. Thus, we need an efficient (iterative) solver, of optimal order if possible. We prove in Section F.2 an optimal MG convergence result for a parameter-dependent $H(\text{curl})$-form corresponding to $\beta = 1$ and α being a constant $\tau > 0$ which can get large. For the purpose of that analysis we need to look at the diagonal of $\mathbf{C}$. It equals $\tau \ \text{diag}((\text{curl} \ \boldsymbol{\varphi}_e, \ \text{curl} \ \boldsymbol{\varphi}_e))_{e \in \mathcal{E}_h^0}$. Because $\text{curl} \ \boldsymbol{\varphi}_e = (1/|e|)(-2\mathbf{b}_e) = 2\nabla\varphi_{\mathbf{x}_k} \times \nabla\varphi_{\mathbf{x}_l}$, where $e = (\mathbf{x}_k, \ \mathbf{x}_l)$ it is clear that $\mathbf{b}_e$ is nonzero (because $\nabla\varphi_{\mathbf{x}_k}$ and $\nabla\varphi_{\mathbf{x}_l}$ are nonparallel constant vectors for $k \neq l$). That is, although $\mathbf{C}$ is only semidefinite with large null space, its diagonal is s.p.d. and it is spectrally equivalent to the scaled mass matrix $h^{-2}\mathbf{G}$. This shows the following result.

Proposition B.18. *The diagonal of* $\mathbf{A}$ *coming from the parameter-dependent* $H_0(\mathit{curl})$ *bilinear form* $(\mathbf{u}, \ \mathbf{v}) + \tau \ (\mathit{curl} \ \mathbf{u}, \ \mathit{curl} \ \mathbf{v})$ *and the lowest-order Nédélec space* $\mathbf{Q}_h$, *is spectrally equivalent to the weighted mass matrix* $(1 + \tau h^{-2}) \ \mathbf{G}$.

C

Computable Scales of Sobolev Norms

C.1 H^s-stable decompositions

This chapter describes a simple construction of H^s-stable computable decompositions of functions based on easily computable quasi-interpolants $\widetilde{Q}_k$. The main results are found in [BPV99].

The quasi-interpolants $\widetilde{Q}_k$ we use in practice are inexpensive to realize because they are based on local projections associated with locally supported basis functions. The stability of the decompositions we prove is important in several applications because we can use them to construct optimal-order preconditioners and stable extension mappings, tools that we frequently explored throughout the book.

C.2 Preliminary facts

Consider a given Hilbert space V, $(.,.)$ and let $\{V_k\}$ be given nested subspaces of V (i.e., $V_1 \subset V_2 \subset \cdots \subset V_k \subset V$). In this section we first show that the norm $|\langle v \rangle|_s^2 = \sum_k \lambda_k^s \|(Q_k - Q_{k-1})v\|^2$ which is based on a given orthogonal (in the given inner product $(.,.)$) projections $Q_k \colon V \mapsto V_k$ can be characterized $|\langle v \rangle|_s^2 \simeq \sum_k \lambda_k^s \|(\widetilde{Q}_k - \widetilde{Q}_{k-1})v\|^2$, based on other (in some sense simpler) operators $\widetilde{Q}_k \colon V \mapsto V_k$ such that $\widetilde{Q}_k$ are first uniformly coercive on V_k, the differences $Q_k - \widetilde{Q}_k$ have certain approximation properties, and finally, the following commutativity property $\widetilde{Q}_k Q_k = \widetilde{Q}_k$ holds.

A main application of this result is a characterization of the norms in the Sobolev spaces H^s for real s in the (open) interval $|s| < 3/2$.

A specific choice of $\widetilde{Q}_k$ in the case of nested finite element spaces V_k of continuous piecewise linear functions obtained by successive steps of uniform refinement, is

$$\widetilde{Q}_k v = \sum_{x_i \in \mathcal{N}_k} \frac{(v, \varphi_i^{(k)})}{(e, \varphi_i^{(k)})} \varphi_i^{(k)}, \quad \{\varphi_i^{(k)},\ x_i \in \mathcal{N}_k\} \text{ is a basis of } V_k, \ e = \sum_{x_i \in \mathcal{N}_k} \varphi_i^{(k)}. \tag{C.1}$$

Here, $\mathcal{N}_k$ is the set of nodes (degrees of freedom) associated with the kth-level grid and $\{\varphi_i^{(k)}\}_{x_i \in \mathcal{N}_k}$ form a Lagrangian (nodal) basis of V_k.

We are interested in the possible stability of the decomposition

$$\sum_k (\widetilde{Q}_k - \widetilde{Q}_{k-1})v, \tag{C.2}$$

for functions $v \in H^s$ for all s in an interval. In the case of piecewise linear finite elements we have $|s| < 3/2$. More specifically, we are interested in the stability of the decomposition (C.2) with respect to the following norms,

$$|\langle v \rangle|_s^2 = \sum_k \lambda_j^s \|(Q_k - Q_{k-1})v\|^2,$$

where $\lambda_j < \lambda_{j+1}$ depend on the specific application.

In what follows we prove that the decomposition (C.2) defines an equivalent norm to $|\langle . \rangle|_s$; namely,

$$|||v|||_s^2 = \sum_k \lambda_j^s \|(\widetilde{Q}_k - \widetilde{Q}_{k-1})v\|^2.$$

The latter one, in contrast to $|\langle v \rangle|_s$ is easily computable if the actions of $\widetilde{Q}_k$ are easy to compute. The latter is the case for the quasi-interpolants defined in (C.1).

In practice, we use finite sums for finite element functions; that is

$$v = \sum_{k=1}^{J} (\widetilde{Q}_k - \widetilde{Q}_{k-1})v, \quad \text{letting } \widetilde{Q}_J = I, \quad \text{and} \quad \widetilde{Q}_0 = 0. \tag{C.3}$$

Based on the above finite decomposition of $V = V_J$ we may construct iterative methods (or preconditioners) if a computable basis is available in the coordinate spaces $\widetilde{W}_k = (\widetilde{Q}_k - \widetilde{Q}_{k-1})V$. Other application of the decomposition is to construct bounded extension mappings. That is, we have a function defined on a boundary of a given domain, and then construct its extension in the interior of the subdomains by trivially extending each component $(\widetilde{Q}_k - \widetilde{Q}_{k-1})v$ by zero in the interior kth-level nodes. Note that this involves interpolation to represent the data on the finest-grid.

In both cases, the fact of main importance is to have the decomposition stable in a proper Sobolev norm of interest for the particular application.

We finally mention that to build an additive preconditioner we do not actually need a computable basis in the coordinate spaces $\widetilde{W}_k$. Indeed, for a given bilinear form $a(.,.)$ on $V \times V$, which defines an operator $\mathcal{A} : V \mapsto V$, we consider the preconditioner $\mathcal{B} : V \mapsto V$ based on the bilinear form

$$(\mathcal{B}v, w) = \sum_k \lambda_k^{-1}((\widetilde{Q}_k - \widetilde{Q}_{k-1})v, (\widetilde{Q}_k - \widetilde{Q}_{k-1})w), \qquad v, w \in V.$$

Here, $\lambda_j = \sup_{v \in V_j} (\mathcal{A}v, v)/\|v\|^2$ stands for the spectral radius of $\mathcal{A}$ restricted to the subspace V_k. Based on the symmetry of $\widetilde{Q}_j$ we come up with the following form of $\mathcal{B}$,

$$\mathcal{B} = \sum_k \lambda_k^{-1}(\widetilde{Q}_k - \widetilde{Q}_{k-1})^2. \tag{C.4}$$

It is clear that the actions of $\mathcal{B}$ are computationally available without explicit knowledge of a basis of $\widetilde{W}_k$.

Based on our norm-equivalence result (that we prove next)

$$a(\mathcal{B}\mathcal{A}v, v) \simeq a(v, v), \quad \text{all } v \in V\cdot$$

we obtain that $\mathcal{B}^{-1}$ is spectrally equivalent to $\mathcal{A}$.

As an example, consider the model bilinear form, $a(v, w) = \tau^{-1}(v, w) + (\nabla u, \nabla v), u, v \in H_0^1(\Omega)$, where $\tau > 0$ is a parameter. We can show that the energy norm $(a(v, v))^{1/2}$ is equivalent, uniformly with respect to τ, to $|\langle v\rangle|_1$ with $\lambda_j \simeq \tau^{-1} + 2^{2j}$. The latter is the spectral radius of $\mathcal{A}$ restricted to V_j. Based on the analysis that follows, we are able to consider parameter-dependent norms with

$$\lambda_j = \tau^{-1} + 2^{2j}. \tag{C.5}$$

The estimates we prove $|||v|||_s \simeq |\langle v\rangle|_s$ appear independent of τ. Examples of parameter-dependent bilinear forms arise from discretizing time-dependent Stokes problems. In summary, decomposition (C.2) or (C.3) is stable in H^s for any $s\colon |s| < 3/2$ uniformly with respect to $\tau > 0$.

The main result of the present chapter given in the following section proves stability of the decomposition (C.2) in an abstract Hilbert space setting. We then verify the assumption for uniform coercivity of the quasi-interpolants $\widetilde{Q}_k$ (restricted to V_k) provided the respective mass (or L_2-Gram) matrices are uniformly sparse. The latter holds in the case of uniformly refined meshes.

C.3 The main norm equivalence result

In this section we prove the main norm equivalence result in an abstract Hilbert space setting. Let V, $(.,.)$ be a given Hilbert space and $V_{j-1} \subset V_j \subset V$ be subspaces such that $\mathcal{C} \equiv \sum_j V_j$ is dense in V. Consider $Q_j : V \mapsto V_j$ the $(.,.)$-orthogonal projections. Due to the density $\lim_{j\to\infty} \|Q_j v - v\| = 0$ for any $\mathbf{v} \in V$. We assume that $\bigcup_k V_k$ is contained in a scale of spaces $\mathcal{H}_s$ for any real $s \in (-s_0, s_0)$, a given interval. In our main application, $s_0 = 3/2$. Let $\|.\|_s$ be the norm of $\mathcal{H}_s$.

For a given sequence $\{\lambda_j\}$, $0 < \lambda_j < \lambda_{j+1}$, define the scale of norms

$$|\langle v\rangle|_s^2 = \sum_j \lambda_j^s \|(Q_j - Q_{j-1})v\|^2\cdot \tag{C.6}$$

We notice that $|\langle v\rangle|_0 = \|.\|$ and we assume that $\mathcal{H}_0 = V$; that is, $\|v\|_0 = \|v\|$. Finally we assume that the spaces satisfy

(I) "Inverse" inequality,

$$\|v\|_\sigma \leq C_I h_j^{-\sigma} \|v\|_0, \qquad v \in V_j.$$

To be specific we let $h_j = 2^{-j}$.

We assume now that there exists a sequence of operators $\widetilde{Q}_k : V \mapsto V_k$ which satisfy:

(A) An "approximation property":

$$\|(Q_k - \widetilde{Q}_k)v\|_0 \leq C_A h_k^\sigma \|v\|_\sigma, \qquad \sigma \geq 0.$$

(C) "Uniform coercivity" of $\widetilde{Q}_k$ when restricted to V_k; that is,

$$\delta \|v_k\|^2 \leq (\widetilde{Q}_k v_k, v_k) \text{ for all } v_k \in V_k. \tag{C.7}$$

(P) A "commutativity" property:

$$\widetilde{Q}_k Q_k = \widetilde{Q}_k.$$

Because Q_k is a projection on V_k, we have $\widetilde{Q}_k = Q_k \widetilde{Q}_k$, which shows the commutativity $\widetilde{Q}_k Q_k = Q_k \widetilde{Q}_k$.

Theorem C.1. *Under the assumptions* (A), *uniform coercivity* (C), *and the commutativity property* (P) *for* $\widetilde{Q}_k$, *based on property* (I) *of the spaces* V_j, *the following main norm characterization result holds.*

$$|||v|||_s^2 \equiv \sum_k \lambda_k^s \|(\widetilde{Q}_k - \widetilde{Q}_{k-1})v\|_0^2 \simeq |\langle v \rangle|_s^2,$$

if the unit lower triangular matrix $\mathcal{L} = (\ell_{k,j})$, *with nonzero entries,*

$$\ell_{k,j} = (2^{-\sigma})^{k-j} \left(\frac{\lambda_k}{\lambda_j}\right)^{s/2}, \qquad k \geq j, \tag{C.8}$$

for a $\sigma = \sigma(s) > 0$, *has a bounded spectral norm. That is, if*

$$\|\mathcal{L}\| \equiv \sup_{(\xi_k),\,(\zeta_k)} \frac{\sum_k \sum_{j \leq k} \ell_{k,j}\, \xi_k \zeta_j}{\left(\sum_k \xi_k^2\right)^{1/2} \left(\sum_k \zeta_k^2\right)^{1/2}} < \infty. \tag{C.9}$$

Proof. Let $v_k = (Q_k - \widetilde{Q}_k)v$. Assume that $|\langle v \rangle|_s < \infty$ for a given s (negative, zero, or positive). Then choose $\sigma = \sigma(s) > 0$ such that (C.9). Consider the expression with any finite number of entries,

$$\sum_k \lambda_k^s \|(Q_k - \widetilde{Q}_k)v\|_0^2.$$

Then, based on the commutativity (P), Cauchy–Schwarz inequality, the approximation properly (A), and the inverse inequality (I) used consecutively, letting $v_k = (Q_k - \widetilde{Q}_k)v$, we obtain

$$
\begin{aligned}
\sum_k \lambda_k^s \|(Q_k - \widetilde{Q}_k)v\|_0^2 &= \sum_k \lambda_k^s ((Q_k - \widetilde{Q}_k)v, (Q_k - \widetilde{Q}_k)Q_k v) \\
&= \sum_k \lambda_k^s (v_k, (Q_k - \widetilde{Q}_k) \sum_{j \le k} (Q_j - Q_{j-1})v) \\
&= \sum_k \lambda_k^s \sum_{j \le k} (v_k, (Q_k - \widetilde{Q}_k)(Q_j - Q_{j-1})v) \\
&\le \sum_k \lambda_k^s \sum_{j \le k} \|v_k\|_0 \|(Q_k - \widetilde{Q}_k)(Q_j - Q_{j-1})v\|_0 \\
&\le \sum_k \lambda_k^s \sum_{j \le k} \|v_k\|_0 C_A h_k^\sigma \|(Q_j - Q_{j-1})v\|_\sigma \\
&\le \sum_k \lambda_k^s \sum_{j \le k} \|v_k\|_0 C_A h_k^\sigma \; C_I h_j^{-\sigma} \|(Q_j - Q_{j-1})v\|_0 \\
&= C_A C_I \sum_k \sum_{j \le k} \left(\frac{1}{2^\sigma}\right)^{k-j} \left(\frac{\lambda_k}{\lambda_j}\right)^{s/2} \lambda_k^{s/2} \|v_k\|_0 \lambda_j^{s/2} \\
&\quad \times \|(Q_j - Q_{j-1})v\|_0 \\
&= C_A C_I \sum_k \sum_{j \le k} \ell_{k,j} \lambda_k^{s/2} \|v_k\|_0 \lambda_j^{s/2} \|(Q_j - Q_{j-1})v\|_0 .
\end{aligned}
\tag{C.10}
$$

Therefore, based on the norm bound of the lower triangular matrix $\mathcal{L}$, we get

$$
\begin{aligned}
\sum_k \lambda_k^s \|(Q_k - \widetilde{Q}_k)v\|_0^2 &\le \|\mathcal{L}\| C_A C_I \left[\sum_k \lambda_k^s \|v_k\|_0^2\right]^{1/2} \left[\sum_j \lambda_j^s \|(Q_j - Q_{j-1})v\|_0^2\right]^{1/2} \\
&= \|\mathcal{L}\| C_A C_I \left[\sum_k \lambda_k^s \|v_k\|_0^2\right]^{1/2} |\langle v \rangle|_s .
\end{aligned}
\tag{C.11}
$$

The latter inequality shows that

$$
\left[\sum_k \lambda_k^s \|(Q_k - \widetilde{Q}_k)v\|_0^2\right]^{1/2} \le \|\mathcal{L}\| C_A C_I |\langle v \rangle|_s . \tag{C.12}
$$

The estimate is independent of the number of terms in the above sums, thus by taking the limit, the estimate remains valid for infinite series.

Now, use the fact that $\widetilde{Q}_k$ are uniformly coercive on V_k, and because $\widetilde{Q}_k Q_k = \widetilde{Q}_k$, we arrive at the inequalities ($v_k = (Q_k - \widetilde{Q}_k)v$),

$$\begin{aligned}
&\sum_k \lambda_k^s \|(Q_k - \widetilde{Q}_k)v\|^2 \\
&\quad \le \delta^{-1} \sum_k \lambda_k^s (\widetilde{Q}_k (Q_k - \widetilde{Q}_k)v, (Q_k - \widetilde{Q}_k)v) \\
&\quad = \delta^{-1} \sum_k \lambda_k^s (\widetilde{Q}_k - \widetilde{Q}_k^2)v, v_k) \\
&\quad = \delta^{-1} \sum_k \lambda_k^s ((Q_k - \widetilde{Q}_k)\widetilde{Q}_k v, v_k) \\
&\quad = \delta^{-1} \sum_k \lambda_k^s ((Q_k - \widetilde{Q}_k) \sum_{j \le k} (\widetilde{Q}_j - \widetilde{Q}_{j-1})v, v_k) \\
&\quad = \delta^{-1} \sum_k \lambda_k^s \sum_{j \le k} ((Q_k - \widetilde{Q}_k)(\widetilde{Q}_j - \widetilde{Q}_{j-1})v, v_k) \\
&\quad \le \delta^{-1} \|\mathcal{L}\| C_A C_I \left[\sum_k \lambda_k^s \|v_k\|_0^2\right]^{1/2} \left[\sum_j \lambda_j^s \|(\widetilde{Q}_j - \widetilde{Q}_{j-1})v\|_0^2\right]^{1/2} \\
&\quad = \delta^{-1} \|\mathcal{L}\| C_A C_I \left[\sum_k \lambda_k^s \|v_k\|_0^2\right]^{1/2} |\!|\!| v |\!|\!|_s.
\end{aligned}$$

The last two rows of the above inequality are proved in the same way as the last six rows of (C.10)–(C.11) combined. Therefore,

$$\left[\sum_k \lambda_k^s \|(Q_k - \widetilde{Q}_k)v\|_0^2\right]^{1/2} \le \delta^{-1} \|\mathcal{L}\| C_A C_I |\!|\!| v |\!|\!|_s. \tag{C.13}$$

Based on the decomposition $(\widetilde{Q}_k - \widetilde{Q}_{k-1})v = (\widetilde{Q}_k - Q_k)v - (\widetilde{Q}_{k-1} - Q_{k-1})v + (Q_k - Q_{k-1})v$, we get the estimate from above

$$|\!|\!| v |\!|\!|_s = \left[\sum_k \lambda_k^s \|(\widetilde{Q}_k - \widetilde{Q}_{k-1})v\|_0^2\right]^{1/2} \le (1 + 2\|\mathcal{L}\| C_A C_I) |\langle v \rangle|_s. \tag{C.14}$$

Similarly, using the identity $(Q_k - Q_{k-1})v = -(\widetilde{Q}_k - Q_k)v + (\widetilde{Q}_{k-1} - Q_{k-1})v + (\widetilde{Q}_k - \widetilde{Q}_{k-1})v$, we obtain the following lower bound,

$$|\langle v \rangle|_s = \left[\sum_k \lambda_k^s \|(Q_k - Q_{k-1})v\|_0^2\right]^{1/2} \le (1 + 2\delta^{-1} \|\mathcal{L}\| C_A C_I) |\!|\!| v |\!|\!|_s. \tag{C.15}$$

Thus the proof is complete. □

Corollary C.2. *Consider the following particular cases*
(i) $\lambda_j = 2^{2j}$. *Then the entries of the lower triangular matrix $\mathcal{L}$ read*

$$\ell_{k,j} = \left(\frac{1}{2^\sigma}\right)^{k-j} \left(\frac{\lambda_k}{\lambda_j}\right)^{s/2} = \left(\frac{1}{2^{\sigma-s}}\right)^{k-j}.$$

It is clear then, that if $\sigma > 0$ and $\sigma > s$, that $\|\mathcal{L}\| \le 2^{\sigma-s}/(2^{\sigma-s}-1)$.
(ii) *A parameter-dependent norm: Choose*

$$\lambda_j = \tau^{-1} + 2^{2j}$$

for a given parameter $\tau > 0$. Then, if $s \le 0$, we have $\ell_{k,j} = (1/2^\sigma)^{k-j}$ $(\lambda_k/\lambda_j)^{s/2} \le (1/2^\sigma)^{k-j}$ and it is clear then that $\|\mathcal{L}\| \le (2^\sigma/(2^\sigma-1))$. For $s > 0$ we have $(k \ge j)$:

$$\begin{aligned}
\ell_{k,j} &= \left(\frac{1}{2^\sigma}\right)^{k-j} \left(\frac{\lambda_k}{\lambda_j}\right)^{s/2} \\
&= \left(\frac{1}{2^\sigma}\right)^{k-j} \left(\frac{\tau^{-1}+2^{2k}}{\tau^{-1}+2^{2j}}\right)^{s/2} \\
&= \left(\frac{1}{2^\sigma}\right)^{k-j} \frac{(2^s)^k}{(2^s)^j} \left(\frac{2^{-2k}+\tau}{2^{-2j}+\tau}\right)^{s/2} \\
&\le \left(\frac{1}{2^{\sigma-s}}\right)^{k-j}.
\end{aligned}$$

It is clear then that $\|\mathcal{L}\| \le (2^{\sigma-s}/(2^{\sigma-s}-1))$ if $\sigma > s$.
In conclusion, the spectral norm of $\mathcal{L}$ is bounded uniformly in $\tau > 0$ for any s if σ is appropriately chosen; namely, for $s < 0$ any positive σ is appropriate, whereas for positive s it is sufficient to choose $\sigma > s$ to bound the norm of $\mathcal{L}$.

C.4 The uniform coercivity property

We show at the end that the quasi-interpolants $\widetilde{Q}_k$ defined in (C.1) satisfy the uniform coercivity bound (C.7). The assumption is that any subspace V_k admits $(.,.)$-stable Riesz basis $\{\varphi_i^{(k)},\ x_i \in \mathcal{N}_k\}$, for a given set of degrees of freedom $x_i \in \mathcal{N}_k$. In the case of finite element spaces, $\mathcal{N}_k$ is the set of nodes associated with a standard Lagrangian basis $\{\varphi_i^{(k)}\}$. Also, as it is well known, the nodal Lagrangian basis is an L^2-stable Riesz basis. The latter means that the Gram matrices $G_k = \{(\varphi_j^{(k)}, \varphi_i^{(k)})\}_{x_i,x_j \in \mathcal{N}_k}$ are uniformly well conditioned,

$$(v, v) = \mathbf{v}^T G_k \mathbf{v} \simeq \theta_k \sum_{x_i \in \mathcal{N}_k} v_i^2 = \theta_k \mathbf{v}^T \mathbf{v}, \quad \text{all } v = \sum_{x_i \in \mathcal{N}_k} v_i \varphi_i^{(k)}, \qquad \mathbf{v} = (v_i)_{x_i \in \mathcal{N}_k}. \tag{C.16}$$

In other words the scaled inner product $\theta_k^{-1}(.,.)$ is bounded above and below by the coefficient vector inner product $\mathbf{v}^T\mathbf{v}$ uniformly w.r.t. $k \geq 0$.

Finally, we assume that G_k are uniformly sparse, namely, that the number of nonzero entries per row is bounded by a number m_0 independent of $k \geq 0$.

The quasi-interpolants $\widetilde{Q}_k$ of interest read

$$\widetilde{Q}_k v = \sum_{x_i \in \mathcal{N}_k} \frac{(v, \varphi_i^{(k)})}{(e, \varphi_i^{(k)})} \varphi_i^{(k)}, \qquad v \in V, \ e = \sum_{x_i \in \mathcal{N}_k} \varphi_i^{(k)}. \tag{C.17}$$

In what follows we assume that $(\varphi_i^{(k)}, \ \varphi_j^{(k)}) \geq 0$. This makes the operators in (C.17) well defined.

We first remark that because $(v, \ \varphi_i^{(k)}) = (Q_k v, \ \varphi_i^{(k)})$ (by the definition of the projection Q_k) we immediately get that $\widetilde{Q}_k v = \widetilde{Q}_k Q_k v$; that is, the commutativity property (P) holds.

Consider the coordinate unit vectors $\mathbf{e}_i = (\delta_{i,j})_{x_j \in \mathcal{N}_k}$, $x_i \in \mathcal{N}_k$. It is clear then, that the following matrix–vector representation holds,

$$(\widetilde{Q}_k v, v) = \sum_{x_i \in \mathcal{N}_k} \frac{(\mathbf{v}^T G_k \mathbf{e}_i)^2}{\mathbf{1}^T G_k \mathbf{e}_i}, \qquad \mathbf{1} = \sum_{x_i \in \mathcal{N}_k} \mathbf{e}_i.$$

Based on the decomposition $\mathbf{v} = \sum_{x_i \in \mathcal{N}_k} ((G_k \mathbf{v})^T \mathbf{e}_i) G_k^{-1} \mathbf{e}_i$, we get

$$\begin{aligned}
(v, v) &= \mathbf{v}^T G_k \mathbf{v} \\
&= \Big(\sum_{x_i \in \mathcal{N}_k} (\mathbf{v}^T G_k \mathbf{e}_i) \mathbf{e}_i \Big)^T G_k^{-1} \Big(\sum_{x_i \in \mathcal{N}_k} (\mathbf{v}^T G_k \mathbf{e}_i) \mathbf{e}_i \Big) \\
&\leq \lambda_{\max}[G_k^{-1}] \sum_{x_i \in \mathcal{N}_k} (\mathbf{v}^T G_k \mathbf{e}_i)^2.
\end{aligned}$$

Therefore, the following estimate is obtained,

$$\frac{(\widetilde{Q}_k v, v)}{(v, v)} \geq \lambda_{\min}[G_k] \min_{x_i \in \mathcal{N}_k} \frac{1}{\mathbf{1}^T G_k \mathbf{e}_i}.$$

Based on the assumption of uniform sparsity of the Gram matrices G_k, that is, that the number of nonzero entries per row of G_k is bounded by an integer $m_0 = \mathcal{O}(1)$, uniformly in $k \to \infty$, the expression $1/(\mathbf{1}^T G_k \mathbf{e}_i)$ is estimated below by $(1/m_0)(1/(\lambda_{\max}[G_k]))$. Indeed, because at most m_0 terms $\mathbf{e}_j^T G_k \mathbf{e}_i$ in the first sum below are nonzero (these indices j define the set $\mathcal{I}(i)$), we get

$$\begin{aligned}
\mathbf{1}^T G_k \mathbf{e}_i &= \sum_j \mathbf{e}_j^T G_k \mathbf{e}_i \\
&\le \sum_{j \in \mathcal{I}(i)} \left(\mathbf{e}_j^T G_k \mathbf{e}_j\right)^{1/2} \left(\mathbf{e}_i^T G_k \mathbf{e}_i\right)^{1/2} \\
&\le \lambda_{\max}[G_k] \sum_{j \in \mathcal{I}(i)} \|\mathbf{e}_j\| \|\mathbf{e}_i\| \\
&\le \lambda_{\max}[G_k]\, m_0.
\end{aligned}$$

That is, the desired uniform coercivity estimate (C.7) takes the final form

$$\frac{(\widetilde{Q}_k v, v)}{(v, v)} \ge \frac{1}{m_0 \mathrm{Cond}(G_k)} = \mathcal{O}(1).$$

D

Multilevel Algorithms for Boundary Extension Mappings

In the case of matrices A coming from a finite element discretization of second-order elliptic PDEs, on a sequence of uniformly refined meshes the following construction can provide a bounded extension mapping in the A-norm. We derive the matrix–vector form of the resulting extension mapping E suitable for actual computations. It becomes evident that E^T also has computable actions. These are found in expressions (D.3) and (D.4) that represent the main result of the present section.

Note that both

$$E = \begin{bmatrix} E_{0,b} \\ I \end{bmatrix} \quad \text{and} \quad E^T = \left[E_{0,b}^T, \quad I \right]$$

are needed explicitly to get the stable two-by-two block form of the transformed matrix $[J, \ E]^T A [J, \ E]$, $J = \left[\begin{smallmatrix} I \\ 0 \end{smallmatrix} \right]$ (cf., Section 3.4.1).

We assume that a finite element function ψ is defined on a boundary Γ of a domain Ω. We extend ψ in the remaining part of Ω to a finite element function v, achieving certain norm-boundedness of the extension. We assume that there is a sequence of easily computable boundary operators $\widetilde{q}_k : \ V|_\Gamma \mapsto V_k|_\Gamma$ corresponding to the restrictions (traces) of nested spaces $V_{k-1} \subset V_k \subset \cdots \subset V_L = V$. Let E_k^0 be the trivial extension of a kth-level function given on $V_k|_\Gamma$ to a function that vanishes at the remaining dofs in Ω. The coefficient vector of a function v is denoted $\mathbf{v}$ and if $v \in V_k$ its kth-level coefficient vector is denoted $\mathbf{v}_k$. The coefficient vectors $\mathbf{v}_k$ restricted to the kth-level dofs on Γ, Γ_k, are denoted $\underline{\psi}_k = \mathbf{v}_k|_{\Gamma_k}$. The intergrid transfer mappings for vectors in the domain Ω are denoted I_k^{k+1}, $I_{k+1}^k = (I_k^{k+1})^T$, whereas their restrictions to Γ are i_k^{k+1}, $i_{k+1}^k = (i_k^{k+1})^T$. Note that here, $k+1$ is a fine-level and k is a next coarse-level. Finally, let $\mathcal{N}_k$ stand for the kth-level degrees of freedom at the kth-level grid. We have $\Gamma_k \subset \mathcal{N}_k$. We need an inner product $(.,.)$ defined for functions on Γ and let $\{\varphi_i^{(k)}\}_{x_i \in \mathcal{N}_k}$ be a Lagrangian basis of the space V_k. This means $\varphi_i^{(k)}(x_j) = \delta_{i,j}$ for $x_i, \ x_j \in \mathcal{N}_k$. The restrictions of $\varphi_s^{(k)}$ for $x_s \in \Gamma$ to Γ will then span the trace space $V_k|_\Gamma$. Finally note that the coefficient vector of basis function $\varphi_s^{(k)}$ is simply the coordinate unit vector

$$\mathbf{e}_s^{(k)} = \begin{bmatrix} 0 \\ \vdots \\ 0 \\ 1 \\ 0 \\ \vdots \\ 0 \end{bmatrix}$$

with the only nonzero entry 1 at position s. Finally, let

$$\mathbf{1}_k = \sum_{x_s \in \Gamma_k} \mathbf{e}_s^{(k)} = \begin{bmatrix} 1 \\ \vdots \\ 1 \\ 1 \\ 1 \\ \vdots \\ 1 \end{bmatrix}.$$

To be specific we consider the following boundary operators,

$$\widetilde{q}_k \psi = \sum_{s:\, x_s \in \Gamma_k} \frac{(\psi,\ \varphi_s^{(k)})}{(1,\ \varphi_s^{(k)})}\ \varphi_s^{(k)}.$$

In a matrix–vector form, we have

$$\underline{\widetilde{q}}_k \underline{\psi} = \sum_{s:\, x_s \in \Gamma_k} \frac{\underline{\psi}^T g_L \left(i_{L-1}^L \dots i_k^{k+1}\right) \mathbf{e}_s^{(k)}}{\mathbf{1}_k^T g_k \mathbf{e}_s^{(k)}}\ \mathbf{e}_s^{(k)}.$$

Here, $g_k = ((\varphi_s^{(k)},\ \varphi_l^{(k)}))_{x_s,\, x_l \in \Gamma_k}$ is the kth-level boundary Gram matrix corresponding to the basis $\{\varphi_s^{(k)}\}_{x_s \in \Gamma_k}$.

Also, we let $d_L = g_L$ and for $k < L$ introduce the diagonal matrices $d_k = \text{diag}\{(g_k \mathbf{1}_k)_s : x_s \in \Gamma\}$.

The extension mapping E is defined based on the following decomposition of any function ψ defined on Γ,

$$\psi = \sum_{k=1}^{L} (\widetilde{q}_k - \widetilde{q}_{k-1})\ \psi,$$

where $\widetilde{q}_L = i$ is the identity on Γ and $q_0 = 0$. Because E_k^0 is the trivial extension (by zero at the kth-level grid) of boundary data in the interior of the domain, then $(E_k^0)^T$

represents restriction to the domain boundary, at the kth-level grid. The extension mapping E of our main interest is defined by

$$E\psi = \sum_{k=1}^{L} E_k^0(\widetilde{q}_k - \widetilde{q}_{k-1})\psi.$$

For $k = L$ we let $\underline{\psi}_L = \underline{\psi}$, $d_L = i =$ identity. For $k < L$, we let

$$\underline{\psi}_k = i_L^k g_L \underline{\psi} = i_{k+1}^k \cdots i_L^{L-1} g_L \underline{\psi}. \tag{D.1}$$

Let $\widetilde{q}_L = i$ (the identity). The vector representation of $(\widetilde{q}_k - \widetilde{q}_{k-1})\psi$ is $d_k^{-1}\underline{\psi}_k - i_{k-1}^k d_{k-1}^{-1}\underline{\psi}_{k-1} = (d_k^{-1} - i_{k-1}^k d_{k-1}^{-1} i_k^{k-1})\underline{\psi}_k$. Finally, introduce for $k = L$

$$\mathbf{w}_L = (g_L^{-1} - i_{L-1}^L d_{L-1}^{-1} i_L^{L-1}) g_L \underline{\psi},$$

and for $k < L$,

$$\mathbf{w}_k = (d_k^{-1} - i_{k-1}^k d_{k-1}^{-1} i_k^{k-1}) i_{k+1}^k \cdots i_L^{L-1} g_L \underline{\psi}. \tag{D.2}$$

Then, the matrix–vector representation of the extension mapping takes the form

$$\begin{aligned} E\underline{\psi} &= \sum_{k=0}^{L} I_{L-1}^L \cdots I_k^{k+1} E_k^0 \mathbf{w}_k \\ &= \sum_{k=0}^{L} I_{L-1}^L \cdots I_k^{k+1} E_k^0 (d_k^{-1} - i_{k-1}^k d_{k-1}^{-1} i_k^{k-1}) i_{k+1}^k \cdots i_L^{L-1} g_L \underline{\psi}. \end{aligned} \tag{D.3}$$

Note that the matrix vector form of E_k^0 has the simple form,

$$E_k^0 \mathbf{w}_k = \begin{bmatrix} 0 \\ \mathbf{w}_k \end{bmatrix} \begin{matrix} \mathcal{N}_k \setminus \Gamma_k \\ \Gamma_k \end{matrix}.$$

Therefore the adjoint to E takes the form

$$E^T = g_L \sum_{k=0}^{L} i_{L-1}^L \cdots i_k^{k+1} (d_k^{-1} - i_{k-1}^k d_{k-1}^{-1} i_k^{k-1}) (E_k^0)^T I_{k+1}^k \cdots I_L^{L-1}. \tag{D.4}$$

The boundedness of E in the A-norm can be proved by assuming that the energy norm based on A can be characterized by a norm induced by certain projections $\{Q_k\} : V \mapsto V_k$ with respect to an inner product $(.,.)_0$, in the sense that we first have

$$\begin{aligned} \sum_{k=1}^{L} \mathbf{w}_k^T (E_k^0)_k^T A_k E_k^0 \mathbf{w}_k &\leq \sum_{k=1}^{L} \lambda_k \, \|E_k^0 \mathbf{w}_k\|_0^2 \\ &\simeq \inf_{v=\sum_k v_k:\ v|_\Gamma = \psi} \sum_{k=1}^{L} \lambda_k \|v_k\|_0^2 \simeq \inf_{v:\ v|_\Gamma=\psi} \mathbf{v}^T A \mathbf{v}. \end{aligned}$$

Here, $\lambda_k = \varrho(A_k) \equiv \max_{\mathbf{v}_k}(\mathbf{v}_k^T A_k \mathbf{v}_k / \|\mathbf{v}_k\|_0^2)$. That is, we have assumed the norm equivalence, $\mathbf{v}^T A \mathbf{v} \simeq \sum_k \lambda_k \ \|(Q_k - Q_{k-1})\mathbf{v}\|_0^2$ and similarly, we have assumed that the norm introduced by the Schur complement $\mathcal{S}_k$ of A_k to Γ_k, is characterized by

$$\underline{\psi}^T \mathcal{S}_k \underline{\psi} \simeq \sum_k \theta_k \ \|(\widetilde{q}_k - \widetilde{q}_{k-1})\underline{\psi}\|^2.$$

Here, $\theta_k \simeq \lambda_k \ \sup_{\underline{\psi}_k}(\|E_k^0 \underline{\psi}_k\|_0^2 / \|\underline{\psi}_k\|^2)$. For finite element matrices coming from second-order elliptic PDEs, we have $\lambda_k = h_k^{-2}$ and $\theta_k = h_k^{-1}$ where the $\|.\|_0$ norm stands for the integral $L_2(\Omega)$-norm and $\|.\|$ stands for the boundary integral $L_2(\Gamma)$-norm. See the next section, for characterizing the H_0^1 Sobolev norm naturally associated with the weak form of the Poisson problem. The motivation of using the computable boundary operators $\widetilde{q}_k$ is that the trace norm on Γ is typically characterized (for finite element matrices A coming from second-order elliptic PDEs) as the Sobolev space $H^{1/2}(\Gamma)$-norm, and the latter has a computable counterpart as described in the present chapter.

Multilevel extension mappings were considered in [HLMN], [0s94], and [Nep95]. The decomposition based on the quasi-interpolants $\widetilde{q}_k$ was analyzed in [BPV99].

E

H_0^1-norm Characterization

In this short appendix we present in a constructive way an $H_0^1(\Omega)$-norm characterization. First the result is proven for a domain Ω which implies full regularity for the Poisson problem,

$$-\Delta u = f(x), \qquad x \in \Omega,$$

subject to $u = 0$ on $\partial\Omega$. Full regularity means that

$$\|u\|_2 \leq C \, \|f\|_0.$$

Such a result is available in the literature for Ω being a convex polygon. Then, we extend it to more general domains by using overlapping decomposition of Ω into convex subdomains.

E.1 Optimality of the L_2-projections

We assume that Ω is triangulated on a sequence of uniformly refined triangulations with characteristic mesh-size $h_k = h_0 2^{-k}$, $k \geq 0$, and it is well known that the respective finite element spaces of piecewise linear functions $V_k = V_{h_k}$ satisfy $\overline{\cup V_k} = H_0^1(\Omega)$. Define the L_2-projections $Q_k : \; L_2(\Omega) \mapsto V_k$. Then, we can prove the following main result,

$$\sum_k h_k^{-2} \|(Q_k - Q_{k-1})v\|_0^2 \simeq \|v\|_1^2. \tag{E.1}$$

More generally, we have the following main characterization of $H_0^1(\Omega)$,

$$\|v\|_1^2 \simeq \inf_{v=\sum_k v_k, \; v_k \in V_k} \; \sum_k h_k^{-2} \|v_k\|_0^2. \tag{E.2}$$

First, we prove the result for convex domain Ω. To this end let us define the elliptic projections $\pi_k : \; H_0^1(\Omega) : \mapsto V_k$ in the standard way as

$$(\nabla \pi_k v, \; \nabla \varphi) = (\nabla v, \; \nabla \varphi), \quad \text{for all } \varphi \in V_k.$$

Because Ω is convex, we have full regularity for the Laplacian. Therefore, π_{k-1} exhibits an optimal L_2-error estimate; that is, we have $\|v - \pi_{k-1}v\|_0 \leq Ch_k\|v\|_1$. Based on this estimate used for $v := (\pi_k - \pi_{k-1})v$ and using the H_0^1-orthogonality of the projections π, we have

$$\sum_k h_k^{-2}\|(\pi_k - \pi_{k-1})v\|_0^2 \leq C \sum_k \|(\pi_k - \pi_{k-1})v\|_1^2$$
$$= C\sum_k \left(\|\pi_k v\|_1^2 - \|\pi_{k-1}v\|_1^2\right) = \|v\|_1^2.$$

This shows that the r.h.s. of (E.2) is bounded in terms of $\|v\|_1^2$ (for the convex domain). For a more general domain Ω, we assume that it can be split into overlapping convex subdomains Ω_m, $m = 1, \ldots, m_0$ for a fixed number m_0. The decomposition is such that for any $v \in H_0^1(\Omega)$, we can find an H_0^1-stable decomposition $v = \sum_m v_m$ with each term supported in the convex subdomain $\overline{\Omega}_m$. Stability here means that we have the estimate $\sum_m \|v_m\|_1^2 \leq C\ \|v\|_1^2$. Because for every component v_m (which is supported in the convex domain) we can find a stable multilevel decomposition (also supported in $\overline{\Omega}_m$), thus a stable decomposition of the finite sum $v = \sum_m v_m$ is constructed which proves that the r.h.s. of (E.2) is bounded in terms of $\|v\|_1^2$, now for the case of more general domains Ω.

We show next that the decomposition based on the L_2-projections Q_k is quasioptimal (for a general, not necessarily convex domain). This follows from the following chain of inequalities, using the fact that Q_k are L_2-symmetric and that $(Q_k - Q_{k-1})^2 = Q_k - Q_{k-1}$, letting $v_j = (Q_j - Q_{j-1})v$,

$$\sum_k h_k^{-2}\|(Q_k - Q_{k-1})v\|_0^2 = \sum_k h_k^{-2}((Q_k - Q_{k-1})v,\ v)$$
$$= \sum_k h_k^{-2}\left((Q_k - Q_{k-1})v,\ \sum_{j\geq k} v_j\right)$$
$$\leq \sum_k h_k^{-2}\|(Q_k - Q_{k-1})v\|_0 \sum_{j\geq k} \|v_j\|_0$$
$$= \sum_k \sum_{j\geq k} \frac{h_j}{h_k}\ (h_k^{-1}\|(Q_k - Q_{k-1})v\|_0)h_j^{-1}\|v_j\|_0$$
$$\leq C \sum_k \sum_{j\geq k} \frac{1}{2^{j-k}}\ (h_k^{-1}\|(Q_k - Q_{k-1})v\|_0)h_j^{-1}\|v_j\|_0.$$

That is,

$$\sum_k h_k^{-2}\|(Q_k - Q_{k-1})v\|_0^2 \leq C\left(\sum_k \sum_{j\geq k} \frac{1}{2^{j-k}}\ h_k^{-2}\|(Q_k - Q_{k-1})v\|_0^2\right)^{1/2}$$
$$\times \left(\sum_j \sum_{k\leq j} \frac{1}{2^{j-k}}\ h_j^{-2}\|v_j\|_0^2\right)^{1/2}$$

$$\leq C\left(\sum_k h_k^{-2}\|(Q_k - Q_{k-1})v\|_0^2\right)^{1/2}$$
$$\times\left(\sum_k h_k^{-2}\|v_k\|_0^2\right)^{1/2}.$$

That is, the decomposition $v = \sum_j (Q_j - Q_{j-1})v$ is quasioptimal. This (together with (E.3) below) shows the well-known norm characterization (E.2) of $H_0^1(\Omega)$ originally proven by Oswald [0s94]; see also, [DK92].

To prove the other direction of (E.2), we proceed as follows. For any decomposition $v = \sum_j v_j$, $v_j \in V_j$, and for a fixed $\alpha \in (0, \frac{1}{2})$, using the inequality $(p, q) \leq \|p\|_\alpha \|q\|_{-\alpha}$ and appropriate inverse inequalities, we have

$$
\begin{aligned}
\|v\|_1^2 &= \sum_k \left(\nabla(\pi_k - \pi_{k-1})v, \sum_{j\geq k} \nabla v_j\right) \\
&\leq \sum_k \sum_{j\geq k} \|(\pi_k - \pi_{k-1})v\|_{1+\alpha}\ \|v_j\|_{1-\alpha} \\
&\leq C\sum_k \sum_{j\geq k} h_k^{-\alpha}\|(\pi_k - \pi_{k-1})v\|_1 h_j^{-1+\alpha}\ \|v_j\|_0 \\
&= C\sum_k \sum_{j\geq k} \left(\frac{1}{2^\alpha}\right)^{j-k} \|(\pi_k - \pi_{k-1})v\|_1 \left(h_j^{-1}\ \|v_j\|_0\right) \\
&\leq C\left(\sum_k \sum_{j\geq k} \left(\frac{1}{2^\alpha}\right)^{j-k} \|(\pi_k - \pi_{k-1})v\|_1^2\right)^{1/2} \\
&\quad \times \left(\sum_j \sum_{k\leq j} \left(\frac{1}{2^\alpha}\right)^{j-k} h_j^{-2}\|v_j\|_0^2\right)^{1/2} \\
&\leq C/(1 - 2^{-\alpha})\ \|v\|_1 \left(\sum_j h_j^{-2}\|v_j\|_0^2\right)^{1/2}.
\end{aligned}
\tag{E.3}
$$

This shows the remaining fact that $\|v\|_1^2$ is bounded in terms of the r.h.s. of (E.2).

Explicit construction of a continuous H_0^1-stable decomposition with components supported in convex polygons was shown in Lions [Li87] for a model L-shaped domain Ω with $m_0 = 2$. We present this example next.

Example E.1. Given the L-shaped domain Ω shown in Figure E.1. Consider the following cut-off function

$$
\chi = \begin{cases}
1, & x \leq 0, \\
1 - \dfrac{bx}{ay}, & (x, y) \in T = \left\{1 \geq y \geq \dfrac{b}{a}x,\ 0 \leq x \leq a\right\}, \\
0, & y \leq \dfrac{b}{a}x,\ x \in [0, a].
\end{cases}
$$

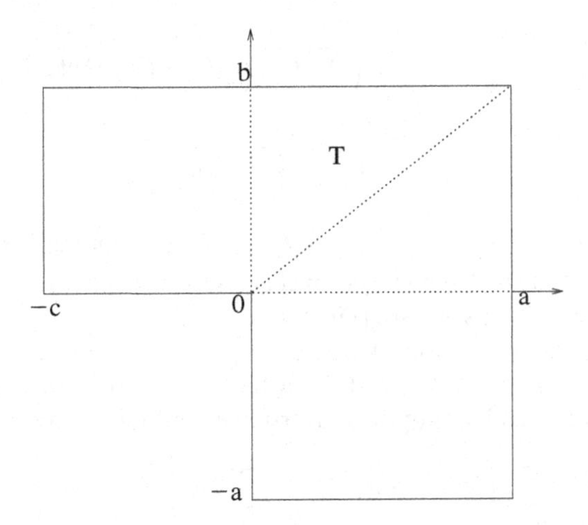

Fig. E.1. *L-shaped domain Ω partitioned into two overlapping rectangles $\Omega_1 = (-c, a) \times (0,\ b)$ and $\Omega_2 = (0,\ a) \times (-a, b)$.*

Its gradient is nonzero only on T and it equals

$$\nabla \chi = \frac{b}{a} \begin{bmatrix} -\frac{1}{y} \\ \frac{x}{y^2} \end{bmatrix}.$$

On T, we have

$$\frac{x^2}{y^2} \leq \frac{a^2}{b^2} \quad \text{and} \quad \frac{1}{x^2 + y^2} \geq \frac{1}{\frac{a^2}{b^2} y^2 + y^2}.$$

This shows that

$$|\nabla \chi|^2 = \frac{b^2}{a^2} \frac{1}{y^2} \left[1 + \frac{x^2}{y^2}\right] \leq \frac{b^2}{a^2} \frac{1}{x^2 + y^2} \left[1 + \frac{a^2}{b^2}\right] \left[1 + \frac{a^2}{b^2}\right].$$

The decomposition of our main interest reads

$$v = \chi v + (1 - \chi) v.$$

Note that $v_1 = \chi v$ is supported in the convex domain (rectangle) $\Omega_1 = (-c, a) \times (0, b)$ and $v_2 = (1 - \chi) v$ is supported in the convex domain (rectangle) $\Omega_2 = (0, a) \times (-a, b)$. To show the desired H_0^1-stability, we have to estimate $|v_1|_1$ in terms of $|v|_1$. We have

$$\begin{aligned} \int_\Omega |\nabla v_1|^2 \, d\mathbf{x} &\leq 2 \int_\Omega v^2 |\nabla \chi|^2 \, d\mathbf{x} + 2 \int_\Omega \chi^2 |\nabla v|^2 \, d\mathbf{x} \\ &\leq 2 \int_\Omega \chi^2 |\nabla v|^2 \, d\mathbf{x} + C \int_\Omega \frac{v^2(\mathbf{x})}{\text{dist}^2\ (\mathbf{x},\ \partial\Omega)} \, d\mathbf{x}. \end{aligned}$$

The stability follows due to a classical inequality

$$\int\limits_{\Omega} \frac{v^2(\mathbf{x})}{\text{dist}^2\,(\mathbf{x},\,\partial\Omega)}\,d\mathbf{x} \leq C\,|v|_1^2,$$

valid for $H_0^1(\Omega)$ functions.

A more general approach of establishing H_0^1-stable decompositions supported in convex subdomains is based on the following simple construction. Let Ω be a polyhedral domain and let $\Omega_1 \cup \Omega_2$ provide an overlapping partition of Ω. In the application, when Ω comes with a triangulation $\mathcal{T}_h$ we assume that Ω_1 and Ω_2 are exactly covered by elements from $\mathcal{T}_h$. Let $\widehat{\Omega}_1 = \Omega_1 \cap \Omega_2$ and $\widehat{\Omega}' = \Omega_1 \setminus \widehat{\Omega}_1$. Then, $\Omega_1 = \widehat{\Omega}_1 \cup \widehat{\Omega}_1'$ is separated by an interface Γ. Given a $\varphi \in H_0^1(\Omega)$ then $g = \varphi|_\Gamma$ as a trace of H_0^1 function will belong to $H_{0,0}^{1/2}(\Gamma)$. Therefore, $\widehat{g}$, the zero extension of g on $\partial\widehat{\Omega}_1 \setminus \Gamma$ will belong to $H^{1/2}(\partial\widehat{\Omega}_1)$. To construct a stable component v_1 of v supported in $\overline{\Omega}_1$, we extend $v|_{\Omega_1 \setminus \Omega_2}$ through Γ into the remaining part of Ω_1, $\widehat{\Omega}_1 = \Omega_1 \cap \Omega_2$, by solving the following Dirichlet boundary value problem,

$$-\Delta\psi = 0 \text{ in } \widehat{\Omega}_1 \quad \text{subject to } \psi|_{\partial\widehat{\Omega}_1} = \widehat{g}.$$

The following a priori estimate holds

$$|\psi|_1 \leq C\,\|\widehat{g}\|_{1/2,\,\partial\widehat{\Omega}_1} \leq C\,\|g\|_{H_{0,0}^{1/2}(\Gamma)}.$$

Then, based on a trace inequality

$$\|g\|_{H_{0,0}^{1/2}(\Gamma)} \leq C\,|v|_{1,\,\widehat{\Omega}_1} \leq C|v|_1,$$

we find that the harmonic extension is stable in H^1. The function v_1 defined as v on $\widehat{\Omega}_1'$, as ψ on $\widehat{\Omega}_1$ and zero outside Ω_1 belongs to $H_0^1(\Omega)$ and by construction is stable; that is, we have $|v_1|_1 \leq C\,|v|_1$. Assuming that Ω_1 is convex, the problem is reduced to eventually further decompose $v_2 = v - v_1$ which is now supported in a smaller domain Ω_2. The process can be repeated several times until all "eliminated" subdomains are convex and they cover Ω. The constants in the stability estimates will depend on the size and shape of the subdomains, which to a certain extent is under our control.

E.1.1 H_0^1-stable decompositions of finite element functions

Assume now that v is a finite element function from a f.e. space V_h vanishing on $\partial\Omega$ where Ω is a polygonal (in a general nonconvex) domain. Let $\{\Omega_i\}$ be a finite set of convex polygons that cover Ω. We assume that $v = \sum_i v_i$ is a $H_0^1(\Omega)$-stable continuous decomposition such that the components v_i are supported in $\overline{\Omega}_i$. Stability means that

$$\sum_i |v_i|_1^2 \leq C\,|v|_1^2.$$

We want to construct an $H_0^1(\Omega)$-stable finite element decomposition $v = \sum_i v_i^h$ with v_i^h supported in $\overline{\Omega}_i$ as well. We assume that the given triangulation $\mathcal{T}_h$ is aligned with the boundaries of the polygons Ω_i. Let $\{\varphi_j\}$ be a nodal basis of V_h. Denote the set of interior (to Ω) nodes by $\mathcal{N}_h$. Then the nodal basis function φ_j is associated with the (interior) node $\mathbf{x}_j \in \mathcal{N}_h$. Consider the quasi-interpolant $\widetilde{Q}_h$ and let $\widetilde{Q}_h^{(i)}$ be the quasi-interpolant associated with the finite element space V_i of functions from V_h that are supported in $\overline{\Omega}_i$. The finite element H^1-stable decomposition of v then is constructed from the representation

$$v = v - \widetilde{Q}_h v + \sum_i (\widetilde{Q}_h^{(i)} - \widetilde{Q}_h) v_i + \sum_i \widetilde{Q}_h^{(i)} v_i.$$

We note that $w \equiv v - \widetilde{Q}_h v + \sum_i (\widetilde{Q}_h^{(i)} - \widetilde{Q}_h) v_i$ is small in L_2 due to the approximation properties of $\widetilde{Q}_h$ and $\widetilde{Q}_h^{(i)}$. We have $\|v - \widetilde{Q}_h v\|_0 \le Ch\ |v|_1$, and $\|(\widetilde{Q}_h^{(i)} - \widetilde{Q}_h) v_i\|_0 \le \|\widetilde{Q}_h^{(i)} v_i - v_i\|_0 + \|v_i - \widetilde{Q}_h v_i\|_0 \le Ch\ |v_i|_1$. That is, based on the assumed stability of $\{v_i\}$, we have $\|w\|_0^2 \le Ch^2(|v|_1^2 + \sum_i\ |v_i|_1^2) \le Ch^2\ |v|_1^2$. Finally, we decompose $w = \sum_i w_i$ where $w_i = \sum_{x_j \in \Omega_j} (1/d_j) w(\mathbf{x}_j) \varphi_j$ so that each w_i is supported in Ω_i and d_j stands for the number of subdomains that contain the node $\mathbf{x}_j$. It is clear that $\{w_i\}$ is H_0^1 stable because based on a standard inverse inequality, and the equivalence of the discrete ℓ_2 and integral L_2 norms, we have

$$\sum_i |w_i|_1^2 \le Ch^{-2} \sum_i \|w_i\|_0^2 \le Ch^{-2} \sum_j w^2(x_j) h^d \le Ch^{-2}\ \|w\|_0^2 \le |v|_1^2.$$

The final decomposition reads

$$v = \sum_i (w_i + \widetilde{Q}_h^{(i)} v_i),$$

which is H_0^1-stable. We already proved that $\{w_i\}$ are stable. Based on the H_0^1-stability of $\widetilde{Q}_h^{(i)}$ and because $\{v_i\}$ come from a continuous H_0^1-stable decomposition the overall stability follows. Finally, note the ith finite element function $w_i + \widetilde{Q}_h^{(i)} v_i$ is supported in Ω_i.

To summarize:

Theorem E.2. *Under the assumption that the overlapping subdomains $\{\Omega_i\}$ are mesh domains, that is, they are covered exactly by the elements from a given quasiuniform triangulation $\mathcal{T}_h$, the existence of continuous H^1-stable decomposition $v = \sum_i v_i$ with functions v_i supported in Ω_i implies the existence of a similar discrete H^1-stable decomposition.*

F

MG Convergence Results for Finite Element Problems

In this chapter, we apply the MG analysis in general terms from Section 5.3 for three particular examples, namely, to finite element problems corresponding to the weighted Laplacian bilinear form $a(u,\ \varphi) = \tau\ (\nabla u,\ \nabla\varphi) + (u,\ \varphi)$, to the weighted $H(\text{curl})$ bilinear form $a(\mathbf{u},\ \boldsymbol{\chi}) = \tau\ (\text{curl}\ \mathbf{u},\ \text{curl}\ \boldsymbol{\chi}) + (\mathbf{u},\ \boldsymbol{\chi})$, and to the weighted $H(div)$-bilinear form $a(\mathbf{u},\ \boldsymbol{\chi}) = \tau\ (\text{div}\ \mathbf{u},\ \text{div}\ \boldsymbol{\chi}) + (\mathbf{u},\ \boldsymbol{\chi})$. In all cases τ is a positive parameter that can get large. The bilinear forms are associated with respective f.e. spaces S_h (the H^1_0-conforming space of nodal piecewise linear functions), $\mathbf{Q}_h$ (the lowest-order Nédélec space) and $\mathbf{R}_h$ (the lowest-order Raviart–Thoma space). We substantially use the fact that the triple $(S_h,\ \mathbf{Q}_h,\ \mathbf{R}_h)$ provides an "exact" sequence which means that ∇S_h equals the null space of the curl-operator restricted to $\mathbf{Q}_h$, and similarly curl $\mathbf{Q}_h$ equals the null space of the div-operator restricted to $\mathbf{R}_h$. For a proof of this result, we refer to [Mo03].

We follow the "recipes" of Theorem 5.7. Given the stiffness matrices A_k, the smoothers M_k, and interpolation matrices P_k, we need to find a multilevel decomposition of any fine-grid vector $\underline{\mathbf{y}}$; that is, starting with $\underline{\mathbf{y}}_0 = \underline{\mathbf{y}}$, for $k \geq 0$, we find $\underline{\mathbf{y}}_k = \underline{\mathbf{y}}_k^f + P_k \underline{\mathbf{y}}_{k+1}$ such that

(i) The (symmetrized) smoothers $\overline{M}_k$ are efficient on the components $\underline{\mathbf{y}}_k^f$ so that the following estimate holds,

$$\sum_k (\underline{\mathbf{y}}_k^f)^T \overline{M}_k \underline{\mathbf{y}}_k^f \leq C\ \underline{\mathbf{y}}^T A \underline{\mathbf{y}}.$$

(ii) The smoothers M_k are efficient on the coarse components $P_k \underline{\mathbf{y}}_{k+1}$ so that the following estimate holds,

$$\sum_k \| (M_k^T + M_k - A_k)^{-(1/2)} A_k P_k \underline{\mathbf{y}}_{k+1} \|^2 \leq C\ \underline{\mathbf{y}}^T A \underline{\mathbf{y}}.$$

(iii) The coarse-grid component $\underline{\mathbf{y}}_\ell$ is stable in energy; that is, we have,

$$\underline{\mathbf{y}}_\ell^T A_\ell \underline{\mathbf{y}}_\ell \leq C\ \underline{\mathbf{y}}^T A \underline{\mathbf{y}}.$$

In all three applications, we assume that M_k is the (forward) Gauss–Seidel smoother coming from A_k (or coming from A_k restricted to a proper subspace). Then, we know (from Proposition 6.12) that $\overline{M}_k$ is spectrally equivalent to the diagonal of A_k. Moreover, $M_k + M_k^T - A_k$ is actually equal then to the diagonal of A_k.

In what follows we often use the following result (originally found in [Yhb]).

Proposition F.1. *Consider the finite element space of our main interest, the standard nodal-based H^1-conforming Lagrangian space S_h, the lowest-order Nédélec space $\mathbf{N}_h$, and the lowest-order Raviart–Thomas space $\mathbf{R}_h$ all associated with a triangulation $\mathcal{T}_h$ obtained by refinement of a coarser triangulation $\mathcal{T}_H$. Let S_H, $\mathbf{N}_H$, and $\mathbf{R}_H$ be coarse counterparts of the fine-grid spaces S_h, $\mathbf{N}_h$, and $\mathbf{R}_h$. Then, the following strengthened inverse inequalities hold.*
(i) For any $\varphi_h \in S_h$ and $\varphi_H \in S_H$, we have

$$(\nabla\varphi_h,\ \nabla\varphi_H) \leq Ch^{-(1/2)}\ \|\varphi_h\|_0\ H^{-(1/2)}\|\nabla\varphi_H\|_0.$$

(ii) For any $\mathbf{\Phi}_h \in \mathbf{N}_h$ and $\mathbf{\Psi}_H \in \mathbf{R}_H$, we have

$$(\text{curl}\ \mathbf{\Phi}_h,\ \mathbf{\Psi}_H) \leq Ch^{-(1/2)}\ \|\mathbf{\Phi}_h\|_0\ H^{-(1/2)}\|\mathbf{\Psi}_H\|_0.$$

(iii) For any $\mathbf{\Psi}_h \in \mathbf{R}_h$ and $\mathbf{\Psi}_H \in \mathbf{R}_H$, we have

$$(\text{div}\ \mathbf{\Psi}_h,\ \text{div}\ \mathbf{\Psi}_H) \leq Ch^{-(1/2)}\ \|\mathbf{\Psi}_h\|_0\ H^{-(1/2)}\|\text{div}\ \mathbf{\Psi}_H\|_0.$$

We first comment that (i)–(iii) are indeed strengthened versions of the directly obtained inverse inequalities. For example, we can proceed as follows. Apply the Cauchy–Schwarz inequality to arrive at $(\nabla\varphi_h,\ \nabla\varphi_H) \leq \|\nabla\varphi_h\|_0\|\nabla\varphi_H\|_0$, Then, after using a standard inverse inequality, we arrive at $(\nabla\varphi_h,\ \nabla\varphi_H) \leq Ch^{-1}\|\varphi_h\|_0\|\nabla\varphi_H\|_0$ which is a much weaker estimate than (i) if $h \ll H$.

Proof. We prove only inequality (i). The remaining two inequalities (ii) and (iii) are similarly proved. We only mention that to prove (ii), we notice that for the lowest-order Raviart–Thomas space $\mathbf{R}_H$, the elementwise curl of $\mathbf{\Psi}_H \in \mathbf{R}_H$ on every coarse element $T \in \mathcal{T}_H$ is zero. For every coarse element $T \in \mathcal{T}_H$ use integration by parts to reduce the integration to ∂T. We have

$$(\nabla\varphi_h,\ \nabla\varphi_H)_T = \int\limits_T \nabla\varphi_h \cdot \nabla\varphi_H\ d\mathbf{x} = \int\limits_{\partial T} \varphi_h(\nabla\varphi_H \cdot \mathbf{n})\ d\varrho.$$

We used the fact that $\nabla\varphi_H$ is constant on T. Now, use the Cauchy–Schwarz inequality and standard inverse inequalities to bound boundary integrals in terms of volume integrals (valid for f.e. functions), to arrive at the local strengthened inverse inequality

$$(\nabla\varphi_h,\ \nabla\varphi_H)_T \leq \|\varphi_h\|_{0,\ \partial T}\|\nabla\varphi_H\cdot\mathbf{n}\|_{0,\ \partial T} \leq Ch^{-(1/2)}\ \|\varphi_h\|_{0,\ T}H^{-(1/2)}\ \|\nabla\varphi_H\|_{0,\ T}.$$

The global strengthened inverse inequality (i) is obtained by summation over $T \in \mathcal{T}_H$ and using the Cauchy–Schwarz inequality. □

F.1 Requirements on the multilevel f.e. decompositions for the MG convergence analysis

As a first step of the analysis, we reformulate the items (i)–(iii) in terms of bilinear forms and f.e. functions, because we exploit specific multilevel decomposition of the respective finite element spaces used, V_k. Here $V_0 = V_h$ stands for the f.e. space on the finest-mesh (or triangulation) $\mathcal{T}_h$, and $V_\ell = V_H$ is the f.e. space on the coarsest-mesh (triangulation) $\mathcal{T}_H$. The triangulations $\mathcal{T}_k$ are obtained by uniform refinement from the initial coarse triangulation $\mathcal{T}_H$. We comment that our notation is a bit non-standard to conform with the notation we used for AMG: level 0 is finest, and level ℓ is coarsest. Thus, the kth-level mesh-size is $h_k = h2^k = H2^{k-\ell}$, $k = 0, \ldots, \ell$, and $h_0 = h = 2^{-\ell} H$ is the finest mesh-size, whereas $h_\ell = H$ is the coarsest one. That is, V_{k+1} is a coarse subspace of V_k.

We use the following convention (unless specified otherwise), for a f.e. function y_k, its coefficient vector is $\underline{\mathbf{y}}_k$ (w.r.t. the given basis of V_k).

We introduce next the mass (Gram) matrices G_k associated with the f.e. space V_k and the L_2-bilinear form $(.,\ .)$. The stiffness matrices A_k are computed from the weighted bilinear forms (introduced earlier) and the respective f.e. spaces V_k (specified later on). A main observation then is that the diagonal of A_k is spectrally equivalent to $\lambda_k\ G_k$ where $\lambda_k = 1 + \tau\ h_k^{-2}$. For the particular $H(\text{curl})$ and $H(\text{div})$ examples, see Propositions B.16 and B.18, respectively.

From the definition of stiffness and mass matrices, we have for any $y_k \in V_k$, $\|y_k\|_0^2 = (y_k,\ y_k) = \underline{\mathbf{y}}_k^T G_k \underline{\mathbf{y}}_k$ and $a(y_k,\ y_k) = \underline{\mathbf{y}}_k^T A_k \underline{\mathbf{y}}_k$. Let $y = \sum_{k=0}^{\ell-1} y_k^f + y_\ell$. Similarly, for $k \geq 0$, let $y_k = \sum_{j=k}^{\ell-1} y_k^f + y_\ell \in V_k$. Then, the sum in (i) can be replaced by

$$\sum_k \lambda_k \|y_k^f\|_0^2 \leq C\ a(y,\ y). \tag{F.1}$$

Similarly, sum (ii) can be replaced first by

$$\sum_k \lambda_k^{-1} \|G_k^{-(1/2)} A_k P_k \underline{\mathbf{y}}_{k+1}\|^2 \leq C\ a(y,\ y). \tag{F.2}$$

Introduce now $\underline{\boldsymbol{\psi}}_k = G_k^{-1} A_k \underline{\mathbf{y}}_{k+1}$. Let $\Psi_k \in V_k$ have coefficient vector $\underline{\boldsymbol{\psi}}_k$. We have

$$\underline{\boldsymbol{\psi}}_k^T G_k \underline{\boldsymbol{\psi}}_k = \|\Psi_k\|_0^2.$$

We also have, noting that $P_{k+1}\underline{\mathbf{y}}_{k+1}$ is the coefficient vector of y_{k+1} viewed as an element form V_k (because $V_{k+1} \subset V_k$),

$$\underline{\boldsymbol{\psi}}_k^T G_k \underline{\boldsymbol{\psi}}_k = \underline{\boldsymbol{\psi}}_k^T A_k P_{k+1} \underline{\mathbf{y}}_{k+1} = a(y_{k+1},\ \Psi_k).$$

That is, we have the identity

$$\|\Psi_k\|_0^2 = a(y_{k+1},\ \Psi_k),$$

which used in (F.2) leads to the next equivalent form of (ii)

$$\sum_k \lambda_k^{-1} \|\Psi_k\|_0^2 = \sum_k \lambda_k^{-1} a(y_{k+1}, \Psi_k) \leq C\, a(y, y). \tag{F.3}$$

MG convergence for weighted H^1-forms

We apply the above technique to the parameter-dependent H_0^1-bilinear form

$$a(y, \varphi) = (y, \varphi) + \tau\, (\nabla y, \nabla \varphi), \qquad y, \varphi \in S_h.$$

Here $\tau > 0$ can be a large parameter. We use here the standard H_0^1-conforming finite element spaces $V_k = S_k$ of continuous piecewise linear functions associated with the vertices of the elements from corresponding triangulations $\mathcal{T}_k = \mathcal{T}_{h_k}$ that are obtained by uniform refinement of an initial coarse triangulation $\mathcal{T}_H$.

We use the multilevel decomposition $y = \sum_{k<\ell}(Q_k - Q_{k+1})y + Q_\ell y$ based on the L_2-projections onto the f.e. spaces S_k. We have the major stability estimate (proven in the previous Chapter E)

$$\sum_{k<\ell} h_k^{-2s} \|(Q_k - Q_{k+1})y\|_0^2 + \|Q_\ell y\|_s^2 \leq C\, \|y\|_s^2, \tag{F.4}$$

for $s = 0, 1$. Let $y_k^f = (Q_k - Q_{k+1})\mathbf{y}$ and $y_k = Q_k y$ for $k \geq 0$. Then estimate (F.1) reads

$$\sum_{k<\ell} \lambda_k \|y_k^f\|_0^2 = \sum_k (1 + \tau h_k^{-2}) \|(Q_k - Q_{k+1})y\|_0^2 \leq C\, a(y, y),$$

which follows directly from (F.4) uniformly in $\tau \geq 0$. The next estimate is (F.3), which for some $\Psi_k \in S_k$ reads

$$\sum_k \lambda_k^{-1} \|\Psi_k\|_0^2 = \sum_k \lambda_k^{-1} a(y_{k+1}, \Psi_k) \leq C\, a(y, y).$$

We have, based on the strengthened inverse inequality (cf. Proposition F.1)

$$\begin{aligned}
\sum_k \lambda_k^{-1} \|\Psi_k\|_0^2 &= \sum_k \lambda_k^{-1} a(y_{k+1}, \Psi_k) \\
&= \sum_k \lambda_k^{-1} \sum_{j>k} \left((y_j^f, \Psi_k) + \tau\, (\nabla \Psi_k, \nabla y_j^f)\right) \\
&\leq \sum_k \lambda_k^{-1} \sum_{j>k} \left(\|y_j^f\|_0 \|\Psi_k\|_0 + C\tau\, h_k^{-(1/2)} \|\Psi_k\|_0 h_j^{-(1/2)} \|\nabla y_j^f\|_0\right).
\end{aligned}$$

The sums involving L_2-terms are bounded as follows.

$$\sum_k \lambda_k^{-1} \sum_{j>k} \|y_j^f\|_0 \|\Psi_k\|_0$$

$$\leq C \left(\sum_k \lambda_k^{-1} \|\Psi_k\|_0^2\right)^{1/2} \left(\sum_k \lambda_k^{-1} \left(\|y_\ell\|_0^2 + \sum_{j=k+1}^{\ell-1} h_j^2 \sum_{j=k+1}^{\ell-1} h_j^{-2} \|y_j^f\|_0^2\right)\right)^{1/2}$$

$$\leq C \frac{H}{\sqrt{\tau}} \left(\sum_k \lambda_k^{-1} \|\Psi_k\|_0^2\right)^{1/2} \left(\|y\|_0^2 + H^2\, |y|_1^2\right)^{1/2}. \quad \text{(F.5)}$$

For the remaining part, using $\lambda_k^{-(1/2)} \leq h_k/\sqrt{\tau}$, we have

$$\sum_k \lambda_k^{-1} \sum_{j>k} \tau\, h_k^{-(1/2)} \|\Psi_k\|_0 h_j^{-(1/2)} \|\nabla y_j^f\|_0$$

$$\leq C\sqrt{\tau} \sum_k \lambda_k^{-(1/2)} \|\Psi_k\|_0 \sum_{j>k} \left(\frac{h_k}{h_j}\right)^{(1/2)} \|\nabla y_j^f\|_0$$

$$= C\,\sqrt{\tau} \sum_k \lambda_k^{-(1/2)} \|\Psi_k\|_0 \sum_{j>k} \left(\frac{1}{\sqrt{2}}\right)^{j-k} \|\nabla y_j^f\|_0$$

$$\leq C\,\sqrt{\tau} \left(\sum_k \lambda_k^{-1} \|\Psi_k\|_0^2\right)^{1/2} \left(\sum_j \|\nabla y_j^f\|_0^2\right)^{(1/2)}$$

$$\leq C \left(\sum_k \lambda_k^{-1} \|\Psi_k\|_0^2\right)^{(1/2)} \sqrt{\tau} |y|_1. \quad \text{(F.6)}$$

In the last line, we used (F.4). Combining (F.5) and (F.6) leads to the estimate

$$\sum_k \lambda_k^{-1}\, \|\Psi_k\|_0^2 = \sum_k \lambda_k^{-1}\, a(y_{k+1},\ \Psi_k) \leq C\, \left(1 + \frac{H^2}{\tau}\right)\, a(y, y),$$

which gives the desired result if $H^2 \tau^{-1} = \mathcal{O}(1)$. In general, we can use decomposition with zero components below a level ℓ_τ. Here, ℓ_τ is the maximal coarse mesh h_{ℓ_τ} for which $h_{\ell_\tau}^2 \tau^{-1} \leq \text{const} < h_{\ell_\tau+1}^2 \tau^{-1}$. We can choose $y_{\ell_\tau} = Q_{\ell_\tau} y$ and at coarse levels $k > \ell_\tau$, we can let $y_k^f = 0$ and hence $y_k = \sum_{j \geq k} y_j^f = 0$.

Note that at level ℓ_τ the corresponding stiffness matrix A_{ℓ_τ} is well conditioned w.r.t. its diagonal (because then $h_{\ell_\tau}^{-2} \tau \simeq \mathcal{O}(1)$). Then, the above estimates still hold. The coarse component $Q_\ell y$ (or $Q_{\ell_\tau} y$) is energy stable, thus we can finally state the following corollary to Theorem 5.7.

Corollary F.2. *The symmetric* $V(1, 1)$*-cycle MG based on Gauss–Seidel smoothing applied to the stiffness matrices* A_k *computed from the weighted bilinear form* $a(y,\ \varphi) = (y,\ \varphi) + \tau\ (\nabla y,\ \nabla \varphi)$ *and standard piecewise linear* H^1*-conforming* *f.e. spaces S_k corresponding to triangulations $\mathcal{T}_k$ obtained by uniform refinement of an initial coarse triangulation $\mathcal{T}_H$, has a convergence factor bounded independently of both the number of refinement steps ℓ as well as the parameter $\tau > 0$. The MG cycle can be stopped (but does not have to) at level ℓ for which $H = h_\ell$ satisfies $H^2\tau^{-1} = \mathcal{O}(1)$.*

Conditions for weighted H(curl) and H(div) forms

For the applications involving $H(curl)$ and $H(div)$, the MG method will exploit intermediate subspaces $V_{k+(1/2)}$(related to the null spaces of the curl and div operators). In other words, for two other f.e. spaces S_k and $\mathbf{N}_k$ (specified later on) we have, either $V_{k+(1/2)} = \nabla S_k \subset V_k$ or $V_{k+(1/2)} = \text{curl } \mathbf{N}_k \subset V_k$. The resulting MG method will exploit (explicitly) the matrices $P_{k+(1/2)}$ that transform the coefficient vector of a f.e. function ϕ, either from S_k or from $\mathbf{N}_k$, to the coefficient vector, of either $\nabla\phi$ or curl ϕ, considered as an element of the respective f.e. space V_k. The stiffness matrix $A_{k+(1/2)}$ then corresponds to the subspaces S_k or $\mathbf{N}_k$ and the bilinear form $(\nabla\cdot, \nabla\cdot)$ or (curl $\cdot$, curl $\cdot$), respectively. The remaining part of the bilinear form $a(.,.)$ vanishes because ∇S_k or curl $\mathbf{N}_k$ represent the null space components of that part of the bilinear form $a(\cdot,\cdot)$. Therefore, we have $A_{k+(1/2)} = P^T_{k+(1/2)} A_k P_{k+(1/2)} = P^T_{k+(1/2)} G_k P_{k+(1/2)}$. We use the (forward) Gauss–Seidel smoother $M_{k+(1/2)}$ coming from $A_{k+(1/2)}$. Its symmetrized version $\overline{M}_{k+(1/2)}$ is spectrally equivalent to the diagonal of the mass matrices $G_{k+(1/2)}$ computed from the f.e. spaces S_k or $\mathbf{N}_k$ (and the L_2-bilinear form).

We are now in a position to define the MG algorithm of interest that exploits the intermediate (fractional order) spaces $V_{k+(1/2)}$ representing the null space components of the curl or div-components respectively.

Algorithm F.1.1 (A MG algorithm for H(curl) or H(div) problems). *Let $A = A_0$ and $\underline{\mathbf{f}}$ be given. Consider the fine-grid discrete problem*

$$A\underline{\mathbf{u}} = \underline{\mathbf{f}}.$$

For a current iterate $\underline{\mathbf{u}}$, compute the corresponding residual $\underline{\mathbf{d}} = \underline{\mathbf{f}} - A\underline{\mathbf{u}}$, and in order to find a MG correction $B^{-1}_{MG}\underline{\mathbf{d}}$, letting $\underline{\mathbf{r}}_0 = \underline{\mathbf{d}}$ and $\underline{\mathbf{y}}_0 = 0$, perform the following steps starting with $j = 0$.

1. *Presmooth with $M_{j+(1/2)}$; that is, solve*

$$M_{j+(1/2)}\mathbf{x}_{j+(1/2)} = P^T_{j+(1/2)}\underline{\mathbf{r}}_j.$$

2. *Update current level iterate $\underline{\mathbf{y}}_j := \underline{\mathbf{y}}_j + P_{j+(1/2)}\mathbf{x}_{j+(1/2)}$ and compute the next intermediate residual*

$$\underline{\mathbf{r}}_j := \underline{\mathbf{r}}_j - A_j P_{j+(1/2)}\mathbf{x}_{j+(1/2)}.$$

3. *Presmooth with M_j; that is, solve*

$$M_j\underline{\mathbf{y}}_j = \underline{\mathbf{r}}_j.$$

4. *Update the current level residual*

$$\underline{\mathbf{r}}_j := \underline{\mathbf{r}}_j - A_j \underline{\mathbf{y}}_j.$$

5. *Restrict the residual to the coarse level* $j+1$*; that is, compute* $\underline{\mathbf{r}}_{j+1} = P_j^T \underline{\mathbf{r}}_j$.
6. *If* $j+1 = \ell$ *solve the coarse problem* $A_\ell \underline{\mathbf{y}}_\ell = \mathbf{r}_\ell$ *and after letting* $j := j-1$ *go to Step (7). Otherwise, with* $j := j+1$ *go to Step (1).*
7. *Interpolate coarse-grid approximation; that is, compute* $\underline{\mathbf{y}}_j := \underline{\mathbf{y}}_j + P_j \underline{\mathbf{y}}_{j+1}$ *and update the residual,* $\underline{\mathbf{r}}_j := \underline{\mathbf{r}}_j - A_j P_j \underline{\mathbf{y}}_{j+1}$.
8. *Postsmooth with* M_j^T*; that is, solve*

$$M_j^T \underline{\mathbf{x}}_j = \underline{\mathbf{r}}_j.$$

9. *Update the current approximation* $\underline{\mathbf{y}}_j := \underline{\mathbf{y}}_j + \underline{\mathbf{x}}_j$ *and the corresponding residual* $\underline{\mathbf{r}}_j := \underline{\mathbf{r}}_j - A_j \underline{\mathbf{x}}_j$.
10. *Postsmooth with* $M_{j+(1/2)}^T$*; that is, solve*

$$M_{j+(1/2)}^T \mathbf{x}_{j+(1/2)} = P_{j+(1/2)}^T \underline{\mathbf{r}}_j.$$

11. *Update the current approximation* $\underline{\mathbf{y}}_j := \underline{\mathbf{y}}_j + P_{j+(1/2)} \mathbf{x}_{j+(1/2)}$ *and compute the corresponding intermediate residual* $\underline{\mathbf{r}}_j := \underline{\mathbf{r}}_j - A_j P_{j+(1/2)} \mathbf{x}_{j+(1/2)}$. *If* $j > 0$ *set* $j := j-1$ *and go to Step (7).*
12. *The next MG iterate equals* $\underline{\mathbf{u}} := \underline{\mathbf{u}} + \underline{\mathbf{y}}_0$.

The output of the above algorithm defines a mapping $B_{MG}^{-1} : \mathbf{d} \mapsto \mathbf{y}_0$ *which is the (inverse) of the MG preconditioner* B_{MG}*; that is, we have* $B_{MG}^{-1}\underline{\mathbf{d}} = \underline{\mathbf{y}}_0$.

The above algorithm can be viewed as a combination of recursive "two-level" and "two-grid" preconditioners (see Definitions 3.12 and 3.13). More specifically, we can define B_k from

$$B_k^{-1} = [P_{k+(1/2)},\ I]\overline{B}_k^{-1}[P_{k+(1/2)},\ I]^T,$$

where

$$\overline{B}_k = \begin{bmatrix} M_{k+(1/2)} & 0 \\ A_k P_{k+(1/2)} & I \end{bmatrix} \begin{bmatrix} \left(M_{k+(1/2)}^T + M_{k+(1/2)} - A_{k+(1/2)}\right)^{-1} & 0 \\ 0 & B_{k+(1/2)} \end{bmatrix} \times \begin{bmatrix} M_{k+(1/2)}^T & P_{k+(1/2)}^T A_k \\ 0 & I \end{bmatrix}.$$

Whereas the fractional step preconditioner $B_{k+(1/2)}$ is defined as a two-grid preconditioner (with inexact coarse solution corresponding to B_{k+1}) from

$$B_{k+(1/2)}^{-1} = \left[I,\ P_k\right]\overline{B}_{k+(1/2)}^{-1}\left[I,\ P_k\right]^T,$$

whereas $\overline{B}_{k+(1/2)}$ is defined as

$$\overline{B}_{k+(1/2)} = \begin{bmatrix} M_k & 0 \\ P_k^T A_k & I \end{bmatrix} \begin{bmatrix} \left(M_k^T + M_k - A_k\right)^{-1} & 0 \\ 0 & B_{k+1} \end{bmatrix} \begin{bmatrix} M_k^T & A_k P_k \\ 0 & I \end{bmatrix}.$$

To analyze the above MG algorithm we use Theorem 5.7 modified accordingly to take into account the intermediate (fractional order) preconditioners $B_{k+(1/2)}$. For any decomposition of $\underline{\mathbf{y}}$, defined recursively by first letting $\underline{\mathbf{y}}_0 = \underline{\mathbf{y}}$, and then for $k = 0, \ldots, \ell - 1$,

$$\begin{aligned} \underline{\mathbf{y}}_k &= P_{k+(1/2)} \underline{\mathbf{x}}_k^{(f)} + \underline{\mathbf{y}}_{k+(1/2)}, \text{ and} \\ \underline{\mathbf{y}}_{k+(1/2)} &= \underline{\mathbf{u}}_k^{(f)} + P_k \underline{\mathbf{y}}_{k+1}, \end{aligned} \tag{F.7}$$

we have first the following inequalities (apply Theorem 3.15 for $\mathcal{J} = P_{k+(1/2)}$, $P = I$ and $\mathcal{D} = B_{k+(1/2)}$)

$$\begin{aligned} \underline{\mathbf{y}}_k^T B_k \underline{\mathbf{y}}_k \le \underline{\mathbf{y}}_{k+(1/2)}^T & B_{k+(1/2)} \underline{\mathbf{y}}_{k+(1/2)} \\ &+ \left(M_{k+(1/2)}^T \underline{\mathbf{x}}_k^{(f)} + P_{k+(1/2)}^T A_k \underline{\mathbf{y}}_{k+(1/2)}\right)^T \\ &\times \left(M_{k+(1/2)}^T + M_{k+(1/2)} - A_{k+(1/2)}\right)^{-1} \\ &\times \left(M_{k+(1/2)}^T \underline{\mathbf{x}}_k^{(f)} + P_{k+(1/2)}^T A_k \underline{\mathbf{y}}_{k+(1/2)}\right). \end{aligned}$$

We also have (apply now Theorem 3.15 for $\mathcal{J} = I$, $P = P_k$, and $\mathcal{D} = B_{k+1}$),

$$\begin{aligned} \underline{\mathbf{y}}_{k+(1/2)}^T B_{k+(1/2)} \underline{\mathbf{y}}_{k+(1/2)} \le \underline{\mathbf{y}}_{k+1}^T B_{k+1} \underline{\mathbf{y}}_{k+1} &+ \left(M_k^T \underline{\mathbf{u}}_k^{(f)} + A_k P_k \underline{\mathbf{y}}_{k+1}\right) \\ &\times \left(M_k^T + M_k - A_k\right)^{-1} \left(M_k^T \underline{\mathbf{u}}_k^{(f)} + A_k P_k \underline{\mathbf{y}}_{k+1}\right). \end{aligned}$$

It is clear then, by applying recursion and the Cauchy–Schwarz inequality in the same way as in the proof of Theorem 5.7, that if we can bound the following sums, for a particular multilevel decomposition (F.7) of $\underline{\mathbf{y}}$,

(A)

$$\sum_k \underline{\mathbf{x}}_k^{(f)^T} \overline{M}_{k+(1/2)} \underline{\mathbf{x}}_k^{(f)},$$

(B)

$$\sum_k \|(M_{k+(1/2)}^T + M_{k+(1/2)} - A_{k+(1/2)})^{-(1/2)} P_{k+(1/2)}^T A_k \underline{\mathbf{y}}_{k+(1/2)}\|^2$$

(C)

$$\sum_k \underline{\mathbf{u}}_k^{(f)^T} \overline{M}_k \underline{\mathbf{u}}_k^{(f)},$$

(D)

$$\sum_k \|(M_k^T + M_k - A_k)^{-(1/2)} A_k P_k \underline{\mathbf{y}}_{k+1}\|^2,$$

all in terms of $\underline{\mathbf{y}}^T A \underline{\mathbf{y}}$, and if the coarse component $\underline{\mathbf{y}}_\ell$ is stable in energy, that is, (E)

$$\underline{\mathbf{y}}_\ell^T A_\ell \underline{\mathbf{y}}_\ell \leq C\, \underline{\mathbf{y}}^T A \underline{\mathbf{y}},$$

we then have a uniform upper bound for $B = B_0$ in terms of $A = A_0$.

For this we need to come up with a decomposition for any f.e. function $y \in V_h$ with some properties that lead to respective multilevel vector decomposition of the coefficient vector $\underline{\mathbf{y}}$ of y. In what follows, we list the conditions (A)–(E) on the multilevel decomposition of interest in terms of finite element functions.

The multilevel decomposition of any f.e. function $y \in V$ starts with $y_0 = y = y^{(\text{null})} + u$, where $y^{(\text{null})} = y'_{1/2}$ belongs either to ∇S_h or to curl $\mathbf{N}_h$. For $k \geq 0$, we have

$$y_k = y'_{k+(1/2)} + u_k.$$

Here $y'_{k+(1/2)}$ belongs either to ∇S_k or to curl $\mathbf{N}_k$; that is, there is a Φ_k in either S_k or $\mathbf{N}_k$ such that $y'_{k+(1/2)} = D\Phi_k$ where either $D = \nabla$ or $D = \text{curl}$. Each of the components $y'_{k+(1/2)}$ and u_k has an "f" part (to be handled by the respective smoother) and a coarse component. More specifically, we have

$$\begin{aligned} y'_{k+(1/2)} &= y'^{(f)}_{k+(1/2)} + y'_{k+(3/2)}, \\ u_k &= u_k^{(f)} + u_{k+1}. \end{aligned} \tag{F.8}$$

In terms of coefficient vectors, we have

$$P_{k+(1/2)}\underline{\mathbf{x}}_k = P_{k+(1/2)}\underline{\mathbf{x}}_k^{(f)} + P_k P_{k+(3/2)}\underline{\mathbf{x}}_{k+1}.$$

Here, $P_{k+(1/2)}\underline{\mathbf{x}}_k$ is the coefficient vector of $y'_{k+(1/2)} = D\Phi_k \in V_{k+(1/2)} \subset V_k$. That is, $\underline{\mathbf{x}}_k$ is the coefficient vector of Φ_k (as an element of either S_k or $\mathbf{N}_k$). Similarly, $P_{k+(1/2)}\underline{\mathbf{x}}_k^{(f)}$ is the coefficient vector of $y^{(f)}_{k+(1/2)} \in V_{k+(1/2)} \subset V_k$. Because $y^{(f)}_{k+(1/2)} = D\psi_k^{(f)}$ for some $\psi_k^{(f)}$ in either S_k (with $D = \nabla$ then) or in $\mathbf{N}_k$ (with $D = \text{curl}$), we have that $\underline{\mathbf{x}}_k^{(f)}$ is the coefficient vector of $\psi_k^{(f)}$ (in the respective space, S_k or $\mathbf{N}_k$). We also have,

$$\underline{\mathbf{u}}_k = \underline{\mathbf{u}}_k^{(f)} + P_k \underline{\mathbf{u}}_{k+1}.$$

The overall recursive two-level vector decomposition reads

$$\underline{\mathbf{y}}_k = \underline{\mathbf{y}}_k^{(f)} + P_k \underline{\mathbf{y}}_{k+1},$$

where

$$\begin{aligned} \underline{\mathbf{y}}_k^{(f)} &= P_{k+(1/2)}\underline{\mathbf{x}}_k^{(f)} + \underline{\mathbf{u}}_k^{(f)}, \\ \underline{\mathbf{y}}_{k+1} &= P_{k+(3/2)}\underline{\mathbf{x}}_{k+1} + \underline{\mathbf{u}}_{k+1}. \end{aligned}$$

Equivalently, we can let $\underline{\mathbf{y}}_k = P_{k+(1/2)}\underline{\mathbf{x}}_k^{(f)} + \underline{\mathbf{y}}_{k+(1/2)}$ and $\underline{\mathbf{y}}_{k+(1/2)} = \underline{\mathbf{u}}_k^{(f)} + P_k\underline{\mathbf{y}}_{k+1}$ thus ending up with the multilevel vector decomposition (F.7)

For any $\underline{y}_0 = \underline{\mathbf{y}}$ of a f.e. function $y \in V_h$ and its particular recursive multilevel decomposition (F.8), based on conditions (A)–(E), it is clear that we need to prove the following estimates,

(a) Let $\underline{\mathbf{x}}_k^{(f)}$ be the coefficient vector of a f.e. function $\psi_k^{(f)}$ from either S_k or $\mathbf{N}_k$ such that $y_{k+(1/2)}^{(f)} = D\psi_k^{(f)} \in V_{k+(1/2)} \subset V_k$ is from the decomposition (F.8). That is, $y_{k+(1/2)}^{(f)}$ equals either $\nabla\psi_k^{(f)}$ or curl $\psi_k^{(f)}$. Then, for the chosen particular decomposition (F.8) of any given $y \in V_h$, using the fact that the symmetrized smoother $\overline{M}_{k+(1/2)}$ is spectrally equivalent to the scaled mass matrix $h_k^{-2}G_{k+(1/2)}$, we need to prove the estimate (see (A))

$$\sum_k \left(\underline{\mathbf{x}}_k^{(f)}\right)^T \overline{M}_{k+(1/2)}\underline{\mathbf{x}}_k^{(f)} \simeq \sum_k h_k^{-2}\left(\underline{\mathbf{x}}_k^{(f)}\right)^T G_{k+(1/2)}\underline{\mathbf{x}}_k^{(f)}$$
$$= \sum_k h_k^{-2}\|\psi_k^{(f)}\|_0^2 \le C\, a(y,\ y).$$

That is, we have to prove the estimate

$$\sum_k h_k^{-2}\|\psi_k^{(f)}\|_0^2 \le C\, a(y,\ y) \tag{F.9}$$

for $\psi_k^{(f)}$ (from either S_k or $\mathbf{N}_k$) defining $y_{k+(1/2)}^{(f)} = D\psi_k^{(f)}$ ($D = \nabla$ or $D =$ curl) and $y_{k+(1/2)}^{(f)}$ are from the chosen particular decomposition (F.8).

(b) Let $\underline{\boldsymbol{\psi}}_k = G_{k+(1/2)}^{-1}P_{k+(1/2)}^T G_k\underline{\mathbf{y}}_{k+(1/2)}$ where $\underline{\mathbf{y}}_{k+(1/2)} = \underline{\mathbf{u}}_k^{(f)} + P_k\underline{\mathbf{y}}_{k+1}$. Let the f.e. function ψ_k from either $\mathbf{N}_k$ or S_k have $\underline{\boldsymbol{\psi}}_k$ as the coefficient vector. Using the fact that the diagonal of $A_{k+(1/2)}$ (which equals $M_{k+(1/2)} + M_{k+(1/2)}^T - A_{k+(1/2)}$ for the Gauss–Seidel smoother $M_{k+(1/2)}$ for $A_{k+(1/2)}$) is spectrally equivalent to $h_k^{-2}G_{k+(1/2)}$, the sum (B) takes the equivalent form,

$$\sum_k \left\|\left(M_{k+(1/2)} + M_{k+(1/2)}^T - A_{k+(1/2)}\right)^{-(1/2)} P_{k+(1/2)}^T G_k\underline{\mathbf{y}}_{k+(1/2)}\right\|^2$$
$$\simeq \sum_k h_k^2 \left\|G_{k+(1/2)}^{-(1/2)} P_{k+(1/2)}^T G_k\underline{\mathbf{y}}_{k+(1/2)}\right\|^2$$
$$= \sum_k h_k^2\underline{\boldsymbol{\psi}}_k^T P_{k+(1/2)}^T G_k\underline{\mathbf{y}}_{k+(1/2)}$$
$$= \sum_k h_k^2\underline{\boldsymbol{\psi}}_k^T P_{k+(1/2)}^T G_k\left(\underline{\mathbf{u}}_k^{(f)} + P_k\underline{\mathbf{y}}_{k+1}\right)$$
$$= \sum_k h_k^2\|\psi_k\|_0^2$$
$$= \sum_k h_k^2\left(D\psi_k,\ y_{k+1} + u_k^{(f)}\right).$$

Here $D = \text{curl}$ or $D = \nabla$. That is, we need to prove the following estimate,

$$\sum_k h_k^2 \|\psi_k\|_0^2 = \sum_k h_k^2 \,(D\psi_k,\; u_k^{(f)} + y_{k+1})$$
$$= \sum_k h_k^2 \,(D\psi_k,\; y_{k+(1/2)}) \leq C\, a(y, y). \tag{F.10}$$

(c) Condition (C) leads to the following estimate, using the fact that $\overline{M}_k$ is spectrally equivalent to $\lambda_k\, G_k$,

$$\sum_k \lambda_k \,\|u_k^{(f)}\|_0^2 \leq C\, a(y, y). \tag{F.11}$$

Recall, that $\lambda_k = 1 + \tau h_k^{-2}$.

(d) We also need the estimate (ii) in the form (F.3) (see (D)) for the coarse components $y_{k+1} = u_{k+1} + y'_{k+(3/2)} = u_{k+1} + D\Phi_{k+1}$; that is,

$$\sum_k \lambda_k^{-1} \,\|\Psi_k\|_0^2 = \sum_k \lambda_k^{-1}\, a(y_{k+1},\, \Psi_k)$$
$$= \sum_k \lambda_k^{-1}\, (a(u_{k+1},\, \Psi_k) + (D\Phi_{k+1},\, \Psi_k)) \leq C\, a(y,\; y), \tag{F.12}$$

where Ψ_k is the f.e. function from V_k that has coefficient vector $\underline{\psi}_k = G_k^{-1} A_k P_k \underline{\mathbf{y}}_{k+1}$.

(e) Finally, we need for the chosen particular decomposition (F.8) of y to have the coarsest component $y_\ell = u_\ell + D\Phi_\ell$ be bounded in energy in terms of y (see (E)); that is, it is sufficient to have

$$a(u_\ell,\; u_\ell) + \|D\Phi_\ell\|_0^2 \leq C\, a(y,\; y). \tag{F.13}$$

F.2 A MG for weighted $H(\text{curl})$ space

Given a polyhedral domain $\Omega \subset \mathbb{R}^3$, consider the space of vector functions $\mathbf{u}$ that have in the $L_2(\Omega)$-sense a well-defined curl $\mathbf{u}$. That is, $\|\text{curl}\,\mathbf{u}\|_{0,\,\Omega}^2 = (\text{curl}\,\mathbf{u},\;\text{curl}\,\mathbf{u})$ is well defined in addition to $\|\mathbf{u}\|_{0,\,\Omega}^2$. This space is denoted by $H(\text{curl},\;\Omega)$ and if $\mathbf{u} \times \mathbf{n} = 0$ on $\partial\Omega$ by $H_0(\text{curl},\;\Omega)$. As used many times, $\mathbf{n}$ stands for a unit vector normal to $\partial\Omega$.

For a given parameter $\tau > 0$, we are interested in the weighted $H(\text{curl})$ bilinear form

$$a(\mathbf{u},\; \mathbf{w}) = (\mathbf{u},\; \mathbf{w}) + \tau\; (\text{curl}\,\mathbf{u},\; \text{curl}\,\mathbf{w}), \qquad \mathbf{u},\; \mathbf{v} \in H_0(\text{curl},\; \Omega). \tag{F.14}$$

For the purpose of defining the MG method of interest, we restrict the above bilinear form to f.e. spaces $\mathbf{N}_h \subset H_0(\text{curl},\; \Omega)$. That is, we assume that Ω is triangulated by a triangulation $\mathcal{T}_h$ of tetrahedral elements obtained by $\ell \geq 1$ successive steps of refinement of an initial coarse triangulation $\mathcal{T}_H$. Then, $a(\cdot, \cdot)$ gives rise to a sequence of stiffness matrices $\mathbf{A}_k$ computed from the sequence of nested Nédélec spaces $\mathbf{N}_k$, $0 \leq k \leq \ell$. Let $\mathbf{N}_0 = \mathbf{N}_h$ be the Nédélec space associated with the fine-mesh $\mathcal{T}_0 = \mathcal{T}_h$ and $h_k \simeq h2^k$, $k = 0, 1, \ldots, \ell$ be the kth-level mesh-size. The coarsest-mesh is $h_\ell = H \simeq 2^\ell h$.

For the multilevel analysis to follow, we concentrate on the lowest-order Nédélec spaces, however, the approach is general and can be extended to higher-order Nédélec elements as well.

The MG method of interest exploits also the finite element spaces $\mathbf{N}_{k+(1/2)} \equiv \nabla S_k \subset \mathbf{N}_k$ which represent the null space of the curl-operator restricted to $\mathbf{N}_k$. Let $\{\varphi_i^{(k)}\}$ be the standard nodal basis of S_k. Here, i runs over the set of vertices $\mathcal{V}_k$ of the elements from $\mathcal{T}_k$. Because we consider essential boundary conditions, $\mathcal{V}_k$ stands for the set of interior vertices (w.r.t. to the polyhedral domain Ω). The mapping that transfers a function $\varphi \in S_k$ expanded in terms of the standard nodal basis $\{\varphi_i^{(k)}\}$ of S_k into the vector f.e. function $\nabla\varphi \in \mathbf{N}_k$ expanded in terms of the edge basis $\{\mathbf{\Phi}_e^{(k)}\}$ of $\mathbf{N}_k$ is denoted by $P_{k+(1/2)}$. It is commonly referred to as a discrete gradient mapping and is actually needed (explicitly) in the MG algorithm below.

For any $h = h_k$, we need the natural interpolant $\mathbf{\Pi}_h$, which for the lowest-order Nédélec space is defined as follows. Let $\{\mathbf{\Phi}_e\}$ be the edge-based basis of $\mathbf{N}_h$. Here e runs over the set of edges $\mathcal{E}_h$ of the elements from $\mathcal{N}_h$. Because we consider essential boundary conditions, $\mathcal{E}_h$ stands here for the set of interior edges (w.r.t. to the polyhedral domain Ω). Then, for any piecewise polynomial function $\mathbf{z}$ (i.e., $\mathbf{z}$ restricted to the elements of $\mathcal{T}_h$ is polynomial, and $\mathbf{z}$ has uniquely defined tangential components $\mathbf{z} \cdot \tau_e$ over the edges $e \in \mathcal{E}_h$), we define

$$\mathbf{\Pi}_h \mathbf{z} = \sum_e \left(\int_e \mathbf{z} \cdot \tau_e \, ds \right) \mathbf{\Phi}_e \in \mathbf{N}_h.$$

It is clear that $\mathbf{\Pi}_h$ is well defined for any sufficiently smooth vector functions $\mathbf{z}$. For higher-order Nédélec spaces, in addition to edge integrals the respective $\mathbf{\Pi}_h$ involves face integrals and even volume integrals over the elements for sufficiently high-order Nédélec spaces. Details are found in Section 5.5 of [Mo03]. We let $\mathbf{\Pi}_k = \mathbf{\Pi}_{h_k}$ and $\mathcal{E}_k = \mathcal{E}_{h_k}$.

Also, let P_k be the interpolation mapping that implements the natural embedding $\mathbf{N}_{k+1} \mapsto \mathbf{N}_k$. It transfers the coefficient vector $\underline{\mathbf{v}}_c = (v_e^c)_{e \in \mathcal{E}_{k+1}}$ of the f.e. function $\mathbf{v} \in \mathbf{N}_{k+1}$ expanded in terms of the edge basis $\{\mathbf{\Phi}_e^{(k+1)}\}_{e \in \mathcal{E}_{k+1}}$ of $\mathbf{N}_{k+1}$, that is, $\mathbf{v} = \sum_{e \in \mathcal{E}_{k+1}} v_e^c \mathbf{\Phi}_e^{(k+1)}$, to the coefficient vector $\underline{\mathbf{v}} = P_k \underline{\mathbf{v}}_c = (v_e)_{e \in \mathcal{E}_k}$ where $\mathbf{v} = \sum_{e \in \mathcal{E}_k} v_e \mathbf{\Phi}_e^{(k)}$. That is, in the second expansion, $\mathbf{v}$ is viewed as an element from $\mathbf{N}_k$ (which contains $\mathbf{N}_{k+1}$).

Introduce the mass (Gram) matrices $\mathbf{G}_k$ computed from the L_2-form $(\cdot, \cdot)$ using the edge basis $\{\mathbf{\Phi}_e^{(k)}\}$ of $\mathbf{N}_k$. Finally, let $\mathbf{A}_{k+(1/2)}$ be the stiffness matrix computed from $a(\cdot, \cdot)$ using the set $\{\nabla\varphi_i^{(k)}\}_{i \in \mathcal{V}_k}$. Because the curl term vanishes on gradients, we actually have that $\mathbf{A}_{k+(1/2)} = P_{k+(1/2)}^T \mathbf{G}_k P_{k+(1/2)}$ which is nothing but the stiffness matrix computed from the Laplace bilinear form $(\nabla\cdot,\ \nabla\cdot)$ from the nodal basis $\{\varphi_i^{(k)}\}_{i\in\mathcal{V}_k}$ of S_k. Let $M_{k+(1/2)}$ be a given smoother for $\mathbf{A}_{k+(1/2)}$, for example, the forward Gauss–Seidel. Similarly, let M_k be also a given smoother for $\mathbf{A}_k$, for example, the forward Gauss–Seidel. Then, we are in a position to define the MG method of interest implemented as in Algorithm F.1.1.

F.2.1 A multilevel decomposition of weighted Nédélec spaces

Let $\{\mathbf{N}_k\}_{k=0}^{\ell}$ be a sequence of nested Nédélec spaces associated with uniformly refined tetrahedral triangulations $\mathcal{T}_k$ of a 3D polytope Ω. Here, $\mathcal{T}_0 = \mathcal{T}_h$ is the finest-mesh whereas $\mathcal{T}_\ell = \mathcal{T}_H$ is the coarsest-mesh. Associated with $\mathcal{T}_k$ consider H^1-conforming Lagrangian finite element spaces $\mathbf{S}_k$, a vector one, and S_k, a scalar one. To be specific, we assume that the vector functions from $\mathbf{N}_k$ have vanishing tangential components on $\partial\Omega$. Similarly, we assume that vector functions from $\mathbf{S}_h$ and the scalar functions from S_h vanish on $\partial\Omega$.

The starting point of our multilevel decomposition is the HX decomposition (see [HX06], or if no essential boundary conditions were imposed, see [Sch05]). It states, that for any $\mathbf{u} \in \mathbf{N}_h$ there is $\mathbf{v} \in \mathbf{N}_h$, $\mathbf{z} \in \mathbf{S}_h$, and a $\varphi \in S_h$ such that

$$\mathbf{u} = \mathbf{v} + \mathbf{\Pi}_h \mathbf{z} + \nabla\varphi,$$

with all components being stable in the following sense,

$$h^{-1}\ \|\mathbf{v}\|_0 \leq C\ \|\text{curl}\ \mathbf{u}\|_0,$$

$$\|\mathbf{z}\|_0 \leq C\ \|\mathbf{u}\|_0, \qquad |\mathbf{z}|_1 \leq C\|\text{curl}\ \mathbf{u}\|_0,$$

and

$$\|\nabla\varphi\|_0 \leq C\ \|\mathbf{u}\|_0.$$

Because we established stable multilevel decompositions for $\mathbf{z} \in \mathbf{S}_h$, $\mathbf{z} = \sum_{k=0}^{\ell}(\mathbf{z}_k - \mathbf{z}_{k+1})$ and $\varphi \in S_h$, $\varphi = \sum_{k=0}^{\ell}\psi_k^{(f)}$, we can consider the following one for $\mathbf{u}$.

Definition F.3 (Multilevel decomposition of Nédélec spaces). *For any* $\mathbf{u} \in \mathbf{N}_h$ *define*

$$\mathbf{u} = \sum_{k=0}^{\ell-1}\mathbf{u}_k^{(f)} + \sum_{k=0}^{\ell-1}\nabla\psi_k^{(f)} + \mathbf{u}_\ell,$$

where, letting $\mathbf{\Pi}_k = \mathbf{\Pi}_{h_k}$,

$$\begin{aligned} \mathbf{u}_0^{(f)} &= \mathbf{v} + \mathbf{\Pi}_h\mathbf{z} - \mathbf{\Pi}_1\mathbf{z}_1, && \textit{for } k = 0,\\ \mathbf{u}_k^{(f)} &= \mathbf{\Pi}_k\mathbf{z}_k - \mathbf{\Pi}_{k+1}\mathbf{z}_{k+1}, && \textit{for } 0 < k < \ell, \end{aligned}$$

and $\mathbf{u}_\ell = \mathbf{\Pi}_\ell\mathbf{z}_\ell + \nabla\psi_\ell$. *To be specific, we let* $\psi_k^{(f)} = (Q_k - Q_{k+1})\varphi$, *for* $0 \leq k < \ell$ *and* $\psi_\ell^{(f)} = Q_\ell\varphi$. *Similarly, we let* $\mathbf{z}_k = \mathbf{Q}_k\mathbf{z}$. *Here,* $Q_k : S_h \mapsto S_k$ *and* $\mathbf{Q}_k : \mathbf{S}_h \mapsto \mathbf{S}_k$ *are the corresponding scalar and vector* L_2*-projections.*

The components $\mathbf{z}_k$ satisfy the stability estimates,

$$\sum_k |\mathbf{z}_k - \mathbf{z}_{k+1}|_1^2 \leq C \sum_k h_k^{-2} \|(\mathbf{Q}_k - \mathbf{Q}_{k+1})\mathbf{z}\|_0^2 \leq C|\mathbf{z}|_1^2, \tag{F.15}$$

and also

$$\sum_k \|\mathbf{z}_k - \mathbf{z}_{k+1}\|_0^2 \leq \|\mathbf{z}\|_0^2.$$

For the terms involving $\psi_k^{(f)}$, we have

$$\sum_k |\psi_k^{(f)}|_1^2 \leq C\ |\varphi|_1^2 \ \text{ and } \ \|\psi_k^{(f)}\|_0 \leq Ch_k\ |\psi_k^{(f)}|_1, \quad \text{for } k < \ell.$$

To estimate the terms involving $(\mathbf{\Pi}_k\mathbf{Q}_k - \mathbf{\Pi}_{k+1}\mathbf{Q}_{k+1})\mathbf{z}$, we observe that

$$\begin{aligned}(\mathbf{\Pi}_k\mathbf{Q}_k &- \mathbf{\Pi}_{k+1}\mathbf{Q}_{k+1})\mathbf{z} \\ &= (\mathbf{\Pi}_k\mathbf{Q}_k - \mathbf{Q}_k)\mathbf{z} + (\mathbf{Q}_k - \mathbf{Q}_{k+1})\mathbf{z} + (\mathbf{Q}_{k+1} - \mathbf{\Pi}_{k+1}\mathbf{Q}_{k+1})\mathbf{z}.\end{aligned} \tag{F.16}$$

Because the middle terms give rise to a stable decomposition in L_2 and their curls also give rise to a stable decomposition in L_2 which is seen due to an inverse inequality, we have

$$\sum_k \|\mathrm{curl}(\mathbf{Q}_k - \mathbf{Q}_{k+1})\mathbf{z}\|_0^2 \leq C \sum_k h_k^{-2} \|(\mathbf{Q}_k - \mathbf{Q}_{k+1})\mathbf{z}\|_0^2 \leq C\ |\mathbf{z}|_1^2, \tag{F.17}$$

it is sufficient to estimate the terms involving the deviation $(\mathbf{\Pi}_k\mathbf{Q}_k - \mathbf{Q}_k)\mathbf{z}$ and $\mathrm{curl}(\mathbf{\Pi}_k\mathbf{Q}_k - \mathbf{Q}_k)\mathbf{z}$.

The following result can be proved similarly to the major estimate (C.12), for $0 \leq s < 3/2$.

Lemma F.4. *We have, for any $s \in [0, 3/2)$, the multilevel deviation norm estimate*

$$\sum_{k=0}^{\ell-1} h_k^{-2s} \|(\mathbf{\Pi}_k - \mathbf{Q}_k)\mathbf{Q}_k\mathbf{z}\|_0^2 \leq C\ |\mathbf{z}|_s^2.$$

Proof. Here we need the L_2-approximation property of $\mathbf{\Pi}_k$. Notice that $\|(\mathbf{\Pi}_k\mathbf{Q}_k - \mathbf{Q}_k)\mathbf{z}\|_0 = \|(\mathbf{\Pi}_k - I)\mathbf{Q}_k\mathbf{z}\|_0 \leq Ch_k^\sigma\ \|\mathbf{Q}_k\mathbf{z}\|_\sigma$ for any $\sigma \in [0, 3/2)$. Then the proof proceeds in the same way as for (C.12), □

Remark F.5. For the lowest-order spaces that we focus on, we have

$$\mathrm{curl}(\mathbf{\Pi}_k\mathbf{z}_k) = \mathrm{curl}\ \mathbf{z}_k \quad \text{for all } \mathbf{z}_k \in \mathbf{S}_k.$$

For higher-order spaces, the decomposition based on the curl $(\mathbf{\Pi}_k - \mathbf{Q}_k)\mathbf{Q}_k\mathbf{z}$ can be estimated by a modification of estimate (C.12).

Lemma F.6. *For any $s \in [0, \frac{1}{2})$ the following multilevel deviation estimate holds,*

$$\sum_k h_k^{-2s} \|curl(\mathbf{\Pi}_k \mathbf{Q}_k - \mathbf{Q}_k)\mathbf{z}\|_0^2 \leq C \sum_k h_k^{-2(1+s)} \|(\mathbf{Q}_k - \mathbf{Q}_{k+1})\mathbf{z}\|_0^2 \leq C\|\mathbf{z}\|_{1+s}^2.$$

Here $\mathbf{z} \in \mathbf{S}_h \subset H_0^{1+s}(\Omega)$, $s \in [0, \frac{1}{2})$.

Proof. Here, we need the commutativity property of $\mathbf{\Pi}_k$ and the Fortin projection $\mathbf{\Pi}_k^{RT}$ associated with the respective Raviart–Thomas space $\mathbf{R}_k$ on the mesh $\mathcal{T}_k$,

$$\text{curl}\ \mathbf{\Pi}_k = \mathbf{\Pi}_k^{RT} \text{curl}. \tag{F.18}$$

This equality holds applied to smooth functions, for example, applied to the continuous vector piecewise polynomials in $\mathbf{S}_k$.

Due to the commutativity property (F.18), we have $\mathbf{\Psi}_k = \text{curl}(\mathbf{\Pi}_k \mathbf{Q}_k - \mathbf{Q}_k)\mathbf{z} = (\mathbf{\Pi}_k^{RT} - I)\text{curl}\ \mathbf{Q}_k \mathbf{z}$. Using the L_2-approximation property of the projections $\mathbf{\Pi}_k^{RT}$, for any $\sigma < \frac{1}{2}$, we have

$$\|\mathbf{\Psi}_k\|_0 \leq C h_k^{\sigma} \|\text{curl}\ \mathbf{Q}_k \mathbf{z}\|_{\sigma}.$$

Below, we choose $0 \leq s < \sigma < \frac{1}{2}$. Proceeding similarly to (C.12), with $\mathbf{Q}_{\ell+1} = 0$, and letting $h_k = 2^{-\ell+k}H$ be the kth-level mesh-size, we derive the estimates,

$$\begin{aligned}
\sum_k h_k^{-2s} \|\mathbf{\Psi}_k\|_0^2 &= \sum_k h_k^{-2s} \|\mathbf{\Psi}_k\|_0 \|\mathbf{\Psi}_k\|_0 \\
&\leq C \sum_k h_k^{-2s} \|\mathbf{\Psi}_k\|_0\, h_k^{\sigma} \left\| \sum_{j=k}^{\ell} \text{curl}(\mathbf{Q}_j - \mathbf{Q}_{j+1})\mathbf{z} \right\|_{\sigma} \\
&\leq \sum_k h_k^{-2s} \|\mathbf{\Psi}_k\|_0 h_k^{\sigma} \sum_{j=k}^{\ell} h_j^{-1-\sigma} \|(\mathbf{Q}_j - \mathbf{Q}_{j+1})\mathbf{z}\|_0 \\
&\leq C \sum_k \sum_{j \geq k} \frac{h_k^{\sigma}}{h_j^{\sigma}} \frac{h_j^s}{h_k^s}\, h_k^{-s} \|\mathbf{\Psi}_k\|_0 h_j^{-1-s} \|(\mathbf{Q}_j - \mathbf{Q}_{j+1})\mathbf{z}\|_0 \\
&= C \sum_k \sum_{j \geq k} \left(\frac{1}{2^{\sigma - s}} \right)^{j-k} h_k^{-s} \|\mathbf{\Psi}_k\|_0 h_j^{-1-s} \|(\mathbf{Q}_j - \mathbf{Q}_{j+1})\mathbf{z}\|_0.
\end{aligned}$$

This shows the desired estimate,

$$\begin{aligned}
\sum_k h_k^{-2s} \|\mathbf{\Psi}_k\|_0^2 &\leq C \left[\sum_k h_k^{-2s} \|\mathbf{\Psi}_k\|_0^2 \right]^{1/2} \left[\sum_j h_j^{-2-2s} \|(\mathbf{Q}_j - \mathbf{Q}_{j+1})\mathbf{z}\|_0^2 \right]^{1/2} \\
&\leq C \left[\sum_k h_k^{-2s} \|\mathbf{\Psi}_k\|_0^2 \right]^{1/2} |\mathbf{z}|_{1+s}^2.
\end{aligned}$$

□

We are now ready to prove estimates (a)–(e) from Section F.1. To conform to the notation from Section F.1, letting $\mathbf{y}_0 = \mathbf{y} \equiv \mathbf{u}$, we introduce

$$\begin{aligned}
\mathbf{y}_k &= \mathbf{\Pi}_k \mathbf{Q}_k \mathbf{z} + \nabla Q_k \varphi, \\
\mathbf{u}_k^{(f)} &= \begin{cases} \mathbf{\Pi}_k \mathbf{Q}_k \mathbf{z} - \mathbf{\Pi}_{k+1} \mathbf{Q}_{k+1} \mathbf{z}, & \text{for } k > 0, \\ \mathbf{v} + \mathbf{\Pi}_k \mathbf{Q}_k \mathbf{z} - \mathbf{\Pi}_{k+1} \mathbf{Q}_{k+1} \mathbf{z}, & \text{for } k = 0, \end{cases} \\
\psi_k^{(f)} &= (Q_k - Q_{k+1})\varphi, \\
\mathbf{y}_{k+(1/2)}^{(f)} &= \nabla \psi_k^{(f)}, \\
\mathbf{y}_{k+(1/2)} &= \mathbf{u}_k^{(f)} + \mathbf{y}_{k+1} = \mathbf{\Pi}_k \mathbf{Q}_k \mathbf{z} + \nabla Q_{k+1} \varphi.
\end{aligned}$$

To verify (F.9) of (a), we use the norm equivalence provided by the L_2-projections and the stability of the HX decomposition

$$\sum_k h_k^{-2} \|\psi_k^{(f)}\|_0^2 = \sum_k h_k^{-2} \|(Q_k - Q_{k+1})\varphi\|_0^2 \le C\ \|\nabla \varphi\|_0^2 \le C\ \|\mathbf{u}\|_0^2 \le C\ a(\mathbf{u},\ \mathbf{u}).$$

Consider now condition (b). We have to estimate

$$\sum_k h_k^2\ \|\psi_k\|_0^2 = \sum_k h_k^2\ (\nabla \psi_k,\ y_{k+(1/2)}) = \sum_k h_k^2\ (\nabla \psi_k,\ \mathbf{\Pi}_k \mathbf{Q}_k \mathbf{z} + \nabla Q_{k+1} \varphi).$$

Here $\psi_k \in S_k$ is such that its coefficient vector (w.r.t. the nodal basis of S_k) equals $\underline{\boldsymbol{\psi}}_k = G_{k+(1/2)}^{-1} P_{k+(1/2)}^T G_k \underline{\mathbf{y}}_{k+(1/2)}$. We have $\mathbf{y}_{k+(1/2)} = \nabla Q_{k+1} \varphi \mathbf{\Pi}_j \mathbf{Q}_j \mathbf{z} = (\mathbf{\Pi}_j \mathbf{Q}_j - \mathbf{Q}_j)\mathbf{z} + (\mathbf{Q}_j \mathbf{z}) + \nabla Q_{k+1} \varphi$. Using integration by parts and an inverse inequality for $\psi_k \in S_k$, we obtain

$$\begin{aligned}
(\mathbf{y}_{k+(1/2)},\ \nabla \psi_k) &= ((\mathbf{\Pi}_k \mathbf{Q}_k - \mathbf{Q}_k)\mathbf{z},\ \nabla \psi_k) - (\text{div}\ (\mathbf{Q}_k \mathbf{z}), \psi_k) + (\nabla Q_{k+1} \varphi,\ \nabla \psi_k) \\
&\le \left[C h_k^{-1} \|(\mathbf{\Pi}_k \mathbf{Q}_k - \mathbf{Q}_k)\mathbf{z}\|_0 + \|\text{div}\ (\mathbf{Q}_k \mathbf{z})\|_0\right] \|\psi_k\|_0 \\
&\quad + (\nabla Q_{k+1} \varphi,\ \nabla \psi_k).
\end{aligned}$$

Use now the representation $\nabla Q_{k+1} \varphi = \sum_{j>k} \nabla (Q_j - Q_{j+1})\varphi$, $(Q_{\ell+1} = 0)$ and the strengthened inverse inequality (see Proposition F.1) to bound the sum

$$\begin{aligned}
\sum_k h_k^2\ (\nabla Q_{k+1} \varphi,\ \nabla \psi_k) &\le C\ \sum_k h_k^2 \sum_{j>k} h_k^{-(1/2)} \|\psi_k\|_0 h_j^{-(1/2)} \|\nabla (Q_j - Q_{j+1})\varphi\|_0 \\
&\le C\ \sum_k h_k \|\psi_k\|_0\ \sum_{j>k} \left(\frac{h_k}{h_j}\right)^{1/2} \|\nabla (Q_j - Q_{j+1})\varphi\|_0 \\
&= C\ \sum_k h_k \|\psi_k\|_0\ \sum_{j>k} \left(\frac{1}{\sqrt{2}}\right)^{j-k} \|\nabla (Q_j - Q_{j+1})\varphi\|_0
\end{aligned}$$

$$\leq C\left(\sum_k h_k^2\|\psi_k\|_0^2\right)^{1/2}\left(\sum_j \|\nabla(Q_j - Q_{j+1})\varphi\|_0^2\right)^{1/2}$$

$$\leq C\left(\sum_k h_k^2\|\psi_k\|_0^2\right)^{1/2}\|\nabla\varphi\|_0.$$

That is, we have the estimate

$$\sum_k h_k^2\ \|\psi_k\|_0^2 = \sum_k h_k^2\ (\mathbf{y}_{k+(1/2)},\ \nabla\psi_k)$$
$$\leq C\sum_k \|(\boldsymbol{\Pi}_k\mathbf{Q}_k - \mathbf{Q}_k)\mathbf{z}\|_0(h_k\ \|\psi_k\|_0)$$
$$+\ C\sum_k h_k^2\|\psi_k\|_0|\mathbf{z}|_1 + C\left(\sum_k h_k^2\|\psi_k\|_0^2\right)^{1/2}\|\nabla\varphi\|_0.$$

Therefore, we proved the following bound

$$\sum_k h_k^2\ \|\psi_k\|_0^2 \leq C\left[\sum_k \|(\boldsymbol{\Pi}_k\mathbf{Q}_k - \mathbf{Q}_k)\mathbf{z}\|_0^2 + CH^2\ |\mathbf{z}|_1^2 + \|\nabla\varphi\|_0^2\right].$$

Using Lemma F.4 for $s = 0$ and the stability of the HX decomposition, we finally have

$$\sum_k h_k^2\ \|\psi_k\|_0^2 \leq C\ \left[\|\mathbf{z}\|_0^2 + H^2\ |\mathbf{z}|_1^2 + \|\nabla\varphi\|_0^2\right] \leq C\left(1 + \frac{H^2}{\tau}\right)\ a(\mathbf{u},\ \mathbf{u}),$$

which verifies estimate (F.10) if $H^2/\tau = \mathcal{O}(1)$, which we assume.

Consider next condition (d). We have to bound the expression

$$\sum_k \lambda_k^{-1}\|\Psi_k\|_0^2 = \sum_k \lambda_k^{-1}a(\mathbf{y}_{k+1},\ \Psi_k), \tag{F.19}$$

where $\Psi_k \in \mathbf{N}_k$ has coefficient vector $\underline{\boldsymbol{\psi}}_k = \mathbf{G}_k^{-1}\mathbf{A}_k P_k \underline{\mathbf{y}}_{k+1}$. Recall that $\lambda_k = 1 + \tau h_k^{-2}$ and $\mathbf{y}_{k+1} = \boldsymbol{\Pi}_{k+1}\mathbf{Q}_{k+1}\mathbf{z} + \nabla Q_{k+1}\varphi$.

We let $\mathbf{z}_{k+1} = \mathbf{Q}_{k+1}\mathbf{z}$ and $\varphi_{k+1} = Q_{k+1}\varphi$. Then the following estimates hold,

$$\|\Psi_k\|_0^2 = (\Psi_k,\ \boldsymbol{\Pi}_{k+1}\mathbf{z}_{k+1} + \nabla\varphi_{k+1}) + \tau(\text{curl}\ \Psi_k, \text{curl}(\boldsymbol{\Pi}_{k+1}\mathbf{z}_{k+1} + \nabla\varphi_{k+1})$$
$$= (\Psi_k,\ \boldsymbol{\Pi}_{k+1}\mathbf{z}_{k+1} + \nabla\varphi_{k+1}) + \tau(\text{curl}\ \Psi_k, \text{curl}(\boldsymbol{\Pi}_{k+1}\mathbf{z}_{k+1} - \mathbf{z}_{k+1}))$$
$$+\ \tau(\text{curl}\ \Psi_k, \text{curl}\ \mathbf{z}_{k+1})$$
$$\leq (\Psi_k, \boldsymbol{\Pi}_{k+1}\mathbf{z}_{k+1} + \nabla\varphi_{k+1}) + \tau\|\Psi_k\|_0 C h_k^{-1}\|\text{curl}(\boldsymbol{\Pi}_{k+1}\mathbf{z}_{k+1} - \mathbf{z}_{k+1})\|_0$$
$$+\ \tau\ (\text{curl}\ \Psi_k,\ \text{curl}\ \mathbf{z}_{k+1}). \tag{F.20}$$

Now, decompose $\mathbf{z}_{k+1} = \sum_{j\geq k+1}(\mathbf{z}_j - \mathbf{z}_{j+1})$ $(\mathbf{z}_{\ell+1} = 0)$. We use the following strengthened inverse inequality (cf. Proposition F.1) valid for $j \geq k$,

$$(\text{curl}\ \Psi_k,\ \text{curl}(\mathbf{z}_j - \mathbf{z}_{j+1})) \leq Ch_k^{-(1/2)}\|\Psi_k\|_0 h_j^{-(1/2)}\|\text{curl}(\mathbf{z}_j - \mathbf{z}_{j+1})\|_0. \tag{F.21}$$

Then, we have the estimate

$$
\begin{aligned}
&\frac{\tau}{\lambda_k}\,(\text{curl }\Psi_k,\ \text{curl }\mathbf{z}_{k+1})\\
&\quad\le C\sum_{j\ge k+1}\frac{1}{\sqrt{\lambda_k}}\,\|\Psi_k\|_0\left(\frac{\tau h_k^{-2}}{\lambda_k}\right)^{1/2}\left(\frac{h_k}{h_j}\right)^{1/2}\left(\sqrt{\tau}\|\text{curl }(\mathbf{z}_j-\mathbf{z}_{j+1})\|_0\right)\\
&\quad\le C\frac{1}{\sqrt{\lambda_k}}\,\|\Psi_k\|_0\sum_{j\ge k+1}\frac{1}{\sqrt{2}^{(j-k)}}\left(\sqrt{\tau}\|\text{curl }(\mathbf{z}_j-\mathbf{z}_{j+1})\|_0\right).
\end{aligned}
$$

This implies the crucial estimate

$$
\begin{aligned}
&\sum_k\frac{\tau}{\lambda_k}\,(\text{curl }\Psi_k,\ \text{curl }\mathbf{z}_{k+1})\\
&\quad\le C\left(\sum_k\lambda_k^{-1}\,\|\Psi_k\|_0^2\right)^{1/2}\left(\tau\sum_j\|\text{curl }(\mathbf{z}_j-\mathbf{z}_{j+1})\|_0^2\right)^{1/2}.
\end{aligned}
$$

Combining (F.20), the last estimate and (F.19), we end up with the main estimate needed in (d):

$$
\begin{aligned}
\sum_k\frac{1}{\lambda_k}\,\|\Psi_k\|_0^2&=\sum_k\lambda_k^{-1}a(\mathbf{y}_{k+1},\ \Psi_k)\\
&\le C\sum_k\frac{1}{\lambda_k}\,\|\mathbf{\Pi}_{k+1}\mathbf{z}_{k+1}+\nabla\varphi_{k+1}\|_0^2\\
&\quad+C\tau\sum_k\frac{\tau h_k^{-2}}{\lambda_k}\,\|\text{curl }(\mathbf{\Pi}_{k+1}\mathbf{z}_{k+1}-\mathbf{z}_{k+1})\|_0^2\\
&\quad+C\tau\sum_j\|\text{curl }(\mathbf{z}_j-\mathbf{z}_{j+1})\|_0^2.
\end{aligned}
$$

Use now that $\left(\tau h_k^{-2}/\lambda_k\right)<1$, Lemma F.6 for $s=0$, estimate (F.17), and the stability of the L_2-projectors in H^1, and the stability of the HX decomposition, we end up with the following counterpart of (F.12)

$$
\begin{aligned}
\sum_k\frac{1}{\lambda_k}\,\|\Psi_k\|_0^2=\sum_k\lambda_k^{-1}a(\mathbf{y}_{k+1},\ \Psi_k)&\le C\left[\frac{H^2}{\tau}\left(\|\mathbf{z}\|_0^2+\|\nabla\varphi\|_0^2\right)+\tau\ |\mathbf{z}|_1^2\right]\\
&\le C\left(1+\frac{H^2}{\tau}\right)a(\mathbf{u},\ \mathbf{u}),
\end{aligned}
\tag{F.22}
$$

Recall that we assumed $H^2\tau^{-1}=\mathcal{O}(1)$.

Condition (c) requires the following estimate

$$
\begin{aligned}
\sum_k\lambda_k\ \|\mathbf{u}_k^{(f)}\|_0^2&=(1+\tau h^{-2})\ \|\mathbf{v}\|_0^2+\sum_k(1+\tau h_k^{-2})\|(\mathbf{\Pi}_k\mathbf{Q}_k-\mathbf{\Pi}_{k+1}\mathbf{Q}_{k+1})\mathbf{z}\|_0^2\\
&\le C\left(a(\mathbf{u},\ \mathbf{u})+\tau\ |\mathbf{z}|_1^2\right)\le C\ a(\mathbf{u},\ \mathbf{u}).
\end{aligned}
\tag{F.23}
$$

Here, we used the representation (F.16), Lemma F.4 for $s = 1$ and the stability of the $\mathbf{z}_k$ terms (F.15), and finally the stability of the HX decomposition. Estimate (F.23) is the counterpart of (F.11).

Finally, condition (e) is simply the energy stability of the coarse component $\mathbf{y}_\ell = \mathbf{\Pi}_\ell \mathbf{Q}_\ell \mathbf{z} + \nabla Q_\ell \varphi$ which is the case due to the properties of the projections $\mathbf{\Pi}_\ell$, $\mathbf{Q}_\ell$, and Q_ℓ, and the stability of the HX decomposition.

Because we have estimated all the necessary terms by $\|\underline{\mathbf{u}}\|^2_{\mathbf{A}}$ uniformly in ℓ and $\tau > 0$ (if $H^2\tau^{-1} = \mathcal{O}(1)$), we have the following main result valid for general polygonal domains (nor necessarily convex).

Theorem F.7. *Consider the parameter-dependent bilinear form* $(\mathbf{u},\ \mathbf{w}) + \tau($*curl* $\mathbf{u}$, *curl* $\mathbf{w})$. *Given a 3D polytope* Ω, *let* $\mathcal{T}_k$ *be a sequence of uniformly refined triangulations of* Ω. *Let* $\mathbf{A}_k$ *be the stiffness matrices computed from the nested Nédélec f.e. spaces* $\mathbf{N}_k =$ *span* $\{\mathbf{\Phi}_e^{(k)}\}$ *where e runs over the interior edges of elements from* $\mathcal{T}_k$. $\mathcal{T}_0 = \mathcal{T}_h$ *is the finest-mesh and* $\mathcal{T}_\ell = \mathcal{T}_H$ *is the coarsest-mesh. Consider also the nodal-based Lagrangian spaces* $S_k =$ *span* $\{\varphi_i^{(k)}\}$ *where* $\mathbf{x}_i$ *runs over the (interior) vertices of the elements in* $\mathcal{T}_k$. *Consider the symmetric* $V(1,1)$*-cycle geometric MG for the Nédélec stiffness matrix* $\mathbf{A} = \mathbf{A}_0$ *as described in Algorithm F.1.1. It requires explicitly the matrices* $P_{j+(1/2)}$ *(discrete gradients) that transform the coefficient vector of a f.e. function* φ *from* S_j *into the coefficient vector of* $\nabla\varphi$ *viewed as an element of* $\mathbf{N}_j$. *It also uses the Gauss–Seidel smoothers* $M_{j+(1/2)}$ *(referred to as the multiplicative Hiptmair smoother) for the Laplace matrices* $\mathbf{A}_{j+(1/2)} = P^T_{j+(1/2)}\mathbf{A}_j P_{j+(1/2)}$ *at intermediate (fractional) levels. The method also utilizes standard Gauss–Seidel smoothers* $\mathbf{M}_j$ *for the Nédélec stiffness matrices* $\mathbf{A}_j$ *as well as the interpolation matrices* P_j *that implement the embedding* $\mathbf{N}_{j+1} \subset \mathbf{N}_j$ *in terms of coefficient vectors. The thus-defined geometric* $V(1,1)$*-cycle MG for the (lowest-order) Nédélec spaces has a convergence factor bounded independently of both the mesh-size* $h \mapsto 0$ *and the parameter* $\tau > 0$ *as long as* $H^2\tau^{-1} = \mathcal{O}(1)$.

We conclude with the comment that originally the MG methods for $H(\mathrm{curl})$ were analyzed in [AFW] and [H99]. Our presentation is based on the HX decomposition found in [HX06].

F.3 A multilevel decomposition of div-free Raviart–Thomas spaces

Given is a 3D polyhedral domain Ω triangulated into a tetrahedral mesh $\mathcal{T}_h$ obtained by $\ell \geq 1$ levels of uniform refinement. Let $\mathbf{R}_h$ be the lowest-order Raviart–Thomas space associated with $\mathcal{T}_h$ and to be specific assume that for $\mathbf{v} \in \mathbf{R}_h$ essential boundary conditions $\mathbf{v}\cdot\mathbf{n} = 0$ are imposed on $\partial\Omega$. Let $k = k(\mathbf{x})$ be a given s.p.d. three-by-three coefficient matrix with spectrum bounded above and below uniformly in $\mathbf{x} \in \Omega$.

In this section, we are interested in solving the saddle-point problem: for a given $\mathbf{f} \in L_2(\Omega)$ find $\mathbf{v} \in \mathbf{R}_h$ such that

$$(k(\mathbf{x})^{-1}\,\mathbf{v},\ \chi) + (p,\ \mathrm{div}\ \chi) = (\mathbf{f},\ \chi) \quad \text{for all } \chi \in \mathbf{R}_h,$$

subject to div $\mathbf{v} = 0$. Here, p is an unknown piecewise constant function with respect to $\mathcal{T}_h$ (Lagrange multiplier or "pressure" in some applications). It can be determined up to an additive constant (if needed). In order to apply the constrained minimization method from Section 9.5 we need to select divergence-free subspaces of $\mathbf{R}_h$ of small dimension and, in order to prove a convergence, we need to establish the existence of a stable decomposition w.r.t. these subspaces. For this, we use the fact that any div-free function $\mathbf{v} \in \mathbf{R}_h$ is actually a curl of a function from the Nédélec space $\mathbf{N}_h$.

Let $\mathbf{S}_h$ be the vector nodal based $\mathbf{H}_0^1 = (H_0^1)^3$-conforming piecewise linear f.e. space associated with the vertices of the elements from $\mathcal{T}_h$. A result in [PZ02] states that for simply connected polyhedral domains Ω for any $\mathbf{u} \in \mathbf{N}_h$ there is a $\mathbf{z} \in (H_0^1(\Omega))^3$ such that curl $\mathbf{u}$ = curl $\mathbf{z}$ = curl $\boldsymbol{\Pi}_h\mathbf{z}$, and $|\mathbf{z}|_1 \le C\|\text{curl } \mathbf{u}\|_0$. Then, for a proper interpolant $\mathbf{z}_h \in \mathbf{S}_h$ of $\mathbf{z}$, the following decomposition

$$\text{curl } \mathbf{u} = \text{curl } \boldsymbol{\Pi}_h(\mathbf{z} - \mathbf{z}_h) + \text{curl } \boldsymbol{\Pi}_h\mathbf{z}_h,$$

is also stable. All components are (obviously) divergence-free. Letting $\boldsymbol{\psi}_h = \boldsymbol{\Pi}_h(\mathbf{z} - \mathbf{z}_h)$, we have $\boldsymbol{\psi}_h \in \mathbf{N}_h$, $\|\boldsymbol{\psi}_h\|_0 \le Ch\ |\mathbf{z}|_1 \le Ch\ \|\text{curl } \mathbf{u}\|_0$ and $\|\text{curl } \boldsymbol{\psi}_h\|_0 \le C\|\text{curl } \mathbf{u}\|_0$. Because now $\mathbf{z}_h \in \mathbf{S}_h$, where $\mathbf{S}_h$ is the vector nodal-based $\mathbf{H}_0^1$-conforming f.e. space, the following multilevel decompositions are available.

$$\text{curl } \boldsymbol{\Pi}_h\mathbf{z}_h = \sum_{k=0}^{\ell-1} \text{curl } (\boldsymbol{\Pi}_k\mathbf{z}_k - \boldsymbol{\Pi}_{k+1}\mathbf{z}_{k+1}) + \text{curl } \boldsymbol{\Pi}_\ell\mathbf{z}_\ell, \qquad \mathbf{z}_k = \mathbf{Q}_k\mathbf{z}_h.$$

Here, $\mathbf{Q}_k$ are the vector L_2-projection onto the (vector) f.e. spaces $\mathbf{S}_k$ of continuous piecewise linear functions $\mathbf{S}_k$. The components $\boldsymbol{\psi}_k = \boldsymbol{\Pi}_k\mathbf{z}_k - \boldsymbol{\Pi}_{k+1}\mathbf{z}_{k+1}$ for $0 < k < \ell$, and $\boldsymbol{\psi}_0 = \boldsymbol{\psi}_h + (\boldsymbol{\Pi}_k\mathbf{z}_k - \boldsymbol{\Pi}_{k+1}\mathbf{z}_{k+1})$ for $k = 0$, as already pointed out in the preceding section (proven similarly to (F.23) where $\mathbf{u}_k^{(f)} = \boldsymbol{\psi}_k^{(f)}$), satisfy for $s = 0, 1$, the major estimate

$$\sum_{k<\ell} h_k^{-2s}\ \|\boldsymbol{\psi}_k^{(f)}\|_0^2 \le C|\mathbf{z}_h|_s^2 \le C\big((1-s)\|\mathbf{u}\|_0^2 + s\|\text{curl } \mathbf{u}\|_0^2\big).$$

Also their curl is stable (due to Lemma F.6 for $s = 0$ and estimate (F.17)); that is, we have

$$\sum_{k<\ell} \|\text{curl } \boldsymbol{\psi}_k^{(f)}\|_0^2 \le C|\mathbf{z}_h|_1^2 \le C\ \|\text{curl } \mathbf{u}\|_0^2.$$

In other words, letting $\boldsymbol{\psi}_\ell = \boldsymbol{\psi}_\ell^{(f)} = \boldsymbol{\Pi}_\ell\mathbf{z}_h$, because $\|\text{curl } \boldsymbol{\Pi}_\ell\mathbf{z}_\ell\|_0 \le C\ |\mathbf{z}_\ell|_1 \le C\ \|\text{curl } \mathbf{u}\|_0$, we have the following main stability result.

Lemma F.8. *For any $\mathbf{u} \in \mathbf{N}_h$, there is a multilevel decomposition $\mathbf{u} = \sum_k \boldsymbol{\psi}_k^{(f)}$, $\boldsymbol{\psi}_k^{(f)} \in \mathbf{N}_k$ for $k = 0, \ldots, \ell$, for which the following stability estimate holds,*

$$\sum_{k<\ell} h_k^{-2}\|\boldsymbol{\psi}_k^{(f)}\|_0^2 + \sum_k \|\text{curl } \boldsymbol{\psi}_k^{(f)}\|_0^2 \le C\ \|\text{curl } \mathbf{u}\|_0^2.$$

For the constrained saddle-point problem the decomposition of interest reads as follows. Let $\{\boldsymbol{\Phi}^h_e\}$ be the edge-based basis of the Nédélec space $\mathbf{N}_h$. Let $\mathbf{N}_H$ be a coarse Nédélec space which is a subspace of $\mathbf{N}_h$. Then, for any divergence-free function $\mathbf{w} \in \mathbf{R}_h$, we can use the following two-level decomposition

$$\mathbf{w} = \text{curl } \boldsymbol{\psi}_h = \text{curl } (\boldsymbol{\psi}_h - \boldsymbol{\psi}_H) + \text{curl } \boldsymbol{\psi}_H.$$

Its first component can be expanded in terms of the basis of $\mathbf{N}_h$. That is, for some coefficients $\{w_e\}$, we have

$$\mathbf{w}^{(f)}_h = \text{curl } (\boldsymbol{\psi}_h - \boldsymbol{\psi}_H) = \sum_e \text{curl } \boldsymbol{\psi}_e, \quad \text{where } \boldsymbol{\psi}_e = w_e \, \boldsymbol{\Phi}^h_e.$$

In other words, $\mathbf{w}^{(f)}_h$ is decomposed as a sum of div-free functions curl $\boldsymbol{\psi}_e$ and each one belongs to the one-dimensional space Range (curl $\boldsymbol{\Phi}^h_e$). If $\boldsymbol{\psi}_H$ approximates $\boldsymbol{\psi}_h$ so that $\|\boldsymbol{\psi}_h - \boldsymbol{\psi}_H\|_0 \le CH \, \|\text{curl } \boldsymbol{\psi}_h\|_0$, assuming that $H \simeq h$, it is easy to show the stability estimate

$$\sum_e \|\text{curl } \boldsymbol{\psi}_e\|_0^2 \le C \sum_e h^{-2} \, \|\boldsymbol{\psi}_e\|_0^2 \le Ch^{-2} \|\boldsymbol{\psi}_h - \boldsymbol{\psi}_H\|_0^2 \le C \, \|\mathbf{w}\|_0^2.$$

As demonstrated earlier, we can choose $\boldsymbol{\psi}_H = \boldsymbol{\Pi}_H \mathbf{z}_h$ for proper $\mathbf{z}_h \in \mathbf{S}_h$, so that the estimate $\|\boldsymbol{\psi}_h - \boldsymbol{\psi}_H\|_0 \le CH \, \|\text{curl } \boldsymbol{\psi}_h\|_0$ holds. Finally, it is clear that for a coefficient $k(\mathbf{x})$ that has a uniformly bounded spectrum (w.r.t. $\mathbf{x} \in \Omega$), a similar stability estimate holds in terms of the $k(\mathbf{x})^{-1}$-weighted L_2-norm. The above process can be applied recursively to any two consecutive meshes $h = h_k$ and $H = h_{k+1}$.

To solve the constrained saddle-point problem, we can use the constrained minimization algorithm from Section 9.5. It involves a sequence of 1D subspace minimization problems (at a given level) plus a coarse constrained minimization problem (referring to the mesh-size H). We do not have to form any saddle-point problems because we have an explicit basis for the divergence-free functions in $\mathbf{R}_h$. The overall method involves recursion over the levels. The following observation is then in order. The solution provided by the resulting multilevel constrained minimization algorithm is obtained in terms of the curl of a function $\mathbf{u}_h$ from $\mathbf{N}_h$, and what is important, $\mathbf{u}_h$ is explicitly available. That is, at the end we have a particular solution of the following problem,

$$(k^{-1}(\mathbf{x}) \text{ curl } \mathbf{u}_h, \text{ curl } \boldsymbol{\psi}_h) = (\mathbf{f}, \text{ curl } \boldsymbol{\psi}_h) \quad \text{for all } \boldsymbol{\psi}_h \in \mathbf{N}_h.$$

In other words, if $\mathbf{Q}^{RT}_h$ is the L_2-projection onto the Raviart–Thomas f.e. space $\mathbf{R}_h$, we have a solution $\mathbf{u}_h \in \mathbf{N}_h$ to the equation

$$\mathbf{Q}^{RT}_h (\mathbf{f} - k^{-1}(\mathbf{x}) \text{ curl } \mathbf{u}_h) = 0.$$

In particular, if $k(\mathbf{x}) = 1$ we have a particular solution to the equation curl $\mathbf{u}_h = Q^{RT}_h \mathbf{f}$. Note that all other solutions are given by $\mathbf{u}_h + \nabla \varphi_h$ for any $\varphi_h \in S_h$ where S_h is the standard nodal f.e. space of continuous piecewise linear functions.

Alternatively, we can implement the above-described multilevel constrained minimization method as in Algorithm F.1.1 where the integer indexed smoothers M_j and M_j^T are omitted and the coarse solution replaced by a divergence-free coarse-grid solution. That is, the resulting MG algorithm is based on the div-free subspaces curl $\mathbf{N}_k \subset \mathbf{R}_k$ and utilizes a Hiptmair-like smoother. Its level-independent convergence can be proved by modifying the analysis from Section F.4.

The case of more general mixed f.e. saddle-point problem

We consider here the more typical saddle-point problem corresponding to the mixed f.e. discretization of $-\text{div}\, k(\mathbf{x})\nabla p = f$ which takes the following form.

Introduce the space W_h of piecewise constants w.r.t. $\mathcal{T}_h$. The mixed finite element problem in question looks for $\overline{\mathbf{v}} \in \mathbf{R}_h$ and $p \in W_h$ such that

$$\begin{aligned}
(k^{-1}\overline{\mathbf{v}},\ \boldsymbol{\chi}) + (p,\ \text{div}\ \boldsymbol{\chi}) &= 0, && \text{for all } \boldsymbol{\chi} \in \mathbf{R}_h,\\
(\text{div}\ \overline{\mathbf{v}},\ q) &= -(f,\ q), && \text{for all } q \in W_h,
\end{aligned}$$

We need to have $(f,\ 1) = 0$ (a necessary condition for solvability of the original second-order PDE if homogeneous Neumann boundary conditions are imposed on p). The above mixed f.e. problem can be reduced to one with a div-free constraint by finding a particular solution $\mathbf{v}_0 \in \mathbf{R}_h$ of the second equation. Assume that we have found one such solution $\mathbf{v}_0$; that is, we have $(\text{div}\ \mathbf{v}_0,\ q) = -(f,\ q)$ for all $q \in W_h$. Equivalently, we have $\text{div}\ \mathbf{v}_0 = -Q_h f$ where Q_h is the L_2-projection onto the space of piecewise constants W_h. Then $\mathbf{v} = \overline{\mathbf{v}} - \mathbf{v}_0 \in \mathbf{R}_h$ solves the div-free saddle-point problem

$$\begin{aligned}
(k^{-1}\mathbf{v},\ \boldsymbol{\chi}) + (p,\ \text{div}\ \boldsymbol{\chi}) &= -(k^{-1}(\mathbf{x})\mathbf{v}_0,\ \boldsymbol{\chi}), && \text{for all } \boldsymbol{\chi} \in \mathbf{R}_h,\\
(\text{div}\ \mathbf{v},\ q) &= 0, && \text{for all } q \in W_h.
\end{aligned}$$

A multilevel procedure, based on solving local saddle-point problems at every discretization level $k \leq \ell$, for finding a particular solution of $\text{div}\ \mathbf{v}_0 = -Q_h f$ was proposed in [VW92]. To describe it, let Q_k be the L_2-projection onto the space W_k of discontinuous piecewise polynomials w.r.t. the kth-level triangulation $\mathcal{T}_k$ of certain degree (zero for the lowest-order Raviart–Thomas space $\mathbf{R}_k$). For any coarse element $T \in \mathcal{T}_{k+1}$ viewed as a kth-level (fine-grid) domain, solve (in some way) the local problem

$$\text{div}\ \mathbf{v}_{0,\ T}^{(k)} = -(Q_k - Q_{k+1})f, \quad \text{on } T,$$

with boundary conditions $\mathbf{v}_{0,\ T}^{(k)} \cdot \mathbf{n} = 0$ on ∂T. Because $(Q_k - Q_{k+1})f$ is orthogonal to the constant functions on T, the above problem is solvable. Note that we need only a particular solution. For example, we can solve the well-posed local saddle-point problem,

$$\begin{aligned}
\left(k^{-1}(\mathbf{x})\mathbf{v}_{0,\ T}^{(k)},\ \boldsymbol{\chi}\right)_T + (p_T,\ \text{div}\ \boldsymbol{\chi})_T &= 0, && \text{for all } \boldsymbol{\chi} \in \mathbf{R}_k|_T : \quad \boldsymbol{\chi}\cdot\mathbf{n}|_{\partial T} = 0,\\
(\text{div}\ \mathbf{v}_{0,\ T}^{(k)},\ q)_T &= -((Q_k - Q_{k+1})f,\ q)_T, && \text{for all } q \in W_k|_T.
\end{aligned}$$

Here, $(.,.)_T$ stands for the L_2-inner product restricted to T. For each T, we find a $\mathbf{v}^{(k)}_{0,\,T} \in \mathbf{R}_k|_T$ such that $\mathbf{v}^{(k)}_{0,\,T} \cdot \mathbf{n}|_{\partial T} = 0$. There is no conflict on the interfaces between any two neighboring elements T_1 and T_2, thus the set $\{\mathbf{v}^{(k)}_{0,\,T}\}_{T\in\mathcal{T}_{k+1}}$ actually defines a global $\mathbf{v}^{(k)}_0 \in \mathbf{R}_k$ which satisfies div $\mathbf{v}^{(k)}_0 = -(Q_k - Q_{k+1})f$.

Note that because we have assumed $(f,\ 1) = 0$, hence $(Q_\ell f,\ 1) = 0$. The latter is a necessary condition for solvability of the last global coarse problem we solve to determine $\mathbf{v}^{(\ell)}_0$ which reads

$$
\begin{array}{lll}
(k^{-1}(\mathbf{x})\mathbf{v}^{(\ell)}_0,\ \chi) + (p^{(\ell)}_0,\ \text{div}\ \chi) = 0, & & \text{for all } \chi \in \mathbf{R}_\ell, \\
(\text{div}\ \mathbf{v}^{(\ell)}_0,\ q) & = -(Q_\ell f,\ q) = (-f, q), & \text{for all } q \in W_\ell.
\end{array}
\tag{F.24}
$$

The desired particular solution then equals

$$
\mathbf{v}_0 = \sum_{k=0}^{\ell} \mathbf{v}^{(k)}_0.
$$

By construction then, we have div $\mathbf{v}_0 = -\sum_{k=0}^{\ell-1}(Q_k - Q_{k+1})f - Q_\ell f = -Q_0 f = -Q_h f$, which is the desired result.

F.4 A multilevel decomposition of weighted $H(\text{div})$-space

In some applications it is of interest to solve s.p.d. problems coming from f.e. discretizations of the weighted $H(\text{div})$-bilinear form

$$
a(\mathbf{v},\ \chi) = (\mathbf{v},\ \chi) + \tau\ (\text{div}\ \mathbf{v},\ \text{div}\ \chi),
$$

using Raviart–Thomas spaces $\mathbf{R}_k$ corresponding to triangulations $\mathcal{T}_k$ of tetrahedral elements obtained by successive refinement. We recall that $\mathcal{T}_\ell = \mathcal{T}_H$ is the coarsest triangulation and $\mathcal{T}_0 = \mathcal{T}_h$ is the finest one. The parameter τ is assumed either $\mathcal{O}(1)$ or large. If it happens that $\tau \le Ch_k^2$ for $\ell \ge k \ge k_0$ at coarse levels k the respective stiffness matrix $\mathbf{A}_k$ coming from $a(\cdot,\ \cdot)$ and the space $\mathbf{R}_k$ will be well conditioned because due to a standard inverse inequality for $\mathbf{v} \in \mathbf{R}_k$ we will have $\tau\|\text{div}\ \mathbf{v}\|_0^2 \le C\tau h_k^{-2}\|\mathbf{v}\|_0^2 \le C\|\mathbf{v}\|_0^2$. That is, the mass term will dominate the spectrum of $\mathbf{A}_k$. Then we do not need to have a full multilevel cycle involving all ℓ levels; we can instead stop at coarse-level k_0 and use only smoothing at that final coarse-level and end up with an optimal order k_0-level MG method.

The MG method that we are interested in relies on a stable multilevel decomposition $\mathbf{v} = \mathbf{v}_0 + \text{curl}\ \mathbf{u}$, where $\mathbf{v}_0 \in \mathbf{R}_h$ is such that div $(\mathbf{v} - \mathbf{v}_0) = 0$, hence because $\mathbf{v} - \mathbf{v}_0$ being a div-free Raviart–Thomas function there is a function $\mathbf{u}$ from the Nédélec space $\mathbf{N}_h$ such that curl $\mathbf{u} = \mathbf{v} - \mathbf{v}_0$. For the div-free Nédélec component we already derived a stable (div-free) multilevel decomposition. At every level k we need to smooth based on one-dimensional space spanned by curl $\boldsymbol{\Phi}_e^{(k)}$, where $\{\boldsymbol{\Phi}_e^{(k)}\}$ is the edge-based basis of the Nédélec f.e. space $\mathbf{N}_k$.

The multilevel procedure presented in the previous section that defines the needed $\mathbf{v}_0$ is actually "energy"-stable. We now give a brief description of the proof from [VW92].

Given $\mathbf{v} \in \mathbf{R}_h$, consider $f = -\text{div } \mathbf{v} \in W_h$ and compute $\mathbf{v}_0^{(k)}$ for $k = 0, \ldots, \ell - 1$ as explained in Section F.3 and let at the coarsest level $\mathbf{v}_0^{(\ell)}$ be such that div $\mathbf{v}_0^{(\ell)} = Q_\ell \text{div } \mathbf{v}$. Then, div $\sum_{k=0}^{\ell} \mathbf{v}_0^{(k)} = \text{div } \mathbf{v}$. By construction, because Q_k are L_2-orthogonal projections, we have the identity

$$\sum_{k=0}^{\ell} \|\text{div } \mathbf{v}_0^{(k)}\|_0^2 = \sum_{k=0}^{\ell-1} \|(Q_k - Q_{k+1})\text{div } \mathbf{v}\|^2 + \|Q_\ell \text{div } \mathbf{v}\|_0^2 = \|\text{div } \mathbf{v}\|_0^2.$$

Use the fact that the local saddle-point problems are stable. The following a priori estimate holds

$$\|\mathbf{v}_0^{(k)}\|_0 \leq C\, h_k\, \|(Q_k - Q_{k+1})\text{div } \mathbf{v}\|_0 \quad \text{for } k < \ell. \tag{F.25}$$

The factor h_k comes from a scaling argument because the diameter of the local domains $T \in \mathcal{T}_{k+1}$ is $\mathcal{O}(h_k)$. The constants C depend on the topology of the elements in $\mathcal{T}_k$ which are geometrically similar to a finite set of coarse ones (by assumption). Hence, C can be considered to be a fixed constant.

Let us now consider the case $k = \ell$ which appears to be somewhat more involved. We need to find a $\mathbf{v}_0^{(\ell)} \in \mathbf{R}_\ell$ such that div $\mathbf{v}_0^{(\ell)} = Q_\ell \text{div } \mathbf{v}$ and $\mathbf{v}_0^{(\ell)}$ be stable in terms of $\mathbf{v}$ in the parameter-dependent norm. Denote $\mathbf{R}_H = \mathbf{R}_\ell$, $W_H = W_\ell$, and $Q_H = Q_\ell$. Consider the following homogeneous Neumann problem posed on Ω,

$$\Delta p = Q_H \text{div } \mathbf{v} \in L_2(\Omega) \quad \text{with } \nabla p \cdot \mathbf{n} = 0 \text{ on } \partial\Omega.$$

Because $(1,\ Q_H \text{div } \mathbf{v}) = (1,\ \text{div } \mathbf{v}) = \int_{\partial\Omega} \mathbf{v} \cdot \mathbf{n}\, d\varrho = 0$, the above Neumann problem is solvable. Use now the following regularity result valid for Lipschitz polyhedral domains Ω, that for some $\delta \in (0, \frac{1}{2}]$ the following a priori estimate holds,

$$\|p\|_{(3/2)+\delta} \leq C \|Q_H \text{div } \mathbf{v}\|_{-(1/2)+\delta}.$$

Such a result is formulated in Lemma A.53 in [TW05] and originates in [D88].

The Fortin projection $\boldsymbol{\Pi}_H \nabla p$ (cf., Definition B.12) is well defined because $\nabla p \cdot \mathbf{n} \in H^\delta(F)$ for any face F of the elements in $\mathcal{T}_H$. Define $\mathbf{v}_0^{(\ell)} = \boldsymbol{\Pi}_H \nabla p$. We have, div $\nabla p = Q_H \text{div } \mathbf{v}$, hence due to the commutativity div $\boldsymbol{\Pi}_H = Q_H \text{div}$, we also have

$$\text{div } \mathbf{v}_0^{(\ell)} = \text{div } (\boldsymbol{\Pi}_H \nabla p) = Q_H \text{div } \mathbf{v}.$$

Also, based on $\boldsymbol{\Pi}_H$'s L_2-approximation property (because ∇p is sufficiently smooth, cf., e.g., Theorem 5.25 in [Mo03]), the a priori estimate for p in terms of $Q_H \text{div } \mathbf{v}$, and an inverse inequality for $Q_H \text{div}\, \mathbf{v}$, we arrive at,

$$\begin{aligned} \|\mathbf{v}_0^{(\ell)} - \nabla p\|_0 &= \|(I - \mathbf{\Pi}_H)\nabla p\|_0 \\ &\le CH^{(1/2)+\delta} \|p\|_{(3/2)+\delta} \\ &\le CH^{(1/2)+\delta} \|Q_H \text{div}\, \mathbf{v}\|_{-(1/2)+\delta} \\ &= CH^{(1/2)+\delta} \|Q_H \text{div}\, \mathbf{v}\|_{-1+((1/2)+\delta)} \\ &\le C\|Q_H \text{div}\, \mathbf{v}\|_{-1} \\ &\le C\|(I - Q_H)\text{div}\, \mathbf{v}\|_{-1} + C\|\text{div}\, \mathbf{v}\|_{-1} \\ &\le C(H\|\text{div}\, \mathbf{v}\|_0 + \|\mathbf{v}\|_0). \end{aligned}$$

We used the approximation property of the discontinuous (at least) piecewise constant projection Q_H in $H^{-1}(\Omega)$; that is, for $g = \text{div}\, \mathbf{v} \in L_2(\Omega)$, we used the estimate

$$\begin{aligned} \|(I - Q_H)g\|_{-1} \equiv &\sup_{\varphi \in H^1(\Omega)} \frac{((I - Q_H)g,\ \varphi)}{\|\varphi\|_1} \\ = &\sup_{\varphi \in H^1(\Omega)} \frac{((I - Q_H)g,\ (I - Q_H)\varphi)}{\|\varphi\|_1} \\ \le &\ CH\ \|(I - Q_H)g\|_0 \le CH\|g\|_0. \end{aligned}$$

Finally, use the estimates $\|\nabla p\|_0^2 = (Q_H \text{div}\ \mathbf{v},\ p) = (\text{div}\ \mathbf{v},\ Q_H p - p) + (\mathbf{v},\ \nabla p) \le C\|\nabla p\|_0 (H\|\text{div}\, \mathbf{v}\|_0 + \|\mathbf{v}\|_0)$. That is, $\|\nabla p\|_0 \le C(H\|\text{div}\, \mathbf{v}\|_0 + \|\mathbf{v}\|_0)$. This shows the L_2-stability of $\mathbf{v}_0^{(\ell)}$,

$$\|\mathbf{v}_0^{(\ell)}\|_0 \le \|\mathbf{v}_0^{(\ell)} - \nabla p\|_0 + \|\nabla p\|_0 \le C\ [H\|\text{div}\, \mathbf{v}\|_0 + \|\mathbf{v}\|_0]. \tag{F.26}$$

To conclude, we can say that there is a $\mathbf{v}_0^{(\ell)} \in \mathbf{R}_\ell$ such that div $\mathbf{v}_0^{(\ell)} = Q_\ell \text{div}\, \mathbf{v}$ and $\mathbf{v}_0^{(\ell)}$ satisfies the stability estimate in the parameter–dependent norm

$$\|\mathbf{v}_0^{(\ell)}\|_0^2 + \tau\|\text{div}\, \mathbf{v}_0^{(\ell)}\| \le C\left[\|\mathbf{v}\|_0^2 + \left(1 + \frac{H^2}{\tau}\right)\ \tau\|\text{div}\, \mathbf{v}\|_0^2\right]. \tag{F.27}$$

So far, we have shown the following multilevel energy stability result,

$$\sum_{k=0}^{\ell} \left(\|\mathbf{v}_0^{(k)}\|_0^2 + \tau\ \|\text{div}\, \mathbf{v}_0^{(k)}\|_0^2\right) \le C\left(\|\mathbf{v}\|_0^2 + \left(1 + \frac{H^2}{\tau}\right)\tau\ \|\text{div}\, \mathbf{v}\|_0^2\right). \tag{F.28}$$

Recall, that we have assumed that $H^2\tau^{-1} = \mathcal{O}(1)$.

Because $\mathbf{v} - \mathbf{v}_0 \in \mathbf{R}_h$ is divergence-free there is a $\mathbf{u} \in \mathbf{N}_h$ (the lowest-order Nédélec space) such that curl $\mathbf{u} = \mathbf{v} - \mathbf{v}_0$. Based on our analysis in the previous sections (Lemma F.8), we can find a multilevel decomposition $\mathbf{u} = \sum_k \boldsymbol{\psi}_k^{(f)}$ where $\boldsymbol{\psi}_k^{(f)} \in \mathbf{N}_k$ such that the following stability estimate holds:

$$\sum_{k<\ell} h_k^{-2}\|\boldsymbol{\psi}_k^{(f)}\|_0^2 + \sum_k \|\text{curl}\ \boldsymbol{\psi}_k^{(f)}\|_0^2 \le C\ \|\text{curl}\ \mathbf{u}\|_0^2 = C\ \|\mathbf{v} - \mathbf{v}_0\|_0^2. \tag{F.29}$$

We let $\boldsymbol{\psi}_\ell = \boldsymbol{\psi}_\ell^{(f)}$. This is the last coarse component of $\mathbf{u}$.

Our goal in this section is to prove the following MG convergence result.

Theorem F.9. *Consider the parameter-dependent $H(\text{div})$ bilinear form*

$$a(\mathbf{v},\ \boldsymbol{\chi}) = (\mathbf{v},\ \boldsymbol{\chi}) + \tau\ (\text{div}\ \mathbf{v},\ \text{div}\ \boldsymbol{\chi}).$$

It gives rise to a sequence of stiffness matrices $\mathbf{A}_k$ coming from the (lowest-order) Raviart–Thomas spaces $\mathbf{R}_k$, $\ell \geq k \geq 0$ associated with tetrahedral triangulations $\mathcal{T}_k$ obtained by uniform refinement of a polyhedral (not necessarily convex) domain $\Omega \subset \mathbb{R}^3$. Consider the geometric MG method for $\{\mathbf{A}_k\}$ with (forward) Gauss–Seidel smoothers $\{\mathbf{M}_k\}$ (of the original Raviart–Thomas face degrees of freedom) plus smoothing involving the curl of the basis functions of the respective (lowest-order) Nédélec space $\mathbf{N}_k$ (associated with $\mathcal{T}_k$). To do this, we need (explicitly) the matrix representation of the mapping (discrete curl) $P_{k+(1/2)}$ that transforms a coefficient vector of a function $\mathbf{u}$ from the Nédélec space $\mathbf{N}_k$ to the coefficient vector of curl $\mathbf{u}$ viewed as an element of the Raviart–Thomas space $\mathbf{R}_k$. We also need the (forward) Gauss–Seidel smoothers $M_{k+(1/2)}$ coming from the matrices $\mathbf{A}_{k+(1/2)} = P^T_{k+(1/2)}\mathbf{A}_k P_{k+(1/2)}$. The resulting MG of interest can be implemented as in Algorithm F.1.1. It gives rise to a symmetric $V(1,1)$-cycle that has the convergence factor bounded independently of the number of levels ℓ as well as of the parameter $\tau > 0$ if $H^2\tau^{-1} = \mathcal{O}(1)$, where $H = h_\ell$ is the coarsest mesh-size.

Proof. We follow the main steps (a)–(e) from Section F.1.

Given a f.e. vector function $\mathbf{v} \in \mathbf{R}_h$, let $\mathbf{v}_0 = \sum_k \mathbf{v}_0^{(k)}$ be its stable decomposition (constructed in the beginning of this section) such that div $(\mathbf{v} - \mathbf{v}_0) = 0$ with components $\mathbf{v}_0^{(k)}$ that satisfy, for $k < \ell$ (F.25), and a coarse component $\mathbf{v}_0^{(\ell)}$ that satisfies (F.27). Let $\mathbf{u} \in \mathbf{N}_h$ be such that curl $\mathbf{u} = \mathbf{v} - \mathbf{v}_0$. We use its multilevel decomposition $\mathbf{u} = \sum_k \boldsymbol{\psi}_k^{(f)}$, $\boldsymbol{\psi}_k^{(f)} \in \mathbf{N}_k$ which is stable in the sense of (F.29) (its existence was shown in Section F.3).

Starting with $\mathbf{y}_0 = \mathbf{y} \equiv \mathbf{v}$, we define (in order to conform with the notation in Section F.1)

$$\begin{aligned}
\mathbf{y}_k &= \sum_{j=k}^{\ell} \left(\mathbf{v}_0^{(j)} + \text{curl}\ \boldsymbol{\psi}_j^{(f)}\right) \in \mathbf{R}_k,\\
\mathbf{u}_k^{(f)} &= \mathbf{v}_0^{(k)},\\
\mathbf{u}_k &= \sum_{j \geq k} \mathbf{u}_k^{(f)} = \sum_{j \geq k} \mathbf{v}_0^{(j)},\\
\mathbf{y}_{k+(1/2)}^{(f)} &= \text{curl}\ \boldsymbol{\psi}_k^{(f)},\\
\mathbf{y}_{k+(1/2)} &= \mathbf{u}_k^{(f)} + \mathbf{y}_{k+1} = \sum_{j=k}^{\ell} \mathbf{v}_0^{(j)} + \sum_{j=k+1}^{\ell} \text{curl}\ \boldsymbol{\psi}_j^{(f)}.
\end{aligned}$$

Condition (a) leads to the verification of following estimate,

$$\sum_k h_k^{-2} \|\boldsymbol{\psi}_k^{(f)}\|_0^2 \leq C\ a(\mathbf{y},\ \mathbf{y}) = C\ a(\mathbf{v},\ \mathbf{v}). \tag{F.30}$$

Because the components $\boldsymbol{\psi}_k^{(f)}$ satisfy (F.29), we have

$$\sum_k h_k^{-2} \|\boldsymbol{\psi}_k^{(f)}\|_0^2 \leq C \; \|\mathbf{v} - \mathbf{v}_0\|_0^2.$$

That is, we need to bound $\mathbf{v}_0 = \sum_{k=0}^{\ell} \mathbf{v}_0^{(k)}$ in terms of $\mathbf{v}$. We have, based on estimates (F.25) and (F.26),

$$\begin{aligned}
\|\mathbf{v}_0\|_0^2 &\leq \left(\|\mathbf{v}_0^{(\ell)}\|_0 + \sum_{k=0}^{\ell-1} \|\mathbf{v}_0^{(k)}\|_0 \right)^2 \\
&\leq C \left(\|\mathbf{v}\|_0^2 + H^2 \; \|\mathrm{div}\; \mathbf{v}\|_0^2 + \left(\sum_k h_k^2 \right)^{1/2} \left(\sum_{k=0}^{\ell-1} h_k^{-2} \|\mathbf{v}_0^{(k)}\|_0^2 \right)^{1/2} \right)^2 \\
&\leq C \left[\|\mathbf{v}\|_0^2 + \left(\frac{H^2}{\tau} \right) \tau \; \|\mathrm{div}\; \mathbf{v}\|_0^2 \right].
\end{aligned} \tag{F.31}$$

By assumption $H^2 \tau^{-1} = \mathcal{O}(1)$, thus we proved (F.30).

Condition (b) leads to the following estimates (for some $\boldsymbol{\psi}_k \in \mathbf{N}_k$)

$$\begin{aligned}
\sum_k h_k^2 \|\boldsymbol{\psi}_k\|_0^2 &= \sum_k h_k^2 \; (\mathbf{curl}\; \boldsymbol{\psi}_k, \; \mathbf{y}_{k+(1/2)}) \\
&= \sum_k h_k^2 \left(\mathbf{curl}\; \boldsymbol{\psi}_k, \; \sum_{j \geq k} \mathbf{v}_0^{(j)} + \sum_{j>k} \mathbf{curl}\; \boldsymbol{\psi}_j^{(f)} \right).
\end{aligned}$$

We use now the following strengthened inverse inequalities (cf. Proposition F.1)

$$(\mathbf{curl}\; \boldsymbol{\psi}_k, \; \mathbf{v}_0^{(j)}) \leq C h_k^{-(1/2)} \|\boldsymbol{\psi}_k\|_0 h_j^{-(1/2)} \|\mathbf{v}_0^{(j)}\|_0$$

and

$$(\mathbf{curl}\; \boldsymbol{\psi}_k, \; \mathbf{curl}\; \boldsymbol{\psi}_j^{(f)}) \leq C h_k^{-(1/2)} \|\boldsymbol{\psi}_k\|_0 h_j^{-(1/2)} \|\mathbf{curl}\; \boldsymbol{\psi}_j^{(f)}\|_0.$$

The last three estimates combined give,

$$\begin{aligned}
\sum_k h_k^2 \|\boldsymbol{\psi}_k\|_0^2 &= \sum_k h_k^2 \; (\mathbf{curl}\; \boldsymbol{\psi}_k, \; \mathbf{y}_{k+(1/2)}) \\
&\leq C \; \sum_k h_k^2 \sum_{j \geq k} h_k^{-(1/2)} \|\boldsymbol{\psi}_k\|_0 h_j^{-(1/2)} [\|\mathbf{v}_0^{(j)}\|_0 + \|\mathbf{curl}\; \boldsymbol{\psi}_j^{(f)}\|_0] \\
&= C \; \sum_k h_k \|\boldsymbol{\psi}_k\|_0 \sum_{j \geq k} \left(\frac{1}{\sqrt{2}} \right)^{j-k} [\|\mathbf{v}_0^{(j)}\|_0 + \|\mathbf{curl}\; \boldsymbol{\psi}_j^{(f)}\|_0] \\
&\leq C \left(\sum_k h_k^2 \|\boldsymbol{\psi}_k\|_0^2 \right)^{1/2} \left(\sum_j \|\mathbf{v}_0^{(j)}\|_0^2 + \sum_j \|\mathbf{curl}\; \boldsymbol{\psi}_j^{(f)}\|_0^2 \right)^{1/2}
\end{aligned}$$

$$\leq C \left(\sum_k h_k^2 \|\boldsymbol{\psi}_k\|_0^2 \right)^{1/2}$$
$$\times \left(H^2 \sum_{j<\ell} h_j^{-2} \, \|\mathbf{v}_0^{(j)}\|_0^2 + \|\mathbf{v}_0^{(\ell)}\|_0^2 + \sum_j \|\text{curl } \boldsymbol{\psi}_j^{(f)}\|_0^2 \right)^{1/2}.$$

Based on estimates (F.25), (F.26), (F.29), and (F.31), we arrive at the counterpart of estimate (F.10)

$$\sum_k h_k^2 \|\boldsymbol{\psi}_k\|_0^2 \leq C \left(1 + \frac{H^2}{\tau}\right) \, a(\mathbf{v}, \, \mathbf{v}).$$

Condition (c) deals with the expression $\sum_{k<\ell} \lambda_k \, \|\mathbf{u}_k^{(f)}\|_0^2$, where $\lambda_k = 1 + \tau h_k^{-2}$. Because $\mathbf{u}_k^{(f)} = \mathbf{v}_0^{(k)}$, using (F.25), we have

$$\sum_{k<\ell} \lambda_k \, \|\mathbf{u}_k^{(f)}\|_0^2 = \sum_{k<\ell} (1 + \tau h_k^{-2}) \, \|\mathbf{v}_0^{(k)}\|_0^2 \leq \left(1 + \frac{H^2}{\tau}\right) \, \tau \, \sum_{k<\ell} h_k^{-2} \|\mathbf{v}_0^{(k)}\|_0^2$$
$$\leq C \left(1 + \frac{H^2}{\tau}\right) \, \tau \, \|\text{div } \mathbf{v}\|_0^2.$$

That is, we have proved the counterpart of (F.11) from condition (c).

Condition (d) deals with the following expression, for some $\boldsymbol{\Psi}_k \in \mathbf{N}_k$,

$$\sum_k \lambda_k^{-1} \, \|\boldsymbol{\Psi}_k\|_0^2 = \sum_k \lambda_k^{-1} a(\mathbf{y}_{k+1}, \, \boldsymbol{\Psi}_k) = \sum_k \lambda_k^{-1} a\left(\boldsymbol{\Psi}_k, \, \sum_{j \geq k+1} (\mathbf{v}_0^{(j)} + \text{curl } \boldsymbol{\psi}_j^{(f)})\right).$$

We first estimate the L_2-part of $a(.,.)$. Expanded it reads

$$\sum_k \lambda_k^{-1} \left((\boldsymbol{\Psi}_k, \, \mathbf{v}_0^{(\ell)}) + \sum_{j>k}^{\ell-1} (\boldsymbol{\Psi}_k, \, \mathbf{v}_0^{(j)}) + \sum_{j>k} (\boldsymbol{\Psi}_k, \, \text{curl } \boldsymbol{\psi}_j^{(f)}) \right).$$

The first term is estimated as follows.

$$\sum_k \lambda_k^{-1} (\boldsymbol{\Psi}_k, \, \mathbf{v}_0^{(\ell)}) \leq \left(\sum_k \lambda_k^{-1} \|\boldsymbol{\Psi}_k\|_0^2 \right)^{1/2} \left(\sum_k \lambda_k^{-1} \right)^{1/2} \|\mathbf{v}_0^{(\ell)}\|_0$$
$$\leq \left(\sum_k \lambda_k^{-1} \|\boldsymbol{\Psi}_k\|_0^2 \right)^{1/2} \left(\frac{1}{\tau} \sum_k h_k^2 \right)^{1/2} \|\mathbf{v}_0^{(\ell)}\|_0$$
$$\leq C \, \frac{H}{\sqrt{\tau}} \left(\sum_k \lambda_k^{-1} \|\boldsymbol{\Psi}_k\|_0^2 \right)^{1/2} \|\mathbf{v}_0^{(\ell)}\|_0. \tag{F.32}$$

Similarly, for the second term, using Cauchy–Schwarz inequalities and (F.25), we have

$$
\begin{aligned}
&\sum_k \lambda_k^{-1} \sum_{j>k}^{\ell-1} (\boldsymbol{\Psi}_k,\ \mathbf{v}_0^{(j)}) \\
&\le \left(\sum_k \lambda_k^{-1} \|\boldsymbol{\Psi}_k\|_0^2\right)^{1/2} \left(\sum_k \lambda_k^{-1} \left(\sum_{j>k}^{\ell-1} \|\mathbf{v}_0^{(j)}\|\right)^2\right)^{1/2} \\
&\le \left(\sum_k \lambda_k^{-1} \|\boldsymbol{\Psi}_k\|_0^2\right)^{1/2} \left(\sum_k \lambda_k^{-1} \sum_{j>k}^{\ell-1} h_j^2 \left(\sum_{j>k}^{\ell-1} h_j^{-2} \|\mathbf{v}_0^{(j)}\|^2\right)\right)^{1/2} \\
&\le C\left(\sum_k \lambda_k^{-1} \|\boldsymbol{\Psi}_k\|_0^2\right)^{1/2} \left(\frac{1}{\tau} \sum_k h_k^2\, H^2\right)^{1/2} \left(\sum_{j>k}^{\ell-1} h_j^{-2} \|\mathbf{v}_0^{(j)}\|^2\right)^{1/2} \\
&\le C\left(\sum_k \lambda_k^{-1} \|\boldsymbol{\Psi}_k\|_0^2\right)^{1/2} \frac{H^2}{\sqrt{\tau}} \|\operatorname{div} \mathbf{v}\|_0.
\end{aligned}
\tag{F.33}
$$

To estimate the last L_2-term, in addition to Cauchy–Schwarz inequalities, we also use inverse inequality for the curl terms for $j < \ell$. This leads to the estimates

$$
\begin{aligned}
\sum_k \lambda_k^{-1} \sum_{j>k} (\boldsymbol{\Psi}_k,\ \operatorname{curl}\ \boldsymbol{\psi}_j^{(f)}) &\le C \sum_k \lambda_k^{-1} \|\boldsymbol{\Psi}_k\|_0 \sum_{j=k+1}^{\ell-1} h_j^{-1} \|\boldsymbol{\psi}_j^{(f)}\|_0 \\
&\quad + \sum_k \lambda_k^{-1} \|\boldsymbol{\Psi}_k\|_0 \|\operatorname{curl}\ \boldsymbol{\psi}_\ell\|_0 \\
&\le C \sum_k \lambda_k^{-\frac{1}{2}} \|\boldsymbol{\Psi}_k\|_0 \frac{1}{\sqrt{\tau}} \sum_{j>k} \left(\frac{1}{2}\right)^{j-k} \|\boldsymbol{\psi}_j^{(f)}\|_0 \\
&\quad + \sum_k \lambda_k^{-1} \|\boldsymbol{\Psi}_k\|_0 \|\operatorname{curl}\ \boldsymbol{\psi}_\ell\|_0. \\
&\le C \left(\sum_k \lambda_k^{-1} \|\boldsymbol{\Psi}_k\|_0^2\right)^{1/2} \left(\frac{1}{\tau} \sum_j \|\boldsymbol{\psi}_j^{(f)}\|_0^2\right)^{1/2} \\
&\quad + \sum_k \lambda_k^{-1} \|\boldsymbol{\Psi}_k\|_0 \|\operatorname{curl}\ \boldsymbol{\psi}_\ell\|_0.
\end{aligned}
$$

The term $\sum_k \lambda_k^{-1} \|\boldsymbol{\Psi}_k\|_0 \|\operatorname{curl}\ \boldsymbol{\psi}_\ell\|_0$ above can be estimated as follows

$$
\begin{aligned}
\sum_k \lambda_k^{-1} \|\boldsymbol{\Psi}_k\|_0 \|\operatorname{curl}\ \boldsymbol{\psi}_\ell\|_0 &\le \left(\sum_k \lambda_k^{-1} \|\boldsymbol{\Psi}_k\|_0^2\right)^{1/2} \left(\sum_k \lambda_k^{-1}\right)^{1/2} \|\operatorname{curl}\ \boldsymbol{\psi}_\ell\|_0 \\
&\le C \frac{H}{\sqrt{\tau}} \left(\sum_k \lambda_k^{-1} \|\boldsymbol{\Psi}_k\|_0^2\right)^{1/2} \|\operatorname{curl}\ \boldsymbol{\psi}_\ell\|_0.
\end{aligned}
$$

Combining the last two estimates, based on (F.29), gives the desired bound

$$\sum_k \lambda_k^{-1} \sum_{j>k} (\boldsymbol{\Psi}_k,\ \text{curl}\ \boldsymbol{\psi}_j^{(f)})$$
$$\leq C \left(\sum_k \lambda_k^{-1} \|\boldsymbol{\Psi}_k\|_0^2\right)^{1/2} \frac{H}{\sqrt{\tau}} \left(\sum_j h_j^{-2} \|\boldsymbol{\psi}_j^{(f)}\|_0^2\right)^{1/2}$$
$$+ C\, \frac{H}{\sqrt{\tau}} \left(\sum_k \lambda_k^{-1} \|\boldsymbol{\Psi}_k\|_0^2\right)^{1/2} \|\text{curl}\ \boldsymbol{\psi}_\ell\|_0$$
$$\leq C\, \frac{H}{\sqrt{\tau}} \left(\sum_k \lambda_k^{-1} \|\boldsymbol{\Psi}_k\|_0^2\right)^{1/2} \|\mathbf{v} - \mathbf{v}_0\|_0. \tag{F.34}$$

Now, we turn to the div part of $a(.,.)$. It reads

$$\tau \sum_k \lambda_k^{-1} \sum_{j>k} (\text{div}\ \boldsymbol{\Psi}_k,\ \text{div}\ \mathbf{v}_0^{(j)}).$$

Here, we use the strengthened inverse inequality (cf. Proposition F.1) valid for $j \geq k$,

$$(\text{div}\ \boldsymbol{\Psi}_k,\ \text{div}\ \mathbf{v}_0^{(j)}) \leq C h_k^{-(1/2)}\ \|\boldsymbol{\Psi}_k\|_0 h_j^{-(1/2)}\ \|\text{div}\ \mathbf{v}_0^{(j)}\|_0,$$

This leads to the estimate

$$\tau \sum_k \lambda_k^{-1} \sum_{j>k} (\text{div}\ \boldsymbol{\Psi}_k,\ \text{div}\ \mathbf{v}_0^{(j)})$$
$$\leq C\sqrt{\tau} \sum_k \lambda_k^{-\frac{1}{2}}\ \|\boldsymbol{\Psi}_k\|_0 \| \sum_{j>k} \left(\frac{1}{\sqrt{2}}\right)^{j-k} \|\text{div}\ \mathbf{v}_0^{(j)}\|_0$$
$$\leq C\sqrt{\tau} \left(\sum_k \lambda_k^{-1} \|\boldsymbol{\Psi}_k\|_0^2\right)^{1/2} \left(\sum_j \|\text{div}\ \mathbf{v}_0^{(j)}\|_0^2\right)^{1/2}$$
$$\leq C \left(\sum_k \lambda_k^{-1} \|\boldsymbol{\Psi}_k\|_0^2\right)^{1/2} \sqrt{\tau}\ \|\text{div}\ \mathbf{v}\|_0. \tag{F.35}$$

Combining (F.35), (F.34), (F.33), and (F.32) based on (F.31) and (F.26), we arrive at the desired counterpart of estimate (F.12) of (d)

$$\sum_k \lambda_k^{-1}\ \|\boldsymbol{\Psi}_k\|_0^2 \leq C \left(1 + \frac{H^2}{\tau} + \left(\frac{H^2}{\tau}\right)^2\right)\ a(\mathbf{v},\ \mathbf{v}).$$

Because the coarse component $\mathbf{y}_\ell = \mathbf{v}_0^{(\ell)} + \text{curl}\ \boldsymbol{\psi}_\ell$ is "energy"-stable, (due to (F.27) and (F.29)), we also have that (e) is satisfied. Thus the proof is complete. □

MG methods for $H(\text{div})$-problems were analyzed in [VW92] (2D only), in [H97] (3D), as well as in [AFWii] and [AFW]. The presentation in this section extends the approach from [VW92] based on the results available for $H(\text{curl})$ presented earlier in Section F.2.

G

Some Auxiliary Inequalities

G.1 Kantorovich's inequality

Another result that can be proved by looking at the sign of the discriminant of an appropriate quadratic form, is the popular Kantorovich's inequality.

Proposition G.1. *For any s.p.d. A, let* $\lambda_{\min} = \lambda_{\min}[A]$ *and* $\lambda_{\max} = \lambda_{\max}[A]$ *stand for its extreme eigenvalues, and let* $\kappa = (\lambda_{\max}[A])/(\lambda_{\min}[A])$ *be the condition number of A. Then, the following Kantorovich's inequality holds,*

$$\frac{\mathbf{v}^T A\mathbf{v}\ \mathbf{v}^T A^{-1}\mathbf{v}}{(\mathbf{v}^T\mathbf{v})^2} \le \left(\frac{\kappa+1}{2\sqrt{\kappa}}\right)^2.$$

Proof. Consider the quadratic form

$$Q_0(\lambda) \equiv (\lambda - \lambda_{\max})(\lambda - \lambda_{\min}).$$

We have $Q_0(\lambda) \le 0$ if $\lambda \in [\lambda_{\min},\ \lambda_{\max}]$. Similarly,

$$Q(\lambda) \equiv (\lambda - \lambda_{\max})(1 - \lambda^{-1}\lambda_{\min}) \le 0$$

for $\lambda \in [\lambda_{\min},\ \lambda_{\max}]$. Therefore, the matrix

$$Q(A) \equiv (A - \lambda_{\max} I)(I - A^{-1}\lambda_{\min}),$$

is symmetric negative semidefinite; that is,

$$\mathbf{v}^T A\mathbf{v} - (\lambda_{\min} + \lambda_{\max})\mathbf{v}^T\mathbf{v} + \lambda_{\min}\lambda_{\max}\mathbf{v}^T A^{-1}\mathbf{v} \le 0$$

for all $\mathbf{v}$.

Finally, consider the quadratic form (for any real t)

$$\varphi(t) = \mathbf{v}^T A\mathbf{v}\, t^2 - (\lambda_{\min} + \lambda_{\max})\mathbf{v}^T\mathbf{v}\, t + \lambda_{\min}\lambda_{\max}\mathbf{v}^T A^{-1}\mathbf{v}.$$

Because $\varphi(0) > 0$ and $\varphi(1) \leq 0$ it follows that the quadratic equation $\varphi(t) = 0$ must have a real root. Therefore its discriminant D must be nonnegative; that is,

$$D \equiv (\lambda_{\min} + \lambda_{\max})^2 (\mathbf{v}^T \mathbf{v})^2 - 4\lambda_{\min}\lambda_{\max} \mathbf{v}^T A \mathbf{v} \mathbf{v}^T A^{-1} \mathbf{v} \geq 0,$$

which is the Kantorovich's inequality. □

G.2 An inequality between powers of matrices

Lemma G.2. *Let A and B be two given s.p.d. matrices such that*

$$\mathbf{v}^T A \mathbf{v} \leq \mathbf{v}^T B \mathbf{v}.$$

Then,

$$\mathbf{v}^T A^{1/2} \mathbf{v} \leq \mathbf{v}^T B^{1/2} \mathbf{v}.$$

Proof. Let $X = A^{1/2} + B^{1/2}$ and $Y = B^{1/2} - A^{1/2}$. Note that X is s.p.d. and Y is symmetric. We also have the identity,

$$YX + XY = 2(B - A). \tag{G.1}$$

Consider the generalized eigenvalue problem,

$$Y\mathbf{q} = \lambda X^{-1} \mathbf{q}.$$

Because X^{-1} is s.p.d., and Y is symmetric, the eigenvalues λ are real. Using the symmetry of X and Y, we have $\mathbf{q}^T YX\mathbf{q} = \mathbf{q}^T (YX)^T \mathbf{q} = \mathbf{q}^T XY\mathbf{q} = \lambda\ \mathbf{q}^T \mathbf{q}$. The latter, together with (G.1), imply

$$0 \leq 2\mathbf{q}^T (B - A)\mathbf{q} = \mathbf{q}^T (XY + YX)\mathbf{q} = 2\lambda\ \mathbf{q}^T \mathbf{q}.$$

Thus, $\lambda \geq 0$. Hence, $X^{1/2} Y X^{1/2}$ is symmetric positive semidefinite, and therefore $Y = B^{1/2} - A^{1/2}$ is symmetric positive semidefinite, which is the desired result. □

The following more general result holds which can be proved by using theory of interpolation spaces.

Lemma G.3. *Let A and B be two given s.p.d. matrices such that*

$$\mathbf{v}^T A \mathbf{v} \leq \mathbf{v}^T B \mathbf{v}.$$

Then, for any $\alpha \in [0, 1]$, we have

$$\mathbf{v}^T A^{\alpha} \mathbf{v} \leq \mathbf{v}^T B^{\alpha} \mathbf{v}.$$

Proof. Following Appendix B of Bramble [Br93]), we can use the following construction. Define the K_A functional

$$K_A^2(t,\ \mathbf{v}) = \inf_{\mathbf{w}} \left(\|\mathbf{w}\|^2 + t^2(\mathbf{v}-\mathbf{w})^T A(\mathbf{v}-\mathbf{w})\right)$$
$$= \inf_{\mathbf{w}} \begin{bmatrix} \mathbf{v} \\ \mathbf{w} \end{bmatrix}^T \begin{bmatrix} t^2 A & -t^2 A \\ -t^2 A & I + t^2 A \end{bmatrix} \begin{bmatrix} \mathbf{v} \\ \mathbf{w} \end{bmatrix}.$$

Using the main minimization property of the Schur complement

$$t^2 A - t^2 A(I + t^2 A\,)^{-1} t^2 A = t^2 A(I + t^2 A)^{-1},$$

of the s.p.d. matrix

$$\begin{bmatrix} t^2 A & -t^2 A \\ -t^2 A & I + t^2 A \end{bmatrix},$$

we get

$$K_A^2(t,\ \mathbf{v}) = \inf_{\mathbf{w}} \begin{bmatrix} \mathbf{v} \\ \mathbf{w} \end{bmatrix}^T \begin{bmatrix} t^2 A & -t^2 A \\ -t^2 A & I + t^2 A \end{bmatrix} \begin{bmatrix} \mathbf{v} \\ \mathbf{w} \end{bmatrix} = t^2 \mathbf{v}^T A(I + t^2 A)^{-1} \mathbf{v}.$$

Let the eigenvectors $\{\mathbf{q}_k\}$ of $A\mathbf{q}_k = \lambda_k \mathbf{q}_k$, $k \geq 1$, define an orthonormal basis. Then, $\mathbf{v} = \sum_{k\geq 1} \mathbf{q}_k^T \mathbf{v}\ \mathbf{q}_k$, and hence

$$K_A^2(t,\ \mathbf{v}) = \sum_{k\geq 1} t^2 \frac{\lambda_k}{1 + t^2 \lambda_k} (\mathbf{q}_k^T \mathbf{v})^2. \tag{G.2}$$

Let

$$C_\alpha = \left(\int_0^\infty \frac{t^{1-2\alpha}}{1+t^2}\, dt \right)^{-(1/2)} = \left(\frac{2}{\pi} \sin \pi\alpha \right)^{1/2}.$$

Define now the norm

$$|||\mathbf{v}|||_{A,\ \alpha} = C_\alpha \left(\int_0^\infty t^{-2\alpha} K_A^2(t,\ \mathbf{v}) \frac{dt}{t} \right)^{1/2}.$$

Based on identity (G.2), the following result is easily seen (see [Br93], Theorem B.2),

$$|||\mathbf{v}|||_{A,\ \alpha} = \left(\mathbf{v}^T A^\alpha \mathbf{v}\right)^{1/2}.$$

Indeed,

$$\int_0^\infty t^{-2\alpha} K_A^2(t,\ \mathbf{v}) \frac{dt}{t} = \sum_{k\geq 1} (\mathbf{q}_k^T \mathbf{v})^2 \int_0^\infty t^{-2\alpha} t^2 \frac{\lambda_k}{1 + t^2\lambda_k} \frac{dt}{t}$$
$$= \sum_{k\geq 1} (\mathbf{q}_k^T \mathbf{v})^2 \lambda_k^\alpha \int_0^\infty (\sqrt{\lambda_k} t)^{-2\alpha} \frac{\sqrt{\lambda_k} t}{1 + (t\sqrt{\lambda_k})^2}\, d(\sqrt{\lambda_k} t)$$

$$= \sum_{k\geq 1} (\mathbf{q}_k^T \mathbf{v})^2 \lambda_k^\alpha \int_0^\infty x^{-2\alpha} \frac{x}{1+x^2}\, dx$$
$$= C_\alpha^{-2} \sum_{k\geq 1} (\mathbf{q}_k^T \mathbf{v})^2 \lambda_k^\alpha.$$

Now, using the fact that $\mathbf{v}^T A^\alpha \mathbf{v} = \sum_{k\geq 1} (\mathbf{q}_k^T \mathbf{v})^2 \lambda_k^\alpha$, the desired identity follows,

$$|||\mathbf{v}|||_{A,\,\alpha}^2 = C_\alpha^2 \int_0^\infty t^{-2\alpha} K_A^2(t,\ \mathbf{v}) \frac{dt}{t} = \sum_{k\geq 1} (\mathbf{q}_k^T \mathbf{v})^2 \lambda_k^\alpha = \mathbf{v}^T A^\alpha \mathbf{v}.$$

Finally, because $\mathbf{x}^T A \mathbf{x} \leq \mathbf{x}^T B \mathbf{x}$, we have

$$\begin{aligned} K_A^2(t,\ \mathbf{v}) &= \inf_{\mathbf{w}} \left(\|\mathbf{w}\|^2 + t^2 (\mathbf{v}-\mathbf{w})^T A (\mathbf{v}-\mathbf{w})\right) \\ &\leq \inf_{\mathbf{w}} \left(\|\mathbf{w}\|^2 + t^2 (\mathbf{v}-\mathbf{w})^T B (\mathbf{v}-\mathbf{w})\right) \\ &= K_B^2(t,\ \mathbf{v}). \end{aligned}$$

The latter, implies that

$$\mathbf{v}^T A^\alpha \mathbf{v} = |||\mathbf{v}|||_{A,\,\alpha}^2 \leq |||\mathbf{v}|||_{B,\,\alpha}^2 = \mathbf{v}^T B^\alpha \mathbf{v},$$

which is the desired result. □

G.3 Energy bound of the nodal interpolation operator

Let T be a convex polygon. In our application, T is a simple finite element, such as a tetrahedron. Then, there is an extension mapping that transforms any continuous function v from $H^1(T)$ to a function $\widehat{v}$ from $H_0^1(\widehat{\Omega})$, where $\widehat{\Omega}$ is a larger domain (with diameter proportional to the diameter of T) which preserves the norm. That is, we have

$$\widehat{v}|_T = v, \qquad |\widehat{v}|_{1,\,\widehat{\Omega}}^2 \leq C_T \left(|v|_{1,\,T}^2 + H^{-2}\ \|v\|_{0,\,T}^2\right).$$

The constant C_T may depend on the shape of T but not on its size (or diameter). Here, H stands for a characteristic size of the diameter of T. For norm-preserving extension mappings we refer to Grisvard ([Gri85], p. 25, and the references given therein).

All the estimates below are first derived for domains with unit size and the corresponding result is obtained by transformation of the domain.

Let $\mathcal{T}_h$ be a triangulation with characteristic mesh-size h which covers T exactly. Choose a node $x_0 \in T$ from the mesh associated with $\mathcal{T}_h$. Let $G(x,\ x_0) = 1/r$, $r = |x - x_0|$. We have

$$-\Delta\ G(x,\ x_0) = 0, \qquad \text{for all } x \in \mathbb{R}^3 \setminus \{x_0\}.$$

Now, choose an $\epsilon > 0$ such that $K_\epsilon(x_0) \equiv \{x : |x - x_0| < \epsilon\} \subset \widehat{\Omega}$. The Green's formula then gives

$$0 = -\int\limits_{\widehat{\Omega}\setminus K_\epsilon(x_0)} \widehat{v}\Delta G(x,\ x_0)\,dx = \int\limits_{\widehat{\Omega}\setminus K_\epsilon(x_0)} \nabla G \cdot \nabla \widehat{v}\,dx - \int\limits_{\partial K_\epsilon(x_0)} \widehat{v}\frac{\partial G}{\partial r}\,d\Gamma.$$

Schwarz inequality, the fact that $(\partial G/\partial r) = -(1/\epsilon^2)$ on $\partial K_\epsilon(x_0)$, imply

$$\frac{1}{\epsilon^2}\left|\int\limits_{|x-x_0|=\epsilon} \widehat{v}\,d\Gamma\right| \le \left(\int\limits_{\widehat{\Omega}\setminus K_\epsilon(x_0)} |\nabla G|^2\,dx\right)^{1/2} |\widehat{v}|_{1,\,\widehat{\Omega}}$$

$$\le C_T\left(|v|^2_{1,\,T} + H^{-2}\,|v|^2_{0,\,T}\right)^{1/2}\left(\int\limits_{\widehat{\Omega}\setminus K_\epsilon(x_0)} |\nabla G|^2\,dx\right)^{1/2}.$$

Also, because $|\nabla G| \le 1/r^2$, and after introducing spherical coordinates in the last volume integral, we obtain,

$$\frac{1}{\epsilon^2}\left|\int\limits_{|x-x_0|=\epsilon} \widehat{v}\,d\Gamma\right| \le C_T\left(|v|^2_{1,\,T} + H^{-2}\,|v|^2_{0,\,T}\right)^{1/2}\left(\int_\epsilon^\infty r^2/r^4\,dr\right)^{1/2},$$

which implies the inequality,

$$\frac{1}{\epsilon^2}\left|\int\limits_{|x-x_0|=\epsilon} \widehat{v}\,d\Gamma\right| \le \frac{1}{\sqrt{\epsilon}}\,C_T\left(|v|^2_{1,\,T} + H^{-2}\,|v|^2_{0,\,T}\right)^{1/2}. \tag{G.3}$$

Apply now estimate (G.3) for a function $v + v_0$ where v_0 is a piecewise linear finite element function associated with the triangulation $\mathcal{T}_h$, such that it vanishes at the vertices of T. Here, we view T as an element from a coarse triangulation $\mathcal{T}_H$. Also, v is a piecewise linear function associated with $\mathcal{T}_H$, hence linear on T. Let $\epsilon \simeq h$ and use Taylor's expansion (noting that we may assume that the extension $\widehat{v + v_0}$ is linear on $T \cap K_\epsilon(x_0)$),

$$v(x_0) = v(x) + v_0(x) + (x_0 - x)\cdot\nabla(\widehat{v + v_0})(x), \quad \text{for } x \in K_\epsilon(x_0).$$

Then, because $|\widehat{v + v_0}|_{1,K_\epsilon(x_0)} \simeq \epsilon^{3/2}|\nabla(\widehat{v + v_0})|$, $|x - x_0| \le \epsilon$, and $\int_{|x-x_0|=\epsilon} 1 d\Gamma \simeq \epsilon^2$, we have

$$|v(x_0)| \le C\left(\epsilon^{-2}\left|\int\limits_{|x-x_0|=\epsilon} \widehat{v + v_0}\,d\Gamma\right| + \epsilon\epsilon^{-(3/2)}\,|\widehat{v + v_0}|_{1,\,T}\right).$$

Using the fact that v is linear on T, we have that its maximal value on T is attained at a vertex x_0 of T, based on (G.3), we get

$$\max_{x\in T} |v(x)| = |v(x_0)| \leq C \left(\epsilon^{-2} \left| \int\limits_{|x-x_0|=\epsilon} \widehat{v+v_0}\, d\Gamma \right| + \epsilon^{-(1/2)} |\widehat{v+v_0}|_{1,\,T} \right)$$
$$\leq \epsilon^{-(1/2)} C_T \left(|\widehat{v+v_0}|^2_{1,\,T} + H^{-2} |\widehat{v+v_0}|^2_{0,\,T} \right)^{1/2}$$
$$\leq h^{-(1/2)} C_T \left(|v+v_0|^2_{1,\,T} + H^{-2} |v+v_0|^2_{0,\,T} \right)^{1/2}.$$

Use now the expansion of v in terms of the nodal basis φ_i of the finite element space associated with the vertices x_i of the triangulation $\mathcal{T}_h$. We have

$$v = \sum_{x_i} v(x_i)\varphi_i,$$

and

$$\nabla v = \sum_{x_i} v(x_i)\, \nabla\varphi_i.$$

Since $|\nabla\varphi_i| \simeq 1/h$ and the fact that only finite number of entries take part in the above sum for any x restricted to an element T, we get

$$|v|_{1,\,T} \leq C H^{1/2} \max_{x\in T} |v(x)|$$
$$\leq C_T \left(\frac{H}{h}\right)^{1/2} \left(|v+v_0|^2_{1,\,T} + H^{-2} |v+v_0|^2_{0,\,T} \right)^{1/2}.$$

Now use Poincaré's inequality,

$$H^{-3} |w|^2_{0,\,T} - \left(H^{-3} \int\limits_T w\, dx \right)^2 \leq C H^{-1} |w|^2_{1,\,T},$$

for $w = v + v_0 + \text{const}$. By an appropriate choice of the const we can achieve $\int_T w\, dx = 0$. Therefore, we proved the local estimate,

$$|v|_{1,\,T} \leq C_T \left(\frac{H}{h}\right)^{1/2} |v+v_0|_{1,\,T}. \tag{G.4}$$

In 2D, we use $G(x,\, x_0) = \log\, 1/r$. Repeating the above analysis with appropriate changes, we end up with the following local estimate,

$$|v|_{1,\,T} \leq C_T \left(1 + \log\frac{H}{h}\right)^{1/2} |v+v_0|_{1,\,T}. \tag{G.5}$$

References

[AFWii] D. N. Arnold, R. S. Falk, and R. Winther, "*Preconditioning in $H(div)$ and applications*," Mathematics of Computations **66**(1997), pp. 957–984.

[AFW] D. N. Arnold, R. S. Falk, and R. Winther, "*Multigrid in $H(div)$ and $H(curl)$*," Numerische Mathematik **85**(2000), pp. 197–217.

[AHU] K. Arrow, L. Hurwicz, and H. Uzawa, *Studies in Nonlinear Programming*, Stanford University Press, Stanford, CA, 1958.

[A78] G. P. Astrahantsev, "*Methods of fictitious domains for a second order elliptic equation with natural boundary conditions*," USSR Computational Math. and Math. Physics **18**(1978), pp. 114–121.

[Ax94] O. Axelsson, "*Iterative Solution Methods*" Cambridge University Press, Cambridge, 1994.

[Ax87] O. Axelsson, "*A generalized conjugate gradient, least–square method*," Numerische Mathematik **51**(1987), pp. 209–227.

[AG83] O. Axelsson and I. Gustafsson, "*Preconditioning and two–level multigrid methods of arbitrary degree of approximation*," Mathematics of Computation **40**(1983), pp. 214–242.

[AN03] O. Axelsson and M. Neytcheva, "*Preconditioning methods for linear systems arising in constrained optimization problems*," Numer. Linear Algebra with Applications **10**(2003), pp. 3–31.

[AP86] O. Axelsson and B. Polman, "*On approximate factorization methods for block–matrices suitable for vector and parallel processors*," Lin. Alg. Appl. **77**(1986), pp. 3–26.

[AV89] O. Axelsson and P. S. Vassilevski, "*Algebraic multilevel preconditioning methods*," I, Numerische Mathematik **56**(1989), 157–177.

[AV90] O. Axelsson and P. S. Vassilevski, "*Algebraic multilevel preconditioning methods II*," SIAM J. Numer. Anal. **27**(1990), pp. 1569–1590.

[AV91] O. Axelsson and P. S. Vassilevski, "*A black box generalized CG solver with inner iterations and variable-step preconditioning*," SIAM J. Matr. Anal. Appl. 12(1991), 625–644.

[AV92] O. Axelsson and P. S. Vassilevski, "*Construction of variable-step preconditioners for inner-outer iterative methods*," Proceedings of IMACS Conference on Iterative methods, April 1991, Brussels, Belgium, *Iterative Methods in Linear Algebra*, (R. Beauwens and P. de Groen, eds.), North Holland, 1992, pp. 1–14.

[AV94] O. Axelsson and P. S. Vassilevski, "*Variable-step multilevel preconditioning methods problems*," Numer. Linear Algebra with Applications **1**(1994), pp. 75–101.

[ABI] O. Axelsson, S, Brinkkemper, and V.P. Il'in, "*On some versions of incomplete block-matrix factorization methods*," Lin. Alg. Appl. **38**(1984), pp. 3–15.

[BTW] L. Badea, X.-C. Tai, and J. Wang, "*Convergence rate analysis of a multiplicative Schwarz method for variational inequalities*," SIAM J. Numerical Analysis **41**(2003), pp. 1052–1073.

[Ba78] A. Banegas, "*Fast Poisson solvers for problems with sparsity*," Mathematics of Computation **32**(1978), pp. 441–446.

[BD80] R. E. Bank and T. F. Dupont, "*Analysis of a two–level scheme for solving finite element equations*," Report CNA–159, Center for Numerical Analysis, The University of Texas at Austin, May 1980.

[BH03] R. E. Bank, and M. J. Holst, "*A new paradigm for parallel adaptive meshing algorithms*," SIAM Review **45**(2003), pp. 291–323.

[BaS02] R. E. Bank, and R. K. Smith, "*An algebraic multilevel multigraph algorithm*," SIAM J. Sci. Comput. **23**(2002), pp. 1572–1592.

[BV06] R. E. Bank and P. S. Vassilevski, "*Convergence analysis of a domain decomposition paradigm*," Lawrence Livermore National Laboratory Technical Report UCRL-JRNL-222227, June 2006.

[BW99] R. E. Bank and C. Wagner, "*Multilevel ILU decomposition*," Numerische Mathematik **82**(1999), pp. 543–576.

[BDY88] R. E. Bank, T. F. Dupont, and H. Yserentant, "*The hierarchical basis multigrid method*," Numerische Mathematik **52**(1988), pp. 427–458.

[Ba02] R. E. Bank, P. K. Jimack, S. A. Nadeem, and S. V. Nepomnyaschikh, "*A weakly overlapping domain decomposition preconditioner for the finite element solution of elliptic partial differential equations*," SIAM J. Sci. Comput. **23**(6)(2002), pp. 1817–1841.

[BWY] R. E. Bank, B. D. Welfert, and H. Yserentant, "*A class of iterative methods for solving saddle–point problems*," Numerische Mathematik **56**(1990), pp. 645–666.

[BGL] M. Benzi, G. H. Golub, and J. Liesen, "*Numerical Solution of Saddle Point Problems*," Acta Numerica **14**(2005), pp. 1–137.

[BW97] J. Bey and G. Wittum, "*Downwind numbering: Robust multigrid for convection diffusion problems*," Applied Numerical Mathematics **23**(1997), pp. 177–192.

[Bl86] R. Blaheta, "*A multi-level method with correction by aggregation for solving discrete elliptic problems*," Aplikace Matematiky **31**(1986), pp. 365–378.

[Bl87] R. Blaheta, "*Iterative methods for numerical solution of boundary value problems in elasticity*," Charles University, Faculty of Mathematics and Physics, 1987, Prague, (in Czech).

[Bl02] R. Blaheta, "*GPCG-generalized preconditioned CG method and its use with nonlinear and non-symmetric displacement decomposition preconditioners*," Numerical Linear Algebra with Applications, 9 (2002), pp. 527–550.

[H05] S. Börm, L. Grasedyck, and W. Hackbusch, "Hierarchical Matrices," Max-Planck-Institut, Leipzig, Lecture note no **21**, revised version 2005.

[BD96] F. A. Bornemann and P. Deuflhard, "*The cascadic multigrid method for elliptic problems*," Numer. Math. **75**(1996), pp. 135–152.

[DB95] D. Braess, "*Towards algebraic multigrid for elliptic problems of second order*," Computing **55**(1995), pp. 379–393.

[B01] D. Braess, "*Finite elements. Theory, Fast Solvers, and Applications in Solid Mechanics*," Cambridge University Press, 2001, 2nd edition.

[BH83] D. Braess and W. Hackbusch, "*A new convergence proof of the multigrid method including the V–cycle*" SIAM J. Numerical Analysis **20**(1983), pp. 967–975.

[BS97] D. Braess and R. Sarazin, "*An efficient smoother for the Stokes problem*," Appl. Numer. Math. **23**(1997), pp. 3–19.

[Br93] J. H. Bramble, "*Multigrid Methods*," Pitman Research Notes in Mathematics Series, No 294, Longman Scientific and Technical, 1993.

[BP87] J. H. Bramble and J. E. Pasciak, "*New convergence estimates for multigrid algorithms*," Mathematics of Computation **49**(1987), pp. 311–329.

[BP88] J. H. Bramble and J. E. Pasciak, "*A preconditioning technique for indefinite systems resulting from mixed approximations of elliptic problems*," Mathematics of Computation **50**(1988), pp. 1–17.

[BP93] J. H. Bramble and J. E. Pasciak, "*New estimates for multigrid algorithms including the V–cycle*," Mathematics of Computation **60**(1993), pp. 447–471.

[BP97] J. H. Bramble and J. E. Pasciak, "*Iterative techniques for time dependent Stokes problems*," The International Journal on Computers and Mathematics with Applications **33**(1997) pp. 13–30.

[BEPS] J. H. Bramble, R. E. Ewing, J. E. Pasciak, and A. H. Schatz, "*A preconditioning technique for the efficient solution of problems with local grid refinement*," Comp. Meth. Appl. Mech. Eng. **67**(1988), pp. 149–159.

[BPL] J. H. Bramble, J. E. Pasciak, and R. D. Lazarov, "*Least squares for second order elliptic problems*," Comput. Methods in Appl. Mech. Engrg. **152**(1998), pp. 195–210.

[BPV97] J. H. Bramble, J. E. Pasciak, and A. T. Vassilev, "*Analysis of the inexact Uzawa algorithm for saddle–point problems*," SIAM J. Numer. Anal. **34**(1997), pp. 1072–1092.

[BPV99] J. H. Bramble, J. E. Pasciak, and P. S. Vassilevski, "*Computable scales of Sobolev norms with application to preconditioning*," Mathematics of Computation **69**(1999), pp. 463–480.

[BPX] J. H. Bramble, J. E. Pasciak, and J. Xu, "*Parallel multilevel preconditioners*," Mathematics of Computation **55**(1990), pp. 1–22.

[BPX91] J. H. Bramble, J. E. Pasciak, and J. Xu, "*The analysis of multigrid algorithms with nonnested spaces or noninherited quadratic form*," Mathematics of Computation **56**(1991), pp. 1–34.

[BPZ] J. H. Bramble, J. E. Pasciak, and X. Zhang. "*Two-level preconditioners for 2m'th order elliptic finite element problems*," *East-West Journal on Numerical Mathematics* **4**(1996), pp. 99–120.

[BPWXii] J. H. Bramble, J. E. Pasciak, J. Wang, and J. Xu, "*Convergence estimates for product iterative methods with application to domain decomposition*," Mathematics of Computation **57**(1991), pp. 1–21.

[BPWXi] J. H. Bramble, J. E. Pasciak, J. Wang, and J. Xu, "*Convergence estimates for multigrid algorithms without regularity assumptions*," Mathematics of Computation **57**(1991), pp. 23–45.

[AB77] A. Brandt, "*Multilevel adaptive solutions to boundary-value problems*," Mathematics of Computation **31**(1977), pp. 333–390.

[B86] A. Brandt, "*Algebraic multigrid theory: The symmetric case*," Applied Mathematics and Computation **56**(1986), pp. 23–56.

[B00] A. Brandt, "*General highly accurate algebraic coarsening*," Electronic Transactions on Numerical Analysis **10**(2000), pp. 1–20.

[BC83] A. Brandt and C. W. Cryer, "*Multigrid algorithms for the solution of linear complementarity problems arising from free boundary problems*," SIAM J. Sci. Stat. Comput. **4**(1983), pp. 655–684.

[BD79] A. Brandt and N. Dinar, "*Multigrid solutions to flow problems*," Numerical Methods for Partial Differential Equations, (S. Parter, ed.), 1979, Academic Press, New York, pp. 53–147.

[BL97] A. Brandt and I. Livshitz, "*Ray–wave multigrid method for Helmholtz equation*," Electronic Transactions on Numerical Analysis **6**(1997), pp. 162–181.

[BR02] A. Brandt and D. Ron, "*Multigrid solvers and multilevel optimization strategies*," in: J. R. Shinerl and J. Cong (eds), Multilevel Optimization and VLSI CAD, Kluwer Academic Publishers, Boston, 2002, pp. 1–69.

[BMcRr] A. Brandt, S. F. McCormick, and J. W. Ruge, "*Algebraic Multigrid (AMG) for automatic multigrid solutions with application to geodetic computations*," Report, Inst. for Computational Studies, Fort Collins, Colorado, 1982.

[BMcR] A. Brandt, S. F. McCormick, and J. W. Ruge, "*Algebraic multigrid (AMG) for sparse matrix equations*," Sparsity and its Applications, (David J. Evans, ed.), 1985, Cambridge University Press, Cambridge, pp. 257–284.

[BZ06] J. Brannick and L. T. Zikatanov, "*Algebraic multigrid methods based on compatible relaxation and energy minimization*," in Domain Decomposition Methods in Science and Engineering XVI, Lecture Notes in Computational Science and Engineering, Springer-Verlag, Berlin - Heidelberg **55**(2007), pp. 15–26.

[Br02] S. C. Brenner, "*Convergence of the multigrid V-cycle algorithms for second order boundary value problems without full elliptic regularity*," Mathematics of Computation **71**(2002), pp. 505–525.

[BS96] S. C. Brenner and L. R. Scott, "*The Mathematical Theory of Finite Elements*," Springer-Verlag, Berlin - Heidelberg - New York, 1996.

[AMGe] M. Brezina, A. J. Cleary, R. D. Falgout, V. E. Henson, J. E. Jones, T. A. Manteuffel, S. F. McCormick, and J. W. Ruge, "*Algebraic multigrid based on element interpolation (AMGe)*," SIAM Journal on Scientific Computing **22**(2000), pp. 1570–1592.

[aSA] M. Brezina, R. Falgout, S. MacLachlan, T. Manteuffel, S. McCormick, and J. Ruge, "*Adaptive smoothed aggregation (αSA)*," SIAM J. Sci. Comput. **25**(2004), pp. 1896–1920.

[aAMG] M. Brezina, R. Falgout, S. MacLachlan, T. Manteuffel, S. McCormick, and J. Ruge, "*Adaptive algebraic multigrid*," SIAM J. Sci. Comp. **37**(2006), pp. 1261–1286.

[Br99] M. Brezina, C. Heberton, J. Mandel, and P. Vaněk, "*An iterative method with convergence rate chosen a priori*," UCD CCM Report **140**(1999), available at: http://www-math.cudenver.edu/~jmandel/papers/.

[BF91] F. Brezzi and M. Fortin, *Mixed and Hybrid Finite Element Methods*, Springer-Verlag, New York - Berlin - Heidelberg, 1991.

[MGT00] W. L. Briggs, V. E. Henson, and S. F. McCormick, "*A Multigrid Tutorial*," 2nd edition, SIAM, 2000.

[BVW03] P. N. Brown, P. S. Vassilevski, and C. S. Woodward, "*On mesh–independent convergence of an inexact Newton–multigrid algorithm*," SIAM J. Scientific Computing **25**(2003) pp. 570–590.

[CDS03] X.-C. Cai, M. Dryja, and M. Sarkis, "*Restricted additive Schwarz preconditioner with harmonic overlap for symmetric positive definite linear systems*," SIAM J. Numerical Analysis **41**(2003), pp. 1209–1231.

[CGP] Z. Cai, C. I. Goldstein, and J. E. Pasciak, "*Multilevel iteration for mixed finite element systems with penalty*," SIAM J. Scientific Computing **14**(1993), pp. 1072–1088.

[CLMM] Z. Cai, R. D. Lazarov, T. Manteuffel, and S. McCormick, "*First order system least–squares for second order partial differential equations: Part I*," SIAM J. Numerical Analysis **31**(1994), pp. 1785–1799.

[CPV] G. F. Carey, A. I. Pehlivanov, and P. S. Vassilevski, "*Least–squares mixed finite element methods for nonselfadjoint elliptic problems, II: Performance of block–ILU factorization methods*," SIAM J. Sci. Comput. **16**(1995), pp. 1126–1136.

[ChV95] T. F. Chan and P. S. Vassilevski, "*A framework for block–ILU factorizations using block–size reduction*," Mathematics of Computation **64**(1995), pp. 129–156.

[Ch78] R. Chandra, "*Conjugate Gradient Methods for Partial Differential Equations*," Ph.D. dissertation, Yale University, New Haven, CT (1978).

[ChG05] S. Chandrasekaran, P. Dewilde, M. Gu, T. Pals, X. Sun, A.-J. van der Veen, and D. White, "*Some fast algorithms for sequentially semi–separable representations*," SIAM J. Matrix Anal. Appl. **27** (2005), pp. 341–364.

[Ch03] T. Chartier, R. Falgout, V. E. Henson, J. Jones, T. Manteuffel, S. McCormick, J. Ruge, and P. S. Vassilevski, "*Spectral AMGe (ρAMGe)*," SIAM J. Scientific Computing **25**(2003), pp. 1–26.

[Ch05] T. Chartier, R. Falgout, V. E. Henson, J. Jones, T. Manteuffel, S. McCormick, J. Ruge, and P. S. Vassilevski, "*Spectral element agglomerate AMGe*," in Domain Decomposition Methods in Science and Engineering XVI, Lecture Notes in Computational Science and Engineering, Springer-Verlag, Berlin - Heidelberg **55**(2007), pp. 515–524.

[ChV99] S.-H. Chou and P. S. Vassilevski. "*A general mixed covolume framework for constructing conservative schemes for elliptic problems*," Math. Comp. **68**(1999), pp. 991–1011.

[ChV03] E. Chow and P. S. Vassilevski, "*Multilevel Block Factorizations in Generalized Hierarchical Bases*," Numerical Linear Algebra with Applications **10**(2003), pp. 105–127.

[Ci02] P. G. Ciarlet, "*The Finite Element Method for Elliptic Problems*," SIAM Classics in Applied mathematics, vol. 40, SIAM, Philadelphia, 2002.

[CGM85] P. Concus, G. H. Golub, and G. Meurant, "*Block preconditioning for the conjugate gradient method*," SIAM J. Sci. Stat. Comput. **6**(1985), pp. 220–252.

[Co99] T. H. Cormen, C. E. Leiserson, and R. L. Rivert, "*Introduction to Algorithms*," MIT Press, Cambridge, MA, 1999.

[DK92] W. Dahmen and A. Kunoth, "*Multilevel preconditioning*," Numerische Mathematik **63**(1992), pp. 315–344.

[D92] I. Daubechies, "*Ten Lectures on Wavelets*," SIAM, Philadelphia, 1992.

[D88] M. Dauge, "*Elliptic Boundary Value Problems on Corner Domains*," Springer-Verlag, New York, 1988.

[DES] R. S. Dembo, S. C. Eisenstat, and T. Steihaug, "*Inexact Newton methods*," SIAM J. Numer. Anal. **19**(1982), pp. 400–408.

[DMS84] S. Demko, W.F. Moss, and P.W. Smith, "*Decay rates of inverses of band matrices*," Mathematics of Computation **43**(1984), pp. 491–499.

[DT91] Y.-H. De Roeck and P. Le Tallec, "*Analysis and test of a local domain decomposition preconditioner*," in: Fourth International Symposium on Domain Decomposition Methods for Partial Differential Equations (R. Glowinski, Y. Kuznetsov, G. Meurant, J. Périaux, and O. Widlund, eds.) SIAM, Philadelphia, 1991.

[PMISi] H. De Sterck, U. M. Yang, and J. J. Heys, "*Reducing complexity in parallel algebraic multigrid preconditioners*," SIAM J. on Matrix Analysis and Applications **27**(2006), pp. 1019–1039.

[PMISii] H. De Sterck, R. D. Falgout, J. W. Nolting, and U. M. Yang, "*Distance–two interpolation for parallel algebraic multigrid*," Numerical Linear Algebra with Applications 15(2008) to appear.

[Dfl94] P. Deuflhard, "*Cascadic conjugate gradient methods for elliptic partial differential equations. Algorithm and numerical results*," in: D. Keyes and J. Xu (eds), Proceedings of the 7th International Conference on Domain Decomposition Methods 1993, pp. 29–42. AMS, Providence.

[DLY89] P. Deuflhard, P. Leinen, and H. Yserentant, "*Concepts of an adaptive hierarchical finite element code*," IMPACT Compt. Sci. Engrg. **1**(1989), pp. 3–35.

[Do07] C. Dohrmann, "*Interpolation operators for algebraic multigrid by local optimization*," SIAM Journal on Scientific Computing (2007) (to appear).

[DJ05] Z. Dostal and J. Schoeberl, "*Minimizing quadratic functions over non-negative cone with the rate of convergence and finite termination*," Computational Optimization and Applications **30**(2005), pp 23–44.

[DW87] M. Dryja and O. B. Widlund, "*An additive variant of of the Schwarz alternating method method for the case of many subregions*," Technical Report TR–339 (Ultracomputer Note 131), Department of Computer Science, Courant Institute, 1987.

[DW95] M. Dryja and O. B. Widlund, "*Schwarz methods of Neumann–Neumann type for three–dimensional elliptic finite element problems*," Comm. Pure Appl. Math. **48**(1995), pp. 121–155.

[EV91] V. L. Eijkhout and P. S. Vassilevski, "*The role of the strengthened C.-B.-S.-inequality in multilevel methods*," SIAM Review 33(1991), 405–419.

[EES83] S. C. Eisenstat, H. C. Elman, and M. H. Schultz, "*Variational iterative methods for nonsymmetric systems of linear equations*," SIAM J. Numer. Anal. **20**(1983), pp. 345–357.

[EG94] H. Elman and G. Golub, "*Inexact and preconditioned Uzawa algorithms for saddle–point problems*," SIAM J. Numer. Anal. **31**(1994), pp. 1645–1661.

[ESW06] H. Elman, D. Silvester, and A. Wathen, "*Finite Elements and Fast Iterative Solvers with applications in incompressible fluid dynamics*," Oxford University Press, New York, 2005.

[EG04] A. Ern, J.-L. Guermond, "*Theory and Practice of Finite Elements*," Springer, New York, 2004.

[ELV] R. E. Ewing, R. D. Lazarov, and P. S. Vassilevski, "*Local refinement techniques for elliptic problems on cell-centered grids, II: Optimal order two–grid iterative methods*," Numer. Lin. Alg. with Appl. **1**(1994), pp. 337–368.

[ELLV] R. E. Ewing, R. D. Lazarov, Peng Lu, and P. S. Vassilevski, "*Preconditioning indefinite systems arising from mixed finite element discretization of second order elliptic problems*," in: Springer Lecture Notes in Mathematics, vol. 1457, 1990, pp. 28–43.

[Fe64] R. P. Fedorenko, "*A relaxation method for solving elliptic difference equations*," USSR Comput. Math. Math. Phys. **1**(1961), pp. 1092–1096.

[Fe64] R. P. Fedorenko, "*The speed of convergence of one iterative process*," USSR Comput. Math. Math. Phys. **4**(1964), pp. 227–235.

[FV04] R. D. Falgout and P. S. Vassilevski, "*On generalizing the AMG framework*," SIAM Journal on Numerical Analysis **42**(2004) pp. 1669–1693.

[FVZ05] R. D. Falgout, P. S. Vassilevski, and L. T. Zikatanov, "*On two-grid convergence estimates*," Numerical Linear Algebra with Applications **12**(2005), pp. 471–494.

[FETI] C. Farhat and F.-X. Roux, "*An unconventional domain decomposition method for an efficient parallel solution of large–scale finite element systems*," SIAM Journal on Scientific Computing **13**(1992), pp. 379–396.

[FETIdp0] C. Farhat, M. Lesoinne, and K. Pierson, "*A scalable dual–primal domain decomposition method*," Numerical Linear Algebra with Applications **7**(2000), pp. 687–714.

[FETIdp] C. Farhat, M. Lesoinne, P. Le Talec, K. Pierson, and D. Rixen, "*FETI–DP: a dual primal unified FETI method – part I: A faster alternative to the two–level FETI method*," Intern. J. Numer. Meth. Eng. **50**(2001), pp. 1523–1544.

[F77] M. Fortin, "*An analysis of the convergence of mixed finite element methods*," RAIRO Anal. Numér. **11**(1977), pp. 341–354.

[GL81] A. George and J. W. Liu "*Computer Solution of Large Sparse Positive Definite Systems*," Prentice-Hall, Englewood Cliffs, NJ, 1981.

[GR86] V. Girault and P. A. Raviart, "*Finite element methods for the Navier-Stokes equations*," Springer Series in Computational Mathematics, vol. 5, Springer-Verlag, New York, 1986.

[GY99] G. H. Golub and Q. Ye, "*Inexact preconditioned conjugate gradient method with inner–outer iterations*," SIAM J. Scientific Computing **21**(1999), pp. 1305–1320.

[GK03] J. Gopalakrishnan and G. Kanschat, "*A multilevel discontinuous Galerkin method*," Numerische Mathematik **95**(2003), pp. 527–550.

[Gr97] A. Greenbaum, "*Iterative Methods for Solving Linear Systems*," vol. 3 of Frontiers in Applied Mathematics, SIAM, Philadelphia, 1997.

[Gr94] M. Griebel, "*Multilevel algorithms considered as iterative methods on semidefinite systems*," SIAM J. Scientific Computing **15**(1994), pp. 547–565.

[GO95] M. Griebel and P. Oswald, "*On the abstract theory of additive and multiplicative Schwarz algorithms*," Numerische Mathematik **70**(1995), pp. 163–180.

[Gri85] P. Grisvard, "*Elliptic Problems in Nonsmooth Domains*," Pitman, Boston - London - Melbourne, 1985.

[HLMN] G. Haase, U. Langer, A. Meyer, and S. V. Nepomnyaschikh, "*Hierarchical extension operators and local multigrid methods in domain decomposition preconditioners*," East–West J. Numer. Math. **2**(1994), pp. 173–193.

[H82] W. Hackbusch, "*Multi-grid convergence theory*," In: "*Multi-Grid Methods* (W. Hackbusch and U. Trottenberg, eds.) Springer Lecture Notes in Mathematics **960**(1982), pp. 177–219.

[H85] W. Hackbusch, "*Multigrid Methods and Applications*," Springer-Verlag, Berlin - Heidelberg - New York - Tokyo, 1985.

[H94] W. Hackbusch, "*Iterative Solution of Large Sparse Systems of Equations*," Springer-Verlag, New York - Berlin (1994).

[HP97] W. Hackbusch and T. Probst, "*Downwind Gauss-Seidel smoothing for convection dominated problems*," Numerical Linear Algebra with Applications **4**(1997), pp. 85–102.

[H] Don Heller, "*Some aspects of the cyclic reduction algorithm for block–tridiagonal linear systems*," SIAM Journal on Numerical Analysis **13**(1976), pp. 484–496.

[HV01] V. E. Henson and P. S. Vassilevski, "*Element-free AMGe: General algorithms for computing interpolation weights in AMG*," SIAM Journal on Scientific Computing **23**(2001), pp. 629–650.

[H97] R. Hiptmair, "*Multigrid method for H (div) in three dimensions*," Electronic Transactions in Numerical Analysis **6**(1997), pp. 133–152.

[H99] R. Hiptmair. "*Multigrid method for Maxwell's equations*," SIAM Journal on Numerical Analysis **36**(1999), pp. 204–225.

[HX06] R. Hiptmair and J. Xu, "*Nodal auxiliary space preconditioning in $H(curl)$ and $H(div)$*," Research Report No. 2006-09, May 2006, Seminar für Angewandte Mathematik, Eidgenössische Technische Hochschule, CH-8092 Zürich, Switzerland.

[HHNL] I. Hlaváček, J. Haslinger, J. Nečas, and J. Lovišek, "*Solution of Variational Inequalities in Mechanics*," Springer-Verlag, New York, 1988.

[IoV04] A. H. Iontcheva and P. S. Vassilevski. "*Monotone multigrid methods based on element agglomeration coarsening away from the contact boundary for the Signorini's problem*," Numer. Linear Algebra Appl. **11**(2004), pp.189–204.

[JV01] J. E. Jones and P. S. Vassilevski, "*AMGe based on element agglomerations*," SIAM J. Scientific Computing **23**(2001), pp. 109–133.

[Kato] T. Kato, "*Estimation of iterated matrices, with application to the von Neumann condition*," Numerische Mathematik **2**(1960), pp. 22–29.

[K82] R. Kettler, "*Analysis and comparison of relaxed schemes in robust multigrid and preconditioned conjugate gradient methods*," in: *Multigrid Methods, Proceedings* (W. Hackbusch and U. Trottenberg, eds.) Lecture Notes in Mathematics, vol. 960, Springer Verlag, Berlin and New York, 1982, pp. 502–534.

[KLPV] C. Kim, R. D. Lazarov, J. E. Pasciak, and P. S. Vassilevski, "*Multiplier spaces for the mortar finite element method in three dimensions*," SIAM J. Numer. Anal. **39**(2001), pp. 519–538.

[KPV] T. Kim, J. E. Pasciak, and P. S. Vassilevski, "*Mesh independent convergence of the modified inexact Newton method for a second order nonlinear problem*," Numerical Linear Algebra with Applications **13**(2006), pp. 23–47.

[KV06] T. V. Kolev and P. S. Vassilevski, "*AMG by element agglomeration and constrained energy minimization interpolation*," Numerical Linear Algebra with Applications **13**(2006), pp. 771–788.

[KV06i] T. V. Kolev and P. S. Vassilevski, "*Some Experience With a H1-Based Auxiliary Space AMG for H(curl) Problems*," LLNL Technical Report UCRL-TR-221841, June 2006.

[KV06ii] T. V. Kolev and P. S. Vassilevski, "*Parallel H1-Based Auxiliary Space AMG Solver for H(curl) Problems*," LLNL Technical Report UCRL-TR-222763, July 2006.

[KPVa] T. V. Kolev, J. E. Pasciak, and P. S. Vassilevski, "H(curl) auxiliary mesh preconditioning," Numerical Linear Algebra with Applications **14**(2007), to appear.

[KPVb] T. V. Kolev, J. E. Pasciak, and P. S. Vassilevski, "*Algebraic construction of mortar finite element spaces with application to parallel AMGe*," Lawrence Livermore National Laboratory Technical Report UCRL-JC-150807.

[JK02] J. K. Kraus, "*An algebraic preconditioning method for M–matrices: linear versus nonlinear multilevel iteration*," Numer. Linear Alg. with Appl. **9**(2002), pp. 599–618.

[Ku85] Yu. A. Kuznetsov, *Numerical methods in subspaces*, in: Vychislitel'nye Processy i Sistemy II, G. I. Marchuk, ed., Nauka, Moscow, 1985, pp. 265–350 (in Russian).

[KM78] Y. A. Kuznetsov and A. M. Matsokin, "*On partial solution of systems of linear algebraic equations*," Vychislitel'nye Metody Lineinoi Algebry (Ed. G.I. Marchuk). Vychisl. Tsentr Sib. Otdel. Akad. Nauk SSSR, Novosibirsk (1978), pp. 62–89. In Russian, English translation in: Sov. J. Numer. Anal. Math. Modelling **4**(1989), pp. 453–467.

[KR96] Yu. A. Kuznetsov and T. Rossi, "*Fast direct method for solving algebraic systems with separable symmetric band matrices*," East–West J. Numer. Math. **4**(1996), pp. 53–68.

[LQ86] U. Langer and W. Queck, "*On the convergence factor of Uzawa's algorithm*," J. Comput. Appl. Math. **15**(1986), pp. 191–202.

[BL00] B. Lee, T. Manteuffel, S. McCormick, and J. Ruge, "*First-order system least squares (FOSLS) for the Helmholtz equation*," SIAM Journal on Scientific Computing **21**(2000), pp. 1927–1949.

[BP04e] B. Lee, S. F. McCormick, B. Philip, and D. J. Quinlan, "*Asynchronous Fast Adaptive Composite–grid methods: Numerical results*," SIAM Journal on Scientific Computing **25**(2003), pp. 682–699.

[BP04] B. Lee, S. F. McCormick, B. Philip, and D. J. Quinlan, "*Asynchronous Fast Adaptive Composite–grid methods for elliptic problems: Theoretical foundations*," SIAM Journal on Numerical Analysis **42**(2004), pp. 130–152.

[LWXZ] Y.-J. Lee, J. Wu, Jinchao Xu, L. T. Zikatanov, "*A sharp convergence estimate of the method of subspace corrections for singular system of equations*," Report # AM259(2002), CCMA, Department of Mathematics, Penn State University.

[Li87] P.-L. Lions, "*On the Schwarz alternating method. I*," in: R. Glowinski, G. H. Golub, G. A. Meurant, and J. Periaux, eds., *1st International Symposium on Domain Decomposition Methods for PDEs, held in Paris, France, January 7–9, 1987.* SIAM, Philadelphia, 1988, pp. 1–42.

[Li04] O. Livne, "*Coarsening by compatible relaxation*," Numerical Linear Algebra with Applications **11**(2004), pp. 205–227.

[MM82] J.-F. Maitre and F. Musy, "*The contraction number of a class of two–level methods; an exact evaluation for some finite element subspaces and model problems*," In: *Multigrid Methods, Proceedings of the 1st European Conference on Multigrid Methods*, Cologne 1981. Lecture Notes in Mathematics, vol. 960, 1982, pp. 535–544. Springer Verlag, Heidelberg.

[Man93] J. Mandel, "*Balancing domain decomposition*," Communications in Applied Numerical Methods **9**(1993), pp. 233-241.

[MMR] J. Mandel, S. F. McCormick, and J. Ruge, "*An algebraic theory for multigrid methods for variational problems*," SIAM J. Numer. Anal. **25**(1988), pp. 91–110.

[MN85] A. M. Matsokin and S. V. Nepomnyaschikh, "*A Schwarz alternating method in subspace*," Soviet Math. **29**(1985), pp. 78–84.

[Mc01] S. MacLachlan, T. Manteuffel, and S. F. McCormick, "*Adaptive reduction-based AMG*," Numerical Linear Algebra with Applications **13**(2006), pp. 599–620.

[M92i] T. P. Mathew, "*Schwarz alternating and iterative refinement methods for mixed formulations of elliptic problems, part I: Algorithms and numerical results*," Numerische Mathematik **65**(1993) pp. 445–468.

[M92ii] T. P. Mathew, "*Schwarz alternating and iterative refinement methods for mixed formulations of elliptic problems, part II: Convergence theory*," Numerische Mathematik **65**(1993) pp. 469–492.

[Mc84] S. F. McCormick, "*Multigrid methods for variational problems: further results*," SIAM J. Numer. Anal. **21**(1984), pp. 255–263.

[SMc84] S. F. McCormick, "*Fast adaptive composite–grid (FAC) methods: Theory for the variational case*," in: *Defect Correction Methods: Theory and Applications*, K. Böhmer and H. J. Stetter, eds., Computing Suppl. 5, Springer Verlag, Vienna, 1984, pp. 115–122.

[Mc85] S. F. McCormick, "*Multigrid methods for variational problems: general theory for the V-cycle*," SIAM J. Numer. Anal. **22**(1985), pp. 634–643.

[McT86] S. F. McCormick and J. Thomas, "*The fast adaptive composite grid (FAC) method for elliptic equations*," Mathematics of Computation **46**(1986), pp. 439–456.

[MvF77] J. A. Meijerink and H. A. van der Vorst, "*An iterative solution method for linear systems of which the coefficient matrix is a symmetric M–matrix*," Mathematics of Computation **31**(1977), pp. 148–162.

[Mo03] P. B. Monk, "*Finite Element Methods for Maxwell's Equations*," Oxford University Press, New York, 2003.

[Nep91a] S. V. Nepomnyaschikh, "*Method of splitting into subspaces for solving boundary value problems in complex form domain.*" Sov. J. Numer. Anal. and Math. Modelling **6**(1991), pp. 151–168.

[Nep91b] S. V. Nepomnyaschikh, "*Mesh theorems of traces, normalizations of function traces and their inversion*," Sov. J. Numer. Anal. and Math. Modelling **6**(1991), pp. 1–25.

[Nep95] S. V. Nepomnyaschikh, "*Optimal multilevel extension operators*," Report SPC 95-3, Jan. 1995, Technische Universität Chemnitz–Zwickau, Germany.

[Not98] Y. Notay, "*Using approximate inverses in algebraic multilevel methods*," Numerische Mathematik **80**(1998), pp. 397–417.

[Not0a] Y. Notay, "*Optimal order preconditioning of finite difference matrices*," SIAM Journal on Scientific Computing **21**(2000), pp. 1991–2007.

[Not0b] Y. Notay, "*Flexible conjugate gradients*," SIAM Journal on Scientific Computing **22**(2000), pp. 1444–1460.

[No05] Y. Notay, "*Algebraic multigrid and algebraic multilevel methods; a theoretical comparison*," Numerical Linear Algebra with Applications **12**(2005), pp. 419–451.

[NV07] Y. Notay and P. S. Vassilevski, "*Recursive Krylov-based multigrid cycles*," Numerical Linear Algebra with Applications 15(2008) to appear.

[O97] M. E. G. Ong, "*Hierarchical basis preconditioners in three dimensions*," SIAM Journal on Scientific Computing **18**(1997), pp. 479–498.

[0s94] P. Oswald, "*Multilevel Finite Element Approximation. Theory and Applications*," Teubner Skripten zur Numerik, Teubner, Stuttgart, 1994.

[PS75] C. C. Paige and M. A. Saunders, "*Solution of sparse indefinite systems of linear equations*," SIAM J. Numer. Anal. **12** (1975), pp. 617–629.

[Pa61] S. Parter, "*The use of linear graphs in Gaussian elimination*," SIAM Review **3**(1961), pp. 119–130.

[PZ02] J. E. Pasciak and J. Zhao, "*Overlapping Schwarz methods in H (curl) on polyhedral domains*," Journal of Numerical Mathematics **10**(2002), pp. 221–234.

[RT] R. Rannacher and S. Turek, "*Simple nonconforming quadrilateral Stokes element*," Numer. Methods for Partial Differential Equations **8**(1992), pp. 97–111.

[RTa] T. Rossi and J. Toivanen, "*A nonstandard cyclic reduction method, its variants and stability*," SIAM J. Matrix Anal. Appl. **20** (1999), pp. 628–645.

[RTb] T. Rossi and J. Toivanen, "*A parallel fast direct solver for block–tridiagonal systems with separable matrices of arbitrary dimension*," SIAM Journal on Scientific Computing **20**(1999), pp. 1778–1793.

[RS87] J. W. Ruge and K. Stüben, "*Algebraic multigrid (AMG)*," in *Multigrid Methods*, S. F. McCormick, ed., vol. 3 of Frontiers in Applied Mathematics, SIAM, Philadelphia, 1987, pp. 73–130.

[RW92] T. Rusten and R. Winther, "*A preconditioned iterative method for saddle point problems*," SIAM J. Matrix Anal. Appl. **13** (1992), pp. 887–904.

[RVWa] T. Rusten, P. S. Vassilevski, and R. Winther, "*Interior penalty preconditioners for mixed finite element approximations of elliptic problems*," Mathematics of Computation **65**(1996), pp. 447–466.

[RVWb] T. Rusten, P. S. Vassilevski, and R. Winther, "*Domain embedding preconditioners for mixed systems*," Numer. Linear Alg. Appl. **5**(1998), pp. 321–345.

[Sa93] Y. Saad, "*A flexible inner–outer preconditioned GMRES algorithm*," SIAM Journal on Scientific Computing **14**(1993), pp. 461–469.

[Sa03] Y. Saad, "*Iterative Methods for Sparse Linear Systems*," 2nd edition, SIAM, Philadelphia, 2003.

[SS86] Y. Saad and M. H. Schultz, "*GMRES: A generalized minimal residual algorithm for solving nonsymmetric linear systems*," SIAM Journal on Scientific Computing **7**(1986), pp. 856–869.

[SS98] S. Schaffer, "*A semicoarsening multigrid method for elliptic partial differential equations with highly discontinuous and anisotropic coefficients*," SIAM Journal on Scientific Computing **20**(1998), pp. 228–242.

[SW96] A. H. Schatz and J. Wang, "*Some new error estimates for Ritz-Galerkin methods with minimal regularity assumptions*," Mathematics of Computation **65**(1996), pp. 19–27.

[Sch98] J. Schöberl, "*Solving the Signorini's problem on the basis of domain decomposition techniques*," Computing **60**(1998), pp. 323–344.

[Sch01] J. Schöberl, "*Efficient contact solvers based on domain decomposition techniques*," Computers and Mathematics with Applications **42**(2001), pp. 1217–1228.

[Sch05] J. Schöberl, "*A multilevel decomposition result in $H(curl)$*," Proceedings of the 5th European Multigrid Conference, 2005, to appear.

[SZ03] J. Schöberl and W. Zulehner, "*On Schwarz–type smoothers for saddle–point problems*," Numerische Mathematik **95**(2003), pp. 377–399.

[Sh94] V. V. Shaidurov, "*Some estimates of the rate of convergence for the cascadic conjugate gradient method*," Preprint No. 4(1994), Otto-von-Guericke-Universität Magdeburg, Germany.

[Sh95] V. V. Shaidurov, "*Multigrid Methods for Finite Elements*," Kluwer Academic, Publ., Dordrecht, Boston, 1995.

[SW93] D. S. Silvester and A. J. Wathen, "*Fast iterative solution of stabilized Stokes systems, part II: using general block preconditioners*," SIAM J. Numer. Analysis **31**(1994), pp. 1352–1367.

[SSa] V. Simoncini and D. B. Szyld, "*Flexible inner-outer Krylov subspace methods*," SIAM J. Numer. Analysis **40**(2003), pp. 2219–2239.

[SSb] V. Simoncini and D. B. Szyld, "*Theory of inexact Krylov subspace methods and applications to scientific computing*," SIAM J. Sci. Comput. **25**(2003), pp. 454–477.

[DD] B. F. Smith, P. E. Bjorstad, and W. D. Gropp, "*Domain Decomposition Methods, Parallel Multilevel Methods for Elliptic Partial Differential Equations*," Cambridge University Press, New York, 1996.

[So06] P. Šolin, "*Partial Differential Equations and the Finite Element Method*," Wiley-Interscience, John Wiley and Sons, Hoboken, NJ, 2006.

[SLC] D.-K. Sun, J.-F. Lee, and Z. Cendes, "*Construction of nearly orthogonal Nedelec basis for rapid convergence with multilevel preconditioned solvers*," SIAM J. Sci. Comput. **23**(2001), pp. 1053–1076.

[Sz06] D. B. Szyld, "*The many proofs of an identity on the norm of oblique projections*," Numerical Algorithms **42**(2006), pp. 309–323.

[TW05] A. Toselli and O. Widlund, "*Domain Decomposition Methods - Algorithms and Theory*," Springer-Verlag Berlin, Heidelberg, 2005.

[TOS] U. Trottenberg, C. Oosterlee, and A. Schüller, "*Multigrid*," Academic Press, San Diego, 2001.

[VV94] H. A. van der Vorst and C. Vuik, "*GMRESR: A family of nested GMRES methods*," Numerical Linear Algebra with Applications **1**(1994), pp. 369–386.

[VSA] P. Vaněk, "*Acceleration of convergence of a two-level algorithm by smoothing transfer operator*," Applications of Mathematics **37**(1992), pp. 265–274.

[SA] P. Vaněk, M. Brezina, and J. Mandel, "*Convergence of algebraic multigrid based on smoothed aggregation*," Numerische Mathematik **88**(2001), pp. 559–579.

[Van86] S. Vanka, "*Block–implicit multigrid calculation of two–dimensional recirculating flow*," Computer Methods Appl. Mech. Eng. **59**(1986), pp. 29–48.

[Var62] R. S. Varga, "*Matrix Iterative Analysis*," Prentice–Hall, Englewood Cliffs, NJ, 1962.

[V84] P. S. Vassilevski, "*Fast algorithm for solving a linear algebraic problem with separable variables*," Compt. rend. de l'Acad. bulg. Sci. **37**(1984) No 3, 305–308.

[V90] P. S. Vassilevski, "*On some ways of approximating inverses of banded matrices in connection with deriving preconditioners based on incomplete block-factorizations*," Computing **43**(1990), pp. 277–296.

[V92a] P. S. Vassilevski, "*Preconditioning nonsymmetric and indefinite finite element matrices*," J. Numer. Linear Alg. with Appl. **1**(1992), 59–76.

[V92b] P. S. Vassilevski, "*Hybrid V–cycle algebraic multilevel preconditioners*," Mathematics of Computation **58**(1992), pp. 489–512.

[V98] P. S. Vassilevski, "*A block–factorization (algebraic) formulation of multigrid and Schwarz methods*," East–West J. Numer. Math. **6**(1998), # 1, pp. 65–79.

[V96] P. S. Vassilevski, "*On some applications of the H^σ–stable wavelet–like hierarchical finite element space decompositions*," in: *The Mathematics of Finite Elements and Applications, Highlights 1996*, (J. R. Whiteman, ed.), Chichester, 1997, J. Wiley & Sons, pp. 371–395.

[V02] P. S. Vassilevski, "*Sparse matrix element topology with application to AMG and preconditioning*," Numerical Linear Algebra with Applications **9**(2002), pp. 429–444.

[VA94] P. S. Vassilevski and O. Axelsson, "*A two–level stabilizing framework for interface domain decomposition preconditioners*," in: Proceedings of the Third International Conference $O(h^3)$, Sofia, Bulgaria, August 21–August 26, Sofia, Bulgaria, "*Advances in Numerical Methods and Applications*," (I. T. Dimov, Bl. Sendov and P. S. Vassilevski, eds.), World Scientific, Singapore, 1994, pp. 196–202.

[VL96] P. S. Vassilevski and R. D. Lazarov, "*Preconditioning mixed finite element saddle–point elliptic problems*," Numer. Linear Alg. Appl. **3**(1996), pp. 1–20.

[VW92] P. S. Vassilevski and J. Wang, "*Multilevel iterative methods for mixed finite element discretizations of elliptic problems*," Numerische Mathematik **63**(1992), 503–520.

[VW97] P. S. Vassilevski and J. Wang, "*Stabilizing the hierarchical basis by approximate wavelets, I: Theory*," Numer. Linear Alg. Appl. **4**(1997), pp. 103–126.

[VW99] P. S. Vassilevski and J. Wang, "*Stabilizing the hierarchical basis by approximate wavelets, II: Implementation and Numerical Results*," SIAM J. Sci. Comput. **20**(1999), pp. 490–514.

[VZ05] P. S. Vassilevski and L. T. Zikatanov, "*Multiple vector preserving interpolation mappings in algebraic multigrid*," SIAM J. Matrix Analysis and Applications **27**(2005–2006), pp. 1040–1055.

[Wab03] M. Wabro, "*Algebraic multigrid methods for the numerical solution of the incompressible Navier–Stokes equations*," Universitätsverlag Rudolf Trauner, Johannes–Kepler–Universität Linz, 2003.

[Wag97] C. Wagner, "*Frequency filtering decompositions for symmetric matrices*," Numerische Mathematik **78**(1997), pp. 119–142.

[Wag96] C. Wagner, "*On the algebraic construction of multilevel transfer operators*," Computing **65**(2000), pp. 73–95.

[WW97] C. Wagner and G. Wittum, "*Adaptive filtering*," Numerische Mathematik **78**(1997), pp. 305–328.

[Wal88] H. F. Walker, "*Implementation of the GMRES method using Householder transformations*," SIAM J. Sci. Comput. **9**(1988), pp. 152–163.

[WCS00] W. L. Wan, T. F. Chan, and B. Smith, "*An energy–minimizing interpolation for robust multigrid methods*," SIAM Journal on Scientific Computing **21**(2000), pp. 1632–1649.

[JW94] J. Wang, "*New convergence estimates for multilevel algorithms for finite element approximations*," Journal of Computational and Applied Mathematics **50**(1994), pp. 593–604.

[WX94] J. Wang and R. Xie, "*Domain Decomposition for Elliptic Problems with Large Jumps in Coefficients*," in: *The Proceedings of Conference on Scientific and Engineering Computing*, National Defense Industry Press, Beijing, China, 1994, pp. 74–86.

[Wi89] G. Wittum, "*Multi-grid methods for Stokes and Navier Stokes equations. Transforming smoothers: Algorithms and numerical results*," Numerische Mathematik **54**(1989), pp. 543–563.

[Wi90] G. Wittum, "*On the convergence of multi-grid methods with transforming smoothers*," Numerische Mathematik **57**(1989), pp. 15–38.

[Wi92] G. Wittum, "*Filternde Zerlegungen - Schnelle Löser für grosse Gleichungssysteme*," Teubner Skripten zur Numerik Band 1, (1992), Teubner Verlag, Stuttgart.

[Wo00] B. I. Wohlmuth, "*A mortar finite element method using dual spaces for the Lagrange multiplier*," *SIAM Journal of Numerical Analysis* **38**(3)(2000), pp. 989–1012.

[Wo00] B. I. Wohlmuth. "*Discretization Methods and Iterative Solvers Based on Domain Decomposition*," Lecture Notes in Computational Science and Engineering **17**(2001), Springer, Berlin.

[SW97] S. J. Wright, "*Primal–Dual Interior–Point Methods*," SIAM, Philadelphia, 1997.

[Xu92a] J. Xu, "*Iterative methods by space decomposition and subspace correction*," SIAM Review **34**(1992), pp. 581–613.

[Xu92b] J. Xu, "*A new class of iterative methods for nonselfadjoint or indefinite problems*," SIAM J. Numer. Anal. **29**(1992), pp. 303–319.

[Xu96a] J. Xu, "*Two–grid discretization techniques for linear and nonlinear problems*," SIAM Journal Numerical Analysis **33**(1996), pp. 1759–1777.

[Xu96b] J. Xu, "*The auxiliary space method and optimal multigrid preconditioning technique for unstructured girds*," Computing **56**(1996), pp. 215–235.

[XZh07] J. Xu and Y. Zhu, "*Uniform convergent multigrid methods for elliptic problems with strongly discontinuous coefficients*," preprint, Penn State University, 2007.

[XZ02] J. Xu and L. T. Zikatanov, "*The method of alternating projections and the method of subspace corrections in Hilbert space*," J. Amer. Math. Soc. **15**(2002) pp. 573–597.

[XZ04] J. Xu and L. T. Zikatanov, "*On an energy minimizing basis in algebraic multigrid methods*," Computing and Visualization in Science **7**(2004), pp. 121–127.

[YJ80] D. M. Young and K. C. Jea, "*Generalized conjugate gradient acceleration of nonsymmetrizable iterative methods*," Linear Algebra and Appl. **34**(1980), pp. 159–194.

[Yhb] H. Yserentant, "*On the multilevel splitting of finite element spaces*," Numerische Mathematik **49**(1986), pp. 379–412.

[Y86] H. Yserentant, "*On the multilevel splitting of finite element spaces for indefinite elliptic boundary value problems*," SIAM J. Numer. Anal. **23**(1986), pp. 581–595.

[Y93] H. Yserentant, "*Old and new convergence proofs for multigrid methods*," Acta Numerica (1993), pp. 285–326.

[Zh92] X. Zhang, "*Multilevel Schwarz methods*," Numerische Mathematik **63**(1992), pp. 521–539.

Index